Global Ecodynamics

A Multidimensional Analysis

Springer-Verlag Berlin Heidelberg GmbH

Kirill Ya. Kondratyev, Vladimir F. Krapivin,
Victor P. Savinykh and Costas A. Varotsos

Global Ecodynamics

A Multidimensional Analysis

Springer

Published in association with
Praxis Publishing
Chichester, UK

PRAXIS

Professor Kirill Ya. Kondratyev
Counsellor of the Russian Academy of Sciences
Scientific Research Centre for Ecological Safety
Nansen Foundation for Environment and
Remote Sensing
St Petersburg
Russia

Professor Dr Vladimir F. Krapivin
Institute of Radioengineering and
Electronics
Russian Academy of Sciences
Moscow
Russia

Associate Professor Dr Costas A. Varotsos
University of Athens
Faculty of Physics
Department of Applied Physics
Laboratory of Upper Air
Athens
Greece

Professor Dr Victor P. Savinykh
Moscow University of Geodesy and
Cartography
Moscow
Russia

SPRINGER–PRAXIS BOOKS IN ENVIRONMENTAL SCIENCES
SUBJECT *ADVISORY EDITOR*: John Mason B.Sc., M.Sc., Ph.D.

ISBN 978-3-642-62213-7

Library of Congress Cataloging-in-Publication Data
Global ecodynamics : a multidimensional analysis / Kirill Ya. Kondratyev ... [et al.].
p. cm.
Includes bibliographical references (p.).
ISBN 978-3-642-62213-7 ISBN 978-3-642-18636-3 (eBook)
DOI 10.1007/978-3-642-18636-3
1. Global environmental change. 2. Global environmental change–Mathematical models.
I. Kondrat'ev, K. IA. (Kirill IAkovlevich)

GE149.G545 2004
363.7–dc22 2004045333

Originally published by Praxis Publishing Ltd.Chichester, UK, in 2004
Softcover reprint of the hardcover 1st edition 2004

Cover design: Jim Wilkie
Project Management: Originator Publishing Services, Gt Yarmouth, Norfolk, UK

Printed on acid-free paper

Contents

Preface

During recent decades the stirring up of the processes of globalization practically in all spheres of present-day civilization activities has aggravated and brought forth numerous problems resulting from the nature–society (N–S) interaction. It has become apparent that to solve these problems it is necessary to develop new concepts and approaches to the interpretation of global environmental changes that would enable one to select the first-priority directions in studies and to reliably assess the state of the nature–society system (NSS). One of these priorities is to predict global climate change. The growing interest in the problem of global climate change, determined by its practical importance and by available contradictory estimates of the anthropogenic contribution to climate change, necessitates a systematization of knowledge of and data on the observed climate change and causes of this change. Despite an enormous amount of projects and programmes of studies of past and present climatic trends, the problem of reliable prediction of future climate change remains far from being solved. Emissions to the atmosphere of greenhouse gases (GHGs), mainly carbon dioxide, is considered as one of the main causes of an expected climate warming resulting in sufficiently negative consequences for humankind. Therefore, an attempt has been made in this book to construct a formalized technology to assess the level of the greenhouse effect due to anthropogenic sources of carbon dioxide as well as the effects of other gas components.

To understand the factors that determine the feedbacks in the global natural–technogenic system of the cycles of carbon and other chemicals, a hierarchy of model units is constructed which parameterize a totality of physical and biogeochemical processes which are responsible for transport of various substances. Elements of these units are substantiated based on the synthesis of partial models taking into account balance relationships at the boundaries of different media.

The book contains nine chapters. In Chapter 1 present-day global ecodynamics are studied, the concept of the greenhouse effect is criticized, and the problem of parameterization of anthropogenic processes is discussed. Chapter 2 gives data on the sources and sinks of carbon, considers some conceptual schemes of the global CO_2 cycle, and proposes a new approach to studying the problem of the greenhouse effect. In Chapter 3 an attempt has been made to assess the role of soil–plant formations in assimilation of carbon dioxide from the atmosphere, to describe the models of production processes in surface biocenoses, and to analyse the characteristics of forest ecosystems. Chapter 4 generalizes the earlier models of atmosphere–ocean gas exchange in order to synthesize an efficient model of the CO_2 cycle in the atmosphere–ocean system. Emphasis is directed to the role of the Arctic Basin in CO_2 assimilation from the atmosphere. Chapter 5 considers the problems of pollution and dynamics of the high-latitude environment. Chapter 6 presents a method of parameterization of the biospheric budgets of carbon, nitrogen, sulphur, phosphorus, oxygen, ozone, and water as a single biogeochemical system. Chapter 7 describes a global model of NSS functioning, and with the use of this model, assesses various scenarios of the anthropogenic forcing on the environment. The emphasis is directed towards the analysis of the existing contradictions in the estimates of the observed climate change. Results of measurements and of modelling the aerosol radiative forcing (ARF) are considered, the concept of the geoinformation monitoring is formulated, and the results of global simulation experiments are given. Chapter 8 considers a set of problems related to geoinformation monitoring. Different questions of global warming energetics, geopolitics, and radiative forcing are discussed. Chapter 9 describes important aspects of using the monitoring systems to solve the intermediate problems of global ecodynamics. The problem of the search and identification of catastrophic disasters is discussed, as well as problems with decision making – schemes are proposed to accomplish the monitoring of environmental dynamics (Hauglustaine, 2002).

The book further develops methods discussed in the publications of Kondratyev (1999a), Krapivin and Kondratyev (2002), and Kondratyev *et al.* (2002a, 2003a), in which models have been proposed to simulate regional and global processes in the NSS. The simulation experiments have shown that an increase of reliability of estimates can be ensured due to a consideration of correlations in a multidimensional phase space of NSS, reflecting an interaction of the key natural and anthropogenic components. The selection of these components is one of the important problems of present-day ecodynamics. Its solution is connected with the use of new information technology called geographical information monitoring system (GIMS) technology (geographical information system (GIS) + model) and oriented towards a combination of the empirical and theoretical components of ecoinformatics. The main principles of the GIM technology are (Krapivin and Marenkin, 1999; Krapivin and Potapov, 2002):

1 Combination, integration, and coordination of the existing governmental, departmental, and other systems of collecting primary information about the environment, on a single organizational and scientifically methodical basis.
2 Optimization of material and financial expenditure for creation, functioning, and improvement of the environmental control system.

3 Coordination and compatibility of information fluxes in the system based on the use of a single system of classification, coding, formatting and structuring of data.
4 Centralization of the access to information through international information networks with a maximum list of users.
5 Provision of the international character of the global geoinformation monitoring independent of the fact that country boundaries and the system boundaries do not coincide.

While analysing the present trends of civilization development, and assessing the global ecodynamics, the book considers the problem of global environmental changes as an interdisciplinary problem. The complex approach to solving this problem is based on the GIMS-technology, and its application is connected with the selection of information levels in the NSS and of criteria for assessing its state.

The problem of present-day civilization and the prospects of its development is so broad and multifaceted that the aspects considered here are only a small part of a wider field of scientific and methodical studies of the processes of the N–S interaction. The proposed adaptive evolutionary scheme of combining of the data from monitoring and the results of simulation modelling enables one to hope that it can be a mechanism which will ensure a transition to sustainable development.

The book will be useful for specialists dealing with the development of constructive information technologies in the field of nature protection, with global modelling, climate change, the problems inherent in the relationships between human society and nature, as well as geopolicy.

Figures

Tables

Abbreviations

ABL	Atmospheric Boundary Layer
ACE	Advanced Composition Explorer
ACIA	Arctic Climate Impact Assessment
ACSYS	Arctic Climate System Study
ADOM	Arctic Dispersion and Oxidants Model
AERONET	Aerosol Robotic Network
ALOMAR	Arctic Lidar Observatory for Middle Atmospheric Research
AMAP	Arctic Monitoring and Assessment Programme
AMS	American Meteorological Society
ANWAP	US Arctic Nuclear Waste Programme
AO	Arctic Oscillation
AOD	aerosol optical depth
AOT	aerosol optical thickness
APARE	Asia/North Pacific Regional Experiment
APDA	Arctic Precipitation Data Archive
APS	atmosphere–plant–soil
APT	advanced picture transmission
ARCSS	Arctic System Science Programme
ARDB	Arctic Runoff Database
ARF	aerosol radiative forcing
ARM	Atmospheric Radiation Measurement Programme
AS	Air Standard
ATMOS	Atmospheric Trace Molecule Spectroscopy
AURAMS	A Unified Regional Air-Quality Modelling System
AVHRR	Advanced Very-High Resolution Radiometer
BC	black carbon
BIBEX	Biomass Burning Experiment

BU	biotic unit
CA	carbon aerosol
CACTUS	Chemistry, Aerosols and Climate Tropospheric Unified Simulation
CAM	Canadian Aerosol Model
CANEX	Canadian Experiment
CART	Cloud and Radiation Testbed
CCCMA	Canadian Centre for Climate Modelling and Analysis
CCD	Corrective Cloud Differential
CCSR	Centre for Computing and Social Responsibility
CEPS	Canadian Emissions Processing System
CERES	Clouds and Earth's Radiant Energy System
CF	cloud feedback
CFC	chloroflourocarbons
CGCM	Canadian Global Climate Model
CIESIN	Consortium for International Earth Science Information Network
CliC	Climate and Cryosphere (Project)
CLIVAR	International Research Programme on Climate Variability and Predictability
CM	climate model
CNN	cloud condensation nucleii
COADS	Comprehensive Ocean–Atmosphere Data Set
COS	carbonyl sulphide
CPV	circumpolar vortex
CR	Club of Rome
CRA	counter-radiation of the atmosphere
CREAMS	Chemical Reactions Modelling System
CRF	cloud radiative forcing
CSIRO	Australian National Scientific Centre
CSU	Charles State University
CTM	chemical tracer model
DIC	dissolved inorganic carbon
DMS	dimethylsulphide
DMSO	dimethylsulphoxide
$DMSO_2$	dimethylsulphone
DNVERSITAS	International Biodiversity Programme
DOC	dissolved inorganic carbon
DOM	dissolved inorganic matter
DTR	diurnal course of temperature
EAAS	European Arctic Aerosol Study
EASOE	European Arctic Stratosphereic Ozone Experiment
EC	emission coefficients
ECHAM	European Centre Hamburg
ECMWF	European Centre for Medium-Range Weather Forecasts

ENSO	El Niño Southern Oscillation
EOF	empiricle orthogonal function
EOS	Earth Observing System
ERB	Earth radiation budget
ERBE	Earth Radiation Budget Experiment
ERCD	efficient radius of cloud droplets
EROC	Ecological Rates of Change Programme
EUCREX	European Cloud-Radiation Experiment
EXPRESSO	Experiment for Regional Sources of Sinks and Oxidants
F	forcings
FAO	Food and Agriculture Organization
FCCC	Framework Convention in Climate Change
FGGE	First GARP Global Experiment
FIRE	First ISCCP Regional Experiment
FPAR	fraction of PAR absorbed by the green portion of the vegetation canopy
GACP	Global Aerosol Climatology Project
GAME	GEWEX-related Asian Monsoon Experiment
GARP	Global Atmospheric Research Programme
GCM	general circulation model
GCP	global carbon project
GCRP	US Global Change Research Programme
GDP	gross domestic product
GEF	Global Ecological Foundation
GEODAS	Geophysical Data System
GEWEX	Global Energy and Water Cycle Experiment
GFDL	Geophysical Fluid Dynamics Laboratory
GGD	Greenhouse Gas Dynamics (Programme)
GHG	greenhouse gas
GIMS	geographical information monitoring system
GIS	geographical information system
GISP2	Greenland Ice Sheet Project 2
GLOBE	Global Backscatter Experiment
GLOBEC	Global Ocean Ecosystems Dynamics Programme
GME	Global Meteorological Experiment
GMNSS	NSS global model
GNP	gross national product
GOES	Geostationary Observing Environment Satellite
GOME	Global Ozone Monitoring Experiment
GPP	gross primary productivity
GSM	global simulation model
GST	ground surface temperature
GTE	Global Tropical Experiment
GTOP	Global Tropospheric Ozone Project
GWP	global warming potential

GWP	global warming potential
GWP	gross World product
HALOE	Halogen Occultation Experiment
HARS	Human Dimensions of the Arctic System
HCFC	hydrochloroflourocarbons
HRPT	high-resolution picture transmission
HU	hydrological unit
IAOF	International Arctic Ocean Expedition
IASC	International Arctic Science Committee
ICGGM	International Centre on the Global Geoinformation Monitoring
ICLIPS	Integrated Assessment of Climate Protection Strategies
IFN	ice-forming nucleii
IGAC	International Global Atmospheric Chemistry (Project)
IGBP	International Geosphere–Biosphere Programme
IGOS-P	Integrated Global Observing Strategy-Partnership
IHBL	Internal heat boundary layer
IHDP	International Human Dimensions Programme of Global Environmental Change
IIASA	International Institute of Applied Systems Analysis
INDOEX	Indian Ocean Experiment
IPCC	Intergovernmental Panel on Climate Change
IR	infrared
ISCCP	International Satellite Cloud Climatology Project
ISS	International Space Station
ITCZ	intertropical convergence zone
ITEX	International Tundra Experiment
JGOFS	Joints Global Ocean Flux Study
KP	Kyoto Protocol
LAI	leaf area index
LAII	Land/Atmospher/ICE Interactions
LORE	Limb Ozone Retrieval Experiment
LRTAP	Long-Range Transport or Airborne Pollutants
LSM	land surface modles
LSR	leaf stomatal leaf stomatal resistance
LUAB	level of the upper atmosphere boundary
LWRF	long wave radiative forcing
MAGG	models of atmospheric general circulation
MBWB	model of the biospheric water balance
MC2	Mesoscale Compressible Community Model
MEA	Millenium Ecosystem Assessment
MEOS	Meteorological Earth-Observing System
MGC	minor gaseous component
MGCDC	model of the global carbon dioxide cycle
MGNC	model of the global nitrogen cycle
MGOC	model of the global oxygen cycle

MGPC	model of the global phosphorus cycle
MGSC	Model of the Global Sulphur Cycle
MISR	Multi-angle Imaging Spectroradiometer
MLS	Microwave Limb Sounder
MMB	mathematical model of the biosphere
MMIA	Methods and Models for Integrated Assessment
MODIS	Moderate-Resolution Imaging Spectroradiometer
MOZART	Model for Ozone and Related Chemical Tracers
MSA	methanosulphonic acid
MSC	Meteorological Service of Canada
MSD	mass deficiency
MUE	model for the upwelling ecosystem
N–S	nature–society
NAM	northern hemisphere annual mode
NAO	North Atlantic Oscillation
NARE	North Atlantic Regional Experiment
NBP	net biome productivity
NCAR	National Centre for Atmospheric Research
NEP	Net ecosystem productivity
NMHC	non methane hydrocarbons
NMHC	non-methane hydrocarbons
NOAA	National Oceanic and Atmospheric Administration
NOVI	normalized difference vegetation index
NPJS	northern polar jet stream
NPP	net primary productivity
NSF	US National Science Foundation
NSS	nature–society system
OAII	Ocean/Atmosphere/Ice Interactions
ODE	ozone depletion
ODS	ozone depleting substances
OSEM	Okhotsk Sea Ecosystem Model
OSWR	outgoing short-wave radiation
PAGES	Pilot Analysis of the Global Ecosystems
PAH	polycyclic aromatic compounds
PALE	Paleoclimates of Arctic Lakes and Estuaries
PAN	peroxyacetylinitrate
PAR	photosynthetic active radiation
PARCS	Paleoenvironmental Arctic Sciences
PBL	planetary boundary layer
PCB	polychlorinated biphenyls
PCEM	Peruvian Current Ecosystem Model
PCL	permissible concentration level
PCM	parallel climate model
PCRF	polar cloud radiative forcing
PDUS	primary Data User Station

PEM	Pacific Exploratory Mission
PGT	Pasquill–Gifford–Turner
PIC	products of incomplete combination
PNJS	polar night jet stream
PNJ	polar jet stream
POAM	Polar Ozone and Aerosol Measurement
POC	particulate organic carbon
POLARIS	Photochemistry of Ozone Loss in the Arctic Region in Summer (Mission)
POP	persistent organic pollutant
PSC	polar stratospheric clouds
PV	potential vorticity
QBO	quasi-biennial oscillation
RAISE	Russian–American Initiative on Shelf–Land Environments in the Arctic
RAS	Russian Arctic Seas
RF	radiative forcing
RNG	Renormalization Group
RS	remote sensing
SAFARI	The Southern Africa Fire-Atmosphere Research Initiative
SAGE	Stratospheric Aerosal and Gas Experiment
SAM	southern hemisphere annual mode
SAS	surface active substances
SAT	surface air temperature
SBI	Shelf–Basin Interactions
SBL	surface boundary layer
SBUV	solar backscatter ultraviolet
SCAR-B	smoke/sulphate clouds and radiation
SCATE	Sulphur Chemistry in the Antarctic Troposphere Experiment
SCE	snow cover extent
SCM	single-column model
SCW	sensitivity of cloud water
SEARCH	Study of Environmental Arctic Change
SESAME	Second Stratospheric Arctic and Mid-latitude Experiment
SHEBA	Surface Heat Budget of the Arctic Ocean
SHF	super high frequency
SiB2	simple biosphere model-2
SIMS	Synthesis, Integration, and Modelling Studies
SLP	sea level pressure
SMOW	Saint Mary Of the Woods
SOLSE	Satellite Ozone Limb Sensing Experiment
SOLVE	Sage-III Ozone Loss and Validation Experiment
SOP	Special Observing Periods
SPARC	Stratospheric Processes and their Role in Climate
SR	synthesized retrieval

SRB	surface radiation budget
SRVI	simple ratio vegetation index
SRXFA	synchrotron radiation X-ray fluorescence analysis
SSAPP	Simulation System for the Atmospheric Polution Physics
SSCRO	simulation system to control the regional ozonosphere
SSM/I	Special Sensor Microwave Images
SSMAE	Spatial Simulation Model of the Arctic Ecosystem
SST	sea surface temperature
SSW	sudden stratospheric warming
STARE	Southern Atlantic Region Experiment
SUCCESS	Subsonic Aircraft: Cloud and Contrail Effects Special Study
SVI	spectral vegetation index
SWRF	short-wave radiative forcing
TADAP	Transport and Deposition of Acidifying Pollutants
TARFOX	Tropospheric Aerosol Radiative Forcing Observational Experiment
TCO	terrestrial carbon observations
TES	Tropospheric Emission Spectrometer
THD	Technische Hochschule Darmstand
THESEO	Third European Stratosphere Experiment on Ozone
TO	tropospheric ozone
TOC	total ozone content
TOMS	Total Ozone Mapping Spectrometer
TOR	Tropospheric Ozone Research
TRACE-A	Transport and Atmospheric Chemistry near the Equator – Atlantic
TsuBiM	Tsukaba Biosphere Model
UNCED	UN Conference on Environment and Development
UNEP	UN Environmental Program
UNGASS	UN General Assembly Special Session
UQL	upper quasi-homogeneous layer
VOC	volatile organic compounds
WCRP	World Climate Research Programme
WSOC	water-soluable organic compounds
WSSD	World Summit on Sustainable Development

About the authors

Kirill Ya. Kondratyev received his PhD in 1956 at the University of Leningrad where he also became Professor in 1956 and was appointed Head of the Department of Atmospheric Physics in 1958. From 1978–1982 he was Head of the Department of the main Geophysical Observatory and from 1982–1991 Head of the Institute for Lake Research – both in Liningrad. Since 1992 he has been Councillor of the Russian Academy of Sciences. Kirill has published more than 100 books in the fields of atmospheric physics and chemistry, remote sensing, planetary atmospheres, and global change.

Vladimir F. Krapivin received his PhD in geophysics from the Moscow Institute of Oceanology in 1973. He became Professor of Radiophysics in 1987 and Head of the Applied Mathematics Department at the Moscow Institute of Radioengineering and Electronics in 1972. Since 1996 he has also been Director of the Institute of Ecoinformatics. He is a full member of the Russian Academy of Natural Sciences, and has specialized in investigating global environmental change by the application of modelling technology. He has published 12 books in the fields of ecoinformatics, theory of games, and global modelling.

Victor P. Savinikh, PhD, professor, space pilot, and rector of the Moscow State University of Geodesy and Cartography has performed three spacewalks. His first space flight was realized in 1981 on the spacestation Salut-6 lasting 75 days. In 1985 he took part in a space mission lasting 169 days on the space station Salut-7. His third space expedition was in 1988 on *Mir* as a member of the Soviet–Bulgar space mission lasting 10 days. He is the author of many books and papers on the problems of remote sensing and global environmental change. His scientific works involve satellite remote sensing studies and data processing.

Costas A. Varotsos received his BSc in Physics at Athens University in 1980, and PhD in Atmospheric Physics in 1984. He was appointed Assistant Professor in 1989 at the Laboratory of Meteorology in the Physics Department of the Athens University, where he also established the Laboratory of Upper Air. In 1999 he became Associate Professor of the Department of Applied Physics at Athens University. He has had published more than 250 papers and 15 books in the fields of atmospheric physics, atmospheric chemistry, and environmental change.

Summary

The basic global problems of the nature–society system (NSS) dynamics have been considered as well as analysis of the significance of ecological, socio-economic, and political factors in the formation of global changes to the environment. The problems of modelling global ecodynamics have been discussed and a new concept of the synthesis of geoinformation monitoring systems has been suggested based on the technology of evolutionary eco-informatics and open systems. Emphasis has been given to the problem of estimating the greenhaouse effect considering most significant natural and anthropogenic processes in the environment. Within the global model, a spatial model of the biospheric carbon budget is suggested which describes the fluxes between the atmosphere, surface biocenoses, and ocean ecosystems. Based on simulation experiments, an assessment has been made of the role of various parts of the biosphere in the assimilation of carbon dioxide from the atmosphere, and on this basis the concentration of carbon dioxide in the atmosphere has been predicted. The problem of transition of the NSS to long-range sustainable development has been studied.

1

Global ecodynamics

1.1 TRENDS IN CIVILIZATION DEVELOPMENT

1.1.1 Introduction

A globalization of man's impact on the environment with the continuing growth of population and a great uncertainty of the prospects of civilization development have raised numerous principal problems of economic, political, and social character. One of them is an assessment and prediction of the state of the global carbon cycle emphasizing the role of anthropogenic processes. Clearly, this problem cannot be selected and discussed separately from the other problems of the society–nature (NS) interaction. Nevertheless, to substantiate a set of relevant problems, it is necessary from the broad spectrum of natural–anthropogenic processes to select the key aspects, the knowledge of which will enable one to answer the principal questions surrounding the sustainable development of the global nature–society system (NSS). Though the notion of sustainable development has not been clearly defined, the term "sustainable development" is widely used in scientific literature. Various discussions on the origin and meaning of this term have long taken place among specialists in various fields. They include the widely known decisions made at the international conferences in Stockholm (1972), in Rio de Janeiro (1992), and in Johannesburg (2002), as well as an accomplishment of several international programmes, such as the International Geosphere–Biosphere Programme (IGBP), International Human Dimensions Programme of Global Change (IHDP), the World Climate Research Programme (WCRP), International Biodiversity Programme DIVERSITAS, etc. (Braswell *et al.*, 1996).

The principal conclusion of the July 2001 Amsterdam Conference "Challenges to Changing Earth" is that: "the Earth System behaves as a single, self-regulating system of physical, chemical, biological and human components. The interactions and feedbacks between the component parts are complex and exhibit multi-scale temporal and spatial variability." In reality, diverse important impacts of human activities on a global scale can be traced both on the land and in the World Ocean. In some respects, the anthropogenic changes to the environment have become comparable or have even exceeded the limits of the natural environmental changes. "Human-driven changes cause multiple effects that cascade through the Earth System in complex ways. These effects interact with each other and with local- and regional-scale changes in multidimensional patterns that are difficult to understand and even more difficult to predict." Therefore, human activities have the potential to switch the Earth System into irreversible modes of operation that may prove unfavourable and even impossible for other life.

> In terms of some key environmental parameters, the Earth System has moved well outside the range of natural variability exhibited over the last half million years at least. The nature of changes now occurring simultaneously in the Earth System, their magnitudes and rates of change are unprecedented. The Earth is currently operating in a non-analogue state.

These extracts from the decisions of the Amsterdam Conference suggest that "an ethical framework for global stewardship and strategies for Earth System management are urgently needed. The accelerating human transformation of the Earth's environment is not sustainable." This problem can be solved by developing new scientific directions in global ecology, for example, ecoinformatics (Arsky *et al.*, 1990, 1992; Krapivin and Kondratyev, 2002; Krapivin and Potapov, 2002; Savinykh, 2001; Savinykh and Tsvetkov, 2000, 2001, 2002). No doubt, this scientific direction should comprise a wide spectrum of natural and social sciences, creating an efficient mechanism for using available methods and technologies and ensuring their goal-directed development. A certain basis for this unification has already been created (Gorshkov, 1990; Kondratyev, 1990, 1998b; Gorshkov *et al.*, 2000; Gerstengarbe, 2002).

Following the work of Steffen and Tyson (2001), we shall analyse some formulations aimed at defining the Earth System as a complex system of interacting physical, chemical, and biological components connected with human activity, which determines the processes of transportation and transformation of substances and energy, and thus keeps the planet habitable.

> The Earth is a system that life itself helps to control. Biological processes interact strongly with physical and chemical processes to create the planetary environment, but biology plays a much stronger role than previously thought in keeping Earth's environment within habitable limits.

It should be mentioned that this thesis has undoubtedly a fundamental meaning, but in this context it would be rather appropriate to refer to the conception of the biotically regulated environment extensively elaborated by Gorshkov (1990) and Gorshkov *et al.* (2000).

> Global change is much more than climate change. It is real, it is happening now and it is accelerating. Human activities are significantly influencing the functioning of the Earth System in many ways; anthropogenic changes are clearly identifiable beyond natural variability and are equal to some of the great forces of nature in their extent and impact.

We have to emphasize that in the contemporary world, when the scientific discussion on the extent of anthropogenic pressure on climate has been transformed into a political discussion and reached the level of heads of state, it should be pointed out that global climate change deserves much keener attention in the context of the Kyoto Protocol, which has been dealt with in a number of publications by Russian scientists (Kondratyev and Demirchian, 2001b; Kondratyev *et al.*, 2001a,b, 2003a,b; Kondratyev, 2002c). However, we agree with Steffen and Tyson (2001) in the sense that climate is not the first priority in the totality of global change problems. It is of key importance that:

- Over a few generations humankind has been exhausting fossil fuel reserves that were generated over several hundred million years.
- Nearly 50% of the land surface has been transformed with significant negative consequences for biodiversity, nutrient cycling, soil structure and biology, and climate.
- More nitrogen is now fixed synthetically (anthropogenically) and applied as fertilizers in agriculture than is fixed naturally in all terrestrial ecosystems.
- More than half of all the accessible freshwater is used by humankind, and underground water resources are being depleted rapidly.
- The concentrations of greenhouse gases (GHGs), including CO_2 and CH_4, have substantially increased in the atmosphere.
- Coastal and marine habitats are being dramatically altered; the areas of wetlands have almost halved.
- About 22% of marine fisheries are overexploited, and 44% more are at the limit of exploitation.
- Both on land and in the World Ocean intensive processes of biodiversity reduction take place.
- The human activity causes numerous and intensive effects on the Earth System in complex ways. Therefore global changes cannot be understood in terms of a simple cause–effect paradigm. Effects of human activities propagate in the Earth System interacting with each other and in the processes of various scales.
- The Earth's dynamics is characterized by critical thresholds and abrupt changes. Human activity can unintentionally intensify changes pregnant with catastrophic consequences for the Earth System. Apparently, such changes could be avoided in the case of depletion of the stratospheric ozone layer. The Earth System has operated in different quasi-stable states, with abrupt changes occurring between them over the last half million years. Human activities clearly have the potential to switch the Earth System to alternative modes of operation that may prove irreversible.

Though the last two points are very important, we must keep in mind that until now there are only preliminary quantitative assessments and only fragmentary data exist, there is not enough information for concrete judgement about the potential changes. This is true, for instance, in the case of the impact on the global ozone content in the stratosphere (Kondratyev and Varotsos, 2000). In this context, the problem of the anthropogenic impact on the global biogeochemical cycles of carbon, sulphur, nitrogen, and other chemical compounds deserves serious attention (Ledley *et al.*, 1999; Alverson, 2000; Kondratyev, 2000b; Irion, 2001; Kondratyev and Krapivin, 2001d; Zavarzin and Kolotilova, 2001; Krapivin and Kondratyev, 2002).

> The Earth is currently operating in a non-analog state. In terms of key environmental parameters, the Earth System has recently moved well outside the range of the natural variability exhibited over at least the last half million years. The nature of changes now occurring simultaneously in the Earth System, their magnitudes and rates of change are unprecedented.

Such general assessments are important for developing an efficient technology to assess global environmental changes (Kontorovich *et al.*, 1999; Zerchaninova and Potapov, 2001). In the context of perspectives of the contemporary civilization development a key factor is undoubtedly the continuing global population increase: during the second half of the 20th century the global population doubled, at the same time the grain crops production tripled, the energy consumption increased four-fold and economic activity increased five-fold. Gorshkov (1990) has clearly demonstrated that *Homo sapiens* has long moved out of his ecological niche, which has brought forth the first signs of a developing ecological catastrophe.

Taking into account these conceptual circumstances, Steffen and Tyson (2001) have formulated two important conclusions: that society should respond to the emerging ecological threat and work out respective scientific developments aimed at an adequate understanding of the processes that govern the variability of the Earth System:

- Ethics of global stewardship and strategies for Earth System management are urgently needed. The inadvertent anthropogenic transformation of the planetary environment is, in effect, already a form of management, or rather dangerous mismanagement. This transformation is not sustainable. Therefore the business-as-usual way of dealing with the Earth has to be replaced as soon as possible by planned strategies of an adequate management.
- A new system of global environmental science is gradually emerging. The largely independent efforts within the framework of various international programmes form the basis for the Earth System science capable of solving these problems. This new science will employ innovative integration methodologies of its organization into a global system with transnational infrastructures, and a continuing dialogue with stakeholders around the world.

Analysis of the published suggestions, opinions, and decisions of international governmental and non-governmental organizations concerning the strategy of studies and development of environmental sciences in conditions of increasing anthropogenic forcings suggest the conclusion that there is no complete and objective list of priorities regarding the problems of global changes. This means that humankind has no scientifically grounded strategy of interaction with the environment.

1.1.2 Contemporary global ecodynamics

The accumulated data on global environmental changes enable one to formulate a general pattern of these changes and characterize possible trends for the decades to come (Losev, 2001; Räisänen, 2001; Prentice *et al.*, 2001; Potter *et al.*, 2001; Potosnak *et al.*, 1999; Philander, 2001; Peng *et al.*, 2002; Pennise *et al.*, 2001; Kondratyev *et al.*, 2003a–c; Kondratyev and Krapivin, 2003a):

- The area of natural land ecosystems shrinks at a rate of 0.5–1.0%/yr. By the beginning of the 1990s about 37% of these ecosystems remained. This trend will lead to the total abolition of natural ecosystems.
- From the estimates for 1985 the rate of human consumption of pure primary biological production constituted 40% on land and 25% over the whole biosphere. In the future these estimates will reach 80–85% and 50–60%, respectively.
- Between 1972 and 1995 the GHGs concentrations grew at a rate of tenths of percent to several percent annually. Concentrations of CO_2 and CH_4 are expected to grow at the expense of the increasing rate of biota destruction.
- An annual 1–2% depletion of the ozone layer, the growth of the ozone hole in the Antarctic, and the growth of the areas of other ozone holes. This trend will remain even with the complete cessation of chlorofluorocarbon (CFC) emissions.
- Shrinking areas of forests, especially tropical forests. During 1990–1995 the forest areas shrunk at a rate of 13 Mha/yr, and the ratio of "aforestation/forest cutting" was 1:10. By 2030 the area of the moderate-belt forests will decrease, and that of the tropical forests will halve.
- An areal increase of the arid areas consisting 40% of land, the growth of technogenic desertification caused by the broken moisture cycle, forest cutting, and soil contamination. Reduction of agricultural lands per capita, fertility reduction, growing acidity, and salinity of soil.
- During the last decades of the 20th century the level of the World Ocean has risen by 1–2 mm/yr. This rate may increase to 7 mm/yr.
- Natural disasters and technogenic accidents in the late 20th century were characterized by an annual increase of 5–7% with an increasing in damage caused by such events of 5–10% and growth of the amount of human victims of 6–12%. Preservation and intensification of these trends is a natural result of global processes.

- Biological species in the second-half of the 20th century disappeared 100–1000 times faster than at any time in the past. The trend may increase as the biosphere is further destroyed.
- There was a qualitative depletion of land water caused by increasing volumes of sewage, a growing number of the point and areal sources of pollution, and a broadening spectrum of pollutants and their increasing concentration. With this trend preserved, by 2030 more than 60% of the population will suffer a water shortage.
- There is a persistent growth of the mass and number of pollutants accumulated in media and organisms. A migration of pollutants, especially radionuclides in food chains, intensifies.
- Worsening quality of life, growing number of diseases connected with the destruction of man's ecological niche, environmental pollution (genetic included), and an appearance of new diseases. Poverty, food shortage, high infant mortality, inadequate provision of the developing countries with pure water, high accident rate, growing consumption of medicine, an increase of allergic diseases in the developed countries, and a decrease of immune status. In the 21st century these trends will remain with an increasing number of infectious diseases and the appearance of new ones.
- The global distribution, including humans, of supertoxicants through the trophic chains, resulting in damaged human endocrine systems worsening the reproduction system, the brain, and other vital organs. All this will result in the growing number of childless couples.
- Artificial introduction and accidental invasion of alien species into the ecosystems, damaging the ecosystems. An increased transport of pests and diseases of plants, animals, and humans, leading to a reduction of biodiversity. These trends will remain in the 21st century.
- The changes observed in the World Ocean: destruction of reefs, reduction of the mangrove ecosystems, depletion of fish supplies as a result of intensive fishing, reduction of the whale herds, pollution of inland and coastal waters, and "red tides".

These trends suggest two principal conclusions: (1) in the second half of the 20th century there was a continued worsening of environmental characteristics; (2) relevant expected trends will not only remain negative in the future but may even worsen. Thus the trends of unsustainable development will intensify. To confirm this, we shall give some quantitative estimates.

Table 1.1 illustrates the dynamics of the global population, the principal features of which are the continuing growth of an absolute size of population, with its rate decreasing. Of course, the mean tendencies mask the strongest regional differences. Practically, in all industrial countries the size of the population is either slowly increasing or decreasing. The USA is an exception, where the increment of population is 1%/yr partly due to immigration constituting one-third of the increase. 95% of the global increment of population in 2000 fell on the developing countries including Asia (57% with an absolute increment of 45 mn),

Table 1.1. World population (1950–2000).

Year	G	ΔG	Year	G	ΔG
1950	2.555	38	1984	4.774	81
1955	2.780	53	1985	4.855	83
1960	3.039	41	1986	4.938	86
1965	3.346	70	1987	5.024	87
1970	3.708	78	1988	5.110	86
1971	3.785	77	1989	5.196	87
1972	3.862	77	1990	5.284	83
1973	3.939	76	1991	5.367	83
1974	4.015	74	1992	5.450	81
1975	4.088	72	1993	5.531	80
1976	4.160	73	1994	5.611	80
1977	4.233	72	1995	5.691	78
1978	4.305	75	1996	5.769	78
1979	4.381	76	1997	5.847	78
1980	4.457	76	1998	5.925	78
1981	4.533	80	1999	6.003	78
1982	4.613	81	2000	~6.080	77
1983	4.694	80			

Note: $G =$ size of population (bn); $\Delta G =$ population increment (mn/yr).

Africa (23%), Latin America (9%), and the Middle East (5%). About half the global increment in 2000 fell on India, China, Pakistan, Nigeria, Bangladesh, and Indonesia. A number of factors favour a reduction of birth rate and the rate of population growth, including: improvement of the economic situation and the public health system in many countries; and progress in women's education in line with their increasing status (easier access to contraceptives, etc). So, for instance, in Iran the rate of population growth decreased from 3.2% in 1986 to 0.8% in 2000. The decrease of birth rate has been caused by a negative factor, such as the spread of AIDS (about 3 million people died in 2000 of AIDS, and the fatal outcomes of all diseases totaled 22 million people).

Naturally, during the second half of the 20th century the global gross domestic product (GDP) continued growing (though slowing down) (Table 1.2). Relative growth in the scales of the global economy in 2000 constituted 4.7% and exceeded the 1999 level (3.4%). The global GWP reached 43 billion dollars with the USA providing an average GWP of $7102 per capita. This progress has been achieved due to a successful development of the economy in the USA and Western Europe, the recovery of the economy in Asia after the 1997 financial crisis, and the economy in Latin America after the 1998 crisis, as well as a marked improvement in the situation of the countries with transition economies. Note that the growth of the economy in China in 2000 constituted 7.5% and was the most significant in Asia. Figures for India, Pakistan, and Bangladesh are equal, respectively, to 6.7, 5.6, and 5%, which is

Table 1.2. Gross world product (GWP) (1950–2000).

Year	U	U_G	Year	U	U_G
1950	6.4	2,502	1984	26.2	5,485
1955	8.1	2,921	1985	27.1	5,582
1960	10.0	3,306	1986	28.0	5,673
1965	12.8	3,822	1987	29.0	5,778
1970	16.3	4,407	1988	30.3	5,938
1971	17.1	4,505	1989	31.2	5,996
1972	17.8	4,599	1990	31.9	6,031
1973	19.0	4,810	1991	32.0	5,957
1974	19.4	4,829	1992	32.4	5,941
1975	19.7	4,816	1993	33.2	6,000
1976	20.7	4,977	1994	34.5	6,150
1977	21.5	5,083	1995	35.8	6,295
1978	22.4	5,210	1996	37.3	6,475
1979	23.1	5,282	1997	39.0	6,666
1980	23.6	5,306	1998	39.9	6,732
1981	24.2	5,329	1999	41.2	6,871
1982	24.4	5,280	2000	~43.2	~7,102
1983	25.1	5,341			

Note: U = net GWP (10^9 US\$, 1999); U_G = per capita GWP (US\$, 1999).

especially important for the region with a population exceeding 1.2 billion people, where the per capita income is about 1 dollar/day or even less.

Fossil fuel consumption is a significant indicator of the development of both the economy and technology (Table 1.3). As seen from Table 1.3, the year 2000 was the second consecutive year of decrease in the global level of total fossil fuel consumption after a long period of its growth by a factor of 3.5 starting from 1950. Now the contribution of the fossil fuel consumption in the commercial production of energy is about 90%, the share of coal being 25%. This mean-global share is gradually decreasing, but in the USA the consumption of coal (~25% of the global use) increased in 2000 by 1.6% due to the development of coal powered electricity stations. Contrary to this, the coal consumption in China (which is also 25% of the global use) decreased by 3.5% (from 1996 the decrease has reached 27%). In India, the third largest consumer of coal (7%), the level of consumption in 2000 increased by 5.4%.

Global oil consumption (its share in energy production is 41%) has increased by 1.1%. In the USA (the leading petroleum user at 26%), the increase is only by 0.1%, whereas in the Asian countries of the Pacific region (representing 27% of the global oil consumption) the use of oil has increased by 2.6%. The respective figures for Western Europe are 22% and 0.2%. As for natural gas (supplying 24% of the commercial energy production), the level of its consumption has increased by 2.1%. In the USA (where 27% of the global natural gas is consumed), the scale of

Table 1.3. Global fossil fuel consumption (1950–2000) (10^6 t of oil equivalent).

Year	C	O	G	Year	C	O	G
1950	1,043	436	187	1983	19,181	2,632	1,463
1955	1,234	753	290	1984	2,001	2,670	1,577
1960	1,500	1,020	444	1985	2,100	2,654	1,640
1965	1,533	1,485	661	1986	2,135	2,743	1,653
1970	1,635	2,189	1,022	1987	2,197	2,789	1,739
1971	1,632	2,313	1,097	1988	2,242	2,872	1,928
1972	1,629	2,487	1,150	1989	2,272	2,921	1,904
1973	1,668	2,690	1,184	1990	2,244	2,968	1,938
1974	1,691	2,650	1,212	1991	2,189	2,967	1,970
1975	1,709	2,616	1,199	1992	2,179	2,998	1,972
1976	1,787	2,781	1,261	1993	2,171	2,969	2,012
1977	1,835	2,870	1,283	1994	2,186	3,027	2,019
1978	1,870	2,962	1,334	1995	2,218	3,069	2,075
1979	1,991	2,998	1,381	1996	2,298	3,150	2,170
1980	2,021	2,873	1,406	1997	2,285	3,224	2,155
1981	1,816	2,781	1,448	1998	2,243	3,241	2,181
1982	1,878	2,656	1,448	1999	2,034	3,332	2,277
				2000	~2,034	~3,332	~2,277

Note: C = coal; O = oil; G = gas.

consumption has increased by 2.4%. The consumption of natural gas in the Baltic countries has also strongly increased: Lithuania – 29%; Estonia – 30%; and Latvia – 45%. Leaders of the 16% growth of natural gas consumption in Asia and West Europe turned out to be South Korea and Spain. The last several years have been characterized by high prices of oil, with a maximum (US$34.5 for a barrel) in 1985, which has drawn attention to analysis of prospects for oil and gas extraction on the Arctic shelf of Russia and Alaska. Both of these are problematic since the Alaskan resources are comparatively small, and significant investments would be required for Russia.

According to the International Energy Agency, the total fossil fuel consumption may increase by 57% (2%/yr) by the year 2020, preserving a 90% share of consumption in total energy production. The supposed global trend of coal consumption increase will constitute 1.7%/yr, with two-thirds of this increase due to India and China. Petroleum will remain the main source of energy (increasing by 1.9%/yr) and its share in primary energy production will reach 40%, but the highest rate of consumption increase (2.7%/yr) is characteristic of natural gas (mainly due to the gas operating electric power stations).

There is a prospect for changes of energy sources, though the available trends in this direction are characterized by low energy production rates. Nevertheless, the rate of increasing generation of energy by nuclear power stations in 2001 constituted 0.4%, and generally, this increase reached 7% during the last decade of the 20th

Table 1.4. Global CO_2 emissions due to fossil fuel (1950–2000) and its atmospheric concentration (1960–2000).

Year	H_1	C_A	Year	H_1	C_A
1950	1,612	–	1984	5,098	344.2
1955	2,013	–	1985	5,271	345.7
1960	2,535	316.7	1986	5,453	34.0
1965	3,087	319.9	1987	5,574	348.7
1970	3,997	325.5	1988	5,789	351.3
1971	4,143	326.2	1989	5,892	352.7
1972	4,305	327.3	1990	5,931	354.0
1973	4,538	329.5	1991	6,020	355.5
1974	4,545	330.1	1992	5,879	356.4
1975	4,518	331.0	1993	5,861	357.0
1976	4,776	332.0	1994	6,013	358.9
1977	4,910	333.7	1995	6,190	360.9
1978	4,950	335.3	1996	6,315	362.6
1979	5,229	336.7	1997	6,395	363.8
1980	5,155	338.5	1998	6,381	366.6
1981	4,984	339.8	1999	6,340	368.3
1982	4,947	341.0	2000	~6,299	~369.4
1983	4,933	342.6			

Note: $H_1 = CO_2$ emissions into the atmosphere (10^6 t C); $C_A = CO_2$ concentration in the atmosphere (ppm).

century. The energy generation from wind power and use of photogalvanic solar arrays grows more rapidly.

Data of Table 1.4 reflect the dynamics of CO_2 emissions into the atmosphere and its increase in atmospheric concentration. Starting from 1950, the atmosphere has gained an amount of CO_2 equivalent to 217 Gt C with the annual four-fold increase of emissions. An important fact is that the fossil fuel use per unit production, calculated with CO_2 emissions taken into account, decreases: in 2000 this decrease constituted 3.6% (this figure is equivalent to 148 t C per 1 million US$ GWP), and during 50 years it reached 41%. Contrary to the Kyoto Protocol, CO_2 emissions in the western industrial countries increased during the period from 1990 by 9.2%, and in the USA – by 13% on average. The respective contribution of the developing countries has reached 22.8%. According to the Intergovernmental Panel on Climate Change (IPCC), by 2020 the annual global CO_2 emissions may reach 9–12.1 Gt C and by 2050 they will be within 11.2–23.1 Gt C. Therefore the problem of global climate change due to CO_2 emissions is discussed in detail and studied in this book. As far as possible, a complete analysis of the biogeochemical CO_2 cycle is given in the chapters below.

The most important component of the economic development is grain production (Table 1.5), which has recently started decreasing from 1869 Mt in 1999 to 1840 Mt in 2000. The corn crop in 2000 turned out to be 2% below the maximum

Table 1.5. Global grain production (1950–2000).

Year	P	P_G	Year	P	P_G
1950	631	247	1984	1,632	342
1955	759	273	1985	1,647	339
1960	824	271	1986	1,665	337
1965	905	270	1987	1,598	318
1970	1,079	291	1988	1,549	303
1971	1,177	311	1989	1,671	322
1972	1,141	295	1990	1,769	335
1973	1,253	318	1991	1,708	318
1974	1,204	300	1992	1,790	328
1975	1,237	303	1993	1,713	310
1976	1,342	323	1994	1,760	314
1977	1,319	312	1995	1,713	301
1978	1,445	336	1996	1,872	324
1979	1,411	322	1997	1,881	322
1980	1,430	321	1998	1,872	316
1981	1,482	327	1999	1,869	311
1982	1,533	332	2000	~1,840	~303
1983	1,469	313			

Note: P = net grain production (10^6 t), P_G = per capita consumption (kg).

grain production reached in 1997 (1881 Mt). The main reason of this decrease was a reduced corn crop in China from 391 Mt in 1998 to 353 Mt in 2000 (i.e., by about 10%). This has resulted from low grain prices that discouraged the farmers from increasing grain production, as well as drought and water deficiency in the northern region of the country. Grain production in the USA, second to China, increased during the same period from 332 to 343 Mt, mainly due to the increased yield of maize.

The mean global grain yield decreased slightly in 2000 (2.75 t/ha) compared to 1999 (2.77 t/ha), when it reached its multiyear maximum. During the past years there were only slight fluctuations about an average of 2.75 t/ha, but in 2000 the consumption per capita decreased to 303 kg, which was 13% below the maximum level of grain consumption per capita in 1984. This decrease was mainly due to the Western European nations, the former USSR, and Africa. In 2000 and during the previous two years the maize production (588 Mt) for the first time exceeded the level of wheat production (580 Mt), with the production of rice constituting 401 Mt. About 43% of the global maize yield has been spent by the USA on stock breeding. The league table for wheat production is as follows: China, India, and USA. China also dominates in rice production. The USA exports more than 75% of its maize. The main exporter of wheat is also the USA, followed by France, Canada, and Australia. The leading exporter of rice is China, followed by Thailand, the USA, and Vietnam. Until recently, Japan had been the main importer of wheat, but during the last years has been overtaken by Brazil, Iran, and Egypt.

Table 1.6. Foreign debt of the countries of FSU and Eastern Europe (1970–1999).

Year	*V*	Year	*V*
1970	0.26	1985	1.47
1971	0.29	1986	2.57
1972	0.32	1987	2.73
1973	0.37	1988	1.68
1974	0.42	1989	1.70
1975	0.51	1990	1.77
1976	0.59	1991	1.80
1977	0.74	1992	1.85
1978	0.86	1993	1.97
1979	0.98	1994	2.15
1980	1.07	1995	2.30
1981	1.18	1996	2.35
1982	1.23	1997	2.40
1983	1.33	1998	2.61
1984	1.35	1999	2.57

Note: V = sum of debt (10^9 US$, 1999).

One of the key indicators of the level of socio-economic development is the amount of foreign debt. Table 1.6 totals the data for the developing and Eastern European countries, as well as for the former USSR republics, showing a continuously growing debt, at a maximum in 1998, which has stabilized somewhat in 1999 (US$2.57 bn dollars USA) due to inflation. Serious changes in the financial trends in Brazil, Indonesia, Russia, and South Korea after 1996 are largely explained by the financial crises in 1997 and 1998.

To understand and predict the global changes in the environment, of great assistance and importance is the data on human health. In Table 1.7 a sample is given of the most demonstrative trend in AIDS propagation. Only this disease is able to reduce the size of the able-bodied population of the planet, especially as it concerns the nations of Africa and the Caribbean Basin. Expenses on health services reaching approximately US$3 bn vividly demonstrates the urgency of the problem. These expenses have been distributed rather unevenly. The nations with a low and average income and with 84% of global population have only got 11% of the global expenses on health services, despite the fact that 93% of the damage caused by diseases falls within these countries. If the expenses on health services in the world average 5% of the GWP, in the USA they reach 13.7%, and in Somalia – only 1.5%. Respectively, the annual expenses per capita range from US$50 in poor countries to US$4100 in the USA.

Another of the important indicators of the socio-economic trends is the level of militarization. Table 1.8 characterizes the dynamics of nuclear armaments as an indicator of technical progress. Here the cause and effect feedbacks of the world NSS are directly regulated by political aspects.

Table 1.7. Global data on the number of HIV-infected people and deaths resulting from AIDS (1980–2000).

Year	G_C	M_G	Year	G_C	M_G
1980	0.1	0.0	1990	10.0	1.7
1981	0.3	0.0	1991	12.8	2.4
1982	0.7	0.0	1992	16.1	3.3
1983	1.2	0.0	1993	20.1	4.7
1984	1.7	0.1	1994	24.5	6.2
1985	2.4	0.2	1995	29.8	8.2
1986	3.4	0.3	1996	35.3	10.6
1987	4.5	0.5	1997	40.9	13.2
1988	5.9	0.8	1998	46.6	15.9
1989	7.8	1.2	1999	52.6	18.8
			2000	~57.9	~21.8

Note: G_C = HIV-infected population (10^6); M_G = deaths resulting from AIDS (10^6).

Table 1.8. World nuclear arsenal (1945–2000).

Year	Y	Year	Y
1945	2	1983	66,979
1950	303	1984	67,585
1955	2,490	1986	69,478
1960	20,368	1987	68,835
1965	39,047	1988	67,041
1970	39,691	1989	63,645
1971	41,365	1990	60,236
1972	44,020	1991	55,772
1973	47,741	1992	52,972
1974	50,840	1993	50,008
1975	52,323	1994	46,542
1976	53,252	1995	43,200
1977	54,978	1996	40,100
1978	56,805	1997	37,535
1979	59,120	1998	34,535
1980	61,480	1999	31,960
1981	63,054	2000	~31,535
1982	64,769		

Note: Y = number of nuclear warheads.

To summarize the results of global civilization development during the second half of the 20th century, it is necessary first of all to underline that this was a period of unprecedented rapid change in global population, the biosphere, global economy, and society as a whole. The world has become more rich economically but poorer from an ecological point of view. The following most important trends have been observed during this period:

- The global population has exceeded 6 bn, increasing by 3.5 mn during the last half century (i.e., more than doubling). Most of the growth has taken place in the developing countries, which were already overpopulated. The growth was especially rapid in the urban areas where the population increased almost four-fold. A remarkable feature of the demographic dynamics in the industrial countries is an increase in the share of old people.
- The growth of the world economy has exceeded (approximately seven-fold during 50 years) the rate of population growth, which provided (on average) a significant increase in living standards. However, at the same time 1.2 bn people still live in severe poverty and 1.1 bn do not have access to pure drinking water.
- The global grain yield has nearly tripled since 1950, which allowed people to enrich their diet, but led to dangerous ecological consequences: lowering the level of subsoil waters and intensifying natural water pollution due to excessive overuse of fertilizers and pesticides.
- Only a small part of primary forests (boreal and tropical) have been preserved on the planet, more than one-half of wetlands and over one-quarter of coral reefs have been lost. This was accompanied by considerable damage to biodiversity and, seriously, the mechanisms for the biotic regulation of the environment have been broken (Gorshkov *et al.*, 2000; Gorshkov, 1995).
- The anthropogenic impacts on the global climate and ozone layer are of great concern, though there are more uncertainties than established trends.

It is especially important that socio-economic and ecological dynamics during the second half of the 20th century was that, as a rule, the observed trends were unpredictable. As the world becomes more and more complicated and inhomogeneous, environmental prediction becomes more difficult but necessary, since an adequate planning aimed at minimizing the risk and maximizing favourable conditions for life depends on its accuracy.

1.1.3 Sustainable development

The paradox of the contemporary situation is that despite the unprecedented scale of discussion on sustainable development, the world is continuing to follow the path of unsustainable development. The definition of the term "sustainable development" proposed by the International Commission on Environment and Sustainable Development (Bruntland Commission) as, "a development that provides the needs of the present generations of mankind, and at the same time, gives the opportunity to the coming generations to satisfy their needs as well," is too general and non-constructive. This term should, of course, reflect the key role of biospheric dynamics, which has been mentioned in the works by Gorshkov (1990, 1995),

Kondratyev (1998b, 1999c), Gorshkov *et al.* (2000), Losev (2001), and Strich (2002). In these publications it has been shown that the communities of the organisms within the biocenosis and the whole biosphere regulate the chemistry of the most important media – atmosphere, hydrosphere, and soil. A necessity has also been emphasized to determine the limits of permissible disturbances in these media. Losev (2001) points to the following four problems to be solved to move on to sustainable development.

The first problem is a clash of interests between our civilization and nature. The modern civilization is the result of 10 thousand years of spontaneous human development, when people could develop without any restrictions defined by the biosphere. By the beginning of the 20th century, as a result of the continuing expansion, humanity had gone beyond this limit, with subsequent global environmental changes. As a result, the 20th century has become the century affected by a severe ecological crisis, which has not been realized yet by many people and politicians, unlike the global nature-protection infrastructure (created over the last 25 years), environmental protection expenses, and application of resource-preserving technologies.

The ecological crisis is a signal that humanity cannot continue at random its development but must coordinate its history with the laws of the biosphere from which man is inseparable. First of all, the development should be coordinated with the law of energy flow distribution in the biosphere that determines the limits of the corridor of ecologically safe civilization development.

The signals from the anthropogenically destroyed biosphere either do not yet affect the majority of the people with enough force or do not connect these signals with the rapidly developing environmental crisis. At the same time, as long as the virgin ecosystems exist on our planet, there is hope that irreversible changes have not yet begun. These signals should be adequately perceived as a guide for action. None of the national programmes or strategies on sustainable development contain an adequate response to such signals because they have no theoretical scientific basis. Some ideas on the development of such a basis and technology to perceive signals from the biosphere have been discussed by Kondratyev *et al.* (2000).

The second problem is economical. It is the development of economy that has led to the destruction of the natural mechanisms of environmental stabilization and to the development of the present-day ecological crisis. The ecological costs of civilization have begun to convert into economical, social, and demographic ones. They have reduced the efficiency of the world economy and its investing activity, have slowed down the rate of growth, stimulated poverty, broken genetic programmes, undermined human health, etc. In its interaction with nature, humankind still remains an appropriator, not only using natural resources free of charge but also thoughtlessly destroying the biosphere – the basis of life itself. In the 21st century humankind will have to radically change its attitude towards the organization of production cycles and excessively extracting resources free of charge, producing unlimited waste.

The third problem of sustainable development is the social sphere. The existing social–economic system not only has led civilization to a clash with nature but has also not solved many social problems. The gap between poor and rich people is widening and the difference between the living standards in poor and rich countries is growing. It means that the global socio-economic system does not correspond to the requirements for the sustainable development of humankind.

The fourth problem is the demographic crisis, leading to a decrease in the areas of arable lands and in the volume of food production per capita. It intensifies the ecological crisis, which, in turn, deteriorates an important social indicator – human health. The social inequality of people destabilizes the situation in the developing countries, causes conflicts within and migration of the population, gives rise to extremism and terrorism, and finally evokes global scale conflicts.

Thus the modern civilization is in the state of system crisis, which has been repeatedly discussed at a number of international conferences. The way out of this crisis is in the transition to sustainable development, requiring a real change of the vector of civilization development. This crisis is somehow connected with the problem of growing atmospheric CO_2 concentration, which worries both the economists and politicians. It is important to predict consequences of global climate change to work out the strategies of regional and global economic and political initiatives in the future.

In this regard, US scientists are accomplishing a complex programme to study the global carbon cycle. The main goals of this programme are as follows:

- To develop techniques to measure, monitor, and model components of carbon cycle in nature in order to synthesize an integrated control system for these components.
- To work out a scientific basis to assess the ability of land and oceanic ecosystems to extract and accumulate CO_2, with the existing anthropogenic activity taken into account.
- To develop techniques to identify and assess the sources of CO_2 and other GHGs on regional and global scales.
- To search for an algorithm to describe the functions of these sources in the future and to provide the prognostic estimates of the climate trends.
- To assess the potential strategies of carbon extraction from the environment.
- To develop an efficient technology of the information fields for an objective planning of environmental studies in the future and realization of the decisions when managing the land ecosystems.
- To create a unified information base for the available and planned environment research programmes.

1.2 GLOBAL CHANGES: REAL AND POSSIBLE CHANGES IN THE FUTURE

1.2.1 Global models by Club of Rome (CR) and Forrester (1971)

The problem of global environmental change is the subject of discussions by many authors (Krapivin and Kondratyev, 2002; Savinykh, 1993; Krapivin, 1993;

Kondratyev and Krapivin, 2001c; Nitu *et al.*, 2000a,b; Pielke, 2001; Sellers *et al.*, 1996a,b; Svirezhev, 2002). Though the notion of global change can be considered established enough despite the preserved numerous terminological differences (it refers especially to the definition of sustainable development), note that we mainly discuss here the interaction between society (socio-economic development) and nature. Most substantial features of global change appear multicomponental in character, interactivity, and non-linearity. These features complicate the prognostic estimates to such an extent that the notion of prediction has recently been ousted by more vague concepts of "scenarios" or "projections." The uncertainty of the scenarios grows due to the fact that, as a rule, there are no probabilistic estimates for various scenarios (this completely refers, for instance, to the problems of global climate change).

One of the possible mechanisms to overcome these uncertainties is a new scientific direction intensively developed during the last years – global ecoinformatics – within the framework of which information technologies have been created, which ensure a combined use of various data on the past and present state of the NSS. The creation of a model of the NSS which is based on knowledge and available data and blends with an adaptive evolutionary concept of the geoinformation monitoring, which enables one to realize an intercorrection of the NSS model and the regime of the global data collection, can be considered an important step in global ecoinformatics. As a result, the NSS structure can be optimized to achieve sustainable interaction between nature and human society and to create an international strategy for the coordinated use of natural ecosystems.

One of the important scientific directions of ecoinformatics is the development of models of various processes taking place in the NSS. Modelling refers to spheres of knowledge. It is connected with the fact that the model has ensured an understanding of correlations between the NSS fragments making it possible to see the whole picture from the mosaic of the processes isolated at first sight. An especially important property of the model is its ability to reflect the present duality of the anthropogenic component. On the one hand, a human is an element of nature whose behaviour is determined by nature, but on the other hand, he can project his behaviour and through this transform the natural environment into an artificial one.

Three decades ago (in 1972) a first report of the Club of Rome (CR) (Meadows *et al.*, 1972) was published which raised worldwide interest. In 1972 an International Conference on the Environment was held in Stockholm, which also caused wide international resonance. Later on, developments of the problems of global change were mainly concentrated within the framework of the IGBP and the WCRP, which then were supplemented with the IHDP. These and attendant international programmes of environmental studies have enabled one to create extensive global data bases on the various NSS components and, in particular, to fix their trends during the last decades.

Let us briefly analyse the results of developments in the problems of global observations for the last three decades and (which is most important) answer why, despite huge efforts and enormous expense (reaching many US$bn), the global ecological situation not only has not improved but has become worse. First of all, we shall briefly discuss the models developed by Forrester and the CR.

The titles of the monographs *Limits to the Growth* (Meadows *et al.*, 1972) and *The World Dynamics* (Forrester, 1971) reflected an appearance of a new fundamental concept: development of human society (population size and increasing scales of human activity) cannot be boundless and have already approached certain limits – initially from the viewpoint of the levels of the use of natural resources. The main goals of the CR developments consisted in an analysis of global demographic dynamics and estimates of natural resources, as well as in substantiation of the model of global ecodynamics and possible scenarios of the ecodynamics of the future. Meadows *et al.* (1972) formulated the principal goals: "The goal of the project is to study a complex of problems of concern for all nations: poverty among abundance; environmental degradation; loss of confidence in public institutions; uncontrolled expansion of cities; unreliable employment; estranged young people; neglect of traditional values; inflation and other economically destructive phenomena."

Instead of an introduction, the authors of *Limits to the Growth* quoted the address by U. Tang, the UN Secretary-General: "I do not want to overdramatize the events, but I only want, based on information I have got as the UN Secretary-General, to make the conclusion that the UN participants have possibly ten years at their disposal to regulate their old disagreements and to start a global cooperation in order to curb the arms, to improve the environment, to restrain the explosive growth of population, and to stimulate efforts in the sphere of the socio-economic development. If the global cooperation does not become a first-priority goal, I am very much afraid that all these problems will reach stunning levels, beyond our capabilities to control them."

Now, three decades later, conclusion might be drawn that U. Tang (like Meadow *et al.* (1972)) adhered to the concepts of catastrophism: at first sight, it seems that with all the problems of the development of present civilization, the world survived and did not reach a critical point of irreversible negative trends. Analysis of realities of the present global ecodynamics does not permit such an optimistic conclusion. As for the initial CR developments and U. Tang's speech, one can only admire their foresightedness. It is corroborated by the evolution of resolutions carried by many different international forums (IPCC, 1996, 2001a–d; Baliunas, 2002; Kondratyev, 2002c; Young and Wefering, 2002). Below we shall briefly discuss the basic content of the CR report, whose authors begin with the analysis of the global demographic dynamics.

From 1650 onwards, when the global population size was about 0.5 billion people, the population increment constituted 0.3% annually, which corresponds to the period of population doubling equal to 250 years. By 1970, the size of population had reached 3.6 billion people, and the rate of increment increased to 2.1%/year (respectively, the period of doubling decreased to 36 years): the increase of population became "superexponential." An important demographic indicator is the human

Table 1.9. The rate of the growth of the population and the economy in different countries.

Country	G	ΔG	Ψ (1968)	$\Delta\Psi$	Ψ (2000)
China	730	1.5	90	0.3	100
India	524	2.5	100	1.0	140
USSR	238	1.3	1,100	5.8	6,330
USA	201	1.4	3,980	3.4	11,000
Pakistan	123	2.6	100	3.1	250
Indonesia	113	2.4	100	0.8	130
Japan	101	1.0	1,190	9.9	23,200
Brazil	88	3.0	250	1.6	440
Nigeria	63	2.4	70	−0.3	60
FRG	60	1.0	1,970	3.4	5,850

Note: G = population in 1968 (bn); ΔG = rate of population growth between 1961–1968 (%); Ψ = per capita GDP (US$); $\Delta\Psi$ = per capita annual rate of GDP growth (%).

lifetime, which in 1650 constituted about 30 years. By 1970 it reached 53 years and continued increasing. Meadows *et al.* (1972) pointed out the lack of prospects for stabilizing the global population size by the year 2000.

The second important parameter is the rate of industrial development; between 1965 and 1968 the rate of growth averaged 7%/year (the respective per capita indicator was 5%). It is important that this growth mainly concentrated in industrial countries where the rate of population growth was, however, relatively low (as a result tremendous contradictions have formed between developed and developing countries characteristic of the present-day). The data of Table 1.9 illustrate the rate of growth of the GDP in different countries. The CR report justly emphasizes the low probability that this dynamic of the growth of both population and GDP will remain even until the end of the 20th century, since many factors of this dynamic will suffer change. It is of interest, however, that the example of Nigeria, where one should expect the end of interethnic conflicts and, respectively, the growth of the economy, proved lame, reflecting a continuation and deepening of the socio-economic conflicts on the African continent (convincingly demonstrated in the continent's growth of foreign debt). Of interest are estimates (Meadows *et al.*, 1972) of possible levels of per capita GDP in 2000 (Table 1.9). Though the reliability of such extrapolative estimates cannot be high, they are in shocking disagreement with reality, which testifies to extremely complex political and socio-economic predictions. One of the key aspects of socio-economic dynamics has turned out to be quite predictable: the process of economic growth being followed by an increased contrast between rich and poor countries.

Answering the question of what is necessary to provide a persistent growth of an economy and population size up to the year 2000 and further, Meadows *et al.* (1972)

separately analysed the aspects of physical (material) and social needs. Satisfaction of material needs requires, first of all, a solution to the problems of sufficiency of food, drinking water, and other natural resources.

Although reliable data on feeding conditions are absent, it is supposed in the CR report that about 50–60% of the population of the developing countries (about one-third of global population) are underfed. The area of arable soils constitutes about 3.2 Gha, of which about 50% are cultivated. According to data from the Food and Agriculture Organization (FAO, data from the international organization on the problems of food and health), the cultivation of the remaining soils is economically inexpedient. It follows from the available estimates that even if all arable soils are used, whilst preserving the rate of population growth, an acute shortage of agricultural soils will occur by 2000 (below shows that this prediction has proved abortive). In this connection the CR report emphasizes that humankind has limited capabilities to prevent catastrophic consequences resulting from exponential population growth in conditions of limited resources.

Meadows *et al.* (1972) obtained various prognostic estimates using the global model, the structural description of which causes no objections. But as follows from subsequent publications (Krapivin and Kondratyev, 2002), the model does not take into account numerous undisputable feedbacks (both direct and indirect) between society and nature, and of principal importance, their spatial heterogeneity. Nevertheless, the predictions made have brought forth many problems for specialists in the field of global modelling, prompting them to develop efficient technologies for environmental control. Moreover, it is apparent that the problems discussed in Meadows *et al.* (1972) cannot be solved without using the systems of global monitoring of the environment.

1.2.2 The present-day state of the NSS

It should be mentioned that during the last 30 years, due to the efforts of many scientists, the key priorities of global ecodynamics have been formulated and perspective trends in solutions of numerous new problems have been outlined. Clearly, to work out a global strategy of sustainable development a constructive formalized approach to the description of the NSS is needed, taking into account its multidimensional and multicomponential nature, as well as non-linearity and interactivity of the processes taking place within it. Many feedbacks in the NSS have strengthened, prevailing over other feedbacks. So, from the available estimates, about 1.2 billion people now live on less than one US dollar per day; about 3 million people die every day of AIDS; 100–150 million people suffer from asthma; 2.4 billion people need better sanitary living conditions; 150–300 Mha of cultivated soils (representing 10–20% of all agricultural soils) have become degraded; and more than 2 billion people suffer from a shortage of drinking water, food, and dwellings. Finally, there is a trend towards increasing population mortality due to intensified terrorism and various technogenic catastrophes. All this changes the concept of the global model and requires a search for new information technologies to assist the control of trends in the NSS.

The CR model reflects only a narrow spectrum of feedbacks (levels) in the NSS (population, finance, pollution, food production, and mineral resources). Besides, the model neither reflects a direct role of biospheric feedbacks nor considers the spatial heterogeneity of these feedbacks. Therefore, it could not be objectively predictive even within the framework of successfully formulated scenarios, which, nevertheless, introduce into the model numerous uncertainties with a broad spectrum of possible issues. A comparison of models developed by Forrester (1971) and Meadows *et al.* (1972) reveals their conceptual and functional identity. Depending on variations of initial suppositions (scenarios) on limited or unlimited irreversible resources as well as on stabilization of the population size, the results of prognostic estimates of the state of the NSS components coincide qualitatively for both models, but differ substantially from the reality which faced us at the end of the 20th century. The basic difference occurs in the estimation of population size whose rate of growth in the 1990s became stable at about 80 million people per year, passing a maximum of 87 million people per year in the late 1980s and returning to the level of the 1970s at the beginning of the 21st century. It is explained by the existence of uncertainties in the models of the CR and Forrester. These uncertainties are defined by the unpredictable changes owing to the manifestation of activity of new feedbacks in the NSS. Understanding the structure and importance of heterogeneous and complex feedbacks in the present world has changed rapidly together with their difficult to predict dynamics. So, duc to an extension of trade relations, transport, and information networks during the last 30 years, the spatial correlation between ecological, demographic, political, and economic events has grown.

One of the substantial differences between the CR model and our present-day world is connected with concepts of the use of mineral resources and food production. It is clear now that such components of the global NSS functioning as alternative energy sources, energy-saving systems and technologies, birthrate and mortality, global structures of population and migration, response of nature to anthropogenic forcings, and many other key indicators need a parameterization and reflection in the model scheme. For instance, in the sphere of food production the role of aquaculture in the period 1984–1999 grew distinctly. During this period, production from global aquaculture grew by almost 400%, from 6.9 Mt in 1984 to 33.3 Mt in 1999. This growth was heterogeneous both in space and in production components. For instance, the share of fish in the global food market increased from 19% in 1990 to 31% in 2001, 68% of which fell on the aquaculture companies in China. The global model component responsible for fossil fuels is also rather complex. There are about ten well-studied alternative mechanisms for their economic use and substitution. This aspect is present in the CR model in the form of several primitive scenarios.

A general idea about the present state of the NSS is demonstrated in Table 1.10. It shows that the decrease or stabilization of the per capita food production as of the early 21st century predicted by the CR model has proved to be inaccurate. This is

Table 1.10. The present state of the basic components of the global NSS.

Global NSS component	Estimate of the component as of the end of the 20th century
Grain production	
Total (Mt/yr)	1,836
Per capita (kg/yr)	302
Meat production	
Total (Mt/yr)	232
Per capita (kg/yr)	38.2
Area of irrigated soils	
Total (Mha)	274
Area per 1000 people (ha)	45.7
Fossil fuel expenditure (Mt of oil equivalent)	
Coal	2,186
Oil	3,504
Gas	2,164
Energy production	
Nuclear power stations (GW/yr)	348
Wind-powered systems (MW/yr)	18,100
Mean global temperature (°C)	14.3
Carbon emission due to fossil fuel burning (Mt C/yr)	6,480
Partial pressure of CO_2 in the atmosphere (ppm)	370.9
Production of metals (Mt/yr)	902
Production of round timber (Mm^3/yr)	3,336
Oil spills due to anthropogenic activity 1000 t/yr	48.6
Gross product	
Total (US$bn/yr)	44.9
Per capita (US$/yr)	7,392
Foreign debt of developing countries and countries of the former Soviet Block (US$bn)	2.53
Global population	
Total (bn)	6.08
Annual increment (mn)	77

natural because food production is characterized by more complicated cause and effect feedbacks than those considered in the models by CR and Forrester (1971). It has to be pointed out that although production of grain, meat, and other components of human food suffered increases and decreases, it gives no reasons for pessimistic predictions for decades to come. As has been mentioned above, production of fish produce sharply increased over the last years. Clearly, the global model should consider the bioproductive processes in the World Ocean and inland water basins in order to have a possibility of estimating the limits of their capabilities to produce food. As for grain, at the turn of the century its production decreased and its consumption increased. So, for instance, the per capita grain production in 2001 constituted 299 kg, which was below 1984 values by 14%. At the same time, if we consider the long-range trend of grain production, between 1950 and 1984 it increased by 38%. The observed imbalance of grain consumption and production and the respective decrease in global scale supplies are only a short-term fluctuation of the process of food production. In fact, other constituents of human food are characterized by positive gradients in production, although distribution is non-uniform over countries and continents. For instance, meat production (beef, pork, and poultry) in 2001 grew at a rate of 2%. Beginning in 1950 the per capita meat production doubled from 17.2 kg/yr to 38.2 kg/yr in the year 2000. In 2001, a maximum of pork and poultry production was observed.

Finally, as for one of the key indicators of the NSS state, namely, non-renewable natural resources, here the CR concept of their continuous reduction and their limiting role in the development of other NSS levels differs from real trends at the turn of the century. The real global scale extraction of coal, oil, and natural gas grows at a rate of 1–2%/yr (in 2001 it was 1.3%). The volumes of fossil fuel consumption are also growing, albeit non-uniformly amongst countries and correspondingly in differing types of fuel. On the whole, the global increase of consumption of oil, coal, and gas constitutes 0.2%, 3.2%, and 1.2%, respectively. Besides, the use of nuclear, solar, and wind energy is growing, and energy-saving technologies are rapidly developing. At the end of the 20th century there was a rapid growth of industrial production of photoelectric cells transforming sunlight into electricity. In this manner, the world gets more than 1140 MW. There is also an increasing trend in the use of solar energy. Therefore the dependence of global dynamics on energy resources should be parameterized not on the basis of simplified models but taking into account the whole spectrum of available information on the nature of multiple feedbacks in the NSS and especially the trends in scientific–technical progress.

1.2.3 Perspectives of the global model development

Over the 30 years after the appearance of CR predictions and the respective global model, a serious step forward has been made in the field of global modelling. The new approach proposed in Krapivin and Kondratyev (2002) was based on the idea of the NSS as a self-organizing and self-structuring system, the correlation of whose elements in time and space is ensured by the process of natural evolution. The anthropogenic constituent in this process is aimed at breaking this integrity. Attempts to formally parameterize the process of coevolution of nature and humankind as elements of the biosphere are connected with a search for a single description of all the processes in the NSS, which would unite efforts of various branches of knowledge about the environment. Such synergism serves as the basis of many studies on global modelling.

Let us cover the Earth's surface Ω with the geographical grid $\{\varphi_i, \lambda_j\}$ with the digitization steps $\Delta\varphi_i$ and $\Delta\lambda_j$ by latitude and longitude, respectively, so that within a cell of land surface $\Omega_{ij} = \{(\varphi, \lambda) : \varphi_i \leq \varphi \leq \varphi_i + \Delta\varphi_i; \lambda_j \leq \lambda \leq \lambda_j + \Delta\lambda_j\}$ all the processes and elements are considered as homogeneous and are parameterized by point models. In the case of a water surface in the territory of the cell Ω_{ij} the water masses are stratified into layers Δz_k thick, that is, 3-D volumes are selected $\Omega_{ijk} = \{(\varphi, \lambda, z) : (\varphi, \lambda) \in \Omega_{ij}, z_k \leq z \leq z_k + \Delta z_k\}$, inside of which all elements of the ecosystem are distributed uniformly. Finally, the atmosphere over the site Ω_{ij} by height h is digitized either by levels of atmospheric pressure or by layers Δh_s thick.

Interactions in the NSS are considered as interactions between natural and anthropogenic components within these spatial structures and also between them. The complex model of the NSS realizes the spatial hierarchy of hydrodynamic, atmospheric, ecological, and socio-economic processes with the division of the whole volume of the environment into structures Ω_{ij} and Ω_{ijk}. The cells of this division are the supporting grid in numerical schemes for solutions based on dynamic equations or in synthesis of data series in evolutionary learning procedures.

The cells Ω_{ij} and Ω_{ijk} are heterogeneous in parameters and functional characteristics. Through this heterogeneity the global model is referenced to databases. Moreover, to avoid an excess structure of the global model, it is supposed a priori that all its elements taken into account in the model and the NSS processes have a characteristic spatial digitization. Ambiguity of spatial digitizations in various units of the global model is removed at an algorithmic level of agreement of data fluxes from the system of monitoring. As a result, the model's structure is independent of the structure of the database. A similar independence between the model's units is also provided. This is realized by data exchange between them only through inputs and outputs under control of the basic the parameters. In case of turning off one or several units their inputs are identified with the corresponding inputs to the database. Then the model's use is reduced to the formation of a spatial image for the modelled medium and to the setting of a control regime for the simulation experiment. Of course, in this case the user should have a certain knowledge base and understand the method of its structuring. For instance, one can use a list of key problems of global ecology or lists of the NSS elements recommended for studies.

The character of the spatial structure of the global model is determined by the database. The simplest version of the point model is realized with the initial information in the form of averaging over the land surface and the whole World Ocean. The spatial heterogeneity is considered through various forms of space digitization. The base form of the spatial division of land and oceans is a heterogeneous grid $\Delta\varphi \times \Delta\lambda$. A realization of a real version of the use of the model is provided by integration of the cells Ω_{ij} so that various forms of the spatial structure of the considered elements and biospheric processes can be present in each unit. Such a flexible setting of the spatial structure of the biosphere makes it possible to easily adapt the model to heterogeneities in databases and to perform simulation experiments with an actualization of individual regions.

Depending on special features of the considered natural processes, the structure of regional division can be identified with climatic zones, continents, latitudinal belts, socio-administrative structure, and natural zones. For climatic processes, many scientists are oriented towards regions with dimensions $\Delta\varphi = 4°$ and $\Delta\lambda = 5°$, the biogeocenotic processes are studied at $\Delta\varphi = \Delta\lambda = 0.5°$, the socio-economic structure is represented by nine regions, the atmospheric processes in the biogeochemical cycles of long-lived elements are approximated by point models, functioning of the ocean ecosystems are described by heterogeneous digitization of the shelf zone into the cells Ω_{ij} with the selection of four parts of the World Ocean pelagians.

The structure of division of the Earth's surface into regions Ω_{ij} covers all the enumerated versions. It means that the general scheme of digitization of the processes in the NSS foresees a hierarchy of levels including global, continental, regional, landscape, local levels, etc. The scheme of independent inclusion of units at all these levels with their combination through parametric interfaces does not prevent an increase in the amount of the model's units due to the introduction of new components, which specify the models of the processes under consideration. The model of the upper level can serve as an information base for the model of the lower level and vice versa. The results of modelling at the lower level can be used to form the information base for the models of higher levels. This mechanism of the information exchange between the models of various levels reduces the requirements of the global database and broadens capabilities of the NSS model.

The NSS model has the support units which provide a means of interaction for the user with the model and database. In particular these are units which realize the algorithms of the spatial–temporal interpolation or coordinate the user's actions with the bank of scenarios. Note that some scenarios can be transformed by the user into the rank of the model's units. Such a duality (excessiveness) is characteristic of scenarios of climate, demography, anthropogenic activity, scientific–technical progress, and agriculture. The user's interface makes it possible to select the structure (Ω_{ij}) in the default mode or to allow the required spatial structure to be formed from the base elements by averaging and interpolation.

Thus the synthesis of the global model version requires a preliminary analysis of the present situation with the global databases and knowledge base. Here the specialists face principal difficulties and, first of all, the absence of an adequate knowledge about climatic and biospheric processes as well as an uncoordinated database on the global processes on land, in the atmosphere, and in the oceans.

Another principal difficulty is connected with the inability of present-day science to formulate the requirements for global databases needed to reliably assess the state of the environment and to give a reliable forecast of its development for a sufficiently long time period. Moreover, there is no technology to form databases aimed at creating a global model.

Many scientists have made attempts to answer these questions (e.g., Tianhong *et al.*, 2003). One of the efficient ways to solve these problems is a single planetary adaptive geographical information monitory system (GIMS), which has a hierarchic structure of data collection and forms a multilevel global database. An adaptive character of this system is provided by correcting the regime of data collection and by changing the parameters and structure of the global model.

The global GIMS can be created taking account of the existing structure of databases whose formation continues within the framework of the IGBP and numerous national ecological and nature-protection programmes. The developed system of world data centres favours rapid use of accumulated information about the global processes and simplifies the GIMS synthesis. However, significant progress in this direction connected with large economic expense cannot lead to a successful solution of the problem of global environmental control.

According to Kondratyev (1998b), to control the global geobiosystem of the Earth, regular observations of specified key variables are needed. With an increasing probability of drastic global change, the spectrum of these variables will vary, and the global prediction system should be constantly modernized. The methodological substantiation of an adequate information content of variables for the monitoring system can be objective only in the case of GIMS functioning. Many of the enumerated variables can be calculated using the respective models, and there is no need to measure them. However, so far, measurements are planned in parallel to model development, and there are no results in the field of planning the global experiment that would raise hopes. As follows from some studies (Kondreatyev and Galindo, 2001; Kondratyev and Krapivin, 2002), the bases of global knowledge and data make it possible to synthesize and develop GIMS series (Kondratyev *et al.*, 2000).

An inclusion of the global model into the GIMS structure enables one to consider it as an expert system. It means that there is a possibility of the complex analysis of numerous elements of the NSS in conditions of realization of hypothetic situations, which can appear for natural or anthropogenic reasons. The scheme in Figure 1.1 reflects basic elements taken into account in the NSS global model (GMNSS). A concrete realization of each unit of the GMNSS is determined by the level of knowledge of the processes reflected in the unit. The units responsible for modelling the biogeochemical and biogeocenotic processes are described with balanced equations.

Let $\psi_s(t)$ be the information content of an element ψ in a medium S at a moment t. Then, following the law of preservation of matter and energy, we write the following balanced equation:

$$\frac{d\psi_s}{dt} = \sum_j H_{js} - \sum_i H_{si}$$

where fluxes H_{js} and H_{si} are, respectively, incoming and outgoing fluxes with respect to the medium S. Summing up is made by external media i and j interacting with S. In fact, the medium S implies elements of digitization of the environment by latitude φ, longitude λ, depth z, and height h. A variety of functional parameterizations of fluxes H_{pq} are determined by the level of knowledge of physical, chemical, and biological features of the element ψ.

A parameterization of the processes of photosynthesis, dying-off, and respiration of plants in land ecosystems is based on knowledge of phytocenology, which includes information about external and internal system connections of the vegetation community. These are temperature dependences of photosynthesis and evapotranspiration of plants, gas exchange processes between plants and the

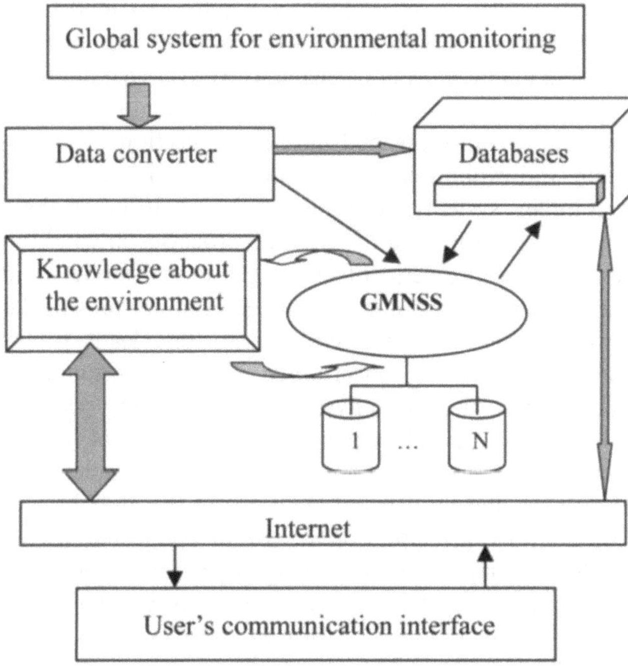

Figure 1.1. The principal scheme of the use of GIMS-technology to synthesize the global system of control of the environment with the use of the standardized means of telecommunications and GMNSS.

atmosphere, impacts of the solar radiation energy on the processes of growth and exchange, relationships between plants and processes in the soil, and interaction of vegetation cover with the hydrological cycle.

The GMNSS units responsible for the parameterization of climatic and anthropogenic processes are of complex character (i.e., partially described by equations of motion and balance, and with a partially evolutionary model constructed for them, based only on the observational data).

1.2.4 Preliminary conclusions

Having considered approaches to an assessment of the NSS dynamics, from Forrester (1971) and Meadows *et al.* (1972) to recent publications (Krapivin *et al.*, 1982; Krapivin and Kondratyev, 2002; Kondratyev *et al.*, 2003d), one can draw a conclusion that an appreciable progress in a search of ways to reach global sustainable development can be made with a systems approach to the multifunctional monitoring of the NSS. Thanks to the CR authors, more than 30 years ago the contradiction was first emphasized between the growth of population size and limited natural resources, and for the first time after the works by

Vernadsky (1944) an attempt was made to use numerical modelling to study the NSS evolution. Of course, the CR model oversimplifies real internal connections in the NSS, describing interactions of its elements by averaged indirect relationships, without direct account of economic, ecological, social, and political laws. The possibility of such an account appeared later in connection with the studies in the fields of simulation and evolutionary modelling and the theory of optimization of interaction for complex systems, resulting in the creation of the methods and algorithms of prognostic estimation of dynamic processes in conditions of a priori uncertainty. However, the problems of creating a global model which adequately matches the real world still cannot be solved even in present-day conditions. A complete consideration of all the NSS parameters leads to insurmountable multivariance and information uncertainty with unsurmountable problems. Besides, in the spheres such as physics of the ocean, geophysics, ecology, medicine, sociology, etc., an adequate parameterization of real processes will always be problematic because it is impossible to have a complete database. Nevertheless, a search of new efficient ways of synthesis of the global system of the NSS control, based on adaptive principles of the use of the global model and renewed databases, seems to be perspective and raising hopes for the generation of reliable forecasts of the NSS dynamics. Preliminary calculations with the use of the GMNSS have shown that the role of biotic regulation in the NSS has been underestimated, and the forecasts, for instance, of the levels of the greenhouse effect have been overestimated (Streets *et al.*, 2001). Therefore in this book an attempt has been made to synthesize the GMNSS taking into account the earlier experience and accumulated databases as well as knowledge about the environment and human society.

Further improvement of the GMNSS will be connected with the balanced development of studies both in the sphere of parameterizations in the NSS and in modernization of the Earth observation systems, covering the whole thematic space of the NSS (Kramer, 1995):

- Sun–land interactions (physical mechanisms of the transport of mass, momentum, and energy in the geosphere).
- Atmospheric dynamics (atmospheric chemistry, atmospheric physics, meteorology, hydrology, etc.).
- Dynamics of the World Ocean and coastal zones (winds, circulation, sea surface roughness, colour, photosynthesis, trophic pyramides, pollution, and fisheries).
- Litosphere (geodynamics, fossil fuel and other natural resources, topography, soil moisture, and glaciers).
- Biosphere (biomass, soil–plant formations, snow cover, agriculture, interactions at interfaces, river runoff, sediments, erosion, biodiversity, and biocomplexity).
- Climatic system (climate parameters, climate-forming processes, radiation balance, global energy balance, greenhouse effect, long-range climate forcings, and delay of climate effects).
- Socio-political system (demography, geopolicy, culture, education, population migration, military doctrines, religion, etc.).

1.3 ANTHROPOGENIC PROCESSES AND THEIR MODELLING

1.3.1 Anthropogenic factors in global ecodynamics

The human (H)–nature (N) interaction is a function of a broad complex of factors functioning both in human society and the environment (Kondratyev, 1999b; Demirchian and Kondratyev, 1998, 1999; Kondratyev *et al.*, 1997; Marchuk and Kondratyev, 1992; Gorshkov *et al.*, 2002; Jenkins *et al.*, 2002). The principal problem of this interaction is the globalization of the human effect on the environment with a rapid development of megalopolises acting as the sources of anthropogenic impacts on the environment. Almost half the population of the Earth live in cities occupying only about 3% of the land area. This trend of concentration within cities and the transformation of the latter into megalopolises will intensify, judging from numerous expert estimates. The number of large cities (with populations of 3–8 mn) and megapolises (>8 mn) in the 21st century will increase in the developing countries, with limited resources for a search and realization of optimal urban infrastructures. Therefore the ecological condition of the territories with an 80% global population increment will cause worry and demand additional investment (Kondratyev, 1996).

Many cities are contaminated by waste, huge territories have been transformed into waste heaps, the discharge of contaminated sewage to the hydrosphere and the emission of polluting matter and gases to the atmosphere have increased. Processes of transformation and even vegetation cover destruction have intensified. Of course, human economic activity has always taken place during historical development. Since the prehistorical epochs until now people have gradually increased their influence on nature, intensified the use of mineral resources, contaminated the environment, and broken the connections between various natural events established during the processes of evolution, leading to, in particular, a biodiversity reduction, deforestation of huge territories, soil salination, depletion of biological resources of the seas and oceans, all of which are a real threat to the existence of life itself on the Earth. In the period from Palaeolithic epoch to the epoch of feudalism these forcings had been first local and then regional. The technical progress in the 18th and 19th centuries had contributed to broadening the scale of human impacts on nature, causing an extermination of some species of animals, wholesale destruction of forests, and rapacious exploitation of other natural resources. In the 20th century these forcing have become of global scale and dangerous for the biosphere. The problem arose of the search for the strategy of an optimal interaction between the systems H and N. In this regard, due to accomplishment of numerous international and national programmes of environmental studies, databases have been accumulated that make it possible to assess the level and direction of anthropogenic processes (Tables 1.11 and 1.12).

The principal regularity noted by many authors (Kondratyev, 1998a; Watson *et al.*, 2000) is that the rate of the impact of human civilization on the biosphere depends directly on population growth. Such categories as energy production, amount of consumed mineral resources, and investments depend directly on the

Table 1.11. Dynamics of the global water consumption.

	Municipal services		Industry		Agriculture		Total	
Years	W_G	P_G	W_G	P_G	W_G	P_G	W_G	P_G
1900	20	25.0	30	6.7	350	74.3	430	67.5
1940	40	20.0	120	5.0	660	72.7	870	61.0
1950	60	18.3	190	4.7	860	73.3	1,190	59.1
1960	80	17.5	310	4.8	1,510	76.7	1,990	63.2
1970	120	16.7	510	3.9	1,930	78.9	2,630	61.5
1975	150	16.7	630	4.0	2,100	76.2	3,080	60.0
1985	250	15.2	1,100	4.1	2,400	79.2	3,970	56.4
1995	320	15.1	1,560	3.9	2,760	78.4	4,750	54.9
2000	440	14.8	1,900	3.7	3,400	76.5	6,000	50.0

Note: W_G = water consumption (km^3/yr); P_G = irretrievable water loss (km^3/yr).

Table 1.12. Characteristics of regional water consumption in the 20th century.

	Population		Industry		Agriculture			
Region	*a*	*b*	*a*	*b*	*a*	*b*	Σ	*b*
Europe	29	11	160	5	125	67	314	24
Asia	40	15	60	6	1,400	79	1,500	74
Africa	4	18	3	7	110	82	117	78
North America	41	21	270	30	210	62	521	28
South America	4	15	8	10	55	82	67	74
Australia	1	10	8	8	13	77	22	49

Note: a = net water consumption (km^3); b = per cent of irretrievable losses; Σ = total amount of water consumption (km^3).

size of the population (Demirchian and Kondratyev, 1998). The result is that the global scale forcing of society on natural processes has reached a level where the survival of humankind as a biological species raises doubts. To survive, humankind should learn to coevolve with nature, and to do that, it is necessary to thoroughly study it (Kondratyev, 1992b, 1993; Kondratyev *et al.*, 1992).

According to the general opinion of experts in various sciences, the basic trends in human economic activity are to regulate the global processes in the biosphere, such as production of energy, industrial materials, and food. On average, the efficiency of these processes calculated per capita tends to increase. For instance, in the USA the per capita amount of energy increases annually by ~2.5%. The mean global energy increment exceeds that of population by 2%. So, for the territory of the FSU the average population increment constituted 1.4%/yr, and the annual growth of energy production is estimated at 7%. On the whole, the energy potential has been, so far, one of the characteristics of the level of the human society development. Other parameters determining the state of the production and the economy of the region, territory, country, and globe, depend directly on the energy potential (Nakicenovic *et al.*, 1998).

The atmosphere, hydrosphere, and soil get more and more contaminated. Practically, every principal geospheric components within which there is life, suffers from growing anthropogenic forcing. For instance, during the last years more than 4 Mt of oil have been discharged into the World Ocean, with the concentration of mercury in the oceans almost doubling.

Since all the domains of the geosphere are interconnected, a destructive forcing on one of them threatens all the others. Therefore, for example, burning of the final product of sewage purification cannot be considered a solution to the problem of pollutant control, since this technology only transfers the load from the hydrosphere to the atmosphere. Equally inefficient is the cleaning of soil of food organic discharge by pumping into the subsoil waters, with the result that rivers get overloaded with bacteria and deficient in oxygen. In short, transporting the pollutants into any domain of the geosphere damages all spheres of life. Although, of course, all spheres react differently to contamination, and a redistribution of anthropogenic forcings among various geospheres can be an element governing the environment. Therefore while analysing the processes of pollution generation, it is necessary to determine the complex characteristics of the anthropogenic processes. For this purpose we shall consider each geosphere and formally describe the processes of their pollution.

One of the aspects of the human impact on the Earth's climate is atmospheric pollution. Along with the GHG biogeochemical cycles that break the natural heat balance of the Earth, a relevant global model should take into account other pollutants that change the optical properties of the atmosphere, such as industrial dust, combustion gases, radioactive substances, smoke, molecular admixtures, and vapours of water and other liquids. For instance, molecular admixtures getting to the atmosphere can markedly change its temperature (up to 5°C), with the most dangerous being admixtures whose absorption spectrum occurs in the atmospheric "transparency windows". Atmospheric molecular pollutants are, methane, water vapour, sulphur dioxide, hydrogen sulphide, ozone, carbon monoxide, nitrogen oxides, etc. These components are divided into stable and short-lived (Table 1.13). The short-lived pollutants, such as SO_2 or mercury vapour are unstable in the atmosphere and cannot propagate for long distances, being unable to transform local anomalies into global ones. On the contrary, the long-lived pollutants like CO_2 tend to accumulate in the atmosphere and propagate over large territories. Therefore their concentration during long time periods becomes uniform throughout the global atmosphere.

Dust and aerosols get to the atmosphere from many sources. Small solid particles are emitted to the atmosphere together with gases from the internal-combustion engines and industrial smokes from the metallurgical works and power stations. About half of the solid particles ejected with combustion gases contain lead which tends to propagate for long distances. For example, anthropogenic lead was

Table 1.13. Lifetimes of some atmospheric components.

Component	Atmospheric residence time
Carbon dioxide	5 years
Carbon monoxide	0.1–3 years
Water vapour	10 days
Sulfur dioxide	3 days
Ozone	10 days
Hydrogen chloride	3–5 days
Nitrogen oxide	5 days
Nitrogen dioxide	5 days
Nitrous oxide	120 years
Ammonia	2–5 days
Methane	3 years
Freons	50–70 years

discovered in the Greenland mainland ice. A dangerous component of atmospheric pollution is mercury whose dispersed liquid particles are harmful to all living beings. From the available estimates the total amount of mercury emitted annually into the atmosphere reaches 1100 t. Mercury sulphate with a concentration of 10 ml/l kills living organisms within several days, and with a lower concentration it substantially violates life cycles. For example, tests on *Monodonta articulata* with a concentration of mercury salts (sulphate, acetate or chloride) of 0.01 mg/l reduced O_2 consumption. A discharge of mercury and copper into the sea in California killed gigantic focus algae which were only exposed to negligible concentrations of these metals (0.05 g mercury and 0.1 g copper per 1 t water) over four days. Mercury in concentrations dangerous for humans is found in fish and mollusks in more and more water bodies of the World Ocean (the coastal waters of the USA, Sweden, Norway, Canada, the Mediterranean) (McIntyre, 1999).

Also dangerous are radioactive substances – some of which can remain in the atmosphere for a long time. Of special danger are accidents at atomic reactors and atomic bombs explosions which represent the most intensive sources of radioactive elements.

Both technogenic and natural processes affect changes in the chemical and optical properties of the atmosphere. Table 1.14 lists some sources of atmospheric pollution. Each source is special. This list is constantly renewed due to the appearance of new spheres of human activity.

The atmospheric pollution changes human and animal habitats and causes climatic variations. The spectrum of human activity is broadening towards a higher comfort of living, and this requires an increased energy consumption and productivity of natural and man-made ecosystems (Table 1.15). Feedbacks appear

Table 1.14. Sources of atmospheric pollution.

Sources of pollution	Pollutant
Natural	
Volcanoes, fumaroles, solfataras	Gases, volcanic dust, and mercury vapour
Natural emissions of natural gas and oil	Hydrocarbons
Deposits	
Mercury	Mercury vapour
Sulphides	Sulphur dioxide
Radioactive ore	Radon
Sea and ocean surfaces	Chlorides, oil, and sulphates
Underground coal fires	CO_2, CO, SO_2, hydrocarbons
Natural forest and steppe fires	Smoke
Plant transpiration	Water vapour, aromatic and other volatiles
Anthropogenic	
Burning of solid and liquid organic matter	CO_2, CO, SO_2, lead, hydrocarbons, mercury vapour, cadmium, and nitrogen oxides
Metallurgy of black, non-ferrous, and rare metals	Dust, SO_2, mercury vapour, and metals
Atomic industry	Radioactive substances
Nuclear explosions	Radioactive isotopes
Cement industry	Dust
Construction explosions	Dust
Forest and steppe fires	Smoke
Gas- and oilfields	Hydrocarbons
Automotive transport	CO, soot, and nitrogen oxides

From Brenninkmeijer and Röckmann (1998, 1999), Grigoryev and Kondratyev (2001a,b), and Demirchian *et al.* (2002).

that require changes in technologies of natural resource exploitation. On the other hand, the increase in atmospheric pollution grows linearly with growing population.

The next important sphere of anthropogenic activity manifestation is the pollution of the World Ocean with oil products, organic and inorganic compounds, and various chemicals. From the available estimates, the total amount of oil products getting into the oceans annually varies between 32 and 60 Mt (Klubov *et al.*, 2000). Oil pollution is not uniform in the oceans, and in many water bodies it exceeds the admissible norm thousands of times (0.5 mg/l). From the estimates of many authors, the concentrations of oil hydrocarbons in the surface microlayer of the World Ocean exceeds maximum permissible norms by a factor of 19.8 in the Atlantic Ocean, 3.6 in the Pacific Ocean, 4.4 in the Indian Ocean, 29.5 in the North Sea, 25 in the Norwegian Sea, and 4.8 in the Caribbean Sea. The property of oil is such that after a leakage it spreads over the water surface forming a film 1.02×10^{-3} mm thick. The areal extent of an oil spot from a 1 t leakage of oil is 12 km^2. At present about 30% of the World Ocean surface is covered with oil film. This film hinders the atmosphere–ocean gas exchange as well as decreases the

Table 1.15. Some characteristics of the energy parameters in the biosphere.

Characteristic	Estimate
Energy of the Earth's mass (Joule)	10^{46}
Global consumption of all kinds of energy (kJoule/yr)	$(2.1–3) \times 10^7$
Energy utilized by the biosphere (Joule/yr)	10^{20}
Energy used by humans (Joule/yr)	10^{21}
Energy of tides and ebbs (kW-hr/yr)	6.4×10^{14}
Energy of rivers (kW-hr/yr)	23×10^{12}
Energy of wind (kW-hr/yr)	15×10^{23}
Solar energy consumption (%)	
Falling on the Earth surface	100
Accumulated in photosynthesis	0.1
Food source	0.001
Energy needs satisfied due to burning of fossil products of photosynthesis (%)	95
Consumption of photosynthesis products as food (%)	1
Coefficient of performance	
Stockbreeding	0.2
Agriculture	
Extensive	20
Intensive	2
Greenhouse	0.02
Energy spent in agriculture on nitric fertilizer production by chemical fixation of atmospheric nitrogen (%)	30
Annual rate of the energy consumption increase (%)	4
Volume of the burnt organic fuel (Gt/yr)	6–7
Annual expenditure of natural gas (10^{12} m^3)	1.4
Geological supplies of coal and brown coal (Gt)	$(14–15) \times 10^3$
Energy content in these supplies (10^9 Joule)	201×10^{12}
Natural gas supplies (10^{12} m^3)	145–340
Explored oil deposits (Gt)	65
Extracted oil (Gt)	147
Global oil output (Gt/yr)	7.4
Share of the sea oil-products in the global oil output (%)	19–20
Extraction of mineral fuel from the ocean bottom	
Oil (10^6 t/yr)	450
Gas (10^9 m^3/yr)	180
Known supplies of combustibles in units of conditional fuel (Gt)	27×10^3
Volume of extracted conditional fuel (Gt/yr)	6
Total energy of fossil fuels (kJoule)	10^{20}
Share of fossil fuel energy (%)	
Oil	10
Gas	10
Coal	80

The ratio of the amount of explored deposits of fossil combustibles to their annual extraction	
Oil	37
Gas	41
Coal	720
Energy emitted by burning timber (kJoule/yr)	2×10^6
Power of all energy constructions over the globe (kW)	10^{10}
Share of total anthropogenic energy emissions over the Earth in its radiation budget	0.02
Methods of energy expenditure (%)	
Heat power stations and central heating	30–35
Industry	30–35
Transport	25–30
Household needs	5–10

coefficient of water transparency. Deposited oil is accumulated in bottom sediments, which violates the natural process of the vital functions of benthos animals and bacteria. Important trophic and energetic feedbacks in the structure of the marine ecosystem get broken. Damage by oil pollution is transported through trophic chains of marine ecosystems (Stephen *et al.*, 2000; Patin, 1999).

Along with the oil pollution of the World Ocean in the late 20th century, dangerous toxic pollutants of the marine domain have become appreciable, such as lead, mercury, arsenic, and other heavy metals, which damage the marine organisms in which they accumulate. The latter makes these organisms unfit for consumption. Quantitatively, the background natural concentrations of heavy metals in numerous water bodies of the oceans exceed five to tenfold the admissible levels and in some places even reach dangerously high levels. So, with the total input of mercury to the oceans at 5000 t/yr due to its local use in chemical and paper industries the highest concentrations of mercury are observed in the coastal waters of Japan, Scandinavia, the Netherlands, and Canada.

By approximate estimates the annual input of lead into the World Ocean constitutes 40,000 t, half of which is deposited from the atmosphere. This half contains a substance especially dangerous for marine organisms – tetraethyl lead, usually added to petroleum. Lead, like mercury, can accumulate in marine organisms.

The World Ocean pollution spectrum is rather broad. Solid wastes and sewage play a marked role in the total pollution of the oceans. From rivers and ships a huge amount of rubbish gets into the oceans. All this leads to the lumbering of the oceanic domain and can worsen the living conditions for marine animals.

Along with pollutants directly harmful to marine domains, huge amounts of biogenic elements are introduced into the oceans together with everyday sewage and fertilizers washed out from the fields (e.g., various compounds of nitrogen, phosphorus, and silicon). In some cases such inflows are harmless or even useful, but they can often lead to negative consequences causing eutrophication, or secondary biological pollution. In this case an excessive reproduction and development of algae takes place that consume all the oxygen dissolved in the water, resulting in the death of masses of living organisms. The eutrophication processes can take place without anthropogenic interference (e.g., in the region of the Peru upwelling). These natural processes occur in a balanced regime developed in the process of evolution and, on the whole, are not dangerous for the ocean ecosystems dynamics.

Bearing in mind that the annual global input of sewage has exceeded 450 km^3 and in the future it may increase 10 to 15-fold due to population growth, the eutrophication processes will become all-embracing with unpredictable consequences. Therefore one of the goals of ecoinformatics is to assess these processes and their consequences. Clearly, the technogenic pollution processes and natural self-cleaning ability of the oceanic waters are in a constant fight. A complex of physical, chemical, and biological processes of pollutant decomposition and transformation with the resulting non-toxic substances provide the oceans with a powerful barrier to anthropogenic forcing. However, the rivers, for example, have already lost this ability since the water medium does not have time to clean itself completely, and the pollutants therefore accumulate. As a result, many rivers have been transformed into ditches (Vega *et al.*, 1998; Ragas and Rsew Leuven, 1999).

The World Ocean is capable of self-cleaning of such stable pollutants as oil. Due to the atmosphere–water surface interaction, oil gradually evaporates: during the first 24 hours by 15%, during the next ten days by ~20%. Storm winds forming foam crests on the water surface favour an input of water dust into the atmosphere also carrying oil particles. Solar UV radiation, chemical oxidation, and bacterial decomposition also destroy the oil products. Microorganisms use oil hydrocarbons as a source of carbon and energy and transform these hydrocarbons into bacterial cells, carbonic acid, and water. Besides, the microorganisms and many hydrobionts promote the transport of oil into the ground.

Anthropogenic forcings on the environment should be assessed together with natural processes, in order to work out a technology to reliably predict the consequences of human activity. For this purpose it is necessary to analyse the sphere of anthropogenic activity – utilization of natural resources. On the whole, there are the following groups of resources: fossil (geological and mineral); water; soil; vegetation; fauna; atomic; climatic; planetary; cosmic, including the chemical compounds of the Earth's crust elements; deposits containing oil, gas, coal, various salts, forests, etc. Natural resources are divided into renewable and non-renewable, exhaustible and inexhaustible. The renewable resources are soil, vegetation, and fauna. The non-renewable resources are fossil fuels. Solar radiation and wind energy are inexhaustible resources. Water resources and atomic energy are practically globally inexhaustible but can become scarce in some regions. Depending on time and scientific–technical progress, the anthropogenic processes can change both their scale and direction.

Assessments of fossil fuel supplies make it possible to predict, to some extent, the levels of their anthropogenic utilization. Numerous valuable elements are contained in the Earth's crust in relatively small amounts. For instance, the share of aluminum and iron is, respectively, 1/24 and 1/30. Almost half of all chemicals contain oxygen. Land is the basic source of minerals and raw materials for humankind. Chemical elements from the World Ocean are used in small amounts.

At the dawn of its industrial activity humankind had used only 19 chemical elements and their compounds, in the 18th century 28, and in the early 20th century more than 100. With the development of science and technology the use of minerals and their extraction has intensified. So, during the last 100 years the annual consumption (and therefore, extraction) of coal, iron, manganese, and nickel has increased 50 to 60-fold, with that of tungsten, aluminum, molybdenum, and potassium increasing 200 to 300-fold. Annually about 100 Gt of ore and mineral fuel are extracted, as well as more that 300 Mt of mineral fertilizers.

The per capita amount of extracted fuel is the principal indicator of energy supply. In 1972 the per capita amount of conditional fuel constituted about 2 t, but in the early 1970s the annual output of conditional fuel constituted almost 6 Gt, spent as follows: transport – 25/30%; heat power stations – 30/35%; industry – 30%; and everyday needs – 5/10%.

The basic types of minerals are fuel-energy resources, ores of black and alloy metals, non-ferrous and noble metals, and non-metal raw materials. According to an assumed classification, the trends of extracting these resources are as follows. Fuel-energy raw materials are oil, gas, and coal. In the 20th century the output of oil and gas increased from 10 to 1800 Mt and from 5 to 1300 billion m^3, respectively. At the same time, the output of coal increased only four to fivefold. The development trends of the output of other minerals such as iron and manganese ores, chromites, copper, lead, tin, molybdenum, etc., are growing. According to the available estimates, this pace of extraction will completely exhaust the acceptable supplies of aluminum in 570 years, iron in 250 years, zinc in 23 years, lead in 19 years, and by the mid-21st century most of the metals will disappear completely. The danger of the possible realization of this prediction intensifies the extraction and use of fossil non-metals, such as asbestos, black lead, mica, melted spar, etc.

Noble metals (platinum, gold, and silver) are special among minerals. Their production is small by volume but they are important not only in the development of industry and scientific instrument making but also in the formation of the socio-economic relationships in human society.

Consumption of natural resources also includes the use of fresh water and forest. The global water balance has been discussed by Krapivin and Kondratyev (2002). It includes a component that affects the natural water cycle. In particular, though the water used for household needs returns to the cycle but in quite a different way than natural exchange between water and individual biospheric elements. Water consumption for household needs is followed by irreversible losses, which, for instance, in the USA and Russia constitute 10–20%. Water supply in agriculture is characterized by 20–40% of irretrievable losses. On the whole, the volume of irretrievable water consumption by the population exceeds 20 km^3/yr or 17%. Industrial water consumption constitutes 500 km^3/yr on a global scale.

Especially large fluctuations in the water cycle are caused by irrigation. The total amount of water used for irrigation is determined by physico-geographical conditions, agricultural crops, and the technical equipment of irrigation systems and their application technology. The general characteristic of the water consumption dynamics is given in Tables 1.11 and 1.12.

Along with the quantitative change of the water balance in the 20th century there has developed the process of the exhaustion of water resources connected with the increasing pollution of natural waters. Most sewage containing harmful chemicals is discharged into the hydrophysical network without preliminary purification, causing a change in the water resources quality and making them unfit for further consumption without purification. This circumstance has led to a preferential use of underground, less polluted waters. As a result, conditions have been created for a sharp violation of the natural water cycle in the biosphere.

Forest is one of the natural resources being catastrophically exhausted. Forests are in a process of being exterminated over the globe. For example, in the USA only 18 Mha of 365 Mha of forested area remains (Watson *et al.*, 2000). The area of tropical forests has decreased by two-thirds. In many countries the forested areas occupy a small part of their territory (India – 18%, Greece – 15%, Spain – 12.5%, China – 9%, Cuba – 8%). On Madagascar the forests have been exterminated almost completely.

Soils as an independent natural body are one of the natural resources of great importance for the biosphere. Mother rock, living organisms (plants, animals, and micro-organisms), climate, and relief are all factors of soil formation. The anthropogenic forcing on soil is manifested through its impoverishment. The impoverished ploughed soil is easily subjected to erosion. Sub-ground and open cast mining is a factor promoting soil destruction. Fertile soils become barren if deprived of vegetation areas transforming them into the so-called "industrial deserts".

Thus the anthropogenic factor in the present-day conditions is versatile and affects practically all the natural processes. Clearly, a complete account of the whole spectrum of anthropogenic forcings on the environment is impossible because of lack of required data. In modelling the global carbon cycle a unit is needed in the models responsible for the parameterization of anthropogenic processes and providing their consideration in simulation experiments.

1.3.2 Modelling the demographic processes

The demographic situation determines the dynamics of anthropogenic processes. Therefore attempts to develop models of demographic processes are justified. The existing predictions of changes of the size of the population and variations in its spatial distribution enable one to synthesize the scenarios to be used in a global model as well as to try to solve the problem of its verification. An adequate model of the global demographic process requires an extensive database covering the characteristics of changes in the standards of demographic behaviour, detailed information on intensities of demographic processes in various regions of the planet, assessments and criteria of demographic policy, and especially important, the structural indicators of human society (Demirchian and Kondratyev, 1999).

Many authors have considered the problem of parameterization of demographic processes. The demographic processes are considered part of the planetary biospheric processes in which *Homo sapiens* plays the role of a user of biocenoses products and a regulator of energy and matter fluxes among them.

1.3.2.1 Matrix model of the population size dynamics

In accordance with the needs of the global model, the unit of population dynamics includes the impact of the following factors (Krapivin and Marenkin, 1999; Logofet, 2002):

- per capita food provision F (calculated as the sum of several shares of vegetation and animal population of the region as well as fish catch);
- the share A of animal protein in the human diet (determined from the contribution to F made by animals and fish);
- the level of public medical service M (in this model version it is in proportion to per capita financing);
- genetic load for human population G (grows slowly with population development and depends on the level of the environmental pollution).

The sex structure of the population and the processes of population migration between regions are ignored. The age structure contains three groups (0–14 years, 15–64 years, and 65 years and older). There is also a fourth group of population (invalids and disabled persons out of these three age groups). For terminology convenience the fourth group is considered as not affecting the age structure.

Let S_t be the age structure at a time moment t, then the population size dynamics can be described by the following matrix equation:

$$S_{t+1} = D \times S_t$$

where D is the demographic matrix 4×4 including the effect of the factors F, A, M, and G:

$$D = \begin{Vmatrix} d_{11} & d_{12} & 0 & 0 \\ d_{21} & d_{22} & 0 & 0 \\ 0 & d_{32} & d_{33} & 0 \\ d_{41} & d_{42} & d_{43} & d_{44} \end{Vmatrix}$$

Matrix D first differs from the traditional matrix for the model of the population with the age structure by Leslie (1945) by the inequality to zero of its diagonal elements, explained by the overlapping of the next generations, and, second, by the last line that reflects the non-age character of the fourth group. The diagonal elements of matrix D are determined from apparent balanced relationships:

$$d_{ii} = 1 - \mu_i - \sum_{j>i}^{4} d_{ji} \qquad (i = 1 \div 4) \tag{1.1}$$

which include the coefficients of mortality μ_i of the ith group, descending functions of the per capita food provision, and the medical service level, $\mu_i = \mu_i(F, M)$ with:

$$\lim_{F,M \to \infty} \mu_i(F, M) = \mu_{i,\min} > 0$$

where $\mu_{i,\min}$ characterizes a minimum of physiological mortality with optimal food provision and medical service.

It is supposed that the reproductive potential of the human population is entirely concentrated in the second age group. The d_{12} coefficient value is considered to be a regional constant. Note that the birth rate coefficient is a complex and poorly studied function of many variables, the ethnic traditions being one of the important variables. The birth rate is affected by religion, for instance, the Moslems consider children to be their wealth. And in South Asia the religious norms order each family to have at least one son. Catholicism also markedly affects the birth rate index. Therefore if the function μ_i is well parameterized by the statistical data, the coefficient d_{12} is still an unidentified parameter. Nevertheless, the d_{12} value for different global regions can be estimated from the birth rate data.

The coefficients of transition to the age groups d_{21} and d_{32} are determined from the duration of the respective age group and the hypothesis of a uniform distribution of ages within the group, namely: $d_{21} = 1/15$ and $d_{32} = 1/50$.

Genetically stipulated diseases and a deficit of protein in the food of children are the main causes of transitions from a younger age group to the invalid group. Thus $d_{41} = d_{41}(G, A)$ is the age function of its arguments. Coefficients d_{42} and d_{43} are small compared to d_{41} and determined by the sum of the vital parameters of the environment, the level of medical service, and other indicators of the anthropogenic medium (investments). In a first approximation, suppose $d_{4k} = \Delta_k \mu_k (k = 23)$, where $\Delta_k \ll 1$.

Demographic matrix D has a certain set of properties which make it possible to reveal the typical trends of the demographic process. A maximum eigenvalue is:

$$\lambda_{\max}(D) = \max\{\lambda_1; d_{33}; d_{44}\} > 0$$

where

$$\lambda_1 = 0.5(d_{11} + d_{22}) + 0.25[(d_{11} + d_{22})^2 + d_{12}d_{22} - d_{11}d_{22}]^{1/2}$$

It follows from Equation (1.1) that if $\lambda_{\max}(D) \geq 1$, then $\lambda_{\max}(D) = \lambda_1$ (i.e., that is, the rate of *Homo sapiens* population growth in the neighbourhood of stationary age distribution is determined by the λ_1 value). The stationary age structure is calculated as an eigenvector corresponding to $\lambda_{\max}(D)$. The coefficients and modifying dependences of matrix D calculated, as a rule, from the data of Stempell (1985), give the dominating eigenvector $P_D = (0.369; 0.576; 0.05; 0.005)$. The first three components of vector P_D coincide, to within the second sign, with the global data on the age complement of the global population.

The matrix version of the global model demographic unit has a time step of one year. An inclusion of this unit in the chain of other global model units with an arbitrary time step $\Delta t < 1$ year is needed as an adjusting procedure. This is possible, for instance, if the demographic unit is included only at time moments multiple to 1 year. In this version the spasmodic changes of the population size will cause some disbalance in continued trajectories of the global system. Another procedure free of this drawback consists in the use of a prognostic equation:

$$S_{t+\Delta t} = S_t + \Delta t(D - I)S_t \tag{1.2}$$

where I is an identity matrix. One can demonstrate that for the age structure vectors coinciding in direction with the eigenvector of matrix D, the main term of a relative error accumulated during one year with a time step $\Delta t = 1/n$ constitutes $(1 - 1/n)(\eta - 1)^2/2$. Let the right-hand part of Equation (1.1) be denoted by $f(S, \Delta t)$. Then if e is the eigenvector of matrix D with an eigenvalue η, then with the time step $\Delta t = 1/n$:

$$f(e, 1/n) = e + (1/n)(D - I)e = (1 + [\eta - 1]/n)e$$

It is apparent that $S_{t+n\Delta t} = f^{(n)}(S_t, 1/n)$ and $S_t = ef^{(n)}(e, 1/n) = (1 + [\eta - 1]/n)^n e$. But if $\Delta t = 1$, then $S_{t+1} = f(e, 1) = \eta e$, so that a relative error resulting from a subdivision of the 1-year time interval is:

$$(1 + [\eta - 1]/n)^n - \eta = n(n - 1)(\eta - 1)^2/(2n^2) + 0((\eta - 1)^3)$$

In other words, the accuracy of an approximation of Equation (1.2) rises with Δt approaching unity.

1.3.2.2 Differential model of population dynamics

The matrix model considered above enables one to use the demographic statistics by the age groups but requires knowledge of many parameters, which leads to uncertainties in the global model. Therefore a second version of the demographic unit is suggested which simulates the dynamics of only the total size of the population. This version assumes that an impact of numerous environmental factors and social aspects on the population size dynamics G_i in the ith region is manifested through birth rate R_{Gi} and mortality M_{Gi}:

$$dG_i/dt = (R_{Gi} - M_{Gi})G_i \qquad (I = 1, \ldots, m) \tag{1.3}$$

Birth rate and mortality depend on food provision and its quality, as well as environmental pollution, living standards, energy supply, population density, religion, and other factors. Within the global model all these factors will be taken into account following the principle:

$$R_G = (1 - h_G)k_G G H_{GV} H_{GO} H_{GG} H_{GMB} H_{GC} H_{GZ}$$

$$M_G = \mu_G H_{\mu MB} H_{\mu G} H_{\mu FR} H_{\mu Z} H_{\mu C} H_{\mu O} G + \tau_{GO} \tau_{GC}^{\omega(G)}$$

where for simplicity purposes we shall omit the i index attributing the relationship to the ith region; h_G is the coefficient of food quality ($h_G = 0$ when the quality of food is perfect); coefficients k_G and μ_G point to levels of birth rate and mortality, respectively; indices τ_{GO} and τ_{GC} characterize the dependence of the regional population mortality on O and C – indicators of the environmental state (within the global model it is the content of O_2 and CO_2 in the atmosphere) manifesting through human physiological functions; coefficient $\omega(G)$ characterizes an extent of the influence of population density on mortality (in the present conditions $\omega(G) \approx 0.6$); functions $H_{GV}(H_{\mu FR})$, $H_{GO}(H_{\mu O})$, $H_{GC}(H_{\mu C})$, $H_{GMB}(H_{\mu MB})$, $H_{GG}(H_{\mu G})$, and $H_{GZ}(H_{\mu Z})$ describe, respectively, an impact on birth rate (mortality) of the environmental factors, such as food provision, atmospheric O_2

and CO_2 concentrations, living standards, population density, and environmental pollution. Functions $H_{\mu O}$ and $H_{\mu C}$ approximate the medico-biological dependences of mortality on the atmospheric gas composition. Let us consider all these functions in more detail. For this purpose we shall formulate a number of hypotheses concerning the forms of dependences of mortality and birth rate on various factors.

The results of numerous studies with national specific features taken into account enable one to assume the following dependence as an approximation of the function H_{GV}: $H_{GV} = 1 - \exp(-V_G)$, where V_G is an efficient amount of food determined as a weighted sum of the components of the *Homo sapiens* food spectrum:

$$V_{Gi} = k_{G\Phi i}\Phi + k_{GFi}\left(F_i + \sum_{j \neq i} a_{Fji}F_i\right) + k_{Gri}I_i(1 - \theta_{Fri} - \theta_{uri})$$
$$+ k_{GLi}L_i + k_{GXi}\left[(1 - \theta_{FXi})X_i + (1 - \nu_{FXi})\sum_{j \neq 1} a_{Xji}X_j\right]$$

Here Φ is the volume of vegetable food obtained from the ocean; F represents protein food; L represents food produced by forests; X represents vegetable food produced by agriculture; I represents fishery products; coefficients $k_{G\Phi}$, k_{GF}, k_G, k_{GL}, and k_{GX} are determined using the technique described in Section 4.5; a_{Fji} and a_{Xji} are, respectively, shares of the protein and vegetable food in the ith region available for use by the population of the ith region; θ_{FXi} and ν_{FXi} are shares of vegetable food produced and imported by the ith region, respectively, to produce the protein food; θ_{Fri} and θ_{uri} are shares of fishery spent in the ith region on the protein food production and fertilizers, respectively.

With an increasing food provision the population mortality drops to some level determined by the constant $\rho_{1,\mu G}$ at a rate $\rho_{2,\mu G}$, so that $H_{\mu FR} = \rho_{1,\mu G} + \rho_{2,\mu G}/F_{RG}$, where the normalized food provision F_{RG} is described by the relationship $F_{RG} = F_{RG}(t) = V_G/G(F_{RG}(t_0) = F_{RGO} = V_G(t_0)/G(t_0))$. Similarly, we assume that the birth rate depending on the living standard M_{BG} of the population is described by the function with saturation so that a maximum of birth rate is observed at low values of M_{BG}, and at $M_{BG} \to \infty$ the birth rate drops to some level determined by the value of a_{*GMB}. The rate of transition from maximum to minimum birth rate with changing M_{BG} is set by the constants $a_{1,GMB}$ and $a_{2,GMB}$:

$$H_{GMB} = a_{*GMB} + a_{1,GMB}\exp(-a_{2,GMB}M_{BG})$$

where

$$M_{BG} = \{(V/G)[1 - B - U_{MG} - U_{ZG}]/[1 - B(t_o) - U_{MG}(t_0) - U_{ZG}(t_0)]\}$$
$$\times [E_{RG}(t)/E_{RG}(t_0)]$$
$$E_{RG}(t) = 1 - \exp[-k_{EG}M(t)/M(t_0)]$$

The dependence of mortality on the living standard is described by the decreasing function:

$$H_{\mu MB} = b_{1,\mu G} + b_{2,\mu G}\exp(-b_{*\mu G}M_{BG})$$

This function shows that population mortality with an increasing per capita share of capital drops with the rate coefficient $b_{*\mu G}$ to the level $b_{1,\mu G}$.

The birth rate and mortality in certain limits are, respectively, decreasing and increasing functions of population density:

$$H_{GG} = G_{1,G} + G_{*G} \exp(-G_{2,G} Z_{GG}) \qquad H_{\mu G} = \theta_{1,\mu G} + \theta_{2,\mu G} Z_{GG}^{\omega \mu G}$$

where $Z_{GG} = G(t)/G(t_0)$.

Finally, an important aspect of the *Homo sapiens* ecology is the environmental state. In this connection the problems of anthropobiocenology have been widely discussed in scientific literature and many authors try to find out the required regularities. Without going into the details of these studies, most of which cannot be used in the global model, we shall confine ourselves to the following dependences:

$$H_{GZ} = l_{1,G} \exp(-l_{*G} Z_{RG}) \qquad H_{\mu Z} = n_{1,\mu G} + n_{2,\mu G} Z_{RG}$$

$$\tau_{GC} = \begin{cases} \tau_{1,GC} + \tau_{2,GC}(C_a - C_{1,G}) & \text{for } C_a > C_{1,G} \\ \tau_{1,GC} & \text{for } 0 \leq C_a \leq C_{1,G} \end{cases}$$

$$H_{\mu O} = f_{1,\mu G} + f_{2,\mu G}/O(t) \qquad H_{\mu C} = \exp(k_{\mu G} C_a)$$

$$H_{GO} = 1 - \exp(-k_{GO} O) \qquad H_{GC} = \exp(-k_{GC} C_a) \qquad Z_{RG} = Z(t)/Z(t_0)$$

$$\tau_{GO} = \begin{cases} \tau_{1,GO} & \text{for } O > O_{1,G} \\ \tau_{2,GO} - (\tau_{2,GO} - \tau_{1,GO}) O/O_{1,G} & \text{for } 0 \leq O \leq O_{1,G} \end{cases}$$

where C_a is the atmospheric CO_2 concentration, and $C_{1,G}$ and $O_{1,G}$ are safe levels (for humans) of the atmospheric CO_2 and O_2 content.

1.3.3 Parameterization of anthropogenic processes

The biosphere Ω as a complicated unique system is functioning following the laws of coevolution of its subsystems, with the human society H and nature N being the basic ones. The impact of human activity on nature, being of comparatively small scale, can apparently only be assessed with newly developed technology. Clearly, for this purpose a systems approach is needed to formalize the ecological, technological, economic, and geopolitical interactions of the subsystems H and N. In general, the system H has at its disposal technologies, science, economic structure, size of population, etc. The system N has a set of mutually dependent processes, such as climatic, biogeocenotic, biogeochemical, geophysical, etc. (Kondratyev *et al.*, 1983; Kondratyev and Grigoryev, 2000, 2002; Kondratyev and Galindo, 2001; Kondratyev *et al.*, 2001a; Kondratyev, 2000b).

From the viewpoint of the theory of systems, H and N are open systems. Their division is a conditional procedure aimed at selecting controlled and non-controlled components of the environment. Without going into philosophical and methodical aspects of this procedure, we assume that both systems are symmetrical from the viewpoint of their simulation (i.e., we assume that each system has a goal, structure, and behaviour of its own). Let $H = \{H_G, H_S, |H|\}$ and $N = \{N_G, N_S, |N|\}$, where H_G and N_G are the goals of the systems, H_S and N_S are strategies of the systems' behaviour, and $|H|$ and $|N|$ are the system structures, respectively. Then, the *H–N* interaction can be described by the process of (V, W)-exchange in that each of the systems, to reach its goal, spends the resources V and, in exchange, obtains a new resource of an amount W. Each system is aimed at an optimization of the (V, W)-exchange with another system (i.e., to maximize W and minimize V).

Now write the equations for (V, W)-exchange:

$$W_H(H^*, N^*) = \max_{\{H_s,|H|\}} \min_{\{N_s,|N|\}} W(H,N) = \min_{\{N_s,|N|\}} \max_{\{H_s,|H|\}} W(H,N) \tag{1.4}$$

$$W_H(H^*, N^*) = \min_{\{H_s,|H|\}} \max_{\{N_s,|N|\}} W(H,N) = \max_{\{N_s,|N|\}} \min_{\{H_s,|H|\}} W(H,N) \tag{1.5}$$

where H^* and N^* are optimal systems. Here, in contrast to traditional game theory models, there exists a power spectrum of the *H–N* interaction covering the final intervals of changes in the payoffs W_H and W_N depending on the aggressiveness of each of them. A concrete definition of the payoff function requires a certain systematization of the mechanisms of human and natural coevolution. One of the widely used models of the balanced development of the world community and nature subjected to the criterion in Equations (1.4) and (1.5) consists of identification of the system H with a number of large cities with adjacent industrial and recreation zones. There are numerous versions for consideration and formal descriptions of such structures. In particular, there exists a well known method of logic/information modelling of the processes of rational nature use and a simulation method of controlling the ecological/economic systems (Ougolnitsky, 1999). According to these methods, to solve a concrete problem, it is necessary to conceptualize the information base of the model and to select the most general relationships between the elements of the interacting systems. This procedure is completed with enumeration of all functional elements of the systems and determination of the capacity loads on their elements. The whole procedure ends with a synthesis of the simulation model, which within assumed assumptions is an instrument of investigation. In the case considered we assume that the structure of the system H includes population G, pollutants Z, and natural resources M (i.e., $|H| = \{G, Z, M\}$). Similarly, the structure of the system N consists of elements such as the climate parameter (temperature T), environmental quality Q, areas of forests σ_L, and agricultural lands σ_X, (i.e., $|N| = \{T, Q, \sigma_L, \sigma_X\}$).

The behavioural strategy of the system H is formed from the distribution of investments into the retrieval of resources U_{MG}, struggle with pollutions U_Z, and agricultural investments U_{BG} (i.e., $H_S = \{U_{MG}, U_{ZG}, U_{BG}\}$). The behavioral strategy of the system N is identified with the rate of investments ageing T_V, population mortality μ_G, agricultural productivity H_X, the cost of resources retrieval G_{MG}, and time constant of biospheric self-cleaning of pollutants T_B:

$$N_S = \{T_V, \mu_G, H_X, G_{MG}, T_B\}$$

Equations (1.4) and (1.5) are basic equations for the model of survival. In general, this model is formulated in terms of the theory of evolutionary technology of modelling. If all possible states of the biosphere with acceptable conditions for human life constitute the multitude $\Gamma = \{\Gamma_i\}$, then as a result of the effect of C_k on the biosphere, two outcomes are possible: $C_k(\Gamma_i) \rightarrow \Gamma_i \in \Gamma$ and $C_k(\Gamma_i) \rightarrow \Gamma_s \notin \Gamma$. When the sequence of biospheric states $\{C_k(\Gamma_i)\} \in \Gamma$, then we can speak about the persistent coevolution of the system $H \cup N$. No doubt, there is a problem of an adequacy between real processes and their simplified presentation as a model. Nevertheless, despite the philosophical doubt in expedience of mathematical modelling for the perspective assessment of the kinetics of biospheric parameters, the model approach has proved to be profitable. The use of the biospheric model instead of the biosphere itself is convenient, first, because there is more information about the model than about the biosphere, second, because the model is easier to handle, and third, because direct experiments with the biosphere are dangerous. All these aspects are the subject of studies for global ecoinformatics aimed at achieving a sufficient similarity of the observed behaviour of the system $H \cup N$ and the model. This is possible through the constant renewal of databases and the broadening of knowledge in accordance with the technology of the formation of the information multitude of the biospheric parameters suggested by Krapivin and Kondratyev (2002).

Trends in human activity are determined by a great number of factors. All of them are reduced to economy and energetics, whose interconnections can be described by the linear regression $V = k_{Ve}e + b_{Ve}$, where V is the regional budget, and e is the energy produced in this region. The constant coefficients k_{Ve} and b_{Ve} reflect the specific features of economic activity and its efficiency. The general scheme of a possible model of the socio-economic processes includes V_i as the size of the funds in the ith region, which this region can use when planning its anthropogenic activity:

$$dV/dt = G_{VG}G_{VMG} - V/T_{VG}$$

where T_{VG} is the time constant of basic funds deterioration; and coefficients G_{VG} and V_M determine the rate of the funds generation. The function V_M defines the dependence of the rate of the funds generation on the living standard M_B of the population of the region. This dependence is described by a logarithmic function:

$$V_{MG} = k_{MGV} \ln(1 + k_{1,MGV}M_{BG})$$

where the coefficient k_{MGV} is chosen from the condition:

$$k_{MGV} \ln(1 + k_{1,MGV}M_{BG}(t_0)) = 1$$

In this case the coefficient G_{VG} represents the volume of the funds generated per capita at the moment t_0 in the region. The time constant of the basic funds deterioration is a function of the scientific–technical progress and is considered in the model as a controlling parameter.

The basic funds determine the intensity and direction of anthropogenic activity. In particular, the generation of pollutants and their utilization are substantially regulated by this distribution.

Let Z be the concentration of an arbitrary pollutant and Z_{VG} the rate of the pollutant's assimilation due to realization of the technologies of the environmental purification. Then:

$$dZ/dt = Z_{VG} - Z_{TV}$$

where $Z_{TV} = Z_T + Z_V$; Z_T is the natural rate of the pollutant decomposition; and Z_V is the rate of the pollutant assimilation due to the environmental purification.

Assume that the rate of pollution generation in the region is in proportion to population density with the coefficient k_Z and depends on the per capita volume of the funds $V_{RG} - V/G$:

$$Z_{VG} = k_Z G Z_{*VG}$$

where

$$Z_{*VG} = Z_{VG,\max}\,[1 - \exp(-G_{1,ZG} V_{RG})]$$

The rate of an artificial assimilation of pollutants is determined by the capital share U_{ZG} to be used to intensify the cleaning processes in the environment: $Z_V = U_{ZG}V/G_{ZV}$ where G_{ZV} is the cost of environmental cleaning of a unit of pollution.

The rate of natural removal of harmful wastes is in direct proportion to the pollutants concentration and indirect proportion to the decomposition time T_Z: $Z_T = Z/T_Z$. Naturally, the processes of generation and utilization of pollutants are more complicated. Therefore the respective unit of the global model needs further development considering a new knowledge in the field of pollution utilization. The generated pollutant is scattered in the environment getting partially into the atmosphere, soil, and hydrosphere. Possible means of pollutant propagation in these media are described in other chapters. However, for a complex analysis of the pollutant an equation for the global model is required which would reflect the general state of the atmosphere. Let us introduce the atmospheric turbidity index B measured by the weight of admixtures in the atmosphere per unit area of the biospheric surface. A change in this index is determined by: the share N_B in the general pollution Z generated in the region; the amount of wastes from energy-generating enterprises (the coefficient N_A); and the water vapour content in the atmosphere W_A, and the rate of its natural cleaning $1/TB$ assumed to be equal to the inverse value of time TB of sedimentation of dust and smoke particles and water droplets. The effect of water vapour on atmospheric transparency is manifested through the derivative $dW_A/dt = \rho_B\, dT/dt$ where T is the atmospheric temperature and ρ_B is the coefficient. As a result, the dynamic equations for the system variable B will be:

$$dB/dt = N_B Z + N_A b_{GC} G + \rho_B dT/dt - B/T_B + B_n$$

where B_n is the rate of atmospheric pollution resulting from natural processes (rock weathering, erosion, volcanism, etc., at a rate of $\sim$0.78 t/km^2/yr).

The horizontal transport of pollutants and the respective change in B are simulated with an account of wind speed and the spatial grid $\Delta\varphi \times \Delta\lambda$. The parameter T_B is calculated with the formula $T_B = h_e/v_s$ where h_e is the efficient height of the pollutant scattering in the atmosphere and v_s is the rate of particle sedimentation calculated with the Stokes formula: $v_s = 2g\rho_k r^2/(9\zeta)$. Here g is the

Table 1.16. Annual amount of particles emitted into the atmosphere with a radius >20 μm.

Type of particles	Flow of particles (10^6 t/yr)
Natural particles of soil and stone	100–500
Particles of forest fires and timber industry waste burning	3–150
Sea spray	300
Volcanic dust	25–150
Particles generated by gas production	
Sulphates from H_2S	130–200
Ammonium salts from HN_3	80–270
Nitrates from NO_x	60–430
Hydrocarbons from vegetable compounds	75–200
Synthetic particles	10–90

acceleration of gravity, ρ_k is the particle's density, r is the particle radius, and ζ is the coefficient of molecular viscosity. For instance, for particles with $r \in [10^{-5}, 5 \times 10^{-3}]$ cm and $\rho_k \in [1, 4] g/\text{cm}^3$ we obtain $v_s = 10^2$ cm/s, which for the calm atmospheric conditions gives $T_B = 7$ hours.

Some characteristics of atmospheric aerosols are given in Tables 1.13, 1.14, and 1.16.

Most complicated is the scattering process of pollutants in the water domain and in the soil layer. These processes are simulated in the respective chapters in studies of the hydrospheric components and associated processes. The relationships between natural and anthropogenic formation of the air or water quality deduced and applied in simulation experiments contain the parameters determined by the pace of expenditure and renewal of natural resources. The pace dynamics are determined by the rate of population growth and civilization development. On average, the annual global acceleration of the natural resource consumption constitutes about 4%, with a 2% mean annual population increment. Hence, the value of per capita natural resource consumption R_{MG} is an important indicator of present civilization development, since the annual consumption of accessible resources constitutes $\Delta M = R_{MG}G$. The consumption of natural resources increases with the improving living standard but an extent of this growth decreases with increasing investments into science. Clearly, this relationship is complicated and has been poorly studied. Therefore to consider this problem, assume the following parameterization:

$$R_{MG} = m_G \ln(1 + M_{BG})$$

where the parameter m_G is the time function set in simulation experiments as a scenario.

The protection and restoration of natural resources requires certain material expenses. Assume that the effect of these measures is in proportion to the share

U_{MG} of the respective investments and in inverse proportion to the cost G_{MG} of the retrieved conditional unit of resources:

$$dM/dt = \Delta M + U_{MG}V/G_{MG}$$

The investments spectrum also includes the sphere of agricultural production, including expenses on scientific investigations in the field of selection, plant physiology, agrotechnology and reclamation, as well as solution of the technical and social problems. Let qV be the share of capital invested in agriculture. We obtain:

$$dq_V/dt = (U_{BG}B_{FG}B_{qG} - q_V)/T_B$$

where the controlling parameter $U_{BG} = U_{BG}(t)$ in simulation experiments is described by the scenario and the remaining components are determined by the following dependences:

$$B_{FG} = \exp(-b_{BG}F_{RG}) \qquad B_{qG} = b_{1,BG} + b_{2,BG}(q_{MG}/q_{FG})^{\alpha SG}$$

$$q_{MG} = b_{1,qG} + b_{2,1G}(M_{BG})^{\alpha qG} \qquad q_{FG} = a_{qG}(F_{RG})^{\beta qG}$$

1.3.4 Megalopolises

A megalopolis is an Earth's surface area with high urbanization, developed industry, and other human activity attributes concentrated in a limited territory. The number of such territories in the world is constantly growing and their total area increases with the increasing size of global population. Characteristic examples of large megalopolises are Moscow, Tokyo, New York, Ho Chi Minh, etc. Some typical estimates of the environmental parameters in megalopolises are given in Table 1.17. For instance, the Moscow megalopolis is characterized by high concentrations of the sources of anthropogenic pollution over the limited territory (energy enterprises, chemical industry, and automobile transport). Their share in total emissions

Table 1.17. Comparative characteristic of the developed and developing megalopolises.

	Estimate of the parameter	
Parameter	Moscow	Ho Chi Minh
Area (km^2)	100	31
Population size (10^3 people)	8,894	8,563
Planted area (km^2)	14	16
Number of enterprises (10^3)	25	14
Number of pollution sources (10^3)	100	70
Emissions of harmful pollutants into the atmosphere (10^3 t/yr)	1,153	248
Length of the open water bodies within the territory (km)	398	101
Volume of sewage (10^6 m^3/yr)	3,036	226

constitutes 90.4%. The polluted air plume from Moscow can be observed at a distance of about 100 km from the city. The state of the water bodies within the megalopolis is determined by the input of sewage, surface runoffs as well as runoffs from industrial enterprises to the basic rivers Moskwa and Yausa, as well as into 70 rivers and springs within the territory of the megalopolis. The concentration of chemicals in the Moskwa River both in the city and downstream varies widely depending on the season. For instance, the copper ion content varies during the year from 0.004 to 0.013 mg/l (4–13 permissible concentration level (PCL)) with a maximum in the spring. The content of oil products varies within 0.25 to 0.6 mg/l (5–12 PCL). A similar situation exists in the small Yausa River where the concentration of copper varies from 0.007 to 0.12 mg/l and that of oil products from 0.38 to 0.7 mg/l. The ecological service of the megalopolis monitors the local concentrations of pollutants forming a respective time series of data for the state of the environment. There is a global network of megalopolises, consideration of which in the global model results in an increase in its accuracy. Many megalopolises are in the stage of formation. Ho Chi Minh is an example of a young megalopolis as well as some adjacent provinces of South Vietnam – Dong Nai, Bin Zyong, and Baria-Vung Tau. This megalopolis occupies an important place in the economy of Vietnam covering 75% of the industrial production of South Vietnam and 50% for the whole country (Si and Hai, 1997; Bui Ta Long, 1998). The development of the megalopolis infrastructure includes ecological and sanitary control services for the environment at an early stage, which makes possible the perspective planning of the environment.

An analysis of the data on the structure of the environment of megalopolises of the world suggests the conclusion that, for its complex assessment, it is possible to develop a sample system of the models simulating the transport and propagation of pollutants in the atmosphere and in water bodies. The input information for this model can come from both the monitoring systems and the global model.

Characteristic linear dimensions of a megalopolis constitute tens of kilometers. This means that to simulate the processes of atmospheric transport of pollutants a Gaussian-type model can be used. To assess the quality of the atmosphere in the megalopolis it is sufficient to form a composition of the Gaussian streams and, at the points of their intersections, to summarize the concentrations of the respective types of pollutants.

Let the source of pollutant of type s be located at a point (0, 0) in the coordinate system (x, y, z) where the 0_x axis is in the wind direction and the 0_z axis is normal to the Earth surface at a height h. Then the concentration C_s of the pollutant of the type s at any point in (x, y, z) can be calculated with the formula:

$$C_s(x, y, z) = Q(2\pi U\sigma_y\sigma_z)^{-1} \exp\{-y^2/[2(\sigma_y)^2] - (z+H)^2/[2(\sigma_z)^2] - (z-H)^2/[2(\sigma_z)^2]\} \tag{1.6}$$

where $H = h + \Delta h$; $\sigma_y = (U/[2k_y(x)])^{1/2}$; $\sigma_z = (U/[2k_z(x)])^{1/2}$; k_y and k_z are the coefficients of eddy diffusion corresponding to the y and z axes; and U is the wind speed. The parameters Δh is calculated from the Briggs formula, $\Delta h = \Delta_1$ for unstable and neutral conditions of the lower atmospheric layer and $\Delta h = \Delta_2$ for stable conditions of the lower atmospheric layer:

$$\Delta_1 = \begin{cases} 1.6F^{1/3}x^{2/3} & \text{for } x \leq x_p \\ 1.6F^{1/3}(3.5x_p)^{2/3} & \text{for } x > 3.5x_p \end{cases}$$

$$\Delta_2 = \begin{cases} \min\{2.6(FU^{-1}/S_n)^{1/3}, 1.6F^{1/3}(3.5x_p)^{2/3}\} & \text{for } U > 1.4\, m/c \\ 5.3F^{1/4}S_n^{-3/8} - D/2 & \text{for } U \leq 1.4\, m/c \end{cases}$$

where $F = g(T_s - T_a)V/(\pi T_s)$ is the buoyancy coefficient; $S_n = (g/T_a)(\partial T_a/\partial_z + 0.01)$ is the coefficient of stability; $\partial T_a/\partial z$ is the potential temperature gradient at the point of pollutant emission; T_s is the temperature of the emitted gas (°K); T_a is the ambient air temperature (°K); D is the diameter of the pollution source (m); V is the gas expenditure per unit time (m^3/s); g is the gravity acceleration; and x_p is the distance along the wind direction where the turbulent diffusion dominates over pollution transport:

$$x_p = \begin{cases} 14F^{5/8} & \text{for } F \leq 55 \\ 34F^{2/5} & \text{for } F > 55 \end{cases}$$

To calculate the pollutant concentration at an arbitrary point in the megalopolis territory, it is necessary to take into account the impacts of all the sources. For this purpose we choose an arbitrary system of coordinates $x0y$. Assume that the wind direction vector is at an angle α to the $0x$ axis. Denote the coordinates of the pollution sources Q_i $(i = 1, \ldots, n)$ by (a_i, b_i). Then the net concentration of the pollutant of type z at an arbitrary point (u, v) at a height z will be:

$$C_s(u, v, z) = \sum_{i=1}^{n} C_s(u, v, z; a_i, b_i)$$

where $C_s(u, v, z; a_i, b_i)$ is the concentration of the pollutant of type s at the point (u, v) at height z due to the source Q_i. Knowledge of the angle α makes it possible to reach an agreement between the coordinates of the systems $u0v$ and $x0y$.

The hydrological cycle parameters for a megalopolis are calculated using a general model described by Krapivin and Kondratyev (2002). However, simplifications are possible here due to the absence of some elements of the standard scheme or changes in their weightings. The water flow by canals (pipelines) is most characteristic of a megalopolis. In this case we have the following equation:

$$\sigma^{-1}\partial Q/\partial t - Q\sigma^{-2}\partial\sigma/\partial t + Q\sigma^{-1}[\sigma^{-1}\partial Q/\partial x + Q\sigma^{-2}(\partial\sigma/\partial x + \partial\sigma/\partial y \times \partial y/\partial x)] + g(\partial z/\partial x + \partial y/\partial x + p) + E_{IL} = 0 \quad (1.7)$$

where z is the level of the canal; y is the water depth in the canal measured from level z; p is the canal bottom friction; E_{IL} is the impact of inflows and outflows; σ is the effective area of the canal's cross section at the point with the x coordinate; x is the water flow direction; and Q is the water flow at point x.

In addition to Equation (1.7), an equation is needed for the water medium continuity: $\partial\sigma/\partial t + \partial Q/\partial x = q$, where q is the intermediate inflow per unit time per unit length of the canal's section.

A change of temperature T_w of an element of the water volume V is described by the relationship:

$$V\partial T_w/\partial t = (lc^{-1}\rho^{-1})\partial(\sigma_x\Delta_x\partial T_w/\partial x)/\partial x - l\partial(Q^* T_w)/\partial x$$
$$+ [SH_{al} + \theta]c^{-1}\rho^{-1} - T_w\partial V/\partial t$$

where l is the length of the canal section; S is the section surface area on the border with the atmosphere; ρ is the water density; c is the specific heat capacity of water; σ_x is the area of the cross section of the canal section at point x; Q^* is an advective heat flux through the section's boundaries; θ is the outer heat source; Δ_x is the coefficient of eddy diffusion; and H_{al} is the heat flux velocity through a unit surface area.

The heat flux through the surface on the border with the atmosphere is the sum of the following components: $H_{al} = E_1 + E_2 - E_3 - E_4 - E_5$, where E_1 and E_2 are fluxes of the incoming short wave and long wave radiation, respectively; E_3 is the outgoing long wave radiation flux; E_4 is the evaporation-driven outgoing heat flux; and E_5 is the convective heat exchange. The E_1 parameter is a complex function of cloudiness, atmospheric dust load, sun location, and other geophysical parameters. For the E_2 flux we use the dependence $E_2 = 1.23 \times 10^{-16}[1 + 0.17\chi^2](T_a + 273)^6$, where χ is the amount of cloud. The E_3 flux is estimated from the expression $E_3 = \varepsilon\sigma_1(T_a + 273)^4$, where ε is the water emissivity ($\sim$0.07) and σ_1 is the Stephan-Boltzman constant (1.357×10^{-8} cal/m^2/s/°C^4). For $T_a \in [0°, 30°]$C an approximation $E_3 = 7.36 \cdot 10^{-2} + 1.17 \times 10^{-3}\, T_a$ is valid.

Heat losses through evaporation are presented as the dependence $E_4 = \rho L e_a$, where ρ is the water density (kg/m^3), L is the latent heat of evaporation (kcal/kg), and E_a is the rate of evaporation (m/s). The latter parameter is approximated by the relationship $E_a = (a + bW)(p_s - p_a)$, where W is the above-surface wind speed, p_s is the pressure of the saturated vapour over the surface (hPa), p_a is the vapour pressure in the atmosphere (hPa), and a and b are empirical coefficients. The p_s parameter is well approximated by the linear function $p_s = \alpha_p + \beta_p T_a$ whose coefficients are given in Table 1.18.

Finally, the convective heat transport $E_5 = R_4 E_4$, where $R_4 = 6.1 \times 10^{-4} p_a(T_a - T_d)(p_s - p_a)$ is the atmospheric pressure (hPa), and T_d is the dry air temperature (°C). We obtain $H_{al} = \mu_{al} - \lambda_{al} T_a$, where $\mu_{al} = E_1 + E_2 - 7.36 \times 10^{-2} - \rho L(a + bW)(\alpha_p - p_a - 6.1 \times 10^{-4} p_a T_d)$ and $\lambda_{al} = 1.17 \times 10^{-3} + \rho L(a + bW)\ (\beta_p + 6.1 \times 10^{-4} p_a)$.

The atmosphere–water gas exchange is described with the equation $V\partial C/\partial t = SK_C(C^* - C)$, where $K_C = K_2\theta^{(T_a - 20)}$, C is the concentration of the water-dissolved gas, C^* is the concentration of saturated dissolved gas at a local temperature, K_2 is the coefficient of gas re-airing at 20°C, and θ is the temperature

Table 1.18. Coefficients of the temperature dependence of the atmospheric vapour pressure.

T_a(°C)	α_p	β_p
0–5	6.05	0.522
5–10	5.10	0.710
10–15	2.65	0.954
15–20	−2.04	1.265
20–25	−9.94	1.265
25–30	−22.29	2.151
30–35	−40.63	2.761
35–40	−66.90	3.511

coefficient. The K_2 coefficient at known wind speed W (m/s) and coefficient of molecular diffusion D_g is calculated with the formula:

$$K_2 = D_g \times 10^6[200\text{–}60W^{1/2}]^{-1}$$

The coefficient D_g is equal to 2.04×10^{-9} for O_2 and 1.59×10^{-11} for CO_2. The system of geoinformation monitoring in the megalopolis is based on the data of measurements of the environmental parameters and Equations (1.4) and (1.5). It allows us to analyse and assess the most vulnerable areas of the megalopolis's structure to natural catastrophes as well as reactions of the system N to the behaviour of the system H in the realization of the projects of natural complex reconstruction. As a result, an interaction takes place between the external shell of the global model with the details of the description of the environment by the scale $\Delta\varphi \times \Delta\lambda$ and the model of the megalopolis where the spatial resolution is more detailed. The resulting diverse schemes of agreement between the global model and the megalopolis model are determined both by its specific features and by the formal system circumstances. Their study and development are the subject of future investigations into the realization of constructive ideas discussed in the monograph by Kondratyev (1999a).

1.3.5 Scenarios of anthropogenic processes

As experience shows, the development of the strategy of global change monitoring with the present-day development of the science of global processes is impossible using any models that simulate the functioning of the NSS. No doubt the developed methods of modelling the type of evolutionary technology (Bukatova *et al.*, 1991; Bukatova and Makrusev, 2003) give hope to creating a global model which, based on the accumulated knowledge, will make it possible to obtain reliable long-term predictions of global change and to thereby substantiate conditions for the sustainable development of the environment. Obviously, a success in this direction depends on the complex use of the modelling technologies and environmental observations. So far, the global models require units, which would realize a parameterization of some processes in the NSS in the form of scenarios. The construction of the latter is the result of expert knowledge and a certain amount of imagination of an investigator. This problem has been discussed in detail by Alcano *et al.* (2001), Gyalistras (2002), and in a special IPCC report (Watson *et al.*, 2000). To some extent, the Kyoto Protocol can serve as one of the scenarios. Some countries of the EEC, who signed this protocol, agreed to reduce GHG (carbon dioxide, methane, nitrogen oxides, hydrofluorocarbons, perfluorocarbons, and sulphur hexafluoride) emissions by 2008–2012 by 12.5%, with CO_2 emissions decreasing (due to this assumption) by 2010 by 20% compared to 1990 levels. In each country the diverse scenarios of this type bring forth a high level of global uncertainty, the solution of which is impossible without certain generalizations. Therefore we shall dwell upon some generalizing scenarios (Nabuurs *et al.*, 2000).

1.3.5.1 Fishery scenario

The use of World Ocean products in the 20th century was characterized by an increasing trade of fish and other elements of higher trophic levels doubling every 10–15 years. A maximum catch in 1970 was 61 Mt. The per capita consumption of sea products varied between 7.2 to 11.8 kg. Numerous forecasts of the possible per capita use of World Ocean products give about 12 kg/yr. Therefore, we can assume a hypothetical dependence of the intensive fishing of nekton in each *i*th region:

$$\lambda_{GRi} = \begin{cases} \lambda_{G0i} - [\lambda_{G0i}(\lambda_{G\infty i})(t - t_0)/(t_{\lambda Gi} - t_0)] & t_0 \le t \le t_{\lambda Gi} \\ \lambda_{G\infty i} & t > t_{\lambda Gi} \end{cases}$$

where λ_{G0i} is the fishing intensity at the moment t_0 (e.g., in 1970 $\lambda_{G0i} \approx 0.0286$ and in 1999 $\lambda_{G0i} \approx 0.0312$) and $\lambda_{G\infty i}$ is the maximum possible trade intensity reached at the moment $t_{\lambda Gi} \ge t_0$. The λ_{Gri} coefficient characterizes the fishing trade in the *i*th region and is considered in the demographic unit of the global model. A human being is considered one of the higher trophic levels with respect to nekton. It is assumed that in the multitude Γ there is the element s_H, for which the following relationship is valid:

$$C_{rs_H} R_{s_H} = r \sum_{i=1}^{m} \lambda_{GRi}$$

The contribution of the fishing industry I_i in the diet of *Homo sapiens* V_G is determined by the ratio of the volume of products fished out of the World Ocean to the volume of agricultural products.

1.3.5.2 Scenario of the distribution of the soil–plant formation areas

An intensity of food production on land depends on the area covered with agricultural crops and on their productivity. Clearly, within the global model, at the level considering details of all processes and elements it is impossible to predict all the directions of agricultural development. Therefore all the processes of increasing productivity have been generalized, reflecting general trends. Let us introduce for the agricultural formation the identifier k and consider that in any region, part of the territory can be occupied with cultivated land:

$$\sigma_{ki} = \begin{cases} \sigma_{koi} + (\sigma_{k^*i} - \sigma_{koi})(t - t_0)/(t_{kiS} - t_0) & t_0 \leq t \leq t_{kiS} \\ \sigma_{k^*i} & t > t_{kiS} \end{cases}$$

$$H_{kiS} = \begin{cases} 1 + (H_{k^*i} - 1)(t - t_0)/(t_{kBiS} - t_0) & t_0 \leq t \leq t_{kBiS} \\ H_{k^*i} & t > t_{kBiS} \end{cases}$$

where $\sigma_{ki} \leq \sigma_i$ is the area of land cultivated in the ith region; and H_{kiS} is the indicator of changes in the agricultural productivity with respect to the time moment $t_0(H_{kiS} = R_{ki}(\varphi, \lambda, t)/R_{ki}(\varphi, \lambda, t_0))$.

1.3.5.3 Investment scenario

According to our suppositions, in the ith region the area σ_{ki} under agricultural crops starting from the moment t_0, changes following the linear law from σ_{k0i} to σ_{k^*i} in the time interval to t_{kiS}. The time for reaching the σ_{k^*i} level depends on the amount of investment into agriculture with the inverse proportion coefficient t_{k^*i} and on other factors affecting the time constant t_{ki}: $t_{kiS} = t_0 + t_{ki} + t_{k^*i}/(q_{V_i} V_i)$. Possible changes in the agricultural productivity H_{kiS} are also approximated by a linear law where the value H_{k^*i} shows how many times the productivity in the ith region can change from t_0 to t_{kBiS}. The reserves for increasing H_{ki} are rather large. Even the highly productive plants such as sugar cane annually consume only about 2% of the solar energy reaching the Earth's surface, with crops consuming 1%, and other plants even less. The plant physiology and agrotechnology have broad possibilities for increasing agricultural productivity and, hence, the amount of food for *Homo sapiens*. Assume that to reach the level H_{k^*i}, the population of the ith region has the time $t_{kBiS} - t_0$:

$$t_{kBiS} = t_0 + \bar{t}_{kBi} + t_{kBi}/(q_N V_i)$$

where the constituent $\bar{t}_{kBi}$ is independent of the agricultural investments, and t_{kBi} characterizes the efficiency of these investments.

According to the V_{Gi} food spectrum for *Homo sapiens* a certain role is played by forest ecosystems, the general trend in changes of their areas being characterized by a negative derivative. On a global scale, the forest resources are constantly being depleted, so that on the border of the 20th and 21st centuries the total area of forests constitutes $\sigma_L = 4184$ Mha, with a 31% density (about 28.3% of the land area).

Two principal processes are observed in forestry: deforestation and aforestation. Let us describe the totality of these processes using the scenario of forest area changes:

$$\sigma_{Li} = \begin{cases} \sigma_{L0i} + (\sigma_{L^*i} - \sigma_{L0i})(t - t_0)/(t_{Li} - t_0) & t_0 \leq t \leq t_{Li} \\ \sigma_{L^*i} & t > t_{Li} \end{cases}$$

This dependence foresees that the area under forests in the ith region until the moment t_{Li} varies linearly from σ_{L0i} to σ_{L^*i} and then remains constant. Values of the input parameters constitute the freedom of choice in simulation experiments.

To complete the formulation of the global scenario describing the level of food provision for *Homo sapiens* let us consider the law of changing the stockbreeding productivity. Suppose that with the constant constituent k_{FSi}, it increases in proportion to agricultural investments q_{Vi} with the coefficient k_{F^*i} and has a constituent (varying with time) with an exponential law of change:

$$k_{Fi} = k_{F3i} + k_{F^*i}q_{Vi} + (k_{F1i} - k_{F3i})[1 - \exp\{-k_{F2i}(t - t_0)\}]$$

The direction of the anthropogenic activity is determined by the intensity of pollution generation, time of pollution utilization, cost of environmental cleaning, amount of investments into the renewal of resources and into the prevention of the environmental pollution, investments into industry and agriculture, rate of natural resources expenditure, and search for new sources, etc. The hypothetical trends of these processes should be set in order to realize the predictions using the global model.

Let the intensity of pollution vary between k_{Z0i} at the time moment t_0 and k_{Z1i} at the moment t_{Z^*i}. Similarly, vary the time of waste utilization T_{Zi} and the cost of cleaning of unit pollution G_{ZVi}, respectively between the values $T_{Z0i} = T_{Zi}(t_0)$, $T_{Z1i} = T_{Zi}(t_{Zi})$ and $G_{2ZVi} = G_{ZVi}(t_0)$, $G_{1ZVi} = G_{ZVi}(t_{ZVi})$. Since the estimates of these indicators can vary for many reasons and on different timescales, the simplest scenario for the experiment will be the following functional presentations:

$$k_{Zi} = \begin{cases} k_{Z0i} - (k_{Z1i} - k_{Z0i})(t - t_0)/(t_{Z^*i} - t_0) & t_0 \le t \le t_{Z^*i} \\ k_{Z1i} & t > t_{Z^*i} \end{cases}$$

$$T_{Zi} = \begin{cases} T_{Z0i} - (T_{Z1i} - T_{Z0i})(t - t_0)/(t_{Zi} - t_0) & t_0 \le t \le t_{Zi} \\ T_{Z1i} & t > t_{Zi} \end{cases}$$

$$G_{ZVi} = \begin{cases} G_{2ZVi} - (G_{2ZVi} - G_{1ZVi})(t - t_0)/(t_{ZGi} - t_0) & t_0 \le t \le t_{ZGi} \\ G_{1ZVi} & t > t_{ZGi} \end{cases}$$

It is rather difficult to specify the spheres of human activity, therefore the spectrum of investments is confined here to the functions U_{ZG}, q_v, and U_{MG}. The strategy of the investments distribution for each region is determined by two-step functions of time:

$$U_{MGi} = \begin{cases} U_{MG1i} & t_0 \le t \le t_{MGi} \\ U_{MG2i} & t > t_{MGi} \end{cases}$$

$$U_{ZGi} = \begin{cases} U_{ZG1i} & t_0 \le t \le t_{ZV} \\ U_{ZG2i} & t > t_{ZVi} \end{cases}$$

where t_{MGi} and t_{ZVi} are the moments of the economic policy change in the ith region in the field of investments into the renewal of natural resources and prevention of pollutions, respectively. Here the term "renewal of resources" denotes the processes favouring an increase of M. In particular, this represents a change of mineral resources for others, with their significance recalculated for the levels of significance of the previous ones.

The agricultural investments are one of the most important components of investments. To obtain a high productivity of cultural crops, it is necessary to completely satisfy their needs for water and mineral matter. It means that high yields require great amounts of energy. The same applies for protein food production, where an efficiency of the transformation of vegetable food energy into the energy of meat and fat synthesized in the animal organisms is about 10% (i.e., 10 calories of energy are spent on the production of 1 calorie of the protein food). This energy production requires some share of investments and consumption of fossil products of photosynthesis. Hence here the investments and energy expenditures are mutually dependent. The global model unit that describes the agricultural investments is constructed by setting the dependences of agricultural productivity on energy expenditures. The investment parameter of the control U_{BGi} is described by the two-step function:

$$U_{BGi} = \begin{cases} U_{BG1i} & t_0 \leq t \leq t_{BGi} \\ U_{BG2t} & t > t_{BGi} \end{cases}$$

where t_{BGi} is the moment of the change of the investment policy in agriculture in the ith region. The T_{Bi} parameter characterizing the time of assimilation of the agricultural investments in the ith region can also change at the moment t_{Bi}:

$$T_{Bi} = \begin{cases} T_{B1i} & t_0 \leq t \leq t_{Bi} \\ T_{B2i} & t > t_{Bi} \end{cases}$$

At present humankind's need for energy is 90% satisfied by the burning of fossil products of photosynthesis (coal, oil, and gas) and only 10% by hydroelectric power stations. In the future these relationships should change drastically due to mastery of new technologies of energy production and, in particular, by raising the efficiency of solar energy use. Here the processes of accumulation and transformation of solar energy through improving the technology applied to harnessing the World Ocean energy (waves, tides, ebbs, currents, etc.) can play an important role. Of course, the time at which these processes will be harnessed will be different for various regions at different stages of economic development. In the global model this is reflected by substitution of the initial value of the component M_i for a new M_{0GMi} value. The t_{GMi} parameter is the function of the multiplier m_{Gi} determined by the ratio of the natural resource supplies to their annual expenditure at a given time moment. It can be regulated in each region by different means with numerous aspects of human activity taken into account. Without dwelling upon details, let us take the following scenario of a possible change of the multiplier m_{Gi}:

$$m_{Gi}(t) = \begin{cases} m_{G0i} - (m_{G0i} - m_{G\infty i})(t - t_0)/(t_{Mi} - t_0) & t_0 \leq t \leq t_{Mi} \\ m_{G\infty i} & t > t_{Mi} \end{cases}$$

Within the time interval $[t_0, t_{MGi}]$ in each region the cost G_{MGi} of the natural resources renewal can vary linearly from the value G_{MG0i} to the value G_{MG1i}:

$$G_{MGi}(t) = \begin{cases} G_{MG0i} - (G_{MG0i} - G_{MG1i})(t - t_0)/(t_{MGi} - t_0) & t_0 \le t \le t_{MGi} \\ G_{MG1i} & t > t_{MGi} \end{cases}$$

The basic capital $V_i\,(i = 1, \dots, m)$ invested into the development of industry, science, agriculture, construction, and other spheres of human activity reduces with its efficiency decreasing. To parameterize the processes taking place, let us introduce the term "the ageing time" T_{VG} ($\sim$40 years). An assessment of this parameter is a characteristic of each region. As before, assume that T_{VG} varies linearly from the value T_{VG0i} to the value T_{VG1i} within the time interval $t \in [t_0, t_{Vi}]$ and then remains constant:

$$T_{VGi}(t) = \begin{cases} T_{VG0i} - (T_{VG0i} - T_{VG1i})(t - t_0)/(t_{Vi} - t_0) & t_0 \le t \le t_{Vi} \\ T_{VG1i} & t > t_{Vi} \end{cases}$$

1.4 ECOLOGICAL SAFETY

The notion of "ecological safety" appeared in connection with the necessity to estimate the danger to the population of some territory that may cause damage to health, constructions or property as a result of changes in environmental parameters. These changes can be caused by both natural and anthropogenic forcings. In the former case danger occurs due to fluctuations in natural processes connected with changes in the synoptic situation, diseases and epidemics, or to natural disasters. In the second case the danger appears as a result of nature's response to human actions (Kondratyev, 1991b).

In a general case, for a given territory an ecological danger results from a change of the parameters of human habitats beyond limits where after a long residence a living organism suffers changes which do not correspond to the natural process of evolution. As a matter of fact, the notions of "ecological danger" or "ecological safety" are connected with the notions of stability, vitality, and wholeness of the biosphere and its elements (Kondratyev, 1990, 1991b). Moreover, being a self-organizing and self-structuring system and developing following the laws of evolution, the NSS creates within itself a multitude of ecological niches whose degree of acceptability for the population on a given territory is determined, as a rule, by national criteria (a combination of PCL, religious dogmas, national traditions, etc.).

Nevertheless, when considering the prospects for life on Earth, it is necessary to proceed from the human criteria of assessing the levels of environmental degradation, since with time, local and regional changes in the environment become global ones. The amplitudes of these changes are determined by mechanisms of the NSS functioning, which provide optimal changes of its elements (Gorshkov *et al.*, 2000, 2002). Humankind deviates more and more from this optimality in the strategy of interaction with the surrounding inert environment. At the same time, human society as an element of the NSS system tries to understand the character of large-scale relationships with nature, directing efforts of many sciences to study the cause-and-effect feedbacks in this system. The basic item of a global human society is the country, that is why the country is considered as the NSS socio-economic component.

The national safety of any country in present-day conditions should be estimated based on many criteria of military, economic, ecological, and social character. Development of an efficient method of objective analysis of the problem of national safety requires the use of the latest methods to collect and process the data on various aspects of the world system's functioning. Such methods are provided by GIMS technology.

Let us consider the economic–ecological aspect of national safety. From the viewpoint of the system analysis, any country can be considered as an object of the system analysis that functions in the space of other complex systems. An interaction of these systems is connected with the controlled and uncontrolled exchange of elements of economic and ecological categories.

An interaction of countries as the complex systems with their socio-economic structures and natural resources raises the problem of optimal behaviour determination. It is necessary to take into account the heterogeneous scientific–technical level of these systems and, hence, different approaches to the choice of criteria for assessing national safety. The GIMS technology proposes the following to solve these problems.

The global NSS model has now been developed. This model describes the main processes in the NSS with their discretization in time and space. The model is based on available databases and information spaces. It is combined into a single national system of ecological monitoring of a territory of a country and is combined with similar global and national systems which allow interaction.

A combination of the NSS model, the system of collection of data about the environment and economic parameters of the regions in a country, the system of computer cartography, and means of informatics results in a single national system of observation and control for economical–ecological safety. This system has a hierarchic structure of information channels with a respective hierarchy of a problem to be solved. In particular, it can promptly provide regions with information about the ecological and economic conditions at any point on the planet. The system will provide information about:

- the current global changes in the environment;
- expected climate changes and the role of present or planned changes to the environment in a country on the changes to the climate and the biosphere;
- the state of the atmosphere, hydrosphere, and soil–plant formations within a territory of a country;
- the availability of required data on ecological, climatic, economic, and demographic parameters of any region;
- the level of ecological safety in a given territory;

- an appearance of phenomena harmful to humans and the environment;
- trends in changes of the states of forests, marshes, pastures, agricultural crops, river and lake systems, and other natural complexes; and
- risks of measures to change the environment.

At the national level, the system can solve the following problems:

- the long-term and timely planning and management of economic activity taking into account its ecological expedience and development of the strategy for rational nature use;
- prompt notification and warning about the processes beyond and within the territory of a country that can worsen the ecological situation and cause long-term changes of the environment with an increased risk to human health in other regions;
- identification of the causes of undesirable changes to the environment in some territories of a country;
- assessment of the consequences of realization of anthropogenic projects for a country and other surrounding and global territories; and
- working out urgent actions to liquidate the causes of ecological catastrophes and natural disasters.

The development and realization of an efficient technology to assess ecological safety on global scales may be possible through the International Centre on the Global Geoinformation Monitoring (ICGGM). This will make it possible to realize the mechanisms for *H–N* coevolution. The basic mechanism will be geared towards new technologies of data processing based on the progress of evolutionary informatics and global modelling. Here the point is about a realization of an approach developed by some authors (Bukatova *et al.*, 1991; Nitu *et al.*, 2000a,b; Bukatova and Makrusev, 2003) to modelling the processes in conditions of inadequate a priori information about their parameters and the presence of principally unavoidable information gaps.

According to the scheme in Figure 1.2, the ICGGM performs the following functions:

- data collection from national monitoring systems and international centres of environmental studies;
- sorting, primary processing, and accumulation of data on natural processes;
- formation of knowledge bases for the processes in the environment;
- simulation and mathematical and physical modelling of climatic, biospheric, cosmic, social, and economic processes;
- prediction of the environmental state and formation of a constantly renewed bank of scenarios of anthropogenic activity;
- answering inquiries from national and international agencies regarding the protection of the environment; and
- substantiation of recommendations for national and international centres of environmental monitoring.

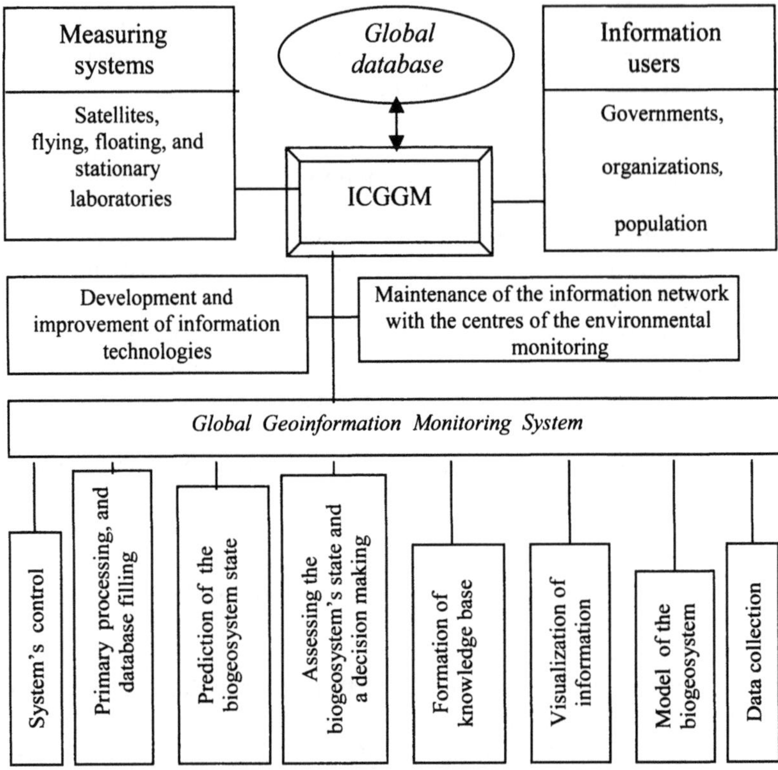

Figure 1.2. Concept of the ICGGM.

Indicators characterizing the GIMS as the base subsystem of ICGGM are grouped by thematic principles of organization of its structure. They are specified in the process of GIMS exploitation and cover the key characteristics of the global topography, synoptic situation in the energy-active zones, content of dangerous atmospheric pollutants in characteristic latitudinal belts, as well as announcements about catastrophes.

A multitude of spatially irregular and fragmentary (in time) data of measurements of geographic, geochemical, ecological, biogeocoenotic, and synoptic characteristics serve as input parameters for the GIMS. Both *in situ* and remote sensing measurements are made in different wavelength regions and with different accuracies. Measurements agree with other GIMS units through algorithmic procedures of primary data processing. The volume of these data will decrease in the process of GIMS functioning. The GIMS input also foresees a possibility to receive signals from scenarios of the anthropogenic development of situations under study.

The GIMS model is demonstrated as a conceptual scheme in Figure 1.3. Relationships between input and output parameters are taken into account through a composition of information fluxes indicated here. The GIMS functions in the adaptive regime, and the final result of the system affects the input

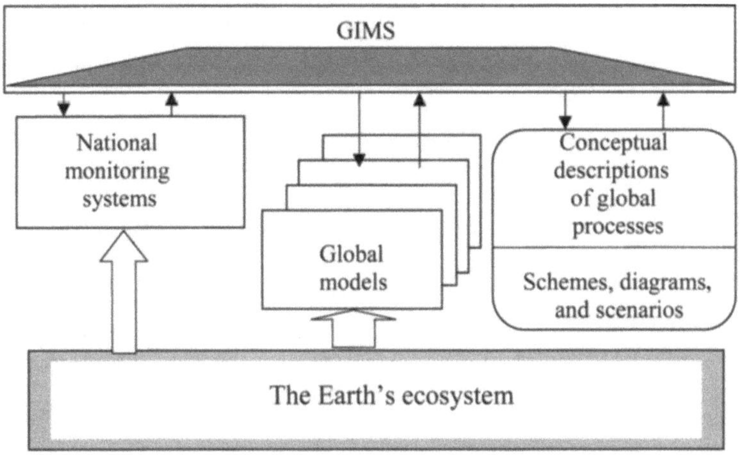

Figure 1.3. The place and role of GIMS technology in studies of the Earth's ecogeosystem.

characteristics of its measuring section. The mathematical aspects of GIMS are shown in Figure 1.4. Here all the biogeochemical and biogeocoenotic processes are described by the systems of balanced equations. However, a considerable part of the not well parametrized (mainly non-stationary) processes are described using the method of the evolutionary modelling oriented towards a parameterization without formulas of very non-stationary processes. The GIMS functioning is presented in Figure 1.5.

The global scale socio-economic structure can be divided into m levels. According to Kondratyev (1993, 1997) $m \geq 9$, giving a structure with three main levels of development of regions: developed, developing, and low-developed. A realization of the ICGGM project will accelerate the process of alignment of this structure, since the planning of the organizational structure of human society will be accelerated, with the provision of purposeful global processes benefitting humankind without damaging nature. International mechanisms will be developed for coordinated nature use by the population of the Earth. The ICGGM will be useful because people will not waste finances on realization of ecologically unacceptable projects thus preserving a balance with nature. With scientific–technical progress, this profit will grow, since the transition to new kinds of resources will become unavoidable (Gorshkov, 1995).

Let us formulate a mathematical problem to estimate the ecological safety of the country with the use of the theory of complex systems and theory of vitality. Let A be the national system of the country. The system A interacts with other similar systems having different spatial location. For simplicity, all other systems will be denoted B. In other words, all other countries will be identified in a first approximation with a single system B. Later on this situation can be complicated by considering many systems, with which system A interacts.

The systems A and B have goals, structures, and behaviours (strategies). The goal $\underline{A}(\underline{B})$ of the system $A(B)$ is its desire to reach certain states preferable for it.

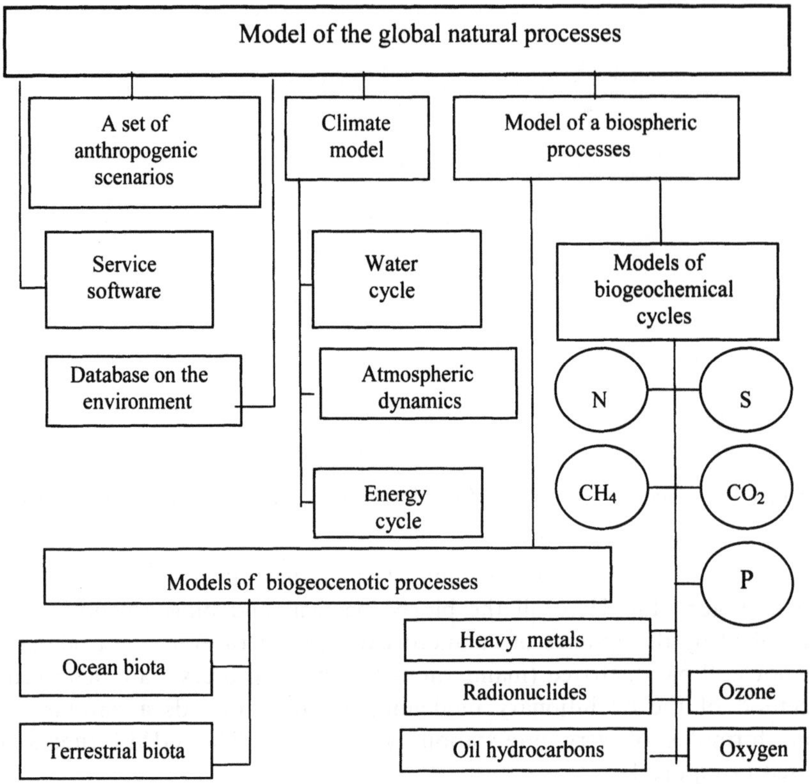

Figure 1.4. Concept of global modelling as a means of studing the cause-and-effect feedbacks in the environment.

These goals can be of multipurpose hierarchic character. The parametric presentation of the goal is one of the important problems. As one possible version, we consider the following components of the goal for the system A: $\underline{A}_1$ is the integral indicator Q of the environmental quality over the territory of the country, it cannot be below the threshold q; $\underline{A}_2$ are maximum permissible concentrations $L(i,j)$ $(j = 1, m_i)$ of substances which cannot be violated over the jth part of the territory in the medium i ($i = 1$ for soil; $i = 2$ for water; and $i = 3$ for the atmosphere); and $\underline{A}_3$ is the economic potential of the country for the time Δt which should increase by s per cent.

The goal $\underline{B}$ of the system A can act towards A as antagonistic, neutral, or cooperative. This is determined by the form of the criteria functions for both sides.

An expediency of the structure $|A|$ $(|B|)$ and the purposeful behaviour $\bar{A}(\bar{B})$ of the system $A(B)$ are estimated by an efficiency with which the system reaches its goal.

The behaviours of the systems can either favour, not prevent, or act against the systems of their goals and the goals of another system. In the first case, a pair of

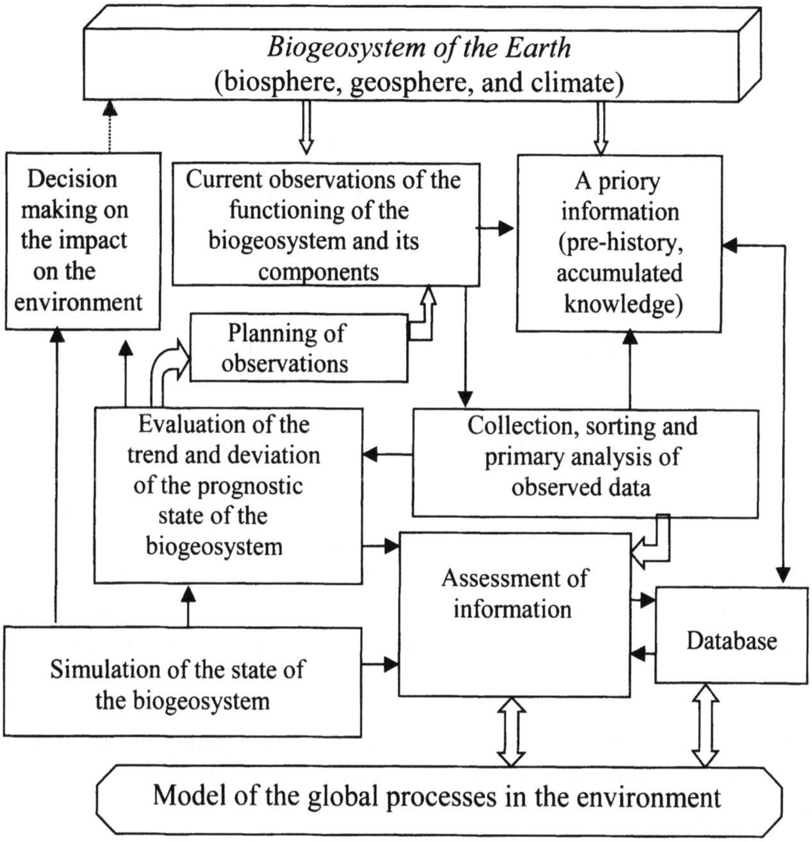

Figure 1.5. Concept of environmental monitoring.

systems can be considered as one system with the common goal interacting with other systems. In other cases, we shall speak about relationships between the systems.

Since the systems imply national ecological systems, it is natural to introduce some statements about ways in which they interact. Such systems are open, and their interaction can be presented in the form of exchange with some resources (finances, technical means, natural resources, etc.). It can be formalized through introducing some resource V spent by the system and some resource W consumed by the system. As a result, a (V, W)-exchange takes place between the interacting systems. Clearly, each system wants to make this process advantageous for itself. This results in the possibility of further formalization of the system's functioning. In other words, the most advantageous (V, W)-exchange can be considered as the goal of each system (i.e., that is, each system wants to get a maximum of W for a minimum of V, which is the function of the structure and behaviour of the interacting systems):

$$W = W(V, |A|, |B|, \overline{A}, \overline{B}) = W(A, B) \tag{1.8}$$

As a result, the interaction of the systems A and B is reduced to the following relationships (models):

$$\underline{W}_a = W_a(V_a, A_0, B_0) = \max_{\{\bar{A},|A|\}} \min_{\{\bar{B},|B|\}} W(V_a, A, B) \tag{1.9}$$

$$\underline{W}_b = W_b(V_b, A_0, B_0) = \max_{\{\bar{B},|B|\}} \min_{\{\bar{A},|A|\}} W(V_b, A, B) \tag{1.10}$$

where A_0 and B_0 are optimal systems.

It is seen from Equations (1.9) and (1.10) that to define its goal, each system should decide what is important for it to become the most profitable (V, W) – to exchange or to prevent another system from getting it. The systems can vary the values of (V, W)-exchanges within some limits $W_1 \leq W_a \leq \overline{W}_1$ and $W_2 \leq W_b \leq \overline{W}_2$, where $\underline{W}_1$ and $\underline{W}_2$ correspond to maximum aggressive states of the systems, and $\overline{W}_1$ and $\overline{W}_2$ to most cautious ones.

If the goals of the systems are known, then we have a definite situation. If each system or one of them hides its intentions, we have a game situation with respect to the choice of the goal. Let $\underline{A}_i$ and $\underline{B}_j$ $(i = 1, \ldots, n; j = 1, \ldots, m)$ be sets of goals for systems A and B, respectively. The goals $\underline{A}_1$ and $\underline{B}_1$ consist of inflicting maximum damage to another system (maximum aggressiveness), and the goals $\underline{A}_n$ and $\underline{B}_m$ correspond to an extreme caution of both systems (maximum favour). All the rest of the goals are located by scales $\{i\}$ and $\{j\}$ as a transition from $\underline{A}_1(\underline{B}_1)$ to $\underline{A}_n(\underline{B}_m)$. Assuming that in the situation $\{\underline{A}_i, \underline{B}_j\}$ the systems get profits $\underline{W}_a = a_{ij}$ and $\underline{W}_b = b_{ij}$, we obtain a bimatrix game to determine an optimal goal with the payoff matrices $\|a_{ij}\|$ and $\|b_{ij}\|$. In a partial case at $\underline{W}_a = -\underline{W}_b$ the game becomes antagonistic.

Note that in a general case the situation should be studied in a probability space (i.e., one should speak about some probability $P(V, W)$ that each system will reach the goal of its own). Moreover, one should consider various manifestations in the systems' behaviour: reliability, information content, controllability, and learning ability. The elements of the systems should have various functions and purposes: both protective and vital.

In addition to Equation (1.8)–(1.10), we should consider dynamic relationships, which describe the time dependence of the system's parameters. In this case the mathematical problem of estimation of the level of the ecological safety of a country's territory is reduced to a differential game.

At the national level, there are many criteria. The government should follow definite sanitary/hygienic and ecological norms in certain climatic situations. These situations should be predicted and serve as initial conditions for the system of estimating ecological safety. The environmental condition is a complex function of temperature T, wind speed U, total content of heavy metals in water E, in the air D, and in soil G; the content of gas of the kth type $(k = 1, \ldots, N)$ in the atmosphere S_k; and vegetation cover biomass M and other parameters:

$$Q = Q(T, M, U, E, D, G, S_1, \ldots, S_N) \tag{1.11}$$

Quite similar is the situation with the functional representation of dependences $L(i,j)$ and other characteristics of the environment on natural and anthropogenic parameters. Moreover, many of these parameters can be presented in the form of functions of the investment policy of the government, for instance, investments into the struggle with pollution, development of agriculture, road construction, new technologies, etc. Indicators of the quality of the environment depend on these parameters, and the problem will be reduced to a search for an optimal investment policy. Several models of the environmental dynamics and the optimizing relationships indicated above determine the problem of synthesis of national policy in the sphere of nature-protection activities taking into account the respective policies of adjacent countries and of the world community.

Hence, the primary priority problem represents specifying the goal functions and their dependences on parameters, with the internal and external national strategies of the ecological monitoring taken into account. In the final form, the numerical problem is reduced to the boundary-value problem for the system of differential equations of the parabolic type. The system of equations will describe the dynamics of pollution on the territory of a country, and the boundary conditions will be determined with behavioural strategies of the adjacent territories taken into account. Solutions for the boundary-value problem will be included in Equations (1.9)–(1.11), which will finally determine the ecological safety.

1.5 BIOCOMPLEXITY

The problem of interaction between various elements and processes in the global NSS has recently attracted the attention of many scientists. Attempts to estimate and predict the dynamics of this interaction have been made in several different scientific spheres. One of these attempts is the "Biocomplexity" programme identified in the USA by the National Science Foundation, within which plans have been made for the 2001–2005 period to study and understand relationships between the dynamics of the complexity of biological, physical, and social systems, and trends in changes of the present environment. Within the framework of this programme, biocomplexity refers to phenomena that result from the dynamic interactions among various components of the Earth's diverse environmental systems.

Biocomplexity is a derivative of biological, physical, chemical, social, and behavioural interactions of the environmental subsystems, including living organisms and the global population. As a matter of fact, the notion of biocomplexity in the environment is closely connected with the rules of biosphere functioning as a unity of its forming ecosystems and natural–economic systems of different scales (local to global). Therefore, to determine and assess the biocomplexity, a combined formalized description is needed of biological, geochemical, geophysical, and anthropogenic factors and processes taking place at a given level of the spatial–temporal hierarchy of units and scales.

Biocomplexity is a characteristic feature of all systems of the environment connected with life. Elements of this manifestation are studied within the framework of the theory of stability and vitality of ecosystems. Note should be taken here that biocomplexity includes indicators of the degree of mutual modification of the interacting systems, and this means that biocomplexity should be studied considering both the spatial and the biological levels of organization. The difficulty is explained by the complicated behaviour of the object under study, especially if the human factor is considered.

Humankind has accumulated sufficient knowledge about environmental systems. Use of this knowledge to study biocomplexity is possible within the framework of the synthesis of a global model that reflects the laws of interactions between environmental components and permits access to an "efficiency" of realization of the scenarios of the human society development, based on the actual data of ground and satellite measurements. It is this problem that serves the basis of all questions set forth by the "Biocomplexity" programme.

Studies of the process of *H–N* interaction are aimed, as a rule, at understanding and assessing the consequences of this interaction. Reliability and accuracy of these assessments depends on criteria, which serve as the basis for conclusions, expert examination, and recommendations. At present, there is no agreed-on method to select such criteria due to a lack of a single scientifically substantiated approach to the ecological normalization of the economic forcings on the environment. The choice of such criteria determines the accuracy of the ecological expertise of the existing and planned productions as well as representativeness of the global geoinformation monitoring data.

The processes taking place in the environment can be represented as a totality of interactions between its subsystems. Since humankind is one of its elements, it is impossible to definitely divide the environment, for instance, into the biosphere and society, since everything on the Earth is correlated and interconnected. The point is to find mechanisms to describe such correlations and interdependences, which would reliably reflect the environmental dynamics and answer the questions formulated in the "Biocomplexity" programme:

1 How does the complexity of biological, physical, and social systems in the environment appear and change?
2 What are the mechanisms of the spontaneous development of numerous phenomena in the environment?
3 How do the systems of the environment with the living components, including those created by humankind, react and adjust themselves to stressful situations?
4 In what ways do information, energy, and matter move within the systems of the environment and though their levels of organization?
5 Is it possible to predict the system's adaptability and to give prognostic estimates of its change?
6 How does humankind affect and respond to biocomplexity in natural systems?

One can add many other no less important questions. For instance, up to what level of complexity should the space-borne observation systems be improved in order that their information be reliable enough to estimate the state of the environment, at least at the moment of receiving information? No less important is the question about an optimal allocation of the means of geoinformation monitoring at different levels of its organization.

The environmental biocomplexity is, to some extent, an indicator of the interconnection of its systems. In this way, one can introduce the unit Ξ of biocomplexity varying with conditions when all interactions in the environment are reduced to the level corresponding to the natural process of evolution. Thus we obtain an integral indicator of the state of the environment on the whole, taking into account biocapability, biodiversity, and biosurvivability. Such an indicator characterizes all kinds of interactions of the environmental components. For instance, at the biological interaction connected with relationships of the type "predator–prey" or "competition for energy resource", there is some minimum level of food availability after which consumer–producer interaction ceases. Chemical and physical processes of the interaction of the environmental elements also depend on a set of certain critical parameters.

All this demonstrates that biocomplexity refers to categories which are difficult to measure and estimate quantitatively. Let us use a formalized estimation. To transfer to gradations of the unit Ξ with a digital scale, we can state that there are relationships of the type $\Xi_1 < \Xi_2$, $\Xi_1 > \Xi_2$, or $\Xi_1 \equiv \Xi_2$ between two values of the unit indicator. In other words, there is always a value of this unit ρ, which determines the level of biocomplexity $\Xi \to \rho = f(\Xi)$, where f is some transformation of the notion of biocomplexity into a number.

Let us try to find a satisfactory model, which would put the verbal portrait of biocomplexity into notions and indicators, following the formalized description and transformation. With this aim in view, we select in the NSS m elements/subsystems of a lower level, the interaction between which we determine from a binary matrix function: $A = \|a_{ij}\|$, where $a_{ij} = 0$, if elements i and j do not interact; and $a_{ij} = 1$, if elements i and j do interact. Then any point $\xi \in \Xi$ is determined as the sum:

$$\xi = \sum_{i=1}^{m} \sum_{j>1}^{m} a_{ij}$$

Here an ambiguity appears, to overcome which it is necessary to complicate the unit Ξ (e.g., by introducing a weighted coefficient for each element of the NSS). The character of these coefficients depends on the nature of the elements. Therefore we select in the NSS two main types of elements: living (including plants) and non-living elements. The living elements are characterized by density estimated in species number per unit area (volume) or in biomass concentration. Vegetation is characterized by the type and share of the covered area. Non-living elements are divided by the level of their concentrations related to the area or volume of space they inhabit. Generally, to each element i some characteristic k_i is ascribed, which corresponds to its significance. As a result, we obtain a specified formula at the transition from the notion of biocomplexity to the unit Ξ of its indicator:

$$\xi = \sum_{i=1}^{m} \sum_{j>1}^{m} k_j a_{ij}$$

Clearly, $\xi = \xi(\varphi, \lambda, t)$, where φ and λ are geographical latitude and longitude, respectively, and t is the current time. For some territory Ω the indicator of biocomplexity is determined as an average:

$$\xi_\Omega(t) = (1/\sigma) \int_{(\varphi,\lambda)\in\Omega} \xi(\varphi, \lambda, t)\, d\varphi\, d\lambda$$

where σ is the area of the territory Ω.

Thus the indicator $\xi_\Omega(t)$ becomes an integral indicator of the NSS complexity, reflecting the individual character of its structure and behaviour at each time moment t in a space of Ω. Following the laws of natural evolution, a decrease (increase) of the ξ_Ω value will follow an increase (decrease) of biodiversity and survivability of the natural–anthropogenic systems. Since a decrease of biodiversity breaks the closed circle of biogeochemical cycles and leads to an increase of the load on non-renewable resources, the binary structure of the matrix A moves to a state, where the level of its survivability lowers.

The NSS consists of the elements/subsystems B_i $(i = 1, \ldots, m)$, whose interaction forms in time depending on many factors. The biocomplexity of the NSS consists of the structural and dynamic complexity of the constituents. In other words, the NSS biocomplexity is formed in the process of interaction of its constituents $\{B_i\}$. With time, the subsystems B_i can change their state and, hence, the topology of connections between them will also change. The evolutionary mechanism of an adjustment of subsystems B_i to their environment makes it possible to bring forth a hypothesis that each subsystem B_i, independent of its type, has the structure $B_{i,S}$, behaviour $B_{i,B}$, and goal $B_{i,G}$. Thus $B_i = \{B_{i,S}, B_{i,B}, B_{i,G}\}$. The goal $B_{i,G}$ of the subsystem B_i is to reach certain states preferable to itself. An expediency of the structure $B_{i,S}$ and purposeful behaviour $B_{i,B}$ of the subsystem B_i is estimated from an efficiency of achieving the goal $B_{i,G}$.

As an example, we take the process of migration of the elements of nekton. Fish migrate towards a maximum of the gradient of nutrients taking into account possible limits of the water basin parameters (temperature, salinity, concentration of oxygen, contamination, etc.). Hence, the elements of nekton have the goal $B_{i,G}$ to increase their ration, and their behaviour $B_{i,B}$ is aimed at determination of the trajectory of movement which would help achieve the goal. There are also the structural changes in the process of formation of shoals, which for each kind of nekton element can be presented in terms of $B_{i,S}$.

Since the interaction of subsystems $\{B_i\}$ is connected with the chemical and energy cycles, it is natural to suppose that each subsystem B_i organizes the geochemical and geographic transformations of matter and energy in order to remain stable. The formalized approach to this process consists in the supposition that in the NSS structure, exchange of some amounts of spent resources V for some amounts of consumed resources W ((V, W)-exchange) take place between subsystems B_i. The goal of the subsystem B_i is the most advantageous (V, W)-exchange (i.e., it tries to obtain maximum W in exchange for minimum V). The quantity W is a complex function of the structure and behaviour of interacting subsystems, $W = W(V, B_i, \{B_k, k \in K\})$, where K is the space of subsystem numbers interacting with the subsystem B_i. Let us designate $B_K = \{B_k, k \in K\}$. Then the following (V, W)-exchange is the result of interactions between the subsystem B_i and its environment B_K:

$$W_{i,0} = \max_{B_i} \min_{B_k} W_i(V_i, B_i, B_k) = W_i(V_i, B_{i,opt}, B_{k,opt})$$

$$W_{k,0} = \max_{B_k} \min_{B_k} W_k(V_k, B_i, B_k) = W_k(V_k, B_{i,opt}, B_{k,opt})$$

Hence, there is some smearing of the goal of the subsystem B_i when determining the levels V_i and V_K. Since in nature the limiting factors are acting, in this case it is natural to suppose the presence of some threshold $V_{i,\min}$, and when reaching this threshold, the energy resource of the subsystem cease to be spent on extraction of external resources (i.e., at $V_i \leq V_{i,\min}$ the subsystem B_i operates in the regime of generation of the internal resource). In other words, at $V_i \leq V_{i,\min}$ the indicator of biocomplexity $\xi_\Omega(t)$ decreases due to a breaking of the connections of the subsystem B_i with other subsystems. In general, $V_{\min}$ is the structural function of the step function (i.e., the changeover of a_{ij} from state $a_{ij} = 1$ to state $a_{ij} = 0$ is not realized for all j at the same time). In any trophic pyramid, the "predator–prey" relationship ceases with decreasing concentrations of victims below some critical level. In other cases the interaction of the subsystems $\{B_i\}$ can cease depending on various combinations of other parameters. The formalized description of possible situations of interaction of the subsystems $\{B_i\}$ can be realized within the framework of a simulation model of the NSS functioning.

2

Greenhouse effect problems

2.1 INTERACTIVITY OF GLOBAL CLIMATE PROBLEMS AND THE CARBON CYCLE

The problem of the carbon cycle and its climatic implications (and respective feedbacks) has attracted attention dating back to the days of S. Arrsenius (Kondratyev, 1992b). However, during the last 2–3 decades the interest in this problem has grown substantially, due primarily to speculative statements with regard to anthropogenically induced global climate warming and apocalyptic predictions of global warming resulting in harmful consequences for humankind (Katz *et al.*, 1999; Borisenkov and Pichugin, 2001; Grigoryev and Kondratyev, 2001a; Wofsy, 2001; Albrecht, 2002; Alverson *et al.*, 2002; Demirchian *et al.*, 2002; Kashapov, 2002). Cihlar *et al.* (2002) have justly emphasized that the problems connected with carbon flux variability are in the focus of not only science but also politics.

2.1.1 Global warming: facts and illusions

Since the reality of global climate warming and the related exaggerations have recently been analysed in detail (Kondratyev, 2002c), we shall only consider related conceptual circumstances.

The failure of the International COP-6 Conference (Nilsson *et al.*, 2002) held in the Hague in November 2000, at which representatives of the states signed the UN Framework Convention on Climate Change (FCCC), cannot be considered unexpected since the participants of COP-6 were not adequately prepared, scientifically. The most important problem, that of the of uncertain assessments of the reservoirs and fluxes of greenhouse gases (GHG) was almost neglected.

After COP-6 the problem became more acute because of President Bush's refusal to support the Kyoto Protocol (KP). On the other hand, the Third IPCC Report published in 2001 (IPCC, 2001b; Cihlar *et al.*, 2002) contains the conclusion that the available information enables one to convincingly speak about the GHG contribution to the formation of climate change and predict climate changes more reliably.

In this regard, of key importance are assessments of uncertainties in the estimates, which serve as the basis for conclusions about climate changes and the required measures to prevent them (Mejer, 2001; Reilly *et al.*, 2001; Newell and Pizer, 2002; Kerr, 2002; Soon *et al.*, 2003). Of special importance is the problem of evaluation of the level of GHG emissions into the atmosphere connected, first of all, with the solution of the problem of the global carbon cycle. Clearly, without a reliable verification of the available estimates of emissions, all discussions with respect to ecological profits of various measures and related expenditures are of abstract character (Nilsson *et al.*, 2002). How, for instance, can imposed fines for not following KP recommendations for GHG emission reductions be justified if it is impossible to prove that emissions in 2012 will differ from those in 1990? So far, discussions on the KP have ignored in particular the quantitative estimates of uncertainties of the levels of GHG sinks (this specifically concerns the biosphere). However, uncertainties in assessments of CO_2 fluxes are rather substantial (exceeding 100% in Russia). Calculations of errors in estimation of total GHG fluxes vary within $\sim\pm5$–25%, whereas the KP levels of GHG emissions reduction average about 5%. The mean global situation can be illustrated – uncertainties in evaluation of GHG emissions due to energy production systems are approximately equal to those in the estimates of CO_2 assimilation by the biosphere and land (Monahan and Dam, 2001; Monnin *et al.*, 2001).

In this situation, solution of the problems of uncertain estimates (initially concerning reliable information on the carbon cycle) and verification become of vital importance. A solution for the verification problem requires an agreement on its mechanisms, which is also very important from the financial point of view. Results of simulation modelling show, for instance, that if a 5.2% confidence level for GHG reduction is raised from 50 to 95%, this entails an increase of expenditure on measures to reduce GHG emissions by a factor of 3–4. The principal conclusion is that science should recommend measures in the field of ecological policy.

A serious uncertainty of the estimates of possible anthropogenic climate change and their determining factors has brought forth within scientific literature and mass media hot discussion (Houghton *et al.*, 2001; Wofsy, 2001; Baliunas, 2002; Ellsaesser, 2002; Hartmann, 2002; Kondratyev and Grigoryev, 2002; Pielke, 2002; Pittock, 2002; Schönwiese, 2002; Wuebbles, 2002). So, for example, in the interview with a journalist from *American Scientific* R. Lindzen (Professor of the Massachusetts Institute of Technology, USA) said that he started worrying about speculative discussions on climate change in the summer of 1988, when J. Hansen (Director of the Goddard Institute for Space Studies, New York) stated that the global climate warming resulted from growing CO_2 concentration due to emissions to the atmosphere of products of fossil fuel burning. This statement prompts R. Lindsen (see Lindsen *et al.*, 2001; Lindsen and Giannitsis, 2003) to explain that climate modelling is at its initial stage of development and, in particular, there is no consensus on the causes of climate change. In the beginning of 2001 he reported on the problem of climate change at a meeting of the US Government. The fact that the global surface air temperature (SAT) has increased by about 0.5°C and the atmospheric CO_2 concentration by about 30% during the last 100 years is far from reflecting any cause-and-effect relation between increasing CO_2 concentrations and growing temperatures. Lindzen believes that the most reliable estimate of climatic sensitivity (SAT increase with a doubling CO_2 concentration) is 0.4°C, reflecting no basis for worrying about catastrophic changes in the future global climate.

According to the IPCC mandate, its task is to prepare an overview of "any climate changes in time, both natural and anthropogenic." Pielke (2001, 2002) observed, however, that at least two climate-forming factors turned out to be either unreliable or not considered at all: (1) impact on global climate of anthropogenically induced changes in the characteristics of land surface; and (2) biospheric impacts of the growth of CO_2 concentration in the atmosphere (including the "fertilization effects"). If it is justified that both these factors exist, the conclusion suggests itself that an agreement of the results of the global climate numerical modelling (in case of mean-global mean-annual SAT) with observations is accidental.

In this connection, Pielke (2002) discussed information that confirms the importance of these two climate-forming factors, and spoke about the possibility to check the validity of such a conclusion. For this purpose data can be used on the impact of anthropogenic changes in land surface characteristics on the local, regional, and global climate. These changes illustrate the fact that this impact is as important as that of a doubled CO_2 concentration in the atmosphere (as well as the growth of other GHG concentrations). No less important is the fact that the atmosphere–land surface interaction is characterized by various non-linear feedbacks, and therefore it may be impossible to predict climate changes for longer than a season.

As for possible biological impacts of CO_2 concentration growth, they are manifested through short-term (biophysical), mid-term (biogeochemical), and long-term (biogeographical) impacts of landscape-forming processes on weather and climate. The biophysical impact includes, for instance, the effect of transpiration on the relationship between latent and sensible heat fluxes as surface heat flux components. The biogeochemical impact incorporates the effect of vegetation growth (fertilization effect) on evaporating leaf area, on surface albedo, and carbon storage. One of the manifestations of biogeographical impacts is a temporal change in the species composition of vegetation communities. Results of numerical modelling testify to the fact that without consideration of biophysical/biogeochemical impacts, the assessments of climate change cannot be considered reliable (Bounoua *et al.*, 2002).

Further development of climate models should consider, in particular, the following aspects of climate formation: (1) direct and indirect impacts of landscape dynamics through biophysical, biogeochemical, and biogeographical processes; (2) consideration of anthropogenic changes in land use on various (local, regional, and global) spatial–temporal scales (Houghton, 1999; Houghton *et al.*, 2000; Jepma *et al.*, 2001); and (3) assessments of possibilities of climate prediction for a longer term than a season bearing in mind functioning of numerous non-linear feedbacks that determine the atmosphere–land surface interaction (Hughes *et al.*, 1999; Joos *et al.*, 1999). With these and other problems unsolved, the importance of the IPCC-2001 Report and US National Report as containing only assessments of global climate sensitivity to changes in some climate-forming factors, becomes rather limited.

In the course of time-consuming preparations of the three published (1990, 1996, 2001) IPCC Reports, the analysis of climate-forming factors and various feedbacks has deepened with each report (so, for instance, apart from just GHG, atmospheric aerosols have also been considered). However, the available complex numerical climate models still cannot be considered adequate from the viewpoint of taking into account every important climate-forming factor. A step forward in IPCC (2001a) was an account of forcings (F) on the climate of anthropogenic changes in land use, limited, however, by consideration of only the effect of land use dynamics starting from 1750, and on surface albedo. The estimates obtained gave an average F value of $-0.2\,W/m^2$ with an uncertainty interval of from -0.4 to $0\,W/m^2$. Thus, these estimates are rather uncertain, let alone the sign of F (a more complete consideration of biophysical, biogeochemical, and biogeographical impacts of the evolution of nature use on the climate suggests the conclusion that in this case $F > 0$).

In this regard of great importance is the fact that the IPCC-2001 Report considers a new range of possible global warming by 2100 (1.4–5.8°C) based only on numerical modelling data (and therefore will inevitably change in the future). In addition, the problem is that the new range can be directly compared with similar estimates obtained earlier. From the viewpoint of reliability of estimates of future climate changes, of great importance is the use in the IPCC-2001 Report of the term "projections" instead of "predictions", since the latter means that the factors left out of account will not be of importance in the future. It is unacceptability of the latter supposition that prevents experts in numerical modelling from stating that climate prediction for the next 100 years is possible. This conclusion confirms the results of the numerical modelling accomplished by Andronova and Schlesinger (2001).

The complicated problem of prognostic assessments of climate, and especially selecting the contribution of an anthropogenic component, is illustrated by the remaining controversy of analysis of the climatic impact of clouds. Tsushima and Manabe (2001) have analysed the effect of cloud feedback (CF) on the formation of the annual change of mean global SAT with the use of the ERB satellite measurement data, bearing in mind an assessment of an adequate consideration of CF in numerical climate models by comparing the calculated and observed variations of mean global SAT. It follows from observations that the mean global annual change of SAT is in phase with the annual change of SAT in the northern hemisphere, and its amplitude is 3.3 K (this phase coherence is determined by concentration in the northern hemisphere of the continents contributing significantly to the formation of the amplitude of the SAT annual course).

Analysis of the observational data on the ERB components for the period from February 1985 to February 1990 has shown that the mean global estimates of both short wave and long wave radiative forcings (SWRF and LWRF) depend weakly on the annual course of mean global SAT (the ERB data considered refers only to the range 60°N–60°S). Thus, the cloud cover dynamics neither intensifies nor reduces the SAT annual change. The considered data on SWRF and optical properties of clouds shows that not only albedo but also cloud amount and cloud top height depend weakly on SAT and, hence, CF does not substantially affect the annual course of mean global SAT.

Based on this conclusion, we can say that the CF negligibly affects the annual course of mean global SAT. However, this speculative opinion is dangerous due to the rather complicated spatial field of SAT. Calculations with the use of three interactive climate models considering the dynamics of cloud microphysical properties, suggest the conclusion that the cloud top height albedo grows substantially with increasing SAT, which disagrees with observation data. This situation reflects prospects for comparing the assessments of the role of CF from the data of numerical modelling and observations from the viewpoint of model validation.

Since J. Hansen was one of the principal initiators of the apocalyptic prediction of global warming, it is of interest to mark an evolution in his own opinion in his recent work (Hansen and Sato, 2001), where it was stated that an increase of mean-annual mean-global SAT from the last 100 years by more than 0.5°C has been, at least partially, due to anthropogenic forcing on the climate. The authors calculated the radiative forcing (RF) due to an observed growth of GHG concentration. The estimates testify to a considerable level of RF due to methane, whose contribution reached about half the RF due to carbon dioxide: $1.4 \pm 0.2\,W/m^2$ including $0.5\,W/m^2$ due to direct forcing and $0.28\,W/m^2$ due to indirect forcing. These results show the necessity for special attention to be given to CH_4 as a GHG and a RF agent. New estimates of RF due to tropospheric ozone have given $0.7\,W/m^2$ instead of 0.3–$0.4\,W/m^2$, as supposed in the IPCC-2001 Report. Apparently, an interval of 0.4–$0.8\,W/m^2$ can be assumed for this RF.

Considerably uncertain (due to uncertain input data) calculations of RF due to "black" carbon (soot) gave $0.8 \pm 0.4\,W/m^2$. Very important is the RF due to reflecting aerosols reaching $-1.4 \pm -0.5\,W/m^2$ with an additional contribution by soil aerosols equal to 0.1–$0.2\,W/m^2$. Considerable contributions to total aerosol RF are made by organic aerosol ($-0.3\,W/m^2$ with an accuracy of a factor of 2) and ammonium sulfate ($-0.2\,W/m^2$). The sum of all positive components reaches $4.3 \pm 0.6\,W/m^2$ (i.e., it exceeds three times the RF due to CO_2), whereas the sum of the negative components is $-2.7 \pm 0.9\,W/m^2$, and, thus, the resulting RF is $1.6 \pm 1.1\,W/m^2$ (i.e., close to that due to CO_2 alone).

Hansen and Sato (2001) performed an analysis of the trends of various GHG, from which it follows that the rates of CO_2 concentration increase grew rapidly during the period from the end of World War II to the mid-1970s, reaching a maximum of about 1.57 ppm/yr. Then it decreased to a comparatively stable level of concentration growth of 1.5 ppm/yr. During the same period the rate of CH_4 concentration increase grew rapidly (from 5 to 15 ppb/yr), slowing down markedly after 1980 (the reasons remain unknown). The total trend of the rate of RF growth due to 13 chlorofluorocarbons (CFC) was characterized by a rapid increase until the mid-1980s and subsequent decrease due to reduced CFC emissions in the process of realization of measures stipulated by the Montreal Protocol. Apparently, before 2010 the RF due to CFC will start decreasing. The rate of total RF growth (for a century) reached a maximum of about $5\,W/m^2$. At this rate of growth the level of equivalent RF corresponding to a doubled CO_2 concentration should be reached by 2050. However, during the last two decades the rate of the growth of total RF decreased to $3\,W/m^2$. With an account of a possible reduction of GHG emissions, the global warming during the forthcoming several decades should constitute $+0.15 \pm 0.05°C$ per decade, under the conditions that the level of CO_2 emissions will be "frozen". The remaining considerable uncertainty of the estimates considered is connected, first of all, with the difficulty of the aerosol-induced RF calculations.

Two years ago the G-8 Group of governments decided to organize the World Climate Change Conference, to discuss the changes of climate and the possibility that they are caused by humankind. The Conference was held in Moscow between 29 September and 3 October, 2003 and was attended by over 2000 participants from more than 100 countries, including scientists, the representatives of governments, the private sector, and non-governmental organizations.

The official goal of the Conference was a "discussion of the natural and anthropogenic factors driving the climate; approaches to reducing anthropogenic emissions; impacts and adaptation measures to on-going climate changes; and hence, to achieve a maximum mutual understanding between scientists, governments, business circles and the public." It is interesting that before the Conference it evoked rather eager interest among the media and numerous environmental organizations, but not at its end, when the final conclusions were announced. Perhaps this was because the official goal of a mutual understanding between governments, and expected unanimous support of the KP had not been achieved. Even the basic questions posed by the Chairman of the organizing committee, professor Yuri Izrael, of "What is really going on on this planet – warming or cooling?", and "whether ratifying of the Kyoto Protocol would improve the climate, stabilize it or make it worse", were not answered. It also became clear that without ratification by Russia the KP would topple.

The climate has always been changing since the Earth was formed, is changing now, and will change in the future. The alternating warm and cold climatic cycles extend from tens to many thousands and even millions of years, and depend on many interactive climate-forming mechanisms. Since the formation of an oxygen atmosphere hundreds of millions of years ago the changes in chemical composition of the atmosphere had rather minor influence on climate, with water being the dominant component responsible for most of the greenhouse effect. There were periods in the past when concentrations of carbon dioxide, a trace GHG (which is not a pollutant, but a life building gas for all living organisms), were about 10 to 20 times higher than now, with no catastrophic "run-away" greenhouse effect occurring on the Earth, and with glaciers covering parts of continents and islands. As stated at the Conference by Andrei Illarionov, the chief economic adviser of the Russian President, "according to scientific data, in the past 400,000 years a dramatic rise of temperature on Earth occurred every 100,000 years, and this was not in the least linked with man's activity. In the past millennium considerable changes of temperature were observed in the 11th, 14th, and 17th centuries." In the 11th century the air temperature around the North Atlantic Ocean, in Europe, Asia, South America, Australia, and Antarctica, was about 1.5°C warmer than now. Earlier still, for a long time between 3500 to 6000 years ago, the period of the "Holocene warming" enjoyed temperatures about 2°C higher than now. Illarionov raised ten important questions shattering the shaky edifice of the anthropogenic global warming hypothesis. His litany was followed by presentations by numerous Russian and foreign critics of this hypothesis. They did not receive satisfying answers from its proponents.

If there is nothing unusual in the current climate change, why is such enormous attention being paid to climate problems in the scientific literature, the mass media, and in public opinion? The answer to this question is not at all simple. Besides science, it involves politics, business, industry, a lot of misanthropic ideology, enormous amounts of money, and group interests (Indermühle *et al.*, 1999).

The World Climate Change Conference was opened by the President of Russia Vladimir Putin who stated that "Even 100 percent compliance with the Kyoto Protocol won't reverse climate change." In response to those calling for quick ratification of the KP, Putin mentioned half jokingly that "They often say, either as a joke or seriously, that Russia is a northern country and if temperature gets warmer by 2 or 3 degrees Celcius, it's not such a bad thing. We could spend less on warm coats, and agricultural experts say grain harvests would increase further." Putin also said that Moscow would "be reluctant to make decisions on just financial considerations. Our first concern should be the lofty idea and goals we set ourselves and not short-term economic benefits." He went on to say, "The government is thoroughly considering and studying this issue, studying the entire complex of difficult problems linked with it. The decision will be made after this work has been completed. And, of course, it will take into account the national interests of the Russian Federation."

The US President George W. Bush Jr rejected the KP in March 2001, as "fatally flawed", because: (1) the document did not have an adequate scientific substantiation; and (2) since the use of fossil fuels prevails in energy production, following the KP would result in serious negative economic consequences, without any real environmental improvement (decreasing the temperature by 0.2°C in 2100, or switching the increase from the year 2100 to 2106). What President Putin will decide is still unknown, but from what he said at the Moscow Conference, it seems that he is thinking along the same lines as the American President, and that he probably will succumb neither to the short-term seemingly lucrative proposal of selling spare Russian CO_2 emission quotas for about US$8 billion per year, nor to the sabre rattling during the Conference by the UE Environmental Commissioner Margot Wallström, who warned Russia that it "would lose politically and economically by not ratifying the Kyoto Protocol". The decision makers must recognize that limiting CO_2 emissions will cause a reduction in the world domestic product, which added up across the whole century represents US$1800 trillion. In Eastern Europe and Russia this reduction would reach 3–3.5% by 2050, and certainly would bring a dramatic rise in unemployment. Andrei Illarionov warned that "The Kyoto Protocol will stymie economic growth. It will doom Russia to poverty, weakness and backwardness." His words echoed the statement in 1998 by the great British astronomer Sir Fred Hoyle that implementing restrictions in CO_2 emissions would be "ruining the world's industries and so bringing on a situation ... returning us all to the Dark Ages." It is clear that the only people who would be affected by abandonment of the KP would be several thousand people who made a living attending conferences on global warming in attractive locations.

The most important problems concerning the climate change myths include the following: (1) the observation data do not confirm the presence of a uniform global warming (this is especially true for the surface temperature in rural regions not influenced by the so-called urban "heat island" effect, and in the American, Canadian, Russian, Norwegian, and Danish Arctic, satellite remote sensing results, and balloon measurements); (2) the increase of the atmospheric GHG assumed for the supposed doubling of the CO_2 concentration in the atmosphere, is about 4 W/m^2, but the uncertainties due to the unreliable accounting of the effects of atmospheric aerosols, clouds, and numerous other factors, reach several tens to hundreds of W/m^2; (3) the results of numerical climate models that substantiate the "greenhouse global warming" hypothesis, are nothing but mathematically expressed opinions of their creators on how the climate works; and (4) recommendations concerning levels of reduction of emissions of GHG are senseless from the viewpoint of their impact on climate change – they would be utterly inefficient.

Obsessive concentration on GHGs as an allegedly dominant factor among the multitude of other powerful climatic phenomena is a mistruth, and not only compromises science but may also hamper the socio-economic progress of developing and industrially developed countries. The presentation of these views by numerous scientists at the World Climatic Change Conference, and also the proposed realistic approach to the problem by the Russian government, is why the recent discussions at the Moscow Conference were so important.

2.1.2 Carbon cycle

Based on the data of the Third IPCC Report (IPCC, 2001b), let us consider the dynamics of the global carbon cycle and climate during the last century (some of the conclusions will be specified later on):

- Atmospheric CO_2 concentration in the pre-industrial epoch constituting 280 ± 10 ppm during several thousand years had been gradually increasing and by 1999 reached 367 ppm.
- The present CO_2 concentration exceeds that observed for the last 420 thousand years and possibly 14 million years. The rate of concentration growth is unprecedented, at least for the last 20 thousand years.
- The principal cause of the CO_2 concentration growth is fossil fuel combustion. Global CO_2 emissions into the atmosphere between 1980–1989 constituted

5.4 ± 0.3 Pg(petagram) C/yr, and during 1990–1999 they reached 6.3 ± 0.4 Pg C/yr (the remaining part of the emission dynamics is determined by the impact of land use and by CO_2 assimilation by the World Ocean).

- The annual growth of total atmospheric CO_2 content was 3.3 ± 0.1 Pg C/yr during 1980–1989 and decreased slightly to 3.2 ± 0.1 Pg C/yr during 1990–1999. The difference between the annual level of CO_2 emissions and its content in the atmosphere characterizes the scale of CO_2 assimilation by the World Ocean and land ecosystems. The dynamics of this assimilation determines the presence of substantial interannual variations in the growth of atmospheric CO_2 content from 1.0 Pg C/yr (1992) to 6.0 Pg C/yr (1998).

Between 1991 and 1997 about 6.2 Pg C/yr of CO_2 were emitted to the atmosphere due to fossil fuel combustion. During this period the atmospheric CO_2 content increased only by 2.8 Pg C/yr. Using the observation data on atmospheric CO_2 concentration, Battle *et al.* (2000) have shown that during this period the biosphere of both the land and of the World Ocean assimilated, respectively, 1.4 ± 0.8 and $2.0 + 0.6$ Pg C/yr. A rapid assimilation of carbon by the biosphere during this period contrasts with conditions in the 1980s, when it was practically neutral. As for the World Ocean, it had been a powerful sink for carbon until 1995, but in 1995–1996 the assimilation of carbon decreased. Apparently, the storage of carbon on land is more variable than in the oceans.

The key role of the World Ocean in the formation of the carbon cycle is determined by photosynthesis of phytoplankton, and as a result, CO_2 transformation into organic carbon. This "biological pump" provides the removal of CO_2 from the atmosphere and carbon storage in the oceanic depths. Since, as local field experiments have shown, an addition of iron ("iron carbon") to seawaters stimulates the process of photosynthesis, the possibility of regulating the CO_2 concentration in the atmosphere through an intensification of the biological pump with the use of iron fertilization has been supposed. Chisholm *et al.* (2001) have convincingly proved, however, that this approach is unacceptable. Real possibilities for atmospheric CO_2 concentration reduction are connected only with a reduction of CO_2 emissions into the atmosphere and with regulating the carbon stored in the land vegetation cover and in the soil (the soil is known to contain about two-thirds of the carbon reserves on land) (Sarmiento and Wolfsy, 1999; Kondratyev, 2001c; Kondratyev and Isidorov, 2001; Kondratyev and Krapivin, 2001a; Sarma *et al.*, 2001; Kondratyev *et al.*, 2003a,b).

A characteristic feature of the ratio of concentrations of stable isotopes of lightweight elements such as hydrogen, carbon, and oxygen, is their weak variability in the substances of the Earth's crust. This variability is determined by differences in the chemical and physical "behaviour" of atoms of the same element, but with different masses and can be an important source of information on geological, biological, climate, and hydrological processes in the past, including an intensity of photosynthesis in seawater, information about which is important for a number of reasons, but especially in the context of studies of the global carbon cycle (Hemming *et al.*, 1998). Bender (2000) performed an analysis of data on the concentration of oxygen isotopes in the euphotic zone of the oceans (near the Bermudas, where measurements were made, the depth reaches 100 m) and biological oxygen with account of information about seawater mixing and gas exchange. This analysis gave estimates of the photosynthetic productivity of seawater. Analysis of measurement results has shown that in these conditions the photosynthetic productivity of the upper layer (about 10 m thick) turns out to be sufficiently intensive over all the year, decreasing with depth in the euphotic zone. As a rule, an intensified CO_2 emission into the atmosphere was observed during the years of El Niño events, although during the long El Niño event between 1991 and 1994 the rate of emissions was low. Apparently, an intensified rate of emissions can be explained by the impact of reduced CO_2 assimilation on land resulting from temperature rise, drought, and fires in the tropics. In this regard some conclusions can be drawn:

- Since fossil fuel combustion is followed by consumption of atmospheric oxygen, this causes a slow reduction of O_2 content in the atmosphere, which turns out, however, to be slower than that determined by consumption of O_2 in the processes of burning, because part of the CO_2 emissions are assimilated by the land biosphere with the subsequent emission of oxygen into the atmosphere. With data on concentrations in the atmosphere of CO_2, O_2, and $\delta^{13}C$ taken into account, the ocean–atmosphere CO_2 exchange turns out to be -1.9 ± 0.6 Pg C/yr (1980–1989) and -1.7 ± 0.5 Pg C/yr (1990–1999) (the minus sign reflects the loss of CO_2 by the atmosphere, i.e., CO_2 assimilation by the ocean). In the case of the land–atmosphere system, the respective fluxes constituted -0.2 ± 0.7 Pg C/yr (1980–1989) and -1.4 ± 0.7 Pg C/yr (1990–1999). These estimates characterize the level of CO_2 assimilation by the land biosphere, which exceeded the loss due to changes in land use.
- Estimates of contributions to the formation of the carbon cycle due to fossil fuel combustion, obtained from data on atmospheric CO_2 concentration measurements, and on the ratio of $\delta^{13}C$ isotopes concentration agree with results of observations of CO_2 and O_2 concentrations with rational suppositions with respect to global distributions of vegetation of the C_2 and C_4 type as well as the timescale of the carbon cycle in the ocean and on land (Bergamaschi *et al.*, 2000; Bilek *et al.*, 2001; Shimji *et al.*, 2000).
- The CO_2 emissions into the atmosphere due to changes in land use in 1980 averaged, with various estimates taken into account, 1.7 Pg C/yr, with a possible variability within 0.6–2.5 Pg C/yr. These emissions were determined mainly by deforestation in the tropics (so far, there is no global data on land use changes in 1990).
- The land–atmosphere CO_2 exchange is governed by the balance between CO_2 emissions due to changes in land use and CO_2 assimilation due to various factors (residual sinks on land). The residual sinks constituted -1.9 Pg C/yr (ranging from -3.8 to $+0.3$ Pg C/yr).
- According to numerical modelling results the latitudinal distribution of CO_2 assimilation by the ocean is mainly determined by variations in natural CO_2 fluxes characterized by tropical emissions and assimilation in high latitudes.

Estimates of the land–ocean CO_2 exchange north of 30°N between 1980 and 1989 ranged from −2.3 to −0.6 Pg C/yr, and in the tropics (30°N–30°S) from −0.1 to +1.5 Pg C/yr. Considering these estimates, the conclusion can be drawn that it is necessary to distinguish between substantial residual sinks of carbon on land both in extratropical latitudes of the northern hemisphere and in the tropics. In both regions the spatial distributions of the CO_2 budget in 1980 and in 1990 were similar, with a weak growing trend of CO_2 assimilation between 1990–1999 as compared to 1980–1989.

- The longitudinal gradients of the atmospheric CO_2 concentration were much below the latitudinal ones. Attempts to estimate the CO_2 budget on the scale of continents have given controversial results because of insufficient observational data.

The theoretical analysis of the dynamics of the functioning of the global carbon cycle has demonstrated that it is governed by complicated feedback mechanisms. Below are enumerated the most important factors of the formation of the carbon cycle on timescales from decades to centuries.

2.1.2.1 *Response of the carbon cycle to changing CO_2 concentrations*

Formation of the global carbon cycle is determined by numerous feedback mechanisms and, in particular, by feedbacks that govern the variability on a timescale of decades and centuries. Here are some results:

- An assimilation of anthropogenic CO_2 by the ocean is regulated by the general circulation of the ocean and its carbonate chemistry (Macdonald *et al.*, 2002a,b). In conditions of growing CO_2 concentration in the atmosphere, an assimilation of carbon by the ocean takes place due to the gradient of partial pressure of CO_2 between the atmosphere and the ocean. The efficiency of this assimilation decreases with growing CO_2 concentrations due to the lowering "buffer capacity" of the carbonate system. The efficiency of the oceanic sink of carbon depends partly on the rate of the atmospheric CO_2 concentration growth, since an assimilation of CO_2 is limited to the rate of surface and deep waters mixing.
- An increase of atmospheric CO_2 concentration does not cause a substantial fertilization of marine bioproductivity but reduces the pH. During the last 100 years variations in marine bioproductivity due to intensified "calcification" with a low pH has promoted an assimilation of CO_2 by the ocean intensified by several percent.
- Due to bioproductivity processes, the transport of carbon takes place from surface to deep layers of the ocean, where its remineralization occurs. This process maintains the gradient of inorganic carbon concentration and preserves the atmospheric CO_2 concentration at a level of 100–200 ppm below that which would exist in the absence of bioproductivity.

- An assimilation of CO_2 on land is determined by net biome productivity (NBP), which is a balance between net primary productivity (NPP) and carbon loss due to heterotrophic respiration and processes of combustion. According to observational data for many ecosystems, when CO_2 concentrations exceed the present level, an increase of NPP should take place (the degree of fertilization being rather variable), determined by reactions of photosynthesis and stomatal resistance to increasing CO_2 concentrations. With a high level of CO_2 concentration (800–1000 ppm) the effect of fertilization may be negligible.

The efficiency of a carbon sink on land depends on the transformation of carbon compounds into the forms characterized by long lifetimes (timber or modified soil organic matter). An intensified carbon sink due to increasing NPP can be manifested only as a result of inertia of these "slow" carbon reservoirs in the process of agricultural development of soils.

2.1.2.2 Feedbacks with climate dynamics that determine the carbon cycle

Climate warming reduces CO_2 solubility and therefore weakens an assimilation of CO_2 by the ocean.

- Global warming can be followed by an intensified vertical stratification of the ocean, probable consequences of which are a reduction of CO_2 emissions due to upwelling and transport of excess carbon to deeper layers of the ocean as well as changes in bioproductivity.
- Over shorter timescales, the warming results in an increasing rate of heterotrophic respiration on land, but an extent of the impact of this process on the long-term processes of CO_2 exchange remains unclear. Both the warming and the regional changes in spatial distributions of precipitation and clouds can lead to changes in the structures of land ecosystems, their geographical distribution, and primary production. The total impact of such forcings should depend on special features of the regional structures of climate change.

2.1.2.3 Impact of land use and other global changes on the carbon cycle

- Changes in land use and agriculture can substantially affect the carbon cycle on land. Apart from such large-scale measures as deforestation and aforestation, more local but also substantial events are possible. For example, reduction in biomass combustion, reduces CO_2 emissions into the atmosphere, reduces the formation of inert (black) carbon, and favours an invasion of forests (e.g., in savannahs). Supplies of soil carbon decrease with over use of virgin soils but can be restored in conditions where the agricultural practice utilizes low levels of soil cultivation (Sands and Leimbach, 2003).
- In some regions a deposition of anthropogenic nitrogen causes an increase of NPP on land, whereas an excess of tropospheric ozone leads to a NPP reduction.
- Anthropogenic input of nutrients into the ocean through river runoff and deposition of atmospheric aerosols (especially nitrate as well as iron in the

composition of atmospheric aerosols) can affect the marine bioproductivity, but qualitative estimates of this impact are still insufficiently reliable.

2.1.2.4 *Theoretical assessments of components of the global carbon cycle*

The available developments in the field of study of carbon cycle numerical modelling have given the following estimates:

- In the period 1980–1989 the ocean–atmosphere carbon exchange varied between −1.9 and 2.9 Pg C/yr, which agrees with the earlier model estimates and data on the atmospheric carbon budget.
- Between 1980 and 1989 the land–atmosphere carbon exchange varied from −0.3 to −1.5 Pg C/yr, which agrees with the data on the atmospheric carbon budget. Contributions to fertilization due to CO_2 and depositions of anthropogenic nitrogen turned out to be substantial, reaching in total between −1.5 and −3.1 Pg C/yr. The effect of the climate on the global carbon cycle formation in 1980 was weak and uncertain even in sign.
- With the IPCC-1992 scenario (Houghton *et al.*, 1992) of anthropogenic CO_2 emissions to the atmosphere, by 2100 the carbon flux between the atmosphere and the ocean should reach -5.5 ± 6.0 Pg C/yr. If the effect of possible climate change is taken into account, this interval becomes 3.9 ± 6.0 Pg C/yr (i.e., an assimilation of CO_2 by the ocean weakens).
- According to similar estimates for land, by 2050 the land–atmosphere carbon exchange should increase due to fertilization within -3.6 ± -8.7 Pg C/yr: an assimilation of CO_2 by land decreases when the effect of climate on this process is taken into account. It follows from some models that after 2050 an assimilation of carbon should rapidly decrease. Estimates for 2100 vary widely between +0.4 and −6.7 Pg C/yr.
- First experiments with the use of interactive global 3-D "climate–carbon" cycle models have led to the conclusion that a great uncertainty of prognostic estimates of CO_2 concentrations is connected mainly with inadequate account of the impact of climate change on the processes taking place on the land surface.

The principal conclusion from these developments is that the formation of the level of CO_2 concentration in the 21st century will be mainly affected by CO_2 emissions to the atmosphere due to fossil fuel combustion. An important contribution is also made by changes in land use. According to available estimates, the global scale deforestation will cause CO_2 emissions into the atmosphere in quantities some 2–4 greater than the amount of CO_2 assimilated as a result of aforestation. The World Ocean assimilates a great deal (up to 70–80% of emissions), but since it takes a lot of time for the whole oceanic thickness to mix, the process of CO_2 assimilation lasts for centuries. Lavorel *et al.* (2002) have shown that forest fires significantly contribute to the atmospheric global annual carbon budget (1.7–4.1 Pg C/yr).

After a detailed analysis of present studies of the global carbon cycle, Falkowski *et al.* (2000) emphasized that the following two most important questions demand special attention: Can we distinguish between contributions of anthropogenic and natural factors to the variability of biogeochemical cycles and climate? What is the sensitivity of the global climate to changes in atmospheric CO_2 concentrations? Answers to these questions can be obtained only within a systems approach with consideration of interactivity of the processes responsible for the formation of biogeochemical cycles and climate (Andronova and Schlesinger, 2001; Baldocchi *et al.*, 2001).

Large supplies of carbon (about 400–500 Gt C) in the northern soils and in permafrost determine the urgency of analysis of the dynamics of the carbon budget in these conditions in the context of the problem of global climate changes (Zavarzin, 1999; Brombach *et al.*, 2001; Romankevich and Vetrov, 2001; Krapivin and Kondratyev, 2002). The process of carbon storage by the ecosystems of the Alaskan tundra that had functioned earlier as a sink for carbon (due to a low temperature and a high moisture content of soils, which reduced the rate of organic matter decomposition), changed its direction as a result of climate warming and drying in the early 1980s, leading to a considerable loss of carbon. Recent observations have shown, however, that the persistent climate warming and drying have reduced the carbon dioxide flux to the atmosphere, and in some cases the tundra functioned as a sink for CO_2 in the summer.

In this regard, Oechel *et al.* (2000) discussed the results of observations of the CO_2 exchange between the tundra ecosystems (soil–vegetation) and the atmosphere carried out between 1960 and the late 1980s over two sites on the tundra. The results obtained show that the summertime tundra functioned as a sink for CO_2 in the periods of the warmest and driest climate during the last four decades, and demonstrated the ability of the tundra ecosystems to adapt themselves metabolically to long-term (on a scale of decades and longer) climate change.

Apparently, the mechanisms governing these processes are connected with the dynamics of the nutrient cycles, physiological acclimatization, and with the reorganization of populations and communities of vegetation cover. However, despite the observed acclimatization, the Arctic ecosystems turned out to be a source of CO_2 for the atmosphere, with an average annual amount of not less than $40\,\mathrm{g\,C/m^2/yr}$ determined by wintertime emissions of CO_2 to the atmosphere. On this basis, further climate warming may intensify the CO_2 emissions to the atmosphere by Arctic ecosystems.

The remaining deficit of observational data that would enable one to analyse the laws of formation of the carbon cycle has stimulated a number of field observations aimed at solving several related problems.

Large-scale efforts have been connected with an accomplishment within the Integrated Global Observing Strategy-Partnership (IGOS-P) programme of the partnership strategy in the field of the global observations under the TCO subprogramme of the use of various observational means and numerical modelling to monitor the carbon exchange between land ecosystems and the atmosphere or the oceans and to estimate their contribution to the formation of the global carbon cycle (Braswell *et al.*, 1997). The related problem is that the lack of an adequate understanding of the laws of the global carbon cycle formation still does not allow one to "close" the annual budget of the global cycle, for which purpose reliable information is needed on the spatial–temporal variability of both the sources and sinks of carbon on the scales of individual regions and ecosystems.

The data to be obtained within the framework of the TCO should give some answers, in particular, to the following questions: what is the spatial–temporal variability of the carbon sources and sinks on land on scales from landscape to global, and what are the causes of this variability? In this regard it is important to take into account two types of processes determining the carbon cycle formation: (1) a relatively slow large-scale accumulation of carbon by vegetation cover and soils; and (2) relatively rapid emissions of carbon over small regions under the influence of various "disturbances". To obtain the observational data that would characterize the formation of the carbon cycle, various conventional and satellite observational means should be used, planned in the context of global observational programmes (Gildor and Tziperman, 2001). The programme includes the following sections:

1 Satellite observations

- Land surface and its changes, including seasonal variations and the effect of fires, with the use of various means of remote sensing (RS) in a broad wavelength interval from the visible to microwave spectrum.
- Biomass dynamics including the structure of vegetation cover.

In some cases (in particular, when it is necessary to calibrate the space-borne instruments for remote sensing) the space and surface-based observations should be made simultaneously.

2 Surface (*in situ*) observations (together with satellite data)

- Carbon reservoirs (vegetation cover, soils, the atmosphere).
- The surface–atmosphere gas exchange (carbon fluxes) provided the needed spatial–temporal representation is available.
- Atmospheric chemical composition (including the data from aircraft observations).

This schematic programme does not contain any observational sections that should provide the accomplishments of other programmes. According to the adopted plans, by 2005 it should be possible to demonstrate using the TCO data to estimate the surface–atmosphere carbon exchange on a subcontinental scale ($\sim 10^7 km^2$) with an error of $\pm 30\%$. By 2008 the error should be reduced to $\pm 20\%$ for the global database.

Thematic developments in the context of studies of the cycle have been planned within the Global Carbon Project (GCP) (Crutzen and Ramanathan, 2000; Young and Wefering, 2002) including the following four principal directions:

- Specifying the laws of the spatial–temporal variability of the present carbon cycle (Dolman *et al.*, 2001; Eastman *et al.*, 2001; Elderfield, 2002).

- Analysis of anthropogenic impacts on the carbon cycle (Zavarzin, 1999; Borisenkov and Pichugin, 2001).
- Simulation modelling of the interactive system of carbon cycle–climate–human activity to reveal instabilities, threshold levels, and possible "surprises" (Danilov-Danilyan and Losev, 2000; Cox *et al.*, 2002).
- Prognostic assessments of the evolution of carbon sources and sinks in the 21st century (Elliott *et al.*, 2001).

Of course, the space-borne observations in the interests of TCO should be part of a broad programme of global monitoring of the environment (Ervin *et al.*, 2002; Engelen *et al.*, 2001). In this regard Habib (2000) described the prospects for NASA scientific instruments developed for the remote sensing of the Earth. In connection with limited financing, new developments are aimed at substantiation of small (and more economic) scientific complexes to solve specific problems instead of developing heavy satellites carrying huge multipurpose observational complexes.

Many programmes of these types of developments are being accomplished and/or at the preparatory stage. In the interests of the 2005–2010 perspective, the strategy and specific satellite programmes are being worked out. These developments should improve the available methods for measuring parameters (the content of ozone and minor gaseous component (MGC), temperature, humidity, precipitation, ocean colour, etc.), which are part of the programmes of the existing observations, as well as assisting with the search for possibilities to retrieve the parameters inaccessible to RS (ocean water salinity, atmospheric CO_2 content (carbon cycle), atmospheric properties at the tropopause, etc.) but which can be accomplished with a breakthrough in the field of present-day instrument-making technologies. New technologies should provide answers to the following general questions: How is the Earth system changing? What are principal causes of these changes? How does the Earth system react to natural and anthropogenic forcings? What are the possible consequences of changes in the Earth system for the development of civilization? Can we predict changes in the Earth system?

Key directions of perspective developments include the following problems (ACECHEM, 2001):

1 seasonal and interannual climate changes;
2 the global carbon cycle (including biology and biogeochemistry of the ecosystems);
3 the chemistry and dynamics of the atmosphere (including problems of atmospheric ozone);
4 the global hydrological cycle; and
5 natural disasters.

Habib (2000) believes that it is possible to use RS to retrieve soil moisture (microwave remote sensing), atmospheric CO_2 concentration (laser sensing), ocean water salinity (passive and active microwave sensing), precipitation rate (microwave sensing), oceanic mixing layer thickness (laser sensing), and the wind field in the troposphere (Doppler lidar sensing). Concrete objectives are: climate predictions (one month, one year, and ten years ahead), studies of changes in the biosphere and on the land surface, precise meteorological predictions for two weeks, 12-hour predictions of dangerous weather events, assessments of air and natural water quality, and predictions of disasters and assessments of anthropogenic forcings.

We shall now mention some concrete developments which are available to study the carbon cycle.

Within the framework of the March–April 1999 Pacific Field Observation Experiment in the tropics (PEM-Tropics B) as a part of the NASA Global Tropical Experiment (GTE) to study further transport of polluting MGC in the atmosphere of a remote (background) region of the Pacific (35°N–35°S), Staudt *et al.* (2001) performed aircraft measurements of the tropospheric chemical composition at altitudes up to 12 km. Data from these measurements have been analysed together with the 3-D numerical modelling results (with the fields of meteorological parameters prescribed from the observational data) to study the laws of the polluting MGC transport from the northern hemisphere. They considered the results of this analysis for carbon monoxide (the CO lifetime is about several months, which makes it possible to use CO as a tracer for long-range transport).

Calculations of the 3-D field of CO concentration over the Pacific showed that in the northern hemisphere tropical Pacific the CO concentration in a free troposphere increased mainly due to biomass burning in south-eastern Asia, which was particularly intensive in the spring of 1999. However, practically, there was no CO increase in the atmospheric boundary layer, with the dominating pollution duc to fossil fuel combustion in western Europe and Asia (Molemaker and de Arellano, 1998). North of 35°N an input of pollutants from western Europe prevailed compared to long-range transport from Asia and was particularly high in the spring due to CO accumulated in the winter at high latitudes. The CO sources located in North America played a secondary role in the formation of the CO concentration field in the atmosphere over the Pacific.

The transport of the Eurasian industrial pollutants from the region of large-scale circulation around the Pacific maximum to the zone of the trade winds caused the formation of a "pollutants river" in the lower troposphere stretching from the north-eastern to western sectors of the equatorial Pacific to the South Pacific intertropical convergence zone. However, this process of long-range transport contributed little to the interhemispherical exchange.

A clear indicator of an intensive interhemispherical exchange through a narrow region of the west to east transport in the upper troposphere over the eastern sector of the equatorial Pacific ("western passage") is an increase of concentrations of CO and other MGC entering the upper troposphere over the south-eastern Pacific from the northern hemisphere. This transport turned out to be the most important way for the global interhemispherical exchange during PEM-Tropics B, especially in the period of La Niña cooling.

Almost a quarter of moist tropical forests are located in Brazil, with the Brazilian vegetation containing carbon 5 times more than the forests of Zair, second to Brazilian forests in the carbon content. The Brazil America Basin covers a territory, 3,940,000 km^2, including 3,648,000 km^2 of forested areas and 292,000 km^2 of non-forested lands. Estimates of total surface biomass for dense forests are 151 ± 39 t/ha (Kabat and Nobre, 2000).

Carvalho *et al.* (2001) calculated biomass consumption in the process of forest burning (to transform them into cultivated lands) in six test sites. The results obtained refer to Canabi Farm (near Alto Floresta) and the Brazil state Mato Grossa where the five sites were chosen (A, B, C, D, E), each with an area of 1 ha after the forests were cut and burnt at different times. Site A bordered on pasture (its three sides being fenced off by forest). The forest was cut and burnt in 1997. The other four sites were located within the forest. On site B the forest was cut and burnt in 1997, and on site C (located on a deforested territory of 9 ha) in 1998. Site D was located within a deforested area of 4 ha, where the forest was cut in 1998 and burnt in 1999. The relative share of biomass in the process of the cut forest burning constituted 22.7%, 19.5%, 47.5%, 61.5%, and 41.8%, respectively. For sites A, B, C, D, and E the surface biomass and the amount of carbon for each site were, respectively, 496 t/ha and 138 t/ha. With these data taken into account, the amount of carbon emitted to the atmosphere in the form of various gases reached 69 t/ha and the amount of CO_2 and CO emitted in the process of burning was 228 t/ha and 15.9 t/ha, respectively.

From the data for Kenya and Brazil, estimates were obtained of the emissions to the atmosphere of GHG and other MGC resulting from charcoal production widely used as a fuel not only in Kenya and Brazil but also in many countries of Africa and the South East Asia. The emission coefficients (EC) defined as an amount of pollutants in grams per kilogram of produced charcoal, were estimated for carbon dioxide, methane, nitrogen monoxide, carbon monoxide, non-methane hydrocarbons (NMHC), nitrogen oxides (NO_x), and total aerosol. Depending on the type of furnaces used in charcoal production the share of wood carbon transformed into charcoal carbon varies from 37% to 69%. With the data for eight types of furnaces the EC varied within 543–3027 (CO_2), 32–26 (CH_4), 143–373 (CO), 24–124 (NMHC), 0.011–0.30 (N_2O), 0.0054–0.13 (NO_x), and 13–14 (TPS). The typical relative distribution of wood carbon among various components was characterized by the following values: 51% (charcoal), 27% (CO_2), and 13% (products of incomplete combustion, PIC). Since the global warming potential (GWP) estimates for PIC exceed substantially the GWP estimates for CO_2 (calculated per carbon atom), in the process of charcoal production the atmosphere obtains a considerable amount of GHG. Calculations based on the use of the published GWP data for CO_2, CH_4, and N_2O showed that the atmosphere obtains from 0.77 to 1.63 kg C-CO_2 (carbon per equivalent amount of CO_2) per one kilogram of charcoal produced. The total contribution to the formation of GWP due to charcoal production in Kenya and Brazil reaches, respectively, about 2.7 and 7.5 mt C-CO_2. These values are negligible, however, compared, for instance, to the GWP level (almost 1700 mt C-CO_2) due to fossil fuel combustion in the USA.

According to the results of some recent studies, to balance the global sources and sinks for carbon dioxide, it is necessary to assume the presence of a CO_2 sink in the northern hemisphere. A considerable part of this sink should be connected with the biosphere–atmosphere interaction in the northern hemisphere mid-latitudes (Barford *et al.*, 2001). To check grounds for this supposition, Haszpra *et al.* (2001) carried out regular measurements of CO_2 concentrations at various latitudes using instruments mounted at heights 10, 48, 82, and 115 m on a radio-TV tower in western Hungary (46°57′N, 16°39′E). The area around the tower was mainly agricultural with small forests and very small (100–400 people) settlements, without industrial enterprises.

Measurements of the vertical profiles of the CO_2 mixing ratio, temperature, humidity, and wind speed (the instruments used have been described in detail by Haszpra *et al.*, 2001) were started in September 1994, and direct (correlative) measurements of the CO_2 flux (at a height of 82 m) began in 1997. Processing of weekly air samples for the content of CO_2, CH_4, CO, H_2N_2O, and ^{13}C and ^{18}O isotopes was carried out in two American laboratories. An analysis of the measurements revealed that in the boundary layer of the Karpaty Basin were found occasional accumulations of CO_2 in winter and a diurnal course of the vertical profile of the CO_2 mixing ratio in summer.

Sorokhtin and Ushakov (1999) substantiated an approximate technique to calculate the CO_2 flux from the data of gradient measurements that can be used to fill in the gaps in the results of direct measurements.

An important source of information on the carbon dioxide fluxes (and MGC in general) between the surface (especially in the presence of forests) and the atmosphere is widely practiced through observations of tower-mounted instruments. In this regard, however, of great importance is an assessment of the representativeness of these observations, which can be obtained, based on the conception of "imprints" (photoprint-F). The F is defined as a conditional probability $f(r - r_{\text{tower}})$ determining the contribution to the observed short-term changes in CO_2 concentrations at the site of the tower r, with a CO_2 flux equal to $\varphi(r)$.

Variations are called short-term when they occur independent of the annual course or interannual trends. They are connected, as a rule, either with the synoptic-scale processes or with near-surface CO_2 flux changes. In this context the observed changes in the mixing ratio are determined by coevolution of the imprints near the surface:

$$\chi(r_{\text{tower}}) = \int f(r - r_{\text{tower}})\varphi(r)\,dx\,dy + \chi_{\text{trends}}$$

where x and y are the horizontal coordinates. Using the processed data of measurements made at the TV tower in Wisconsin (USA), Gloor *et al.* (2001) showed that changes in the near-tower CO_2 mixing ratio used to calculate the CO_2 flux are affected by the input of air parcels with a typical time of transport of up to 1.5 days. With this typical time and special trajectory of transport taken into account, the area of an imprint may be assumed in this case to reach $10^5\ \text{km}^2$, or one-tenth the size of the USA. This areal representativeness determines the high information content of the data of the tower-based observations used to estimate the CO_2 fluxes over large territories. To obtain reliable information for the whole USA, data of observations on about ten towers would be sufficient.

One of the principal factors of the formation of the global carbon cycle is an atmosphere–ocean CO_2 exchange. Averaged over the time interval t, the CO_2 flux is defined as:

$$F_t = (ks(\Delta pCO_2))$$

where k is the exchange coefficient, s is the rate of CO_2 solution in the seawater, and ΔpCO_2 is the difference between the CO_2 partial pressures of the seawater and the atmosphere. The coefficient k depends on wind speed, intensity of the air bubble formation, air surface layer stability, and wind shear. The short-term variability of CO_2 exchange remains almost unknown at present. Bates and Mervilat (2001) showed that an account of this variability is very important, however, being determined by two circumstances: (1) non-linear dependence of k on wind speed; and (2) importance of consideration of the averaged short-term wind speed variability. The respective quantitative estimates have been obtained from the data of observations of wind speed and pCO_2 with the use of an autonomous sea buoy. A comparison of the results of hourly and daily observations showed that differences in CO2 fluxes with this repeatability of observations reached a factor of 3. Such differences are determined by a non-linear relationship between wind speed and exchange coefficient k. This result is of great importance for correct calculations of the averaged regional and global estimates of the atmosphere–ocean CO_2 exchange.

One of the principal problems of present-day dynamics and forecasts of the climate system evolution is a pre-calculation of the RF determined by various factors, including changes in GHG concentrations, which require, in particular, a reliable account of their exchange between the atmosphere and the surface. Possible solutions to this problem include a retrieval of spatially averaged GHG emissions to the atmosphere from the data of their local observations. This approach is usually called a synthesizing retrieval (SR) that consists in the calculations of fluxes as a linear combination of local values with their spatial structure taken into account.

Kaminski *et al.* (2001) discussed a solution to the problem for carbon dioxide which was to assess the effect of the spatial resolution of fluxes on the results of calculations of their spatially averaged values (in SR, the Earth's surface is divided into several vast regions for which averaged CO_2 fluxes are calculated). Calculations showed that the results of aggregation of local flux values with their averaging for large regions depended substantially on the level of spatial resolution. The related errors may be the same as calculated fluxes. An algorithm has been suggested that would reduce such errors by a factor of 2. The remaining considerable uncertainty in the assessments of the global carbon cycle dynamics necessitates further intensive studies. Some ideas in this regard have been expressed by Ciais (1999a,b; 2002).

2.1.3 The carbon cycle and climate

We have already mentioned the interactivity of the carbon cycle and climate variability. Therefore serious efforts have recently been undertaken to carry out a numerical interactive modelling of the global carbon cycle and climate change. Since this problem is very difficult to solve, it is necessary to search for and substantiate adequate schemes of parameterization of the processes that determine the dynamics of the carbon cycle and climate, initially within the framework of reproducing the processes in the atmosphere–ocean system.

In the earlier widely used interactive 1-D models of the atmosphere–ocean system one of the most important parameters is the coefficient of vertical diffusion k_v in the ocean, characterizing the intensity of the heat and mass exchange between the ocean and the atmosphere. In 3-D models (as 1-D models) this parameter is prescribed a priori using an "adjustment" by comparing the results of numerical modelling with observational data. However, during this adjustment there are substantial differences in the k_v values. In this regard, Harvey and Huang (2001) and Harvey (2001) developed a relatively simple interactive model of the atmosphere–ocean–carbon cycle system, which can be used to numerically model the processes of assimilation of heat and CO_2 by the ocean and to assess changes in the level of the ocean due to thermal expansion.

The development of the model was mainly motivated by the desire to study the atmosphere–ocean interaction in the context of the formation of the carbon cycle and further to apply the model to analyse the climatic consequences to various scenarios of anthropogenic emissions of GHGs and aerosols. On the other hand, an important problem of the use of the model arises when retrieving the natural climate change and carbon cycle on geological timescales.

This model is considered diagnostic but not prognostic, with an emphasis on the global-scale processes. The latter determines a selection of three global regions (two polar regions and one non-polar) with special vertical resolution. The model makes it possible to reproduce the vertical distributions of oceanic parameters, such as the concentration of dissolved inorganic carbon, phosphates, dissolved oxygen, alkalinity, and temperature, and to describe the function of the "biological pump" that causes the formation of organic matter, calcite, and argonite. The formation of bottom waters takes place in one of the polar regions due to interaction with the atmosphere and convective mixing in the ocean leading to a subsequent transport of water to the lower part of the non-polar region. The bottom waters formed in the region of polar downwelling then suffer upwelling in the non-polar region, though some water gets from intermediate depths to the second polar region.

The work discusses in detail specific features of the climate in conditions of this interactive model, and in this regard a definition of the relative efficient coefficient of diffusion k_v was substantiated for various tracers. Calculations showed that the k_v values are at a minimum for temperature, at an intermediate level for carbon, and at a maximum for dissolved oxygen. A much lower value of k_v was obtained in the upper layer of the ocean compared to that used earlier in 1-D models ($\sim 0.2\,cm^2/s$ instead of 0.6–$1.0\,cm^2/s$) as well as exhibiting a lower maximum velocity of upwelling (2 m/yr instead of 4 m/yr). Consideration of convective mixing resulted in considerable changes in the thermohaline circulation intensity on the time-dependent reaction of surface temperature and rising level of the ocean. For this reason the reaction of surface temperature and ocean level to decreasing intensity of the thermohaline circulation appeared to be quite different (faster) compared with that obtained using an earlier classical 1-D model. This means, in particular, that climatic sensitivity to external forcing, in agreement with observational data for the last 140 years, is about 10% lower than in the case of the classical model.

A comparison of the results of the numerical modelling of the vertical profiles of various tracers (including the ratios of isotope concentration) with the use of an interactive 1-D model of the atmosphere–ocean–carbon cycle system and observational data revealed the necessity to reproduce the following characteristics: (1) the rate of upwelling in a non-polar region reaching a maximum of about 2 m/yr at a depth of 1 km with a constantly decreasing rate below and above this depth; (2) different values for the coefficient of vertical diffusion for temperature and various tracers in the upper 0.5-km layer of the ocean and (3) the ratio of the rates of formation of carbonate carbon and organic carbon equal to only 0.9. From the observational data (together with information on mixing along isopycnic surfaces and the results of numerical modelling) it follows, in particular, that the effective coefficient of the vertical mixing in the upper layer of the ocean should be at a minimum for temperature and at a maximum for dissolved oxygen, with the values of alkalinity and concentration of phosphate somewhat lower than for dissolved inorganic carbon (DIC). The model parameters adjusted by the data from pre-industrial observations of tracer distributions provided the best agreement with the results of observations of changes in the isotope concentrations.

The interaction between alkalinity (earlier not considered, as a rule, in such models) and DIC smoothes down the impact of changes in the model parameters on the stationary level of pCO_2 in the atmosphere. Calculations showed that calculated estimates of the anthropogenic CO_2 assimilation by the ocean and of the evolution of the CO_2 content in the atmosphere for the period up to 2200 are insensitive to the mixing parameters prescribed for the model (with these parameters being constant during this period). Estimates of sensitivity of the calculated values of pCO_2 to changes in the temperature of the warm surface of the ocean agree well with similar results obtained with the use of a 3-D model of the carbon cycle in the ocean.

Numerical modelling of the interactive dynamics of the atmosphere–ocean–carbon cycle system is of special interest in the context of studies of the laws of the palaeoclimate formation (Barker *et al.*, 2001; Beck *et al.*, 2001). The data about the relatively warm climate on ancient Earth have always been an enigma, since the extra-atmospheric insolation was then much lower than now. According to geological data (which cannot be considered reliable enough), at an initial stage of evolution the Earth's surface had been warmer than now. An explanation of this paradox consisted of the supposition of a very high CO_2 concentration in the atmosphere at that time and, respectively, an exclusively intensive atmospheric greenhouse effect. Sleep and Zahnle (2001) have analysed a possible global CO_2 cycle on ancient Earth and justified the supposition of a high CO_2 concentration.

In this regard, estimates have been obtained of a gigantic amount of CO_2 in the Earth's crust. According to approximate estimates, about 6×10^{21} mol CO_2 are presently contained in the Earth's crust and a comparable or greater amount in its mantle. The present-day estimation of CO_2 degassing by the mantle gives values of about 2.5×10^{12} mol/yr, from which it follows that over a period of about 2.4 billion years the content of carbonates in the surface layer of the Earth's crust could have doubled. In the conditions of early Earth, in the presence of geothermal sources which were 2–3 times more powerful than today, the formation of the ocean's crust and its subduction could have taken place 4–9 times faster than now. Consequently, the degassing was more intensive, and the time for doubling of the carbonate content in the Earth's crust could have decreased down to 300–700 million years.

Apparently, when considering the long-term processes one should not leave out an account of the effect of the mantle in the formation of the CO_2 cycle. The possible related processes include: subduction of carbonates directly deposited onto the sea bottom (pelagial carbonates); subduction of continental carbonates; and subduction of carbonates resulting from changes in oceanic basalt (the latter process depends on the concentration of free CO_2 in the ocean and therefore can function as a buffer regulating the level of CO_2 concentration in the atmosphere). Sleep and Zahnle (2001) analysed the data for CO_2 content in the mantle and for CO_2 exchange between various reservoirs and the mantle within the conception of the tectonic plates. In this regard, the present global CO_2 cycle has been considered first, since even now the CO_2 content in the ocean and in the atmosphere depends on the processes connected with the mantle. There appears to be a feedback determined by the dependence of the level of carbonization of the ocean crust on the partial pressure of CO_2.

At an early stage in the Earth's evolution powerful processes could have dominated in the mantle, with an intensive greenhouse effect due to CO_2 being impossible (it is the product of other GHG). Sleep and Zahnle (2001) suggested a model of the CO_2 cycle in which the ocean crust functions as a carbonate trap subject to subduction, which closes the mantle-connected CO_2 cycle. The factors that confine the subduction of carbonates could promote a concentration of CO_2 in the near-surface reservoirs. The weakest point of the model is, apparently, that although the capacity of the oceanic basalts or CO_2 reservoirs are high ($\sim 3 \times 10^{21}$ mol), it is still only part of the global CO_2 reservoir (25×10^{21} mol). Therefore it is possible that at an early stage in the Earth's evolution the effect of the buffer determined by carbonization of the ocean crust could have been dominated by the contribution of degassing of the Earth's crust subducting to the surface at island arcs and, according to the considered model, it could have taken place, if the share of subducted carbonate that reached the mantle, $C_{deep} < 20\%$ (the present $C_{deep} = 64$–84%). In the case of $C_{deep} < 0.2\%$, a massive CO_2 atmosphere could have formed, and then the surface temperature could have risen to 200°C, making the Earth uninhabitable. The Earth could not become inhabited until C_{deep} became so great that CO_2 fluxes from the mantle to the surface became balanced by the subduction flux to deep layers.

Goddéris *et al.* (2001) studied the carbon cycle in the period of the Early Palaeozoic Era using for this purpose the data on the isotopic composition of seawater, which is one of the most useful sources of information about the Earth's evolution on geological timescales. Diverse information about the evolution and interaction of the lithosphere, biosphere, atmosphere, and hydrosphere is contained in the data reflecting changes in tectonics ($^{87}Sr/^{85}Sr$, $^{187}Os/^{186}Os$, $^{44}Ca/$

^{40}Ca) and/or the redox state ($\delta^{34}S$, $\delta^{13}C$) as well as physical properties of the ocean ($\delta^{18}O$).The earlier review of data on the trends of concentrations of isotopes O, C, Sr, and S during the whole Phanerozoic period showed that the present values of isotopic "signals" are similar to those observed in the late Proterozoic period. The Neoproterozoic period was characterized by $\delta^{18}O$ ~26‰ SMOW, after which (during the transition from Proterozoic to Cambrian period) there was a decrease down to ~22.5‰ SMOW with a subsequent (approximately) linear increase up to present values of about 30‰ SMOW. To a first approximation, oscillations with respect to this rising trend correlate strongly with the Earth's evolution, but the interpretation of this rise itself raises doubts.

A comparison with other isotope data suggested the conclusion that changes in $\delta^{18}O$ are controlled by tectonic processes, especially by the relative intensity of relationships between the hydrosphere and oceanic and continental silicate formations. Therefore large-scale tectonic events could have broken the isotopic balance in the oceans. This scenario has been suggested, for example, as the cause of the high concentration ratio $^{87}Sr/^{86}Sr$ about 65 million years ago, which was ascribed to the effect of the lifting and erosion of the Himalayan and Tibetan Plateaus, determining the modulation of the global carbon cycle and climate (Kump *et al.*, 2000). During the Neoproterozoic Era, similar processes had taken place in Africa, and similarity of the present-day and Neoproterozoic isotopic characteristics of seawaters can reflect a similarity of tectonic processes responsible for their formation.

In this regard Goddéris *et al.* (2001) discussed the results of the numerical modelling of interactive cycles C-alkalinity-S-Sr, from which it follows that during the early Palaeozoic period (from the early Cambrian to the late Devonian period) the evolution of the isotopic composition of seawaters (Sr, O, C, and S) could have been determined by gradual oxidation of large reservoirs of carbon released during orogenic processes in Africa. Based on a supposition that any changes in the isotopic composition of oxygen in seawaters results from breaking an equilibrium of the organic carbon cycle due to the interaction of the cycles of the oxygen and carbon isotopes, changes of $\delta^{18}O$ in brachiopod shells are assumed to be responsible for the forcings. Calculated values of isotopic ratios of $\delta^{18}C$, $^{87}Sr/^{86}Sr$, and $\delta^{34}S$ agree well with the observational data, like the predicted evolution of pCO_2 in the atmosphere and its connection with the global climate. An important feature of the suggested model is that it admits the possibility of $\delta^{18}C$ fluctuations in seawaters in the Palaeosoic period within several ppm.

To solve the problem of the interactive palaeodynamics of the carbon cycle and climate, of great importance are estimates of CO_2 concentrations in the ancient atmosphere. Using a reverse relationship between indices of leaf particles and CO_2 partial pressure for the present-day plants *Ginkgo biloba* and *Metasequoia glyptostroboides*, Royer *et al.* (2001) retrieved the CO_2 concentration in the atmosphere from the middle Palaeocene to early Eocene and in the period of the middle Miocene. The results obtained showed that, as a rule, the CO_2 concentration had not passed the limits of 300–400 ppm (by volume) (i.e., it had been close to the present-day concentration with an exception of some high concentrations near the border of Palaeocene (Eocene)). Thus it follows from these data that an increasing

CO_2 concentration could not have led to an intensification of the atmospheric greenhouse effect that had compensated the low levels of the extra-atmospheric insolation.

2.1.4 Conclusion

The principal conclusion to be drawn from this information is that studies of such a complicated problem as global carbon dynamics are still at an early stage of development, with the main difficulty being the lack of adequate observational data that characterizes the formation of the carbon cycle. One of the possible ways to remove these contradictions and to overcome the available information uncertainties is to adopt the adaptive-evolutionary technology described by Kondratyev *et al.* (2002a). This technology consists in a complex usage of satellite and surface observations of the environment together with the global model of the NSS. It is necessary to improve the available versions of the global model, providing an agreement between the model and accumulated databases. A regular assessment of the differences between the results of modelling and observational data will enable one to determine an efficient regime of global geoinformation monitoring and to reliably assess the role of various regions and water basins in the formation of CO_2 fluxes in the "atmosphere–vegetation–soil–ocean" system.

2.2 GLOBAL CLIMATE AND THE KP

There is no doubt that the main result of the Second UN Conference on Environment and Development (UNCED) held in Rio de Janeiro in 1992 and of the UN General Assembly Special Session (UN GASS) which followed five years later in New York (1997) ("Rio + 5"), was that these two events drew the attention of governments and the general public to the problems of global changes and sustainable development. Unfortunately, both worldwide forums were not adequately prepared. Perhaps the most important negative result was the failed attempt to work out the "Earth Charter", intended as a conceptual document substantiating priorities. Instead, the most amorphous and declarative document – the "Rio Declaration" – was approved. The aim of the World Summit on Sustainable Development (WSSD – "Rio-10") in Johannesburg (26 August–6 September, 2002) has not been achieved either.

Nowadays, the following three global environmental problems have attracted general attention: (1) climate change (global warming); (2) the global dynamics of the stratospheric ozone layer; and (3) the closed nature of the global biogeochemical cycles (the concept of biotic regulation of the environment).

The sad paradox is that despite the UNCED documents and numerous international forums, a principal scientific understanding of real global ecodynamics is

Table 2.1. Recommendations of the KP for CO_2 emissions reductions by various countries by 2008–2012 compared to the 1990 level.

From Kok and Verweij (1999), Watson *et al.* (2000), and Kondratyev and Demirchian (2001a).

L	Country	*Y*	*D*	*Q*	*G*
108	Australia	288,965	16,516	16.91	0.98
94	Canada	462,643	19,705	17.44	0.81
92	Czech Republic	165,792	3,049	16.00	5.25
92	France	366,536	20,966	6.49	0.31
92	Germany	1,014,155	21,861	12.76	0.59
92	Ireland	30,719	11,349	8.77	0.68
92	Japan	1,155,000	24,205	9.35	0.39
92	Luxemburg	11,343	34,614	30.41	1.10
92	The Netherlands	174,000	18,939	11.22	0.61
100	New Zealand	25,476	12,192	7.61	0.59
101	Norway	35,514	26,387	8.37	0.31
94	Poland	414,930	1,459	10.87	7.04
92	Portugal	42,148	6,774	4.00	0.63
100	Russia	2,388,720	3,897	16.11	4.13
92	England	577,012	16,723	10.08	0.59
93	USA	4,957,022	22,046	19.83	0.90

Note: $L = CO_2$ emissions (%); $Y =$ amount of CO_2 emissions (10^3 tC); D = per capita gross national product ($); $Q =$ per capita emissions of CO_2 (tC); $G = CO_2$ emissions per unit gross national product (GNP) (tC/10^3 $).

absent. It is most important to formulate the sequence of events in the NSS, taking into consideration their dependence on the tendencies in industrial development and population growth.

A result of this situation brought to the forefront the problem of "global warming" and hence, an adoption of the UN FCCC, a document, which is lame, confusing and unjust for the developing world. It is focused, regardless of scientific substantiation, on the anthropogenic origin of global climate warming (observed mean-global mean-annual surface air warming) recommending that developed countries reduce GHG emissions to the atmosphere (primarily carbon dioxide emissions).

In December 1997, in Kyoto (Japan) was held the third conference of the representatives of the countries (over 160), which had signed the FCCC. This conference was accompanied by lengthy and heated debates on the necessity to recommend a 5% CO_2 emissions reduction by 2008–2012 for industrially developed countries (relative to the 1990 level, see Table 2.1) despite the absurdity of this recommendation (it has been convincingly shown that such a reduction would not practically affect the global climate, with the CO_2 concentration in the atmosphere continuing to grow). Naturally, the position of the developing countries was primarily concerned with the consideration of conditions of socio-economic development (in overcoming the poverty and its consequences). Of course, the developing countries are not prepared to make reductions in GHG emissions, being more interested in further intensive development of both industry and agriculture. This has not prevented them from adopting the KP with the recommendation to reduce GHG emissions.

One might think that this absurd situation is a consequence of scientific groundlessness regarding the problems of global change. In this regard it should be borne in mind that as far back as 1990 some publications appeared devoted to discussions of key aspects of global change. Gorshkov (1990) and Gorshkov *et al.* (2000, 2002) proposed and substantiated the fundamental concept of the biotic regulation of the environment, while publications by Kondratyev (1990, 1992b, 1998b, 1999c) revealed the inadequacy in the greenhouse hypothesis of global warming and emphasized the necessity to study the processes taking place in the atmosphere–hydrosphere–lithosphere–cryosphere–biosphere climatic system, analysing their complex interactivity and non-linearity. The problem of a global observing system has been thoroughly analysed (Kondratyev, 1998b, 1999c) as well as the problem of atmospheric ozone variability (Kondratyev and Varotsos, 2000).

Unfortunately, such a document as the third IPCC report (IPCC, 2001a–c) although encyclopaedically complete, lacks references to many important publications including numerous monographs containing critical analysis of some sections of the IPCC-2001 report.

An unprecedented increase in interest in the problems of the climate observed during the last decades (this refers, in particular, to mass media) has stimulated the resolution of both scientific and applied developments (which provided a considerable advance in understanding the causes of present climate change), the laws of the paleoclimate, and in substantiation of scenarios of possible changes in the climate in the future (this is concerned with scenarios and not with predictions whose possibilities should be assessed as doubtful). Unfortunately, the growing interest in the problems of the climate are partly explained by the important role of various speculative exaggerations and apocalyptic predictions (e.g., complete thawing of the Arctic sea ice in the first-half of this century) as a result of which the problems of climate change formulated as a concept of anthropogenic global warming have become the focus of geopolicy (Moe and Tanger, 2000). It is a paradox that the Presidents and Prime Ministers of various countries discuss whether the KP should be considered as a scientifically grounded document. The confusion of the situation is determined, in particular, by the lack of sufficiently clear and agreed-on terminology. Leaving out of the account a very complicated situation regarding the definition of the notion of climate (this subject needs a separate discussion), one should remember, for instance, that in the UN FCCC the notion of climate change was defined as being anthropogenically induced. One of the main unsolved problems consists in the absence of convincing quantitative estimates of the contribution of anthropogenic factors to the formation of the global climate (there is no doubt that anthropogenic forcings on climate do exist). The international documents containing analyses of the present-day ideas of climate use the prevalent notion of consensus with respect to scientific conclusions drawn in these documents, as if the development of science was determined not by different views and respective discussions but by general agreement (and even voting) on some concrete problem. Apart from definitions, of further importance is the problem of uncertain conceptual estimates concerning various aspects of climate problems.

This refers, in particular, to the main conclusion in the summary of the IPCC-2001 report (IPCC, 2001c): "An increasing body of observations gives a collective picture of a warming world and most of the observed warming over the last fifty years is likely to have been due to human activities."

It is a pity that in a recent (2003) article in the British newspaper *The Guardian* the former chairman of the IPCC Working Group 1, Professor J. Houghton (Houghton *et al.*, 1996), compared the threat of anthropogenic climate change with weapons of mass destruction, and accused the USA for their refusal to support the concept of global warming and the KP as being the basic cause of this threat. No matter how paradoxical it may be, such a statement has been made against a background of an increasing understanding of imperfection in present global climate models and the absence of their adequate verification, which makes the predictions on the basis of numerical modelling no more than conditional scenarios.

As for the USA, their huge efforts to support climate studies should be encouraged. These efforts have a special emphasis on the improvement of the observing system and on developments in the field of climate problems in general. The US expenditure on these problems (planned for 2004) is to reach US$4.5 bn.

The statement published on behalf of the intergovernmental group G-8 on 27 July, 2003 emphasizes that in the years to come, efforts will be undertaken in three areas: (1) coordination of strategies of global observations; (2) provision of the pure, stable, and efficient use of energy; and (3) provision of stable agricultural production and biodiversity.

The most important facts testifying to the delusion regarding the problems of global climate change are the following: (1) observational data (so far, inadequate from the viewpoint of its completeness and reliability) do not confirm global warming (especially concerning the data of surface observations in the USA (Harris *et al.*, 1992), in the Arctic (Sturm *et al.*, 2001), as well as the results of satellite remote sensing); (2) if the atmospheric greenhouse effect enhancement due to the supposed doubling of CO_2 concentration in the atmosphere constitutes about $4\,W/m^2$, then the uncertainties connected with an account of the climate-forming role of atmospheric aerosols and clouds as well as a number of other factors, amount to tens and even hundreds of W/m^2; (3) the results of numerical climate modelling, that substantiate the hypothesis of the greenhouse global warming and ostensibly agree with the observational data, are nothing more than data made to fit observations; and (4) recommendations on the levels of the GHG emission reductions based on these results are senseless (their realization can have far-reaching negative socio-economic consequences.

In the context of global change, the problems connected with the assessments of the present and future global climate changes are, of course, of primary importance. The concept of global warming still dominates in such assessments, judging from the third IPCC report published in 2001 (Kondratyev and Demirchian, 2001b). These assessments should be treated as only an inertia of the earlier speculative opinions, a suggestion convincingly demonstrated by Boehmer-Christiansen (2000). Clearly, for an objective analysis of all the cause and effect feedbacks and processes in the Earth's climatic system a constructive mechanism is needed to collect and analyse the data. Here we shall briefly discuss the well known data (Kondratyev and Demirchian, 2001b).

2.2.1 Atmospheric temperature

According to surface air temperature (SAT) observations for the period from 1860, the mean-annual mean-global value has increased by $0.6 \pm 0.2°C$. This exceeds the value given in the IPCC-1996 report by about 0.15°C (Kondratyev, 1999a), which is due to a relatively high SAT between 1995 and 2000. The observational data reveal a very strong spatial–temporal variability of the mean-annual SAT on a global scale. For example, the climate warming in the 20th century took place during two periods: 1910–1945 and 1976 to present-day. It follows from new data on the global climate diagnostics that the climate warming in the northern hemisphere in the 20th century was, probably, the strongest during the last 1000 years; the 1990s were the warmest decade, with 1998 being the warmest year. An important feature of the climate dynamics consisted in that, on average, the rate of the nocturnal (minimum) SAT increase on land was nearly twice that of the diurnal (maximum) SAT increase, starting from 1950 (0.2°C against 0.1°C/10 years). This favoured the long duration of frostless periods in many regions at moderate and high latitudes.

The IPCC-2001 report does not mention an expected enhanced climate warming at the northern hemisphere high latitudes, which has been emphasized earlier as a characteristic indicator of the anthropogenically induced global warming. An analysis of the data of direct SAT measurements made at the "North Pole" stations over 30 years and the indirect dendroclimatic data for the last 2–3 centuries, performed by Adamenko and Kondratyev (1999), has not revealed any enhancement of climate warming; the climate changes observed during the last century and the last decade were characterized by a strong spatial–temporal variability (in the Arctic, regions of both warming and cooling of the climate formed simultaneously).

Data from the aerological observations network (starting from the 1950s, when this network became more or less adequate) show the trends of the mean global SAT and of the lower troposphere temperature were almost the same (about 0.1°C/10 years). According to data from satellite microwave remote sensing (starting from 1979), the mean global lower troposphere temperature has been increasing at a rate of 0.06°C/10 years, much lower than the SAT increase (~0.15°C/10 years). This difference in climate warming manifested itself mainly in the regions of the oceans in the tropics and subtropics, the reasons for these differences being unknown (Christy *et al.*, 1998).

The SAT increase was mainly caused by an increase of minimum temperature T_{min}, which almost doubled compared to the maximum temperature T_{max}. Therefore the SAT increase was followed by the decreasing amplitude of the diurnal course of temperature, DTR = $T_{max} - T_{min}$, which turned out to be comparable (–0.8°C/100 years) with the warming trend. Although the DTR trend can serve as an indicator of the level of anthropogenic forcing on the climate, there are no respective quantitative estimates. In this regard, based on the use of the Canadian global climate model (CGCM1), Stone and Weaver (2002) carried out a numerical modelling of SAT changes, considering both the observed growth of GHG concentrations and the prescribed increase of the content of sulphate aerosol in the troposphere, whose effect is estimated by an equivalent change in surface albedo. The DTR constituted –0.2°C and, as the analysis of the numerical modelling results showed, was determined mainly by changes of cloud conditions and soil humidity. A considerable difference from the observed data testifies to the necessity of further improvement of numerical modelling.

The difference of temperature trends near the surface and in the troposphere has caused a heated discussion in the scientific literature (Waple and Lawrimore, 2003). Since the reliability of the satellite remote sensing data raises no doubts, and their spatial representativeness (on global scales) is more reliable than that of the data of ground measurements, this difference should be interpreted as necessitating further analysis of the SAT and sea surface temperature (SST) data adequacy.

Data on changes of the tropopause height have recently attracted attention. As Santer *et al.* (2002, 2003) noted, starting from 1979, the height of the tropopause increased by several hundred meters, agreeing with the results of numerical climate modelling, taking into account the growth of GHG concentrations, whose contribution prevails.

2.2.2 Snow and ice cover extent

Starting from the end of the 1960s, a 10% decrease in the snow cover extent (SCE) and a 2-week reduction of the annual duration of lake and river ice cover in the northern hemisphere middle and high latitudes have been observed, while in the non-polar regions mountain glaciers retreated. In 2002 the northern hemisphere snow cover extent constituted 25,400,000 km^2, on average 200,000 km^2 less than during the preceding 30 years. The annual trend changed from 2,700,000 km^2 (August) to 46,900,000 km^2 (January) (Waple and Lawrimore, 2003).

Starting from the 1950s, the northern hemisphere ice cover extent in spring and summer had been decreasing by 10–15%. Probably, during the last decades (in the periods of late summer to early autumn) the Arctic sea ice cover thickness decreased by about 40%, while in winter the decrease was less substantial. During regular satellite observations (starting from the 1970s) there was no marked trend in the ice cover extent in the Antarctic.

Numerical modelling based on the use of the global climate models has shown (considering the growing concentration of GHGs and aerosols) that in the Arctic the climate warming should increase due to a feedback determined by the melting of sea ice and snow cover, which causes a decrease in surface albedo. On the other hand, it follows from the observed data that during the last decades the SAT has increased over most of the Arctic. One of the regions where a warming took place is northern

Alaska (especially in winter and spring). In this regard, Stone *et al.* (2002) performed an analysis of the data on climate change in the north of Alaska to reveal their impact on the annual trend of SCE and the impact of SCE changes on the surface radiation budget (SRB) and SAT.

This consideration is focused on the most adequate information for the region of Fort Barrow and on the retrospective analysis of data on early springtime snow thawing both at Barrow and in other places. These data show that compared with the 1960s, the complete thawing of snow cover in 2002 took place eight days earlier than the same period in the 1960s. This situation is explained by a reduction of snowfall in winter and by a climate warming in spring, which can be determined by the changing regional atmospheric circulation. During the last decades, the frequency of occurrence of the wintertime northern air currents increased, which could favour a reduction of snowfall in northern Alaska. However, in spring intrusions of humid warm air masses from the northern Pacific Ocean prevail favouring an accelerated ablation of snow at the northern slope of Alaska.

One of the consequences of the early snowmelt consists in increasing SRB. At Barrow the associated RF can exceed $150\,W/m^2$ during the five days after complete snowmelt, and the annual average RF constitutes $2\,W/m^2$. It means that the early disappearance of snow in Alaska contributes into the northern hemisphere climate warming. Also of importance is a high sensitivity of land ecosystems to variations in the SCE. In this regard, there is concern that the discussed forcings are of anthropogenic origin, and an adaptation to them will be followed by considerable socio-economic consequences. Although the observed decrease of the SCE agrees, on the whole, with results of the numerical modelling of anthropogenic warming, natural changes of atmospheric circulation are also important. All this determines the necessity for further observations and numerical modelling.

2.2.3 Surface level and the ocean upper layer heat content

During the 20th century the World Ocean surface level rose by 0.1–0.2 m. Apparently this was caused by the thermal expansion of seawater and ice melting on land due to global warming. The rate of the World Ocean level rise in the 20th century exceeded 10 times that observed during the last 3000 years. Beginning from the end of the 1950s (when SST changes became of mass character), the heat content of the upper ocean layer has been increasing.

Levitus *et al.* (2001) analysed data on the warming of some components of the climatic system during the second-half of the 20th century. These data were derived from the growth of the heat content of the atmosphere and ocean as well as from estimates of the heat losses on melting of some components of the cryosphere. The results under discussion have led to the conclusion that the heat content of the atmosphere and ocean is growing. The growth of the heat content in the 3-km ocean layer between 1950 and 1990 exceeded, at least by an order of magnitude, the increase of the heat content in other components of the climate system. If the observed increase of the ocean heat content between 1955 and 1996 reached $18.2 \times 10^{22}\,J$, then in the case of the atmosphere it constituted only $6.6 \times 10^{21}\,J$.

As for the values of latent heat due to water phase transformations, they were: 8.1×10^{21} J (a decrease of the mass of glaciers on land); 3×10^{21} J (a decrease of the sea ice cover extent in the Antarctic); 1.1×10^{21} J (melting of mountain glaciers); 4.6×10^{19} J (a decrease of the snow cover extent in the northern hemisphere); and 2.4×10^{19} J (melting of permafrost in the Arctic).

The observed data were compared (Levitus *et al.*, 2001) with results of numerical modelling using the Geophysical Fluid Dynamics Laboratory (GFDL) interactive model of the atmosphere–ocean system: (1) taking into account the RF determined by the observed growth of GHG concentrations, changes of the content of sulphate aerosol in the atmosphere, and extraatmospheric insolation as well as volcanic aerosol; and (2) taking into account only GHGs and sulphate aerosols. The results of comparisons have led to the conclusion that the observed changes in the ocean heat content can be explained, mainly, by the growth of GHG concentrations in the atmosphere, though one should bear in mind a substantial uncertainty in the estimate of RF due to sulphate aerosols and volcanic eruptions. The latter reduces the reliability of Levitus *et al.* (2001) with respect to recognition of the anthropogenic warming.

Noting the presence of the interannual variability of the World Ocean heat content, Levitus *et al.* (2001) emphasized: "... we draw attention to the fact that an extreme heating of the World Ocean during the 1990s was partly connected with a multi-decadal warming of the Atlantic and Indian Oceans as well as with the positive polarity of possible two-year oscillations of the Pacific Ocean heat content ... changes of the World Ocean heat content we observe can be connected with modes of hemispherical or global variability of the atmosphere from the ocean level up to the stratosphere. Understanding the nature of such possible connections plays the important role in understanding the mechanisms governing the global climate." As has been mentioned above, recent developments to recognize the anthropogenic climate change have been confined to analyses of comparatively long data series on SAT, although smaller volumes of data on changes of the sea ice cover extent, vertical temperature profiles (data from radiosondes), as well as results of satellite microwave sensing (Christy *et al.*, 1998, 2003; Christy and Spencer, 2003) were also considered. On the other hand, it follows from the numerical modelling results that scarce data on the amplitude of the annual trend and those on the diurnal course of the winter air temperature should be more representative (than SAT).

Since the World Ocean is the most inertial component of the global climate system, analysis of its variability is a first priority, especially since Levitus *et al.* (2001) revealed that an increase of the heat content of the upper layer of all oceans has taken place during the last 45 years.

Comparison of the numerical modelling results of the heat content of the upper 3-km layer of various oceans, realized with the use of the "parallel" climate model (PCM) and observational data, has shown that the calculated anomalies of the heat content (as deviations from data of the control integration) do not differ from the observed values (between 1950 and 1990) at a 0.05 level of statistical significance. An exception (in the case of global averaging) is only found in data for 1970–1979, when the model does not reproduce the heat content anomaly observed during this decade. On the whole, the probability that the considered anomalies in the heat content are explained only by the internal variability of the climate system, does not exceed 5%, which makes it possible to suppose the anthropogenic nature of the climate change.

The nature of the warming of various oceans can be quite different. An intensive vertical mixing and rapid penetration of the warming into deep layers are typical of the Atlantic Ocean (especially in its southern sector). In other sectors this process is much slower. An important conclusion from the obtained results is that it is necessary to simulate with the climate model not only SAT changes but also the ocean heat content. Some weak points of numerical modelling have been noted and, in particular, the internally induced climate variability estimated only from the data of the control numerical modelling.

Cai and Whelton (2002) drew attention to the fact that in the future the ocean can considerably affect the global scale precipitation. Developments in the fields encountering these difficult problems, based on the use of both observed data and results from numerical modelling, have led to quite different conclusions. The climate warming of the last decades was characterized by a spatial structure similar to that of the El Niño/Southern Oscillation (ENSO) event. But since there are no data on such a structure for the entire century, the observed structure of warming is assumed to be a manifestation of the multidecadal natural variability of climate, but not the result of greenhouse forcing.

Initial results of numerical modelling using the interactive models of the atmosphere–ocean system have shown that the structure of the warming characterized by the zonal SST gradient in the equatorial band should be similar to El Niño, but some theoretical developments suggest the conclusion of a similarity of this structure to La Niña. To resolve this controversy, Cai and Whelton (2002) performed a numerical climate modelling with the use of an interactive model developed at the Australian National Scientific Centre (CSIRO Mark 2), which has shown that initially the spatial structure of the warming is similar to La Niña (the strongest warming in the extratropical latitudes with a weak La Niña-like structure in the tropics) which later on (after 1960) is transformed into a structure similar to that of El Niño. Such results were obtained using three versions of numerical modelling (apart from the control integration for the 1000 year period) in which the growth of GHG concentrations in the atmosphere were prescribed from the observed data (1880–1990) and according to the 1992a scenario (1990–2100). The climate-forming role of aerosols was left out of the account.

In the context of the problem of atmosphere–ocean interaction, the role of the Arctic and North Atlantic oscillations should be analysed. By definition, the Arctic oscillations (AO) are characterized by the first empirical orthogonal function (EOF-1) for the monthly mean field of atmospheric pressure at sea level (SLP) in the northern hemisphere. The presence of a "swing" in the sign of the SLP between the Arctic and middle latitudes, with a high zonal symmetry, is typical of their spatial structure. From the time when the notion of AO was introduced (in 1998), heated discussions have been taking place regarding the relationships between the AO and the North Atlantic oscillations (NAO). In this regard it was noted that the temporal coherence between high and middle latitudes is most vividly expressed in the Atlantic sector. Between the pressure fields in the Atlantic and Pacific sectors there is only a very weak correlation. On this basis a conclusion was drawn about the absence of any difference between AO and NAO.

To test the validity of this conclusion, it was critically important to analyse the coherence of variations of the SLP fields in the Arctic (N), regions of the Atlantic Ocean (A), and Pacific Ocean (P) as well as to answer the question whether the AO is the result of two independent "swings" in these three regions, representing thereby an apparent (imaginary) AO. An analysis of monthly mean data of the pressure field observations performed by Itoh (2002) for 53 years (1948–2000) for the period from November to April has shown that SLP variations in the regions A and P (without data for the region N) are independent. With the region N taken into account, there appears a negative correlation between A and P as well as two temporal series, which depend on the structure of the AO in A and P. The correlation between these two series is negligible, but their sum is characterized by a high correlation with the coefficient of temporal variability for the AO. All these results suggest the conclusion that the observed AOs are almost imaginary, since a combined consideration of two almost independent "swings" based on analysis of the observed data enables one to reproduce the AO. The observed AO turns out to be determined by the NAO and by enhanced fluctuations of the Pacific/North Atmosphere (PNA; the structures with wave disturbance in the European-Atlantic sector), which do not correlate.

The transformation of the climate warming spatial structure is determined by warm extratropical waters, which descend downward and through the subtropics reaching the tropical belt where an upwelling occurs. This is the cause of a climate change. The obtained results can be considered as supporting the conclusion that the warming with the spatial structure of El Niño observed during the last decades can be, at least partially, explained by the effect of anthropogenic enhancement of the atmospheric greenhouse effect. However, although the structure of the observed warming is similar to that calculated, observations made before 1950 are less reliable. Besides, conditions similar to La Niña have been recently observed (in 1995–1996 and 1998–2000).

2.2.4 Other climatic parameters

Data on ground surface temperature (GST) are important for climate diagnostics. As Majorovicz *et al.* (2002) noted, an analysis of the GST data obtained in different regions of Canada by measuring the ground temperature in boreholes revealed considerable spatial differences both in GST increase observed in the 20th century and in the time of the onset of the warming. So, for instance, from measurements in 21 boreholes covering a period of about 1,000 years a warming was detected (within 1–3°C) taking place over the last 200 years. The warming was preceded by a long cooling trend in the region 80°W–96°W, 46°N–50°N, which continued until the beginning of the 19th century. According to data for ten boreholes in central Canada, the temperature had reached a minimum in 1820 with a subsequent warming by about 1.5°C. In western Canada, during the last 100 years the warming reached 2°C.

An analysis has been made by Majorovicz *et al.* (2002) of more adequate information on GST from data of measurements in 141 boreholes at a depth of several hundred metres. The holes were drilled in 1970–1990. The results obtained revealed an intensive warming that started in the 18th and 19th centuries, which followed a long period of cooling (especially during the Little Ice Age) continuing during the rest of the millennium. The time of the onset of the present-day warming differs among the regions. An analysis of the spatial distribution of the GST changes over the territory of Canada revealed a substantial delay of the onset of the present warming in the east-to-west direction, with a higher level of GST increase in the 20th century in western Canada. This conclusion is confirmed by the data of SAT observations. It should be borne in mind that the GST increase in eastern Canada had begun about 100 years before the industrial era.

Characteristics of the atmospheric general circulation are important components of climate diagnostics. As Wallace and Thompson (2002) noted, the western–eastern zonal wind component averaged over the 55°N latitudinal belt can be a representative indicator of the primary mode of the surface air pressure anomalies – the annual mode of the northern hemisphere (NAM). Both NAM and a similar index SAM for the southern hemisphere are typical signatures of symbiotic relationships between the meridional profiles of the western–eastern transport in the respective hemisphere and wave disturbances superimposed on this transport. Their index determined (using a respective normalization) as a coefficient for the first term of NAM expansion in empirical orthogonal functions can serve as the quantitative characteristic of the modes. The presence of a positive NAM (or SAM) index denotes the existence of a relatively strong western–eastern transport.

The indices of modes (IM) are characterized by irregular variability together with long-term trends with different signs. If the IM changes its sign, it means a change of polarity. The observed swinging of atmospheric pressure both in polar zones and in the middle latitudes, most vividly manifested in the Atlantic sector, is the NAO. There is a trend of enhancement or attenuation of the wintertime stratospheric northern polar jet stream (NPJS) due to NAM fluctuations. In the autumn, there is a regular enhancement of NPJS due to radiative cooling taking place in the polar night zone.

Data have been accumulated which show that in the periods between January and March the NAM can be predicted for longer periods than it was supposed, due to connections between the NAM and variations of NPJS intensity in the lower stratosphere. Although for a long time sceptic views have been expressed on a possible effect of changes of the stratospheric circulation on changes of weather and climate near the Earth's surface, at present these views are changing. There are convincing grounds to believe that during the last decades, winters in the northern hemisphere have become less severe. In part this could be connected with the effect of greenhouse global warming, but changes of the atmospheric circulation also play an important role.

During the last years, it was recognized that dynamic factors contribute significantly to the observed temperature trends. For instance, in 1995 a marked similarity was observed between the spatial distributions of the SAT field and NAM fluctuations over the last 30 years, with a clear increase in the NAM index. The increasing trend of the index was accompanied by mild winters, changes of the spatial distribution of precipitation in Europe, NPJS increase, and ozone layer depletion in the latitudinal belts >40°N. Similar data are available for the southern hemisphere. The main conclusion is that along with the ENSO event, both NAM and SAM are the leading factors of the global atmosphere variability. In this regard, attention should be focused on the problem of the 30-year trend of NAM towards an IM increase (after 1995 the index lowered). It is still not clear whether this trend is a part of long-term oscillations.

The observation data show that during the 20th century an increase in precipitation constituted 0.5–1%/10 years over most of the land surface in the middle and high latitudes of the northern hemisphere, but a decrease (by about 0.3%/10 years) took place over most of the land surface in subtropical latitudes (an observation which has recently weakened, however). As for the World Ocean, the lack of adequate observational data has not revealed any reliable trends for precipitation. Probably, during the last decades, intensive and extreme precipitation in the middle and high latitudes of the northern hemisphere has become more frequent. Beginning from the mid-1970s, the ENSO events have been frequent, stable, and intensive. This ENSO dynamics was reflected in specific regional variations of precipitation and SAT in most of the zones of the tropics and subtropics. Data on the intensity and frequency of occurrence of the tropical and extratropical cyclones as well as local storms still remain fragmentary and inadequate and do not permit conclusions to be drawn on any trends (Kondratyev and Grigoryev, 2002).

Changes in the biosphere are important indicators of climate. One of them is the bleeching of coral. It is important that enhanced atmospheric forcings on coral reefs do not lead to their disappearance but to their transformation into species more resistant to external forcings (Hughes *et al.*, 2003). Changes in seawater properties are one further indicator of climate changes (Bryden *et al.*, 2003).

2.2.5 Concentration of GHGs and anthropogenic aerosols in the atmosphere

During the period from 1850 to the present-day the atmospheric CO_2 concentration has increased by one-third, reaching a maximum during the last 420 thousand years (and, possibly, during the last 20 million years) judging from paleoclimatic data. About two-thirds of the CO_2 concentration increase for the last 20 years is determined by emissions into the atmosphere from fossil fuel combustion (the rest of the share is determined by deforestation and, to a lesser extent, the cement industry). It is of interest that by the end of 1999 the CO_2 emissions in the USA exceeded the level of 1990 by 20%, and by the year 2008 their increase should exceed this level by a further 10%. Nevertheless, according to the KP, the emissions should be reduced by 7% by 2008 compared to the level of 1990, which requires a total reduction in emissions of about 25%.

According to the available observational data both the World Ocean and land surface are now global sinks for CO_2, in the ocean it is governed by chemical and biological processes, and on land it is connected with an intensified "fertilization" of vegetation due to increasing concentrations of CO_2 and nitrogen in addition to changes in land use. More details have been given in other sections of this book and in numerous publications by other authors (Cox *et al.*, 2000; Braunlich *et al.*, 2001; Cihlar *et al.*, 2002).

There is no doubt that fossil fuel combustion will remain the principal factor in the CO_2 concentration growth in the 21st century, with the biosphere of both the ocean and land as a barrier to concentration growth diminishing with time. According to the IPCC (2001a), by the end of the century, CO_2 concentrations will possibly range between 540 and 970 ppm (pre-industrial and present values are, respectively, 280 and 367 ppm). An important factor of the global carbon cycle is a change in land use but even if total carbon emitted to the atmosphere due to land use is assimilated by land biota, this will only reduce CO_2 concentrations by 40–70 ppm. As for the prognostic estimates of the concentrations of other GHGs, they vary widely. So, for instance, it follows from some estimates that the role of tropospheric ozone as a greenhouse gas can be equal to the contribution of methane and will be substantial, being the factor of deterioration of air quality over most of the northern hemisphere.

Since 1850 atmospheric methane concentration has increased, compared to the observed one (from indirect data), by a factor of 2.5 and continues to grow. The annual rate of CH_4 concentration growth slowed down and became more variable in the 1990s compared to the 1980s. Starting from 1850 the concentration of nitrous oxide has increased by 16%. As a result of implementation of the recommendations of the Montreal Protocol and its subsequent supplements, the concentrations of several halocarbons acting as GHGs and ozone-depleting gases either increased more slowly than before or started to decrease. On the other hand, concentrations of their substitutes and some other synthetic compounds (e.g., perfluorocarbons, PFC, and sulphur hexafluoride, SF_6) started rapidly increasing.

Estimates of changes of the RF that characterizes an enhancement of the atmospheric greenhouse effect and is determined by the growing concentration of MGC in the atmosphere totalled $2.42\,W/m^2$, with the following contributions of various MGCs: CO_2 ($1.46\,W/m^2$), CH_4 ($0.48\,W/m^2$), halocarbons ($0.33\,W/m^2$), and N_2O ($0.15\,W/m^2$). A decrease of the total ozone content observed during the last two decades could lead to a negative RF constituting $0.15\,W/m^2$, which could drop to zero in this century if the measures taken to protect the ozone layer are successful. The tropospheric ozone content increase since 1850 (by about one-third) could have led to a positive RF of about $0.35\,W/m^2$.

Since the publication of the IPCC-1996 report substantial changes have taken place in assessments of RF determined not only by purely scattering sulphate aerosols but also by other types of aerosols, especially carbon (soot) characterized by a considerable absorption of solar radiation as well as organic sea salt and mineral aerosols. A strong spatial–temporal variability of the atmospheric aerosol content and its properties seriously complicates the assessments of the climatic implications of aerosols (Kondratyev, 1999b). The results of new calculations made by Hansen *et al.* (2000) have radically changed the ideas of the role of various factors of the RF formation. According to these calculations, there exists an approximately mutual compensation of climate warming due to the growing CO_2 concentration and cooling due to anthropogenic sulphate aerosols. In these conditions the anthropogenically induced emissions of methane (mainly due to rice paddies) and carbon (absorbing) aerosols should play a more important role.

The climate-forming factors to be taken into account are changes in the extra-atmospheric solar radiation. Contributions of these changes into RF since 1850 could have reached $\sim 20\%$, compared to the CO_2 contribution, mainly due to an intensified insolation during the second half of the 20th century (of importance is an account of an 11-year cycle of insolation). However, possible mechanisms for enhancing the climatic implications of solar activity are far from being understood.

All this confirms the conclusions drawn by many authors that recommendations like those of the KP might have some sense provided they were made based on a complex analysis of every climate-forming process, which is impossible using present-day numerical climate modelling techniques. The recent considerable progress in the development of more complete numerical climate models with an account (interactive, as a rule) of principal components of the atmosphere–hydrosphere–cryosphere–biosphere system raises no doubts, however. An interactive simulation of the global carbon cycle within the framework of climate theory has started, at last. The extreme complexity of climate models and the large number of applied empirical parameterization schemes for various (subgrid, in particular) processes hinders the analysis of adequate models, especially from the viewpoint of their use in future climate predictions. Therefore, so far, the undertaken attempts to compare the results of numerical climate modelling with the observational data have been rather schematic, contradictory, and unconvincing.

Also unconvincing were, for example, the conclusions concerning the secular trend of the mean-annual mean-global SAT for the last one and a half centuries. If according to the IPCC-1996 report, the observed and calculated changes (with the growth of CO_2 and sulphate aerosol concentrations taken into account) of SAT agree well, then, following Hansen *et al.* (2002), an account of methane and carbon aerosols is more important. Unfortunately, in both these cases the conclusions are based on arbitrary opinions, and an agreement with observations is nothing else than forced fitting of the data. Also it is clear that a meaningful comparison of the theory with observations should contain a consideration of regional climate changes (not only SAT) and not only average values of climate parameters, but also their variability characterized by higher order moments.

The IPCC-1996 report contains a disputable conclusion: "The balance of evidence suggests a discernable human influence on global climate" (Kondratyev, 1998a). It also states that the "anthropogenic signal" is already manifested against the background of natural climate change. According to the IPCC-2001 report, "the detection and attribution studies regularly indicate the presence of the anthropogenic signal in the climate observation data for the last 35–55 years .... The nature-induced impacts could play a role in the warming observed in the first-half of the 20th century, but they cannot explain the warming in the second-half of the century." This report also indicates, however, that, "the reconstruction of the climate for the last 1000 years, as well as model estimates demonstrate a low probability that the observed climate changes in the second-half of the 20th century could be of completely natural origin." Also, the report emphasizes a high uncertainty in the quantitative estimates for anthropogenic warming, especially from the viewpoint of contributions from various factors of warming (atmospheric aerosols). The contradictory and unconvincing character of the cited opinions is apparent and does not require any further comment. Of course, the leading role in the substantiation of future climate predictions should belong to integral simulation models describing the dynamics of interactions between the maximum possible number of factors of global natural–anthropogenic system (Tol, 2000). It is unclear, however, how real climate predictions can be made in the case of inadequate input information. In the near future the integral models can only give rather conditional scenarios. The development of detailed models of the climate system individual components, such as biogeochemical cycles of greenhouse gases, will make it possible to reduce the existing uncertainties in assessing their role in the dynamics of climatic characteristics.

From new data for various scenarios of the growing concentrations of GHGs and aerosols, the mean-global mean-annual SAT should rise between 1990 and 2100 within 1.4 to 5.8°C, whereas according to the IPCC-1996 report, this interval was 1.0–3.5°C. In this regard it is symptomatic that an improvement and increasing number of the models have brought forth not narrowing but broadening of the divergence of this process. It is important that the divergence of calculated SAT values using various models with one scenario of GHG emissions and one model with different scenarios of emissions are nearly similar. As for regional climates, they are still statistically unreliable. However, the following conclusion may be considered reliable: the climate in almost all terrestrial regions will warm up faster compared to mean global warming, especially in high latitudes in the cold half of the year. Climate warming pre-calculated for the northern regions of North America and for northern and central Asia, where it exceeds the mean global value by about 40%, turns out to be especially marked. On the contrary, in the summertime for southern and south-eastern Asia and in the wintertime for southern South America the warming should be weaker than the mean global one. The numerical modelling testifies to a forthcoming increase of atmospheric moisture content and enhanced precipitation. In particular, there can be a precipitation enhancement in the northern hemisphere middle and high latitudes as well as in the winter Antarctic (this conclusion is of special importance in the context of the problem of glacier dynamics). In low latitudes there may be regions both with intensified and reduced precipitation (depending on the choice of GHG emissions scenarios).

Calculations of anthropogenic (greenhouse) climate change reveals the possibility of future weakening of the thermohaline circulation (THC) in the northern hemisphere oceans. However, even the models showing such attenuation still reflect the preserved greenhouse warming in Europe. But it is not clear whether the THC collapse is irreversible and what threshold conditions correspond to such a collapse. None of the available models predicts a complete THC collapse in the next 60 years.

According to the results of the numerical modelling of the global warming process, there must be a further reduction of the snow and sea ice cover extent in the northern hemisphere. Further retreat of glaciers is expected in the 21st century (except in the ice sheets of Greenland and the Antarctic). With a chosen scenario of the GHG concentration increase between 1990 and 2100 the World Ocean level can rise by 0.14–0.8 m ($\sim$0.47 m, on average), which exceeds two to fourfold the rate of the ocean level rise in the 20th century.

The anthropogenic impact on the global climate will remain for a long time, which determines the following specific features of the respective processes:

- The supposed stabilization of the level of atmospheric CO_2 concentration requires a considerable reduction of CO_2 emissions into the atmosphere as well as a greater decrease of other GHG emissions.
- The effect of CO_2 emissions on the CO_2 concentration in the atmosphere is of long duration. Even over several centuries after cessation of emissions the share of CO_2 remaining in the atmosphere can reach 20–30% of the whole volume of emissions.
- The rise of the mean-global SAT and the World Ocean level (due to thermal expansion) can also continue for several centuries after the stabilization of the CO_2 concentration level because of the gigantic inertia of the ocean.
- The response of ice sheets to the earlier climate changes can continue for several thousands of years after the climate stabilizes. According to model calculations, maintenance of the local mean global warming at more than 3°C during thousands of years can lead to complete melting of the Greenland ice sheet. With a 5.5°C warming the World Ocean level should rise by 3 m during 1000 years (due to the Greenland ice sheet melting). The present models of the ice sheet dynamics for the western Antarctic show that in the process of its melting the ocean level could rise by no more than 3 m during 1000 years. It should be taken into account that the possible long-term dynamics of the western Antarctic cryosphere has not been properly studied.

Of course, one of the important components of the climatic system is the global biogeochemical CO_2 cycle, specified characteristics of which could remove many uncertainties and reduce the contradiction in the estimates of the greenhouse effect.

2.3 SOURCES AND SINKS FOR CARBON DIOXIDE IN THE BIOSPHERE

Carbon dioxide is circulating in the environment among the reservoirs listed in Table 2.2. The carbon fluxes in the form of CO_2 can be roughly described by the scheme shown in Figure 2.1. In general, the diverse compounds of carbon continuously form, change, decompose, and among all this diversity, natural and anthropogenic CO_2 fluxes are formed in the processes of respiration and decomposition of vegetation and humus, in the burning of carbon-containing substances, in rock

Table 2.2. Global carbon reservoirs (Gt).

Reservoir	Carbon
Atmosphere	720
World Ocean	38,400
Total inorganic carbon	37,400
Surface layer	670
Deep layers	36,730
Total organic carbon	1,000
Lithosphere	
Carbonate sedimentary rocks	>60,000,000
Kerogens	15,000,000
Land biosphere	2,000
Living biomass	600–1,000
Dead biomass	1,200
Biosphere of inland waters	1–2
Burnt fuel	4,130
Coal	3,510
Oil	230
Natural gas	140
Others (peat, etc.)	250

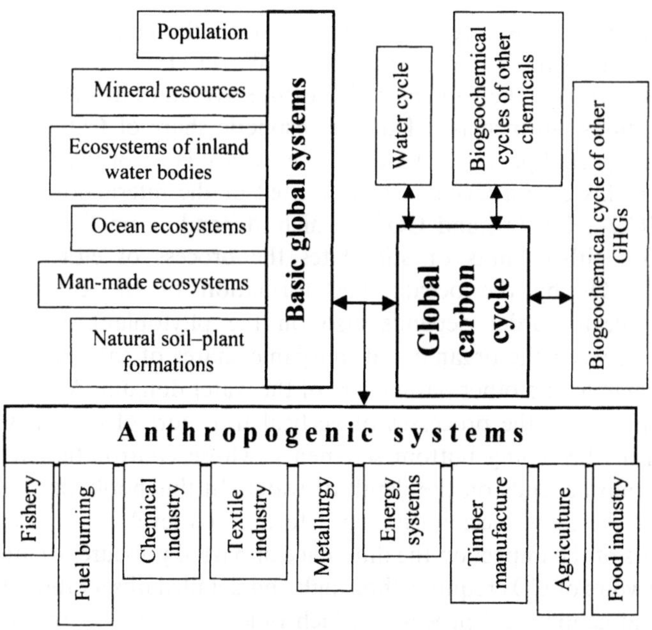

Figure 2.1. The conceptual scheme of the carbon cycle in the environment.

weathering, etc. An amount of CO_2 dissolves in the World Ocean giving carbonic acid and the products of its dissociation. The content of carbon in its reservoirs and estimates of its fluxes among them is the most important problem facing the analysis of the global CO_2 cycle. Numerous schemes of this cycle drawn from analysis of global interactions of living organisms and their physical and chemical media as well as estimates of carbon supplies accumulated during the historical period serve as the basis for predictions of the CO_2 concentration dynamics in the Earth's atmosphere, which has recently been the subject of a heated dispute in regards to assessments of the role of CO_2 in climate warming (Houghton *et al.*, 2001).

An important stage in understanding the processes of the CO_2 exchange between the biospheric reservoirs is the study of the laws of development of various ecosystems during the preindustrial epochs (i.e., without any anthropogenic factors) (Katz, 2002). Natural carbon fluxes between the atmosphere, ocean, surface ecosystems, and inland water bodies are strongly variable both in space and time (from year to year and seasonally). Analyses of the ice cores from Greenland and the Antarctic have reliably shown variations in the atmospheric CO_2 in the past. Eight-thousand years ago the CO_2 concentration in the atmosphere constituted 200 ppmv. By the beginning of the preindustrial epoch this estimate had varied between 275 ppmv and 285 ppmv ($\pm$10 ppmv). By the year 1985 the concentration of CO_2 in the atmosphere reached $\sim$345 ppmv. But in 1998 it was already 366–367 ppmv (Bolin and Sukumar, 2000). The total amount of carbon in the atmospheric CO_2 is estimated at about 700×10^9 tC. The natural CO_2 budget is estimated at $\sim 150 \times 10^9$ tC emitted annually in the processes of respiration and decomposition and assimilated in photosynthesis both on land and in the ocean, as well as in CO_2 dissolving in the World Ocean.

Special emphasis has been recently placed on the circulation of organic and inorganic carbon in the water domain of the World Ocean whose mechanisms are closely connected with the CO_2 partial pressure dynamics in the atmosphere. With some assumptions, an increase of the atmospheric share of CO_2 is followed by an increase of the partial pressure of CO_2 in the surface layer of the ocean at about the same rate. In this case a mechanism is started of the interaction of carbonic acid elements (HCO_3^-, $CaCO_3$, and CO_3^{2-}) which, depending on the relationship of its characteristics, either limits or stimulates the process of either assimilation or emission of CO_2 by the ocean. The formation of the carbon flux on the atmosphere–ocean border depends also on the phytoplankton production, the relationship between the organic and inorganic shares of carbon, the temperature, and hydrodynamic and other parameters of the water domain. At a certain level this flux also depends on the processes of vertical transport of carbon via descending dead organisms becoming bottom sediments, whose contribution to the so-called biological pump depends on the depth of water. In this book an attempt has been made to describe these mechanisms on a formalized level. Of course, a search for critical factors of the impact on the difference of partial pressures of atmospheric and oceanic reservoirs of CO_2 requires thorough and detailed observational studies of the ecological and geophysical processes which function in the oceanic system. In particular, the role of marine organisms in the process of transformation of calcium

Table 2.3. Data on carbon supplies in the surface vegetation and in the 1-m soil layer.
Bolin and Sukumar (2000).

Biome	Area (10^6 km^2)	Carbon supplies (Gt C)		
		Vegetation	Soil	Total
Tropical forests	17.6	212	216	428
Temperate zone forests	10.4	59	100	159
Boreal forests	13.7	88	471	559
Tropical savannahs	22.5	66	264	330
Temperate zone meadows	12.5	9	295	304
Deserts and semi-deserts	45.5	8	191	199
Tundra	9.5	6	121	127
Heavily moistened soils	3.5	15	225	240
Ploughed soils	16.0	3	128	131
Total	*151.2*	*466*	*2,001*	*2,477*

Table 2.4. CO_2 budget and its exchange activity in the biosphere.

Reservoir	Carbon content as CO_2 (10^9 t C)	Component of process in the biosphere	Exchange rate (10^9 t C/yr)
Atmosphere	697	Forests	10
Hydrosphere	35,420	Cultivated soils	4
Geosphere		Grass ecosystems	1
Inorganic carbonates	18.3×10^6	Deserts	0.2
Inorganic non-carbonates	6.8×10^6	Photosynthesis (north of 45°)	15
Coal, gas, oil	7,400		
Substratum, detritus		Human breathing	0.14
In the ocean	3,220	Domestic animal breathing	0.5
On land	710		
Biosphere		Wild animal breathing	0.4
In the ocean	10	Soil respiration	8.4
On land	124		

carbonate and its accumulation (i.e., in the change of the domain acidity) has been poorly studied. All these processes have different directions and power in the open ocean, estuaries, shallow waters, coastal zones, and river deltas.

To analyse the CO_2 dynamics in the biosphere, it is important to take into account a maximum possible number of its reservoirs and fluxes as well as their spatial distributions. It is here that numerous global models of the carbon cycle differ. The present level of these studies does not allow one to answer the principal question about an extent of details of the database concerning the supplies and fluxes of carbon. Therefore many authors analysing the dynamic characteristics of the global CO_2 cycle, rather arbitrarily utilize the fragments of databases on the distribution of the carbon sinks and sources (Fung, 2000). These fragments are exemplified in Tables 2.2–2.7.

Table 2.5. Quantitative characteristics of the types of land vegetation formations.
Bazilevich and Rodin (1967).

Vegetation formation type (see Table 2.6)	Formation area (10^6 km^2)	Annual increment of plants (kg/m^2/yr)	Phytomass (kg/m^2)	Dead organic matter (kg/m^2)
A	2.55	0.17	0.4	1.3
C	2.93	0.36	1.9	5.6
M	2.33	0.38	1.9	5.5
L	1.55	0.65	3.8	9.0
F	5.45	0.54	10.0	8.1
D	5.73	0.63	22.5	10.8
G	6.60	0.65	23.5	14.5
R	2.12	0.87	25.0	25.1
+	7.21	1.25	45.0	24.8
P	5.75	1.72	43.0	22.2
U	3.91	0.56	3.8	15.0
X	3.72	0.74	1.9	38.0
W	4.29	0.79	1.9	33.0
E	1.66	1.11	3.8	21.0
V	2.66	0.38	0.8	12.6
#	2.08	0.45	0.4	12.1
S	2.69	0.25	0.2	8.1
&	1.99	0.35	0.8	8.8
H	7.16	0.12	0.1	1.4
B	1.15	0.47	0.8	16.9
Q	3.54	0.76	1.9	24.0
Z	10.4	3.17	60.0	21.6
Y	7.81	2.46	60.0	20.5
N	9.18	1.42	10.0	15.1
J	17.1	1.35	0.1	2.0
T	11.5	0.18	0.4	4.9
K	0.38	0.18	45.0	21.6
I	0.9	1.96	45.0	21.6
*	14.6	0	0	0
@	2.08	0.45	0.4	12.52
O	0.21	0.49	9.1	11.2

2.4 ANTHROPOGENIC SOURCES OF CARBON

A key component of the global CO_2 cycle is its anthropogenic emissions into the environment. The principal problem studied by most investigators occurs in the assessment of the ability of the biosphere to neutralize an excessive amount of CO_2. It is here that all predictions of the consequences of the greenhouse effect are wide open to criticism. All the models of the global CO_2 cycle are based on scenarios that describe the dynamics of the extraction and burning of fossil fuels

Table 2.6. Identifier of the types of soil–plant formations following the classification after Bazilevich and Rodin (1967).

Type of soil–plant formation	Symbol (see Table 2.5)
Arctic deserts and tundras	*A*
Alpine deserts	*B*
Tundras	*C*
Mid-taiga forests	*D*
Pampas and grass savannahs	*E*
Northern taiga forests	*F*
Southern taiga forests	*G*
Subtropical deserts	*H*
Subtropical and tropical thickets of grass–tugai type	*I*
Tropical savannahs	*J*
Saline lands	*K*
Forest tundra	*L*
Mountain tundra	*M*
Tropical xerophytic open woodlands	*N*
Aspen birch subtaiga forests	*O*
Subtropical broad leaved and coniferous forests	*P*
Alpine and subalpine meadows	*Q*
Broad leaved coniferous forests	*R*
Sub-boreal and saltwort deserts	*S*
Tropical deserts	*T*
Xerophytic open woodland and shrubs	*U*
Dry steppes	*V*
Moderately arid and arid (including mountain) steppes	*W*
Forest steppes (meadow steppes)	*X*
Variably humid deciduous tropical forests	*Y*
Humid evergreen tropical forests	*Z*
Broad leaved forests	+
Subtropical semi-deserts	&
Sub-boreal and wormwood deserts	@
Mangrove forests	#
No vegetation	*

(Rosa and Ribeiro, 2001). All the models require an energy–economy system which requires a detailed parameterization of the geopolitical structure of the world. So far, among the most widely used models of this type is one developed by NASA, in which the globe is divided into nine regions differing in levels of per capita energy consumption and other parameters, the regional structure of which is shown in Table 2.8. With this scenario of the socio-economic structure one can attribute to it strategies of development for each region and assume the possible consequences for the environment of the future behaviour of individual regions. Most similar scenarios use an indicator such as the ratio of acceleration of energy consumption. This parameter varies from 0.2% to 1.5% per year. Various combinations are considered when

Table 2.7. Estimates of some characteristics of land and ocean ecosystems assuming that the content of carbon in a dry substance constitutes 45%.

Ecosystem	σ	MPP	TPP	MPB
Tropical humid forests	17.0	900	15.3	20
Tropical seasonal forests	7.4	675	5.1	16
Temperate zone evergreen forests	5.0	585	2.9	16
Temperate zone deciduous forests	7.0	540	3.8	13.5
Boreal forests	12.0	360	4.3	9.0
Woodland and shrubs	8.0	270	2.2	2.7
Savannah	15.0	315	4.7	1.8
Temperate zone grass systems	9.0	225	2.0	0.7
Tundra and alpine meadows	8.0	65	0.5	0.3
Desolate dense shrubs	18.0	32	0.6	0.3
Stone, ice, and sand	24.0	1.5	0.04	0.01
Cultivated land	14.0	290	4.1	0.5
Marshes and boggy land	2.0	1,125	2.2	6.8
Lakes and rivers	2.5	225	0.6	0.01
All land	*149*	*324*	*48.3*	*5.55*
Open ocean	332	57	18.9	0.0014
Upwelling zones	0.4	225	0.1	0.01
Continental shelf	26.6	162	4.3	0.005
Algae and reefs	0.6	900	0.5	0.9
Estuaries	1.4	810	1.1	0.45
The whole ocean	*361*	*69*	*24.9*	*0.0049*
The whole biosphere	*510*	*144*	*73.2*	*1.63*

Note: σ = area of the ecosystem ($10^6\,km^2$); MPP = mean pure primary production ($g\,C/m^2/yr$); TPP = total pure primary production ($10^9\,t\,C/yr$); MPB = mean plant biomass ($kg\,C/m^2$)

choosing a source of energy among oil, gas, nuclear, solar, hydroelectric, or solid waste. Naturally, one has to take into account the demographic, technical, political, and macroeconomic factors. The size of the population in most of the scenarios is assumed to grow at a rate that would provide 7.9 bn people in 2025 and 10.5 bn people in 2075. If all these suppositions in the scenario are assumed to be true, one can calculate the anthropogenic emissions of CO_2 and other GHGs. Then it is necessary to determine the total temperature impact AT_Σ of these gases.

The anthropogenic constituent in the global CO_2 cycle causes changes in the reservoirs of the CO_2 sink. Maximum changes are connected with urbanization, deformed structures of the soil–plant communities, and hydrospheric pollution. The rates of the change of forest masses to pasture and cultivated land are estimated at $0.05 \times 10^6\,km^2/yr$. Dense tropical forests are substituted with plantations at the rate of $10^5\,km^2/yr$. This process increases the rate of desertification ($\sim 5 \times 10^4\,km^2/yr$), which increases the amount of emitted carbon ($\sim$0.1 Gt C/yr).

Table 2.8. Characteristics of the rates of growth in the economic efficiency and population of world regions following the International Institute of Applied Systems Analysis (IIASA) scenario (Mintzer, 1987).

		Size of population (mn)		
		Year		
Region	P	2025	2050	2075
Australia and Japan	2.3	160	150	150
Africa	1.6	1,600	2,200	2,700
Canada and Europe	1.6	520	540	540
China	1.9	1,600	1,700	1,700
Latin America	1.9	720	850	900
Russia and FSU countries	1.3	470	500	510
Middle Asia	1.9	280	360	410
USA	1.2	290	290	290
South East Asia	1.8	2,600	3,100	3,400

Note: P = annual increment of labour productivity (%); FSU = Former Soviet Union.

Table 2.9. Distribution of the scales of biomass burning (millions of tonnes of dry matter per year).

	Sources						
Region	T	S	L	B	Y	A	C
Tropical America	590	770	0	0	170	200	7.5
Tropical Africa	390	2,430	0	0	240	160	9.3
Tropical Asia	280	70	0	0	850	990	3.3
Tropical Oceania	0	420	0	0	8	17	0
USA and Canada	0	0	0	0	80	250	0.5
West Europe	0	0	0	0	40	170	0.2
Temperate-zone forests	0	0	224	0	0	0	0
Boreal forests	0	0	0	56	0	0	0
Total	*1,260*	*3,690*	*224*	*56*	*1,438*	*2,017*	*21*

Note: T = tropical forests; S = savannah; L = temperate zone forests; B = boreal forests; Y = household fuel; A = agricultural wastes; C = brown coal.

The general pattern of the present-day level of anthropogenic CO_2 fluxes has been rather well studied. Due to burning of solid and liquid fuels, about 20×10^6 t CO_2 are emitted every year (with the solid to liquid ratio of 1 : 1). Burning of gas fuel emits about 4.5×10^6 t CO_2 into the atmosphere. The contribution from the cement industry is estimated at 750×10^3 t CO_2. Individual regions and countries contribute to these fluxes rather non-uniformly. Table 2.9 gives some estimates of such contributions from biomass burning.

The biomass burning in the tropics is one of the main sources of input of MGCs and aerosol particles to the troposphere (Romashkin *et al.*, 1999; Röckmann *et al.*, 1999). The share of the tropics is about 40% of the global land area and about 60% of the global primary productivity. The types of vegetation in the tropics are much more diverse than found in other regions. However, at present the tropical forests and savannahs are being transformed into agricultural land and pastures at a rate of about 1%/year. This transformation is mainly caused by biomass burning which strongly affects the chemical composition of the atmosphere and, hence, the climate.

In the process of biomass burning, huge amounts of NMHC, NO_x, and many other gas components are released to the atmosphere. As shown by analysis of the data from satellite observations, the share of the tropics constitutes about 70% of the burnt biomass, about half of which is concentrated in Africa, with a maximum of biomass burning in the course of a year (in the period of the dry season) observed north of the equator. Savannahs and forests in the tropics also emit a great amount of biogenic compounds into the atmosphere. In connection with the wide-spread fires in savannahs and their strong impact on the environment, Nielsen (1999) performed an analysis of special features of the spatial–temporal distribution of fires in the region where the field experiment EXPRESSO (central Africa) was carried out from the data of the advanced very-high resolution radiometer (AVHRR) carried by NOAA satellites for the periods of dry seasons, from November 1994 to December 1997. The fires variability can be described using three characteristics: (1) fire probability at a given point and at a given moment in time; (2) probability of repeated fires at a given point during a certain time period; and (3) the spatial extent and the burning savannah temperature affecting the conditions at a given point.

Processing of satellite imagery has shown that a fire is not an accidental process. The fire probability increases, for instance, with the fires occurring in the neighbourhood of the point considered. A combined analysis of the characteristics of the spatial and temporal variability of fires has made it possible to substantiate 12 typical regimes of fires as well as the dependence of the special features of the fires on those of the vegetation cover. Although there is no doubt that, as a rule, the savannah fires are caused by humans and not by other factors, specific causes of fires as a function of human activity remain unclear. From the viewpoint of the temporal variability, it is expedient to classify the fires taking into account the beginning of the fire season, the rate of their development, and duration of the fire season. In this context the following types of fires can be identified by the specific dynamics of their development: fast, late, or long.

The contribution of fires in savannahs constitutes more than 40% to the global level of biomass burning due to which the atmosphere receives MGCs, such as NMHC, carbon monoxide, methane, etc., as well as aerosols. According to the available estimates for the period 1975–1980, 40–70% of savannahs were burnt every year, about 60% of such fires occurring in Africa. In 1990 about 2×10^9 t of vegetable biomass were burnt, and as a result, 145 Tg CO entered the atmosphere, which constitute about 30% compared to the anthropogenic CO emissions.

Forest fires have a serious impact on the global carbon cycle. Alhough forest fires can occur naturally, for example, being caused by lightning strikes, the human contribution to their occurrence is constantly growing. An occurrence of a forest fire due to a lightning strike is only possible if it strikes standing wood or, in the case of a thin forest, it strikes the soil covered with moss or litter (Metting *et al.*, 2001; Metzger and Benford, 2001). The electrical resistance of standing wood is known to be at almost 100 times greater than that of growing trees, and therefore when the lightning strikes a living tree, it does not even become charred. Therefore, the monitoring of danger posed by fire gives reliable estimates of the probability of lightning-induced forest fires. A more complicated problem is predicting the anthropogenic causes of forest fires. More than 90% of the forest fires are known to occur in a 10-km zone of populated areas. Hence, the fire load on forests is strongly correlated with the spatial distribution of population density. Of course, an intensity and frequency of occurrence of the fires depends on the climate dryness in a given territory, the forests' density, and health. A forest fire is dangerous not only because it is a source of pollutants for the atmosphere, but also because their consequences are dangerous. Fire changes the forest microclimate, in particular, an illumination and heating of soil intensifies and the hydrological regime of the territory changes. Moreover, in a territory of forest fires the bioproductive ability of biocenoses deteriorates and, hence, the role of this territory in the biogeochemical cycles changes. It is well known that in a region with a dry climate the fire-destroyed forests cannot restore naturally, and the area must be re-forested. Therefore, it is important to know and understand the laws of interaction between the forest fire and biocenos of its territory. For instance, fires in the boreal forests contribute not more than 2% into carbon emissions into the atmosphere, but seriously affect the chemical processes in the high-latitude troposphere and the atmospheric radiative properties. This can lead to global climatic consequences (McKenzie *et al.*, 1998).

In general, for different reasons, biomass burning is a complex anthropogenic source of the atmospheric pollution and of global impact on the biosphere on the whole (Table 2.9). Estimates obtained by many authors show that the RF on the climate determined by aerosols from biomass burning constitute about $-1.0\,W/m^2$ (in the case of pure scattering aerosols the uncertainty of the estimates ranges between -0.3 and $-2.2\,W/m^2$).

2.5 RESOURCES OF THE BIOSPHERE AND THE GREENHOUSE EFFECT

The dynamics of the global carbon dioxide flux is determined by natural and anthropogenic factors. Natural factors are formed in the process of evolution of the biosphere, and their dynamics depends on the interaction between natural ecosystems. The level of the anthropogenic forcing on the global CO_2 cycle is determined by the relationship of natural forces with numerous aspects of humankind's development – political, demographic, cultural, religious, economic, etc. All this diversity of anthropogenic origin in the present world is limited by the resources of the biosphere, which eventually determine these aspects. The tabulated estimates of some resources given in Chapter 1 suggest the conclusion that the omnipotence of the human mind, with its urges to destroy nature, is limited by many circumstances. Humankind contaminates the environment either by using mineral resources or by changing the planet through changing one ecosystem for another. In both situations humankind sooner or later becomes at variance with its goals, and therefore the limiting factors exist in the global dynamics of destruction – a depletion of resources, worsening of living conditions, etc.

A human being as an element of nature, forming his or her environment, first of all is interested in the sources of material production – mineral resources. With growing scientific–technical progress the rate of the mineral resources consumption constantly increases, approaching some critical level. Alhough the late 20th and early 21st centuries are characterized by the broadened spectrum of mineral deposits, nevertheless, no alternative to oil, coal, and gas has yet been found. An increase in energy resource consumption is confirmed by the fact that almost half of the conditional fuel used by humankind ($\sim 90 \times 10^9$t) was consumed during the last 25–30 years.

By the end of the 20th century about 200×10^9 t of mineral deposits will have been extracted. There are no reliable estimates of the global supplies of mineral resources. Therefore, scenarios of carbon emissions into the atmosphere used by many authors should be considered conditional and disputable (Watson *et al.*, 2000; Houghton *et al.*, 2001). Nevertheless, there are undisputable relationships between volumes of emitted CO_2 and national production. This relationship changes in the world between developed and developing countries by a factor of 30, which means a considerable reserve in the working out of an optimal strategy for the formation of the profile of the global curve of CO_2 emissions into the atmosphere.

The World Ocean is one of the poorly mastered sources of mineral resources. Humankind is gradually mastering the mineral fuel supplies lying under the bottom of the oceans. Oil, gas, and coal are already being extracted in large volumes from marine deposits, but of course there exist economic and technical limitations. On the whole, from the preliminary estimates, in 180 gas basins discovered by geologists within the World Ocean, the potential supplies of oil and gas are estimated, respectively, at $300 \times 10^9\,m^3$ and $150 \times 10^{12}\,m^3$. Thus, here there also exists limiting parameters for scenarios of CO_2 emissions into the atmosphere.

Human activity resulting in changing land cover is one of the factors of the anthropogenic forcing on the global carbon cycle dynamics which is difficult to assess. During the last century humankind strongly affected the global distribution of vegetation – one of the important sinks for atmospheric CO_2. This was caused by human activity in engineering, construction, mining, creation of new types of land formations, biological recultivation, etc. Especially dangerous for the environment are processes of deforestation and desertification. For instance, after the estimate by Watson *et al.* (2000) with the realization of the IPCC scenario of reforestation, by the year 2050, an additional 60–87 Gt C (70% tropical forest, 25% temperate zone forests, and 5% boreal forests) could be removed from the atmosphere. An important parameter in this scenario is the rate of tree growth. For instance, if the forest grows at a rate of 3 t C/yr/ha, then 1 t C can be attributed to the effect of atmospheric CO_2 assimilation. In general, the control of land biocenoses can markedly affect the biogeochemical carbon cycle. It is important in what way this control is realized – planned or spontaneous. Humans create anthropogenic landscapes on the Earth's surface to improve their own habitat. An appearance of man-made seas, recreation zones, cities, and other anthropogenic landscapes reduces the level of the land covers' natural evolution. In other words, human activities which change the landscapes can strongly affect the dynamics of atmospheric CO_2.

2.6 THE GREENHOUSE EFFECT AND GLOBAL CARBON CYCLE

During the last decade the term "greenhouse effect" has been used in numerous publications on the problems of global climate changes on the Earth (Demirchian and Kondratyev, 1999; Ledley *et al.*, 1999; Sorokhtin and Ushakov, 1999; Danilov-Danilyan and Losev, 2000; Kondratyev, 2000b; Friedlingstein *et al.*, 2001; Reid, 2001). This term implies a sum of results of simulating the effects caused in the climate system and is associated with a number of natural and anthropogenic processes. On the whole, the term greenhouse effect refers to an explanation of changes in the atmospheric thermal regime caused by the impacts of some gases on the process of radiation transport. Many gases are characterized by a high stability and long residence time in the atmosphere (see Table 2.10). Carbon dioxide is one of them. Other GHGs are CH_4, N_2O, CFC-11 (CCl_3F), and CFC-112 (CCl_2F_2). Their observed growing concentration in the atmosphere is characterized by the following values. Since the pre-industrial period to 1997 the CO_2 concentration had grown on average from 280 ppm to 364 ppm. From the estimates by Ledley *et al.* (1999) in 1996 the atmosphere gained 6.5 Gt C due to fuel combustion and the cement industry. From the data in Fan *et al.* (1998) an additional 1.6 ± 1.0 Gt C were emitted to the atmosphere due to change in land use.

Table 2.10. The characteristics of the most important GHGs.

Mintzer (1987) and Ledley *et al.* (1999).

GHG	Lifetime in the atmosphere (years)	Average concentration	Rate of concentration increase (%)
Carbon dioxide	2–10	362 ppmv	0.5
Nitrogen oxide	100–150	308 ppbv	0.25
Methane	10–11	1,815 ppbv	1.0
CFCs			
CFC-11, $CFCl_3$	50–75	0.34 ppbv	7.0
CFC-12, CF_2Cl_2	102–111	0.54 ppbv	7.0

Such a high uncertainty of the last estimate is explained by the poorly studied spatial pattern of land use. Atmospheric CH_4 concentrations increased from 700 ppb in the pre-industrial period to 1721 ppb in 1994 (Houghton *et al.*, 1996). The sources of anthropogenic CH_4 connected with fuel combustion give approximately 70–120 Mt CH_4/yr. At the expense of rice paddies, biomass burning, stockbreeding, and dust heaps, the atmosphere obtains 200–350 Mt CH_4/yr more. Among the non-anthropogenic sources of CO_2 and CH_4 of special importance are marsh biogeocenoses. According to the estimates by Dementyeva (2000), in 1998 and 1999 the rates of CO_2 emissions from the marshes during the vegetation period were 75.8–216.2 mg CO_2 and 94.4–104.5 mg CO_2 per square meter per hour, respectively. These figures demonstrate a wide range of seasonal variations of CO_2 fluxes from the soil to the atmosphere, which necessitates a detailed inventory of land biogeocenoses.

Atmospheric N_2O concentrations increased from 275 ppb in 1800 to 312 ppb in 1994 (Houghton *et al.*, 1996). The anthropogenic N_2O emissions varied within 3 to 8 Mt N/yr. The principal sources of anthropogenic N_2O emissions to the atmosphere were agriculture and industrial production of acids.

Before 1950 there had been practically no CFC-11 and CFC-12 in the atmosphere. Then, with the use of these components in refrigerators and other devices the problem of ozone depletion has appeared which has led to the Montreal Protocol that foresaw a substantial reduction of the CFC-11 and CFC-12 content in the atmosphere of the 21st century.

The stability of some GHGs are estimated from their average lifetime in the atmosphere, which is defined as the time needed to remove 63% of anthropogenic emissions from the atmosphere (Table 2.10). It is very difficult to estimate this indicator for CO_2 because of different timescales of fluxes between biospheric reservoirs. The model estimates of the CO_2 lifetime show that annually 70–85% of anthropogenic CO_2 are removed from the atmosphere.

Numerous long-term observations in different latitudinal belts show a high level of correlation between temperature and CO_2 content. From estimates by numerous authors the largest contribution to this dependence is made by the atmosphere–ocean interaction (Nefedova and Tarko, 1993). Although the atmosphere and the ocean are in a certain balance with respect to the CO_2 exchange, this balance gets broken regularly. The most serious reasons for this breaking are as follows: (1) changing SAT; (2) changing volume of the ocean; and (3) changing regime of the oceanic vertical circulation. In general, these reasons can be characterized, from the viewpoint of their efficiency, by the ratio of their forcing on CO_2 concentration in the atmosphere. The first reason introduces about 65% to changes in the CO_2 partial pressure in the atmosphere (p_a). The remaining 35% fall on the second and third reasons. Quantitatively, this ratio is characterized by a 6% increase of the CO_2 partial pressure per every °C increase in temperature in the upper ocean layer. A 1% decrease of the ocean volume raises the p_a by 3%.

An estimation of the greenhouse effect requires a complex consideration of interactions among processes of energy transformation on the Earth. However, there is a certain hierarchy in the significance of the diverse processes (from astronomical to biological) affecting the climatic system on various timescales. This hierarchy cannot be constant since the role of individual processes can vary significantly for climate changes. A consideration of individual factors simplifies an analysis of its effect on climate. The level of the greenhouse effect is determined by an excess of surface temperature T_s over the effective temperature T_{eff}. The Earth surface temperature T_s depends on emissivity ε. The effective temperature T_{eff} is a function of the emissivity α of the atmosphere–ocean–land system. In general, the values ε and α depend on many factors, including the atmospheric CO_2 concentration. There are a lot of simple and complex mathematical models used for the parametrical description of these dependences. Unfortunately, so far, there is no model which would meet the requirements of adequacy and reliably simulate the pre-history of the climatic trends on the Earth. Nevertheless, one can state that the greenhouse effect depends non-linearly on the difference $T_s - T_{eff}$ (i.e., on atmospheric transparency, especially in the longwave region). The greater the CO_2 content in the atmosphere, the stronger atmospheric attenuation. In the long wave interval 12–18 μm the effect of CO_2 on atmospheric transparency is the strongest. Its weakest effect is in the intervals 0.875, 0.9, 2.0, 2.7, and 4.3 μm. It is clear that with an increasing partial CO_2 pressure in the atmosphere the role of various CO_2 bands will increase, and it means that with an intensified CO_2 absorption the upward long wave radiation flux will decrease. At the same time, on the Earth's surface, the downward long wave flux will increase. From the available estimates a decrease of the upward and increase of the downward fluxes are characterized by the values 2.5 and 1.3 W/m^2, respectively (Ramanathan and Coakley, 1978).

To assess the level of the greenhouse effect due to CO_2, one should be able to predict its concentration with an account of all feedbacks in the global biogeochemical carbon cycle as well as a correlation of this cycle with the cycles of other GHGs. From the estimates by Gale and Freund (2000), the contribution of various gases into the greenhouse effect averages: CO_2 – 63.5%, CH_4 – 20.5%, nitrogen oxides – 4.5%, and CFCs – 11.5%. Temporal changes of various GHGs are one of the most important problems for the global model.

A complete list of CO_2 fluxes on the Earth is given in Table 2.11, with independent subcycles of CO_2 in the World Ocean and on land. The relationship between these components is realized through the atmosphere and is due to river runoff. From the available estimates, the intensity of the mean global CO_2 circulation constitutes about 6×10^9 tC/yr. Some quantitative characteristics of the CO_2 global biogeochemical cycle have been published by many authors (Bach *et al.*, 1987; Kelley, 1987; Eliasson *et al.*, 1999). The present proportion of the carbon content in various global reservoirs vividly demonstrates the role of each of them in the global CO_2 cycle. This proportion is: atmosphere/land/ocean/geosphere = 1/3/50/10. It follows that the role of the World Ocean in the CO_2 cycle exceeds all the other reservoirs. The law of carbon flow in the ocean is connected with the carbonate system of the ocean, where carbon is divided into organic and inorganic. Carbon dioxide dissolving in the ocean forms carbonic acid. An equilibrium condition of the reaction of carbonic acid formation depends on the partial pressure of gaseous CO_2, concentration of dissolved carbonic acid, temperature, pressure, and salinity of the seawater. Carbonic acid gives a bicarbonate ion and carbonate ion. The latter is the

Table 2.11. Description of the global CO_2 cycle.

CO_2 flux	Processes responsible for the CO_2 flux formation
H_1	Fuel burning
H_2	Desorption
H_3	Sorption
H_4	Erosion of land surface
H_5	Volcanic eruptions
H_6	Photosynthesis in the World Ocean
H_7	Plant respiration
H_8	Vegetation burning
H_9	Humus decomposition
H_{10}	Human activity
H_{11}	Biotic activity in the World Ocean
H_{12}	Human vital functions
H_{13}	Animal vital functions
H_{14}	Plant decay
H_{15}	Animal decay
H_{16}	Deposits
H_{17}	Solution of bottom deposits in the World Ocean
H_{18}	Decay of dead organic matter
H_{19}	Water rise
H_{20}	Water lowering and gravitational sedimentation
H_{21}	Photosynthesis of land vegetation
H_{22}	River sink into the World Ocean

final product of the reaction of dissolved CO_2. The concentration of carbonate ions in the ocean increases until the limit of the calcium carbonate solubility is exceeded.

The process of photosynthesis is a source of organic carbon in the ocean. In the reaction of photosynthesis 106 mol/CO_2, 16 mol/nitrate, and 1 mol/phosphate are used. Under the influence of the trophic pyramid of the ocean, the photosynthesis production changes substantially, and as a result, the concentration of inorganic matter is regulated through deposits of calcium carbonate and organic matter oxidation. In this regulation, the process of vertical mixing of the ocean contributes significantly. The result of all these processes is a CO_2 exchange at the ocean–atmosphere border. The CO_2 cycle in the atmosphere–plant–soil system has been described in detail by Krapivin *et al.* (1982), giving the base fluxes of carbon of land ecosystems. Their photosynthesizing elements assimilate CO_2 from the atmosphere and transform it into substratum and humus; 0.5% being washed out to underground waters and into the oceans. The spatial structure of these fluxes is determined by the distribution of the types of soil–vegetation formations over the globe. Table 2.12 characterizes the role of various types of biocenoses in the global cycle of CO_2. It is clear that an accuracy of the estimates of land vegetation productivity and the reliability and details of its structural classification are important parameters to specify the scheme and parameters of the global CO_2 cycle.

Table 2.12. Biomass and productivity (dry matter) of some biocoenoses.

Watson *et al.* (2000).

Biocoenosis	P	B	τ
Tropical humid forests	2,200	45,000	70
Tropical seasonal forests	1,600	35,000	80
Moderate zone forests			
Evergreen	1,300	35,000	40
Broadleaved	1,200	30,000	40
Arctic forests	800	20,000	90
Bushes	700	6,000	50
Savannah	900	4,000	50
Moderate zone meadows	600	1,600	60
Tundra and alpine vegetation	140	600	100
Desert and semi-desert vegetation	90	700	100
Complete deserts (sand, ice, stones)	3	20	40
Cultivated land	650	1,000	80
Marshes and bogs	3,000	15,000	40

The global CO_2 balance is defined by numerous functional dependences applied by various investigators in their models. Since, on the whole, globally there is no single available database including adequate descriptions of all processes of CO_2 transformation in the ocean and on land, at present all conclusions about catastrophic consequences of the expected warming due to increasing partial pressure of atmospheric CO_2 should be assessed as hypothetical. Of course, not all fluxes given in Table 2.11 are equal in their influence on changes in atmospheric CO_2 concentrations. Such fluxes as CO_2 exchange on the border of the atmosphere with the ocean and land prevail when estimating the atmospheric CO_2 gradients. It is these fluxes that are still poorly assessed and lacking adequate parameterizations. Most of the models proceed from the fact that the CO_2 flux of the ocean–atmosphere border is in proportion to the difference of its partial pressures in these domains. But as shown in Alexeev *et al.* (1992), this flux is described more reliably by the model in which the flux is considered to be in proportion to the square root of this difference.

2.7 SCHEMES OF THE GLOBAL CARBON CYCLE

The reliability of the assessment of the role of CO_2 in greenhouse-effect formation depends on a detailed consideration of the global biogeochemical carbon cycle dynamics in the models and on the accuracy of the assessment of its characteristics. There are dozens of diagrams of the global carbon cycle in the form of CO_2 changes. Let us consider some of them to demonstrate their principal features and to

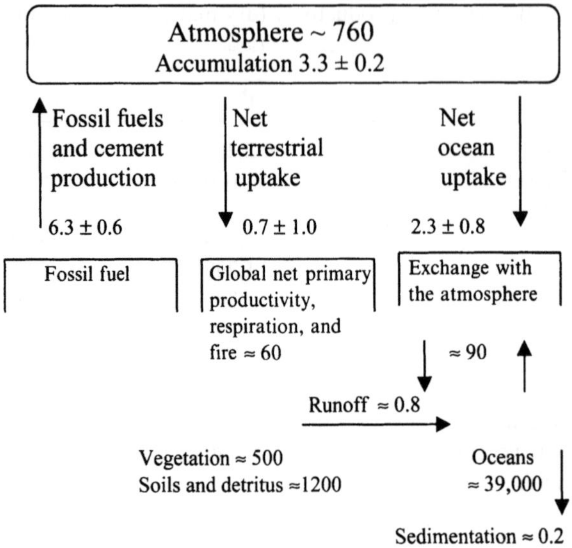

Figure 2.2. Global carbon cycle. Carbon supplies are given in Gt C, and fluxes in Gt C/year. After Bolin and Sukumar (2000).

understand the limits for necessary details of the simulation of carbon cycle compounds, beyond which it is impossible to obtain any additional knowledge about the cycle and, hence, about the greenhouse effect due to CO_2. Note that all known diagrams of the global CO_2 cycle are divided into two classes: pointwise (globally averaged) and spatial (locally averaged). All the diagrams are similar in that the biosphere is divided into the atmosphere, ocean, and land ecosystems. Many diagrams divide carbon into organic and inorganic forms. As a rule, the time step of averaging all processes and reservoirs of carbon is assumed to be one year, and therefore the atmospheric reservoir is considered as homogeneously mixed-up (pointwise). The World Ocean and surface ecosystems are considerably detailed. This detailing is based on global databases for these reservoirs of carbon. As a rule, the final results of the diagram studies are either of methodical character or they predict the atmospheric CO_2 concentrations within the limits of a certain scenario of the anthropogenic activity.

The scheme in Figure 2.2 (Bolin and Sukumar, 2000) gives an idea of the amounts of carbon supplies in basic reservoirs. The estimates shown in this scheme differ drastically from the estimates by other authors. Nevertheless, their relationships and orders of magnitude coincide in most cases. As seen, the largest carbon supply is concentrated in the World Ocean and a minimum of it in the atmosphere.

Natural processes determining the global carbon cycle dynamics have various timescales. Some of them, such as burying of dead organic matter at the bottom of the oceans, have characteristic timescales of hundreds and thousands of years. Others, for instance the biological carbon cycle on land, have a period of several

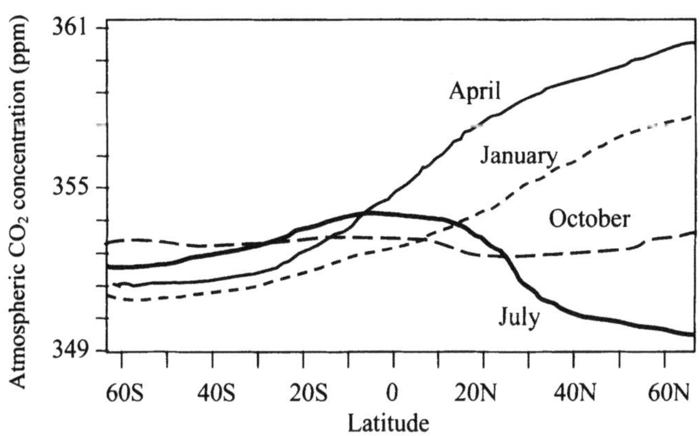

Figure 2.3. The latitudinal distribution of the CO_2 concentration in the atmosphere in 1990 from the TRENFS'93 data. The data have been averaged over the months given in the curves. From Boden *et al.* (1994).

tens of years. Therefore a consideration of the timescales of the carbon cycle in the biosphere is an important stage of studies of the dynamics of the CO_2 content in the atmosphere. Here it is also important to consider the fact that the characteristic time of the complete mixing of the atmosphere constitutes from several months to two years. Regular measurements at different stations of monitoring have shown that the CO_2 concentration varies considerably over one year. The difference between maximum and minimum estimates of the atmospheric CO_2 concentrations varies from 10 ppm at the South Pole to 15.1 ppm at the northern hemisphere high-latitudes. This spatial non-uniformity is explained by the presence in the northern hemisphere of large seasonally photosynthesizing vegetation communities. Figure 2.3 shows some estimates of the latitudinal distribution of CO_2 in the atmosphere. An increased CO_2 content in the northern hemisphere atmosphere is closely connected with the impact of human activity through direct CO_2 emissions and due to the impact on vegetation cover. Almost 90% of the total carbon emissions due to organic fuel burning fall in the 30°N–60°N zone. It follows from the data in Figure 2.3 that the conceptual schemes of the global biogeochemical cycle of carbon dioxide should also consider the spatial non-uniformity of atmospheric processes (Kaminski *et al.*, 2001).

An important constituent of most of the conceptual schemes of the global carbon cycle is the structure of carbon fluxes in the World Ocean. As follows from Table 2.2, there is the possibility to select several layers in the ocean (by depth) and to distinguish between the spatial non-uniformities in the structure of the ocean surface. Most authors consider the vertical structure of the ocean in the form of two or three layers covering the photic layer and deep layers. Within the photic layer, layers can be selected above and below the thermocline. The spatial non-uniformity of the ocean is demonstrated through upwellings and latitudinal zones, with different rates and directions of the ocean–atmosphere CO_2 exchange. More detailed schemes of the global carbon cycle take into account non-uniformities

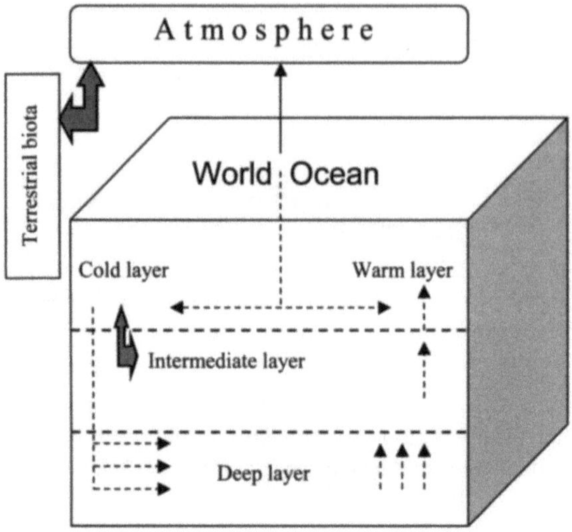

Figure 2.4. A scheme of the carbon reservoirs and fluxes in the models with the vertical structure of the World Ocean.

in the carbonate system of the oceans, which make it possible to considerably increase the accuracy of the respective models. Two of the widely used schemes of this type are shown in Figures 2.4 and 2.5.

The role of the World Ocean in the global CO_2 cycle is mainly manifested through the process of its exchange at the atmosphere–ocean border. The intensity

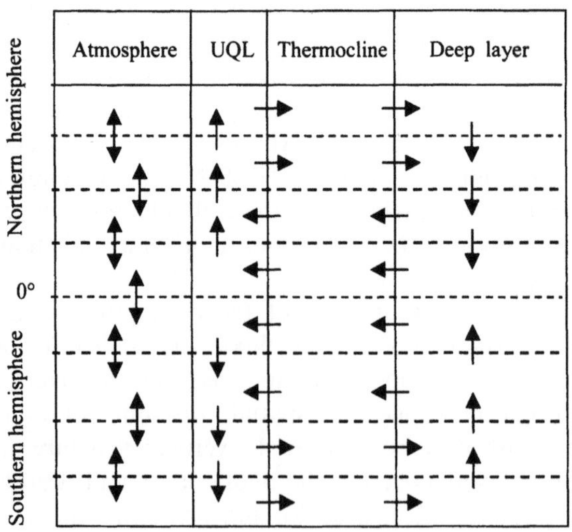

Figure 2.5. A scheme of the World Ocean global circulation as the basis for the description of carbon cycles.

Note: UQL = upper quasi-homogeneous layer.

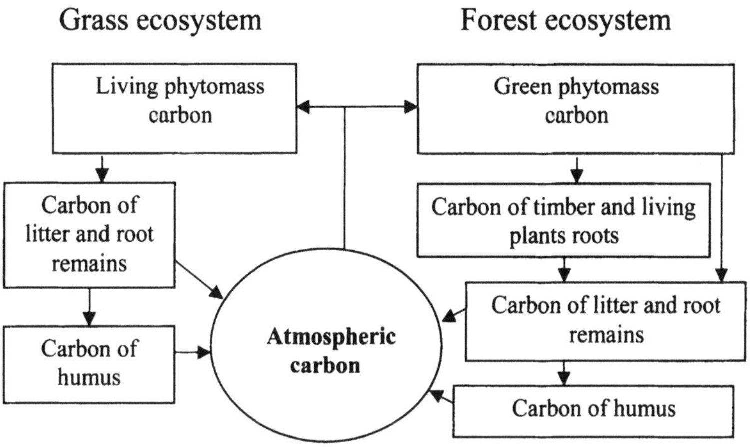

Figure 2.6. Diagram of the carbon fluxes in the atmosphere–plant–soil system with grass and forest ecosystems as an example.

of gas exchange between the ocean and the atmosphere is determined by the dynamic characteristics of the turbulent layers of water and air near the interface. Here, numerous physical schemes appear that reflect the situations of sea wave formation, as well as formation of foam and various films. As a result, carbon dioxide either dissolves in the ocean, providing thereby the inflow of CO_2 needed for photosynthesis, or is emitted from the ocean into the atmosphere. The cause of this binary situation at the air–water border is the difference between the partial pressures of CO_2 in the atmosphere and CO_2 dissolved in the water. Actually, this directed transport of CO_2 at the atmosphere–ocean border is more complicated. Its study requires expensive field experiments and a detailed classification of both synoptic and physico-geographical situations on the ocean surface.

In the programmes of studies of the global carbon cycle emphasis is placed on the role of the surface ecosystems in its formation (Houghton *et al.*, 2001). In the process of photosynthesis, plants assimilate carbon dioxide and emit it into the atmosphere through decomposition of dead plants. Thus, a continuous CO_2 exchange between living and dead organic matter and the atmosphere takes place in the land biosphere. There are many conceptual diagrams which formalize this exchange and serve as the basis for global models of the carbon cycle. Two examples of such diagrams are given in Figures 2.6 and 2.7. Clearly, the accuracy of estimates of carbon fluxes in the terrestrial part of the biosphere is the function of a detailed quantization of the types of soil–plant formations and accuracy of the parameterization of the biocenotic processes. In this regard the world maps of vegetation and soils have been drawn, their areas estimated, and data on the vital functions of soil microorganisms have been accumulated, with technologies of an operational monitoring of landscapes having been worked out. Unfortunately, there are no reliable estimates of the limits for considering the details of the soil–plant formations, which would provide the required accuracy of the carbon fluxes estimation. To solve this problem, a global model of carbon cycle is needed which would reflect a hierarchy of impor-

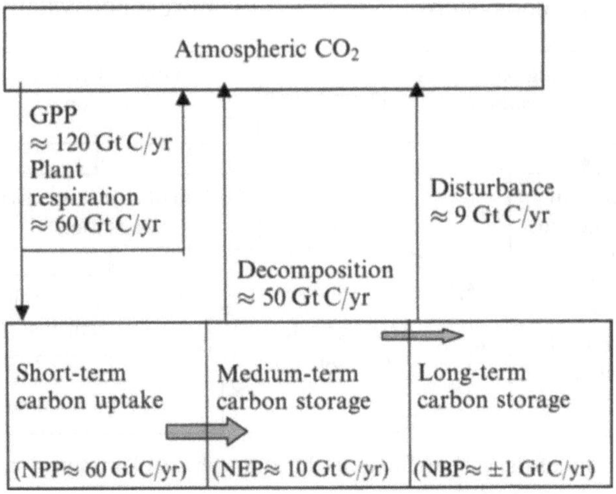

Figure 2.7. The scheme of the global carbon cycle.
From Watson *et al.* (2000).
Note: GPP = Gross Primary Productivity, NEP = Net Ecosystem Productivity.

tance (by expert estimate), of the constituents of the biosphere and the processes of carbon transformation. An exemplary scheme of a carbon flux in this model is characterized in Figure 2.8. Numerical experiments with this model will make it possible to obtain comparative assessments of the consequences of complicating its various constituents and thereby to approach sequentially the level of an optimal spatial quantization.

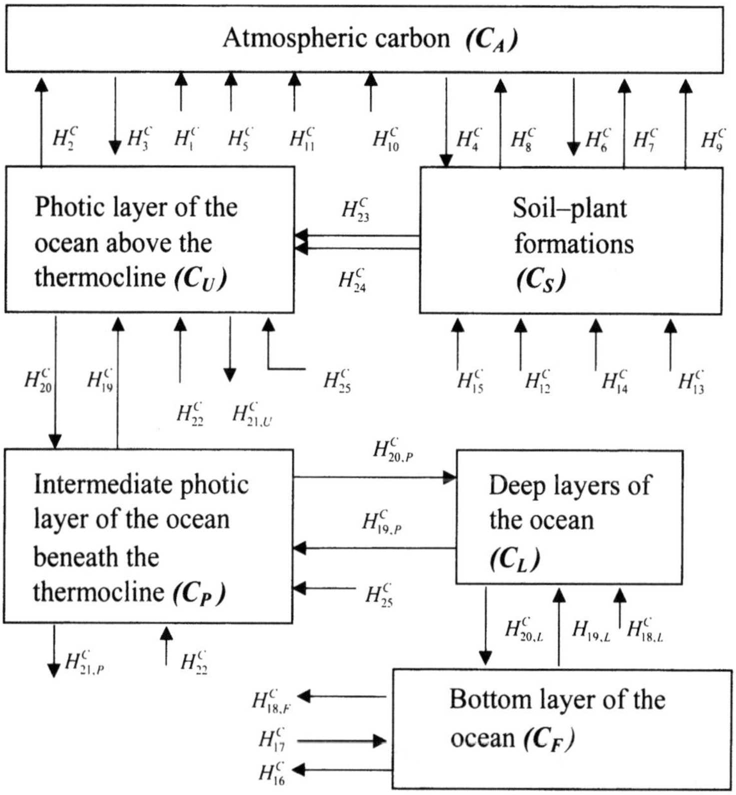

Figure 2.8. The block diagram of the global biogeochemical cycle of carbon dioxide in the atmosphere–land–ocean system (model of the global carbon dioxide cycle (MGCDC-unit)). The CO_2 reservoirs and fluxes are described in Tables 2.9 and 2.13.

From Krapivin and Kondratyev (2002), Krapivin and Potapov (2002), Kondratyev *et al.* (2002), Caspersen *et al.* (2000), Charette *et al.* (2001), Ciais (1999a,b, 2002), Clerbaux *et al.* (2001).

Table 2.13. Reservoirs and fluxes of carbon as CO_2 in the biosphere in a simulation model (MGCDC) of the global biogeochemical cycle of CO_2 shown in Figure 2.8.

CO_2 reservoirs and fluxes	Identifier in the model	Average estimate of the reservoir (10^9 t) and flux (10^9 t/yr)
Carbon		
Atmosphere	C_A	650–750
Photic layer of the ocean	C_U	580–1,020
Deep layers of the ocean	C_L	34,500–37,890
Soil humus	C_S	1,500–3,000
Emissions due to burning		
Vegetation	H_8^C	6.9
Fossil fuel	H_1^C	3.6
Desorption	H_2^C	97.08
Sorption	H_3^C	100
Rock weathering	H_4^C	0.04
Volcanic emanations	H_5^C	2.7
Assimilation by land vegetation	H_6^C	224.4
Respiration		
Plants	H_7^C	50–59.3
People	H_{10}^C	0.7
Animals	H_{11}^C	4.1
Emission		
Decomposed soil humus	H_9^C	139.5
Plant roots	H_{15}^C	56.1
Vital functions		
Population	H_{12}^C	0.3
Animals	H_{13}^C	3.1
Vegetation decay	H_{14}^C	31.5–50
Sedimentation to bottom deposits	H_{16}^C	0.1–0.2
Solution of marine sediments	H_{17}^C	0.1
Decomposition of detritus		
Photic layer	H_{22}^C	35
Deep layers of the ocean	H_{18}^C	5
Upwelling with deep waters	H_{19}^C	45
Downwelling with surface waters and due to gravitational sedimentation	H_{20}^C	40
Photosynthesis	H_{21}^C	69
Underground sink	H_{23}^C	0.5
Surface sink	H_{24}^C	0.5–0.6
Respiration of living organisms in the ocean	H_{25}^C	25

3

Land ecosystems and global ecodynamics

3.1 GLOBAL DYNAMICS OF LAND ECOSYSTEMS

The growing global size of the population and the associated growing forcing of human activity on the environment and ecosystems have become not only the main threat to further sustainable development of our civilization in the context of the global ecological safety but also reflect a dangerous disorder in the normal functioning of various systems of life support (Grigoryev and Kondratyev, 2001a,b). An alarming symptom of barbaric impacts on nature was the bombardment of Yugoslavia by NATO (Kondratyev, 2001a). In connection with the key role of the ecosystems in the processes of natural regulation of environmental properties, of principal importance is an analysis of the available data on global dynamics of ecosystems and an assessment of possible trends. Important information on these problems can be found, in particular, in recent publications by many authors (Sellers *et al.*, 1997; Hackler *et al.*, 1999; Lee *et al.*, 1999; Fang *et al.*, 2001; Kondratyev and Krapivin, 2001a,b; Sedjo, 2001; Gbondo-Tugbawa and Driscoll, 2002; Ito and Oikawa, 2002; Seppelt and Voinov, 2002; Lischke *et al.*, 2002), among which the most informative is the report prepared by the World Resources Institute (USA) supported by the UN Environmental Programme (UNEP), and the World Bank (*A Guide to World Resources*, 2000). This report emphasizes the close relationship between the global ecosystems and the global population – a symbiosis, unique and extremely sensitive to external forcings. These presentations have been thoroughly substantiated by Gorshkov *et al.* (2000) in the form of a concept of biotic regulation of the environment. Monographs by Kondratyev (1998a,b; 1999c) and new publications (Ernst, 2000; Kondratyev *et al.*, 2000; Kondratyev and Krapivin, 2001c,d) contain an analysis of the key aspects of the global changes from the viewpoint of functioning of the global systems of life support and requirements for adequate ecological monitoring.

Global natural and regulated ecosystems play an important role as factors of the environmental dynamics ranging from micro-scales (e.g., soil bacteria) to the whole planet and, are vitally important sources of drinking water, food, timber, paper, and other means of life support. As has been mentioned earlier (Kondratyev, 1998a,b, 2000a; Kondratyev and Demirchian, 2000; Marland *et al.*, 2001), the problem is that the world, on the whole, has already drawn near to exceeding the limits of sustainable impacts on the ecosystem.

An extreme complexity in the problem discussed is that it is necessary to explain (and, as far as possible, to predict) the dynamics of the interactive nature–society system (NSS) (society should be placed first here since its functioning determines its impact on nature) with its numerous feedbacks, non-linear nature, and "surprises". Unfortunately, the present stage of studies of the NSS can be considered not more than initial and preliminary. This refers even to the simple description of the present condition of nature (global ecosystems), which results from the observational data deficit with an apparent abundance of some observational means (especially expensive space-borne means). Therefore the report of the Institute of World Resources (USA) is in many respects incomplete, concentrating only on consideration of five types of ecosystems (the share of land surface is given in brackets): agricultural (28%), coastal regions (22% within a 100-km band), forest (22%), freshwater (<1%), and grass (41%). An abandonment of the World Ocean is, of course, a serious, though justified (in view of information deficit) flaw. Special attention should be also paid to the soil ecosystems (Batijes, 1999; De Lucia *et al.*, 1999; White *et al.*, 1999; La Deau and Clark, 2001; Tangley, 2001; Wofsy *et al.*, 2001; Bettinger *et al.*, 2003).

These ecosystems are very important for the solution of various problems of human life support and regulation of environmental conditions (Table 3.1).

3.1.1 Anthropogenic impacts on land ecosystems

The authors of the report of the World Resources Institute (USA) have briefly summarized the anthropogenic impacts on the ecosystems during the development of civilization, beginning with an intensive application of irrigation systems during the Shumer civilization, which led to soil salination, and to the present-day global processes of atmospheric pollution and depletion of the ozone layer (Kondratyev and Varotsos, 2000). The data in Table 3.2 characterize the most important aspects of anthropogenic impacts on the ecosystems presently.

Examples of destructive (and even catastrophic) impacts on ecosystems and their subsequent economic consequences are numerous (see, e.g., Grigoryev and Kondratyev, 2001a,b). The collapse of cod catches in 1990 in the regions of Canada made about 30,000 fishermen unemployed, and in the region of Newfoundland brought forth serious economic difficulties in 700 settlements. Material losses in China reaching \$US 11.2×10^9 per year have resulted from a deficit of drinking water due to polluted river and subsoil waters. In India, commercial forest felling and the transformation of deforested lands into agricultural lands have not only changed the traditional way of life but also caused a deficit of wood fuel and timber to the detriment of 275 millions of rural populations.

As for the estimates of the consequences of the global anthropogenic impacts, the situation with water resources is a good example: about 28% of the global population have no access to pure drinking water; every year about 5 million people die because of the poor quality of drinking water and sanitary conditions; and about 90% of waste in the developing countries goes into rivers, lakes, and coastal regions. Intensified emissions of CO_2 into the atmosphere have caused considerable changes in the global carbon cycle (Demirchian and Kondratyev, 1999; Kondratyev and Demirchian, 2000).

A most important fact is that the levels of impact on the ecosystems have become global in scale. About 75% of marine fish populations have either decreased due to violation of permitted levels of catch or come close to the threshold of extinction. Intensive forest felling has almost halved the forested areas, and the construction of various economic infrastructures has caused a fragmentation of the forest cover (Kaiser, 2000; Ramall, 2000; Katul and Albertson, 1999; Keller *et al.*, 2001). About 58% of coral reefs are seriously affected by fishing, tourism, and pollution. Almost 65% of arable lands have partially lost their fertility. The scales of economic use of groundwater exceeds their natural rate of recovery by at least $160 \times 10^9\,m^3$/year. In most cases the anthropogenic load on ecosystems has intensified. It is especially concerning in pasture ecosystems where the interference of humankind has reached a level of absolute control (Schulte, 2003; Schulte *et al.*, 2003).

It is well known that the main causes of ecosystem degradation are the growing size of the population, increased demand for natural resources, and enhanced loads on the environment (Table 3.2). Concrete detailed data characterizing the present global situation can be found in numerous publications (Danilov-Daniljan and Losev, 2000; Kondratyev, 1998a, 1999b, 2000a; *A Guide to World Resources*, 2000; Ernst, 2000; Watson *et al.*, 2000). Note that an extremely important feature of the growing scales of consumption is the strong geographical non-uniformity in consumption, reflecting the socio-economic contrasts in the world (Table 3.3).

3.1.2 The state of soil–plant formations

Now we shall discuss the principal results obtained within the framework of the pilot analysis of the global ecosystems (PAGES) (*A Guide to World Resources*, 2000). The main difficulties of this analysis are connected with a deficit of available information. The PAGES project is concentrated on the consideration of the three types of indicators of impacts on ecosystem dynamics: (1) anthropogenic loads (increasing size of population, growing level of resource consumption, pollution, and overexploitation of natural and controlled ecosystems); (2) the spatial extent of ecosystems (size, shapes, localization, geographical distribution); and (3) production of economically important products such as agricultural crops, wood, fish, etc. A considerable drawback of each of these indicators is that they do

Table 3.1. Ecosystems and their functions of life support.

Ecosystem	Human life support	Regulation of the state of the environment and human life support
Agricultural	• Food (agricultural crops). • Fibre (agricultural crops). • Genetic diversity of agricultural crops.	• Maintaining the functions of watersheds (infiltration, regulation of runoff, partial protection of soils). • Provision of habitat for birds, pollination processes, functioning of soil organisms. • Formation of soil organics. • Accumulation of carbon from the atmosphere. • Employment of population.
Coastal	• Fish and mollusks (food). • Fish as food for animals. • Algae (as food and raw material for industry). • Salt. • Genetic resources.	• Some impact on storms (mangrove thickets, "barrier" islands). • Habitat for wild fauna and flora. • Biodiversity maintenance. • Solution and processing of waste. • Provision of harbours and routes for water transport. • Habitats for humans. • Provision of population employment. • Provision of places for recreation and aesthetic satisfaction.
Forest	• Wood. • Wood fuel. • Water for drinking and reclamation. • Non-wood products (wine, bamboo, leaves, etc.). • Food (honey, mushrooms, fruit, etc.). • Genetic resources.	• Removal of atmospheric pollutants and emission of oxygen. • Nutrients cycles. • Maintaining the functions of watersheds (infiltration, water cleaning, regulation of runoff, stabilization of soil properties). • Biodiversity maintenance. • Accumulation of carbon from the atmosphere. • Some weakening of extreme impacts of the weather. • Soil formation. • Provision of habitat for humans and wild animals. • Provision of places for recreation and aesthetic satisfaction.

Freshwater	• Water for drinking and reclamation. • Fish. • Hydroelectric energy. • Genetic resources.	• Regulation of water regime. • Solution and removal of waste. • Cycles of nutrients. • Biodiversity maintenance. • Provision of aquatic habitat. • Provision of water transport routes. • Provision of population employment. • Provision of places for recreation and aesthetic satisfaction.
Grass	• Domestic animals food. • Water for drinking and reclamation. • Genetic resources.	• Maintaining the functions of watersheds (infiltration, water cleaning, regulation of runoff, stabilization of soil properties). • Cycles of nutrients. • Removal of atmospheric pollutants and formation of oxygen. • Biodiversity maintenance. • Soil formation. • Accumulation of carbon from the atmosphere. • Provision of habitat for humans and animals. • Provision of places for recreation and aesthetic satisfaction.

	Agro	Coasts	Forests	Water	Grass
Food/fiber	↘ 2	↘ 3	↗ 2	↑↓ 2	↘ 3
Water quality	↘ 4	↓↑ 3	↘ 3	↘ 4	6
Water amount	↘ 3	6	↘ 3	↘ 3	6
Biodiversity	↘ 4	↘ 3	↘ 4	↘ 5	↘ 3
Carbon supplies	↑↓ 3	6	↘ 3	6	↘ 2
Recreations	6	? 2	6	6	↘ 2
Protection of coasts	6	↘ 4	6	6	6
Wood fuel	6	6	? 3	6	6

Figure 3.1. Expert estimates of the levels of anthropogenic forcing on ecosystems and their consequences.

not contain information about the thresholds of an ecosystems ability to perform their functions of life support.

Figure 3.1 presents an expert assessment of the state of ecosystems and their functioning as life support systems, including these respective trends: increasing ability of ecosystems to perform life supporting functions (↗), a reduced ability (↘), mixed trends (↑↓), unknown trends (?). The figures characterize the levels of ecosystem functioning as a function of life support compared with the situation

Table 3.2. Present-day anthropogenic impacts on ecosystems.

Ecosystem	Anthropogenic impacts	Causes of impacts
Agricultural	• Use of agricultural land for urban construction or industrial purposes. • Pollution of inland waters due to nutrients, runoff, and silting. • Impoverishment of water resources as a result of irrigation. • Reduction of soil fertility due to erosion, change of agricultural practice, or decreased input of nutrients. • Weather change.	• Increasing size of population. • Increasing demand for food and industrial products. • Urbanization. • Governmental subsidies to agriculture. • Poverty and unemployment. • Climate change.
Coastal	• Exhaustion of fish resources (exceeding the permissible catch levels). • Transformation of wetlands and coastal habitats. • Water pollution by agricultural and industrial sources. • Fragmentation and destruction of natural barriers by tides and reefs. • Intrusion of alien species. • Possible rise of sea level.	• Increasing size of population. • Increasing needs for food and the growth of tourism. • Development of urbanization and of recreational places. • Governmental subsidies for fishing. • Inadequate information about the state of ecosystems, especially from the viewpoint of fishing conditions. • Poverty and unemployment. • Uncoordinated ecological policy in coastal regions. • Climate change.
Forest	• Transformation or fragmentation of forest ecosystems as a result of either agricultural use or urbanization. • Biodiversity losses due to fires and deforestation, emissions of stored carbon, pollution of the atmosphere and water basins. • Acid precipitation due to industrial pollutions. • Intrusion of alien species. • Overexploitation of water resources economically.	• Increasing size of population. • Growing demand for wood. • Governmental subsidies connected with deforestation and use of timber. • Inadequate estimation of losses caused by industrial pollution of the atmosphere. • Poverty and unemployment.

Freshwater	• Overexploitation of water sources economically. • Excessive fishing of inland water basins. • Building of dams in the interests of irrigation, creation of hydroelectric power stations, and flood control. • Pollution of water basins due to various factors of economic water use. • Intrusion of alien species.	• Increasing size of population. • General deficit of water and natural non-uniform distribution of water resources. • Governmental subsidies to maintain water use. • Inadequate estimates of losses due to water basin pollution. • Poverty and unemployment. • Growing needs for hydroenergetics.
Grass	• Transformation or fragmentation due to either agricultural use or urbanization. • Biodiversity losses due to fires, emissions of stored carbon, and atmospheric pollution. • Reduction of soil fertility and water pollution by domestic animals. • Unacceptable use of animals for entertainment.	• Increasing size of population. • Growing needs for agricultural products, especially for meat. • Inadequate information on the state of ecosystems. • Poverty and unemployment. • Accessibility and easiness of grass ecosystems.

20–30 years ago: 1 – excellent, 2 – good, 3 – satisfactory, 4 – weak, 5 – unsatisfactory, and 6 – no estimate. The results of the expert assessments shown in Figure 3.1 testify on the whole to considerable anthropogenic forcings on ecosystems (especially during the 20th century) that have led to considerable changes in their ability to perform life support functions. Although in some cases (e.g., food production) the ecosystem productivity has substantially increased, providing thereby the required level of production, in other cases (drinking water quality, biomass preservation, etc.) there was a degradation in ecosystem functioning. Let us now consider these results in more detail (see also Kondratyev, 1998b, 1999c; Ernst, 2000).

3.1.2.1 Food production

The state of the present agricultural ecosystems responsible for 95% of protein due to cultivated plants and animals and 99% of calories consumed by humankind is contradictory. On the one hand, agricultural yields have increased, but on the other hand, in most of the countries the quality of the agricultural ecosystems have worsened. So, for instance, a decrease in the natural soil fertility

Table 3.3. Contrasts in per capita consumption.

Country	*S*	*F*	*M*	*G*	*P*	*R*	*A*
USA	21,680	21.0	122.0	975.0	293.0	6,902	489.0
Singapore	16,340	34.0	77.0	159.0	168.0	7,825	120.0
Japan	15,554	66.0	42.0	334.0	239.0	3,277	373.0
Germany	15,229	13.0	87.0	496.0	205.0	3,625	500.0
Poland	5,087	12.0	73.0	696.0	54.0	2,585	209.0
Trinidad/Tobago	4,864	12.0	28.0	237.0	41.0	6,394	94.0
Turkey	4,377	7.2	19.0	502.0	32.0	952	55.0
Indonesia	1,808	18.0	9.0	311.0	17.0	450	12.2
China	1,410	26.0	47.0	360.0	30.0	700	3.2
India	1,166	4.7	4.3	234.0	3.7	268	4.4
Bangladesh	780	11.0	3.4	250.0	1.3	67	0.5
Nigeria	692	5.8	12.0	228.0	1.9	186	6.7
Zambia	625	8.2	12.0	144.0	1.6	77	17.0

Note: S = total per capita consumption (US$, 1997); F = fish (kg, 1997); M = meat (kg, 1998); G = grain (kg, 1997); P = paper (kg, 1998); R = fossil fuels (kg, equivalent oil, 1997); A = cars (per 1,000 people, 1996).

is recorded on 65% of agricultural land. In spite of this, a wide application of fertilizers, irrigation, and various new technologies (seeds, pesticides, etc.) has more than compensated for the impact of deteriorating conditions of agricultural ecosystems, and this trend will remain in the near future (for how long it is unclear).

Much more problematic is the situation of fishing. The present state of the coastal marine ecosystem is only satisfactory (and is worsening). In 28% of the water bodies in the most intensive fishing regions the fish supplies have either decreased due to overfishing or have only just started to recover. In 47% of the water basins the fishing levels are close to maximum permissible biological levels. Unfortunately, the information on the potential supplies of biomass in the ocean is contradictory. The general laws of the distribution of the organic life in the World Ocean and in various seas have been more or less studied. Phytoplankton as the source of productivity for other trophic levels of the ocean is known to develop mainly in the upper 100-m layer; with 65% of the zooplankton biomass and over 90% of nectone being located in the top 500-m layer. The benthos animal habitat is mainly at a depth down to 200 m (about 80%). By productivity, the World Ocean is very non-uniform. Less than 25% of its area is occupied by ecosystems with a plankton biomass above 200 mg/m^3. More than 50% of the area is occupied by unproductive ecosystems (<50 mg/m^3).

The situation in freshwater basins is rather contradictory. On the one hand, there is deprecatory fishing, but on the other hand, the natural fish supplies are replenished through fish breeding. The scale of fish breeding in ponds is constantly growing. On the whole, an increasing dependence of fish catch on the aquacultural production and a decrease of natural fish supplies threaten the population of the developing countries, where aquacultural production is absent.

3.1.2.2 Amount of water

The water resources of the biosphere have been systematized in detail from the viewpoint of both supply and quality. The main water fluxes between the oceans and land as well as volumes of river, lake, subsoil, and soil waters have been estimated. On the whole, global freshwater supplies are not great. Annually all the rivers bring about 50×10^3 m^3 of water to the World Ocean, which is only 1.5 times greater than the volume of Lake Baikal or the American Great Lakes. Water supplies are distributed non-uniformly over the continents and are not in proportion to the population density (Gleick, 1993).

The construction of dams and other forms of river regulation have seriously changed the conditions of use of water resources for both people and water ecosystems. The level of economic use of river water has reached on average about half the size of the available resources (with this situation strongly varying between countries). Building of dams and other engineering constructions has led to a considerable fragmentation of about 60% of the largest river systems in the world. For instance, as a result, the time needed for river water to reach the sea has tripled on average. The regime of inland waters has changes due to deforestation and drying of wetlands. This has especially manifested itself in the tropics where the forests are a key factor of the water regime dynamics. The global area of freshwater wetlands has almost halved, which has placed its toll on water supplies and flood regimes. Unfortunately, at the present level of global model development the role of wetlands as independent elements of the biosphere have not been reflected. As Spieles and Mitsch (2003) have shown, there is a clear-cut connection in the wetlands ecosystems between the state of the trophic structure and carbon cycle. This means that the presence in the global model of the unit simulating the functioning of the wetlands ecosystem will make it possible to increase the accuracy of the estimate of the CO_2 flux at the atmosphere–land boundary.

The main causes of the ongoing deterioration in the quality of natural waters are chemical pollution of water basins and input of nutrients, a reduced ability (indirect) of the ecosystem to perform water filtration, and intense soil erosion due to land use (Watson *et al.*, 2000). The pollution by nutrients resulting from river water runof polluted with fertilizers is a serious problem for all agricultural regions in the world. This kind of pollution takes the form of a eutrophication of waters in lakes and near-shore regions of the sea (this is especially observed in the Mediterranean and Black Seas, in the north-western part of the Gulf of Mexico, and in many large lakes). During the last two decades a dangerous fluorescence of water basins has taken place due to the input of nutrients. In many cases the ability of freshwater and marine coastal ecosystems to maintain the required quality of natural waters has been broken. Although, on the whole, in industrial countries the quality of natural inland waters has somewhat improved, the situation in the developing countries is quite different, especially near large cities and industrial regions.

3.1.2.3 The carbon cycle

To avoid repetition, we remind the reader that only plants and soil micro-organisms favour a removal of carbon dioxide from the atmosphere and storage in their tissues, slowing down the anthropogenic growth of CO_2 concentrations in the atmosphere. It is important to emphasize in this context that an intensified exploitation of ecosystems as sources of food and other means of life support negatively affect the ability of the ecosystem to accumulate carbon. This takes place, in particular, when forests are substituted with agricultural ecosystems (or even with secondary forests) with a much weaker potential for carbon accumulation. This land use provides a considerable source of CO_2 emissions which enter the atmosphere, up to 20% of the total anthropogenic emissions of CO_2.

Despite these circumstances, the ecosystems still remain a considerable reservoir of carbon which plays an important role in the formation of its global cycle. Tropical and boreal forests, where carbon is accumulated in the vegetation cover and in the soil (especially in peatbogs), are the main reservoirs of carbon on land. At present 38–39% of carbon is concentrated in the forests and 38% in the grass ecosystems. The share of carbon in the agricultural ecosystems is 26–28% (Bazzaz, 1998; Stern *et al.*, 1998; Schlesinger, 1999; Rosenberg and Izaurralde, 2001; Scholes and Noble, 2001; Smith, 2002).

Monitoring the dynamics of the biogeochemical cycles of carbon and nitrogen as well as the water cycle is of paramount importance for understanding the conditions of the ecosystem's existence. These cycles are the main indicators of the biospheric metabolism and determine the global function of the ecosystem functioning. The level of anthropogenic forcings on these cycles plays a key role.

Between 1850 and 1998 the concentration of CO_2 in the atmosphere increased from 285 ppm to 366 ppm (i.e., by about 30%). This has mainly resulted from fossil fuel burning, but changes in land use that have led to carbon emissions into the atmosphere have also contributed much. About 33% of carbon accumulated in the atmosphere during the last 150 years has been caused by deforestation and changes in land use. The process of emissions of carbon dioxide and other GHGs into the atmosphere determined by biospheric dynamics is interactive with respect to climate changes due to increasing CO_2 concentrations. So, for instance, the climate warming in the Arctic is fraught with an accelerated decomposition of gigantic masses of permafrost in the ecosystems of tundra and taiga (permafrost melting would be followed by powerful emissions into the atmosphere of an important GHG such as methane). An important aspect of the problem of increasing CO_2 concentrations is the effect of fertilization – stimulating the growth of some species of vegetation in conditions of increased CO_2 concentrations. In particular, in middle and high latitudes one can expect a 5% increase of the crop yield but in low latitudes (mainly in Africa) a 10% decrease of the yield is possible.

As for the global nitrogen cycle, in this case the impact of human economic activity has turned out to be even more substantial and biologically significant. For most of the species of natural vegetation a deficit of nitrogen is the basic factor limiting development (therefore nitrogen fertilizers are used). The increased use of fertilizers, fossil fuel burning, and deforestation have resulted in the excessive use of nitrogen by ecosystems. This has entailed serious problems of eutrophication (especially for freshwater and coastal ecosystems) – an excess of nitrogen stimulates the development of "parasite" algae to such an extent that a deficit of oxygen occurs for various aquatic organisms, including fish populations.

3.1.2.4 Water cycle

As has been mentioned above, the level of anthropogenic impacts on water systems has become rather substantial, which is illustrated, for instance, by taking almost half the river runoff for economic (mainly agricultural) needs, a figure which may reach 70% by the year 2025. The transport of freshwater from rivers and other reservoirs to the agricultural fields ensures an increase in yields, but can also cause various natural ecosystems and water users located downstream serious damage. Although the waters are often returned to the rivers they are strongly polluted and, as a rule, unfit for further use. The best example of such processes is a change of the water regime in the Aral Sea Basin.

3.1.2.5 Biodiversity

The biodiversity losses during the last century have acquired threatening scales and concern practically all ecosystems (mainly as a result of the loss of habitats). So, for instance, the area of forests has decreased by at least 20% and probably (on global scales) by 50%. Some forest ecosystems (like the dry tropical forests of Central America) have disappeared completely. In many countries up to 50% of mangrove plantations have been lost, the area of wetlands has almost halved, and in some regions the size of territories covered with grass has decreased by 90%. Only the tundra, Arctic, and marine deepwater systems have changed little, although in these cases substantial anthropogenic impacts also occur sometimes.

Even when the initial areas of the ecosystems have been preserved, many species are threatened with pollutions, overexploitation, intrusion by "alien" species, and degradation of habitats. From the viewpoint of biodiversity, the freshwater ecosystems have suffered most. For instance, about 20% of the freshwater fish species have been exterminated, and many other species are in a very dangerous condition. The forest, grass, and coastal ecosystems face serious survival problems. Their biodiversity is threatened with frequent diseases of marine organisms, intensive florescence of water bodies, and a considerable reduction in the amphibian population. Apart from the respective losses for medicine (due to a decreasing volume of raw materials for drugs), gene banks, and ecotourism, a reduction of biodiversity also threatens ecosystem productivity, integrity, and resistance to various external forcings.

Finally, it is clear that there are now numerous signs of anthropogenically caused degradation of ecosystems, giving them reduced ability to function as systems of life support and reducing their ability to control environmental properties. The data discussed above show that all the ecosystems considered suffer serious and increasing anthropogenic forcings. Intensive processes of transformation of natural ecosystems take place due to deforestation, pollution, construction of dams, and biological intrusions of alien species. Anthropogenic forcings result in considerable transformations of the global biogeochemical cycles, whose normal functioning is a guarantee of the saftely of the ecosystem.

So far, the negative trends in the state of the global ecosystem do not vividly threaten the high levels of production of various products and services. Food and fibre production has never reached such high levels as present-day levels, but dam construction has ensured an unprecedented regulation of the river runoff to provide the water in the interests of agriculture. However, the progress in creating material well-being is fraught with the long-term prospect of reduction of productivity of the global ecosystems and their reduced ability to regulate the environmental quality. The use of new technologies such as synthetic fertilizers, improved means of fishing, and irrigation systems have all masked a decrease of bioproductivity of natural ecosystems, ensuring achievement of the required growth of food and fibre production. As for the long-term prospects (and it is in this context that sustainable development is defined), one should remember about the reduced functioning of natural systems of life support, which manifests itself, for instance, in the form of dangerous decreases of biodiversity, reduced quality of drinking water, intensive emissions of GHGs into the atmosphere, and many other negative phenomena.

In the light of the World Summit on Sustainable Development (WSSD) in Johannesburg (2002) ("Rio+10"), of key importance is a substantiation of priorities regarding the problem of the functioning of the NSS. The acuteness of the situation becomes especially apparent when considering the limited success of the UN Second Conference on Environment and Development (Rio de Janeiro, 1992), the Special UN Session "Rio+5", and the complete fiasco of the 6-year series of annual conferences of the representatives of the governments signing the UN Framework Convention on Climate Change (FCCC) that served the basis for the Kyoto Protocol (KP). Speculative exaggerations (sometimes reaching the level of apocalyptic global predictions) characteristic of the concept of "global warming" have diverted attention (and to a considerable extent the financial resources) from the first priority problem: the global ecosystem's safety due to increasing size of the global population and intensifying anthropogenic forcing on natural systems of life support. In this regard the authors of *A Guide to World Resources* (2001) draw attention to the availability of the "ecosystem approach" suggesting that the use of natural resources should be based on a consideration of the ability of the global natural systems of life support to function normally (Shulze *et al.*, 2000).

The ecosystem approach has the following main features:

1 complex (systematic) character of analysis of interactions in the NSS;
2 not fragmentary but integral analysis of ecosystem functioning;
3 consideration of all the spatial–temporal diversity of the processes with the indispensable priority of long-term prospects;
4 a detailed consideration of interactive processes in the NSS; and
5 justification and accounting of the limits of the normal functioning of ecosystems, and in this regard, permissible anthropogenic loads.

Table 3.4 illustrates the differences between traditional and ecosystem approaches to the use of natural resources as applied to forestry. Of course, advantages of the ecosystem approach raise no doubts. The problem is, however, to what extent the conditions of the socio-economic development of each country permit a successive realization of the ecosystem approach. The negative manifestations of the dynamics of various ecosystems reflect vividly great (sometimes insurmountable) difficulties in realizing the ecosystem approach. In this regard the participants of the PAGES project have formulated the following general recommendations:

- to stimulate the development of science and observational means;
- to realize and quantitatively estimate the value of functions performed by the ecosystems;
- to stimulate a broad discussion of the goals, scientific foundations, and practical importance of the ecosystem approach; and
- to provide the participation of all required specialists in the solution of problems of ecosystems management.

The scientific understanding of the laws of ecosystem functioning and, moreover, a possibility to predict the ecosystem dynamics are still in the initial phase. To a great extent this is explained by inadequate complex information about the ecosystems and their functioning. So, for instance, the problem of the global carbon cycle attracts serious attention since long ago (Kondratyev, 1998b, 1999c; Demirchian and Kondratyev, 1999; Kondratyev and Demirchian, 2000; Watson *et al.*, 2000). Nevertheless, even the system requirements have not been formulated yet for complex observational data, which would make it possible to also obtain adequate information about the laws of the carbon cycle (of course, this requires a combined use of both usual and satellite means of observations). In this regard, the UN Secretary-General Kofi A. Annan justly noted: "It is impossible to justify an efficient ecological policy if it is not based on reliable scientific information. Though in many spheres a considerable progress has been reached in obtaining the observational data, there remain considerable gaps in our knowledge. In particular, so far, no complete global assessment of the state of the main ecosystems existing over the globe has been obtained. A response to this necessity was the planned Millenium Ecosystems Assessment (MEA) as an important international initiative aimed at mapping the 'health' of our planet (Lu and Li, 2003). This initiative has been backed by many Governments, as well as by UNEP, UNDP, FAO, and UNESCO. I call upon the Governments–UN members to give the needed financial support in assessing the state of ecosystems for the last Millenium and take an active part in the programme."

A realization of the MEA programme started in 2001 represents the first step towards an assessment of the state of the global ecosystem, as well as development and realization of measures required to preserve the ecosystems under demanding conditions when the need for natural resources is increasing.

Table 3.4. Differences between traditional and ecosystem approaches to studies of natural resources using forestry as an example.

	Traditional approach	Ecosystem approach
Goal	• Maximum possible production of consumer goods. • Maximizing the value of produce. • Maintaining the state of the forests or the use of forest products at levels equal to or below those ensuring the growth or restoration of the forests.	• Maintenance of forest ecosystems as an interconnection component of the natural medium with stable production of consumer products. • Preserving the possibility of choice in the future. • Providing persistent productivity of ecosystems, taking an account of forest aesthetics and social acceptability.
Scale	• Works on the level of forest stands within the boundaries of the states or property.	• Works on the levels of ecosystems and landscapes.
Role of science	• Consideration of the problems of forestry management as an applied science.	• Forestry management with scientific data and social factors taken into account.
Role of management	• Special attention to final results (production of goods and services) from the viewpoint of provision of needs for wood, solution of recreational problems, maintenance of wild nature. • Reaching the management that ensures the required industrial production. • Priority of wood production. • Desire to avoid a wood deficit. • Consideration of forests as systems of wood production. • Assessment of the economic efficiency of forestry.	• Emphasis on the safety of forest ecosystems with an account of required soil properties, biological diversity, and ecology on the whole in the interests of raising the level of production of goods and services. • Approach to management that ensures the solution of nature-protection problems. • Complex approach to the solution of the production problems considering similar importance of the problems of nature protection. • Desire to avoid losses of biodiversity and soil degradation. • Approach to forests as safe natural systems. • Assessment of economic profits and social acceptability.

3.2 FOREST ECOSYSTEMS AND THE GREENHOUSE EFFECT

During the last years, the problem of the impact of atmospheric carbon dioxide on the global climate has been discussed both by scientists and politicians. Some people believe that humankind will inevitably change the climatic situation on the Earth due to an enhanced greenhouse effect, which will change living conditions for the worst. Therefore, it is necessary to reduce the industrial emissions of CO_2. Others, agreeing with the consequences of the greenhouse effect, deny the strategy put forward by the KP and believe that the recommended reduction quotas will lead to an aggravation of the economy in many regions of the globe, without solving the problem of the greenhouse effect, but further worsening the global ecological situation. The opponents of the KP strategy think that the greenhouse effect can only be prevented by the correct management of the structure of surface cover and by introducing a strict control on World Ocean pollution (Kondratyev, 1998b; Kondratyev and Demirchian, 2001a,b). In this regard, the Intergovernmental Panel on Climate Change (IPCC) at its 8th session of SBSTA in June 1998 in Bonn and at the 14th session of the IPCC in October 1998 in Vienna prepared a special report on the role of the strategy of using surface cover (forests, in particular) in the global balance of CO_2. This report discusses the problems of interaction between the anthropogenic activity in the field of surface cover reconstruction and the distribution of CO_2 and other GHGs in the biosphere. An assessment is given of various scenarios following from the KP and concerning the problem of the impact of human society on the surface cover structure in general and on forested territories, in particular. A brief analysis of this report is given below.

3.2.1 Forests as sinks for carbon dioxide

Item 3.1 and Appendix 1 of the KP foresee a limitation and then a reduction of GHG emissions during the period up to 2008–2012. Before this time some problems should be solved to assess the role of the use of the Earth's surface. In particular, among these problems is that of the formalized description of the processes of change of the Earth cover structure, such as afforestation, forest reconstruction, deforestation and the associated carbon supplies. Understanding of the meteorological processes as functions of GHGs refers to one of the key problems facing humankind in the first decade of the third millenium. Only an adequate knowledge of the meteorological phenomena on various spatial–temporal scales in conditions of supplies of CO_2 and other GHGs will enable one to make the correct and constructive decisions in the field of global environment protection.

The dynamics of land ecosystems depends on interactions between biogeochemical cycles, which during the last decade of the 20th century suffered anthropogenic modification. This especially refers to the cycles of carbon, nitrogen, and water. The surface ecosystems (in which carbon remains in living biomass), decomposing organic matter, and soil, play an important role in the global CO_2 cycle. Carbon exchanges between these reservoirs and the atmosphere take place through photosynthesis, respiration, decomposition, and burning. Human interference into this

Table 3.5. Carbon stocks in vegetation and soil carbon pools down to a depth of 1 m. From Watson *et al.* (2000).

Biome	Area (10^9 ha)	Carbon stocks (Gt C)		
		Vegetation	Soil	Total
Tropical forests	1.76	212	216	428
Temperate forests	1.04	59	100	159
Boreal forests	1.37	88	471	559
Tropical savannahs	2.25	66	264	330
Temperate grasslands	1.25	9	295	304
Deserts and semideserts	4.25	8	191	199
Tundra	0.95	6	121	127
Wetlands	0.35	15	225	240
Croplands	1.60	3	128	131
Total	*15.12*	*466*	*2,011*	*2,477*

process takes place through changing of the structure of the Earth's covers, pollution of the surface and soil areas of water basins, as well as through direct emissions of CO_2 into the atmosphere (Rojstaczer *et al.*, 2001; Qi *et al.*, 2001; Potter and Klooster, 1999).

The role of various ecosystems in the formation of carbon supplies in the biospheric reservoirs determines the rate and direction of changes in the regional meteorological situations and in the global climate. An accuracy of assessment of the level of these changes depends on reliability of the data on the surface ecosystem inventory. Table 3.5 shows that a considerable scattering of the estimates of carbon supplies in various types of vegetation suggests the conclusion that it is important to specifically classify the surface ecosystems in more detail.

The anthropogenic constituent of the global carbon budget, beginning from the mid-19th century, increases the amplitude of the effect on every natural element. From 1850 until 1998 about 270($\pm$60) Gt C were emitted as CO_2 into the atmosphere due to fuel burning and cement production. About 136($\pm$55) Gt C went to the atmosphere as a result of anthropogenic reconstruction of surface cover. This has led to an increase of atmospheric CO_2 by 176($\pm$10) Gt C (i.e., the partial pressure of carbon dioxide in the atmosphere has increased from 285 to 366 ppm, or 28%). In other words, over 148 years, 48% of emitted carbon has remained in the atmosphere and has not been assimilated by surface or ocean ecosystems (230($\pm$60) Gt C were assimilated).

Some idea about the global carbon budget can be obtained from the data in Table 3.6. This shows that the rates and trends of carbon accumulation in the surface ecosystems are rather uncertain. However, it is clear that the surface ecosystems are important assimilators of excess CO_2. Understanding the details of such assimilation is only possible through modelling the process of plant growth (i.e., considering the

Table 3.6. Characteristic of the mean annual CO_2 budget.
From Watson *et al.* (2000) and Lee *et al.* (1998).

Characteristic	Estimate (Gt C/year) 1980–1989	1990–1998
(1) CO_2 emission due to fossil fuel combustion and cement production	5.5 ± 0.5	6.3 ± 0.6
(2) CO_2 storage in the atmosphere	3.3 ± 0.2	3.3 ± 0.2
(3) Oceans uptake	2.0 ± 0.8	2.3 ± 0.8
(4) Net terrestrial uptake = (1) − [(2) + (3)]	0.2 ± 1.0	0.7 ± 1.0
(5) CO_2 emission due to changes in the use of land resources	1.7 ± 0.8	1.6 ± 0.8
Residual terrestrial uptake = (4) + (5)	1.9 ± 1.3	2.3 ± 1.3

effect of the nutrient elements of soil and other biophysical factors on plant photosynthesis.

According to Table 3.5, the forest ecosystems and associated processes of natural afforestation, forest reconstruction, and deforestation should be studied in detail. The same has been emphasized in Items 3.3 and 3.4 of the KP, where the necessity is emphasized to determine national and international strategies for forest management. In a forest range, the volume of the reservoir for CO_2 coming from the atmosphere is a function of the density of its canopy, and over a specific time period, a change of this volume is determined by the level and character of the dynamic processes of the transition of a given type of forest to that of another. The causes of this transition can be natural, anthropogenic, and/or mixed. Biocenology tries to create a universal theory of such transitions, but so far, there is only a qualitative description of the observed transitions. As mentioned in the KP, of importance is the correct definition of the notions of afforestation, forest reconstruction, and deforestation. Afforestation means to forest a land area used previously (for 20–50 years and longer) for other purposes. Usually this term determines the process of natural succession at the expense of propagation of forest over other territories without human interference (Goldstein *et al.*, 2003). The process of forest reconstruction is defined as planting trees. Deforestation is a substitution of the forest territory for another ecosystem. Thus two opposite processes are possible in the forest ecosystem dynamics that can be controlled by both nature and humans. Each of these processes has its versions characterized by the special dynamics of the vegetation over a given territory. Of special status is the process of foresting a territory where historically trees had never grown. In this case this territory immediately becomes important in the CO_2 dynamics.

Table 3.7 illustrates an impact of the afforestation/deforestation processes on carbon supplies following the FAO scenario (Watson *et al.*, 2000) where the forest represents a land area not less than 0.5 ha, with trees more than 5 m tall and the canopy covering more than 10% of the area. Deforestation is determined as a change of the surface cover with the canopy covering less than 10% of the area, as well as a

Table 3.7. Assessment of the calculated change of the mean annual carbon supply for an afforestation/deforestation scenario.
From Watson *et al.* (2000).

			TR		*FR*	
Region	*RF*	*AF*	*A*	*B*	*A*	*B*
Boreal	35	0.4–1.2	0.5	0.1	−18	−185
Temperate	60	1.5–4.5	2.1	1.9	−90	−501
Tropical	120	4–8	13.7	2.6	−1644	−1352

Note: A = deforestation; B = afforestation; RF = change of average carbon supply after deforestation (tC/ha); AF = average rate of CO_2 assimilation at afforestation (tC/ha/year); TR = change of area (10^6 ha/year) resulting from the deforestation/afforestation transition; FR = forecast of changes in carbon supplies (10^6 tC/year) in 2008–2012 after the FAO scenario.

change of the class of the forest resulting in negative consequences (e.g., a decrease of productivity). Afforestation is the planting of trees over the area where trees had never before grown. Note that "natural broadening" (i.e., propagation of forest over agricultural territories without human interference) due to the FAO scenario is also referred to as afforestation. Finally, forest reconstruction is a direct planting of trees on the territories earlier covered with forest.

The technology of considering the scenarios of the type in Table 3.7 makes it impossible to choose the scenario to be recommended for use. An approach suggested in the KP is too oversimplified for reliable assessments of the CO_2 dynamics as a function of numerous natural and anthropogenic parameters.

3.2.2 Global cycles of GHGs

The fact that differences in the estimates of the global carbon cycle made by various authors reaches 100%, demonstrates the impossibility, based on present-day knowledge, to reliably simulate the global CO_2 dynamics and, hence, identify objective causes to make decisions (recommendations) similar to those of the KP.

Carbon dioxide, which dissolves in the oceans and is assimilated by vegetation, leaves the atmosphere through the complicated mechanisms of the biospheric cycles of substances and energy. In the oceans such a mechanism is the carbonate system which, with a certain combination of environmental parameters, draws-off the atmospheric CO_2 and sends it to the bottom sediments. The fate of CO_2 fixed on land depends on which ecosystem and which reservoir of carbon is its assimilator (e.g., living biomass or soil). Carbon fixed in a reservoir in a year or less (for leaves and roots) either returns to the atmosphere or remains in the reservoirs with characteristic cycle times of decades to centuries (for trunks, stems, branches, and organic matter). As a result, a pure carbon flux forms between the surface ecosystems and the atmosphere, introducing a slight instability between the assimilation via photosynthesis and various processes of carbon return to the atmosphere. Plants, soil micro-organisms, biochemical processes, and animals contribute to this instability

(Nabuurs and Sikkema, 2001; Ni, 2001; Niles and Schwarze, 2001; Pacala *et al.*, 2001; Post *et al.*, 2001). Climate change and human activity contribute to this process by transforming the structure of the surface ecosystems and polluting the environment. From the available estimates, pure assimilation of CO_2 by the surface biosphere constituted 0.1–0.2 Gt C/year between 1980 and 1989 and 0.07–0.6 Gt C/year during 1989–1998 (see Table 3.6). At the same time, because of anthropogenic interference into the structure of surface cover, pure emissions of CO_2 reached 1.7 ± 0.8 Gt C/year and 1.6 ± 0.8 Gt C/year, respectively. On the whole, variations in the global carbon cycle are estimated at ±1 Gt C/year, and a clear correlation is observed between variations and phenomena such as El Niño or volcanic eruptions (Hernandez *et al.*, 2001; Jones and Cox, 2001). Total carbon flux assimilated by plants from the atmosphere reaches approximately 120 Gt C/year. Part of this carbon rapidly returns to the atmosphere due to transpiration ($\sim$60 Gt C/year) and decomposition of dead organic matter ($\sim$10 Gt C/year). Forests contribute most to this balance (see Table 3.5). As seen from Figure 3.2, various forests have quite different internal and external supplies and fluxes of carbon. Hence, the estimates of carbon fluxes averaged over large territories do not even give approximate patterns of changes of its supplies to the biospheric reservoirs. Therefore, numerous carbon circulation models based on these data cannot be an objective instrument for decision making.

A similar conclusion is valid for other GHGs, such as CH_4 and N_2O (see Tables 3.8 and 3.9) (Isidorov and Kondratyev, 2001). The concentration of methane increased from 700 ppbv in the pre-industrial epoch to 1700 ppbv today. About 550×10^6 t CH_4/year are emitted into the atmosphere from various sources. From 1980 until 1990, the CH_4 concentration in the atmosphere increased at a rate of 13 ppbv/year, but in 1990 and in 1996 the rates were 8 ppbv/year and 4 ppbv/year, respectively. The lifetime of methane is about 9 years. The contribution of anthropogenic sources of methane into its total flux in the atmosphere is estimated at 70%. Clearly, this share consists of not only direct emissions of methane from industry and agriculture, but also the contributions from the man-made transformations of surface cover.

The sources of another important GHG, N_2O, are also closely connected with the structure of the surface cover. The global emission of N_2O is estimated at 14×10^6 t N/year, with the anthropogenic contribution to this reaching 50%. This estimate is strongly affected by agricultural land, tropical forest, and especially

Table 3.8. Global estimates of CH_4 sources.
From Prather *et al.* (1994).

CH_4 source	10^6 t CH_4/year	10^9 t C/year
Domestic animals (wastes)	110 (85–130)	0.6 (0.5–0.7)
Rice plantations	60 (20–100)	0.3 (0.1–0.6)
Biomass burning	40 (20–80)	0.2 (0.1–0.5)
Natural heavily moistened soils	115 (55–150)	0.7 (0.3–0.9)

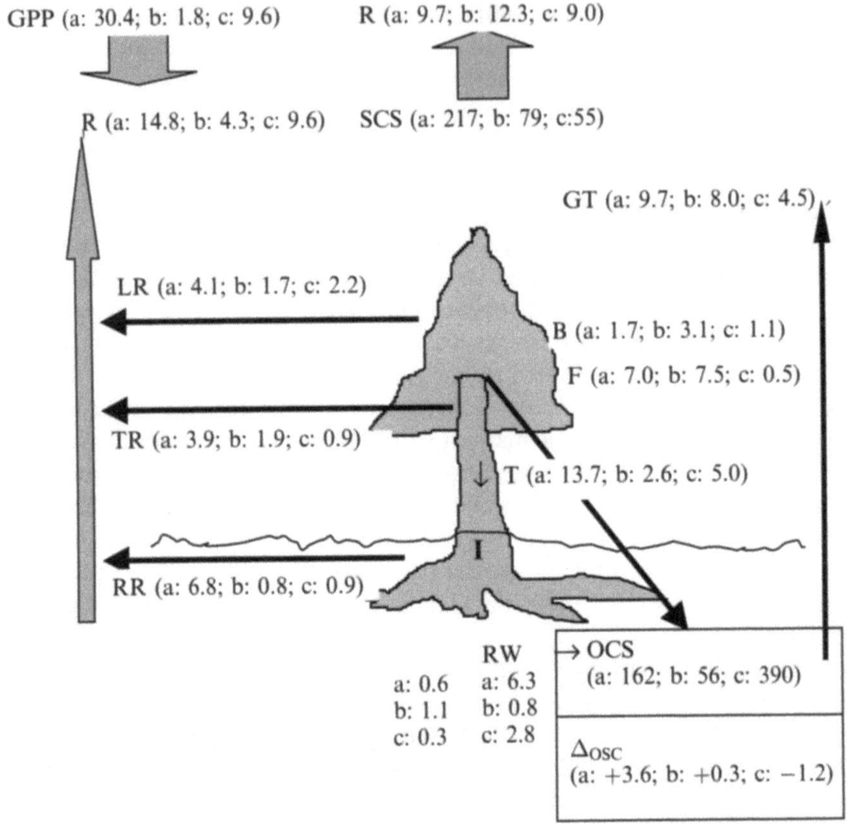

Figure 3.2. Annual supplies (t C/ha) and fluxes (t C/ha/yr) of carbon for tropical (a), temperate (b), and boreal (c) forests.
From Watson *et al.* (2000).
Note: GPP = gross primary production, RA = autotrophic respiration, RH = heterotrophic respiration, RL = foliage respiration, RW = above-ground wood respiration, RR = root respiration, SOC = soil organic carbon stock, Δ_{SOC} = net increment in soil organic carbon, E = below-ground detritus, K = below-ground net biomass carbon increment, B = above-ground net biomass carbon increment, O = above-ground detritus (litter fall and mortality), T = translocation from above to below ground.

by soil moistening. The microbiological processes in soils are primary sources of N_2O.

Land use directly affects the processes of GHGs exchange between the surface ecosystems and the atmosphere. Changes in land use, such as forest clearing to broaden agricultural fields or for settlements cause synchronous (or slightly delayed) changes in the surface cover structure and in the distribution of carbon supplies to the surface ecosystems. These actions are connected with the necessity to increase food production and to create new areas for human habitation and recreation. Therefore, the problem of the balanced regulation of the processes of land use becomes urgent. To solve this problem using the approaches suggested by the KP is impossible. It is necessary to put into practice some methods of

Table 3.9. Global estimates of N_2O sources.
From Prather *et al.* (1994).

N_2O source	10^6 t N_2O/year	10^9 t C/year
Cultivated soils	3.5 (1.8–5.3)	0.9 (0.5–1.4)
Biomass burning	0.5 (0.2–1)	0.1 (0.05–0.3)
Domestic animals	0.4 (0.2–0.5)	0.1 (0.05–0.13)
Natural soils in tropical latitudes		
Wet forests	3 (2.2–3.7)	0.8 (0.6–1)
Dry savannahs	1 (0.5–2)	0.3 (0.1–0.5)
Natural soils		
Forests	1 (0.1–2)	0.3 (0.03–0.5)
Steppes	1 (0.5–2)	0.3 (0.1–0.5)

ecoinformatics which make it possible to study the NSS as an integral system based on a complex global model that describes a totality of biogeochemical, biogeocenotic, climatic, demographic, and socio-economic processes with their binding with spatial and temporal scales. The idea of such a model has been described by Krapivin and Kondratyev (2002). One of the versions of the global model described by Krapivin (1993) uses a spatial geographical grid ($4° \times 5°$) with detailed forest ranges, following the scheme in Figure 3.2. According to this model, the estimates of CO_2 assimilation drastically change the results of the scenario modelling given in Watson *et al.* (2000). Respectively, both predictions and conclusions are changing. Of course, it is necessary to take into account a variety of possible scenarios (Matsumura, 2000; Nabuurs *et al.*, 2000; Pielke, 2001; O'Neil and Oppenheimer, 2002).

3.2.3 Management of the forest ecosystems

Let us return to the problem of anthropogenic changes to the structure of forest ranges. In the KP its solution is connected with the problem of definition of such notations as forest, afforestation, tree planting, deforestation, biome, surface cover, land use, degradation, etc. A question arises here: to what extent is the use of either definition justified for making decisions and how does the result of its realization depend on their quality? The KP not only does not answer this question but doesn't even raise it. In this respect, Table 3.10 is demonstrative and serves as the basis for drawing various conclusions. In fact, a huge range of uncertainty exists here due to which the processes of afforestation and deforestation cannot be estimated uniquely. Clearly, one way to solve the principal problem and the respective partial problems is to make a detailed inventory of the forest ranges and to use the obtained data in the global model. The specific features of this inventory should be determined in an adaptive regime using the model and gradually changing the spatial–temporal scales, starting from the known model with a mean annual time step and the

Table 3.10. Directions of carbon supply changes with different forms of soil cover reconstruction.
From Watson *et al.* (2000).

Type of land cover reconstruction	Direction of changes in carbon supplies					
	Biomass		Forest cover/ wood			
	A	U	S	L	Q	M
Cultivated land → forest	↑	↑	–	↑	↑	↑
Non-cultivated land → forest	↑	↑	–	↑	?	↑
Forest → cultivated land	↓↓	↓	↓	↓	↓	–
Forest → grazing land	↓↓	↓	↑	↓	?	–

Note: A = above ground; U = under ground; S = short-lived; L = long-lived; Q = organic matter; M = wood production.

geographical 4° × 5° grid. Some studies in this direction are being carried out within the framework of the IGBP.

3.2.4 Concluding remarks

The problem discussed above is urgent both from the viewpoint of ecological and economic global changes to the environment. Combining these aspects, it is necessary to find an efficient mechanism for global forestry management. This problem cannot be solved independently of other nature-protection problems. Apparently, even on a local level a decision about changing the Earth's cover should be made on the basis of assessments of the global consequences in the future. The technology of making such well-considered decisions was proposed in Ougolnitsky *et al.* (2000) and was called geographical information monitoring systems (GIMS) technology.

Returning to the KP, we note that despite numerous quite acceptable conclusions and scenarios, recommendations to introduce quotas on GHG emissions cannot be considered scientifically substantiated (Gorshkov and Makaryeva, 1999). After all, even the most optimistic assessments of the role of the forest ecosystems in CO_2 assimilation from the atmosphere vary from 10 to 27%.

To solve the problem of the use of forests, coordinated with the dynamics of global change, is only possible with the use of GIMS-technology including a model of the forest dynamics, which describes changes of the structure of forest areas and temporal variations as a result of natural growth of plantations and a complex of external forcings.

From the viewpoint of humankind it would be worthwhile and advantageous to concentrate the efforts of experts in different sciences dealing with the problem of global modelling, in a far more organized manner that at present (Kondratyev *et al.*, 2000).

3.3 PRIMARY PRODUCTION AND CARBON DIOXIDE

Relationships between the global CO_2 cycle and surface vegetation manifest themselves through the dependence of primary production and the rate of decomposition of dead biomass on atmospheric temperature and CO_2 concentration. The temperature dependence is manifested most in northern latitudes, where mean annual temperature variations can reach high levels, and the vegetation period of plants varies from two to seven months. An impact of atmospheric CO_2 on the growth of plants depends on many factors. There are two basic types of plants which vary in their reaction to changes in the partial pressure of atmospheric CO_2. The first, most widespread type of plants (type C_3) is characterized by photorespiration realized due to ferments, which can simultaneously assimilate and emit CO_2 and O_2. This process has the so-called compensation point Γ, when the balance of all functions of the plant with respect to CO_2 concentration (C_a) is optimal. This point is characterized by the value $\Gamma \cong 50$ mmol/mol at 25°C, grows with increasing temperature and is proportional to the value $C_a - \Gamma$ up to the level 1000 mmol/mol. The efficiency of the initial use of light increases with increasing CO_2 in proportion to $(C_a - \Gamma)/(C_a + \Gamma)$ (Goudrian *et al.*, 1990).

The other type of plants (C_4), such as high tropical grass (maize, sugar cane, millet, and sorghum), assimilate CO_2 from the atmosphere independent of the O_2 concentration, so that Γ remains practically constant and at a low level $\cong$5 mmol/mol. These plants react weakly to changes in the concentrations of carbon dioxide.

Numerous laboratory studies of the reactions of both types of plants to changing values of C_a testify to a wide variance of the quantitative estimates of variations in photosynthesis for the C_3 type. On average, the plants react to changes of CO_2 concentrations with a 1-month lag, and a CO_2 doubling can cause an increase in the photosynthesis rate from 10 to 100%. Further increase of C_a up to 400% leads to an effect of photosynthesis saturation for some plants (i.e., an addition to the rate of photosynthesis reaching 20%), and in some cases (e.g., for *Setaria lutescens*) photosynthesis is oppressed (Bazzaz, 1986). Practically all the plants of the C_4 type at present values of C_a are in a state of photosynthesis saturation. Winslow *et al.* (2003) studied the effect of the seasonal availability of water for the grass ecosystem (C_3/C_4) against the sensitivity of photosynthetic responses and it was shown that the ecosystem production depends on the combination of temperature and precipitation.

There are many effect of CO_2 on plants, being manifested through changes in the nutrients regime of photosynthesis. Table 3.11 exemplifies this effect. Of course, elemental composition of the plant varies. It includes C, O, H, N, S, P, K, Ca, Mg, and Fe, and the exchange processes in the plant–atmosphere system include chemical compounds, such as CO_2, CH_4, H_2O, H_2S, NH_3, and NO_2. A living plant

Table 3.11. Changing content of the nutrient elements in trees as a result of a 2-year impact of changed CO_2 concentrations for *Acer pseudoplatanus* (*A*) and *Fugus sylvatica* (*F*) (Bazzaz, 1986).

Chemical element	CO_2 concentration, mmol/mol					
	370		520		670	
	A	*F*	*A*	*F*	*A*	*F*
N	1	1	1.09	1.11	1.17	1.21
P	1	1	1.10	1.33	1.20	1.50
K	1	1	1.15	1.21	1.27	1.33
Ca	1	1	1.17	1.13	1.25	1.23
Mg	1	1	1.14	1.22	1.29	1.22
Mn	1	1	1.00	1.08	1.07	1.25
Fe	1	1	1.43	1.17	1.54	1.50

consists of 50–95% water, with the remaining part, the so-called dry substance, composed of 70–98% organic substance, which can burn. In other words, each plant on the Earth plays a role of its own in the global biogeochemical cycle of CO_2 and other chemical elements. Therefore, all the existing models of the CO_2 cycle based on rough classifications of the soil–plant formations are incorrect, and their reliability can hardly be assessed from the available databases on vegetation cover and its parameters.

The tundra and forest tundra biocenoses, whose share is about 4% of the land surface, are quite special. Their role in assimilation or emission of CO_2 is seasonal in character. Tundra with its marshes, water basins, and lakes is a source of CO_2 for the Arctic atmosphere (Lafreur *et al.*, 2001). The soils of the arctic tundra play a special role in this process. During the period of snow melting they emit carbon monoxide (CO). Above the soil surface the CO partial pressure can reach 100 ppm with a mean annual value of 0.05 ppm (Kelley, 1987). The CO concentration in air bubbles produced in water bodies and lakes with decaying biomass is estimated at 5–20 ppm. In the spring, near the bottom of the tundra plants the CO partial pressure in the air reaches 40 ppm, markedly decreasing by the end of summer. As a result, in spring, the CO_2 partial pressure in the near-surface atmosphere of the Arctic tundra can reach 2100 ppm. What happens in winter with the CO_2 exchange on the atmosphere–tundra border is practically unknown. However, there are data on the photosynthetic and respiratory activity of the tundra vegetation, from which it follows that this activity is preserved even under insolation of 5–7 W/m^2/day and with negative temperatures (Tiezen, 1973). This means that the tundra vegetation in late summer and early winter can serve as a sink for atmospheric CO_2. The sink of CO_2 from the atmosphere due to assimilation by the tundra vegetation is estimated at 146 g/m^2/day (Kelley, 1987; Eliasson *et al.*, 1999). Therefore, to assess the role of the soil–vegetation formations in the global cycle of CO_2, it is necessary that the model has their sufficiently detailed spatial and objective classification. Unfortu-

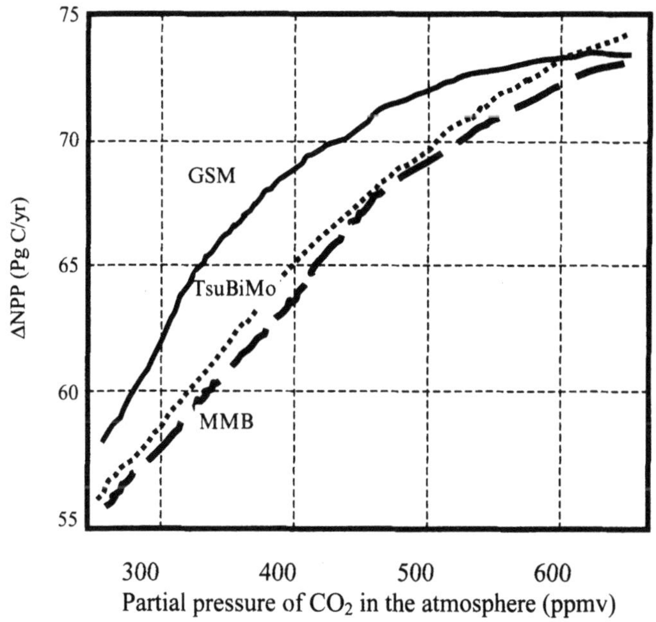

Figure 3.3. Comparison of three models of the global carbon cycle to assess the response of pure primary production (NPP) of vegetation to changing concentrations of atmospheric CO_2. Note: GSM = global simulation model (Kondratyev *et al.*, 2002); TsuBiMo = Tsukaba Biosphere Model (Alexandrov and Oikawa, 2002); MMB = mathematical model of biosphere (Krapivin, 1993).

nately, there is no information of the form that could be introduced into the model, and therefore the published estimates of the role of the soil–vegetation formations considered in the global models of the CO_2 cycle cannot be seen to be reliable. Although, for instance, the estimates of the surface vegetation role in CO_2 assimilation over the territory of the FSU countries given in the monograph by Krapivin and Kondratyev (2002) do seem realistic.

One of the promising results of modelling the effect of CO_2 fertilization has been discussed by Alexandrov and Oikawa (2002) who, based on the Tsukaba Biosphere Model (TsuBiMo), obtained a globally distributed estimate of the contribution of various types of vegetation to the change of CO_2 flux at the atmosphere–land border. To compare this result with the Global Simulation Model (GSM) data, Figure 3.3 presents curves of the dependence of pure primary production of the surface vegetation on changes of the atmospheric CO_2 partial pressure. The difference between these curves shows that a consideration in the GSM of additional feedbacks compared with a simple TsuBiMo specifies this dependence and recommends a more thorough structure of the GSM units. Nevertheless, it should be borne in mind that despite the apparent simplicity of the parametric descriptions assumed in TsuBiMo, the result obtained with its use is reliable due to a detailed description of the spatial structure of vegetation cover (50 × 50 km). Therefore, the strategy for modelling of the global carbon dioxide cycle should follow a paradigm of the complete description of all direct interconnections and feedbacks with the use of a set of functional dependences, the choice of which is made in the process of an adaptive-evolutionary numerical experiment.

3.4 THE ATMOSPHERE–LAND CARBON EXCHANGE

The interaction between two carbon reservoirs – the atmosphere and the land – is expressed by carbon fluxes formed through ecological, geophysical, and geochemical processes, including photosynthesis, respiration of plants and animals, decomposition of dead organic matter, vegetation and fuel burning, volcanic eruptions, rock weathering, etc. Which of these processes prevails, depends on many factors. Therefore in the scheme of Figure 3.2 and in Table 3.10 various processes are taken into account.

The most important aspect in studying the global carbon cycle is the contribution of the interaction between the surface vegetation and the atmosphere to CO_2 exchange. This dependence is based on the fact that all plants create their biomass by assimilation of atmospheric constituents C, O_2, H, N, S, P, K, Ca, Mg, Fe, of which the main constituents are carbon, oxygen, nitrogen, and sulphur. Clearly, in a detailed analysis of the process of photosynthesis one should take into account the kinetics of CO_2, CH_4, H_2O, H_2S, NH_3, and NO_2. A minimum requirement with ensured CO_2 assimilation is the availability of CO_2, H_2O, light, chlorophyll, and proper environmental conditions (temperature and humidity). Therefore the complex assimilation formula can be written as follows:

$$\underset{\substack{(6\,\text{moles})\\(246\,\text{g})}}{CO_2} + \underset{\substack{(12\,\text{moles})\\(216\,\text{g})}}{H_2O} + \underset{(\text{solar energy})}{675\,\text{kcal}} \rightarrow \underset{\substack{(1\,\text{mol})\\(180\,\text{g})}}{C_6H_{12}O_6} + \underset{\substack{(6\,\text{moles})\\(192\,\text{g})}}{O_2} + \underset{\substack{(6\,\text{moles})\\(108\,\text{g})}}{H_2O}$$

This formula can be used to calculate the balance between the plants and the atmosphere for the CO_2 exchange only, but cannot be used for water, since the water is a limiting factor for photosynthesis, and the plants (because transpiration has not been considered in this formula) use much more water. A more detailed pattern of variations of the carbon exchange can be gained from the data in Table 3.8. In the global models the process of carbon assimilation should be carefully detailed to avoid the violation of the balanced description of other processes. Usually this is realized through introducing corrections (Krapivin and Kondratyev, 2002). For instance, possible losses in the balanced relationship for photosynthesis are taken into account. These losses are assumed to be 20–30% (i.e., on average, 6 mol CO_2 give 0.75 mol of glucose).

It is also necessary to consider the spatial heterogeneity of the Earth's cover differing in biomass, density, and intensity of organic matter formation. On average, 90% of the total biomass ($\approx$830 Gt C) are forest ranges ($\approx 50 \times 10^6\,m^2$), 50% of these constituting tropical forests ($\approx 24.5 \times 10^6\,m^2$), with only 10% ($\approx$84 Gt C) refering to bushes, savannahs, meadows, deserts, semi-deserts, marshes, and cultivated land. The process of organic matter formation is highly inhomogeneous: forest ranges 33 Gt C/year, the remaining vegetation 20 Gt C/year. These heterogeneities lead to mosaic patterns of bioproductivity and therefore should be taken into account in the synthesis of the model.

The relationship between the CO_2 global cycle and surface vegetation dynamics is manifested through the dependence of primary production and the rate of dead biomass decomposition on atmospheric temperature and CO_2 concentration taking into account the binary digitization of plants into two classes.

Let $R_\kappa(\varphi, \lambda, t)$ be the photosynthesis production for vegetation of the type κ at a latitude φ and longitude λ at a time moment t. Then the CO_2 flux from the atmosphere into the living biomass can be described with a simple model:

$$H_6^C(\varphi, \lambda, t) = C_{23} R_\kappa(\varphi, \lambda, t)$$

where the coefficient C_{23} reflects the efficiency of the mechanism of the photosynthesis response and, on average, is estimated at $C_{23} \cong 0.549$. Bjorkstrom (1979) proposed to approximate an assimilation of CO_2 by vegetation with the formula:

$$H_6^C = k_b(1 + \beta \ln [C_A/C_A^*] C_\kappa)$$

where the parameter $\beta \in [0, 1]$ is the measure of the ability of the vegetation system to react to an increase of the atmospheric CO_2 partial pressure; C_κ is the carbon content in the biomass of the κth type of vegetation; k_b is the coefficient of proportion depending on the temperature and type of vegetation; and C_A^* is the concentration of carbon dioxide in the atmosphere in the pre-industrial period.

Various authors estimate the flux $H_6^C(\varphi, \lambda, t)$ as ranging from 16.7 to 35 Gt C/year. This negligible scatter of the estimates enables one to reliably estimate the coefficients in approximations for H_6^C. A more detailed description of H_6^C requires construction of an additional unit of the model, taking into account the relationship between CO_2 concentration and functioning of the surface biomes on a given territory. These specifications have been made in the publications by Krapivin and Vilkova (1990), Nefedova and Tarko (1993), and Krapivin *et al.* (1996). To specify the function R_κ, additional empirical dependences are used, exemplified in Tables 3.12 and 3.13. The database of the modelling system contains similar information as well as data on the parameters of the soil–plant formations. Of course, the global data on the CO_2 balance in the biosphere are rather contradictory and formed with a delay and averaging over long time periods (see, e.g., Table 3.6). Such data are exemplified in Table 2.5 and Figure 3.4. In the publication by Vilkova *et al.* (1998) the regression formulae are given which enable one to calculate productivity $F(T_a, W)$, humus supply $H(T_a, W)$, and phytomass supply $B(T_a, W)$ as a function of atmospheric temperature T_a(°C) and precipitation W (mm/year):

$$\begin{aligned} F(T_a, W) &= 4.25 \times 10^{-4} T_a^3 - 8.76 W^3 - 1.99 T_a^2 W + 4.29 T_a W^2 + 2.29 T_a^2 \\ &\quad + 19.05 W^2 - 8.79 T_a W + 4.56 T_a - 14.16 W + 4.18 \\ H(T_a, W) &= -5.16 T_a^3 - 161.4 W^3 - 9.41 T_a^2 W + 6.79 T_a W^2 - 9.47 T_a^2 \\ &\quad + 199.51 W^2 - 4.37 T_a W + 7.47 T_a - 44.17 W + 4.93 \\ B(T_a, W) &= -9.02 T_a^3 + 225.79 W^3 + 1.11 T_a^2 W - 29.39 T_a W^2 - 5.87 T_a^2 \\ &\quad - 511.72 W^2 + 41.29 T_a W - 11.37 T_a + 356.97 W - 62.94 \end{aligned}$$

Table 3.12. Dependence of the annual production (kg/m²/year) on the mean global temperature and total precipitation $F(T_a, W)$.

Precipitation W (mm/year)	Atmospheric temperature (°C) (T_a)											
	−14	−10	−6	−2	2	6	10	14	18	22	26	30
3,125							3.4	3.5	3.7	3.8	3.9	4.0
2,875							3.2	3.3	3.5	3.6	3.7	3.8
2,625							3.0	3.2	3.3	3.4	3.5	3.6
2,375							2.8	2.9	3.0	3.1	3.2	3.3
2,125							2.5	2.6	2.7	2.9	2.9	3.0
1,875						1.6	2.3	2.3	2.4	2.5	2.6	2.7
1,625				0.4	0.6	1.3	2.0	2.1	2.1	2.2	2.3	2.4
1,375		0.2	0.3	0.4	0.7	1.1	1.7	1.9	1.9	2.1	2.1	2.0
1,125	0.2	0.3	0.3	0.4	0.6	1.0	1.6	1.8	1.9	1.8	1.8	1.7
875	0.2	0.3	0.4	0.5	0.8	0.9	1.5	1.4	1.3	1.3	1.2	1.2
625	0.3	0.3	0.5	0.6	0.9	0.9	0.9	0.8	0.8	0.7	0.7	0.7
375	0.4	0.4	0.5	0.7	0.6	0.6	0.6	0.5	0.5	0.5	0.4	0.4
125	0.1	0.3	0.3	0.2	0.2	0.2	0.2	0.2	0.2	0.1	0.1	0.1

Table 3.13. Dependence of the humus content (kg/m²) in a 1-m layer of soil on the mean annual temperature and total precipitation amount $H(T_a, W)$.

Precipitation W (mm/year)	Temperature (°C) (T_a)										
	−14	−10	−6	−2	2	6	10	14	18	22	26
3,000							21.9	21.8	21.2	21.1	19.8
2,750							22.4	22.4	22.4	21.2	20.3
2,500							22.5	22.5	22.4	21.3	20.6
2,250							22.6	22.6	22.5	21.5	20.7
2,000							22.7	22.7	22.7	21.5	20.9
1,750							22.8	22.7	22.7	21.6	21.1
1,500						15.6	22.8	22.8	22.7	21.6	21.3
1,250				6.0	6.1	16.2	23.2	23.0	22.9	21.7	21.4
1,000		5.0	5.0	6.5	6.1	16.9	25.8	24.1	23.1	22.5	22.2
750	1.0	5.5	6.0	11.0	21.5	35.1	25.1	24.0	23.1	23.1	22.3
500	1.0	6.1	7.4	9.1	19.1	58.3	45.1	23.3	19.2	14.3	18.2
250	1.1	6.1	7.5	11.0	13.5	14.3	10.1	5.2	3.4	1.1	1.0

The interaction of vegetation and the atmosphere is characterized by the CO_2 flux H_7^C, resulting from the process of respiration. Therefore, if T_κ denotes the losses of the gross production by vegetation of type κ in the process of respiration, then $H_7^C = C_{26}T_\kappa$. For a first approximation we can take $T_\kappa = \gamma B_\kappa$, where B_κ is the vegetation biomass and the product γC_{26} reflects the share of

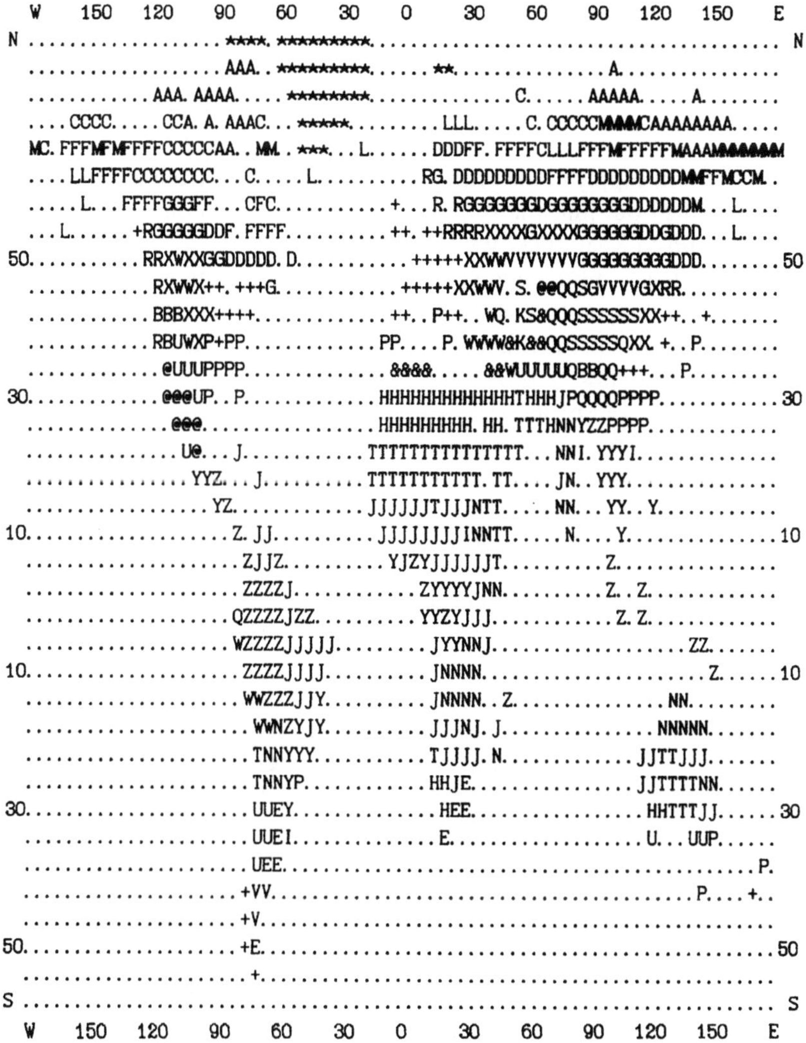

Figure 3.4. The spatial distribution of soil–plant formations over the 4° × 5° geographical grid. Notations are given in Table 2.5.

organic carbon emitted per unit time by the vegetation surface cover into the atmosphere.

The functioning of the atmosphere–land border in the process of CO_2 exchange also includes the fluxes H_9^C, H_{14}^C, and H_{15}^C, which play an important role in the carbon balance of the biosphere. Bjorkstrom (1979) used the following relationships to describe these fluxes in his model of the biospheric balance of CO_2:

$$H_9^C = B_\kappa/\theta_9 \qquad H_{14}^C + H_{15}^C = B_\kappa/\theta_{14,15}$$

where θ_9 is the characteristic time of the carbon cycle in soil and $\theta_{14,15}$ is the characteristic time of the transition of living biomass carbon to the soil (about 1000 years). If R_Q denotes the rate of humus decomposition and M_κ is the rate of vegetation decay, then:

$$H_9^C = C_{30}R_Q \qquad H_{14}^C = C_{18}M_\kappa \qquad H_{15}^C = C_{15}B_\kappa$$

With the carbon supply in the litter equal to 421.1 t/km^2, from which 12.95 t/km^2 are annually washed out into the soil, with 82.5% of this amount remaining in the upper soil layer up to 8 cm thick, we obtain $C_{18} = 0.31$.

The flux H_9^C has a more complicated dependence on the environmental parameters such as soil temperature and humidity. In the global model H_9^C is assumed to be a linear increasing function. The non-linear effects are considered due to heterogeneity of the types of soils manifested through the processes of oxidation of organic matter. The most inert system is peat covering about 4.3×10^6 km^2 and containing 860 Gt of organic carbon. In a stable condition the organic substance in the soil (via a process of oxidation) emits CO_2 and accumulates the same amount of carbon in the process of plant decay:

$$(H_9^C = H_{14}^C + H_{15}^C + H_{12}^C + H_{13}^C)$$

For peat and tropical forests this balance is not observed. The soil in the tropical forests emits CO_2 at a rate almost twice as high as the rate of CO_2 input to the soil from dead plants.

For the model of the atmosphere–soil CO_2 exchange to be complete, it is necessary to take into account geophysical and demographic aspects of the formation of additional carbon fluxes. They include volcanic emanations (H_5^C), rock weathering (H_4^C), vital functions of animals (H_{11}^C, H_{13}^C) and humans (H_{10}^C, H_{12}^C), as well as vegetation burning (H_8^C). Although nowadays some of them do not play a substantial role in the total balance of CO_2, an account of these fluxes is worthwhile for greater adequacy of the model's response in conditions of stress simulation experiments. In the models the H_5^C flux is usually assumed both a function of time and spatial coordinates φ and λ. Fluxes $H_i^C\,(i = 10, \ldots, 13)$ are considered to be proportional to the sizes of population G and animals F:

$$H_{10}^C = C_3G \qquad H_{11}^C = C_2F \qquad H_{12}^C = C_{22}G \qquad H_{13}^C = C_{21}F$$

With respect to the H_4^C flux, during the process of weathering silicate rocks the rate of CO_2 extraction from the atmosphere is negligible compared with a similar process for carbonate rocks. Therefore consider the contribution of such rocks to the flux H_4^C. In equilibrium conditions the relationship $[C_a^{2+}][HCO_3]^2/p_a = \text{const}$ is valid. Usually $2[C_a^{2+}] = [HCO_3]$, and therefore:

$$\Delta[HCO_3]/[HCO_3] = (\Delta p_a/p_a + 1)^{1/3} - 1$$

The flux H_8^C reflects a fragment of the anthropogenic interference to the global cycle of carbon dioxide. The formation of industrial CO_2 can be described rather precisely as follows:

$$H_1^C + H_8^C = \gamma_0[\exp{(rt)}]^6 \qquad \text{where } r \approx 0.029\,\text{yr}^{-1}$$

where γ_0 is the proportionality coefficient.

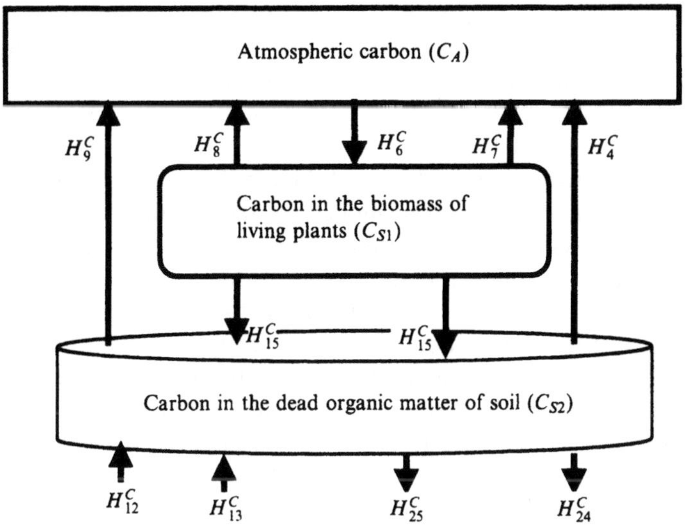

Figure 3.5. The scheme of carbon fluxes in the model of the atmosphere–vegetation–soil system. Notations are given in Table 2.9.

The surface cycle of carbon is, of course, more complicated in real conditions. Therefore, many fluxes of carbon in Figure 2.8 should be considered as a combination of fluxes characterized by the scheme of Figure 3.5. A more detailed surface carbon cycle can be considered by dividing the body of plants and soils into their constituents. This process requires a deeper knowledge of the biogeocenotic parameters, which on global scales is only possible in the form of average estimates of phytomass, biomass of the trunks and roots, and supplies of organic matter in the litter and humus.

3.5 PROBLEMS OF BIOCENOLOGY IN CONNECTION WITH THE GREENHOUSE EFFECT

To solve the problem of modelling the global greenhouse effect with an account of the role of surface ecosystems, it is necessary to use the methods, technologies, methodologies, and algorithms of ecoinformatics (Krapivin and Kondratyev, 2002). Ecoinformatics is the science of natural and anthropogenic systems and their interactions. It is characterized by the processes of differentiation and integration depending on the field of investigation. In the sphere of biocenology these aspects can be clearly manifested, when only those types are selected from a variety of the surface ecosystems, which are typical of the assumed spatial grid of digitization of the Earth's surface. Unfortunately, at present there are no fixed terms which can uniquely identify the subsystems of the environment. Many authors put different meanings to such terms as the biosphere, ecosystem, biogeocenosis, biome, geosystem, and environment. From the viewpoint of informatics, the development of methods of data processing and models representing the functioning of the natural–anthropogenic systems, such definitions are not important since they do not affect the formalized constructions of informatics. They manifest themselves only at the stage of interpretation of the results of data processing or modelling (Yokozawa, 1998; Kondratyev *et al.*, 2002a,b).

Interactions between informatics and biocenology have been demonstrated mainly in spheres such as numerical modelling and monitoring. Since the nature of the biocenology systems are determined by several abiotic and biotic factors, the use of informatics methods in biocenology requires a more detailed classification of these factors (Tikunov, 1997). One important factor is the climatic effect on a specific character of the distribution of plants. Each type of plant reacts with a time delay to temperature changes. The problem of modelling this dependence has been studied by many authors. It is a problem of succession of the vegetation cover depending on both climate change and on anthropogenic forcing resulting from fires and deforestation. Different ideas and methodologies of correlation studies in biocenology are caused by the instability of the climate processes. Therefore, an application of stochastic mechanisms to analyse a hierarchy of the cycles of climate change is one target for ecoinformatics.

Ecoinformatic methodology requires a certain level of formalization. For instance, Walter and Breckle (1985) introduced a zonal classification of the types of climatic situations characterized in Table 3.14. This classification simplifies the procedure of the spatial description of biocenotic processes in the case of their modelling and in the formation of the biocenology database. Of course, this example is one of numerous versions attempting to formalize the environmental structure.

The classification aspect of biocenology is important for the use of quantitative methods in the formalized description of the dynamics of various parameters of the surface ecosystems. The models of the cycles of some chemical constituents of the

Table 3.14. Ecological classification of the types of soil cover found in characteristic climatic zones.

	Specific features of the zone	
Climatic zone	Type of soil	Type of vegetation
1 Equatorial	Brown clay	Evergreen forests
2 Tropical	Red clay	Deciduous forests or savannah
3 Subtropical	Grey soil; stony, desert, and salt soils	Desert vegetation
4 Mediterranean	Brown soils	Wood vegetation
5 Warm temperate climate	Red or yellow forest soils	Temperate evergreen forests
6 Cold temperate climate	Brown and grey forest soils	Broad-leaved deciduous forests
7 Continental	Black, brown and grey soils	Steppes
8 Boreal	Ashen-grey soils	Taiga
9 Polar	Tundra	Undersized vegetation

biosphere described by Krapivin and Kondratyev (2002) would be impossible without a division of the soil–vegetation formations into homogeneous categories. The best known classification of the soil–vegetation formations is the classification by Bazilevich and Rodin (1967) given in Figure 3.4 and Table 2.5. Other classifications can be found on the Internet in the database of the World Resources Institute (www.elsevier.com/locate/worldresources).

The level of detailed description of surface ecosystems is determined by many circumstances. The experts in concrete spheres of biocenology study various facets of the functional behaviour of the ecosystem components. The problem of ecoinformatics is to combine obtained knowledge to generate a new knowledge – a product of the synthesis of information fluxes from various scientific directions. So, for instance, phytocenology studies plant physiology and the processes of competition for light, nutrients, and space. Here one cannot do without methods of ecoinformatics making it possible, at a formal level, to formulate the behaviour of one plant or of a community of plants.

The theory of complex survivability enables us to formulate a model of the organization of phytocenoses, based on a hypothesis of the usefulness of ecologo-cenotic strategies behind vegetation community development. For the first time, the so-called syn-ecological addition introduced by Rabotnov (1987) makes it possible to consider the process of the plant survival not as a mutual adaptation of concrete species but in a more general way, as a coexistence of species with different adaptive strategies. Each plant or group of plants in its competition for life can vary elements of its life strategy: change the distribution of roots over the soil horizons, change the ratio of its structure above and below ground, and change its requirements to fit the ecological niche. Therefore, the use of methods based on the theory of the vitality of complex systems would favour a deeper understanding of the mechanisms of succession in vegetation communities.

The biocenological aspect of ecoinformatics is especially manifested in the spheres of anthropogenic effects on the environment (landscape ecology and agricultural ecosystems (Muller and Steinhardt, 2003)). The methods of ecoinformatics make it possible to formalize the scheme of the interaction of factors that form the landscape, and lead to an appearance of the potential vegetation in place of that destroyed by humans. There are formalizations in the landscape ecology, the use of which can help in the synthesis of the respective models of landscape succession. For instance, there exists a 4-level hierarchy of landscape units: ecotop, micro-, meso-, and macro-flora. However, there is a disagreement in the subject sphere among various authors on typifying the territorial heterogeneities of vegetation communities caused by internal landscape boundaries. Therefore, the use of system methods of ecoinformatics with the possibilities of numerical experiments will help standardize the notions of landscape ecology with an account of human interference.

In agroecology, methods of ecoinformatics are introduced to develop the technologies to assess energy fluxes in controlled ecosystems. It is important that the specific energy expenditures of industrial agriculture are higher (by an order of magnitude) than in natural ecosystems. Humans have considerably reduced species variety in agroecosystems in order to raise the productivity of useful species by managing their energy fluxes. Therefore, respective corrections should be introduced into the biogeochemical cycles (for instance, by introducing corrections in the Table 2.5 data). A variety of agroecosystems are not considered in the global biospheric models, resulting in substantial reduction of the reliability of simulation modelling.

3.6 A SYSTEMS APPROACH IN BIOCOENOLOGY

Practical problems of the rational use and reproduction of biospheric resources require a quantitative study of the internal nature of the organisms constituting its communities (biocenoses). This is necessary in studies of World Ocean productivity, the structure and functioning of the communities residing in it, as well as in assessing the state of recreational zones. The use of only biochemical methods to solve these problems is insufficient in principle. It is necessary to apply biocenotic methods. The latter, in contrast to biogeochemical methods, have no solid quantitative basis provided by physics and chemistry.

The quantitative basis of biocenology is cybernetics, one component of which is called the theory of complex systems. This theory manifests itself in biocenology through numerous models of concrete biocenoses aimed at solving the problems of predicting their dynamics and at finding the technologies to control these dynamics. In this respect the Lotki-Volterra models (Svirezhev, 1987) are rather popular as well as their numerous modifications that introduce a factor limiting some functions of biocenosis into the model through environmental parameters. Unfortunately, all these models are only of theoretical character and serve to study a simplified phase structure of an idealized biocenosis. Attempts to complicate the model in order to consider real processes leads to complex systems of differential equations in partial derivatives, with the coefficients requiring serious field investigations (Bratus *et al.*, 2001; Novozhilov, 2002). Such a situation in biocenology turns out to be the rule rather than the exception. An attempt to express the ecological concepts as formulas or other formalized constructions is closely connected with the necessity of field measurements. Hence, the numerical modelling in biocenology also serves to justify the necessity of measurements.

Biological systems have a hierarchical structure. Unicellular organisms are at its lowest level, and the biosphere its top level. Biosystem A is determined by its structure $|A|$ and behaviour $\bar{A}$. The ecological law would not be broken if we suppose that a biosystem has some goal $\underline{A}$. The is no doubt that the strategic goal of any biosystem is either to survive or favour the survival of a higher biosystem. However, the tactical goals of a biosystem favouring a realization of the strategic goal are not apparent.

Biosystems are open systems. This means they interact with the medium B, which also includes other biosystems. The biosystem–medium interaction consists of an exchange between them of substance, energy, and information. A set of the biosystem's requirements to the medium is called its niche. Biosystems with similar niches can be considered as the same form of life. If the requirements of the biosystem to the medium are described by values of some parameters prescribed within certain ranges, the niche can be determined as some region in the respective multidimensional parametric space. Each element of the biosystem can also be described by some set of parameters, and in this case the bonds between the elements of the biosystem are presented as the dependence between the respective parameters. In other words, the biosystem A can be described by N parameters $\bar{x}(t) = \{x_i(t), i = 1, \ldots, N\}$, depending on time t. The structure $|A(t)|$ and behaviour $\bar{A}(t)$ of the biosystem are functions of these parameters. Therefore, $A(t) = F(\bar{x}(t))$. Hence, the trajectories of the system A in time correspond to some curve in the N-dimensional Euclidean space.

For instance, the model of the system A is called an abstract formation $A_M(t) = F_M(\bar{x}^M(t))$, depending on M components of vectors $\bar{x}(t)\{\bar{x}^M(t) \in \bar{x}(t), M \leq N\}$ without considering all bonds existing between them. Apparently, as $M \to N$ a discrepancy between the trajectories of the system itself and its model should decrease. An extent of the discrepancy is chosen in the form of any natural criterion, for instance, in the form of a maximum difference between respective coordinates of the trajectories.

Note that the modelling of biocenoses and its spatial distribution is the present-day equivalent to classical experimental studies. The principal obstacle here is the multidimensionality of the phase space, which is impossible to overcome. Therefore, one should not expect more from numerical modelling than it can give. The well-developed numerical models perform mainly an explanatory function making it possible to simulate some processes in nature on an accelerated timescale and, without interference into real processes, to assess the effect of variations of individual characteristics on the dynamics of the environmental systems. Important models of biocenoses are similar, as a rule, in complexity to the modelled objects, and therefore often lose their explanatory character. Solution of the problems is connected with the use of the model only as an element of a more complicated construction discussed in Kondratyev *et al.* (2000), GIMS technology.

Let the biosystem's niche $A(t)$ be described by L parameters $\{c_j(t), L \ll N\}$ lying within the limits $a_j \leq c_j(t) \leq b_j (j = 1, \ldots, L)$. Then the biosystem's environment bond can be described by a system of equations:

$$G_j(\{x_i(t)\}) = c_j(t) \qquad (j = 1, \ldots, L)$$

All the systems whose parameters meet these relationships form the class Ξ.

Let us introduce some goal functional on the trajectory of the system $A(t)$:

$$V = f(\{x_i(t)\})$$

Let the parameter V reach a maximum at $\{x_{i,0}(t)\}$ on the trajectory $A_{opt}(t)$. Dependences of N parameters of an optimal system on L of its characteristics is a solution for this extreme problem:

$$x_{i,0} = H_i[c_1(t), \ldots, c_L(t)]$$

which enables one to reduce the model's dimensionality. By forming a multitude of tactical goals of the system $\Xi = \{\underline{A}_1, \ldots, \underline{A}_r, \ldots, \underline{A}_m\}$ and establishing correlations

$A_{r,opt}(t) = g(\underline{A}_r)$, and $\underline{A}_r \in \Xi$ either experimentally or at an expert level, we obtain a limited set of possible optimal models whose trajectories can be compared with the trajectories of $A(t)$, and an inverse problem that can be solved using the given test of proximity:

$$\underline{A}_{r,0} = g^{-1}(A_{r,0,opt}(t))$$

As for the determination of the multitude of tactical goals – at the level of population they are depredation, parasitism, mimicry, etc. Determination of these goals at the level of biocenoses is a more difficult problem, connected with understanding the categories of purposefulness, expediency, and reasonableness, as manifestations of the biosystem's efficiency. The problem is somehow simplified when biocenosis is a part of the economic system.

3.7 MODELS IN BIOCENOLOGY

Numerical biocenology has become an independent scientific field due to the works by many authors who combined in their studies the potentials of ecology and numerical modelling (Friend, 1998; Legendre and Legendre, 1998; Svirezhev and Zavalishin, 2003). This combination has resulted in a reorientation of many principles of traditional ecology, which created a kind of information interface between ecology and other sciences and raised the role of complex interdisciplinary studies. The sphere of interest of this new scientific field includes the problems of complexity and generality in the development of ecological laws.

The recently developed models cover such problems as the parameterization of vegetation cover succession, description of the production process as a function of the environmental factors (Wirtz, 2000), modelling the competition for common resources (Alexeev *et al.*, 1992), modelling the natural–anthropogenic systems (Polishchuk, 1992; Ougolnitsky, 1999), modelling the spatial–temporal dynamics of the ecological systems, and many others (Roo-Zeilinska and Solon, 1997; Perituen, 1998).

Many authors solve the problem of modelling a succession of vegetation communities based on their beliefs on the succession processes. There are two fundamental interpretations of these processes based on the concepts of endo-ecogenesis and exo-ecogenesis. The most stable ideas of the character of the dynamic phenomena in phytocenoses is an idea of a single process of vegetation development under the influence of both external and internal factors. In this case the process of succession covers such factors as soil development, layer by layer division of the vegetation community, change of the height of plants and their accumulated biomass, microclimate formation, change of species diversity, and relative stability of the community. The classification of the forms of the vegetation community dynamics and its affecting factors is important for further synthesis of the numerical model. Vasilevich (1983) introduced the following categories of succession: endogenic (autogenic), exogenic (allogenic), continuous, and post-disruptive (post-catastrophic). In the case of endogenic succession the source of changes is located in the vegetation itself and through changes in the medium affecting it.

There are no stages in continuous successions. In the course of post-disruptive succession, changes in the community take place from unstable to stable conditions. Exo-ecogenesis is characterized by prevailing external factors.

A new development in the modelling of succession is the use of the population dynamics. Hulst (1979) proposed three types of models of succession. The first type is a model of succession with response but without competition (i.e., there is an effect of reaction of the vegetation community to changes in the environmental conditions). Succession is described by the usual differential equation:

$$dN_i/dt = r_i N_i(k_i - N_i)/k_i$$

where N_i is the number of individuals of the ith species; r_i is the rate of exponential growth; and k_i is a maximum of population density.

The second type of succession model is based on suppression of one species by another via the process of intrusion. In this case the coefficient of competition is used:

$$\alpha_{ij} = \partial[dN_i/dt]/\partial N_j/\partial[dN_i/dt]/\partial N_i$$

and the model itself is written as:

$$dN_i/dt = r_i N_i\left(k_i - \sum \alpha_{ij} N_i\right)/k_i$$

The third type of model of succession is alternative in character. If in the first two models the population density is a function of the size of a previous species of vegetation, here a maximum population density k_i depends on time.

Production of organic matter by a vegetation community depends on physiological processes in the organisms of the plants, resources of solar radiation, moisture, elements of mineral nutrition, temperature, and many other internal and external factors (Svirezhev *et al.*, 2003). Modelling of the production process takes into account photosynthesis, respiration, matter motion, transpiration, division of cells, etc. There are numerous correlations between these factors and production of plants. The formulas from Lieth (1985) are used most often. To estimate the annual production of plants R_p (g/cm^2/year) as a function of mean annual temperature T(°C) and mean annual precipitation r (mm), the following estimate can be used:

$$R_p = 3000 \min\{(1 + \exp(1.315 - 0.119T))^{-1} \quad (1 - \exp(-0.000664r))\}$$

In a more general case the vegetation biomass density P on a given territory is modelled by the balanced equation:

$$dP/dt = \min\{\rho_P, R_{PP}\} - M_P - T_P$$

where ρ_P is a maximum production/biomass (P/B) coefficient of the considered type of vegetation; R_{PP} is the vegetation productivity for the P/B coefficient values less than ρ_P for the present values of the P/B coefficient; and M_P and T_P are parameters of decay and respiration expenditure per unit time, respectively.

The use of parametric descriptions of productivity in biogeocenotic models is an important addition to the monitoring of vegetation communities. Unfortunately, the existing database and knowledge do not give any synthesized information about the huge amount of empirical correlations representing dependences of productivity and other functional components of vegetation communities.

Model descriptions of the competition between individual plants is a special problem in numerical biogeocenology. On the whole, the mutual impact of plants is described by the so-called competition indices (Bogatyrev, 1988). A freely growing plant is a plant whose function is independent of resource and is determined by only species characteristics and external parameters. An interaction between plants starts with their influence on the same resource and with mutual intrusion into territories occupied by free growth. Coenopopulations have more complicated and diverse interrelationships and, as a result, they differentiate niches and habitats. According to Mirkin *et al.* (2002), who introduced five types of strategies for plants and their functioning in the environment (K – violents, S – ecotopic patients, S_k – phytocenotic patients, R – true explerents, R_k – false or phytocenotic explerents), a systems approach to a model description of competition and survival is possible. All these strategies, to a different degree, are inherent in individual species, and therefore the synthesis of the model of survival is determined by a concrete set of requirements of a plant to the niche parameters. Usually a plant's niche is an aggregation of the space it occupies, resources it uses, and additional conditions, for instance, the presence of pollinators, biomass consumers, etc.

Each plant or some totality of plants N interacts with the environment H (including other plants) exchanging some amount of resources V_N for an obtained volume of resources W_N. The structure $|N|$ and the behaviour $\underline{N}$ of plants or their communities are in an equilibrium state with the structure $|H|$ and behaviour $\underline{H}$ of the environment, and this state is described with a minimax criterion:

$$W_{N,opt} = \max_{\{|N|,\underline{N}\}} \min_{\{|H|,\underline{H}\}} W_N(|N|, \underline{N}, |H|, \underline{H})$$

$$W_{N,opt} = \max_{\{|H|,\underline{H}\}} \min_{\{|N|,\underline{N}\}} W_N(|N|, \underline{N}, |H|, \underline{H})$$

The goals N_S and H_S of vegetation continuum N and environment H, respectively, can be opposite, partially coinciding, or indifferent. Due to the developed optimal strategy, plants of different kinds adjust to the levels of (V, W) exchange, which are possible on a given territory. Thus, because of differentiation of niches, there exist multilevel vegetation communities, for instance, in the forest. The factors of niche differentiation are multifaceted, covering location of roots at different depths, different periods of blossoming, different responses of plants to climate change, different needs of species for mineral nutrition, etc. In general, a plant's niche can vary in time both due to the dynamics of the plant itself and in connection with the drifting of adjacent niches. The diversity of possible situations brings forth a series of models of the functioning of vegetation communities.

One of the characteristic examples of the models of vegetation communities is the discrete model by Galitsky (1985). Let us consider it in more detail. The equation that describes a change in the plant's biomass B is written as:

$$dB/dt = k(t)(\alpha B^k + (B/B_f)^\gamma B_f) - \alpha B^k$$

where $k(t) = \min\{A/A_f(t), 1\}$ is the indicator of the plant's provision with a territory; A is the size of the available territory for the plant's growth; A_f is the size of the territory required for free growth; B_f is the biomass of a freely growing plant; α is the characteristic of the item of expenditure in the balance (on the biomass maintenance); and γ is the index of the deficit of territory.

The gist of the Galitsky model's functioning is that the biomass of each plant located in a cell of its own in a common mosaic changes independently of other plants in the community until either this plant or its neighbour dies from lack of territory. At the moment of dying the adjacent plants rush to distribute the vacant territory, and after changing the parameter A, the equation of the model is valid again. As a result, as individual plants die, the original mosaic changes on a local scale, gradually covering larger areas. The criterion of dying-off is simple: a plant is considered dead if its biomass becomes below some threshold level B_{dead} $(B_{\text{dead}}(t) = \varepsilon B_f(t), 0 < \varepsilon < 1)$.

3.8 MODELLING THE FOREST ECOSYSTEMS

The forest ecological system is an important constituent of the global continuum of soil–vegetation formations on the Earth (see Table 2.5). Therefore, the development of models that would describe forest dynamics is a principal stage in the synthesis of the global model. Of course, the importance of the models of the forest ecosystems is greater, since forest resources and their use affect the economic potentials of regions (Ougolnitsky *et al.*, 2000).

3.8.1 Modelling the growth of a tree as an element of the forest biocenosis

Many authors describe the dynamics of the total biomass of an individual plant in the form of a logistic equation:

$$dB/dt = k(B_{\text{lim}} - B)B$$

where B is the biomass; B_{lim} is the threshold amount of biomass; and k is the coefficient reflecting the effect of environmental factors and the type of plant on its growth.

To describe the dynamics of tree growth, the simplest functions are often used, such as $B = a\exp(-b/t), B = ax^b c^{-ct}, B = a\exp(b\ln t + c\ln^2 t)$, etc. Bichele *et al.* (1980) were first to propose a more complicated model for tree growth described by the system of ordinary differential equations:

$$dB_j = K_\Phi - k_R R_j - V_j + K_M$$

where $K_\Phi = k_F \sum_{i=1}^{4} \alpha_{ij}\Phi_i$; $K_M = M\sum_{i=1}^{4} \beta_{ij}$; B_j is the biomass of individual organs of a plant $(j = 1, 2, 3, 4)$; M is the dry biomass of the whole plant; Φ_i is an amount of carbon dioxide assimilated during 24 hours by the ith organ of the plant ($k_F\Phi_I$ is an equivalent dry biomass, k_F is the coefficient of equivalence); R_i is an amount of carbon dioxide emitted during 24 hours by the ith organ of the plant ($k_R R_i$ is an equivalent dry biomass, k_R is the coefficient of equivalence); V_i is the dry biomass of part of the ith organ; α_{ij} is the share of "fresh" assimilates created during 24 hours in the *i*th organ of the plant, moving during the same period of time into the *j*th organ of the plant; β_{ij} is the exchange of "old" assimilates during 24 hours between the *i*th and *j*th organs (per gram of dry biomass of the whole plant); K_Φ is an increment of biomass due to photosynthesis and redistribution of fresh assimilates; and K_M is an increment connected with the redistribution of old assimilates; and $k_R R_j$ is the biomass expenditure on respiration.

Of course, the coefficients of the model described above are functions of the environmental parameters. An account of the effect of the environment on the dynamics of the tree growth is realized, as a rule, within the principle of the limiting factors. Among the factors of the environment are photosynthetic active radiation (PAR), carbon dioxide, temperature, humidity, as well as mineral salts in the soil. By the principle of limiting factors, a plant grows until the "resources" of some factor become limited. An introduction of this principle is connected with many uncertainties, and therefore in concrete situations it must be tested with the use of experimental estimates.

More developed models of tree growth, instead of a set of empirical parameters whose uncertainties are often a great barrier to practical application of the models, include the functional dependences of the elements of the tree growth on the environmental factors. Among the most widespread and often applied is the Libikh criterion – that the growth of a plant is confined to an element, whose concentration is at a minimum. By this criterion a plant, or its organs, grow until the resources of some element, for instance, PAR, moisture, temperature, carbon dioxide, nutritious salts are exhausted. There are other criteria testifying to a weighed influence of all environmental factors, these weights are characteristic indicators of the plant type.

The process of tree growing can be presented as a result of interaction of three processes of growth: change of the leaf (needle) mass, growth of the trunk, and development of the root system. This process is schematically shown in Figure 3.2 and 3.6.

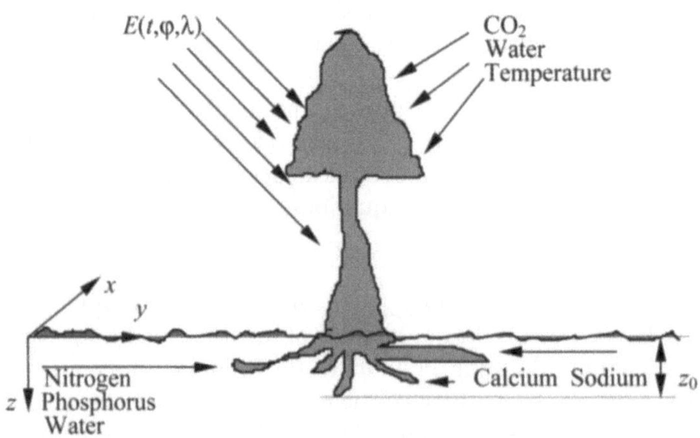

Figure 3.6. Model of the growth of a tree as an element of the forest biogeocenosis.

Note: z_0 = thickness of the root system layer; E = solar radiation.

Let us denote the biomass of the canopy, trunk, and roots of the tree as x_1, x_2, and x_3, respectively. Then the balanced equations of a tree model can be written as:

$$x_i(t+\Delta t) = x_i(t) + \varepsilon_i(t)\{R_x - T_x \Delta t\} \qquad (i = 1, 2, 3)$$

where Δt is the characteristic time step; R_x is the gross productivity of a tree due to photosynthesis; T_x is the biomass expenditure on respiration; ε_i is the share of new biomass moving into the biomass of the ith organ of the tree.

The ε_i parameters are the functions of time as well as other parameters of the environment. Following the principle of maximum survival, a supposition can be made that the principle of maximum primary productivity is valid:

$$\tilde{Y}^*(t+\Delta t) = \max_{\{\varepsilon_i(t)\}} Y\{x_i(t) + \varepsilon_i(t) Y[x_i(t), E(t, \varphi, \lambda), C_A, T_A]\Delta t, E, C_A, T_A\}$$

3.8.2 Modelling the water regime of plants

One of the important limiting factors of tree growth is water, the movement of which in the soil and the tree body determines the dynamics of the tree biomass change. Following Kirilenko (1990), let us denote the soil as a homogeneous porous layer $[0, z_g]$, where z_g is the depth of groundwater. Let us introduce the following notation: W = relative soil moisture (kg H_2O/m^3/[kg soil/m^3]), ρ = water density, and ρ_s = volume density of soil. Let us also introduce a system of coordinates (x, y, z), where the z-axis is directed down from the soil surface ($z = 0$). Then the movement of water through the soil can be described by Darsi's 3-D equation (Bras, 1990):

$$V = -K[\nabla(\Phi_s - \rho g z)]/(\rho g) \tag{3.1}$$

where $V(t, x, y, z)$ is the rate of water flow; $K(t, x, y, z)$ is the hydraulic conductivity of water; g is the acceleration due to gravity, and $\Phi(t, x, y, z)$ is the water potential of soil.

The equation of soil moisture dynamics is:

$$\partial W/\partial t = -(\rho/\rho_s)(f + \operatorname{div} V) \tag{3.2}$$

where $f(t, x, y, z)$ is the sink function.

Dependences of the water potential of soil and its hydraulic conductivity on soil moisture are written as:

$$K = K_s^* W^m \qquad \Phi_s = \Phi_s^* W^{-n}$$

where

$$K_s^* = K_{FM} W_{FM}^{-m} \qquad \Phi_s^* = \Phi_{MG} W_{MG}^n \qquad m = 2n + 1$$

W_{FM} is the total water capacity of soil; K_{FM} is the soil conductivity corresponding to total water capacity; and m and n are constants.

The system of Equations (3.1) and (3.2), with assumed notation, can be rewritten as:

$$\partial\omega/\partial t = (-\rho/\rho_s)[(K_{FM}/W_{FM}) \operatorname{div}\{\mu\omega^n \nabla\omega + i\omega^n\} + f/W_{FM}]$$

$$V = K_{FM}(\mu\omega^n \nabla\omega + i\omega^n)$$

where $\omega = n\Phi_{FM}/(\rho g)$ and i is the unit vector directed along the z-axis.

In the case of homogeneous processes in the (x, y) plane the problem is simplified:

$$\partial\omega/\partial t = (-\rho/\rho_s)[(K_{FM}/W_{FM})\partial/\partial z\{\mu\omega^n\partial\omega/\partial z + \omega^n\} + f(t,z)/W_{FM}]$$
$$V(t,z) = K_{FM}(\mu\omega^n\partial\omega/\partial z + \omega^n)$$

Boundary and initial conditions are given in the form:

$$\omega(t, z_g) = 1 \qquad V(t,0) = P(t) - e(t) \qquad \omega(t_0, z_g) = \omega^0(z)$$

where $P(t)$ is precipitation intensity; $e(t)$ is the intensity of evaporation from the soil surface; and $\omega^0(z)$ is the initial soil moisture.

The function f describes the intensity of water absoption by a tree's root system:

$$f(t,z) = \xi_0[\Phi_s(t,z) - \psi_0(t)]\,ds/dt$$

where $\xi_0 = \xi_k\sigma_0$ is the conductivity of water entry to the root system; $\xi_k = r_k^{-1}$ is the specific conductivity of water entry to the root; r_k is the specific resistance of water entry to the root; σ_0 is the general area of the sucking roots of the tree per unit soil surface area ($\sigma_0 = \beta_k m_k$); m_k is the mass of roots beneath unit soil area; β_k is an empirical constant; ψ_0 is the water potential within the root system; and $s(z)$ is the function of the vertical distribution of the root system $[s(0) = 0, s(z_0) = 1]$.

To complete the synthesis of the tree's water regime, let us parameterize the processes of water flow of the plant and transpiration. Assume that from the root system the water flows via the trunk to the canopy (branches and leaves) and then evaporates into the atmosphere (Lai *et al.*, 2000; Lal *et al.*, 1998). Divide the above ground part of the tree of height H into n equal layers: $H = n\Delta z$. The water moves across the boundaries of the ith layer due to the difference between water potentials ψ_i and ψ_{i-1} overcoming the resistance of the xylem vessel walls:

$$r_{ks}^*(z) = (S_{ks}^*(z)\xi_{ks})^{-1}$$

where ξ_{ks} is the specific conductivity of the xylem; and S_{ks}^* is the cross section of the tree at a height z. As a result, with the area of the cross section supposed to be vertically constant, we obtain the following formula to describe the rate of water flow in the trunk across the ith layer:

$$v_{i-1,i}^* = -\xi_{ks}S_{ks}^*\Delta z^{-1}(\psi_i - \psi_{i-1} + \rho g\Delta z)$$

According to Kirilenko (1990), the intensity of transpiration from the ith layer can be described with the formula:

$$\gamma_i^* = \frac{d^i}{r_L^i}S_L^{*i}$$

where d^i is the deficit of saturation of an absolute air humidity in the atmosphere; S_L^{*i} is the leaf area of the ith layer, $r_L^i = r_{st}^i + r_a$; r_{dt}^i is the stomatal resistance; and r_a is the air boundary layer resistance.

To complete the description of the model after Kirilenko (1990), consider a standing wood of density ρ_F. The model equations are written as:

$$d^2\varphi/d\eta = U\nu(\eta)\varphi(\eta) \qquad \eta \in [0, \alpha] \tag{3.3}$$

the boundary conditions are:

$$\begin{cases} \varphi_0' = -[G(\tilde{\varphi}_s - \varphi_0) + F] \\ \varphi_\alpha' = -F \end{cases} \qquad d\varphi/d\eta = -F \qquad \text{for } \eta \in [\alpha, 1] \tag{3.4}$$

At $\varphi_0 \leq 0$, the system of Equations (3.3) and (3.4) becomes:

$$\begin{cases} \dfrac{d\varphi}{d\eta} = -F \\ \varphi_0 = \tilde{\varphi}_s \end{cases}$$

with the following notation: $\eta = z/H, \varphi(\eta) = 1 + \psi(\eta H)/\psi^*, \psi^*$ is the leaf water potential, $1/D_1$ is a minimum stomatal opening resistance, $U = Hr_{ks}D_1 d\beta m_L/\psi^*$, $F = \rho gH/\psi^*$, $G = Hr_{ks}\xi_0$, $\alpha = a/H$, β is the leaf index, $h = -z$, and

$$a = \begin{cases} H & \text{when } \psi(H) \geq \psi^* \\ h & \text{when } \psi(h) = -\psi^* \qquad \psi(H) < -\psi^* < \psi(0) \\ 0 & \text{when } \psi(0) \leq -\psi^* \end{cases}$$

3.9 MODELLING THE ENERGY FLUXES IN THE ATMOSPHERE–PLANT–SOIL (APS)

The energy exchange in the APS system plays an important role in the formation of climatic processes on every spatial and temporal scale, and therefore it should be considered in the models of the greenhouse effect to increase the reliability of the estimates. Problems with this include the development of methods to measure the energy fluxes in the air–surface layer as well as their modelling. A special aspect of these problems concerns the development of technologies of estimation of energy fluxes in the APS system on global scales to take these estimates into account in the global climate systems. There are well-developed methods for remote sensing a wide set of climatic parameters including energy fluxes (Yakovlev, 2001; Krapivin and Kondratyev, 2002; Marenkin, 2002). However, in the remote sensing of the APS system a complicated problem arises due to the assessment of the screening impact of vegetation over the wavelength regions used in the monitoring system.

Beginning in the 1980s, studies have been carried out on developing a model of photosynthesis for a vegetation bud. It was shown in Farquhar *et al.* (1980) how a biological model of the CO_2 assimilation by leaves is connected with fermentative kinetics and electronic properties of the chloroplast, which made it possible to rather plausibly describe the process of photosynthesis for the plants of class C_3. Then Ball (1988) and Collatz *et al.* (1990, 1992, 2000), using model technology and observation data on the process of the vegetation bud conductivity in photosynthesis, derived a semi-empirical model for the vegetation bud functioning with an account of changes in the environmental temperature, humidity, atmospheric CO_2 concentration, and intensity of assimilation of the vegetation species itself.

The problem of the consequences of the distribution of the photosynthetic potential of leaves and canopy of plants with respect to illumination has also been considered and a criterion has been developed to determine a common potential corresponding to maximum photosynthetic abilities of the plant in CO_2 assimilation.

Sellers (1985, 1987) developed methods of the vertical integration of leaf canopies, models of light scattering and absorption, and photosynthesis and conductivity of the vegetation bud. These studies revealed an interaction between spectral characteristics of a vegetation species (spectral vegetation index – SVI) and important functional relationships that regulate the processes of photosynthesis and evaporation from leaf canopies. Apart from this, an accomplished analysis revealed a connection between the share of the PAR, absorbed by the vegetation canopy with the respective characteristic coefficient of vegetation species (simple ratio vegetation index – SRVI). Here we consider SRVI and relative difference (normalized difference vegetation index – NDVI) as SVI. Spectral characteristics of vegetation species SRVI and NDVI are based on the remote sensing of the reflected emission in the visible and near-infrared (IR) regions and determined as follows:

$$\mathrm{SRVI} = \frac{a_N}{a_V} \qquad \mathrm{NDVI} = \frac{a_N - a_V}{a_N + a_V}$$

where a_N and a_V are hemispheric reflections and radiance in the visible and near-IR wavelength regions, respectively.

The parameters a_N and a_V are measured directly above the leaf canopy. Therefore, the effects of absorption and scattering of emissions by the atmosphere are not considered here. For satellite remote sensing a method has been developed to correct the experimental data by the Amsterdam School of International Relations (ASIR). A theoretical dependence (close to linear) has been established between PAR and SRVI confirmed later by experimental measurements in the region of Kansas, USA (Sellers *et al.*, 1992, 1996a–c). A minimum and maximum of FPAR (the fraction of PAR absorbed by the green portion of the vegetation canopy) are determined experimentally, and the final relationship becomes:

$$\mathrm{PAR} = \frac{(\mathrm{SR} - \mathrm{SRVI}_{i,\min}) \times (\mathrm{PAR}_{\max} - \mathrm{PAR}_{\min})}{(\mathrm{SRVI}_{i,\max} - \mathrm{SR}_{i,\min})} \tag{3.5}$$

where $\mathrm{PAR}_{\max} = 0.950$ and $\mathrm{PAR}_{\min} = 0.001$, but in general, $\mathrm{PAR}_{\max}$ and $\mathrm{PAR}_{\min}$ depend on the type of vegetation; and $\mathrm{SRVI}_{i,\max}$ and $\mathrm{SRVI}_{i,\min}$ can be calculated from NDVI, based on experimental data.

Sellers *et al.* (1996b) use the Π parameter reflecting the amount of vegetation cellulose taking part in the process of photosynthesis. The Π parameter changes from 0 (lack of vegetation) to ≈ 1.5 for bushy species rich in photic vegetation cellulose, approximately estimated as:

$$\Pi \approx \frac{\mathrm{PAR}}{\bar{k}}$$

where $\bar{k}$ is the coefficient of attenuation for PAR averaged in time.

Apart from spectral characteristics, Sellers *et al.* (1996c) considered the photic characteristic (green leaf area index L_g) and the photic share (green part N)

for a concrete vegetation species. Many authors derived dependences of L_g on FPAR for species growing separately and in groups. The result can be presented in the form:

$$L_g = (1 - F_{cl})L_{g,i,\max}\frac{\log(1 - \text{PAR})}{\log(1 - \text{PAR}_{\max})} + F_{cl}\frac{L_{g,i,\max}\text{PAR}}{\text{PAR}_{\max}}$$

where F_d is the share of bushy vegetation determined for each of the vegetation classes L_s.

The photic share N of vegetation species can be obtained from the following considerations. The plant canopy consists both of leaf matter (L_g) participating in the process of photosynthesis and of trunk cellulose L_s (0.076 for forests and 0.05 for meadows and pastures), as well as a dead fraction L_d. The dead cellulose is supposed to remain in the leaf canopy for about a month before leaf fall or other means of extraction. Thus the photic share of vegetation cellulose is determined as:

$$N = \frac{L_g}{L_T}$$

where $L_T = L_s + L_d + L_g$; L_d is the dead fraction of vegetation cellulose $= L_{g_{n-1}} - L_{g_{nl}}$, $L_d \geq 0$; L_g is the photic fraction determined from PAR; L_T is the total territorial characteristic of a vegetation species; and L_s is the fraction of trunk cellulose.

3.9.1 The model of photosynthesis of the leaf canopy

In the model of the biosphere (simple biosphere model-2 – SiB2) derived by Sellers *et al.* (1996b) there are equations to determine temperature, humidity, and evaporation for a vegetation canopy and a 3-layer soil. Apart from this, models of radiative transfer (soil–leaf canopy–atmosphere) and photosynthesis were synthesized. The model of the leaf canopy photosynthesis considers a 1-layer vegetation species, surface and root layers, as well as a zone of recharging. Equations of the SiB2 are based on the concept of flux exchange in the soil–leaf canopy–atmosphere system (Kim *et al.*, 2001). For this purpose the notions and estimates of resistance r and conductivity g in inverse proportion ($r = 1/g$) are introduced. The flux in this case is determined by analogy with the electrostatic form:

$$\text{Flux} = (\text{potential difference})/\text{resistance}$$

Sellers *et al.* (1996b,c) considered the fluxes of sensible and latent heat transport through evaporation of water vapour from a vegetation canopy λE_c and the surface soil layer, as well as the CO_2 flux (A_c and R_{SOIL} for soil). The potential differences connected with the fluxes are presented in the form of differences of temperatures, pressures, and partial pressure of CO_2 for heat transport, latent heat transport, and CO_2, respectively.

In the model of photosynthesis of interest are CO_2 fluxes (A_c and R_{soil}) and the fluxes of latent heat exchange λE_c of the leaf canopy as well as it respiration constituent λE_{ct}.

The model of the leaf canopy photosynthesis derived by Collatz *et al.* (1990) for vegetation species of classes C_3 and C_3 together with the bud model (Ball, 1988) have been used to derive the SiB2. In this model, photosynthesis and conductivity are closely connected (Sellers *et al.*, 1996a–c; Marenkin, 2002).

According to Collatz *et al.* (2000), there are three factors limiting the plant photosynthesis intensity. These are an efficiency of the photosynthetic ferment system w_c (the so-called limitation after Rubusco (Sellers *et al.*, 1996b)), an amount of PAR absorbed by cellulose chlorophyll w_e, and an ability of a vegetation species to assimilate and emit to the environment the products of photosynthesis w_s. As a result, the value of photosynthesis intensity A is a minimum of the three limitations enumerated above:

$$A \leq \min(w_c, w_e, w_s)$$

where A is the photosynthesis intensity (mol/m^2/s); w_c is a Rubusco (ferment) limitation of photosynthesis intensity (mol/m^2/s); w_e is a limitation of assimilation by illumination (mol/m^2/s); and w_s is a limitation of carbon content for a vegetation species (mol/m^2/s).

The physiological limitation of assimilation w_c is a function of the ferment supplies of a species, in other words, it is an ability of cellulose to realize biochemical reactions. The ferment reserve of the plant is described in the model with the $V_{\max}$ parameters, a maximum catalytic supply of ferments (Rubisco) taking part in photosynthesis. The w_c value also depends on the leaf canopy temperature and soil moisture. Collatz *et al.* (1990) suggested the following expression for w_c for vegetation classes C_3 and C_4:

$$w_c = V_m \left[\frac{c_i - \Gamma^*}{c_i + K_c(1 + O_2/K_o)} \right] \qquad \text{for } C_3$$

$$w_c = V_m \qquad \text{for } C_4$$

where V_m is a maximum catalytic supply of ferments (Rubisco) (mol/m^2/s); c_i is the partial pressure of CO_2 in the leaf (Pa); O_2 is the partial pressure of O_2 in the leaf (Pa); Γ^* is the point of CO_2 compensation (Pa) $= O_2/S$; S is a special interaction between CO_2 and O_2; K_c is the Michaelis–Menten constant for CO_2 (Pa); and K_o is the constant of O_2 suppression (inhibition).

The V_m parameter is expressed through $V_{\max}$, the function of dependence on the soil temperature and moisture. The S, K_c, and K_o parameters are functions of temperature.

A limitation of assimilation intensity w_e is described as follows:

$$w_e = (F_\pi n)\varepsilon_3(1 - \omega_\pi) \left[\frac{c_i - \Gamma^*}{c_i + 2\Gamma^*} \right] \qquad \text{for } C_3$$

$$w_e = (F_\pi n)\varepsilon_4(1 - \omega_\pi), \qquad \text{for } C_4$$

where F_π is a PAR vector (W/m^2); n is the normal to the leaf surface; $\varepsilon_{3,4}$ is the quantum efficiency of CO_2 assimilation (mol); and ω_π is the dispersion coefficient for PAR.

The third limitation w_s is written as:

$$w_s = \frac{V_m}{2} \qquad \text{for } C_3$$

$$w_s = \frac{V_m}{2} \qquad \text{for } C_4$$

where p is atmospheric pressure (Pa).

The photosynthesis intensity A as a minimum of functions w_c, w_e, and w_s is estimated in a simple way, but observations show that the transition from one limitation to another is not that spasmodic. Therefore, Collatz *et al.* (1992) proposed a method of combining the contributions of all the of three of the limitations in the form of two equations, to be solved with the least values:

$$\beta_{ce} w_p^2 - w_p(w_c + w_e) + w_e w_c = 0$$

$$\beta_{ps} A^2 - A(w_p + w_s) + w_p w_s = 0$$

where β_{ce} and β_{ps} are coefficients of correlation and w_p is a "smoothed" minimum between w_c and w_e (mol/m^2/s).

Theoretically β_{ce} and β_{ps} can have values from unity (representing a lack of relationships between the limitations) to zero. Real estimates vary from 0.8 to 0.99.

Pure assimilation of CO_2 by a plant A_n is equal to $A_n = A - R_d$, where R_d represents CO_2 losses resulting from respiration connected with the content of carboxylase in cellulose: $R_d = f_d V_m$, where $f_d = 0.015$ for C_3 and $f_d = 0.025$ for C_4.

The connection of the Collatz model of photosynthesis with the Ball semi-empirical model of vegetation bud conductivity is written as:

$$g_s = m\frac{A_n}{c_s} h_s p + b$$

where g_s is the vegetation bud conductivity (mol/m^2/s); $g_s(\text{m/s}) = 0.0224 \dfrac{T}{T_f}\dfrac{p_0}{p} g_s$ (mol/m^2/s); m is an empirical coefficient (9 for vegetation of class C_3, 4 for C_4; 6 for coniferous forests); b is an empirical coefficient (mol/m^2/s or m/s) (0.01 for C_3, = 0.04 for C_4); h_s is the relative moisture of the leaf surface; c_s is the partial pressure of CO_2 in the stem (Pa); p is atmospheric pressure (Pa); p_0 is the normal atmospheric pressure (1.013×10^5Pa); and T_f is absolute zero temperature (273.16°K).

There is a relationship between partial pressure of CO_2 (c_i, c_s) and relative humidity h_s. This dependence can be obtained by analysis of the processes taking place on the vegetation bud–atmosphere border, the respective conductivities g_i and g_s, pure CO_2 flux, and water vapour flux E_{it} as a result of the process of photosynthesis.

This relationship is expressed as:

$$E_{lt} = g_l(e_s - e_a)\frac{\rho c_p}{\lambda\gamma} = g_s(e_i - e_s)\frac{\rho c_p}{\lambda\gamma} \qquad h_s = \left[\frac{e_s}{e_i}\right]$$

$$A_n = \frac{(c_a - c_s)}{p}\frac{g_l}{1.4} = \frac{(c_s - c_i)}{p}\frac{g_s}{1.6}$$

where e_a, e_s, and e_i are partial pressures of water vapour in the open air, on the surface, and in the stem, respectively (Pa); c_p is the specific heat capacity of air (joule/kg/K); γ is a psychometric constant (Pa/K); g_l is the water vapour conductivity on one side of the leaf (mol/m^2/s or m/s) (1/2rl); E_{lt} is respiration intensity (kg/m^2/s); c_a and c_s are partial pressures of carbon dioxide outside and inside the stem, respectively (Pa); and $e_i = e^*(T_c)$ (Pa).

Equations of the photosynthesis–conductivity model describe the processes taking place in a vegetation bud, and the next step is to integrate the relationships with respect to a leaf canopy to obtain total values of A_c and g_c. This was undertaken by Sellers *et al.* (1996a) with a supposition that the vertical distribution of vegetation ferments in the canopy is connected with the time-averaged distribution of PAR:

$$V_{\max_L} = V_{\max_0} e^{-\bar{k}L}$$

where $V_{\max_L}$, $V_{\max_0}$, are the $V_{\max}$ values on the leaf area index (LAI $= L$) level and at the top of the canopy (LAI $= 0$), respectively (mol/m^2/s); and $\bar{k}$ is a time-averaged value of the coefficient of PAR attenuation.

An amount of PAR assimilated by the leaf canopy is:

$$F_\pi n \approx F_{\pi_0} \left[\frac{G(\mu)}{\mu}\right] e^{-\bar{k}L}$$

where F_{π_0} is the value of the PAR flux caught by the leaf canopy (W/m^2); and $G(\mu)$ is a projection of the orientation of the canopy leaves towards the direction of the incoming radiation flux (μ). Sellers *et al.* (1996a) established that the use of the time-averaged $\bar{k}$ value instead of the instant areas k value leads to only negligible uncertainties in calculations.

Since now the w_c, w_e, and w_s parameters are connected (depending on the vertical coordinate) with the LAI of the leaf canopy through an exponent, it is possible to use them to obtain an integral estimate of A_c. We have already discussed the difference of bushy and separately growing species ($V < 1$) and the notion of the photic share ($N < 1$). Therefore bearing in mind that in the case of dense thickets the photosynthesis intensity is reduced due to the lack of "green" (rich-in-ferment) vegetation material, and the integration is realized from 0 to L_T/V:

$$A_c = A_{n_0} \int_0^{L_T/V} VNe^{-\bar{k}L} dL = A_{n_0}\Pi \tag{3.6}$$

where $A_{n_0} = A_n$ for leaves at the top of the canopy (mol/m^2/s):

$$A_{n_0} = f_c(V_{\max_0}, \ldots,), f_e(F_{\pi_0}, \ldots,), f_s(V_{\max_0}, \ldots,),$$

$$\Pi = \frac{VN\left(1 - e^{-\bar{k}L_T/V}\right)}{\bar{k}} \qquad \Pi \approx \mathrm{PAR}/\bar{k}$$

The A_c value is used to determine the respective conductivity of the leaf canopy using the modified expression:

$$g_c = m\frac{A_c}{c_S}h_S p + bL_T$$

where h_S and c_S are the volume analogues to values h_s and c_s relative to the leaf canopy.

The g_c parameter is used to estimate the respiration intensity λE_{ct}:

$$\lambda E_{ct} = \left[\frac{e^*(T_c) - e_a}{1/g_c + 2r_b}\right]\frac{\rho c_p}{\gamma}(1 - W_c)$$

where $e^*(T_c)$ is the saturated water vapour pressure at a temperature T_c (Pa); e_a is the water vapour pressure in an open atmosphere near the leaf canopy (Pa); ρ and c_p are the air density and specific heat capacity, respectively (kg/m^2, joule/kg/K); γ is a psychometric constant; and W_c is the share of the leaf-canopy wet area.

To complete a description of the equations of the photosynthesis–conductivity model, the following expressions are given that estimate the ferment supply of vegetation cellulose V_m (Marenkin, 2002):

$$V_m = V_{\max} f_T(T_c) f_W(W_2) \qquad K_c = 30 f_T(T_c)$$

$$K_o = 30{,}000 f_T(T_c) \qquad S = 2600 f_T(T_c)$$

$$Q_t = (T_c - 298)/10 \qquad f_w(W_2) = \{1 + \exp[0.02(\psi_c - \psi_r)]\}^{-1}$$

$$f_T(T_c) = \begin{cases} 2Q_t/\{1 + \exp[s_1(T_c - s_2)]\} & \text{for } C_3 \text{ and } V_m, \\ 2Q_t/\{1 + \exp[s_1(T_c - s_2)]\} \times \{1 + \exp[s_3(s_4 - T_c)]\} & \text{for } C_4 \text{ and } V_m, \\ 2Q_t/\{1 + \exp[s_5(T_c - s_6)]\} & \text{for } R_d \text{ and } V_m, \\ 2.1Q_t & \text{for } K_c \\ 1.2Q_t & \text{for } K_0 \\ 0.57Q_t & \text{for } S \end{cases}$$

where ψ_c is a critical water potential (m); ψ_s is the value of saturation potential for moisture content in the soil (m); B is an empirical parameter; s_l is the parameter of suppression of high and low temperatures; and $\psi_r = \psi_s W_2^{-B}$ is the potential of moisture content in the root later of the soil (m).

3.9.2 The dynamic photosynthesis–conductivity model

The physiological processes described by the A_c and g_c parameters do not exhibit an instant response to changes in the control derivatives. It takes about one minute to stabilize the photosynthesis process after changes in the external conditions, but for conductivity the time step constitutes several minutes. Hence, for time steps longer than one minute the photosynthesis intensity A_c can be calculated using relationship

(3.6), and a delay in changes of conductivity can be described reliably by the equation:

$$\frac{\partial g_c}{\partial t} = -k_g(g_c - g_{\inf}) \tag{3.7}$$

where k_g is a time constant for the response of vegetation cellulose ($0.00113\,\mathrm{s}^{-1}$); $g_{c_{\inf}} = g_c$ for $t \to \infty$; $g_{c_{\inf}} = m(A_c/c_S)h_S p + b$.

The solution to Equation (3.6) is written as:

$$g_c = e^{-\bar{k}_g t} g_{c_0} + (1 - e^{-\bar{k}_g t}) g_{c_{\inf}}, g_{c_0} = g_c \qquad \text{for } t = 0$$

and a change in the g_c during a time step Δt will be:

$$\Delta g_c = g_c - g_{c_0} = (1 - e^{-\bar{k}_g \Delta t})(g_{c_{\inf}} - g_{c_0})$$

These correlations make it possible to calculate the energy fluxes in the multi-layer structure of the APS system. As a result, one can specify the elements of the climate models, and assess from the space-borne monitoring data the impact of surface vegetation on the synoptic situation. It is especially important in arid and excessively humid zones where the state of vegetation cover depends strongly on small changes of temperature and other meteorological parameters. Another important conclusion from these correlations can be drawn from the study of the effect of vegetation cover on the processes of CO_2 assimilation from the atmosphere. Clearly, the use of such models increases an adequacy of the global carbon cycle modelling.

3.10 MODELLING THE CONIFEROUS FOREST PRODUCTIVITY

Coniferous forests play an important role in the global carbon cycle (Safonov *et al.*, 1999). Therefore, in this section we shall consider a parameterization of the production process in the forest ecosystem with the pine forest stand of one age proposed and developed by Kirilenko (1990) for practical use in climate models. The model reflects correlations of mean annual and seasonal processes without consideration of the succession of vegetation communities. It describes the dynamics of carbon and water in the ecosystem using a system of five ordinary differential equations, one partial differential equation, and many algebraic equations reflecting the biotic and abiotic processes in the coniferous forest.

The model includes the following values: phase derivatives, input (external) and output (internal) derivatives, as well as parameters (constants, coefficients).

The phase derivatives include (kg/m^2 of soil):

- carbon of the structural mass of young pine needles m_1 and old ones m_2, $m_L = m_1 + m_2$;
- carbon of the above ground parts of the trees (branches, boughs, and trunks) and growing roots m_3;
- carbon of thin roots m_4; and
- fund of mobile carbon assimilants m_5.

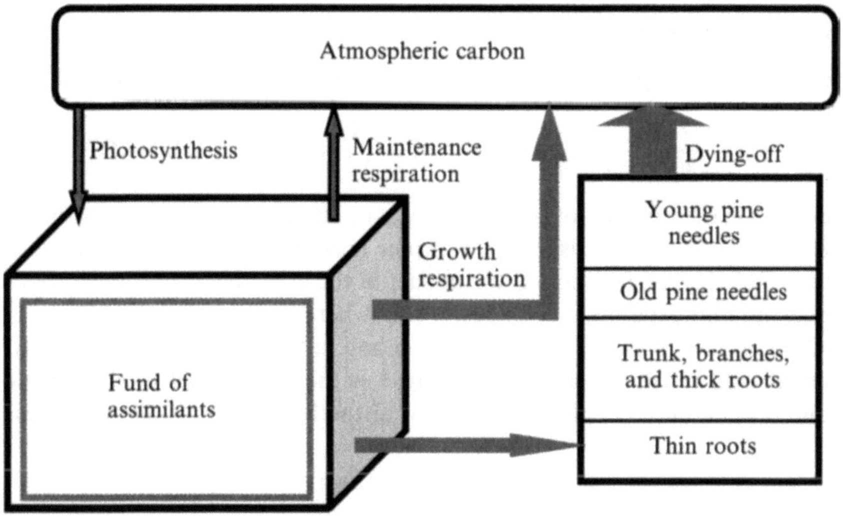

Figure 3.7. The conceptual scheme of the carbon cycle in the model of the production process in the coniferous forest.
From Kirilenko (1990).

The input variables characterize the environmental conditions:

- incoming solar radiation intensity E;
- air temperature T and humidity H;
- length of the light period of day τ;
- daily amount of precipitation P; and
- concentration of carbon dioxide in the atmosphere C_o.

The output variables are photosynthesis, average height and diameter of trees, transpiration, respiration, increment in tree organs, and other characteristics of the coniferous forest ecosystem calculated in the model.

The biomass of the phase variables in the model is measured in carbon units, and the respective scheme of carbon fluxes is shown in Figure 3.7. Notations for dynamic equations are as follows:

- s_3 (m_3) is the area of the trunk per m^2 of soil;
- θ_i (τ_B) is the function of distribution of assimilants (i.e., the share of the fund of assimilants spent on the increment of the ith organ);
- τ_B is the biological time;
- d_i is the share of the dead biomass of the ith organ ($i = 1, \ldots, 4$);
- F is a photosynthesis value;
- r_i is the respiration intensity of the growth of the ith organ; and
- p_i is the respiration intensity of the maintenance of the ith organ.

3.10.1 Photosynthesis

To parameterize the process of photosynthesis we use the Chartier formula:

$$F = \frac{\sigma(T_L)\Omega(\tilde{\omega}_S)}{2(r - r_c)} \frac{S_L}{2} \tau \left(C_0 + r\alpha E - \sqrt{(C_0 - r\alpha E)^2 + 4r_c \alpha E C_0} \right)$$

where $\sigma(T_L)$ is the function of the dependence of photosynthesis intensity on the leaf temperature T_L; $S_L = \beta m_L$ is the area of pine needles; $\tilde{\omega}_S$ is the weighted mean soil moisture; $\Omega(\tilde{\omega}_S)$ is the function of the impact of the excessively moistened root-layer of the soil on photosynthesis; $r = r_a^c + r_{st}^c + r_c$, where r_a^c and r_{st}^c are diffuse resistances for CO_2 molecules in the leaf boundary layer and in the leaf stoma, respectively, and r_c is an efficient resistance of carboxylizing; τ is the duration of the light period of day; and α is an initial angle of inclination of the light curve of photosynthesis:

$$\tilde{\omega}_S = \int_0^{A_0} \omega_S(z)\, ds(z)$$

where $s(z)$ is the function of distribution of the root system along the vertical cross-section of the soil layer and A_0 is a maximum capacity of the root layer of the soil.

For conifers, $r_a^c \ll r_{st}^c$, and along the height of the tree the value of photosynthesis per unit mass of pine needles changes negligibly, and therefore we shall consider some point, averaged over the tree height, for which the intensity of the PAR is determined by the Beer–Lambert law:

$$E = \kappa_E E_e \exp(0,5 K_e h S_L H^{-1})$$

where E_e is the intensity of the incoming solar radiation over the forest; κ_E is the share of PAR in the total solar radiation; and K_e is the coefficient of extinction.

Assuming that the temperature of pine needles coincides with that of the atmosphere ($T_L = T$), we use the following approximation for the $\sigma(T)$ function:

$$\sigma(T) = \max\left\{0, \frac{T - T_1}{T_0 - T_1} \exp\left[l^{-1} - l^{-1}\left(\frac{T - T_1}{T_0 - T_1}\right)^l\right]\right\}$$

where the parameter $l = 1,8$; T_1 is a minimum temperature at which the photosynthesis still takes place; and T_0 is a temperature optimal for photosynthesis.

The Ω function characterizes the dependence of photosynthesis on soil moisture. With a short-term excessive-moistening of soil photosynthesis does not decrease, and therefore $\Omega(\tilde{\omega}_S) = 1 \forall \tilde{\omega}_S \in [0, 1]$. In a general case we use the dependence:

$$\Omega(\tilde{\omega}_S) = \begin{cases} 1 & \tilde{\omega}_S \in (0, \omega_{HB}) \\ (1 - \tilde{\omega}_S)/(1 - \omega_{HB}) & \tilde{\omega}_S \in [\omega_{HB}, 1] \end{cases}$$

3.10.2 Distribution of assimilants

To describe the strategy of distribution of assimilants, let us introduce the function $\theta_i(\tau_B)$, where τ_B is the biological time. Let us determine the ratio of the sum of efficient temperatures T_l to the mean multiyear sum of efficient temperatures T_Σ accumulated during the vegetation period:

$$\tau_B(j) = \sum_{l=1}^{j} T_l / T_\Sigma$$

where j is the day number.

The θ_l function is determined from the expression:

$$\theta_i = a_i^\theta \exp\left[-b_i^\theta (c_i^\theta - \tau_B)^2\right] \qquad (i = 1, \dots, 4)$$

where the a_i^θ coefficient is equal to a maximum of θ_i; c_i^θ is the biological time of a maximum of assimilates' flux to the ith organ; and the b_i^θ coefficient determines the steepness of the θ_i function.

The flux of assimilants at a time moment τ_B into the ith organ of the plant is equal to $m_5\theta_i(\tau_B)$.

3.10.3 Respiration and dying-off

Assume that respiration connected with the growth of the ith organ of a tree is in proportion to the rate of its increment and is equal to $r_i dm_i/dt$, and the respiration for maintenance of leaves and sucking roots is in proportion to their mass and is equal to $p_i m_i$ $(i = 1, 2, 4)$. The p_i coefficient depends on temperature:

$$p_i = a_i^p \exp[k_i^p (T - T^p)]$$

where a_i^p is the respiration for maintenance at a temperature T^p.

The respiration for maintenance of the trunk and thick roots is in proportion to the trunk area s_3 and is equal to $s_3 p_3 \sqrt{\beta_{st}}$, where β_{st} is the share of the trunk mass in m_3. The dying-off of the organs is considered to be in proportion to their masses with the coefficients d_i $(i = 1, \dots, 4)$.

3.10.4 Water cycle

The water cycle in a tree is described by a system of equations that parameterize the water motion over the territory (see Chapter 6), in soil, in a tree, and its elements. Water is one of the limiting factors of the tree growth through a value of the stomatal resistance that affects the photosynthetic activity of plants. The water cycle in the forest territory is the sum of two basic parts – abiotic and biotic, which describe the processes of water motion in soil and its absorption by the root system, its motion in the plant, and transpiration. There are a lot of models of these processes. The difference between them is determined by different suppositions about the character of the soil structure and the location of the root system. Kirilenko (1990) proposed to describe the water motion in the soil with an equation that generalizes the Darsi classical model for the case when the soil is considered a homogeneous porous layer:

$$\frac{\partial \omega}{\partial t} = -\frac{\rho}{\rho_S}\left[\frac{K_{PB}}{\Phi_{PB}}\frac{\partial}{\partial z}\left(\mu\omega^n\frac{\partial \omega}{\partial z} + \omega^n\right) + \frac{f(z,t)}{\Phi_{PB}}\right]$$

$$V(z,t) = K_{PB}\left(\mu\omega^n\frac{\partial \omega}{\partial z} + \omega^m\right)$$

with the boundary conditions:

$$\Phi(z_g, t) = 1 \qquad V(0,t) = W(t) - L(t)$$

and the initial condition:

$$\omega(z_g, t_0) = \omega^0(z)$$

where $W(t)$ is precipitation intensity; $L(t)$ is the intensity of evaporation from the soil surface; ρ is water density; ρ_S is the volume density of soil; Φ is the relative soil moisture; Φ_{PB} is the total water capacity of soil; V is the rate of moisture motion in soil; $\mu = n\Phi_{PB}/(\rho g)$; n is a constant; g is the acceleration due to gravity; $\Phi_{PB} = \phi_S\omega^n$; ϕ_S is the water potential of soil; z_g is the root layer depth; K_{PB} is the conductivity corresponding to soil moisture Φ_{PB}; and z is the vertical coordinate. The f function describes the intensity of water absorption by the root system:

$$f(z,t) = [\xi_0\bar{\phi}\varphi(z,t) + v(t)]dS/dz$$

where $\varphi(z,t) = \varphi_S(z,t) - \tilde{\varphi}_S(t)$ is deviation of the dimensionless water potential from an average value for the root layer of the soil; $\xi_0 = \xi_k\sigma_0$ is the conductivity of the entry of water into the root system; $\xi_k = 1/r_k$ is the specific conductivity of the entry of water into the root, where r_k is the specific resistance of the entry of water to the root; and σ_0 is the total area of the sucking roots of the tree per unit soil area:

$$\varphi_S(z,t) = 1 - \left(\frac{\omega(z,t)}{\bar{\omega}}\right)^{-n} \qquad \tilde{\varphi}_S(t) = 1 + \tilde{\phi}_S/\bar{\phi}$$

where $\tilde{\varphi}_S$ is an average dimensionless water potential for the root layer of the soil. The rate of water absorption by roots from the soil equals:

$$v = \int_0^{z_g} f(z,t)\,dz$$

Water from the root system moves upward along the trunk, through the branches into the leaves and evaporates into the atmosphere. Let us assume that an amount of the transpired moisture is equal to that assimilated by the root system, and the rate of motion is equal to v. The integral effect of the impact of the coniferous forest on water cycle in its territory is the sum of the effects of each tree. The practical application of the developed model needs information about spatial heterogeneities and distributions of the characteristics of the forest stand. An application of the model to deciduous and mixed forest systems requires additional units which would describe the processes of the transformation of water potentials as well as the carbon cycle in leaves. Intensive studies are being carried out in this direction (Papakyriakou and McCaughey, 1991; Esteve, 1998; Perituneṇ, 1998; Ferm and Hultberg, 1999; Kolbe, 1999; Peng, 1999).

3.11 SUCCESSION IN THE TUNDRA–TAIGA SYSTEM

In connection with the growing anthropogenic processes on the Earth and their propagation towards the northern latitudes, some problems arise connected with the prediction of the location of the forests' northern boundary. Connections between the observed propagation of this boundary and forest fires, reindeer grazing, anthropogenic transformations, and climate change have been widely discussed in scientific literature. Various hypotheses have been expressed about correlations between the natural–technogenic processes in the zone of this boundary. For instance, there is an opinion that forest fires can influence the northward propagation of the birch forests. This can lead to considerable changes in the native habitats and the reindeer propagation. In the late 20th century some disagreements were observed between forestry and deer breeding. In Finland the areas of reindeer propagation have moved 400–500 km south from the boundary of the forests (i.e., up to the boreal zone). In other words, a change of the forest boundary is followed by a change of the soil layer, and this, due to shifts in biogeocenotic cycles, can direct the succession of the northern boundary of forests in an unpredicted direction. Moreover, this prediction, based on the pre-history of the development of this boundary in natural conditions, cannot be reliable. Therefore, the use of the methods of ecoinformatics with its models and technologies for monitoring is now a single adequate method for predicted estimates of the dynamics of the northern forest boundary. Here a lot of ecological problems appear among which are the following:

- combining various elements of the forest ecosystems in the zone of the northern boundary of the forest (reasons for the presence or absence of some type of trees);
- the role of various diseases of the northern forests and their sensitivity to climate changes and other external factors;
- the role of fires and their location and power in the change in the structure and configuration of the transfer boundary between forests and tundra;
- the role of insects with an account of the dependence of their propagation and biomass on climate change in the dynamics of the northern forests boundary;
- patterns of seed scattering for various plants in the zone of the forests northern boundary and their effect on its dynamics; and
- the impact of the dynamics of the forests northern boundary on wild animal migration and habitat.

An expected, climate warming can result in the substitution of some vegetation ecosystems for others (exogenic succession). It will be especially manifested in high latitudes where taiga transforms into tundra. Bogatyrev (1988) proposed a model,

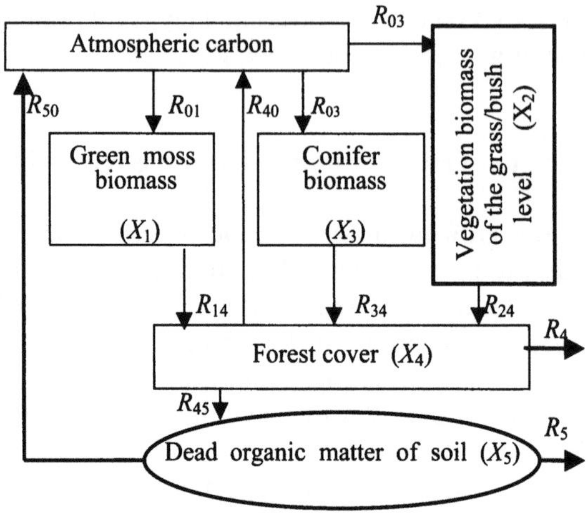

Figure 3.8. The scheme of the carbon cycle in the tundra–taiga system.
From Bogatyrev (1988).

which parameterizes the transformation processes in the tundra–taiga system and makes it possible to reveal some important regularities. The phase variables are the biomass of moss (X_1), vegetation of the grass/bush level (X_2), conifers (X_3), forest cover (X_4), and the dead organic matter of the soil (X_5). The phase variables are measured in tonnes of carbon per unit area (t/ha), and their spatial distribution is assumed to be uniform. The carbon flux scheme is given in Figure 3.7.

The system of balanced equations of the model of the transfer processes on the taiga–tundra border according to the scheme in Figure 3.8 is:

$$dX_i/dt = R_{0i} - R_{i4} \qquad (i = 1, 2, 3)$$
$$dX_4/dt = \sum_{i=1}^{3} R_{i4} - R_{40} - R_{45} - R_4$$
$$dX_5/dt = R_{45} - R_{50} - R_5$$

where the R_{ij} functions depend on many parameters, and their determination is the principal stage of the model formulation. Bogatyrev (1988) proposed the following formulas for these functions:

$$R_{0i} = F_i\varphi_i(T_M)G_i(X_3)f_i(X_5)H_i(W) \quad R_{i4} = k_iX_i \qquad (i = 1, 2)$$
$$R_{03} = F_3\varphi_3(T_M)G_3(X_3)f_3(X_5)H_3(W)\chi(X_3)$$
$$R_4 = m_4X_4$$
$$R_5 = m_5X_5$$
$$R_{34} = k_3X_3[1 + \kappa(T_M - T_A)]^{-1}$$
$$R_{40} = \sigma k_4X_4\varphi_4(T_M)$$
$$R_{50} = k_5X_5\varphi_5(T_M)$$
$$R_{45} = (1 - \sigma)k_4X_4\varphi_4(T_M)$$

where the F_i coefficients express the dependence of the biomass increment on its type (the characteristic indicators of vegetation); the φ_i functions assign the laws of the changes in the biomass increment as a function of atmospheric temperature beneath the canopy T_M; and the G_i, f_i and H_i functions describe deviations of the annual biomass increment from a maximum level at a given temperature T_M due to a deficit of illumination, an amount of mineral resource in the soil, and moisture, respectively. For trees, the annual production also depends on their biomass, and this dependence is represented with the χ function. For vegetation of the grass/bush level and moss, their annual increment is supposed to be independent of their biomass.

Determine the functions φ_i, G_i, f_i, H_i, and χ:

$$\varphi_i(T_M) = 1 + (\delta_i/100)(T_M - T_Y) \qquad (i = 1, 2)$$

$$\varphi_3(T_M) = 1 + (\delta_3/100)(T_M - T_A)$$

$$G_3(X_3) = \exp(-\alpha_i X_3) \qquad (i = 1, 2, 3)$$

$$f_1(X_5) \equiv 1$$

$$f_i(X_5) = \exp(-a_i/X_5) \qquad (i = 2, 3)$$

$$H_i(W) = \exp\left\{-b_i\left(W - \frac{W_A - W_Y}{\sigma_i}\right)^2\right\} \qquad (i = 1, 2, 3)$$

$$\chi(X_3) = \left(\frac{X_3}{X_{3A}}\right)^\gamma$$

where T_Y and T_M are atmospheric temperatures in the tundra and taiga, respectively; $\delta_i\,(i = 1, 2, 3)$ are indicators of variability of the annual biomass increment with an increasing temperature beneath the canopy by 1°C; α_i, a_i, b_i, κ, and $\gamma \in (0, 1)$ are constant coefficients; W_A and W_Y represent the soil moisture in taiga and tundra, respectively; $k_i\,(i = 1, 2, 3)$ are proportions coefficients; and $\sigma \in (0, 1)$ is the share of the forest cover transforming into dead organic matter as a result of decomposition.

To parameterize the atmospheric temperature beneath the tree canopy, suppose that it cannot be below the environmental temperature T, and increases with the mass of trees up to some level $X_3 = 0.5\beta_2/\beta_1$, and then decreases:

$$T_M(X_3, T) = T \cdot \begin{cases} f(X_3) & 1 \le f(X_3) \le T_A/T \\ T_A/T & f(X_3) > T_A/T \\ 1 & f(X_3) < 1 \end{cases}$$

where:

$$f(X_3) = 1 - \beta_1 X_3^2 + \beta_2 X_3$$

where β_1 and β_2 are positive coefficients.

The model described can be used as an independent unit of the global model, providing a transformation of the spatial structure in Figure 3.4 and specifying the local elements of the model of the water balance on the territory described in Krapivin and Kondratyev (2002). For real application of the model, its parameters should be estimated with a definitive attachment to geographical latitude.

3.12 GLOBAL MODEL OF THE CARBON CYCLE

Let us assume that each site of land 4° by latitude and 5° by longitude belongs to one of the types of the soil–vegetation formations enumerated in Table 2.5, or without vegetation. On the ith site of land of the area S_i of the soil–vegetation formation is characterized by an amount of carbon per unit area in the biomass of living plants B_i and in the dead organic matter of the soil D_i. Following Svirezhev *et al.* (1985), let us write a system of balanced equations:

$$\left.\begin{aligned} \partial B_i/\partial t &= F(T_i, P_i)[1 + (\delta/10)(C_a/C_a^0)] - K_i B_i \\ \partial D_i/\partial t &= K_i B_i - E(T_i, P_i) D_i \\ \partial C_a/\partial t &= \sum_{i=1}^{N} \{[F(T_i, P_i)(1 + (\delta/10)(C_a/C_a^0 - 1)) - E(T_i, P_i) D_i] S_i\} + V \end{aligned}\right\} \quad (3.8)$$

where $F(T_i, P_i)$ is the annual production of plants per unit area (see Table 3.12); $E(T_i, P_i)$ is the specific rate of decomposition of dead organic matter of soil per unit area; T_i and P_i are the mean annual atmospheric temperature and the annual precipitation amount over the ith site of land, respectively; δ is the indicator of the annual production increase (%) with a 10% increase of CO_2 content in the atmosphere; K_i is the specific intensity of the biomass dying-off of plants on the ith site of land; $C_a^0 = C_a(t_0)$, where t_0 is an initial time moment (in the global models it is usually assumed that t_0 represents the year 1850); and V is an anthropogenic input of carbon into the atmosphere.

Suppose that in the absence of anthropogenic impacts ($V = 0$) the amount of carbon in the APS system is constant, and in the early industrial period the system was in a state of equilibrium. Then at the moment t_0 the following relationships are valid:

$$K_i = B_t^0/F(T_i^0, P_i^0) \qquad E(T_i^0, P_i^0) = F(T_i^0, P_i^0)/D_i^0(T_i^0, P_i^0) \qquad (i = \overline{1, N})$$

As a result, the model makes it possible to calculate the dynamics of carbon in three reservoirs with their spatial distribution taken into account. Depending on the details of the types of vegetation covers (see Tables 1.12, 2.5, and 2.6), calculations of the concentration of carbon in the vegetation cover, soil, and atmosphere from Equation (3.8) requires a knowledge of the functions F and E, as well as the δ coefficient.

4

Global environmental change and the World Ocean

4.1 THE PRESENT-DAY WORLD OCEAN

To increase the reliability of assessing the role of the World Ocean in the global carbon cycle, a more detailed description is needed of the production processes in the ocean ecosystems. Along with the physical and chemical processes of transformation and motion of carbon in the ocean medium, the biological processes play an important role. In particular, phytoplankton, like the nutrient elements, assimilates dissolved CO_2 from the salt water. As a result, an organic substance is formed that partially goes into the nutrient chains of the trophic pyramids of the World Ocean and partially descends to bottom sediments. A totality of all processes of carbon motion in the ocean medium creates a gradient of CO_2 concentration between surface and deep waters (Klubov *et al.*, 2000; Raupach *et al.*, 2001; Rau and Caldeira, 2002). Therefore, a study of the structure and functioning of the ocean ecosystems becomes one of the most important and rapidly developing directions of marine biology. Its various aspects are being developed in many countries within the framework of the International Biological Programme. In particular, the international programme JGOFS (Joint Global Ocean Flux Study) is dedicated to a study of biochemical processes in the ocean to obtain a deeper knowledge of the response of the ocean to external forcings. One of the goals of the studies is to obtain a possibility to predict the system's behaviour as a result of changes of some of its parameters. However, due to the unique nature and broad spatial extent of the World Ocean, it is difficult to quantitatively estimate all the elements of the system at different moments of development and in different regions of the World Ocean and, moreover, to assess the effect of their change on the functioning of the system on the whole. With this aim in view, regular expeditions are organized, databases of satellite monitoring of the World Ocean basins are accumulated, and methods are developed for using the model approach for these purposes. Of course, an accuracy of assessing the role of water ecosystems in global changes of greenhouse gas (GHG) cycles in general depends much on a thorough and accurate description of photosynthesis as a function of the environmental parameters and especially with consideration of its seasonal variations (Seibel and Walsh, 2001; Thebault and Rabouille, 2003).

The ocean, covering 71% of the planet's surface, provides a substantial amount of food products consumed by humans, with the remaining part being obtained from cultivated land. At the same time, the total amount of organic matter produced in the ocean is approximately equal to that produced by land vegetation. By rough estimates, the total biomass of nekton constitutes 53×10^9 t. The catch of fish and other species from the World Ocean is estimated at 70×10^6 t/yr, which constitutes 20% of the protein consumed by humans. The catch of traditional species is close to its limit ($\sim 90 \div 100 \times 10^6$ t/yr). However, supplies of krill and other species are not utilized.

This disagreement between the role of the land and ocean ecosystems in food production is explained, first of all, by the fact that on land an intensive cultural economy is realized, which is weakly developed in the seas and oceans. Therefore, possible ways of increasing the ocean bioproductivity have not yet been applied. First, humankind uses the final trophic chains of natural communities of the World Ocean – fish and whales. Each successive trophic level obtains about one-tenth the share of energy accumulated at the previous level. On land, two levels of organisms (vegetation and herbivorous animals) are used, but in the ocean and in the seas this increases up to 5 levels. The direct use of non-fish species will make it possible to sharply increase the amount of proteins obtained from the ocean.

Second, a question arises about the transition from free fishing to cultural methods of economy in the World Ocean (i.e., the question of an artificial increase of productivity of the biological communities of the ocean). To do this, it is necessary, first of all, to study the methods of controlling the output of the final produce in the biological systems of the World Ocean. To determine the rational ways of affecting the ocean communities, it is necessary to study their structure and functioning, to understand the production processes, and transformations of matter and energy flux at different trophic levels of the ocean ecosystems. It is necessary to develop a theory of control in the biological systems of the coastal waters and the open ocean, which differ both in natural hydrophysical and biogeochemical parameters as well as in the extent of anthropogenic impacts.

Marine communities are complicated biological systems of populations of individual species. As a result of their interaction, communities are in dynamic development. Their spatial structure is mostly determined by the composition of numerous biotic and abiotic factors, which depend on a totality of the oceanic parameters. The latter are determined by the laws of general circulation of the ocean waters, including tides and ebbs, zones of convergence and divergence, wind and thermohaline currents, etc.

In the late 20th century an urgent problem arose of predicting the dynamics of the ocean systems in conditions of increasing anthropogenic impacts (chemical poisoning, mechanical liquidation of living organisms, and environmental changes) as well as assessing their role in the dynamics of the whole biosphere. So, in recent studies of the climatic impact of GHGs it has been shown that the role of the World Ocean in this process has been underestimated. In particular, data has been given in Kondratyev and Johannessen (1993) on the role of the Arctic basins of the World Ocean in the formation of the global CO_2 cycle, from which it follows that previous assessments of this role were incorrect. This is connected with the fact that an account of biological and gravitational processes playing the combined role of a CO_2 pump from the atmosphere to deep layers of the ocean, was inadequate in the earlier models of the global biogeochemical carbon cycle. Therefore, specified models of the working regime for this pump with climatic zones taken into account may be of importance for predicting estimates of the greenhouse effect (Choudhury and Kukla, 1979; Gorshkov, 1990; Doskey and Gaffney, 1992; Anderson *et al.*, 1998; Bates *et al.*, 1998; Bousquet *et al.*, 2000; Caldeira and Duffy, 2000; Krapivin, 2000b).

The impact of the ocean ecosystems on the biogeochemical cycles is manifested through the atmosphere–water border and is usually parameterized based on observational data. However, in this impact of great importance is the vertical structure of the processes taking place in the ocean medium. The nature of these processes depends significantly on external factors. For instance, according to Legendre and Legendre (1998), in the Arctic zones of the World Ocean the patchy structure of the springtime blossoming of phytoplankton is determined by winter conditions of ice formation and subsequent ice melting. In other zones these external factors are pollution of the atmosphere and ocean surface, changes in phytoplankton living conditions, and functioning of the carbonate system.

Phytoplankton is at one of the initial stages of the trophic hierarchy of the ocean system. As field observations have shown, the World Ocean has a patchy structure formed by a combination of non-uniform spatial distributions of insolation, temperature, salinity, concentration of nutrient elements, hydrodynamic characteristics, etc. The vertical structure of the phytoplankton distribution is less diverse and possesses rather universal properties. These properties are manifested through the existence of 1–4 vertical maxima of phytoplankton biomass.

Variability of the patchy topology and vertical structure is connected with seasonal cycles and has been studied experimentally in many climatic zones of the World Ocean. The typical qualitative and quantitative indicators of this variability have been found. The combined distributions of abiotic, hydrological, and biotic components of the ocean ecosystems have been studied.

The complexity and mutual dependence of all the processes in the ocean substantially hinder uncovering the laws of formation of phytoplankton spots and establishing correlations between various factors regulating the trophic relationship intensity in the ocean ecosystems. For instance, many studies revealed a close relationship between primary production and amount of phytoplankton. At the same time, this relationship suffers a breakdown depending on a combination of the synoptic situation and insolation. It turns out that the extent of the breakdown depends significantly on a combination of the groups of phytoplankton (Legendre and Legendre, 1998).

An analysis of the accumulated observation data on assessments of the produce of seas and oceans and an attempt by many authors to uncover the laws of produce formation characteristics for various water basins have revealed various laws in local relationships between productivity and environmental parameters.

An efficient method of studying the vertical structure of the ocean ecosystems is numerical modelling based on measurements of their characteristics (Kuck *et al.*, 2000). To derive the model, it is necessary to know the structure of trophic relationships in the ecosystems, specific features of hydrological conditions, and information about other characteristics of the environment. Experience in modelling has raised the possibility of an efficient prediction of the dynamics of the World Ocean communities. An example of such a model is a 3-D model of the ecosystem of the Peruvian Current (Krapivin, 1996) and the Okhotsk Sea (Aota *et al.*, 1993). In all these models the unit of parameterization of the vertical structure of the ecosystem is central.

4.2 INTERACTIONS BETWEEN THE ATMOSPHERE AND THE OCEAN

Knowledge of exchange processes between the atmosphere and the ocean is very important in assessing the greenhouse properties of CO_2. In most of the known models of the global CO_2 cycle, the relationship between partial pressure of CO_2 dissolved in water p_c and general concentration C_U of inorganic carbon in the surface waters is calculated based on the buffer coefficient:

$$\xi = (p_c - p_{c,o})C_{U,o}(C_U - C_{U,o})^{-1}/p_{c,o}$$

where the "o" index refers to the pre-industrial period. However, numerous observations show that the use of only this coefficient when calculating the H_2^C and H_3^C fluxes gives rough estimates, without the effect of many additional factors being taken into account.

One of the first attempts to consider the spatial heterogeneity of the World Ocean and to simulate the impact of temperature gradients on the CO_2 exchange between the upper layer of the ocean and the atmosphere was made by Bjorkstrom (1979). The idea of dividing the World Ocean basins into two parts, corresponding to warm and cold waters, was later developed by numerous authors (Nefedova and Tarko, 1993). Pervaniuk (2001) developed this scheme dividing the World Ocean surface into 211 homogeneous basins. This discretization was based on combining basins $4^\circ \times 5^\circ$ in size. The adjacent cells of this grid, exhibiting similar directions for the vertical components of water flow velocities, are combined into one water basin. The average annual field of water flow velocities was considered at a depth of 75 m. The size of the water basins were determined with the dependence of CO_2 fluxes on the climatic factors taken into account. A maximum of latitudinal size of the basin was assumed to be 8°. Above 80°N the World Ocean is presented as one site.

The most complete study of the physical mechanisms of the CO_2 transport for different conditions at the water–air interface was carried out by Alexeev *et al.* (1990, 1992). Here for the first time, detailed measurements of CO_2 fluxes have been made, and their dependence on the parameters of the state of the atmosphere–ocean interface has been analysed, with an account of wind-driven mixing, an appearance of foam on the water surface, breaking waves, and pollution. Parametric descriptions of the process of the ocean–atmosphere gas exchange have been proposed for each of the types of water–air interface conditions (McGillis *et al.*, 2001). For instance, it was shown that with the sea roughness and foam layer increasing, the intensity of gas exchange grows substantially. With the foam layer 5 cm thick the rate of gas exchange exceeds the gas exchange across a free surface by a factor of 2.4. This fact is important for the characteristic of the gas exchange in the tidal zones and strong storm regions, where the foam stretches can be several centimetres thick. Also of interest is the result connected with the availability of surface active substances (SAS) in the upper layer of the ocean. With the SAS concentration about $7.8 \times 10^{-4}\%$ (by volume), all other things being equal, the rate of gas exchange slows down to 60%. However, in the zones of foam formation this effect decreases considerably.

On the whole, over the World Ocean basins the estimates of the CO_2 flux between the atmosphere and the upper layer of the ocean vary between 16 and 1250 mol/m^2/yr. This variability means that the global model of the CO_2 biogeochemical cycle should accurately consider every feature mentioned.

In a stationary state the hydrosphere and atmosphere are in a certain equilibrium with respect to CO_2 exchange, being broken with fluctuations of temperature, ocean surface level, vertical circulation regime, etc. The amount of CO_2 assimilated and emitted in the process of exchange between the ocean and the atmosphere constitutes 55.6×10^9 t C/yr. Algae assimilates 16.7×10^9 t C/yr from the atmosphere. These values are non-uniformly distributed over the water surface and vary strongly over time. A characteristic place in CO_2 exchange is occupied by the Arctic Ocean, whose low temperatures determine a high absolute content of CO_2 dissolved in the surface layer in any season. On average, in the summer, the Arctic waters mainly assimilate CO_2 and emit O_2 to the atmosphere, and in winter, on the contrary, they emit CO_2 and assimilate O_2. This clearly expressed seasonal change is smoothed down in the basins of the North Atlantic seas where a decrease of the ice cover, increase in the period of photosynthetic activity, and the autumn–winter vertical convection lead to an intensification of the gas exchange with the atmosphere towards a prevailing of the H_3^C flux. Of the amount of CO_2 assimilated from the atmosphere, carbon constitutes 0.18×10^9 t C/yr. Compared to total assimilation of CO_2 by the World Ocean estimated at $5.2–6.6 \times 10^9$ t C/yr, the contribution of the Arctic water basins to the removal of excess carbonic acid from the atmosphere is small (Table 4.1). However, this assessment is rather underestimated, considering the results obtained by Kelley (1987). The average estimates show that the partial pressure of CO_2 in the atmosphere exceeds that in the Arctic seas by 110 ppm. The CO_2 deficit in the marine medium happens mainly in the period of the springtime blossoming and is estimated at $\sim$450 g C/m^2. This means that the CO_2 flux from the atmosphere into the marine medium can vary from 1.5 to 4.0 g C/m^2/day. This assessment changes considerably as a function of longitude. For instance, the Norwegian Sea has a deficit of the CO_2 partial pressure between 20 and 50 ppm. For the Bering Sea, $\Delta p_{CO_2} \cong 70$ ppm. Between the CO_2 content in the seawater and

Table 4.1. Annual budget of CO_2 exchange with the atmosphere for water bodies of the Arctic basin and northern seas (10^6 t C/year).

Region of the ocean	Summer			Winter			
	Assimilation	Emission	Balance	Assimilation	Emission	Balance	Year
Arctic basin	4.4	0.0	+4.4	–	–	–	+4.4
Arctic seas							
East Siberian Sea	1.0	2.1	−1.1	0.0	0.7	−0.7	−1.8
Kara Sea	11.4	2.4	+9.0	0.0	0.2	−0.2	+8.8
Laptev Sea	2.4	3.6	−1.2	0.0	2.9	−2.9	−4.1
Chuckchee Sea	11.2	0.0	+11.2	0.0	0.6	−0.6	+10.6
Total	*26.0*	*8.1*	*+17.9*	*0.0*	*4.4*	*−4.4*	*+13.5*
North Atlantic seas							
Barents Sea	71.9	0.0	+71.9	0.0	0.0	0.0	+71.9
Greenland Sea	19.1	0.0	+19.1	8.7	0.0	+8.7	+27.8
Norwegian Sea	64.8	0.0	+64.8	0.0	6.0	+6.0	+5.8
Total	*155.8*	*0.0*	*+155.8*	*8.7*	*6.0*	*+2.7*	*+158.5*
Arctic Ocean	***186.2***	***8.1***	***+178.1***	***8.7***	***10.4***	***−1.7***	***+176.4***

its temperature a clear linear correlation is observed with the proportion coefficient 10 ppm CO_2 per °C.

All this suggests the conclusion that the Arctic waters pump away carbon dioxide from the atmosphere. Therefore, a detailed study of this process is needed to specify the global model of the carbon cycle and to raise the accuracy of estimation of the greenhouse effect due to emissions of anthropogenic CO_2.

On the whole, in the synthesis of the global model of the CO_2 biogeochemical cycle, the unit that simulates the oceanographic part of this cycle must describe the functioning of the ocean carbonate system. Alexeev *et al.* (1990), analysing the CO_2–HCO_3–CO_3^{2-} system and the distribution of pH values of the ocean waters, found that more than 80% of the dissolved carbon dioxide is in the form of hydrocarbonate ions of HCO_3. This means that in the synthesis of the model of the ocean carbonate system only the first stage of dissociation of carbonic acid can be reliably considered. As a result, the H_3^C flux of CO_2 dissolved in the upper layer of the ocean can be calculated with the formula:

$$H_3^C = \psi(T_L)\sqrt{p_a}/(1 + 0.5C_S)$$

where $\psi(T_L)$ is the function of the effect of temperature on CO_2 solubility; and C_S is the water salinity. An expression for the ψ function has been studied inadequately and there are only a few experimental estimates (Alexeev *et al.*, 1992). Of course, a lot of simplified approaches to the determination of the direction of CO_2 fluxes on the atmosphere–ocean border prevail here. In a simplistic case, the fluxes H_2^C and

H_3^C can be considered as mutually exclusive. That is, a valve is supposed to exist on the atmosphere–ocean border, the switching of which is determined by the pH parameter. A critical level of pH is about 8.11. At pH ≤ 8.11 the ocean assimilates CO_2, and at pH > 8.11 the ocean emits CO_2.

On global scales, in the World Ocean regions with cold waters (northern latitudes, upwelling zones) the flux H_3^C prevails, and in the warm waters the flux H_2^C prevails. The dynamics of this regime is maintained by reactions in the hydrosphere between CO_2 and water resulting in formation of carbonic acid:

$$HCO_3 + H_2O \Leftrightarrow H_2CO_3 + OH^-$$

$$CO^{2-}{}_3 + 2H_2O \Leftrightarrow H_2CO_3 + 2OH^-$$

In salt water, there is dissolved carbon dioxide, non-dissociated molecules of H_2CO_3, anions of HCO_3, and CO_3^{2-}. They all are in equilibrium:

$$CO_2 + H_2O \Leftrightarrow H_2CO_3 \Leftrightarrow HCO_3 + H^+ \Leftrightarrow CO^{2-}{}_3 + 2H^+$$

An equilibrium state between carbonic acid bicarbonate and carbonate ions is rapidly established. The solution of CO_2 and formation of H_2CO_3 takes place much more slowly. However, since many authors dispute the availability of H_2CO_3, the summarized concentration of CO_2 and H_2CO_3 should be considered, denoting it as $[CO_2]$. Let us introduce notations to characterize the total concentration of intermediate components:

$$\sum C = [CO_2] + [H_2CO_3] + [HCO_3^-] + [CO_3^{2-}] \tag{4.1}$$

The next characteristic of the hydrosphere is alkalinity:

$$A = [HCO_3^-] + 2[CO_3^{2-}] + [H_2CO_3] + [OH^-] - [H^+] \tag{4.2}$$

where $[Ca^{2+}]$ and $[H^+]$ are concentrations of ions of calcium and hydrogen, and $[HCO_3^-]$ and $[CO_3^{2-}]$ are concentrations of bicarbonate and carbonate ions.

Using these notations, let us formulate the equilibrium conditions:

$$K_0 p_c = [CO_2] \tag{4.3}$$

$$[HCO_3^-] = K_1[CO_2]/[H^+] \tag{4.4}$$

$$[CO_3^{2-}] = K_2[HCO_3^-]/[H^+] \tag{4.5}$$

$$[Ca^{2+}] = L_p[H^+]^2/(K_1K_2[CO_2]) \tag{4.6}$$

where K_0 is the indicator of CO_2 solubility (250 mmol/l atm at 30°C and 640 mmol/l atm at °C); L_p is the solubility indicator for $CaCO_3$; and K_1 and K_2 are first and second apparent coefficients of dissociation of carbonic acid (dependent on temperature and pressure).

From Equations (4.1)–(4.6) we obtain:

$$\sum C = [CO_2](1 + K_1/[H^+] + K_1K_2/[H^+]^2)$$

$$A = [CO_3^{2-}](K_1/[H^+] + 2K_1K_2/[H^+]^2)$$

Eriksson (1963), using a differentiation operator δ, calculated relative changes of $[CO_2]$, $[Ca^{2+}]$, p_c, and A. Let us denote any of these components as U and taking $\Delta U = \delta U/U$, we obtain:

$$\Delta p_c = \Delta[CO_2] - \Delta K_0 \tag{4.7}$$

$$\Delta[Ca^{2+}] = \Delta L_p + 2\Delta[H^+] - \Delta K_1 - \Delta K_2 - \Delta[CO_2] \tag{4.8}$$

$$\Delta\Sigma C = \Delta[CO_2] + F_1\Delta K_1/F_0 + a_2\Delta K_2/F_0 - F_2\Delta[H^+]/F_0 \tag{4.9}$$

$$\Delta A = \Delta[CO_2] + \Delta K_1 + 2a_2\Delta K_2/F_2 - F_3\Delta[H^+]/F_2 \tag{4.10}$$

where $F_0 = 1 + a_1 + a_2$; $F_1 = a_1 + a_2$; $F_2 = a_1 + 2a_2$; $F_3 = a_1 + 4a_2$; $a_1 = K_1/[H^+]$, and $a_2 = K_1K_2/[H^+]^2$.

Based on Equations (4.7) through to (4.10), and considering the experimental data, Eriksson (1963) showed that with a 1°C increase of the temperature of the upper layer of the marine medium the partial pressure of CO_2 in the atmosphere can grow by 4.2–5.8%, and with a 1% decrease of the water volume the partial pressure of CO_2 in the atmosphere grows by 3%, with 1% of carbon dioxide in the water medium precipitating as $CaCO_3$. With variations of the equilibrium state in deep layers of the hydrosphere a 1% output of CO_2 in the form of gas is followed by a 1% sediment of CO_2 in the form of $CaCO_3$. Moreover, variations in the CO_2 partial pressure correlates with changes in the concentration of phosphorus P, so that a decrease of P in deep layers by 1% leads to an increase of p_c by 5.6%.

Eriksson (1963) found out that a 1% increase of alkalinity causes a 2.26% decrease of the CO_2 partial pressure in the atmosphere and a 98% reduction of total CO_2 supply in the hydrosphere. In conditions lacking external input of Ca into the hydrosphere, a 1% increase of alkalinity causes an increase of the rate of $CaCO_3$ deposition (i.e., the total alkalinity decreases by 0.92%). Thus a 1% increase of the water alkalinity is equivalent to an increase in pH by 0.5.

An equilibrium between various components of the hydrospheric carbonate system depends on temperature and pressure, a combination of which correlates with pH so that at a given temperature and pressure the equilibrium is a function of only $\mathrm{pH} = -\lg[H^+]$. The effect of temperature on pH in the first approximation can be described by the Merrey law (Ivanov, 1978): $\Delta\mathrm{pH} = -0.0111\Delta T$, valid at pH $\in [7.5, 8.4]$, $T \in [1, 30]$ °C, and salinity from 10 to 40‰. The dependence of pH on pressure p_c follows the Buch–Grippenberg law: $\mathrm{pH} = d\Delta p_c$, where, on average, $d = -0.0254$. A more accurate presentation of this law is given in Table 4.2.

The connection of the equilibrium condition by CO_2 exchange with pH on the atmosphere–ocean border is such that with the CO_2 pressure in the atmosphere at

Table 4.2. Empirical dependence of pH on atmospheric pressure.

pH at atmospheric pressure	7.5	7.7	7.9	8.1	8.3
pH with pressure increasing by 1000 dbar	−0.035	−0.028	−0.023	−0.021	−0.02

330×10^{-6} atm, an equilibrium occurs at 20°C for pH = 8.16 and at 0°C for pH = 8.11. With a lower pH value the ocean will assimilate CO_2, and with a higher pH it will emit CO_2. Hence, to describe the functions of the fluxes H_2^C and H_3^C on the atmosphere–ocean border, the structure of the ocean carbonate system should be thoroughly studied. A simplified description of these fluxes is usually based on the comparison of the partial pressures of CO_2 in the atmosphere and in the ocean. According to the data in Alexeev *et al.* (1992), the fluxes H_2^C and H_3^C are approximated by the function $H_i^C = k_i(p_a - p_c)^{1/2}$, where p_a and p_c are partial pressures of CO_2 in the atmosphere and in the ocean, respectively. The partial pressure of CO_2 in the atmosphere at the level of the ocean can be calculated with the formula:

$$p_a = 0.421542 \times 10^{-18} M_C(273.15 + T)$$

where M_C is the CO_2 mass in tonnes and T is the air temperature in °C.

According to Bjorkstrom (1979), the functional dependence of p_c on the parameters of the ocean carbonate system can be presented in the form $p_c = [CO_2]/K_0$. From the condition of chemical equilibrium, according to Equations (4.4) and (4.5), it follows:

$$[CO_2] = [H^+]\sum C/a$$

where $a = [H^+]^2 + [H^+]K_1 + K_1K_2$; and $\sum C = C_U/W_U$, where W_U is the volume of an elementary reservoir. Denote:

$$K_W = [H^+][OH^-], K_B = [H^+][B(OH)_4^-]/[B(OH)_3]$$
$$B_T = [B(OH)_3] + [B(OH)_4^-]$$

Then we obtain from Equation (4.2):

$$A = K_W[H^+]^{-1} + K_B B_T(K_B + [H^+])^{-1} + C_U([H^+]K_1 + 2K_1K_2)/(aW_U) - [H^+]$$

or

$$A = [CO_2]\left(\frac{K_1}{[H^+]} + \frac{2K_1K_2}{[H^+]^2}\right) + \frac{B_T}{1 + [H^+]/K_B} + \frac{K_W}{[H^+]} - [H^+] \tag{4.11}$$

The solution of this equation relative to $[H^+]$ enables one to determine p_c as a function of C_U. From the estimates by Bjorkstrom (1979), $K_B = 2 \times 10^{-9}$ and $K_W = 10^{-14}$.

The right-hand part of Equations (4.2) and (4.11) imply an account of all weak acids in the salt water. Ions of other compounds are taken into account through the dependence of the equilibrium constants on salinity or chlorine content. The characteristic parameter of equilibrium of the carbonate system is a variable:

$$X = \frac{(K_1K_2)^{1/2}}{[H^+]} = \left[\frac{[CO_3^{2-}]}{[CJ_2]}\right]^{1/2}$$

In the terms of this parameter we have:

$$\sum C = (1 + KX + X^2)[CO_2] \tag{4.12}$$

$$A = (KX + 2X^2)[CO_2] + W(X) \tag{4.13}$$

where $K = \sqrt{K_1/K_2}$; and $W(X) = B_T/\{1 + [K_1K_2]^{1/2}X^{-1}\}/K_B + K_W X[K_1K_2]^{-1/2} - [K_1K_2]^{1/2}X^{-1}$.

Excluding $[CO_2]$ from Equations (4.12) and (4.13):

$$(2C - A')X^2 - K(A' - C)X - A' = 0 \tag{4.14}$$

where E$A' = A - W(X)$.

Since $W(X) \ll A$, to solve Equation (4.14) the following iterative procedure can be used beginning with an estimation of $X = X_1$ ($\cong 1$):

$$A'_i = A - W(X_i) \qquad X_i = 0.5[-b + (b^2 - 4ac)^{1/2}]/a \qquad (i \geq 2)$$

where $a = 2C - A'_{i-1}$ and $b = K(A'_{i-1} - C)$.

The exchange processes taking place on the atmosphere–ocean border have been experimentally studied by Kiseleva (1990), Zaitsev (1988), and others. It has been shown in their works that at high wind speeds the rate of gas exchange sharply increases. This is connected with the mechanism of foam formation on wave crests as well as an intensive catching of air bubbles and their penetration beneath the water surface. The dependence of the amount of spray Q on the height over the water surface is well approximated by a linear function. For instance, for the wind speed $V = 11.1$ m/s this approximation is (Kiseleva, 1990):

$$Q = \begin{cases} Q' \exp\{-14.5z + 0.684\} & \text{without wave breaking} \\ Q' \exp\{-7.48z + 0.842\} & \text{with wave breaking} \end{cases}$$

where $Q'[\mathrm{cm}^{-2}\, s^{-1}]$ is the dimensional multiplier.

Experimental estimates of the net gas transport due to the droplet mechanism have shown that the CO_2 flux between water and air can vary from 1.44 mg CO_2/m^2/hr in the absence of wave breaking to 26.6 mg CO_2/m^2/hr. The rate of the gas exchange at the atmosphere–ocean border is markedly affected by the foam formation on the ocean surface. In this case the rate of gas exchange can increase up to 28%. All this suggests the conclusion that a detailed modelling of the mechanisms of the atmosphere–water CO_2 transport is needed. Monitoring the World Ocean by satellite allows one to identify the zones of foam formation and, hence, along with the modelling of the processes of foam formation, to perform direct measurements on the areas of foam-covered basins.

The mechanism of air bubble behaviour in the upper layer of the ocean has been poorly studied. The available theoretical results are based, as a rule, on a number of suppositions, which in many cases can drastically distort the ideas about real processes of the gas exchange between the atmosphere and the ocean. Among these suppositions the following have been used most frequently:

- air bubbles in the upper layer of the ocean almost instantly acquire the temperature of the environment, since the molecular coefficient of heat conductivity exceeds that of gases by 2–3 orders of magnitude;
- air bubbles in water do not affect its dynamics; and
- air bubbles do not merge in water.

The size of air bubbles in water varies constantly because of their gradual solution. The gas flux through the bubbles is described by the relationship:

$$Q = 4\pi RD|\Delta p|S \cdot Nu,$$

where $|\Delta p|$ is the difference between the partial pressures of gas in the water and in the atmosphere; R is the bubble's radius;

$$Nu = \begin{cases} 1 + 0.5Pe + 0.5Pe^2 \ln(2Pe) & \text{for } R \leq 20\,\mu\text{m} \\ \sqrt{2/3\pi}Pe^{1/3} & \text{for } 20\,\mu\text{m} < R \leq 200\,\mu\text{m} \\ 0.45Pe^{1/3}\,Re^{1/3} & \text{for } 200\,\mu\text{m} < R \leq 400\,\mu\text{m} \end{cases}$$

Re is the Reynolds number ($Re = 2vR/D$); and Pe is the Pekle number ($Pe = vR/\nu$), where ν is water viscosity, v is the velocity of the bubble's motion, and D is the coefficient of the gas molecular diffusion.

Of course, the ocean surface condition substantially affects its gas exchange with the atmosphere. The size of the basins covered with either foam or white caps depends directly on a combination of the parameters such as wind speed, water temperature, and sea currents. An analysis of the statistical characteristics of the patchy pattern of the ocean surface made by many experts makes it possible to describe the percentage distribution of the areas covered with foam (S_f) and white caps (S_l) with the following binary functions of wind speed V (at a height of 10 m):

$$S_f = \begin{cases} 0 & \text{for } V < 5\,\text{m/s} \\ 0.65\{+4.76 \times 10^{-2}(V - 5^2)\} & \text{for } V \geq 5\,\text{m/s} \end{cases}$$

$$S_l = \begin{cases} 0 & \text{for } V < 5\,\text{m/s} \\ 0.015\{1 + 2.2 \times 10^{-2}(V - 5^3)\} & \text{for } V \geq 5\,\text{m/s} \end{cases}$$

According to Kiseleva (1990), the relationship between S_f and S_l at $V \geq 5\,\text{m/s}$ obeys the following rule:

$$S_f/S_l = 50 - 3.4(V - 5)$$

It is clear that the structure of the atmosphere–ocean border can affect the gas exchange within widely varying values of fluxes H_2^C and H_3^C (see Figure 2.8). Unfortunately, the level of knowledge about the laws of changes of the atmosphere–ocean border structure as a function of synoptic situations does not permit a clear estimation of the limits of detailing the processes taking place on this border, to obtain the values of the fluxes H_2^C and H_3^C close to real ones. It means that at this stage of the synthesis of the global model of the CO_2 cycle some uncertainty still exists.

4.3 THE ZONAL MODEL OF THE GLOBAL CARBON CYCLE IN THE ATMOSPHERE–OCEAN SYSTEM

Consider one of the versions of parameterization of carbon dioxide global dynamics in the atmosphere–ocean system proposed and studied by Nefedova (1994). The spatial heterogeneity of the World Ocean is approximated by the zonal scheme in accordance with the latitudinal dependence of climatic processes as well as processes of mixing in the atmosphere and in the oceans. There are 14 latitudinal zones, 10° in size. In the vertical, there are three layers in the World Ocean: the upper quasi-homogeneous layer (UQL) with a time-dependent thickness, the thermocline, and the deep ocean. As a result, the World Ocean is divided into 42 volume parts. An upwelling is assumed to exist between 40°N and 40°S along with a downwelling at high latitudes. In the UQL the water flows poleward from the equator, and in deep layers the water flows in the opposite direction (Figure 2.5). Over each water basin, the atmosphere is simulated by a point model. The carbon exchange between the zones of the atmosphere takes place due to advection H_i^a and turbulent diffusion H_i^d:

$$H_i = H_i^a + H_i^d (i = \overline{1.14})$$

$$H_i^a = 2\pi R_0 h_a V_i^* (C_i^a - C_{i+1}^a) \cos \phi_i$$

$$H_i^d = [2\pi R_0 A_h h_a / \Delta\phi_i](C_i^a - C_{i+1}^a) \cos \phi_i$$

where $C_i^a = M_i^a / V_i^a$ is the concentration of carbon in the ith zone of the atmosphere; M_i^a is the carbon mass in the ith zone of the atmosphere; V_i^a is the volume of the ith zone of the atmosphere; R_0 is the Earth's radius; ϕ_i is the latitude of the southern boundary of the ith zone; h_a is the altitude of the atmosphere (10 km); V_i^* is the average velocity of the meridian transport of air masses in the atmosphere (0.2–1.0 m/s); and A_h is a coefficient ($10^5\,m^2/s$).

The CO_2 exchange at the atmosphere–ocean border is described by the traditional law:

$$H_i^{a0} = k(u_i)(P_i^o - P_i^o)$$

where

$$P_i^a = k_a M_i^a R T_i^a S_i^{-1} \mu^{-1}$$

$k(u_i)$ is the proportion coefficient depending on wind speed; P_i^o and P_i^o are partial pressures of CO_2 in the ith zone of the atmosphere and the ocean, respectively; k_a is the share of the 100-m air column mass in the mass of a 10-km column (≈ 0.01602); R is the universal gas constant (8.31451 Joule/mol/K); T_i^a is the air temperature at the level of the ocean in the ith zone ($T_i^a = T_i^{as} + \Delta T_i^a$); T_i^{as} is the seasonal temperature component; and ΔT_i^a is the average annual change of air temperature caused by the increased CO_2 content in the atmosphere.

A sea surface temperature (SST) change is assumed to take place in phase with a change of air temperature and by the same value:

$$T_i^o = T_i^{os} + \Delta T_i^a$$

The partial pressure of CO_2 dissolved in the surface waters is proportional to its concentration in the water and inversely proportional to its solubility. This dependence is established by solving the system of Equations (4.12) and (4.13) that describe the functioning of the ocean carbonate system. For the quantitative solution of this system one can use, for instance, the methods of secants. As a result, we obtain $[CO_2]$ and P_i^o. Based on the data on the temperature dependence of the equilibrium constants for the respective chemical reactions, we find:

$$P_i^o = P_i^o(C_i^1, T_i^o) \qquad (i = \overline{1.14})$$

The turbulent fluxes of carbon at the lower boundary of the UQL (H_{UQL}) and the upper boundary of thermocline (H_T) are described by the functions:

$$H_{\text{UQL}} = \begin{cases} C_i^1 - C_i^2)(V_i + dh_i/dt) & \text{for } dh_i/dt + V_i > 0 \\ 0 & \text{for } dh_i/dt + V_i \leq 0 \end{cases}$$

$$H_T = \begin{cases} (C_i^1 - C_i^2)(V_i + dh_i/dt) & \text{for } dh_i/dt + V_i \leq 0 \\ 0 & \text{for } dh_i/dt + V_i > 0 \end{cases}$$

where C_i^1 and C_i^2 are concentrations of inorganic carbon in the UQL and thermocline, respectively; h_i is the UQL depth of the ith zone of the ocean (depends on the season); and V_i is the water lifting (in upwelling) or lowering (in downwelling) (0.001–0.1 cm/s).

The turbulent fluxes of carbon at the thermocline–deep ocean border is considered to be proportional to the coefficient k_T of the difference of carbon concentrations in the bordering layers:

$$H_{TF} = k_T(C_i^3 - C_i^2) \qquad (i = \overline{1.14})$$

where C_i^3 is the concentration of inorganic carbon in the deep ocean at the ith zone.

Production and circulation of organic carbon in the ocean have been studied in detail and parameterized by many authors (Krapivin, 1996; Nitu *et al.*, 2000 a,b; Krapivin and Kondratyev, 2002). In the model by Nefedova and Tarko (1993) the following approximations have been assumed:

$$B_i^p = B_i^o\{1 + (\delta_i - 1)/(\delta_i + 1)\sin(wt - \pi/2)\} \qquad \text{for } \varphi \in [60°\text{N}, 90°\text{N}] \cup [60°\text{S}, 90°\text{S}]$$
$$(i = 1.14)$$

$$B_i^p = B_i^o\{1 + (\delta_i - 1)/(\delta_i + 1)\sin(2wt - \pi/2)\} \quad \text{for } \varphi \in [10°\text{N}, 60°\text{N}] \cup [10°\text{S}, 60°\text{S}]$$
$$(i = 2 \div 6.9 \div 13)$$

$$B_i^p = B_i^o \qquad \text{for } \varphi \in [0°, 10°\text{N}] \cup [0°, 10°\text{S}], \qquad (1 = 7.8)$$

where B_i^o is the average annual production of phytoplankton ($150–550 \times 10^9$ t/yr); and δ_i is the ratio of maximum and minimum values of the rate of organic matter production.

Organic matter produced in the UQL decomposes and descends to deeper layers. Let B_i^{1d}, B_i^{2d}, and B_i^{3d} be the rates of the organic matter decomposition in the UQL, thermocline, and deep ocean, respectively; γ – the ratio of the average annual amount of organic matter decomposing in the UQL to the average annual production of phytoplankton; τ – the time lag for organic matter decomposition; D – an average depth of the ocean (3653–3795 m); and H – the depth of the thermocline bottom (30–200 m). Then:

$$B_i^{1d} = \gamma B_i^p(t - \tau) \qquad (i = \overline{1.14})$$

$$B_i^{2d} = (1 - \gamma)(H - h_i)/(D - h_i)\langle B_i^{1p}\rangle$$

$$B_i^{3d} = (1 - \gamma)(D - H)/(D - h_i)\langle B_i^{1p}\rangle$$

Here the angular brackets $\langle\ \rangle$ denote the calculation of the average value for the annual cycle.

So, the seasonal model of the global carbon cycle developed in the works of Nefedova and Tarko (1993) and Nefedova (1994) is simulated by a system of 56 ordinary differential equations with periodic coefficients ($i = 1, \ldots, 14$):

$$\frac{d}{dt}C_i^1 h_i - \left(\frac{dh_i}{dt} + V_i\right)C_i^1 = H_i^{ao} - H_{\text{UPL}} + L_i^1 - B_i^p + B_i^{1d}$$

$$\frac{d}{dt}C_i^2(H - h_i) + \left(\frac{dh_i}{dt} + V_i\right)C_i^2 = H_T + H_{TF} + B_i^{2d}$$

$$\frac{d}{dt}C_i^3(D - H) + V_i C_i^3 = L_i^3 - H_{TF} + B_i^{3d}$$

$$\frac{dM_i^a}{dt} = \gamma_i^a + V_i^a(H_i - H_{i-1}) - 0.012 S_i H_i^{ao}$$

where γ_i^a are the anthropogenic carbon emissions in the ith zone of the atmosphere.

4.4 MODELLING OF THE CARBON CYCLE IN THE WORLD OCEAN

Along with the physico-chemical processes of transformation of carbon compounds in the hydrosphere mentioned above, the general circulation in the World Ocean plays an important role in their long-range transport. According to the available estimates, about 80% (~1.7 Gt) of organic matter formed in the process of photosynthesis descends to deeper layers and oxidizes giving CO_2. This process is balanced by a slow lifting of water, and thereby forming a stable CO_2 exchange. However, in local situations there are strong deviations from such stable conditions, which can only be described by the scheme of the spatial–temporal structure of the ocean waters motion. The block-scheme of this structure identifies the surface, intermediate, deep, and bottom waters as well as the currents topography on every horizon. In the scientific literature the simulation models of World Ocean circulation vary widely. Therefore, the development of this part of the global model unit is not difficult.

Following Nitu *et al.* (2000b), with depth z the ocean can be divided into four basic layers: photic to well-heated depths ($\Omega_U = U[z_i,\ z_{i+1}],\ z_0 = _0$;

$i = 0.1, \dots, m-1$); intermediate water ($\Omega_P = U[z_i, z_{i+1}]$, $i = m, \dots, m+n-1$); deep water ($\Omega_L = U[z_i, z_{i+1}]$, $i = m+n, \dots, m+n+l-1$); and bottom water Ω_F. From the hydro-physico-ecological characteristics the layer Ω_U, as a function of latitude φ, longitude λ, and season t, can be attributed to warm or cold waters, the layer Ω_P is photic but always with low water temperatures, and in the Ω_L layer phytoplankton is not produced. Finally, the layer Ω_F plays the role of a boundary layer.

On average, the layer Q_U has an area of $360 \times 10^{12}\,\mathrm{m}^2$ and depth $z_m \cong 75\,\mathrm{m}$, with warm waters covering an area of $240 \times 10^{12}\,\mathrm{m}^2$. The layer Ω_P is located at depths from z_m to $z_{m+n} \cong 200\,\mathrm{m}$. In concrete model realizations given in Krapivin and Kondratyev (2002) the following assumptions have been made: $m = 10$, $n = 2$, $l = 8$.

The vertical CO_2 transport in the ocean is determined by advective fluxes $H^C_{19,ij}$ and gravitational sedimentation of dead organic matter ($H^C_{20,ij}$). An advective transport from the ith into the j-th reservoir of the ocean is considered proportional to the concentration of carbon in the respective reservoirs: $H^C_{19,ij} = \lambda_{2,ij} C_{a,i} (a = U, P, L)$, where $\lambda_{2,ij} = V_{ij}/V_i$; V_{ij} is the water volume transported per unit time from the ith reservoir into the jth reservoir; and V_i is the volume of the ith reservoir.

The following algorithm is widely used for a parameterization of the process of carbon sedimentation. The flux under unit area of the ocean is supposed to decrease exponentially with depth. If we denote the inflow of organic matter in the ith reservoir as g_i and the net outflow of organic matter from the water surface as H_{20T}, we obtain:

$$H^C_{20,1} = H_{20,T} \qquad H^C_{20,i} = g_{i-1}(\sigma_i/\sigma_{i-1}) \exp[(-z_i - z_{i-1})/D_s] \qquad (i = 2, \dots, m+n+l)$$

where σ_i is the area of the ith reservoir; and D_s is the characteristic time of the organic matter particles sedimentation before their decomposition. The rate of decomposition in each reservoir is equal to:

$$R_{D,i} = H^C_{20,i} - H^C_{20,i+1} \qquad (i = 1, \dots, m+n+l); \qquad R_{D,F} = H^C_{20,m+n+l} - H^C_{16}$$

However, if the time of the transition of the organic matter particles from one layer to another is short compared to D_s, then it is better to take $H^C_{20,i} = \lambda_1 C_{a,i}$ and $H^C_{16,i} = \lambda_4 C_{F,i}$. In addition to these fluxes one should take into account the fluxes of detritus decomposition, solution of bottom sediments, and carbon consumption during the process of photosynthesis:

$$H^C_{17,i} = \text{const.} \qquad H^C_{18,i} = \lambda_3 D_{L,i} \qquad H^C_{22,i} = \lambda_3 D_{U,i} \qquad H^C_{21,i} = C_{31} R_{\Phi,j}$$

The estimates for the modelled parameters of the particular oceanic processes of the carbon cycle vary greatly. For instance, from the data of various authors the estimates of assimilation of carbon from the hydrosphere during the process of photosynthesis ranges from 10 to 155 Gt C/yr. The value 127.8 Gt C/yr is the most widely used value. However, because of large dispersions of these estimates the calculation of the C_{31} coefficient is fraught with great uncertainty and, therefore, to be specified, it requires numerical experiments with the use of other, more accurate data.

Finally, suppose that the surface layers of the ocean are filled with carbon due to its runoff from the land $H_{24,i}^C = C_7 W_{sio}$, $H_{23,i}^C = C_8 W_{Hio}$, where W_{sio} and W_{Hio} are the river and underground runoffs into the World Ocean, respectively.

Considering the notations in Figure 2.8, the balanced equations used to describe the global carbon cycle are written as:

$$\frac{\partial C_A}{\partial t} = H_5^C - V_\varphi \frac{\partial C_A}{\partial \varphi} - V_\lambda \frac{\partial C_A}{\partial \lambda} + \begin{cases} -H_2^C + H_3^C & (\varphi, \lambda) \in \Omega_O \\ H_1^C + H_4^C - H_6^C + \sum_{i=7}^{11} H_i^C & (\varphi, \lambda) \in \Omega \backslash \Omega_O \end{cases}$$

$$\frac{\partial C_{S1}}{\partial t} = H_6^C - H_7^C - H_8^C - H_{14}^C - H_{15}^C$$

$$\frac{\partial C_{S2}}{\partial t} = \sum_{i=1}^{4} H_{i+11}^C - H_4^C - H_9^C - H_{23}^C - H_{24}^C$$

$$\frac{\partial C_U}{\partial t} + v_\varphi^U \frac{\partial C_U}{\partial \varphi} + v_\lambda^U \frac{\partial C_U}{\partial \lambda} = Q_U + H_{22}^C + H_{25}^C - H_{21,U}^C - H_3^C + H_2^C + H_{19,U}^C - H_{20,U}^C$$

$$\frac{\partial C_P}{\partial t} + v_\varphi^P \frac{\partial C_P}{\partial \varphi} + v_\lambda^P \frac{\partial C_P}{\partial \lambda} = H_{19,P}^C - H_{19,U}^C + H_{20,U}^C - H_{20,P}^C - H_{21,P}^C + H_{22}^C + H_{25}^C$$

$$\frac{\partial C_L}{\partial t} + v_\varphi^L \frac{\partial C_L}{\partial \varphi} + v_\lambda^L \frac{\partial C_L}{\partial \lambda} = H_{18,L}^C + H_{19,L}^C - H_{19,P}^C + H_{20,P}^C - H_{20,L}^C$$

$$\frac{\partial C_F}{\partial t} + v_\varphi^F \frac{\partial C_F}{\partial \varphi} + v_\lambda^F \frac{\partial C_F}{\partial \lambda} = H_{17}^C - H_{16}^C + H_{18,F}^C - H_{19,L}^C + H_{20,L}^C$$

The flux Q_U is formed from H_{23}^C and H_{24}^C. Let $Q_U = 0$ for the pelagic regions Ω_{OP} of the World Ocean. Let us describe the formation of Q_U on the shelf by a simple algorithm with the supposed uniform distribution of the sink from the Kth region to the Mth water basin:

$$Q_U = \begin{cases} 0 & (\varphi, \lambda) \in \Omega_{OP} \\ (H_{23}^C + H_{24}^C)\sigma_{OP}/\sigma_L & (\varphi, \lambda) \in \Omega_O \backslash \Omega_{OP} \end{cases}$$

where σ_{OP} and σ_L are the areas of the water basins Ω_{OP} and Ω_L, respectively.

4.5 ORGANIC CARBON AND OCEANIC ECOSYSTEMS

4.5.1 Models of the oceanic ecosystems

Each element of the ocean ecosystem A can be described by some number of parameters, and the connection between the elements can be presented as the connection between the respective parameters. Then, on the whole, the ecosystem A can be described by N time-dependent parameters $x^*(t) = \{x_j(t),\ j = 1, \ldots, N\}$. The structure $|A(t)|$ and behaviour $\overline{A}(t)$ of the ecosystem A which can be observed in more or less detail, are the functions of these parameters. Therefore, the ecosystem itself $A(t) = \{|A(t)|, \overline{A}(t)\}$, as a collection of both the structure and behaviour, is the function of these parameters:

$$A(t) = F(x^*(t)) \tag{4.15}$$

Hence, according to Equation (4.15) with a change of time t the ecosystem A will be characterized by the trajectory in the $(N+1)$-dimensional Eucledian space. The following abstract formation will be considered a model of the ecosystem $A(t)$:

$$A_M(t) = F_M(x^{*,M}) \tag{4.16}$$

that depends on $M \leq N$ components of the vector $x^*(t)$ $(\{x^{*,M}(t)\} \in \{x^*(t)\})$ and considers all the connections existing between them.

The nearer M is to N and the more completely the connections between the vector $x^{*,M}(t)$ components are taken into account, the less is the disagreement of the trajectories of the ecosystem $A(t)$ and its model $A_M(t)$. The latter can be measured by any natural measure, for instance, by a maximum of absolute difference of all the respective coordinates of trajectories or by the integral of absolute difference of all respective coordinates for a final time period. In other words, let us introduce some goal:

$$V = Q(\{x_i(t)\}) \tag{4.17}$$

functional on the trajectory of the ecosystem $A(t)$. The form of the Q function is determined by the requirements made by the system A to the environment. The natural evolutionary process is considered to lead to an optimal system and therefore the system $A_{\text{opt},M}(t)$, which ensures an extremum of the functional (4.17), can be considered an optimal model of the ecosystem A.

The value of disagreement between the trajectories of the ecosystem A and an optimal model $A_{\text{opt},M}(t)$ is affected by the degree of correspondence of the goal functional V chosen by the relationship (4.17) to a real goal $\underline{A}$ of the ecosystem A. The set $\underline{G} = \{\underline{A}_1, \ldots, \underline{A}_r, \ldots, \underline{A}_m\}$ of possible reliable goals $\{\underline{A}_r\}$ of the ecosystem can be derived on the basis of the experience accumulated by oceanographers. Then, determining $A_{\text{opt},M,r} = g_r(\underline{A}_r)$, $\underline{A}_r \in \underline{G}$, we obtain a limited assembly of possible optimal systems $A_{\text{opt},M,r}(r = 1, \ldots, m)$, whose trajectories together with the trajectory of the ecosystem A, are within the space of possible trajectories. If we find $A_{\text{opt},M,r0}$, whose trajectory has a minimum disagreement with $A(t)$, then, based on this, we can find the most reliable goal of the ecosystem:

$$\underline{A}_{r0} = (g_r)^{-1}(A_{\text{opt},M,r0}(t))$$

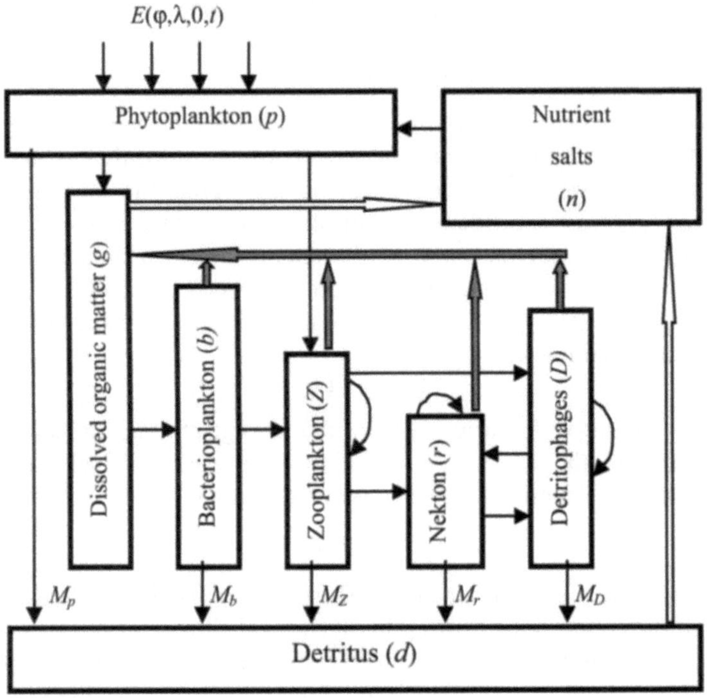

Figure 4.1. The standard block-scheme of the model of the ocean ecosystem. The arrows show the energy flux directions.

According to the principles mentioned above, the derivation of a numerical model of the ocean ecosystem A (Figure 4.1) requires either a detailed description of its states or derivation of an adequate complex of numerical models of energy exchange between the trophic levels, taking place in A, as well as interactions of the biotic, abiotic, and hydrophysical factors. Of course, in this case an availability of a certain set of hypothesis is assumed concerning the character of the balanced relationships in the ecosystem A.

The basic hypothesis is that in the ecosystem A the only original source of energy and matter for all forms of life is solar radiation energy (E). According to numerous theoretical and experimental studies, the sunlight penetration into deep layers of the ocean follows an exponential law:

$$E = uE_0 \exp\left[-\int_0^z \{\delta p(\varphi,\lambda,x,t) + \beta d(\varphi,\lambda,x,t) + \nu Z(\varphi,\lambda,x,t)\}dx - \alpha z \right] + (1-u)E_0 \exp(-\zeta z) \tag{4.18}$$

where $E_0 = E(\varphi,\lambda,0,t)$ is the illumination of the ocean surface; α is the coefficient of light absorption by the filtered seawater; δ, β, and ν are the coefficients of light attenuation due to phytoplankton (p), detritus (d), and zooplankton (Z), respectively; and u and ζ are the parameters chosen in a concrete situation to bring

$E(\varphi, \lambda, z, t)$ closer to a real pattern of illumination changing with depth. Note that here the impact of the biomass of other trophic levels on water transparency is considered to be negligibly small.

Illumination affects the rate of photosynthesis R_p. The R_p parameter as a function of E has a maximum at some optimal value of E_{max}, which drifts from this critical value when the illumination increases or decreases. A maximum of R_p at various latitudes φ is located at depths varying as a function of season (i.e., sun elevation). Thus in the tropical zones this variability with depth is most pronounced. On average, the photosynthesis maximum is located at depths of 10–30 m, and in the open water bodies it can be observed at depths below 30 m. Here $E_{max} = 65$–$85\,\text{cal/cm}^2/\text{day}$. Beginning from the depths where $E = 20$–$25\,\text{cal/cm}^2/\text{day}$, photosynthesis decreases in proportion to E. An apparent suppression of phytoplankton by light is observed at $E > 100\,\text{cal/cm}^2/\text{day}$. These estimates are quite different in northern latitudes, where the photosynthesis maximum is located, as a rule, on the surface.

The rate of photosynthesis at a depth z depends on the water temperature T_W, the concentration of biogenic elements n, and the phytoplankton biomass p, as well as on other factors, which are not considered here. To express this dependence, various equations are used, which in either form reflect the limiting role of the elements E, n, and p. Considering that $\partial p/p\partial z \to 0$ at $n \to 0$ and $\partial p/p\partial z \to$ const. with increasing n, let us take the following function as the basic one to describe the photosynthesis intensity at a depth z:

$$R_p(\varphi, \lambda, z, t) = k_0(T_W)K_T f_2(p) f_3(n) \tag{4.19}$$

where

$$\begin{aligned} &K_T = Af_1(E) \qquad A = kA_{max}/E_{max} \qquad f_1(E) = E\exp[m(1 - E/E_{max})] \\ &f_2(p) = [1 - \exp\{-\gamma_1 p\}] \qquad f_3(n) = [1 - \exp\{-\gamma_2 n\}]^{\theta} \end{aligned} \tag{4.20}$$

where k is the proportion coefficient; $k_0(T_W)$ is the function characterizing the dependence of the rate of photosynthesis on the water temperature T_W; A_{max} is an assimilation number in the region of maximum photosynthesis (an increment per unit weight of the phytoplankton organisms); and γ_1, γ_2, θ, and m are the constants, the choice of which can determine the species characteristics of the phytoplankton elements. For A_{max} the following estimates are valid: $A_{max} = 5.94E_{max}$ in the region of the photosynthesis minimum and $2.69E_{max}$ for other regions. According to these estimates, an assimilation number of the tropical phytoplankton in the region of maximum photosynthesis averages 11–12 mg C/hr. Thus for the Peru upwelling $A_{max} = 6.25$ mg C/hr. The light saturation of photosynthesis in the equatorial regions is reached at $9\,\text{cal/cm}^2/\text{day}$.

As for the dependence of $k_0(T_W)$, the specific intensity of the phytoplankton photosynthesis first increases with the temperature, changing from low to high values, reaching in some interval of temperatures a maximum value, optimal for p, and then, with the temperature further increasing photosynthesis begins to decrease. Near the maximum the following approximation is often used:

$$k_0(T_W) = \exp\{(T_W - T_{W,opt})\ln(\theta_0)\} \qquad 0 < \theta_0 \leq 2$$

The dependence of the rate of photosynthesis on the concentration of nutrient elements $n(\varphi, \lambda, z, t)$ (phosphorus, silicon, nitrogen, and others), expressed in Equation (4.19) by the exponential term, is of course, more complicated. The nutrient elements are one of the most important parts of the ecosystem, since they regulate the energy flux in the ecosystem. The nutrient elements supplies are spent in the process of photosynthesis at a rate R_n, usually approximated by the expression $R_n = \delta R_p$, where δ is the proportion coefficient. The nutrient element supplies are replenished due to upwelling of deep waters, where they are formed as a result of chemical processes of the dead organic matter decomposition. This process is controlled by several abiotic conditions characteristic of various climatic zones of the World Ocean. The vertical flux of nutrient elements is determined by conditions of water mixing. In the tropical zones, where the vertical structure of the water has a clear 3-layer configuration with a layer where the temperature leaps suddenly (the layer of thermocline), the vertical motion of nutrient elements is confined to this layer. In the water bodies where the thermocline is located at depths of 40–100 m, the upper layer is usually poor in nutrient elements, and their input to this layer takes place only in the zones of upwelling. In this case the average rate of the vertical water lifting beneath the thermocline varies between 10^{-3} to 10^{-2} cm/s, and in the zones of upwelling (breaking through the thermocline) it can reach 0.1 cm/s.

4.5.2 Equations of the World Ocean ecosystems dynamics

The entire volume of ocean water is considered as a single biocenosis, in which the flux of organic matter produced at the surface layers and then descending to maximum depths of the ocean, is the main connecting factor. All the model parameters are assumed to be able to change as functions of place and time, and their parametric description is made by average characteristics (i.e., deterministic models).

Suppose the food bonds between the trophic levels are described by the Ivlev model (i.e., the consumption of various kinds of food by the ith trophic level is proportional to their efficient biomasses). Taking into account the diagram of food bonds assumed in Figure 4.1 and the structure of the trophic pyramids of the typical ocean ecosystem, consider each trophic level in detail.

Bacterioplankton b plays an important role in the trophic chains of the ocean. From available estimates, not less than 30% of the bacterioplankton mass are in natural aggregates exceeding 3–5 μm in size, therefore acting as food for filtrates. This fact must be taken into account when deriving a model of the ecosystem, since in many regions of the World Ocean the bacteria production is comparable with the production of phytoplankton (Kolber *et al.*, 2001). Bacteria, taking a special place in the trophic pyramid, differ by variable exchange, strongly decreasing with the deficit of food, followed by a respective decrease in the rate of their growth. The food for bacteria is mainly detritus d and dissolved organic matter g emitted by phytoplankton (Hansell and Carlson, 2002). As a result, the ration for bacteria can be described by the Ivlev formula:

$$R_b = k_b b[1 - \exp(-k_{1,d} d - k_{1,g} g)] \tag{4.21}$$

where k_b, $k_{1,d}$, and $k_{1,g}$ are coefficients determined experimentally.

The equation describing the dynamics of the bacterioplankton biomass is written in the form:

$$\partial b/\partial t + V_{\varphi}\partial b/\partial\varphi + V_{\lambda}\partial b/\partial\lambda + V_z\partial b/\partial z = R_b - T_b - M_b - \sum_{s\in T_b} C_{bs}R_s + k_{2,\varphi}\partial^2 b/\partial\varphi^2 + k_{2,\lambda}\partial^2 b/\partial\lambda^2 + k_{2,z}\partial^2 b/\partial z^2 \quad (4.22)$$

where T_b and M_b are losses of the bacterioplankton biomass due to energy exchange with the environment and dying-off; Γ_b is a set of the trophic subordination of bacterioplankton (in a typical case Γ_b is one element Z); and $C_{b,s}$ is the share of bacterioplankton in the food ration of the sth element of the ecosystem. The parameters T_b and M_b are described by the following relationships:

$$T_b = t_b b \quad (4.23)$$

$$M_b = \max\{0, \mu_b(b - \underline{B}_b)^{\xi}\} \quad (4.24)$$

where t_b is the specific expenditure on the exchange with the environment; μ_b is the rate of bacteria dying-off; and B_b and ξ are the constants that determine the dependence of the intensity of the bacteria dying-off on their concentration. The coefficient $k_2 = (k_{2,\varphi}, k_{2,\lambda}, k_{2,z})$ determines the process of turbulent mixing of the ocean waters.

The dynamic equation for the phytoplankton biomass is:

$$\partial p/\partial t + V_{\varphi}\partial p/\partial\varphi + V_{\lambda}\partial_p/\partial\lambda + V_z\partial p/\partial z = R_p - T_p - M_p - \sum_{s\in T_p} C_{ps}R_s + k_{2,\varphi}\partial^2 p/\partial\varphi^2 + k_{2,\lambda}\partial^2 p/\partial\lambda^2 + k_{2,z}\partial^2 p/\partial z^2 \quad (4.25)$$

where Γ_p is the set of the trophic subordination of phytoplankton; C_{ps} is the share of the phytoplankton biomass in the food ration of the elements of the sth trophic level of the ecosystem; T_p is the expenditure of the energy exchange with the environment; and M_p is the dying-off of phytoplankton cells. The latter parameters are determined by the following relationships:

$$M_p = \max\{0, \mu_p(p - \underline{p})^{\theta}\} \quad (4.26)$$

$$T_p = t_p p \quad (4.27)$$

where t_p is the specific expenditure of the phytoplankton cells; on respiration μ_p is the coefficient of phytoplankton dying-off; and $\underline{p}$ and θ are the coefficients characterizing the dependence of the rate of phytoplankton cells dying-off on their concentration.

Zooplankton is an important trophic element in the ocean ecosystem. In the scheme of Figure 4.1 it is presented by an integral level Z, which implies the presence of a large number of sublevels with intersecting trophic bonds. Zooplankton consumes phytoplankton and bacterioplankton. Zooplankton itself is food for a lot of animals, in the scheme of Figure 4.1 denoted as nekton r and detritus consumers D.

Let us describe the zooplankton ration by the Ivlev formula:

$$R_Z = k_Z(1 - \exp[-\nu\overline{B}]) \tag{4.28}$$

where $\overline{B}$ is the biomass of the accessible food ($\overline{B} = \max\{0, B - B_{\min}\}$); k_Z is a maximum of the ration with an excess of food; and ν is the coefficient characterizing the level of starvation. The ration maximum is assumed to be equal to the needs for food, which in turn, are determined by the exchange intensity T_1 and a maximum possible increment P_1 at a given intensity of exchange. Both the latter parameters are related to the coefficient $q_2 = P_1/(P_1 + T_1)$, so that we obtain:

$$k_Z = T_1 u(1 - q_{2,\max})$$

where $1/u$ is the food assimilation, $q_{2,\max} = \max q_2$.

Equation (4.28) implies that with a small amount of food the ration of zooplankton grows in proportion to the amount of food, then, as the ration approaches a maximum of k_Z, its dependence on B diminishes. Since in fact one trophic level never completely consumes the other, there is a limitation in Equation (4.28) that establishes $R_Z = 0$ at $B \leq B_{\min}$, where $B_{\min}$ is a minimum of the unconsumed food biomass. In Equation (4.26) the $\underline{p}$ parameter plays the same role but in the process of the phytoplankton cell mortality.

Thus a change of the zooplankton biomass follows the law described by the following differential equation:

$$\begin{aligned} \partial Z/\partial t + V_\varphi \partial Z/\partial\varphi + V_\lambda \partial Z/\partial\lambda + V_z \partial Z/\partial z = R_Z - T_Z - M_Z \\ - H_Z \sum_{s\in\Gamma_Z} C_{Zs} R_s + k_{2,\varphi}\partial^2 Z/\partial\varphi^2 \\ + k_{2,\lambda}\partial^2 Z/\partial\lambda^2 + k_{2,z}\partial^2 Z/\partial z^2 \end{aligned} \tag{4.29}$$

where Γ_Z is a set of the trophic subordination of zooplankton; C_{Zs} is the share of the zooplankton biomass in the food ration of the sth trophic level; and H_Z, T_Z, and M_Z are the losses of the zooplankton biomass due to unconsumed food and expenditures on respiration and mortality, respectively. The latter three parameters may be described by the relationships:

$$H_Z = h_Z R_Z \qquad T_Z = t_Z Z \qquad M_Z = (\mu_Z + \mu_{Z,1} Z)Z \tag{4.30}$$

where the coefficients h_Z, t_Z, μ_Z, and $\mu_{Z,1}$ are determined empirically for a concrete species of zooplankton.

As seen from Equation (4.29), zooplankton is considered as a passive element of the ecosystem subject to physical processes of transference in space with the motion of water masses. However, zooplankton is known to migrate mainly in the vertical direction. In the given model one can use a simple mechanism to simulate the process of the vertical migration of zooplankton. For this purpose, let us divide the whole water thickness into two layers: $0 \leq z \leq z_0$ and $z_0 < z \leq H$. Suppose the zooplankton migration between these layers depends on food availability (i.e., some part of the zooplankton from the layer $[z_0, H]$ can satisfy its needs for food in the layer $[0, z_0]$) so that with $B_{\min}$ taken into account the whole vertical profile $B(\varphi, \lambda, z, t)$ is considered.

The coefficients $C_{as}(a = p, Z)$ in Equations (4.25) and (4.29) may be determined from the supposition that the consumption of various kinds of food in the sth trophic level is proportional to their efficient biomasses:

$$C_{as} = k_{sa}\overline{B}_a \left[\sum_{a \in S_s} k_{sa}\overline{B}_a \right]^{-1} \tag{4.31}$$

where $\overline{B}_a$ is the efficient biomass of the ath food; S_s is the set of trophic subordination of the sth component; and k_{sa} is the proportion coefficient that determines the significance of the sth constituent in the food ration of the ath element (an estimation of these parameters is found in Section 4.5.6).

According to the scheme in Figure 4.1, the equations describing the dynamics of the biomass of nekton, detritus consumers, detritus, dissolved organic matter, and nutrient salts will be:

$$\partial r/\partial t = R_r - H_r - T_r - M_r - \sum_{s \in \Gamma_r} C_{rs} R_s \tag{4.32}$$

$$\partial D/\partial t = R_D - H_D - T_D - M_D - \sum_{s \in \Gamma_D} C_{Ds} R_s \tag{4.33}$$

$$\begin{aligned} \partial d/\partial t + V_\varphi \partial d/\partial \varphi + V_\lambda \partial d/\partial \lambda + V_z \partial d/\partial z = {} & M_b + M_D + M_r + M_p + M_Z + H_Z \\ & + H_r + H_D - \mu_d d - C_{dD} R_D + k_{2,\varphi} \partial^2 d/\partial \varphi^2 \\ & + k_{2,\lambda} \partial^2 d/\partial \lambda^2 + k_{2,z} \partial^2 d/\partial z^2 \end{aligned} \tag{4.34}$$

$$\begin{aligned} \partial n/\partial t + V_\varphi \partial n/\partial \varphi + V_\lambda \partial n/\partial \lambda + V_z \partial n/\partial z = {} & \mu_d d - \delta R_p + k_{2,\varphi} \partial^2 n/\partial \varphi^2 \\ & + k_{2,\lambda} \partial^2 n/\partial \lambda^2 + k_{2,z} \partial^2 n/\partial z^2 \end{aligned} \tag{4.35}$$

$$\begin{aligned} \partial g/\partial t + V_\varphi \partial g/\partial \varphi + V_\lambda \partial g/\partial \lambda + V_z \partial g/\partial z = {} & T_p + T_b + T_r + T_D + T_Z - C_{gb} R_b \\ & + k_{2,\varphi} \partial^2 g/\partial \varphi^2 + k_{2,\lambda} \partial^2 g/\partial \lambda^2 \\ & + k_{2,z} \partial^2 g/\partial z^2 \end{aligned} \tag{4.36}$$

Here $H_a = (1 - h_a)R_a$ is the unassimilated food of the ath element $(a = r, D)$; $T_a = t_a a$ is the expenditure on energy exchange; $M_a = (\mu_a + \mu_{a,1} a)a$ is the mortality; ρ_g is the indicator of the rate of replenishing the supplies of nutrient elements due to decomposition of the dissolved organic matter; and δ is the coefficient of consumption of nutrient elements in the process of photosynthesis.

As follows from Equation (4.32) and (4.33), a supposition is made in the model that the elements of nekton and detritus consumers do not move in space with the motion of the water masses. These elements are assumed to migrate independent of the hydrophysical conditions of their environment. Consider two possible versions of modelling the processes of migration. The first version is connected with an addition to the right-hand parts of Equations (4.32) and (4.33) of the terms describing the turbulent mixing with the coefficients $k_2^* > k_2$. In other words, the process of migration is identified with the process of an intensified turbulent mixing (i.e., it is accidental). However, the process of fish migration manifests some expediency in the choice of the direction of their movement. According to the biological principle of adaptation, the migration of fish is subject to the principle of the complex maximization of the efficient food ration, with the environmental parameters remaining within the conditions of their habitat. Hence, the motion of fish in space with characteristic velocities ensures their location in the regions where at the time the food and other abiotic conditions (temperature, salinity, dissolved oxygen, and chemical concentration) are most favourable. This means that fish migrate in the direction of a maximum gradient of efficient food with preserved limitations of the environmental parameters.

4.5.3 The vertical structure of the ocean

To qualitatively assess the role of the World Ocean ecosystems in the global carbon cycle, it is useful to consider some aspects of the laws of formation of the vertical distributions of organic matter. It follows from Equation (4.25) that the number s of maxima in its solution, $p(\varphi, \lambda, z, t)$ at a fixed time at any point (φ, λ) is determined by the number l of changes of the sign of the derivative $\partial p/\partial z$, with $s \leq 0.5(1 + l)$. It is important to know the location of the phytoplankton biomass maxima as a function of the depth in order to understand the whole range of the biogeocenotic processes in the ocean ecosystem. Of course, the amount and the location of the p function maxima are determined by many characteristics of the water medium, among which the hydrological heterogeneity between various layers of the water space is one of the most important. The results of numerous field studies show that there can be four maxima of phytoplankton and by depth these maxima can be located from the surface down to 500 m. Of course, in most cases there are 1–2 maxima down to 50–100 m.

A theoretical analysis of Equation (4.25) is possible after its simplification:

$$dp/dz = F(z)p - \nu N(z) - w(z) \tag{4.37}$$

where F, N, and w are the rate of photosynthesis, the zooplankton density, and the rate of sedimentation, respectively, at a depth z. The concentration of nutrient elements $n(z)$ is supposed to be taken into account in the form of the function $F(z)$.

The Equation (4.37) is linear relative to p, and therefore it will not change if it is multiplied on both sides by the value:

$$\exp\left\{-\int_0^z [F(\xi) - \nu N(\xi)]\right\}$$

We have:

$$\exp\left\{-\int_0^z [F(\xi) - \nu N(\xi)]\right\}\{dp/dz - [F(z) - \nu N(z)]p\} = w(z)\exp\left\{-\int_0^z [F(\xi) - \nu N(\xi)]\right\} \tag{4.38}$$

Note that:

$$\exp\left\{-\int_0^z [F(\xi) - \nu N(\xi)]\right\}\{dp/dz - [F(z) - \nu N(z)]p\} = (d/dz)\left[p\exp\left\{-\int_0^z [F(\xi) - \nu N(\xi)]\right\}\right. \tag{4.39}$$

Substituting Equation (4.39) into (4.38) and integrating both parts with respect to z, we have upon simplification:

$$p(z)\exp\left\{-\int_0^z [F(\xi) - \nu N(\xi)]\right\} = -\int_0^z w(y)\left\{\exp\left[-\int_0^y [F(x) - \nu N(x)]dx\right]\right\}dy + C \tag{4.40}$$

Substituting $z = 0$ into Equation (4.40), we obtain $C = p(0)$. Finally, from Equation (4.40) we obtain the formula for the vertical distribution of the phytoplankton biomass:

$$p(z) = \left\{p(0) - \int_0^z w(y)\exp\left\{-\int_0^y [F(x) - \nu N(x)]dx\right\}dy\right\}\exp\left\{\int_0^z [F(\xi) - \nu N(\xi)]d\xi\right\} \tag{4.41}$$

Based on the dependence (4.41) let us investigate the character of the $p(z)$ distribution from the viewpoint of the number of maxima and their location. It is important for specification of the role of various depths in the functioning of the ocean carbonic system. Let us denote:

$$g(h) = \int_0^h N(z)\,dz \qquad \psi(h) = \int_0^h F(z)dz$$

then the expression (4.41) is rewritten as:

$$p(h) = \left\{p(0) - \int_0^h w(y)\exp[\nu g(y) - \psi(y)]dy\right\}\exp[\psi(h) - \nu g(h)] \tag{4.42}$$

Thus $p(h)$ is the functional of $g(h)$ and $\psi(h)$. The g function has a simple interpretation if $h_2 > h_1$, then the difference $g(h_2) - g(h_1)$ represents the zooplankton biomass located in the layer $h_1 \leq z \leq h_2$. The functions $g(h)$ and $\psi(h)$ are monotonically non-decreasing, since, $T(h)$ and $F(h)$ are not negative.

Before proceeding to a qualitative analysis of the function $p(h)$, let us obtain some auxiliary results. Let us introduce the following sets:

$$H^+ = \{h \geq 0 : F(h)/N(h) = \nu/F'(h)/N'(h) > \nu\}$$

$$H^- = \{h \geq 0 : F(h)/N(h) = \nu/F'(h)/N'(h) < \nu\}$$

and the function:

$$\varphi(h) = \nu g(h) - \psi(h)$$

Lemma 4.1 *The $\varphi(h)$ function is non-decreasing at $\forall h \in \Omega_1^+ \cup \Omega_2^+$ where:*

$$\Omega_1^+ = \{h \geq 0 : h \in [0, h_1^+]/h_1^+ = \inf H^+, F(0)/N(0) < \nu\}$$

$$\Omega_2^+ = \{h \geq 0 : h \in [h_i^-, h_j^+]/h_i^- \in H^-, h_j^+ \in H^+, (h_i^-, h_j^+) \cap H^- = \varnothing\}$$

Lemma 4.2 *The $\varphi(h)$ function is non-increasing at $\forall h \in \Omega_1^- \cup \Omega_2^-$ where:*

$$\Omega_1^- = \{h \geq 0 : h \in [0, h_1^-]/h_1^- = \inf H^-, F(0)/N(0) < \nu\}$$

$$\Omega_2^- = \{h \geq 0 : h \in [h_i^+, h_j^-]/h_i^+ \in H^+, h_j^- \in H^-, (h_i^+, h_j^-) \cap H^+ = \varnothing\}$$

The validity of Lemmas 4.1 and 4.2 follows from the elementary properties of the function $\varphi(h)$. In particular, we have: $\varphi(0) = 0$, $\varphi'(0) = \nu N(0) - F(0), \ldots,$ $\varphi^{(i)}(0) = \nu N^{(i-1)}(h) - F^{(i-1)}(h)$.

Note 1 The conclusion of Lemma 4.1 also remains valid at $F(0)/N(0) = \nu$, provided $(\exists s)[F^{(s)}(0)/N^{(s)}(0) < \nu,\ F^{(j)}(0)/N^{(j)}(0) = \nu, j = 0, 1, \ldots, s1]$.

Note 2 The conclusion of Lemma 4.2 also remains valid at $F(0)/N(0) = \nu$, provided $(\exists s)[F^{(s)}(0)/N^{(s)}(0) > \nu,\ F^{(j)}(0)/N^{(j)}(0) = \nu, j = 0, 1, \ldots, s1]$.

According to Equation (4.42) the function $p(h)$ is the function of $\varphi(h)$:

$$p(h) = \left\{p(0) - \int_0^h w(y) \exp[\varphi(y)]dy\right\} \exp[-\varphi(h)] \tag{4.43}$$

Based on the results of Lemma 4.1 and Lemma 4.2, from Equation (4.43) one can easily formulate the monotonic properties of the function $p(h)$. Since $\varphi(h)$ does not decrease in the set $\Omega^+ = \Omega_1^+ \cup \Omega_2^+$, the function $\exp[-\varphi(h)]$ will be non-increasing at $\forall h \in \Omega^+$. At $w(h) \geq 0$ the function $p(0) - \int w(y) \exp[\varphi(y)]dy$ is non-increasing at any values of $h \geq 0$. Therefore, $p(h)$ is the non-increasing function of $\forall h \in \Omega^+$.

In the set $\Omega^- = \Omega_1^- \cup \Omega_2^-$, $p(h)$ is the product of the non-decreasing and non-increasing functions. Hence, in Ω^-, $p(h)$ is either non-decreasing, or non-increasing or has a maximum. Since $\forall h \in \Omega^- F(h)/N(h) > \nu$, then $p(h)$ will be the non-increasing function at $\forall h \in H_0 = \{h : h \geq h_0\}$, where:

$$p(0) = \int_0^{h_0} w(y) \exp[\varphi(y)]dy$$

Denote by H_1 the multitude of values $h \geq 0$, for which the following condition is valid:

$$w(h)[F(h) - \nu N(h)]^{-1} \exp[\varphi(h)] + \int w(y) \exp[\varphi(y)] dy > p(0)$$

Then $p(h)$ will be the non-increasing function at $\forall h \in H_1 \cap \Omega^-$.

Formulate the obtained results as the following theorem.

Theorem *The function $p(h)$ is non-increasing at $\forall h \in H^+ \cup (H_0 \cap H^-) \cup (H_1 \cap H^-)$ and non-decreasing at $\forall h \in H^- \cap (E\, H_0) \cap (E\, H_1)$, where $E = \{h : h \geq 0\}$.*

In a more general case $\partial p/\partial$ can be presented in the form of the sum:

$$\partial p/\partial z = (\partial p/\partial z)_{EB} + (\partial p/\partial z)_O + (\partial p/\partial z)_\tau + (\partial p/\partial z)_M + (\partial p/\partial z)_Z + (\partial p/\partial z)_T \tag{4.44}$$

where each of the terms on the right-hand side reflects a change of the phytoplankton biomass due to changes in illumination and concentration of nutrient elements (*EB*), sedimentation (*O*), temperature changes (τ), mortality (*M*), consumption by zooplankton (*Z*), and turbulent mixing (*T*). According to Equation (4.43), it can be supposed that:

$$(\partial p/\partial z)_{EB} \approx F(t, p, z) = Af_1(E)f_2(n)[1 - 10^{-\gamma p}]$$

At $\gamma p \ll 1$ we obtain:

$$p^{-1}(\partial p/\partial z)_{EB} \approx Af_1(E)f_2(n)$$

With an increasing depth in other stationary conditions the water temperature is known to monotonically decrease, dropping sharply in the layer of the thermocline (i.e., $(\partial p/\partial z)_\tau \leq 0$). The similar term $(\partial p/\partial z)_M$ has the form of the sectionally continuous non-positive function. However, due to the physical processes of mixing a situation can appear when this monotonicity gets broken.

The values of the constituent $(\partial p/\partial z)_Z$ depend on the distribution of $N(z)$. The term $(\partial p/\partial z)_T$ is totally determined by the hydrological conditions through the water thickness, and its behaviour can change arbitrarily, apparently, in the layer above the thermocline $(\partial p/\partial z)_O \leq 0$.

Studies of all components in totality will make it possible to determine the number of changes of sign of the derivative $(\partial p/\partial z)$, which uniquely determines the number of the phytoplankton maxima. In particular, if $\partial p/\partial z$ changes sign seven times in the layer under study, and $(\partial p/\partial z)_{z=0} > 0$ and $(\partial p/\partial z)_{z\to\infty} < 0$, then the function $p(z)$ has four maxima.

Finally, assuming that zooplankton uniformly consume phytoplankton through the depth of the layer considered, and neglecting in Equation (4.43) all constituents apart from the first one, based on Equation (4.41) we obtain the following expression for the specific rate of the change of phytoplankton biomass:

$$p^{-1}(\partial p/\partial t) \approx b - ap \tag{4.45}$$

where:

$$a = (k\gamma^2/2)E \times 10^J \qquad b = k\gamma E \times 10^J \qquad m = (E_{max}/E_0) \times 10^Y$$

$$J = m(1 - E/E_{max}) \qquad Y = \alpha H_{max}$$

Equation (4.45) has the solution:

$$p(t, h) = [\gamma/2 + \exp\{-tk\gamma E \times 10^S\}]$$

where:

$$S = (E_{max}/E_0) \times 10^c \qquad c = \alpha H_{max}(1 - E/E_{max})$$

Since at $h \to \infty E \to 0$, then $p(t, h) \approx [\gamma/2 + K \exp\{-tk\gamma E\}]^{-1}$, where: $K = \exp\{-10^{-m}\}$.

4.5.4 Model of the Peruvian Current ecosystem

Each region of the World Ocean has characteristic hydrophysical and ecological properties of its own. Therefore, a specification of the model of the carbon cycle in the World Ocean with these properties taken into account will raise the reliability of the model assessment of the role of the ocean systems in the assimilation of excess CO_2 from the atmosphere. With this aim in view, consider the region of the Peruvian Current characterized by high concentrations of nutrient elements (mkg-at/l): nitrites (0.6–0.8), nitrates (5.5–14.9), phosphates (0.6–1.31), silicic acids (3–45); and large values of primary production (cal/m^2/day), 17.9–69.3 for phytoplankton, 3.26 for bacteria, 0.12–0.73 for zooplankton, and 0.7–3 for nekton (mainly, anchovy). Near the Peruvian coastline occurs an upwelling of deep waters that feed the photic layer of the bordering region of the ocean with nutrient elements. Here the main object of the industrial fishing is anchovy whose average production is estimated at 100×10^8 kg/yr. From the available data, anchovy constitutes above 95% of the catch, and therefore, the derivation of the model of the functioning of the Peruvian ecosystem is also important for the economy of the bordering states. Of course, the real limits of the interaction of the natural–anthropogenic system of the concrete state of South America with the global environment are impossible to assess without an account of the role of the ecosystem of the Peruvian Current.

The Peruvian Current refers to the regions of the World Ocean where complicated interconnected and interdependent physical, biological, and chemical processes take place in a relatively small area. Near the coastlines of Peru and Chile the production of organic substance is so large that the available oxygen is insufficient for it to be oxidized, and as a result, a layer with an oxygen deficit appears at depths of 100–800 m. This is accompanied by the underfeeding of fauna.

In the hierarchic structure of the trophic relationships of the Peruvian Current community, the following components can be selected: B_1 – bacterioplankton; B_2 – phytoplankton; B_3 – phytoplankton consumers; B_4 – predatory zooplankton; B_5 – industrial anchovy; B_6 – anchovy grubs; B_7 – predatory fish; B_9 – anchovy young; B_{10} – detritus; B_{13} – dissolved organic matter (cal/m^3); B_8 – birds (birds/m^2); B_{12} – biogenic elements (mg/m^3); B_{14} – dissolved oxygen (ml/l); and B_{11} – volume of anchovy catch (t/km^2). The characteristic feature of the trophic graph of the system is the availability of connections between the components of the ocean biocenoses (B_5, B_7, B_9) and the land biocenoses (B_8). An introduction of this connection into the model has turned out to be essential, since maritime birds (cormorants, pelicans, etc.) in the regions of the Peruvian Current markedly affect the dynamics of the whole system. From the available estimates, anchovy constitutes from 80 to 96% of the food spectrum for birds. Therefore, the characteristic feature of the developed model is an account of the elements of interaction of the land and ocean biocenoses expressed by the introduction of trophic bonds between birds and anchovy. The trophic pressure on anchovy from other fish is expressed by a single connection with the component B_7 – which is mainly Chili pelamyd (*Sarda chiliensis*), although anchovy is in the food spectrum for other animals, also.

The block-scheme of the Peruvian Current Ecosystem Model (PCEM) reflects the elements of the atmosphere–ocean interaction reduced to a mutual exchange with oxygen, with the given rates of invasion and evasion.

The whole system receives solar radiation energy E whose intensity is estimated by the formula (4.18) at $\alpha = 0.013$, $\delta = \beta = \nu = 5 \times 10^{-5}$. A change in the biomass of the ith component ($i = 1$–$7, 9$) is described by a balanced equation similar to Equation (4.29):

$$\begin{aligned}\partial B_i/\partial t + V_\varphi \partial B_i/\partial\varphi + V_\lambda \partial B_i/\partial\lambda + V_z \partial B_i/\partial z = R_i - T_i - M_i - \sum_{j\in\Gamma_i} C_{ij} R_j \\ + k_{2,\varphi}\partial^2 B_i/\partial\varphi^2 + k_{2,\lambda}\partial^2 B_i/\partial\lambda^2 \\ + k_{2,z}\partial^2 B_i/\partial z^2 - \nu_i \partial B_i/\partial z + Q_i - H_i\end{aligned} \tag{4.46}$$

where Q_i is the characteristic of the age transitions between the stages of the anchovy development ($i = 5, 6, 9$), $Q_i = 0$ for $i = 1$–$4, 7$; R_i is the produce; T_i is the expenditure on energy exchange; M_i is an amount of the dead biomass; H_i is the unassimilated food (cal/m^3/day);

$$C_{ij} = k_{ij}\overline{B}_i / \sum_{m\in S_j} k_{jm}\overline{B}_m$$

Γ_i is the set of the trophic subordination of the ith component; S_j is the food spectrum of the jth component; k_{ab} is the indicator of the degree of satisfaction of the needs of food of the ath component at the expense of the biomass of the bth component; $\overline{B}_i = \max\{0, B_i - B_{i,\min}\}$, where $B_{i,\min}$ is the minimum biomass of the ith component consumed by other trophic levels; $k_2 = (k_{2,\varphi}, k_{2,\lambda}, k_{2,z})$ are the constituents of the coefficient of the turbulent diffusion (with the supposed isotropy of the eddy diffusion in the horizontal plane $k_{2,\varphi} = k_{2,\lambda} = \nu_H$; and ν_i is the rate of the gravitational sedimentation of inactive elements.

The rate of photosynthesis as the function of illumination, nutrient elements, temperature, and pollutions θ is modelled by the relationship:

$$R_2 = \min\{\rho_2 B_2, R_{22}\}$$

where

$$R_{22} = 0.07B_2[1 - \exp(-k_{2,12}B_{12})]E\exp[0.2(1 - E/E_{\max})]e^{-10.7\theta} \times (1 + 0.02\Delta t^o)$$

Here t^o is the water temperature (°C).
The produce R_i on the remaining trophic levels is calculated by the formula:

$$R_i = R_{i,\max}\left[1 - \exp\left(-\sum_{j\in S_i} k_{ij}\overline{B}j\right)\right](1 + 0.02\Delta t^o)B_i$$

The function $T_i(i = 2, \ldots, 9)$ is postulated as the dependence $T_i = \tau_i B_i^{\omega_i}$, where $\omega_i = 0.7$–0.95. The coefficients τ_i are functions of the water temperature t^o and oxygen concentration B_{14}:

$$\tau_i(t^o, B_{14}) = \begin{cases} t_i^*(t^o) & \text{for } B_{14} > B_{14,i}^* \\ t_i^*(t^o)B_{14}/B_{14,i}^* & \text{for } 0 \le B_{14} \le B_{14,i}^* \end{cases}$$

$$t_i^*(t^o) = \begin{cases} \tau_i^* & \text{for } t^o \in [t_{1,i}^o, t_{2,i}^o] \\ \tau_i^*[a_i \exp(-b_i t^o)]^s & \text{for } t^o \notin [t_{1,i}^o, t_{2,i}^o] \end{cases}$$

where $s = (t^o - t_i^{oo})/m_i$, $t_{1,i}^o$, and $t_{2,i}^o$ are the lower and upper boundaries of the temperature zone of adaptation of the ith trophic level; $t_i^{oo} = 0.5(t_{1,i}^o + t_{2,i}^o)$; a_i, b_i, and m_i are constant values, selected from the condition of the measurement data approximation.

In contrast to other hydrobionts, bacteria are characterized by a substantial dependence of gas exchange on the food concentration. Therefore, to describe T_1, assume the following relationship:

$$T_1 = t_1 B^{\omega_1}[1 - \exp(-t_1'\overline{B}_{10} - t_1''\overline{B}_{13})]$$

where t_1 is the expenditure of bacteria on energy exchange per unit biomass; ω_1 is the indicator of the dependence of expenditures on energy exchange on the amount of the bacterioplankton biomass; t_1' and t_1'' are indicators of the dependence of expenditures on energy exchange on the efficient concentrations of detritus and dissolved organic matter, respectively.

For phytoplankton, the losses to respiration are characterized by reduced assimilation of carbon in photosynthesis. For the Peruvian Current these losses average 74.6% of assimilated oxygen, which constitutes 1.45 g/cm^2/day. Let us determine the dynamic indicator of mortality M_i based on the assumptions made above in the form of the dependence $M_i = (\mu_i + \mu_i' B_i)B_i$, where $\mu_i = 0.01$–0.02 and $\mu_i' = 0.1\mu_i$. The H_i parameter is calculated by the formula $H_i = h_i R_i$.

Among living components, anchovy takes a special place in the model, and for its description the following three stages have been chosen: grubs, youngsters, and industrial anchovy. It is assumed that the rate of transition from the ith stage to the jth stage is constant and takes one year.

The inert components in the model are described by the following equations:

$$\left.\begin{aligned}
\partial B_{10}/\partial t + (V\nabla)B_{10} &= \sum_{s=1}^{9}(M_s + H_s) - \rho_{10}B_{10} - \nu_{10}\partial B_{10}/\partial z \\
&\quad - \overline{B}_{10}(k_{1,10}R_1/P_1 + k_{3,10}R_3/P_3 + k_{4,10}R_4/P_4) \\
&\quad + (k_2\nabla^2)B_{10} \\
\partial B_{12}/\partial t + (V\nabla)B_{12} &= \rho_{10}B_{10} + \rho_1\left(T_9 + \sum_{s=1}^{7} T_s\right) - \delta_{12}R_2 \\
&\quad + (k_2\nabla^2)B_{12} + \beta_{12}\partial B_{12}/\partial z \\
\partial B_{13}/\partial t + (V\nabla)B_{13} &= k_{13,0}R_2 - \overline{B}_{13}k_{1,13}R_1/P_1 \\
\partial B_{14}/\partial t + (V\nabla)B_{14} &= \delta_{14}R_2 - \sum_{s=1}^{7}\xi_s T_s - \xi_9 T_9 + (k_2\nabla^2)B_{14} \\
&\quad - \delta_{10}\rho_{10}B_{10} + \max\{0, (z_{14} - z)/|z - z_{14}|\}w_{14}
\end{aligned}\right\} \quad (4.47)$$

The equation to describe changes in the population of birds is written as:

$$\partial B_8/\partial t = (1 - h_8)R_8 - M_8$$

where

$$R_8 = \max\{0, (\Delta^* z - z)/|z - \Delta^* z|\}k_{8,0}B_8(1 - \exp[-V_8])$$

$$V_8 = \sum_{s=5}^{7} k_{8,s}\overline{B}_s + k_{8,9}\overline{B}_9 + \gamma_8' \qquad M_8 = \mu_8 B_8$$

and γ_8' is the share of the non-fish food in the food ration of birds.

The Q_i function of the age transitions is presented only for the anchovy elements and written as a linear approximation:

$$Q_6 = \lambda_{5,6}B_5 - \lambda_{6,9}B_6 \qquad Q_5 = \lambda_{9,5}B_9 \qquad Q_9 = \lambda_{6,9}B_6 - \lambda_{9,5}B_9$$

4.5.5 Model of the upwelling ecosystem

The zones of the deep water upwelling in the World Ocean are known to exhibit high productivity. An upwelling results from the surface waters being wind-driven from the shore, dispersing currents, or water retreat from the shore for other reasons. The rates of water lifting and the upwelling stability are determined by several synoptic parameters. A most characteristic value of the vertical velocity of water lifting in the zone of upwelling is 0.77×10^{-3} cm/s. The depths from which water starts lifting varies widely, with depths around 200 m prevailing.

Consider the ocean ecosystem in the zone of upwelling, beginning from the concept of the successive development of a community from the moment of its formation in the region of the deep water lifting to its final state in the oligotrophic region of convergence. Let us derive a model for the upwelling ecosystem (MUE). During these two states the system develops in time and, respectively, moves together with the water flow. In this case changes take place both in the total energy supply in the community and in its structure (spatial, trophic, and specific). The basic moments of these changes have been generally studied in numerous field investigations. Therefore, the correspondence of the model calculations to these ideas can be one of the criteria for model adequacy.

Suppose the ecosystem from the zone of upwelling develops uniformly horizontally, so that changes are observed both through in depth z with a step $\Delta z = 10\,\mathrm{m}$ and in time with a step $\Delta t = 1$ day. The horizontal velocity of water flow from the zone of upwelling is assumed to be constant and equal to $V = V_\varphi = V_\lambda$, so that the distance
of a unit volume of water from the zone of upwelling is equal to $\Delta r = (\Delta\varphi^2 + \Delta\lambda^2) = V\Delta t$.

The state of the ecosystem on each horizon $z =$ const. is determined by illumination $E(z,t)$, concentration of nutrient elements $n(z,t)$ and detritus $d(z,t)$, and the biomasses of phytoplankton $p(z,t)$, bacteria $b(z,t)$, protozoa $Z_1(z,t)$, microzooplankton $Z_2(z,t)$, small filtraters $Z_3(z,t)$, large filtraters $Z_4(z,t)$, small predatory cyclopoids $Z_5(z,t)$, predatory calamoids $Z_6(z,t)$, and large predatory chaetognaths and polychaete $Z_7(z,t)$. Protozoa include infusorian and radiolarian and microzooplankton includes naupliuses of all copepod. Based on studies of the food for tropical plankton animals, small filter feeders (<1 mm in size) include plankton animals such as *Oikopleura*, copepods *Clausocalanus*, *Paracalanus*, *Acartia*, *Lucicutia*, and small ostracode *Conchoecia*. Large filter feeders include copepods larger than 1 mm, such as *Undinula*, *Pleuromamma*, *Centropages*, *Temora*, *Scolecythrix*, and others, baby euphasiids, pteropods, etc. In the trophic pyramid the impact of nekton elements is considered negligibly small. The food bonds between the enumerated elements are shown in the scheme in Figure 4.2.

It is assumed that 30% of the mass of bacteria are in natural aggregates >3–5 μm in size, which can be eaten and assimilated by filter feeders (Z_3, Z_4); nauplii (Z_2), protozoa (Z_1), and small filter feeders (Z_3) can also feed on non-aggregated bacterioplankton.

The trophic bonds between the elements are described based on the energy principle. Biomass, rates of production and exchange, and rations are expressed in energy units (cal/m^3 or cal/m^2).

As in the general model of the ocean ecosystem, the original source of energy and substance in the community is the primary production of phytoplankton (R_p). External to this comes the solar radiation energy (E) and nutrient elements (n). The vertical structure of the water medium is described by a 3-layer model with one layer above the thermocline, one layer of sharp gradient, and one layer of lower gradient beneath the thermocline. The upper (z_b) and lower (z_l) boundaries of the layer with a sharp gradient are assumed to be located (depth in metres):

$$z_b = \begin{cases} 10 + 2.2t & \text{for } 0 \le t \le 50 \text{ days} \\ 120 + 0.6(t - 50) & \text{for } t > 50 \text{ days} \end{cases}$$

$$z_l = \begin{cases} 30 + 2.4t & \text{for } 0 \le t \le 50 \text{ days} \\ 150 + 1.4(t - 50) & \text{for } t > 50 \text{ days} \end{cases}$$

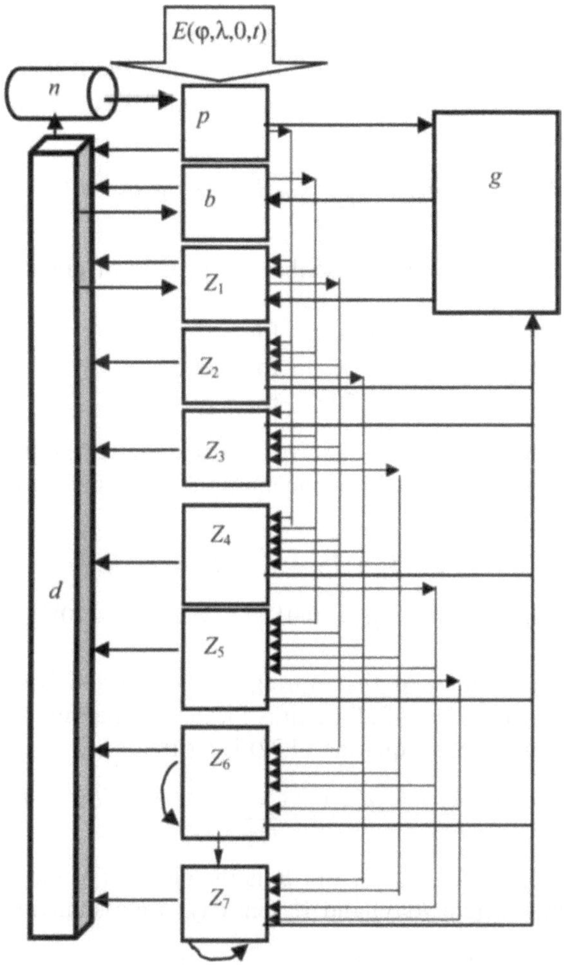

Figure 4.2. The scheme of energy fluxes (cal/m^3/day) through the community of pelagic organisms residing in the 200-m surface layer in the oligotrophic tropical region of the World Ocean.

Thus the thermocline at an initial time moment $t = 0$ is located in the 10–30-m, descending gradually down to 120–150 m on the 50th day and to 150–190 m on the 100th day.

The supply of nutrient elements in the 0–200-m is replenished at the expense of detritus decomposition and emission of unassimilated food by living elements of the community, as well as at the expense of the input from deep layers ($z = 200$ m) due to turbulent mixing (coefficient k_2) and vertical advection (velocity V_Z). The content of nutrients in organic matter is constant and constitutes 10%.

With an account of the assumptions mentioned Equation (4.35) becomes:

$$R_p = k_T[1 - 10^{-0.25p\gamma(t)}][1 - 10^{-0.1n}]^{0.6}$$

where the function $\gamma(t)$ characterizes the time dependence of the production/biomass ratio (P/B-coefficient) for phytoplankton. A maximum of P/B for phytoplankton at the point of upwelling ($t = 0$) is assumed to be 5, then the P/B-coefficient decreases and on the 15th day reaches 1, after which it remains constant. Hence Equation (4.25) is rewritten:

$$\partial p/\partial t = R_p - t_p p - \mu_p P - \sum_{j \in \omega_0/p} C_{pj} R_j + k_2 \partial^2 p/\partial z^2 + (V_z - w_p)\partial p/\partial z$$

where the coefficient w_p (cm/day) describes the process of gravitational sedimentation and is estimated at:

$$wp = \begin{cases} 50 & \text{for } 0 \le z \le z_b \\ 10 & \text{for } z_b < z < z_l \\ 30 & \text{for } z \ge z_l \end{cases}$$

Bacteria feed on detritus and dissolved organic matter emitted by phytoplankton. The allochthonous dissolved organic matter is neglected. In Equation (4.36) an approximation is considered:

$$T_p + T_b + T_Z + T_r + T_D = 0.3R_p$$

It is also assumed that $\mu_o = 0.01$, $t_b = 0.75[1 - 10^{\zeta}]$, $R_p = 3b[1 - 10^{\zeta}]$, $\zeta = -0.2d - 0.3R_p$.

A limitation is established that during a 24-hr period bacterioplankton can consume not more than 10% of all detritus located in the same water layer.

With these assumptions Equation (4.22) is written as:

$$\partial n/\partial t = -0.1R_p + 0.1d + 0.05\sum_{i \in \omega_0} t_i B_i + k_2 \partial_n^2/\partial z^2 + V_z \partial n/\partial z$$

where $\omega_0 = \{p, b, Z_j(j = 1, \ldots, 7)\}$; and B_i is the biomass of the ith element.

The coefficients k_2 (cm^2/day) and V_z (cm/day) are assumed to be:

$$V_z = \begin{cases} 5 & \text{for } 0 \le z \le z_b \\ 0.1 & \text{for } z_b < z < z_1 \\ 1 & \text{for } z \ge z_1 \end{cases} \qquad k_2 = \begin{cases} 200 & \text{for } 0 \le z \le z_b \\ 50 & \text{for } z_b < z < z_1 \\ 150 & \text{for } z \ge z_1 \end{cases}$$

The Equation (4.18) for the case considered is simplified:

$$E(z_n, t) = E_0 \times 10^{-\alpha z_n}$$

where z_n is the depth of the nth layer (m);

$$\alpha = 0.01 + 0.01\sum_{k=0}^{n}[p(z_k, t) + d(z_k, t)]/z_n$$

A marked suppression of phytoplankton is observed at $E > 100\,\text{cal/cm}^2\text{/day}$. Hence, Equation (4.20) is rewritten:

$$K_T = 0.041E \times 10^{0.25[1 - E/E_{\max}]}$$

where $E_{\max} \approx 70\,\text{cal/cm}^2\text{/day}$.

For this case Equation (4.19) is reduced to the form:

$$\partial b/\partial t = R_b - \mu_b b - t_b b - \sum_{j \in \omega_0/(p,b)} C_{bj} R_j + k_2 \partial^2 b/\partial z^2 + (V_z - w_p)\partial b/\partial z$$

The rate of the change of the biomass of each element of zooplankton $\partial Z_i/\partial t (i = 1\text{–}7)$, as in Equation (4.29), is determined by the level of consumption (ration) $R_{Z,i}$, food assimilation $1/u_{Z,i}$, expenditure on energy exchange $T_{Z,i} = t_{Z,i} Z_i$, the rate of dying-off $\mu_{Z,i} = 0.01 Z_i$, consumption of the ith element by the jth element with the coefficient C_{ij}, and the age transition of nauplii to copepod of filter feeders and predators. Thus:

$$\partial Z_i/\partial t = \Omega_{m,i} + (1 - h_{Z,i}) R_{Z,i} - (t_{Z,i} + \mu_{Z,i}) Z_i - \sum_{j \geq i}^{7} C_{ij} R_{Z,j}, (i = 1\text{–}7)$$

where $h_{Z,i} = 1 - 1/u_{Z,i}$; and $\Omega_{m,i}$ characterizes the age transition of nauplii to another category of zooplankton. The transition of nauplii Z_2 to copepod filter feeders (Z_3, Z_4) takes place with an intensity:

$$\Omega_{m,3} = Z_2 Z_3 \{15(Z_3 + Z_4 + Z_5 + Z_6)\}^{-1}$$

$$\Omega_{m,4} = Z_2 Z_4 \{20(Z_3 + Z_4 + Z_5 + Z_6)\}^{-1}$$

and for predatory copepod (Z_5, Z_6) with an intensity:

$$Q_{m.4} = Z_2 Z_5 \{15(Z_3 + Z_4 + Z_5 + Z_6)\}^{-1}$$

$$Q_{m.6} = Z_2 Z_6 \{20(Z_3 + Z_4 + Z_5 + Z_6)\}^{-1}$$

To calculate the ration for the elements of zooplankton using Equation (4.28), the coefficient ν is determined from the condition $\nu = 0.01/B_{\min}$. When calculating the coefficients C_{ij}, it is assumed that the food requirements refer to various food objects in proportion to their biomasses with an account of food selectivity. The coefficients C_{ij} for all elements of the ecosystem are given in Figure 4.3.

The daily migrations of zooplankton are simulated by an addition to the daily ration of the elements $Z_j (j = 4\text{–}7)$, located in the 0–50 m layer, of some share (k_Z) of the total food requirements of the same elements of the community but in the 50–200 m layer. The coefficient k_Z is considered to be time-dependent:

$$k_z = \begin{cases} 0.02 + 0.0016t & \text{for } t \leq 50 \text{ days} \\ 0.1 & \text{for } t > 50 \text{ days} \end{cases}$$

Various scenarios are possible here both for migration and nutrition of the elements of zooplankton. It is only known for certain that some share of the zooplankton elements from deep layers satisfy their needs for food in the upper layers of the ocean.

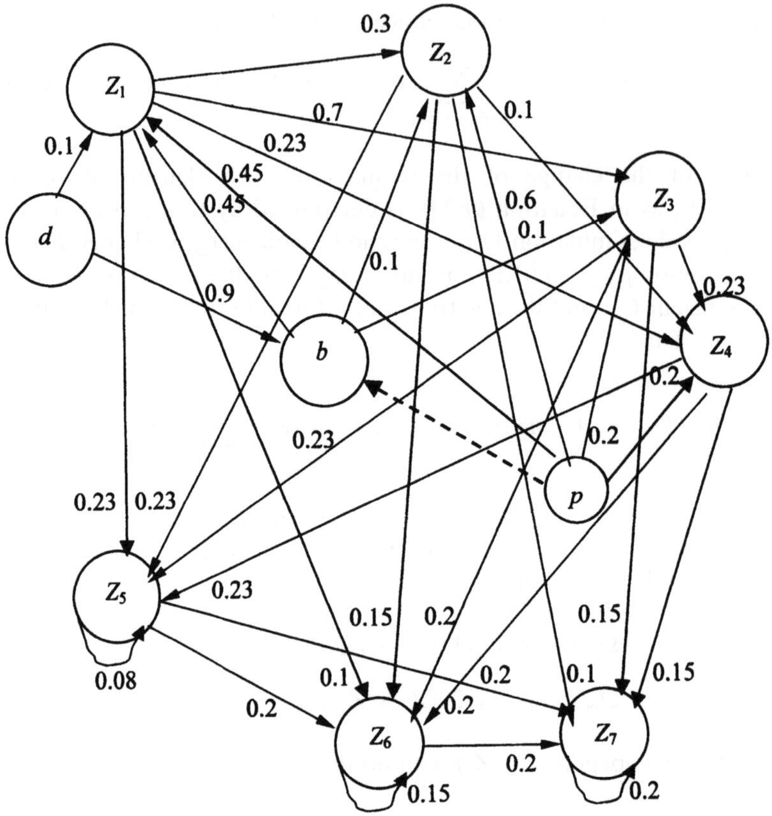

Figure 4.3. The scheme of trophic loads between the elements of the upwelling ecosystem. The letter identifiers correspond to Figure 4.2. The arrows indicate the values of the coefficients C_{ij}.

In this case Equation (4.34) is written as:

$$\partial d/\partial t = \sum_{i \in \omega_0} (H_{Z,i} + M_{Z,i}) - \sum_{j \in \omega_0/p} C_{dj} R_j + k_2 \partial^2 d/\partial z^2 - \mu_d d + (V_z - w_d)\partial d/\partial z$$

where the coefficient of the rate of the gravitational sedimentation of detritus is taken in the form (cm/day):

$$w_d = \begin{cases} 25 & \text{for } z_b \le z \le z_l \\ 50 & \text{for } z < z_b, z > z_l \end{cases}$$

4.5.6 The organic carbon cycle in conditions of freezing seas

The formation of phytoplankton productivity in the northern seas, which for most of the time are covered with ice, is connected with the composition of complicated synoptic, hydrodynamic, geochemical, ecological, and energy variations in the parameters of the marine medium. The complexity of this problem has been mentioned in Legendre and Krapivin (1992), and the problem of the formalized description of the kinetic processes of formation of vertical distributions of phytoplankton biomass have been formulated for the first time in the work by Legendre and Legendre (1998), where a stationary oceanographic criterion has been derived to estimate a new production of phytoplankton.

The ice–water ergocline plays an important role in the formation of the biological productivity of the northern seas as well as the community of algae that infiltrates the snow and ice. Therefore, in deriving a model of the formation of the vertical structure of the marine ecological community it is necessary to take into account the mechanisms of snow cover formation, ice growth, and transformation of hydrodynamic processes. In other words, the dynamic formalization of the marine ecosystem in the northern latitudes covering the season of ice cover formation, should be of a system character and take into account the variability of the synoptic parameters. The latter, apparently, cause the variability of other parameters within the system.

Here a complex model is suggested for marine ecosystem dynamics in the conditions of the northern climate. The model takes into account the seasonal variations of temperature and precipitation, either prescribed or determined by the respective models. The climatic situations in the region considered were parameterized on the basis of a time series of temperature and precipitation. The growth of ice thickness is described by the Stefan model. The formation of the ecological situation in the marine medium is described by the relationships that make it possible to calculate the intensity of light at any depth and to determine the dynamic characteristics of the vertical flux of nutrient elements.

To derive a general model and to obtain a forecast, it is important also to determine the temporal and spatial scales of the processes under study as well as to select the characteristic and special features of the ecosystem's functioning in conditions of the northern climate. Most reliable studies of the stability and characteristic scales have been carried out when studying the atmosphere. Theoretically, it has been proven that the time of predictability of actual weather conditions (temperature, wind speed, cloudiness, and precipitation) is confined to 1–3 days, with the accuracy of the forecast rapidly decreasing when the term of the forecast increases. At the same time, experiments reveal an increase in the predictability of the same parameters (averaged over time and space). On this basis a weather forecast is possible for a period of 10–15 days. Longer term (seasonal and annual) predictions use data averaged over a decade or a month. That is, the ratio of the time of prediction to the time of averaging can be estimated as ≤ 10. The spatial scales are closely connected with the timescales and characterize various phenomena: motions of fronts, cyclones and anticyclones, low-frequency seasonal oscillations, etc. For instance, the daily resolution in time enables one to describe the motion of cyclones and has a characteristic spatial scale $\sim$1000 km.

The characteristic temporal and spatial scales of the ocean and seawater flows have been adequately studied, too, which has made it possible to easily estimate them. The horizontal scales vary from 100 km for synoptic formations to 1000 km for large-scale currents. Oceanographers have identified a substantial link between

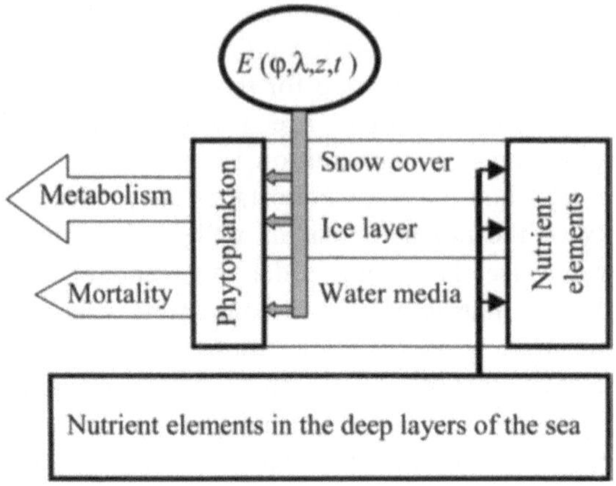

Figure 4.4. The conceptual structure of the model of the organic matter cycle in the ice-covered sea.

primary synoptic hydrophysical processes and elementary circulation mechanisms in the atmosphere with an average duration of 5 days. The seasonal hydrophysical cycles described mainly by mixing of isolines in the meridional circulation have been well estimated. In the interannual variability of the hydrophysical fields the interaction of currents, upwellings, and large-scale atmospheric processes plays an important role. Here the characteristic timescales constitute several months.

Figure 4.4 shows a conceptual scheme of energy fluxes of solar radiation and nutrient elements from deep layers of the sea where their supplies are supposed to be infinite. The nutrient background at a depth z is formed due to dead organic matter decomposition d, and the rate of the vertical advection v. The nutrient elements are absorbed by phytoplankton at a rate inherent to a given kind of algae at a given temperature and illumination. The formation of the phytoplankton biomass is affected by solar radiation intensity, concentration of nutrient elements, dying-off, and expenditures on energy exchange with the media. In principle, to close the trophic relationships, it is worthwhile considering the impact on phytoplankton of the process of consumption by zooplankton and other trophic levels. When time intervals with low temperatures are considered, this effect can be neglected. Therefore, the final realization of all dynamic equations, that describe the interaction of ecological factors in the marine medium, leads to the set of model units shown in Figure 4.4.

The time and space steps also determine the means of parameterization of the processes taken into account. Thus the atmospheric part of the model is of the statistical character and realizes a daily simulation and inertial prediction of the following meteorological parameters: average daily temperature T_a, humidity deficit in the surface air layer D, short wave solar radiation E, and daily precipitation total x.

Consider an application of this approach to model a random vector $Y = \{T_a, E, D\}$. Let $Y_i = \bar{Y}_i + y_i$ $(i = 1, 2, 3)$, where $\bar{Y}$ is the climatic norm and y is a deviation from the climatic norm caused by external accidental forcings. Then $dy_i/dt = A_{ij}y_j + F_i$, where F_i is an accidental forcing, and A_{ij} are the coefficients of relaxation.

Substitute a differentiation by a difference operator following the simplest scheme $dy_i/dt = [y_i(t_{k+1}) - y_i(t_k)]/\tau$, where $\tau = t_{k+1} - t_k = \text{const.}$ and k is the step of discrete time. With account of this scheme and by multiplying the linear equation given above by y_l, after the statistical averaging we obtain: $\langle y_i(t_{k+1})y_l(t_k)\rangle - \langle y_i(t_k)y_l(t_k)\rangle = \tau A_{ij}\langle y_j(t_k)y_l(t_k)\rangle$. The angular brackets denote the statistical averaging over an ensemble of possible weather scenarios. Hence, if the covariance matrix $\langle y_j(t_k)y_l(t_k)\rangle$ is non-singular, the coefficients A_{ij} are easily calculated:

$$A_{ij} = (\langle y_i(t_{k+1})y_l(t_k)\rangle - \langle y_i(t_k)y_l(t_k)\rangle)(\langle y_j(t_k)y_l(t_k)\rangle)^{-1}/\tau$$

Then the covariance of random forcings is calculated:

$$\langle F_iF_j\rangle = \langle (y_i(t_{k+1}) - y_i(t_k) - \tau A_{il}y_l(t_k))(y_j(t_{k+1}) - y_j(t_k) - \tau A_{jm}y_m(t_k))\rangle$$

Usually the sums of precipitation are assumed to follow gamma-distribution:

$$P(x) = x^{\kappa-1}\exp(-\kappa/\beta_0)/[(\beta_0)^{\kappa}\Gamma(\kappa)]$$

where κ and β_0 are empirical parameters for the form and scale of the distribution. Then the daily sums of precipitation obtained are distributed by the dates of the highest probability in accordance with oscillations of other meteorological elements and the value of the probability coefficient:

$$\lambda_P = \{1 + N_2[Y_i - \bar{Y}_{i,2}]C_{ij}[Y_j - \bar{Y}_{j,2}]/(1 + N_2)\}/\{1 + N_1[Y_i - \bar{Y}_{i,1}] \times C_{ij}[Y_j - \bar{Y}_{j,1}]/(1 + N_1)\}$$

where N_1 and N_2 are the number of days with and without precipitation; $\bar{Y}_1$ and $\bar{Y}_2$ are vectors of the averaged meteorological parameters for the days with and without precipitation, respectively; and C_{ij} is the correlation matrix of meteorological parameters.

The input data for the model can be either obtained directly from the meteorological data for the territory considered or calculated with the use of traditional algorithms. For instance, in the absence of short wave radiation measurements one of the empirical connections with cloudiness can be used (Kondratyev, 1999b):

- *The Kimbell formula*: $E = E_y(1 - 0.71n_o)$, where n_o is the average point of cloudiness and E_y is the solar radiation in clear skies.
- *The Budyko formula*: $E = E_y(1 - 0.68n_o)$.

- *The Berliand formula*: $E = E_y(1 - [m_o + 0.38n_o]n_o)$, where m_o is the correction coefficient taking latitude into account.
- *The Matauike formula*: $E = E_y(1 - 0.052[n_o]^{1.3})$.

The results of calculations with these formulae vary most in conditions of continuous cloud. On average, error of calculations are determined by variations of the parameter E/E_y, which constitutes 0.25–0.35 for continuous cloudiness with stratus and cumulostratus clouds and 0.65–0.85 for stratus and cirrostratus clouds.

The ice cover thickness $z_s(t)$ can be described by statistical data with indicated dispersive characteristics: $z_s(t) = G(t) + g(t)$, where the function $G(t)$ characterizes an average thickness of snow cover during a chosen time period, and the function $g(t)$ determines its average variation during the same time period. There are numerous models, which describe the process of the snow cover formation and melting. Within the global model it is expedient to use one of the simplest parameterizations, such as the Sellers model: $S_F = x(27.5 - 0.1T_a)$, $T_a \in [265, 275]$K; $S_A = S_M + Q_0$, $Q_0 = \rho_a C_a U_a[r*(T_a) - r_a]$, $S_M = 0.045(T_a - 273.16)$; where S_F is the share of the snowfall at air temperatures close to zero, S_A is the snow ablation (evaporation + melting), S_M is the amount of snow that has melted per day (water equivalent in cm), $r*$ is the saturated specific humidity, r is the specific humidity, C_a is the resistance coefficient, U_a is the surface wind speed, and ρ_a is the air density.

As a result, the snow cover thickness is calculated by the formula:

$$z_s(t + \Delta t) = z_s(t) + [S_F - S_A]\Delta t$$

We will describe the process of ice formation on the water surface using the Stefan model. Let the ice thickness be described by the function $R(t) = z_I - z_s$, with the part $R_1(t)$ being below the level $z = 0$. Let us denote by ρ_G, ρ_R, and ρ_W the densities of snow, ice, and water, respectively. Then, $R_1 = (z_s\rho_G + R\rho_R)/\rho_W$. The process of ice formation will be subject to the regularity following from the system of equations:

$$\partial T_a/\partial t = a^2(z)\partial^2 T_a/\partial z^2$$

$$\lambda_1 \partial T_a/\partial z|_{z=R_1} = q\rho_R dR/dt + \lambda_2 \partial T_a/\partial z|_{z=R_1}$$

with boundary and initial conditions $T_a(R_1 - R) = f(t) \leq 0°\text{C}$, $T_a(R_1, t) = 0°\text{C}$, $T_a(\infty, t) = P(t)$, $R(0) = 0$, and $T_a(z, 0) = \psi(z)$, where λ_1 and λ_2 are coefficients of the ice and water heat conductivity, respectively, q is the specific heat of ice melting, f is the function of variability of the surface air temperature in the absence of snow either prescribed as a scenario or calculated by the climate model, or snow temperature, P is the prescribed temperature of deep water layers, $a^2(z)$ is the coefficient of the medium heat conductivity:

$$a^2(z) = \begin{cases} a_1^2(z) & \text{for } R_1 - R \leq z \leq R_1 \\ a_2^2(z) & \text{for } z > R_1 \end{cases}$$

Now proceed to a description of the ecological part of the model. The formation of the nutrient background in the case considered, compared to ice-free conditions,

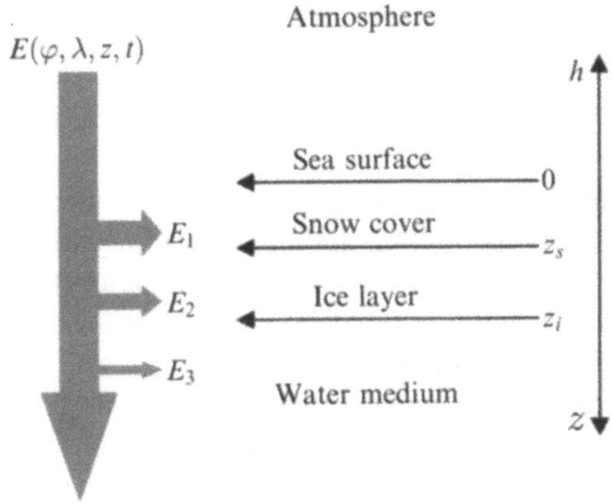

Figure 4.5. The vertical section of the model medium of an arbitrary sea basin in freezing conditions.

has certain specific features manifested through the characteristic process of their motion in the environment indicated in the scheme in Figure 4.5. Select three types of nutrient elements: $n_{AN}(z,t)$ – nitrogen, $n_{AP}(z,t)$ – phosphorus, and $n_{ASi}(z,t)$ – silicon. Symbols $A = (s$ – snow, i – ice, w – water) and $X = \{N, P, Si\}$ are used to make the formulae shorter. The dynamic equations to describe the kinetics of nutrient elements will be:

$$\partial n_{AX}/\partial t = k_1 \partial n_{AX}/\partial z + k_2 \partial^2 n_{AX}/\partial z^2 + Q_{AX} - \delta R_{Ap}$$

where k_1 and k_2 are the coefficients of advection and turbulent diffusion, respectively; R_{Ap} is the rate of photosynthesis in the medium A; Q_{AX} is the inflow of nutrient elements into the medium A from other sources; and δ is the consumption of nutrient elements by unit biomass of pure produce of phytoplankton per unit time. In the simplest case the flux Q_{AX} is identified with the input of nutrient elements due to decomposition of dead organic matter and phytoplankton respiration $\tau_X : Q_{AX} = \delta_X R_d + \lambda_p \tau_X$, where the coefficient δ_X characterizes the content of the nutrient element X in detritus d, and λ_p is the indicator of the rate of replenishing the nutrient element supplies due to respiration of the phytoplankton cells.

To calculate the level of illumination $E(z,t)$, apply an exponential law that reflects the impact of the optical properties of the medium A:

$$E(z,t) = \begin{cases} E_0 & \text{for } z = 0 \\ E_1(z,t) = E_0 \exp\{-\alpha_1 z\} & \text{for } z \in (0, z_s) \\ E_2(z,t) = E_1(z_s,t) \exp\{-\alpha_2 z\} & \text{for } z \in (z_s, z_I) \\ E_3(z,t) = E_2(z_1,t) \exp\{-\alpha_3 z\} & \text{for } z > z_I \end{cases}$$

Let us approximate the rate of photosynthesis in the medium A by the relationship:

$$R_{Ap} = k_p p_A E \exp(1 - E/E_{\text{opt}})(1 + a\Delta T_A)[1 - \exp(-b \min\{n_{sX}, n_{iX}, n_{wX}\})]$$

where k_p is a maximum of phytoplankton productivity at a control temperature T_c; $\Delta T_A = T_A - T_c$; E_{opt} is an illumination optimal for photosynthesis; a is the indicator of a change of the rate of photosynthesis when the temperature of the medium deviates from the control; and b is the degree of limitation of photosynthesis by a biogenic element following the Libich principle.

The dynamic equations for the elements of the sea ecosystem in freezing conditions are written by analogy to the balanced equation given above in other paragraphs of this chapter.

4.5.7 Model of the Okhotsk Sea ecosystem

The Okhotsk Sea is an object most characteristic of the northern latitudes. Numerous investigations have been concentrated on the Okhotsk Sea as a unique ecosystem (Vasiliev, 1991; Terziev *et al.*, 1993). This is connected with the desire to understand the climatic and ecological processes taking place here and to establish their correlations with the global processes, especially in the Arctic Basin. For Russia and Japan this is also important for economic reasons. Therefore, an urgent problem is to derive a model (the Okhotsk Sea Ecosystem Model (OSEM)) that would make it possible to predict the state of the ecosystem as a function of global and regional changes in the environment.

The OSEM structure and the scheme of energy fluxes in the ecosystem of the Okhotsk Sea are shown in Figures 4.6 and 4.7, respectively. The equations of the trophic interconnections are written by analogy to Equations (4.21)–(4.36). Let us approximate a change of the oxygen regime by solving the equations:

$$\partial O/\partial t + V_\varphi \partial O/\partial\varphi + V_\lambda \partial O/\partial\lambda + V_z \partial O/\partial z = \xi_p R_p - \xi_Z T_Z - \xi_r T_r - \xi_D T_D - \xi_d R_d + O_A$$

where O_A is the function describing the oxygen exchange on the atmosphere–sea border; ξ is the indicator of the oxygen production through photosynthesis; $\xi_i (i = Z, r, D)$ is the coefficient of the oxygen expenditure on respiration of hydrobiont i; and ξ_d is the indicator of the oxygen expenditure on the decomposition of detritus.

The spatial and seasonal dynamics of salinity may be described by the balanced relationships:

$$V_i dS_i/dt = \sum_{\substack{j=1 \\ j\neq i}}^{N} [\alpha_{ij} W_j S_j + \mu_{ij}(S_j - S_i)] + (1 - \beta_i) W_i S_i + f_i \qquad (i = \overline{1, N})$$

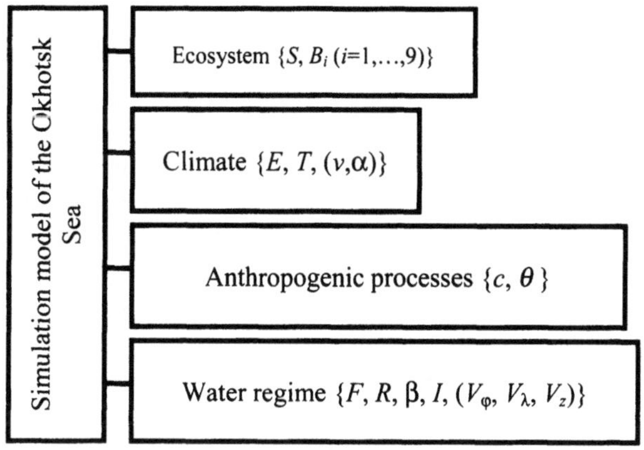

Figure 4.6. The OSEM structure. The ecosystem block includes the following components: bacterioplankton (B_1), phytoplankton (B_2), zooplankton (B_3), benthos (B_4), nekton (B_5), nutrients (B_6), dissolved organic matter (B_7), detritus (B_8), oxygen (B_9), and salts (S). The climate block simulates the temperature (T), illumination (E), and wind (ν, α). The water regime block takes into account the following processes: precipitation (F), evaporation (I), current ($V_\varphi, V_\lambda, V_z$), river runoff, and water exchange with the Pacific ocean (β). The anthropogenic processes block includes the water exchange with Pacific ocean (β) and pollutions (θ).

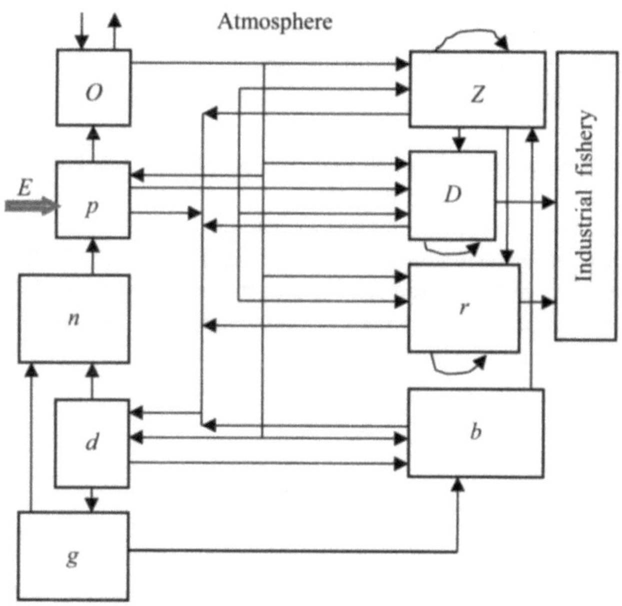

Figure 4.7. Energy fluxes in OSEM. O is the oxygen, p is the phytoplankton biomass, n is the nutrient salts, d is the detritus, g is the dissolved organic matter, b is the bacterioplankton, r is the fishery biomass, D is the benthic animals biomass, and Z is the zooplankton.

where N is the number of segments of the volume $V_i = \Delta\varphi\Delta\lambda\Delta z_i$, after dividing the water body of the Okhotsk Sea in the process of quantization of latitude φ, longitude λ, and depth z; μ_{ij} is the volume coefficient of diffusion equal to $\delta_{ij}\sigma_{ij}/L_i$ (L_i is the average length of the bordering boundary of the ith segment, σ_{ij} is the interface of the segments i and j, and δ_{ij} is the coefficient of turbulent diffusion); f_i is the function of the source describing the external input of salts to the ith segment (with precipitation F, river runoff R from the Pacific Ocean β and from the bottom sediments with water uplifting); W_j is the total water flow from the jth segment; α_{ij} is its share directed to the ith segment ($\alpha_{ij} = 0$ for non-bordering segments); and β_i is the share of the flux W_i, directed outside.

The temperature regime of the Okhotsk Sea is presented by the heat balanced equation:

$$\rho_c V_i dT_i/dt = \sum_j W_{ij,A} + \sum_j W_{ij,D} - W_{i,S} + \Omega_{i,1} + \Omega_{i,2}$$

where T_i is the average temperature of water in the ith segment (°K); $W_{ij,A}$ is the input of heat with the water flow from the jth segment into the ith one; $W_{ij,D}$ is the heat exchange between the ith and jth segments due to turbulent motion of water masses; $W_{i,S}$ is the total heat output from the jth segment; $\Omega_{i,1}$ is the heat input with water fluxes across the external boundary; $\Omega_{i,2}$ is the heat exchange across the external boundaries not connected with water fluxes; ρ is the seawater density (g/cm^3); and c is heat capacity (cal/g/deg).

The scheme to calculate the temperature regime is in fact an approximate simulation. This semi-empirical approach is acceptable for the climatic zone of the Okhotsk Sea and does not require complicated calculations. The exchange processes with water masses between the segments $V_i (i = 1, \ldots, N)$ are simulated with account of information on directions and velocities of the currents contained in the database.

The functioning of the Okhotsk Sea ecosystem is determined by severe climatic conditions. During 6–7 months in a year the sea is almost completely covered with ice 80–100 cm thick. Therefore, in the OSEM the vertical structure of the Okhotsk Sea is presented as a vertical section shown in Figure 4.5. With account of this scheme, in the zones of the frozen sea the energy fluxes in Figure 4.7 get complicated due to partitioning of the component "phytoplankton" into three possible media. Schematically it is shown in Figure 4.4.

5

High-latitude environment and global ecodynamics

5.1 ARCTIC BASIN POLLUTION PROBLEMS

The Arctic region plays a unique role in the global environmental processes, forming numerous adverse correlations with the Earth's climatic system. Low temperatures, a high level of atmospheric circulation, and the availability of large ice-covered water bodies are notable for the high latitudes, quite unlike the other areas of the globe (Kondratyev, 2003a).

The intensive industrial development of the northern territories has led to significant environmental change in these regions, especially in the Russian territories. Oil and gas extraction on the Yamal and Taimir Peninsulas in north-western Siberia, coal and gold extraction in Yakutia and Chukot, as well as the mining industry on the Kola Peninsula have led to a strong anthropogenic intervention into the natural environment. Plant cover has been violated over great territories, the area and productivity of reindeer pastures diminished, and the hydrologic regime of rivers disturbed. The greatest quantities of pollution substances are brought to the northern coast of Russia by rivers, violating the ecosystems of the northern seas.

Further adverse influence on the Arctic ecosystems can lead to rising regional or even global scale negative consequences. Thus, the problem of the development of the northern territories of Russia, the USA, Canada, and the Scandinavian countries, needs to be solved by careful analysis of all types of ecosystem dynamics, using real time data collection and formation of national databases and by defining effective ways to coordinate development of the natural and anthropogenic processes.

The purpose of this chapter is to develop and to investigate a simulation model of the pollution dynamics in the Arctic Basin. There are many experimental and theoretical results giving estimates of the growing dependencies between the pollution dynamics in the World Ocean and the state of the continental environment. The problem of the Arctic Basin pollution causes the most anxiety to investigators (Krapivin and Phillips, 2001; Stephens and Keeling, 2000). It is known that the ecosystems of the Arctic seas are vulnerable to a considerable extent in comparison with the ecosystems of other seas. Processes that clean the Arctic Ocean are slower and marine organisms of the Arctic ecosystem live in the polar climate where the vegetation period is restricted. Some feedback mechanisms operate with significant time delays and the capacity to neutralize the effects of human activity is poor. Apart from these reasons, the Arctic ecosystem has specific boundary conditions connected with the sea-ice ergocline (Legendre and Legendre, 1998). This influences the survivability level of the Arctic ecosystem.

In regards to this circumstance, the Arctic Basin is the object for in vestigations in the frameworks of many national and international environmental programmes, such as the International Geosphere–Biosphere Programme (IGBP), the US Global Change Research Programme (GCRP), the international Arctic System Science Programme (ARCSS), the US Arctic Nuclear Waste Assessment Programme (ANWAP), and the international Arctic Monitoring and Assessment Programme (AMAP). The research strategies of these programmes includes the theoretical and experimental study of the tundra ecosystems, Siberian rivers, and near-shore and open Arctic waters. The main problems arising here are the determination of a set of principle matters that require study in the first place. These include:

1. *Transport modelling of pollutants in the Arctic ecosystems.* Based on the experimental database, it is necessary to prepare a complete set of models and their corresponding computer realizations for describing the processes of transfer and transformation of pollution substances in the natural Arctic ecosystems. This set includes the following models:
   - a model of the transformation of organic pollution in the ecosystems of freshwater basins and streams;
   - models of the self-cleaning processes for oil, radionuclides, heavy metals, and other pollutants;
   - a model of radionuclide and heavy metal accumulation in the river ecosystems of the far north;
   - a model of the transport of radionuclide, heavy metal, and organic pollution with river flow into the near-coastal Arctic waters;
   - a model of pollution wash-out during the spring season in the tundra and forest tundra zones;
   - a model of the kinetics and transformation mechanisms for the biospheric elements in the water systems;
   - a kinetics model of radionuclides and heavy metals in the food chains of the land ecosystems for boreal zones;
   - a model of the surface flow of chemical elements and compounds from territories in mining zones with opencast mines under the climatic conditions of the far north; and

- a model of the seasonal influence of pollution on the phytoplankton and primary production in the northern seas.

2. *Modelling of the exchange processes of carbon dioxide and methane between the tundra ecosystems and the atmosphere.* The global interaction of the Arctic ecosystems with the biosphere and with the Earth's climatic system is carried out, in particular, through the influence upon the biogeochemical cycles of carbon dioxide and methane (Field, 2001; Field and Fung, 1999). Existing models of the global circulation of these greenhouse gases (GHGs) are incomplete in that they do not take into account this interaction. Present estimations of the gas exchange between the Arctic ecosystems and the atmosphere confirm, however, the necessity of making such an account. To create a model set related to the gas exchange in the Arctic reservoirs it is necessary to compile a catalogue of soil-plant formations, ice fields, and land-based and oceanic reservoirs. It is also necessary to put in the database estimations of evapotranspiration, dead vegetation decomposition rate, and productivity of vegetation communities. With the aid of this model set it will be possible to evaluate the role of the tundra ecosystems in forming the greenhouse effect.
3. *Modelling of the hydrological regime and estimation of the pollutant flows in the Arctic Basin.* It is necessary to prepare a set of models to describe the dynamics of the separate aquatories and of the whole hydrosystem of the Arctic Ocean, including:
   - a complex model of the water circulation in the Arctic Basin;
   - regional models of the water circulation in the Arctic seas;
   - a model of the kinetics of radionuclides, heavy metals, and organic pollutants in the trophic structures of the Arctic marine ecosystems;
   - a model of the spread of a pollutant concentration from a point-like source in the near-coastal zone of the Arctic Basin;
   - a model of the transfer of radionuclides, heavy metals, and organic pollutants due to vertical mixing of the Arctic waters; and
   - a model of the conservation and release processes due to freezing and thawing of the ice cover.
4. *Modelling of the Arctic ecosystems under conditions of anthropogenic impacts.* Anthropogenic influence in the Arctic Basin and on adjacent territories is connected with local, regional, and global activities. Therefore it is necessary to produce the following models:
   - models for the influence of radionuclides, heavy metals, and oil hydrocarbons upon the dynamics of the marine ecosystems under Arctic climate conditions;
   - models for the limitation of vegetation cover due to the different types of pollution that are brought to the land ecosystems with precipitation and surface flows;
   - models for the dynamics of vegetation cover subjected to a physical influence;
   - models/scenarios for the town and settlement structures under development;

- models/scenarios for the changes in area of the traditional seasonal regions of activity of the nomadic population; and
- models/scenarios for the social development of the scattered peoples in the far north.

5. *Modelling of the biogeochemical carbon cycle in the "Atmosphere/Arctic Ocean" system.* As has been shown by the experience of modelling the carbon dioxide global cycle, the estimations of the role of the World Ocean in redundant carbon absorption are rather rough (Krapivin and Vilkova, 1990; Nitu *et al.*, 2000a). For the models to be more precise they need to be reinforced by more reliable parameterizations of the physical processes related to the interaction between the bordering layers of the atmosphere and the Arctic aquatories. According to numerous laboratory and natural observations, the directivity of these processes depends considerably on many factors. The most significant of these are the speed of the driving wind, the presence of ice cover, and the vertical distribution profile of the water temperature. A complex composition of these factors determines the variety of possible models and their details. One of the significant problems is to explain the powerful flow of seaweeds during the spring season and hence to construct a parameterization system for the dynamics of photosynthetic processes under conditions of snow and ice cover along with their spring thawing.
6. *Development of a complex model for the functioning of the hydrologic and biogeochemical systems in the Arctic.* In addition to the model set intended for local and fragmentary processes in the Arctic ecosystems, as well as for understanding their global role, it is necessary to synthesize a single model for the whole complex of biogeochemical, biogeocenotic, and hydrologic processes that occur in the boreal systems. Creation of such a model will ease the development of database requirements and will make it possible to obtain a means for estimating the consequences of anthropogenic projects. Using this model consequences can be estimated for forest cutting and fires, for broadening of the zones with disturbed land cover, for land and basin pollution with oil, for hydrogeological changes on separate territories due to the submerging of landscapes, and for territorial pollution with waste materials from the mining industry.
7. *Stability estimation of the Arctic systems under variable global climate conditions.* Human activities in the delicate ecosystems of the far north need to be conducted with great care. But how is the natural balance to be safe from the rising influence of human civilization with its industrial machines? How can the survival of these ecosystems be evaluated in different situations? These and other questions need to be answered as the programme outlined above is realized.

An understanding of the environmental processes in the Arctic regions (a prerequisite for finding scientific solutions to the problems arising there), can be found only by combining many disciplines, including ecology, oceanography, mathematical modelling, and system analysis. This chapter synthesizes many data sources and knowledge from various scientific fields in the form of a Spatial Simulation Model of the Arctic Ecosystem (SSMAE). Separate blocks of the SSMAE were created earlier by many authors (Kondratyev *et al.*, 2002b, 2003a–c). The sequence of these blocks in the SSMAE structure and the adaptation of it to the global simulation model (GSM) provide a technology for computer experiments.

This investigation is a realization of the idea of the ARCSS Programme, which was initiated at the US National Science Foundation as part of their contribution to the US GCRP (McCauley and Meier, 1991). The present chapter describes a simulation system based on sets of computer algorithms for processing data from monitoring of the Arctic regions and for applying mathematical models of natural and anthropogenic processes.

The basic blocks of the SSMAE are oriented towards the description of the dynamics of any given pollutant. For consideration of a specific pollutant it is necessary to include in the SSMAE an additional block with the description of its physical and chemical characteristics. This procedure is demonstrated by examples of blocks that simulate the characteristics of radionuclides, heavy metals, and oil hydrocarbons. The consideration of these pollutants is restricted to elements with averaged properties.

The Arctic Basin aquatory Ω which is studied in this chapter has boundaries which include the peripheral Arctic seas as well as the coastline and southern boundaries of the Norwegian and Bering Seas.

5.2 HIGH-LATITUDE ENVIRONMENTAL SCIENCE

Recent growing attention to the Arctic's environmental problems is motivated by a number of factors, including:

1. a stronger sensitivity of the high-latitude environment to various external forcings;
2. increased understanding of the importance of numerous interactions and feedbacks between components of the Earth's system; and
3. an enhanced necessity to use natural resources located at high latitudes (Arctic shelf especially).

It is fair to say (Vörösmarty *et al.*, 2001) that, "the Arctic system constitutes a unique and important environment with a central role in the dynamics and evolution of the Earth system."

Some of the recent scientific results have been pointed out in the Arctic Climate Impact Assessment (ACIA) Implementation Plan (2000):

- There has been increased coastal erosion in the Bering Sea from storm surges resulting from reduced sea ice.

- Sea ice extent in the Arctic has decreased Arctic-wide by 0.35% per year since 1979. During the summer of 1998, a record reduction of sea ice coverage was observed in the Beaufort and Chukchi Seas.
- Sea ice thickness has also been reduced by between 1–2 m in most parts of the Arctic Ocean and the sub-Arctic seas.
- Streamflow discharge of major Siberian rivers into the Arctic Ocean has increased in recent years and is associated with a warmer climate and enhanced precipitation in the river basins.
- Since 1970, the Arctic Oscillation (AO), which is a measure of the strength of the circumpolar vortex, has strengthened. This has been found to be consistent with temperature change in the Arctic.
- There has been an increased warming of the Arctic Ocean's Atlantic layer and approximately 20% greater coverage of Atlantic water types.
- Record low levels of ozone were measured in 2000 in the Arctic with increasing evidence that these levels are likely to continue for at least the next 20 years.
- Ongoing studies indicate that the current UV levels can have a significant effect on fish larvae survival rates.
- General warming of soils in regions with permafrost, derived primarily from Alaskan data, has been observed over recent years.

It has been emphasized in ACIA (2000) that past assessments indicated that the Arctic is important to global scale processes in at least four important ways:

- The thermohaline circulation dominated by the Arctic Ocean and Nordic Seas is responsible for a considerable part of the Earth's poleward heat transport and may also serve as a sink for CO_2. Alterations of this circulation, as have been observed during climatic changes of the past, can affect global climate and in particular the climate of Europe and North America.
- The melting of the Arctic land ice sheets can cause sea level rise around the world. A compilation of studies suggests that a global warming of 1°C will lead to ~1 mm per year of sea level rise from small ice caps and glaciers. The Arctic will supply over half of this total, with an additional 0.3–0.4 mm per year contribution coming from Greenland although uncertainties remain about the mass balance of the Greenland ice sheet.
- Arctic soils can act as either sinks or sources of GHGs depending on temperature and moisture changes within the Arctic. Moisture has opposing effects on the concentrations of the two major trace gases: with CH the flux declines with soil drying while with CO_2 the flux initially increases. These changes can influence GHG warming globally.
- Our current understanding of the Arctic climate system suggests that positive feedbacks in high-latitude systems, including the snow and ice albedo effect, amplify anthropogenically induced atmospheric changes and that disturbances in the circumpolar Arctic climate may substantially influence global climate.

In the context of the health of the Arctic marine environment, from the viewpoint of proper functioning of economically important ecosystems, a number of key questions can be formulated:

- How was the polar basin formed, where are the plate boundaries?
- What has been the detailed paleoclimatic history of the high Arctic Ocean during the last 1 million years?
- Do decreases in ice extent and upper stratification of the ocean signal a different sea ice regime?
- What is the stability of the sea ice cover, what are the effects of radiative feedback in the Arctic, and how do they modulate the global ocean circulation?
- What is the role of continental shelves in the cycling of C, N, Si and other chemicals?
- What is the productivity of the Arctic Ocean, and what is the structure and diversity of higher trophic levels?
- What are the effects of environmental change, both of climate and of pollutants and contaminants such as the introduction of persistent organic pollutants (POPs) into the food chain?

Of particular interest is the high-latitude climate dynamics. According to Weller and Lange (1999), "While considerable uncertainty still exists about the exact nature of the future impacts of global climate change, there can no longer be any doubt that major changes in the climate have occurred in recent decades in the Arctic, with visible and measurable impacts following the climatic changes. Greater impacts are likely in the future and while some of them will be positive, others will be detrimental to human activities."

Recent analysis of ice cores from the Arctic (Everett and Fitzharris, 2001) has revealed large-scale and rapid paleoclimate changes. Rapid warming took place ~11,500 years ago, at the end of the last glacial period. The coldest parts of ice cores had been as much as 21°C colder than the present temperature in central Greenland, with temperatures increasing by more then 10°C in a few decades. There is evidence of even more rapid change in the precipitation pattern, rapid reorganizations of atmospheric circulation, and periods of rapid warming during the past 20,000 years. Rapid warmings of ~10°C over a few decades during the last glacial period in central Greenland had been followed by periods of slower cooling over a few centuries and then a generally rapid return to glacial conditions. About 20 such intervals, each lasting between 500 and 2,000 years, occurred during the last glacial period.

It has been emphasized (Everett and Fitzharris, 2001; Spero and Lea, 2002; Siani *et al.*, 2001) that the polar systems are extremely sensitive to the variability of temperature, and several aspects of these systems will be affected by any further climate change. The primary impacts will be on the physical environment, including ice, permafrost, and hydrology; on biota and ecosystems, including fisheries and terrestrial systems; and on human activities, including social and economic impacts on settlements, on resource extraction and transportation, and on existing infrastructure. Scenario predictions of potential future global warming indicate a necessity to particularly take into account various phenomena, such as thermocarst erosion in lowland areas and thawing of permafrost accompanied by hydrological and climatic changes. Climate change will affect terrestrial ecological systems through changes in permafrost as well as direct climatic changes, including changes in precipitation, snow cover, and temperature. Terrestrial ecosystems are likely to change from tundra to boreal forests, although vegetative changes are likely to lag behind climatic changes. Major shifts in biomass will be associated with changes in microbiological (bacteria, algae, etc.) and insect communities (some of them may diminish while others prosper).

It has been pointed out (Everett and Fitzharris, 2001) that in the recent geologic past, the tundra was a carbon sink, but recent climatic warming in the Arctic, coupled with the concommitant drying of the active layer and lowering of the water table, has shifted areas of the Arctic from sinks to sources of CO_2 (this problem is, however, far from being solved). An important potential consequence of permafrost thawing is methane emission. Another GHG concentration, tropospheric ozone, may also change due to warming of the troposphere (Kondratyev and Varotsos, 2000).

An interesting illustration of potential future surprises due to interactions and feedbacks has been discussed by Stevenson *et al.* (2000) who obtained future estimates of tropospheric ozone radiative forcing (RF) and methane turnover in the context of the impact of climate change (it should be pointed out that studies of the contribution of tropospheric ozone, O_{3T}, as a GHG as well as assessments of the potential impact of global warming on permafrost melting and methane emissions, are still at the preliminary stage of development). Interactive simulations of climate dynamics and O_{3T} changes during the time period 1990–2100 for scenarios of "high" (A2) or "central" (B2) cases of CO_2 emissions resulted in tropospheric ozone RF equal to +0.27(A2) or +0.09(B2), if climate–ozone coupling was neglected, then relevant RF values were equal to +0.43 (+0.22). When climate change was included, CH_4 lifetime fell by 0–5%. Hence, climate warming exerts a negative feedback on itself by enhancing O_{3T} and CH_4 destruction.

Three principal achievements have stimulated, during recent years, the progress in studying the Arctic environment (Dickson, 1999): (1) further development of observation programmes (including satellites and submarines); (2) declassification of the military Soviet–American archive of ocean "climatology" data; and (3) discovery of the fact that the climatic forcing in the region of Arctic and northern seas in the 1990s has increased in comparison with that observed during the previous century (a similar situation also took place with respect to climate dynamics indicators (such as the AO and North Atlantic Oscillation (NAO)).

Overland and Adams (2001) have pointed out that, "decadal differences between the 1990s and 1980s in winter (JFM) sea level pressure and 300 hPa zonal winds have an Arctic-centered character with nearly equal contributions from the Atlantic and Pacific sectors. In contrast, the differences between positive and negative AO composites defined from monthly values of principal components from the same period have similar magnitudes in the Pacific and Arctic, but have an additional large NAO signature in the Atlantic sector. Thus Arctic changes of decadal scales are more symmetric with the pole than suggested by the standard AO index definition. Change point analysis of the AO shows that a shift in value near 1989 is an alternative hypothesis to a linear trend. Analysis of zonal and meridional winds by longitudinal sectors shows the importance of the standing wave pattern in interpreting the AO, which supplements the view of the AO as a simple zonal average (annular) mode." Thus the AO should be considered as a physical phenomenon connected with the enhancement of the circumpolar vortex and relevant mass and temperature changes in the stratosphere.

By the end of the 1980s and the beginning of the 1990s a very strong NAO enhancement resulted in the powerful transport of warmer and fresher Norwegian Atlantic waters to the north of the Fram Strait and the Barents Sea. Entering the Arctic, the sublayer of Atlantic waters was becoming thinner, warming (by about 2°C) and increasing its horizontal extent (~20%). At shallower depths, the cold haloclyne (which thermally isolates the sea ice cover from the warm Atlantic layer located below) shifted towards the Euroasiatic Basin, which resulted in substantial changes of mass and energy balances of the ice cover surface. This and other phenomena have been studied within a number of recent programmes (Allison *et al.*, 2001). Of particular interest is the climatic impact of polynyas (Holland, 2001).

Alexeev (1998) has emphasized that the Arctic is in many respects a key part of the global climatic system where the strongest natural fluctuations of climatic characteristics develop. The global impact of the Arctic is primarily accomplished through the Arctic Ocean, which is capable of changing its structure circulation regime under the influence of changes in fresh water, salt, and heat exchange with the non-polar parts of the global system. The freshened upper layer and sea ice located above it turn out to be the most active components, with freshwater, heat, and salt transport being the major processes responsible for coupling between the high-latitude environment and its lower-latitude parts.

Specific features of the Arctic atmosphere such as phenomena of Arctic haze, as well as extended cloudiness and radiation, have been studied during the period of the First Global Atmospheric Research Programme (GARP) Global Experiment (FGGE) (Kondratyev *et al.*, 1992; Kondratyev, 1999a–c).

Important progress has been achieved in the field of Arctic climate diagnostics (Adamenko and Kondratyev, 1999; Gillett *et al.*, 2002; Moritz *et al.*, 2002). Basic features of Arctic climate dynamics have been demonstrated, such as strong space and time variability of various scales. It was shown, for instance, that as far as interannual lower troposphere variations are concerned, before the 1980s negative anomalies had prevailed, while later on, for the whole troposphere, positive temperature anomalies were typical. Total polar atmosphere energy (potential plus internal) during the previous 40 years has not changed, however.

A much more difficult situation exists in the field of numerical modelling of high-latitude climate change. It has been mentioned in the Report from the Arctic Climate Impact Assessment Modelling and Scenarios (2001) that current estimates of future changes to the Arctic vary significantly. The model results disagree as to both the magnitude of changes and the regional aspects of these changes.

5.2.1 Climate and cryosphere

An important step forward in studying the Arctic environment is the Climate and Cryosphere (CliC) Project (Allison *et al.*, 2001). The term "cryosphere" describes those portions of the Earth's surface where water is in solid form. This includes all kinds of ice and snow and frozen ground such as permafrost. The cryosphere is an important part of the global climate system. It is strongly influenced by temperature, solar radiation, and precipitation, and, in turn, influences each of these properties. It also has an effect on the exchange of heat and moisture between the Earth's surface (land or sea) and the atmosphere, on clouds, on river flow (hydrology), and on atmospheric and oceanic circulation. Parts of the cryosphere are strongly influenced by changes in climate. The cryosphere may therefore act as an early indicator of both natural and human-induced climate changes.

As a core project of the World Climate Research Programme (WCRP), the CliC Project encourages and promotes research into the cryosphere and its interactions as part of the global climate system. It seeks to focus attention on the most important issues, encourage communication between researchers with common interests in cryospheric and climate science, promote international cooperation, and highlight the importance of this field of science to policy makers, funding agencies, and the general public. CliC also publicizes significant findings regarding the role of the cryosphere on the climate, and recommends directions for future study.

CliC aims to improve understanding of the cryosphere and its interactions with the global climate system, and to enhance the ability to use parts of the cryosphere for the detection of climate change. The scientific goals of CliC are to:

- improve understanding of the physical processes through which the cryosphere interacts within the climate system;
- improve the representation of cryospheric processes in climate models;
- assess and quantify the impacts and consequences of past and future climatic variability on components of the cryosphere; and
- enhance the observation and monitoring of the cryosphere.

To attain these goals, CliC seeks to develop and coordinate national and international activities aimed at increasing the understanding of four main scientific themes:

1. Interactions between the atmosphere and snow and ice on the land surface.
2. Interactions between glaciers and ice sheets and sea level.
3. Interactions between sea ice, oceans, and the atmosphere.
4. Interactions of the cryosphere with the atmosphere and oceans on a global scale.

CliC encourages the use of observations, process studies, and numerical modelling within each of the above topic areas. In addition, CliC promotes the establishment of new cryospheric monitoring programmes.

The cryosphere is also considered as an indicator of climate variability and change. It has been pointed out in Allison *et al.* (2001) that, "Atmosphere–snow/ice–land interactions are concerned with the role of the terrestrial cryosphere within the climate system and with improved understanding of the processes, and of observational and predictive capabilities applicable over a range of time and space scales. Better understanding of the interactions and feedback of the land–cryosphere system and their adequate parameterization within climate and hydrological models are still needed. Specific issues include the interactions and feedback of terrestrial snow and ice in the current climate and their variability, in land surface processes, and in the hydrological cycle. Improved knowledge is required of the amount, distribution, and variability of solid precipitation on a regional and global scale, and its response to a changing climate. Seasonally frozen ground and permafrost modulate water and energy fluxes, and the exchange of carbon, between the land and the atmosphere. How do changes of the seasonal thaw depth alter the land–atmosphere interaction, and what will be the response and feedback of permafrost to changes in the climate system? These issues require improved understanding of the processes and improved observational and modelling capabilities that describe the terrestrial cryosphere in the entire coupled atmosphere–land–ice–ocean climate system.

Over a considerable fraction of the high-latitude global ocean, sea ice forms a boundary between the atmosphere and the ocean, and considerably influences their interaction. The details and consequences of the role of sea ice in the global climate system are still poorly known. Improved knowledge is needed about the broadscale time-varying distributions of the physical characteristics of sea ice, particularly ice thickness and the overlying snow-cover thickness, in both hemispheres, and the dominant processes of ice formation, modification, decay and transport which influence and determine ice thickness, composition, and distribution. We do not know how accurate present model predictions of the sea ice responses to climate change are, since the representation of much of the physics is incomplete in many models, and it will be necessary to improve coupled models considerably to provide this predictive capability.

Key issues on the global scale are: understanding the direct interactions between the cryosphere and atmosphere; correctly parameterizing the processes involved in models; and providing improved data sets to support these activities. In particular, improved interactive modelling of the atmosphere–cryosphere surface energy budget and surface hydrology, including freshwater runoff, is required.

The scientific strategy for the CliC Project is similar in each of the areas of interaction: a combination of measurement, observation, monitoring, analysis, field process studies, and modelling at a range of time and space scales. The CliC modelling strategy must address improved parameterization in models of the direct interactions between all components of the cryosphere, the atmosphere, and the ocean. It will need to do this on a variety of scales from the regional to global, and with a hierarchy of models ranging from those of individual processes to fully coupled climate models. It will also be essential to provide the improved data sets needed for validation of models and parameterization schemes.

Table 5.1. Areal and volumetric extent of major components of the cryosphere.
Allison *et al.* (2001).

Component	Area (10^6 km^2)	Ice volume (10^6 km^3)	Sea level equivalent (m) (a)
Land snow cover (b)			
Northern hemisphere late January	46.5	0.002	
Late August	3.9		
Southern hemisphere late July	0.85		
Early May	0.07		
Sea ice			
Northern hemisphere late March	14.0 (c)	0.05	
Early September	6.0 (c)	0.02	
Southern hemisphere late September	15.0 (d)	0.02	
Late February	2.0 (d)	0.002	
Permafrost (underlying the exposed land surface, excluding Antarctica and southern hemisphere high mountains)			
Continuous (e)	10.69	0.0097–0.0250	0.024–0.063
Discontinuous and sporadic	12.10	0.0017–0.0115	0.004–0.028
Continental ice and ice shelves			
East Antarctica (f)	10.1	22.7	56.8
West Antarctica and Antarctic Peninsula (f)	2.3	3.0	7.5
Greenland	1.8	2.6	6.6
Small ice caps and mountain glaciers	0.68	0.18	0.5
Ice Shelves (f)	1.5	0.66	–

Note: symbols (a)–(f) correspond to the comments in the text.

Table 5.1 data characterize major components of the cryosphere averaged over the last five years of the 20th century. It has been mentioned in Allison *et al.* (2001) that the processes operating in the coupled cryosphere climate system involve three timescales – intraseasonal–interannual, decadal–centennial, and millennial or longer. The longest timescale is addressed through the IGBP Pilot Analysis of the Global Ecosystems (PAGES) Programme, although abrupt climate shifts evidenced in ice core and ocean sediment records (Heinrich events, involving extensive deposition of ice-rafted detritus in the North Atlantic) are also highly relevant to CliC. The other two timescales are commensurate with WCRP interests, as manifest in the Arctic Climate System Study (ACSYS), the Global Energy and Water Cycle Experiment (GEWEX) and the Climate Variability Experiment (CLIVAR). In the space domain, cryospheric processes and phenomena need to be investigated over a wide range of scales from metres to thousands of kilometres, taking into account:

(a) Sea level equivalent does not equate directly with potential sea level rise, as a correction is required for the volume of the Antarctic and Greenland ice sheets

that are presently below sea level. 400,000 km^3 of ice is equivalent to 1 m of global sea level.
(b) Snow cover includes that on land ice, but excludes snow-covered sea ice.
(c) Actual ice areas, excluding open water. Ice extent ranges between approximately 7.0 and 15.4×10^6 km^2.
(d) Actual ice area excluding open water. Ice extent ranges between approximately 3.8 and 18.8×10^6 km^2. Southern hemisphere sea ice is mostly seasonal and generally much thinner than Arctic sea ice.
(e) Data calculated using the Digital Circum-Arctic Map of Permafrost and Ground-Ice Conditions and the Global Backscatter Experiment GLOBE-1 km Elevation Data Set.
(f) Ice sheet data include only grounded ice. Floating ice shelves, which do not affect sea level, are considered separately.

5.2.2 Cryosphere dynamics

Studying the cryosphere dynamics is of importance for many applications, some of which are illustrated in the data of Table 5.2.

Four overarching goals that address major concerns for the WCRP can be identified (Allison *et al.*, 2001). These are:

1. To improve understanding of the physical processes and feedbacks through which the cryosphere interacts within the climate system.
2. To improve the representation of cryospheric processes in models to reduce uncertainties in the simulations of climate and predictions of climate change.
3. To assess and quantify the impacts of past and future climatic variability and change on components of the cryosphere and their consequences, particularly for global energy and water budgets, frozen ground conditions, sea level change, and the maintenance of polar sea ice cover.
4. To enhance the observation and monitoring of the cryosphere in support of the process studies, model evaluation, and detection of change.

Specific questions that help define the primary tasks of the CliC Project are:

(i) How stable is the global cryosphere?
- How well do we understand and model the key processes involved in each cryospheric component of the climate system?
- How do we best determine the rates of change in the cryospheric components?

(ii) What is the contribution of glaciers, ice caps, and ice sheets to changes in global sea level on decadal to century time scales?
- How can we reduce the current uncertainties in these estimates?

(iii) What changes in frozen ground regimes can be anticipated for decadal to century timescales that would have major socio-economic consequences, either directly or through feedback on the climate system?

Table 5.2. Examples of socio-economic sectors affected by changes in the cryosphere. Allison *et al.* (2001).

Socio-economic component	Cryosphere factor
Direct effects	
Loss of coastal land and population displacement	Land ice melt contribution to sea level
Transportation	
Shipping	Iceberg hazard; sea ice extent, thickness
Barge traffic	Freshwater ice season
Tundra roads	Freshwater ice roads; frozen ground thaw
Road/rail traffic	Freeze events; snowfall
Water Resources	
Consumption	Undetermined
Irrigation	Snow/glacier melt runoff
Hydropower	Undetermined
Agriculture	Moisture recharge extremes
Hydrocarbon and mineral resource development	Icebergs and sea ice; frozen ground duration and thickness
Wildlife population	Snow cover; frozen ground and sea ice
Recreation/safety	Snow cover; avalanches
Indirect effects	
Enhanced greenhouse	Thaw of clathrates
Traditional lifestyles (Arctic, sub-Arctic, and high mountains)	Changes in sea ice and freshwater ice, snow cover, and frozen ground
Tourism/local economies	Loss of glaciers; shorter snow season
Insurance sector	Changes in risk factor

(iv) What will be the annual magnitudes, rates of change, and patterns of seasonal redistribution in water supplies from snow and ice fed rivers under climate change?
(v) What will be the nature of changes in sea ice mass balance in both polar regions in response to climate change?
(vi) What is the likelihood of abrupt climate changes resulting from regime changes in ice shelf (ocean and sea ice) – ocean interactions that impact the ocean thermohaline circulation?
(vii) How do we monitor cryospheric components as indicators of change in the climate system?

Monitoring cryosphere dynamics is a key aspect of high-latitude environmental studies, especially because of rather controversial information concerning ice cover dynamics. It is true, for instance, for ice thickness observations. Holloway and Sou (2001) have pointed out that "while submarine records had indicated a stunningly rapid thinning, model results show that the position of submarine observations was exceptionally biased towards regions of thinning."

5.2.3 Arctic environmental pollution

The Arctic region exploration strategy in a broad context of biospheric studies has been discussed in details by Matishov and Matishov (2001), where the necessity of an ecosystem approach to studying land and marine biota has been particularly emphasized, as well as conditions of socio-economic development in high-latitude regions. Aibulatov (2000), and Matishov and Matishov (2001) have discussed the general problems of high-latitude environmental dynamics with special emphasis on radioactive pollution as an echo of the cold war. Aibulatov (2000) has analysed principal sources of artificial radioisotopes in the Russian Arctic seas such as atomic explosions at Novaya Zemlya, the global radionuclide background as a result of the overall nuclear tests conducted on the planet, Russian chemical and mining plants, the Chernobyl accident, western European radiochemical plants, solid and liquid radioactive waste dumping in the Barents and Kara Seas, the Northern Military Marine and its bases, and atomic submarine construction and maintenance facilities and Atomflot (atomic fleet) – Murmansk Shipping Company.

Studies of the distribution of ^{137}Cs, ^{90}Sr, and $^{239,240}Pu$ in the water masses of the North, Norwegian, Barents, Kara, White, and Laptev Seas have resulted in the following conclusions (Aibulatov, 2000):

1. The general level of radioactive contamination of the waters of the Arctic seas, except for several local areas, is characterized at the present time by little difference in comparison with background levels (~6 Bq/kg).
2. The radioactive pollution of the waters of the North and Norwegian Seas is entirely due to the emissions from radiochemical plants located in western Europe.
3. The contamination of waters of the Barents, White, Kara, and Laptev Seas is due to both local (Russian) sources and western European plants.
4. The field observations in the Kara Sea in 1992–1995 have resulted in the conclusion that there have been no substantial radioactive emissions from the burial sites in the area.
5. The contribution of the Ob′ and Yenisey river runoff to the overall radioactive transport is not significant at the present time, except in cases of extremely heavy floods (which happen very rarely).
6. Compared to the open waters of the Arctic Ocean, shelf seas of the Russian Arctic are more heavily contaminated.

Aibulatov (2000) has pointed out that judging from the ^{137}Cs-distribution patterns in the Kara Sea, it becomes evident that the Yenisey and Ob′ rivers (less evident, however, in the latter case) should be considered as transport channels for inputs of technogenic radionuclides to the Arctic Ocean waters. There are radioactive sources in the ocean as well. The ^{137}Cs activity level reached its maximum in 1984 and was equal to 245 Bq/kg (in open sea); during the 1990s (1993) this level was found to be equal to 100 Bq/kg (in the Yenesey estuary).

Arctic fjords have been classified into categories of comparatively clean, contaminated, heavily contaminated, and potentially contaminated. Contaminated areas include, for instance, Kola Gulf and, probably, all the fjords of the northern Kola Peninsula west of Murmansk. The content of radionuclides in phytobenthos, in the coastal zone east of Murmansk, is low. Evidently, there has not recently been any serious radionuclide penetration into this area. A rather low gamma-nuclide level (1–3 Bq/kg) is typical for the zoobenthos of the Barents Sea. This is also true for the Kara Sea.

An impact of all the sources of radioactivity in the zone of the Arctic coast on the local population has not been assessed reliably enough. It was particularly difficult to separately identify natural and anthropogenic components of such an impact. Aibulatov (2000) has discussed the future research of the Russian Arctic radioactive pollution, including:

- Development of a coordinated Russian Arctic Sea Radioactivity Ecological Monitoring Programme.
- Assessments of impacts of different radioactive sources on the contamination of the Arctic marine environment including water basins, land, and atmosphere.
- Studies of detailed space and time variability of various long-lived technogenic radionuclides in bottom sediments.
- A detailed examination of all Novaya Zemlya fjords in connection with the dumping of radioactive waste.
- Research into the impact of radioactive pollution on the Arctic marine ecosystem dynamics.
- Studying medical aspects of environmental pollution in the Arctic.

5.2.4 High-latitude ecodynamics

The fundamental study of radioactivity of the Arctic and sub-Arctic marine ecosystems has been undertaken by Matishov and Matishov (2001), which resulted in the substantiation of a new branch of science – radiational ecological oceanology. Investigations have been conducted looking at the level of artificial radionuclide concentration in both environment and biota of the bays and inlets (the Kola, the Chernaya, the West Litsa), where radioactively dangerous objects are located. In this context, a classification has been suggested for coastal areas (bays, gulfs, and fjords) in accordance with contamination levels for bottom sediments. The important role of biofilters has been discovered for pelagic and coastal zones during the processes of self-purification of water reservoirs and transport of radionuclides from water to bottom sediments. For the first time the levels of ^{137}Cs, ^{90}Sr, $^{239,240}Pu$ concentrations for different types and populations of sea organisms were measured. Migrations of radioisotopes along the trophic chains (from macrophytes and plankton to zoobenthos, fish, birds, seals, and whales) were also studied. The assessments of comparative contributions of global, regional, and local sources of radioactive environmental contamination during the time period from nuclear tests until present-day, have been analysed and used as a source of information for environmental predictions. An important optimistic conclusion concerning the consequences of potential accidents is that for all the prescribed scenarios of radioactive emissions, it is highly improbable that large-scale contamination of the Arctic Ocean will take place with disasterous impacts on marine bioresources. A high biological assimilation capacity in combination with specific features of hydrodynamic and other processes are supposed to serve as a barrier against dangerous pollution of the Arctic Ocean.

Kalabin (2000) has accomplished a study of the environmental dynamics and industrial potential of the Murmansk region, the most urbanized and industrially developed trans-polar region of the planet. Under these conditions, specific features of environmental dynamics result in the enhancement of anthropogenic impacts. In this context, Kalabin (2000) has analysed critical environmental loads for some of the northern ecosystems and emphasized a necessity to investigate their assimilation (buffer) capacity as a principal aspect of sustainable functioning natural systems. The solution of regional problems of sustainable development requires a careful analysis of the interaction between ecodynamics and socio-economic development.

5.2.5 Scientific field programmes

The progress achieved in studying the Arctic environment variability is due to the accomplishment of a number of international research programmes. Of particular importance is the ACSYS developed in 1991 as the WCRP initiative and a practicable programme for the next decade to assess the role of the Arctic in the global climate (Kondratyev *et al.*, 1996; Steffen *et al.*, 2002). Five areas are emphasized: (1) ocean circulation; (2) sea ice climatology; (3) the Arctic atmosphere; (4) the hydrological cycle; and (5) modelling. The scientific goals of ACSYS, which started its main observational phase on January 1994 and will continue for a ten-year period, includes the following three main objectives:

1. understanding the interaction between the Arctic Ocean circulation, ice cover, and the hydrological cycle;
2. initiating long-term climate research and monitoring programmes for the Arctic; and
3. providing a scientific basis for an accurate representation of Arctic processes in global climate models.

The Arctic Ocean Circulation Programme of ACSYS consists of four components:

1. the Arctic Ocean Hydrographic Survey to collect a high-quality hydrographic database representative of the Arctic Ocean;
2. the Arctic Ocean Shelf Studies which are aimed at understanding how the

shelf processes partition salt and freshwater components and at defining the dynamics and thermodynamics of the shelf waters as well as other processes;
3. the Arctic Ocean Variability Project designed to assess the variability of the circulation and density structure of the Arctic Ocean; and
4. the Historical Arctic Ocean Climate Database Project aimed at establishing a universally available digital hydrographic database for the Arctic Ocean, for analysis of climate-related processes and variability, and to provide a data set suitable for initialization and verification of Arctic climate and circulation models.

The ACSYS sea ice programme includes three main components:

1. establishing an Arctic basin-wide sea ice climatology database;
2. monitoring the export of sea ice through the Fram Strait; and
3. the study of Arctic sea ice.

One of the main tasks of the ACSYS Arctic sea ice study is to establish a database of ice thickness and ice velocity. Such data will be supplied by the WCRP Arctic Ice Thickness Project, the International Arctic Buoy Programme, sonar profiling from naval submarines and unmanned vehicles, airborne oceanographic lidars, and polar satellites carrying appropriate instruments.

The Arctic atmosphere provides the dynamic and thermodynamic forcing of the Arctic Ocean circulation and sea ice. Key directions of research include problems such as: cloud–radiation interaction, air–sea interaction in the presence of ice cover (impacts of polynyas are of special interest), Arctic haze, etc.

Primary ACSYS efforts within the project of the hydrological cycles in the Arctic region are aimed at:

- the documentation and intercomparison of solid precipitation measurement procedures used in high latitudes; and
- the development of methodologies for determining areal (regional) distributions of precipitation from station data.

There are two relevant data archiving efforts: Arctic Precipitation Data Archive (APDA) and Arctic Runoff Database (ARDB).

The principal purpose of the ACSYS modelling programme is simulations of climate variations in polar regions which arise from the interaction between atmosphere, sea ice, and ocean.

Apart from the ACSYS projects described above, a number of new research programmes have been developed, such as the Study of Environmental Arctic Change (SEARCH), which is an interdisciplinary, multiscale programme dedicated to understanding the complex of interrelated changes that have been observed in the Arctic environment in the past few decades (Dong *et al.*, 2001; Halloway and Sou,

2001). SEARCH is envisioned as a long-term effort of observations, modelling, process studies, and applications with emphasis on five major thematic areas:

- human society;
- marine/terrestrial biosphere;
- atmosphere and cryosphere;
- ocean; and
- integrated projects/models/assessment.

The ARCSS is an interdisciplinary programme whose principal goals are to (1) understand the physical, geological, chemical, biological, and sociocultural processes of the Arctic system that interacts with the total Earth system and thus contributes to or is influenced by global change, in order to (2) advance the scientific basis for predicting environmental change on a seasons, to centuries, timescale, and for formulating policy options in response to the anticipated impacts of global change on humans and societal support systems. The following four scientific thrusts are included as the central aims of ARCSS:

- to understand the global and regional impacts of the Arctic climate system and its variability;
- to determine the role of the Arctic in global biogeochemical cycling;
- to identify global change impacts on the structure and stability of arctic ecosystems; and
- to establish the links between environmental change and human activity.

ARCSS has five linked ongoing components: Ocean/Atmosphere/Ice Interactions (OAII); Land/Atmosphere/Ice Interactions (LAII); paleoenvironmental studies (including the Greenland Ice Sheet Project Two (GISP2) and Paleoclimates of Arctic Lakes and Estuaries (PALE); Synthesis, Integration, and Modelling Studies (SIMS); and Human Dimensions of the Arctic System (HARS). Aagard (1998) discussed the basic problems with a multidisciplinary look at the Arctic Ocean, including: physical and chemical studies; biological studies; contaminant studies; measurements of the properties and variability of the ice cover and of the surface radiation budget; studies of atmospheric chemistry; and geological observations.

LAII research has three main goals (*Witness the Arctic* 1994):

1. to estimate important fluxes in the region, including the amount of carbon dioxide and methane reaching the atmosphere, the amount of river water reaching the Arctic Ocean, and the radiative flux back to the atmosphere;
2. to predict how possible changes in the Arctic energy balance, temperature, and precipitation will lead to feedbacks affecting large areas; this incorporates changes in water budget, duration of snow cover, extent of permafrost, and soil warming, wetting, and drying; and

3. to predict how the land and freshwater biotic communities of the Arctic will change, and how this change will affect future ecosystem structure and function.

A major LAII research project is the Flux Study; its principal purpose is to provide a regional estimate of the present and future movement of materials between the land, atmosphere, and ocean in the Kuparuk River Basin in northern Alaska.

Of the nineteen LAII projects three are part of the International Tundra Experiment (ITEX), which looks at the response of plant communities to climate change. Three others are concerned with atmosphere processes, including studies of the weather patterns affecting snowmelt, Arctic-wide temperature trends, and water vapour over the Arctic and its relationship to the atmospheric circulation and surface conditions. One project deals with the response of large birds to climate and sea level change at river deltas, and one studies the balance and recent volume changes of the McCall Glacier in the Brooks Range.

Synthesis, integration, and modelling studies are intended to foster linkages and system-level understanding. Research on the past contemporary relationship between humans and global climate change is thought to be critical to understanding the consequences of global change in the Arctic.

There are a number of ARCSS data projects, including: LAII Flux Study Alaska North Slope data sampler CD-ROM; OAII Northeast Water (NEW) Polynya Project CD-ROM; Arctic solar and terrestrial radiation CD-ROM, etc.

A list of the OAII components includes the joint US/Japan cruise, the Western Arctic Mooring Project, and the Northeast Water Polynya Project. Among other OAII projects the most notable are the US/Canada Arctic Ocean Section and the Surface Heat Budget of the Arctic Ocean (SHEBA) Project.

An outstanding effort was accomplished in 1994 within the Canada/US 1994 Arctic Ocean Section when two icebreakers entered the ice in the northern Chukchi Sea on 26 July, 1994, reached the North Pole on 22 August, and left the ice northwest of Spitsbergen on 30 August, thereby completing the first crossing of the Arctic Ocean by surface vessels. This voyage will greatly alter our understanding of biological productivity, the food web, ocean circulation and thermal structure, the role of clouds in the summer radiation balance, the extent of contamination and spreading pathways (especially related to radionuclides and chlorinated organics), and the extent and effects of sediment transport by sea ice (*Witness the Arctic* 1994).

In connection with the SHEBA project, the US Department of Energy's Atmosphere Radiation Measurement (ARM) Programme indicated its intention to develop a Cloud and Radiation Testbed (CART) facility on the north slope of Alaska. The principal focus of this programme will be on atmospheric radiative transport, especially as modified by clouds (which impacts the growth and decay of sea ice), as well as testing, validation, and comparison of radiation transfer models in both the ice pack and arctic coastal environment.

A special place is occupied by the Russian–American Initiative on Shelf–Land Environments in the Arctic (RAISE) with the principal goal of facilitating ship-based research in the Russian Arctic (*RAISE* ..., 2001). Earlier relevant land-based research projects under the RAISE umbrella included the studies of:

- organic material and nutrient fluxes from Russian rivers;
- seasonal flooding dynamics along rivers; and
- reconstruction of late Pleistocene glacial and sea level history on Wrangel Island.

New scientific topics in the near-shore waters of the Russian continental shelf will include a broad range of studies: from the biogeochemical fate of organic materials contributed to the Arctic Ocean by shoreline erosion and river runoff to the social and biological impacts of changes in sea ice distribution.

The western Arctic Shelf–Basin Interactions (SBI) project, sponsored by ARCSS and the US Office of Naval Research, is investigating the Arctic marine ecosystem to improve our capacity to predict environmental change. The SBI Phase II Field Implementation Plan (2002–2006) focuses on three research topics in the core study area:

- northward fluxes of water and bioactive elements through the Bering Strait input region;
- seasonal and spatial variability in the prediction and recycling of biogenic matter on the shelf-slope area; and
- temporal and spatial variability of exchanges across shelf-slope region into the Canada Basin.

5.2.6 Priorities and perspectives

The recent meeting of the International Arctic Science Committee (IASC) has identified the following four science priorities (*Witness the Arctic* 1994):

1. Arctic processes relevant to global systems;
2. effects of global change on the Arctic and its peoples;
3. natural processes within the Arctic; and
4. sustainable development in the Arctic.

The following areas in Arctic global change research have been considered the most significant:

1. terrestrial ecosystem;
2. mass balance of glaciers and ice sheets;
3. regional cumulative impacts; and
4. human dimensions.

An important aspect of studying high-latitude environmental dynamics is an assessment of the impact of potential anthropogenic climate warming. In this context Frederick (1994) has formulated key issues of integrated assessments of the impact of climate change on natural resources. Specific project objectives include:

1. characterizing the current state of natural science and socio-economic modelling of the impacts of climate change and current climate variability on forests, grasslands, and water;
2. identifying what can be done currently with impact assessments and how to undertake such assessments;
3. identifying impediments to linking biophysical and socio-economic models into integrated assessments for policy purposes; and
4. recommending research activities that will improve and remove impediments to the state-of-the-art model integration.

The following questions are supposed to be answered:

- How will the overall system (physical–biological–economic) respond to various imposed stresses?
- How do the uncertainties in the component models add up to give an overall system response uncertainty?
- Is society made more vulnerable to extreme natural events either by changing those events or by reducing the ability to respond with corrective action?
- How likely is it that the consequences of climate change will be severe or catastrophic?
- What is at risk and when is it at risk?
- What are the likely impacts on the landscape and the hydrological system?
- How might the boundary conditions and the overall productivity of the forests, grasslands, and other rangelands be affected?
- How might increasing carbon dioxide levels affect crops and food supplies for humans, livestock, and wildlife?
- What are the socio-economic consequences of these physical and biological changes?
- What are the likely consequences for ecosystems of mitigation actions?
- Can the costs associated with climate change be reduced through natural adaptation of ecosystems or policy-initiated adaptation?

Frederick (1994) has emphasized that the accumulated results of many regional and local climate impact assessments may help provide informed answers to these questions. Nevertheless, the uncertainties surrounding both the nature and the impacts of any future climate change are likely to remain very large, precluding precise estimates of the net benefits associated with alternative policy responses. Even if the range of uncertainty were diminished, it might still be difficult to justify specific measures on narrow economic grounds because (as noted above) the impacts on natural resource systems are apt to be poorly reflected in standard cost-benefit analysis.

Mendelsohn and Rosenberg (1994) have formulated the following questions relevant to the global warming effects in the area of ecological and water resources:

- Do changes in ecosystems provide important feedbacks to the natural carbon, nitrogen, and methane cycles? For example, will the natural sinks or sources be affected by changing precipitation, temperature, and CO_2 levels?
- What are the appropriate output measures of ecosystem component models?
- What are the ecological effects of climate change that policy analysts use to determinate the importance of an ecosystem change?
- What climate change driven shifts in ecosystem boundaries can be predicted?
- Will these effects be subtle and small or large and dramatic and over what time frame and spatial dimensions will they account?
- Will climate change cause a change in the productivity of valuable market or non-market species? For example, to what extent will some forests grow more quickly or more slowly. Will desired non-market species, such as bear, elk, and bald eagles, be more or less plentiful?
- What species could be lost with rapid climate changes? How do the vulnerable species break down by type and geographical distribution? How should conservation policies adapt to a world requiring change?
- How are ecosystems likely to change as the climate evolves over time: will there be a large increase in early succession species and where?
- How will average flows in rivers change with greenhouse warming? How will these flows change over seasons? Will the probabilities of catastrophic events change?
- What values do people assign to the changes in ecosystems by climate change? Which changes are important and which are minor? Can a value be assigned to non-use?
- How much should society be willing to pay to reduce the probability of losing specific species? If different scenarios favour different species, how should society trade between these outcomes?
- What impact do ecosystem changes have upon the economy? For example, how will climate change affect grazing, commercial fishing, timber, or commercial tourism?

It has been suggested that priority programme areas and relevant projects should be as follows:

1. Impacts of global changes on the Arctic region and its peoples.
   - Regional cumulative impacts.
   - Effects of increased UV radiation.
2. Arctic processes of relevance to global systems.
   - Mass balance of glaciers and ice sheets.
   - Terrestrial ecosystems and feedback on climate change.

3. Natural processes within the Arctic.
   - Arctic marine/coastal/river systems.
   - Disturbance and recovery of terrestrial ecosystems.
4. Sustainable development in the Arctic.
   - Sustainable use of living resources.
   - Dynamics of Arctic populations and ecosystems.
   - Environmental and social impacts of industrial development.

Future priorities of the ARCSS include the following research questions (Aagard, 1998): How will the Arctic climate change over the next 50 to 100 years? How will human activities interact with future global change to affect the sustainability of natural ecosystems and human societies? How will changes in Arctic biogeochemical cycles and feedbacks affect Arctic and global systems? How will changes in Arctic hydrological cycles and feedbacks affect Arctic and global systems? Are the predicted changes in the Arctic detectable?

Important perspectives are connected with paleoenvironmental studies by the Paleoenvironmental Arctic Sciences (PARCS) community (*PARCS...*, 2000) which have a principal aim of answering the question: how much do recent observations of climate change in the Arctic reflect natural climate cycles? Relevant major topics include:

- the medieval warm period (apparently, AD 1000–1400) and Little Ice Age (approximately AD 1400–1850);
- high-amplitude Holocene climate cycles; and
- the possible connection of the onset of neoglacial (a mid-Holocene cooling, particularly evident at high northern latitudes) with shifts in the frequency and amplitude of such climate cycles.

According to the PARCS, there are very warm past scenarios that can serve as analogues for future climate warming:

- the early Holocene, when the Arctic had experienced high summer insolation anomalies; and
- the last interglacial period (marine isotope stage 5), which appears as a very strong warming in the paleorecord approximately 125,000 years ago.

Key topics of investigation in relation to these periods are:

- feedbacks and non-linear changes (surprises) as consequences of strong warming – particularly the role of sea ice, ice sheets, and land surface cover; and
- implications of strong warming for Arctic and global carbon budgets.

To summarize what has been mentioned in connection with recent Arctic environmental programmes, it must be emphasized that relevant information was not at all exhaustive (*IASC Project Catalogue,* 2001). An obvious conclusion is that the number of programmes is too great. A clear necessity exists of better coordination of all ongoing efforts and their "regularization".

Vörösmarty *et al.* (2001) are correct in their conclusion that "understanding the full dimension of Arctic change is a fundamental challenge to the scientific community over the coming decades and will require a major new effort at interdisciplinary synthesis. It also requires an unprecedented degree of international cooperation."

5.3 ARCTIC ATMOSPHERE COMPOSITION

The most important feature of the Arctic atmosphere is its high level of variability and complexity of chemical and photochemical elements, being especially dynamic in the stratosphere (Kondratyev *et al.*, 1996). One special problem is that of interaction between ocean and land surfaces and the atmosphere as well as minor gas components (mols) – aerosols – cloud – climate interactions (Kondratyev *et al.*, 1992, 2002a,b). Kane and Reeburgh (1998) have pointed out that large quantities of carbon (10–20% of the Earth's soil carbon inventory) are sequestered in the near-surface soils of this region. This stored carbon represents a significant potential source of carbon dioxide and methane for the atmosphere (see also Kondratyev and Isidorov, 2001). What are the variables and processes controlling the fluxes of CO_2 and CH_4 from the Arctic ecosystems to the atmosphere, and how will these fluxes change in response to variations in climate? Answers to these questions were the goals of the US National Science Foundation (NSF)-sponsored LAII Flux Study, an interdisciplinary study that focused on the Kuparuk River Basin (8140 km^2) and its subwatersheds on the north slope of Alaska. The LAII Flux Study was conducted over a 4-year period (1993–1996) by a group of soil scientists, hydrologists, atmospheric scientists, biogeochemists, modellers, and ecosystem scientists.

The hypotheses addressed by flux study projects and by many other recent studies are as follows:

- Arctic terrestrial and freshwater systems are presently a source of CO_2 and CH_4.
- Climate-induced changes of air temperature and soil moisture will be large enough to provoke changes in trace gas fluxes in the Arctic terrestrial systems.
- Changes in trace gas fluxes will result in climate feedbacks that are comparable to or greater than albedo and vegetation feedbacks.
- In the next 50 years, changes in soil moisture will exercise greater control over regional trace gas fluxes than either air temperature or vegetation; after 50 years, vegetation changes will contribute more significantly to the Arctic climate feedback.
- Climate change and the associated acceleration of soil organic decomposition will increase the nutrient flux to all water bodies.

- Arctic trace gas feedbacks will be sufficient to impact climates beyond the Arctic.

Bottenheim and Barrie (1996) have discussed the central aspects of what drives tropospheric chemical reactions, and have indicated how the unique features of the polar troposphere might be expected to result in specific trends, different from other regions of the globe. It has been shown that under winter conditions (when there is no direct solar radiation) the chemical composition of the troposphere is controlled by chemical reactions at mid-latitude sources followed by transport modulated differences in the concentrations. In the case of the springtime, several measurements are in agreement with predictions from chemical models of the atmosphere (e.g., the seasonal trends in NO_x compounds, the increase in oxidation products of SO_2, and the decrease in hydrocarbons reactive with photochemically generated reactions). Some new features have also been identified, such as halogen atoms driven chemistry, and the apparent efficient formation of NHO (nitrone), which should create much higher OH radical concentrations than anticipated. Heterogeneous chemical reactions on the surface of ice and aerosol particles plays an important role in both the troposphere and stratosphere.

Landers *et al.* (1995) have emphasized that Arctic contaminant research is expensive and current international resources are restricted. It is incumbent upon current and future Arctic research programmes to focus efforts where the greatest and most relevant information can be gained. It has been attempted in this context to help guide future work to focus on the most pressing information needs. The following several summary points relate to environmental research in the Arctic and outside the Arctic as well; some may also relate to the Arctic as a unique region, where continued scientific investigation of airborne contaminants promises to reveal issues of global significance regarding airborne contaminant distribution, the physical and chemical processes involved, and their ecological effects. Arctic contaminant problems are multidisciplinary and the best chance of advancing our scientific information base is through interdisciplinary studies that are integrated from the start and that incorporate traditional knowledge. Human and ecological biomarkers and bioindicators should be developed that are sensitive, relatively inexpensive to measure, and broadly available. The most persistent and deleterious chemical contaminants should now be controlled globally; other chemicals should be substituted that have been adequately screened for their effects on the developing endocrine, immune, and nervous systems, as well as for traditional health endpoints.

5.3.1 MGCs

There are two MGCs which are the most important: carbon dioxide (in the context of its climatic impact and biospheric significance) and ozone (both tropospheric and stratospheric). Since problems of ozone studies will be discussed later, we will first of all consider carbon dioxide as well as the problems of oceanic dimethylsulphide (DMS) emissions, atmospheric pollution of the high-latitude environment, and others problems.

5.3.1.1 Carbon dioxide

The global carbon cycle in general and its Arctic components have been discussed in detail in a number of recent publications (Anderson *et al.*, 1998; Aurela *et al.*, 2001; Houghton *et al.*, 2001; Kondratyev and Demirchian, 2000; Kondratyev and Isidorov, 2001; Kondratyev *et al.*, 2002a; Kondratyev and Krapivin, 2003b; Lafleur *et al.*, 2001; Watson *et al.*, 2000; Wofsy, 2001). Therefore, only a few new results will be mentioned here.

Schindler (1999) has pointed out that due to earlier interpretation of atmospheric CO_2 concentration data the imbalance between carbon released by man's activities and measured CO_2 concentrations in the atmosphere, could be explained by an uptake of carbon at the ocean surface. The rather large seasonal oscillation was regarded as a terrestrial signal, but one that remained in balance from year to year. The results obtained during the 1970s and 1980s have demonstrated, however, that both of these assumptions were incorrect. It has been shown, for instance, that in 1990 the annual carbon flux to the oceans was likely to be less than 0.5×10^{15} g of carbon per year – one quarter the amount necessary to balance the global carbon budgets.

As has been emphasized by Schindler (1999) and Nadelhoffer *et al.* (1999), convincing evidence exists that, just as it cannot be explained by the oceans, neither can the "missing carbon" be rationalized by an uptake by the northern temperate forests – which was until now, the favoured theory – with a maximum contribution to the missing sink for atmospheric CO_2 of about 0.25×10^{15} g of carbon per year.

There are several possibilities explaining where in the biosphere the missing carbon is hiding. In this context the following circumstances are of importance: (1) simple surface fluxes of CO_2 underestimate the role of the oceans; (2) it is possible that northern temperate forests that were stimulated by nitrogen in the past are now saturated with this element; (3) some other northern ecosystems may be responsible for carbon uptake; and (4) northern forests growth due to climate warming could contribute. It is quite probable that a combination of several small components can explain the missing sink.

An important step forward in assessing the contribution of the Russian Arctic Seas (the Barents, White, Kara, Laptev, East Siberian, and Chukchi Seas) into the formation of the global carbon cycle has been made by Romankevich and Vetrov (2001) through the analysis of field observation data obtained in 1993–2000. In this context, primary production of both phytoplankton and phytobenthos; secondary production of zooplankton, bacterioplankton and zoobenthos; vertical carbon fluxes in the photic zone-water column-seaflow system; lateral carbon fluxes from land; and balance and masses of carbon in the Russian Arctic Seas (RAS) have been assessed.

Principal specific features of the RAS, their ecosystem dynamics, and carbon flux changes are strongly pronounced annual variations due to relevant seasonality of all hydrological, physicochemical, and biological processes, as well as low temperatures during the whole year, contrasting insolation (Arctic summer and polar winter), continuous or interrupted presence of ice cover, and a most variable vertical structure of water masses (from strongly expressed stratification to complete homogeneity). Photosynthetic dynamics in the Arctic Seas is determined by the availability of sunlight. This is why in winter, photosynthesis is practically impossible, while in summer, a large part of the RAS is covered by ice with a highly reflective snow cover. Two different groups of seas may be identified within the RAS: the Barents and Chukchi Seas, which are characterized by a stronger influence of the comparatively warm waters of the Atlantic and Pacific Oceans than the Kara, White, Laptev, and East Siberian Seas where powerful river runoff determines the characteristic features of the carbon cycle.

The assessment made by Romankevich and Vetrov (2001) have shown that the main sources of organic carbon (C_{org}) in the RAS are photosynthetic production (160 Mt C_{org}/year), river runoff (23.6 Mt C_{org}/year), particulate and dissolved organic carbon (POC and DOC), and wave abrasion (3.8 Mt C_{org}/year). Underground runoff (1.65 Mt C_{org}/year) and eolian transport (0.4 Mt C_{org}/year) contribute much less C_{org}.

Oechel *et al.* (2000) have also emphasized the presence of a complex dependence of the carbon cycle in the Arctic on highly variable (in space and time) characteristics of phenology, hydrology, and climate, which leads to strong changes of CO_2 exchange between the atmosphere and the ecosystems. This is the reason why these data for Arctic ecosystems are so fragmentary. This was the motivation for simultaneous chamber, tower, and aircraft net CO_2 and energy flux observations during the ARCSS-LAII Flux Study in June–August (vegetation seasons) of 1994–1995 in the Alaska region. The results obtained demonstrate a significant disagreement between various techniques of CO_2 flux measurements.

Lafleur *et al.* (2001) have discussed the results of observations of CO_2 exchange between the atmosphere and sub-Arctic boreal woodland ecosystem with the use of the eddy correlation technique. The observations were conducted during the 1997 vegetation season (when environmental conditions were close to the average climatic situation) in the state of Manitoba (Canada): 58°45′N, 94°04′W. The observation results have revealed the existence of a comparatively high correlation ($r^2 = 0.51$) between the CO_2 flux and photosynthetically active radiation. A rather weak correlation has been discovered, however, between the ecosystem's respiration during the night-time and surface air temperature (SAT). The highest correlation ($r^2 = 0.20$) took place between respiration and SAT within the vegetation cover. Substantial annual variations have been revealed between day-time values with maximum levels of CO_2 assimilation (6.2 g CO_2/m^2/day) at the end of July, and average values of 1.72 g CO_2/m^2/day. Measured values of CO_2 fluxes were much lower than in the points located in central and southern parts of boreal woodlands and only slightly exceeded CO_2 fluxes in the tundra region.

Strongly pronounced space and time variability is also typical of other GHGs, such as methane – as has been shown by observations made by Fridborg *et al.* (1999) in north-eastern Greenland (74°28′N, 20°34′W). The analysis of observational data has led to the conclusion that during the summer of 1997 the Greenland valley studied, functioned as a sink for CO_2 at about 11.5 g C/m^2, while being a source of methane at approximately 0.85 g C/m^2. The driving factors of gas exchange between the surface and atmosphere were soil temperature and water table level.

5.3.1.2 DMS and nitrogen oxides

DMS is the dominating biogenic component emitted by the World Ocean to the atmosphere (Kondratyev, 1999b). Its reaction with hydroxyl (OH) (during the daytime) and NO_3 (during the night-time) leads to formation of sulphate gas compounds as well as aerosols (Sharma *et al.*, 1999). The mechanism of DMS oxidation is due to both homogeneous and heterogeneous chemical reactions with the subsequent formation of sulphur dioxide (SO_2), sulphate (SO_4^{2-}), and methanosulphate acid (MSA).

On a global scale, DMS emissions by the World Ocean reach approximately 15% of the global total sulphur emissions and about 50% of the natural sulphur emissions. An important contribution of DMS emissions in the summertime has been observed in the central part of the Arctic Ocean, which results in the formation of sulphate aerosols in the atmosphere (the relative contribution of biogenic sources can reach 30% and even more at the edge of the ice cover). Since the background number concentration of sulphate aerosol particles in the Arctic is very low (of the order of 10–100 cm^{-3}), a substantial increase of aerosol concentration due to DMS emissions results in the enhancement of the impact of aerosols as cloud condensation nuclei (CCN), on cloud albedo, and conditions of cloud formation.

Sharma *et al.* (1999) have conducted simultaneous gas chromatography measurements of DMS concentrations in both the atmosphere (about 20 m above the sea surface) and in the seawater (samples were taken at depths of about 1 m) during the period 17 July–5 October, 1994 in the course of the Canadian–American expedition on the *Polar Sea* ice-breaker in the region around North America. Measured DMS concentration in the atmosphere varied from 0.25 to 50 $nmol/m^3$ (with the average value equal to 5.1 $nmol/cm^3$ and $\sigma = 8.5\,nmol/m^3$), 222 maximum concentration values were observed near the edge of the ice cover.

Ob'served values of DMS concentration in the seawater varied within the limits of 0.1–12.6 nmol/l (with the average value of 2.2 nmol/l, $\sigma = 2.7$ nmol/l). Maximum DMS concentration took place in the western sector of the Arctic Ocean and near the US east coast (the Sargasso Sea). Maximum DMS concentration in both the atmosphere and the sea were discovered along the edge of the ice cover in the Chukchi Sea. According to observational data from the 70°–76°N (continental shelf of the Chukchi Sea) and 80°–90°N (central part of the Arctic Ocean) latitudinal belts, a decrease of DMS with latitude took place of the order of 68% and 38% per day in the western and eastern sectors of the Arctic Ocean, respectively.

Calculations of DMS flux from the ocean to the atmosphere made by Sharma *et al.* (1999) have given the interval of 0.0017–30 $nmol/m^2/day$ with maximum values occurring in the regions of open water. Due to these calculations, the contribution of the Arctic Ocean into the formation of the sulphur budget in the atmosphere during the summer of 1994 was equal to 0.063 Tg S (in the form of DMS), which is about 0.4% of the global DMS emission from the World Ocean to the atmosphere. Calculations with the use of a 1-D photochemical box model to simulate observational data have demonstrated good enough results and have shown that characteristic DMS life-times in high latitudes of the Arctic varies within 2.5–8 days, while in the latitude belt 16°–30°N it reaches only 1–2 days. The same 1-D model has given a DMS decay time, which reaches only 33% of the measured values. This result indicates the significance of taking into account in halogen chemistry, the reactions with the participation of Br/BrO (not considered in the model), which may be responsible for a substantial DMS sink under conditions of the Arctic atmosphere.

Leck *et al.* (1996) have discussed the broad aim of the Atmospheric Programme of the International Arctic Ocean Expedition (IAOE-91), which was to test the hypothesis that marine biogenically produced DMS gas can exert a significant global climate control. Due to this hypothesis, DMS is transferred to the atmosphere and is oxidized to form aerosol particles, which (partly) can function as CCN and, thus, increase cloud droplet concentration. Leck *et al.* (1996) have pointed out that, in summer, the central Arctic is a specially favourable region for studying the natural sulphur cycle, that the open waters surrounding the pack ice are the only significant sources of DMS, and that there are almost no anthropogenic particle sources. Ob′servation results obtained on board the Swedish ice-breaker *Oden* (1 August to 6 October, 1991, latitudes 75°N to 90°N) show that concentrations of seawater and atmospheric DMS decreased at about the same rate during the period of observations spanning about three orders of magnitude, suggesting that the hypothesis concerning DMS transformation into aerosol particles is true. Elemental analysis of the aerosol also revealed the interesting possibility that debris from Siberian rivers transported on the moving ice, represent a fairly widespread source of supermicrometre crustal material within the pack ice. Measurements of aerosol number size distributions indicate the presence of three distinct model sizes: the "ultrafine", "Aitken", and "accumulation" modes (14, 45, and 175 nm diameter, respectively) within the diameter range 3 nm to 500 nm. The presence of ultrafine particles, implying recent production, was more frequent than has been found in lower latitude remote marine areas. Sudden and often drastic changes in aerosol concentration and size distribution were surprisingly frequent and most common in particles likely to have taken part in cloud formation (>80 nm diameter). Further, more detailed studies of the processes considered are necessary.

Regarding the interpretation of the results of DMS studies, Leck and Persson (1996) have demonstrated the importance of phytoplankton blooms and zooplankton abundance both in the edge zone and in the pack ice during summer ice melt, but seawater DMS showed no simple correlation with phytoplankton standing stock over large areas and seasons. The area-weighed summer and winter DMS fluxes from the Arctic Polar Ocean to the atmosphere were estimated at 2.0 and 0.03 $mmol/m^2/day$, respectively. The total emissions of marine biogenic sulphur from the northern hemisphere high latitudes was estimated at approximately 4 Gmol/yr.

For intercomparison purposes, observation results for Antarctica may be of interest. In January and February 1994, Berresheim *et al.* (1998) accomplished measurements of DMS in air and seawater during the Sulphur Chemistry in the Antarctic Troposphere Experiment (SCATE). DMS concentration values ranged between 6 and 595 pptv (median = 94) and 0.7 and 3.7 nm (median = 2), respectively, and were significantly enhanced during a storm episode. DMS sea-to-air fluxes varied between 0.03 and 0.19 μmol/m^2/day (median = 1.1) based on seawater DMS and wind speed measurements. The atmospheric DMS lifetime was estimated at 9 days based on a 24-hour averaged OH concentration of $1.1 \times 10^{-8}\,cm^{-3}$. Atmospheric dimethylsulphoxide (DMSO) and dimethylsulphone ($DMSO_2$) mixing ratios varied mostly between <0.2 and 15 pptv (with many short time spikes which may be explained by downward mixing of these compounds from the free troposphere). A good correlation was found between DMSO and $DMSO_2$. Aerosol MSA and excess sulphate concentrations also showed a good correlation, with a slope of 0.73. However, DMS was not correlated with any of its measured oxidation products, probably because of its long atmospheric residence time over Antarctica.

Carbonyl sulphide (COS) concentration in polar firn air from one Arctic and two Antarctic locations was measured and the absence of any large trend in COS concentrations over the last fifty or more years was discovered, which argues against COS being the origin of reported increases in stratospheric sulphate aerosol. Also it has discovered that the Arctic firn contained extremely high levels of methyl bromide as well as numerous other organic gases, which are evidently produced *in situ*.

5.3.1.3 *Atmospheric pollution in the Arctic*

Under the specific conditions of the Arctic (being mainly covered by sea ice) natural minor gas and aerosol components come to the Arctic from sources such as volcanic eruptions (Kamchatka, Aleutian Islands and Iceland); soil dust emissions (Greenland, Canada Archipelago and other regions); sea waves (which produce sea salt aerosols); biogenic emissions by oceans (chiefly as DMS); long-range dust transport from Eurasian deserts; and transport of MGC from the stratosphere. In this context, Barrie (1986, 1996) has discussed observation of data obtained in Alert (Canada). In this case the principal sulphur compounds were SO_2 and its oxidation products – sulphate aerosols (SO_4^{2-}). A significant product of DMS oxidation is MSA contained in aerosols. The concentration ratio MSA/SO_4^{2-} varies within 0.2–0.9 (with the average value equal to 0.3). A relevant variability interval for oceans in lower latitudes is 0.04–0.07. Table 5.3 data illustrate basic types of MGCs and aerosols in the Arctic troposphere (Barrie, 1996).

In 1991, the Air Quality Department of the Finnish Meteorological Institute established a monitoring station in the Pallas-Ounastunturi National Park, located north of the Arctic Circle, on the top of the Arctic fell, at a height of 560 m , to make measurements of air chemical composition above the surface inversion (Laurila *et al.*, 1995). This station is located in one of the cleanest areas of the European continent. The measurements of ozone, sulphur dioxide, nitrogen oxides, and peroxyacetyl nitrate were made using computer-controlled automatic analysers. In addition, measurements of volatile organic compounds (VOC), carbon dioxide, methane, and nitrous oxide have been made through the analysis of canister samples taken twice a week (a gas chromatography technique was used). The

Table 5.3. Principal types of MGCs and aerosols in the Arctic troposphere.

Groups of components	Ecological significance of various components
Sulphur compounds	Climate change, visibility, precipitation (acidification of ecosystems), and biogenic tracers
Black carbon	Climate change, fossil fuel burning, and tracers of sources
Heavy metals	Toxicity and tracers of sources
Nitrogen oxides	Oxidants in the atmosphere, acidification of ecosystems, and fertilization
Chemically active organic compounds	Oxidants in the atmosphere, tracers of natural and anthropogenic sources, and indicators of chemical reactions
Ozone	Climate change and chemical stability of the atmosphere
Halogenes (Br, I)	Ozone layer depletion and tracers of marine biogenic components
Policyclic aromatic compounds (PAH)	Toxicity and tracers of sources
Industrial organics	Toxicity and tracers of sources
Pesticides	Toxicity and tracers of sources
GHGs	Climate change

preliminary results indicate that, frequently, long-range transport episodes from continental Europe occur.

The analysis of ice core data for central Greenland (Mount Summit) accomplished by Delmas and Legrand (1998) has obtained information on the high-latitude atmospheric chemical evolution during the last few centuries. Table 5.4 data illustrate the anthropogenically induced changes of the chemical composition of atmospheric aerosols since the beginning of the Industrial Revolution. For

Table 5.4. Atmospheric chemical composition changes during the previous century from data for central Greenland. Hydrogen peroxide data were obtained in 1992–1994.

	Mass concentration (ng/g)	
Component	Pre-industrial	Last decade
Non-seasalt sulphate	26	85
Ammonium	5.3 ± 9.6	–
Nitrate	68	120
Fluoride	0.06	0.19
Formaldehyde	2–3	5
Hydrogen peroxide	125	255
Formate	7.7 ± 2.3	6.3 ± 2.4
Acetate	6.5 ± 1.9	9.8 ± 3.2
Methanosulphonate	2.9 ± 1.9	1.9 ± 0.05

instance, pre-industrial concentration of non-sea sulphates (nss SO_4^{2-}) was equal to 26 ng/g, but after World War II rapid growth in nss SO_4^{2-} concentration started. By 1980–1989 the concentration reached 85 ng/g with a subsequent decrease due to measures to reduce the level of SO_2 emissions. It is curious that during the last glaciation maximum (about 25,000 years ago) sulphate concentration was much higher than in 1960.

No similar trends have been discovered in the cases of Ca and NH_4^+. The background ammonium concentration in the ice core was 5.3 ng/g mainly due to emissions from soils and vegetation and sometimes from forest fires. Ob′served H_2O_2 concentration values exceeded background values of 90 ppb with the subsequent growth of about 20 ppb during the time period 1800–1950. The significant growth of nitrate concentration (Table 5.2) only started in 1950 (i.e., much later than in the case of sulphates).

Very fragmentary data for formaldehyde indicate a doubling of the pre-industrial concentration (2 ng/g) during the last few decades. The principal sources of carboxylic acids (formate, ether and acetate) are natural emissions from vegetation and biomass burning. The decrease of formic acid ether concentration during the previous century is of interest, because it demonstrates the insignificance of an anthropogenic impact in this case. On the other hand, the acetate concentration has substantially increased. The same is true for heavy metals (not shown in Table 5.4), whose concentration strongly increased during the last 20 years by 12.3, 2.7, and 9 times for Zn, Cu, and Cd, respectively. The growth of Pb concentration was even stronger (mainly due to automobile emissions), although violations of natural Pb cycles had already started in the times of ancient Greece and Rome. The present-day level of Pb concentration is approximately the same as at the time of the last glacial maximum.

Kawamura *et al.* (1999) have studied a Greenland ice core (450 years) to obtain data on low molecular weight dicarboxylic acids (C_2–C_{10}) with the use of capillary gas chromatography and a mass spectrometer. The results obtained show a predominance of succinic acid (C_4) followed by oxalic (C_2), malonic (C_5), adipic (C_6), and azelaic (C_6) acids. Data for azelaic acid, which is a specific photochemical reaction product of biogenic unsaturated fatty acids, revealed a characteristic historical trend in the ice core: the concentrations are relatively low during the late 16th to 19th centuries (Little Ice Age) but become very high in the late 19th to 20th centuries (warmer periods) with a large peak in the 1940s. Kawamura *et al.* (1999) have pointed out that lower concentrations of azelaic acid may have been caused by a depressed emission of unsaturated fatty acids from seawater microlayers due to enhanced sea ice coverage during the Little Ice Age. On the contrary, an increased concentration of azelaic acid in the late 19th to 20th centuries are likely interpreted by an enhanced sea-to-air emission of the precursor unsaturated fatty acids due to a retreat of sea ice and/or by enhanced production due to a potentially increased oxidizing capability of the atmosphere.

In order to simulate multiphase chemistry in the Arctic boundary layer, Michalowski *et al.* (2000) have developed a multiphase box model of Arctic halogen chemistry using a PC-based modelling programme called the Chemical Reactions

Modelling System (CREAMS). The model contains 125 gas phase reactions, 19 photolysis reactions, and 16 aqueous reactions occurring in suspended aerosol particles and the quasi-liquid component of snow to simulate mass transfer of species between the gas phase and particles, and between the gas phase and snowpack. A principal purpose of the model was to simulate ozone changes in the boundary layer. It has been shown that O_3 depletion cannot be simulated without the presence of heterogeneous halogen chemistry occurring in the snowpack, and the rate of O_3 depletion is limited by the mass transfer rate of HOBr to the snowpack (a critically important model reaction is $BrO + HCHO \rightarrow HOBr + CHO$).

Since there is a limited number of pollution sources in the Arctic, of special importance is the problem of long-range pollution transport (Kondratyev, 1992a, 1999b). In this context Iversen (1989, 1996) has considered the problem of passive tracer (which is gravitationally neutral and may be in the form of minor gases or aerosol particles) transport under Arctic conditions for distances of more than 1000 km. The speed of air particle motion is determined by atmospheric motions within the range of spatial scales from millimetres up to 10^4 kilometers and characteristic timescales from seconds to months. Relevant atmospheric motions may be classified into two basic categories: (1) waves which are characterized by certain amplitudes, wavelengths, and periods; and (2) closed circulations which have certain amplitudes, diameters, and periods. Both motion categories are usually considered as vortices. A spatial scale of a vortex (L and H represent vertical and horizontal scales, respectively) is determined by a quarter of the wave-length and by a radius of a circulation cell, while a characteristic time is determined by a quarter of the period or by the time of change (T) of an amplitude e-times.

Scales of atmospheric motions may be approximately subdivided into three classes: (1) $L \leq$ km, $H \leq 1$ km, $T \leq 10^3$ s (microscale motions); (2) $10\,\text{km} \leq L \leq 10^3$ km, $H \leq 10$ km, $10^3\,\text{s} \leq T \leq 10^4$ s (mesoscale motions); and (3) $L > 10^3$ km, $H = 10$ km, $1\,\text{day} \leq T \leq 1\,\text{month}$ (macroscale motions). These scales are usually closely connected with the spectrum of kinetic energy (because of the dependence of vortex kinetic energy on their sizes and periods). In the case of microscales, kinetic energy is generated by hydrodynamic instability, while for macroscales (in mid-latitudes) by baroclinic instability of jet stream systems. An important difference between these two classes of motions is that microscale motions are 3-D, while macroscale motions are basically 2-D (horizontal). Taking into account the specific features mentioned, Iversen (1996) has analysed typical features of atmospheric circulation in the Arctic, and in this context, atmospheric transport pathways of the Arctic haze anthropogenic components into the Arctic region as well as to Greenland (the latter problem is of special interest for the interpretation of ice core data).

5.4 ATMOSPHERIC OZONE

The special significance of atmospheric ozone (both in the stratosphere and troposphere) dictates the necessity to discuss ozone variability in the Arctic separately. It is well known that the ozone layer in the stratosphere (about 90% of ozone resides in a layer of the stratosphere approximately between 10 and 50 km above the Earth's surface, which is called the "ozone layer") protects life on the Earth against harmful UV solar radiation (Kondratyev and Varotsos, 2000, 2001; Varotsos, 1997). The ongoing depletion of the ozone layer due to anthropogenic impacts results in the enhancement of surface UV radiation, which may lead to dangerous consequences for man and the biosphere.

In 1970 Professor Paul J. Crutzen (Crutzen, 1970) pioneered the discovery that nitrogen oxides from fertilizers and supersonic aircraft may have adverse impacts on the ozone layer. In 1974, Professors Frank Sherwood Rowland and Mario J. Molina identified CFCs (chlorofluorocarbons) as ozone destroyers, which has been confirmed by subsequent studies. In recognition of their pioneering work, these scientists were awarded the Nobel Prize for Chemistry in 1995 (Katzenstein *et al.*, 2003; Rowland, 1991; Molina and Rowland, 1974).

Tropospheric ozone (which represents about 10% of the total ozone content – TOZ) is a significant GHG. Stratospheric and tropospheric ozone changes contribute differently to climate change as well as have different effects on humans and ecosystems. An important specific phenomenon relevant to tropospheric ozone change is photochemical smog formation.

It is thus obvious that atmospheric ozone is a component of the atmosphere, which deserves special attention. This is why the Montreal Protocol on substances that deplete the ozone layer (in 1987), the Vienna Convention for the Protection of the Ozone Layer (in 1985) and subsequent Amendments and Adjustments to the Protocol have been signed (Sarma, 1998). In accordance with the Helsinki Declaration on the Protection of the Ozone Layer (May 2, 1989), it has been agreed to phase-out the production and consumption of CFCs controlled by the Montreal Protocol as soon as possible, but not later than the year 2010.

The total phase out has been recommended by the London Amendment to the Montreal Protocol on the basis of the 1989 assessment. The discovery of the Antarctic "ozone hole" in 1985 and conclusions of the 1991 assessment, have led to the recommendations of the Copenhagen Amendment (in 1992) to broaden the list of the ozone depleting substances (ODS) by adding HCFCs (hydrochlorofluorocarbons) and methyl bromide. The subsequent 1997 adjustments and amendment marked ten years of the Protocol (Sarma, 1998) and required the phasing out of the following categories of substances (at present there are 95 chemicals controlled by the Protocol):

1. Chlorofluorocarbons: CFCs, halons, hydrobromofluorocarbons; HCFCs, other fully halogenated CFCs, carbon tetrachloride, 1,1,1 trichloroethane; Methylchloroform.
2. Hydrochlorofluorocarbons: HCFCs, hydrobromofluorocarbons; HCFCs and methyl bromide.

Control measures for chemicals include (in the case of developed countries): complete phasing out of halons by 1994, of CFCs, carbon tetrachloride, methyl chloroform and HCFCs by 1996, of methyl bromide by 2005, and of HCFCs by 2030.

It was mentioned (UNEP, 1994) that the data of recent years clearly shows that the growth rates of CFC-11, CFC-12, halon-1301, and halon-1211 are reducing. According to existing data the total combined abundance of ODS in the lower atmosphere peaked in about 1994 and then started to slowly decline, the combined abundance of stratospheric chlorine and bromine having reached the peak at the very end of the 20th century. The increase of total tropospheric organic chlorine in 1992 was only about 60 ppt/year (1.6%), compared to 110 ppt/year (2.9%) in 1989. The abundance of carbon tetrachloride is also decreasing. On the other hand, the atmospheric abundance of several CFC substitutes (HCFCs, HFCs) was increasing, as anticipated.

The result of the recommendations mentioned has been a slowing down of the growth of halocarbon concentrations in the stratosphere (the first example of an advertent positive influence of man on the atmosphere). In accordance with the observations from the UARS satellite, discussed by Russel and Lastovicka (1998), global average (for the 70°N–70°S latitude band) total Cl and F ratio trends at the 55-km altitude have decreased during the time period 1991–1997, from 124 to 116 ppt/yr (Cl) and from 92 to 82 ppt/yr (F).

Bojkov (1999) emphasized that the long period of increasing total chlorine abundance in the troposphere – primarily from CFCs, carbon tetrachloride and methyl chloroform – has ended: chlorine from major CFCs is still increasing slightly and most halons continue to increase substantially (e.g., halon-1211, 5% per year up to the present time). The abundance of HCFCs and HFCs (hydrofluorocarbons) is increasing as a result of their CFC substitutes. Stratospheric halogen loading lags behind tropospheric loading by up to 6 years: the maximum concentrations of chlorine and bromine are expected to peak at ~3.7 ppb around the year 2000 and keep close to this over the next one or two decades. Even with a complete and immediate global elimination of all ODS, emissions would result in stratospheric halogen loading returning to pre-1980 values by the year 2033. Some major scientific findings are as follows:

- The rate of decline in stratospheric ozone at mid-latitudes has slowed, hence, the projections of ozone reduction made in the 1994 UNEP Assessment are larger than has actually occurred. In the northern polar latitudes, in six out of the last nine boreal winter–spring seasons, ozone has declined in some months by 25% to 30% below the 1960s average. Bojkov *et al.* (1998) have shown that during the 1990s over the 35°–50° mid-latitude belts the ozone deficiency in the southern hemisphere was less than over the northern hemisphere by about 39%. The observed TOZ losses from 1979 to 1997 were about 5.4% and 2.8% for northern mid-latitudes in winter/spring and in summer/autumn, respectively, and 5.0% in southern mid-latitudes all the year round. There are no statistical TOZ trends in the equatorial region (20°S to 20°N).

- The springtime Antarctic ozone hole continues unabated. Although the extent of ozone depletion was strongest in 1998, it has over the last 3–4 years remained generally similar to that in the early 1990s, with monthly TOZ values in September and October continuing to be 40–55% below the pre-ozone hole values (with up to a 70% decrease for periods of about 1 week).
- The late winter/spring ozone values in the Arctic were unusually low in six out of the last nine years. The six years were characterized by unusually cold and protracted stratospheric winters with minimum Arctic circumpolar vortex (CPV) temperatures near the threshold for large chlorine activation. Ozone has declined during some spring months by 25–35% below the pre-1976 average. Elevated stratospheric halogen abundance over the next decade or so would imply the Arctic's continuing vulnerability to large ozone losses.
- Over northern mid-latitudes, the downward TOZ trend is largest between 40 and 15 km (>7% per decade) and is smallest near 30 km (2% per decade). The bulk of TOZ decline is between the tropopause and 25 km.
- Stratospheric ozone losses may have caused part of the observed cooling of the lower stratosphere in the polar and upper mid-latitudes and global average negative radiative forcing of the climate system. Much of the observed downward trend in lower stratospheric temperatures (about 0.6°C per decade from 1979 to 1994) is attributed to the ozone loss in the stratosphere. The stratospheric ozone losses since 1980 may offset about 30% of the positive GHG effect over the same time period.
- The understanding of the reaction between increasing surface UV radiation and decreasing column ozone has been further strengthened by ground-based observations. Newly developed satellite methods show promise for establishing global trends in UV radiation. The largest UV increases occur during spring at high latitudes in both hemispheres.
- Based on past emissions of ODS and the projection of the maximum allowances under the Montreal Protocol into the future, the maximum ozone depletion is estimated to lie within the current decade or the next two decades, but its identification and the evidence for the recovery of the ozone layer lie still further ahead. It has been pointed out in *Scientific*... (1999) that, "a full recovery of the Earth's protective ozone shield could occur by the middle of the next century, but it would require that the Protocol is fully implemented."
- The increase of ozone in the troposphere since pre-industrial times is estimated to have contributed to 10 to 20% of the warming due to the increase in long-lived GHGs during the same period.

An important new phenomenon in the Antarctic, discovered in 2002, was the split of the ozone hole in September 2002.

Sarma (1998) emphasized the role of the Montreal Protocol, as well as subsequent amendments and adjustments, in avoiding negative environmental consequences such as: (1) at least 50% ozone depletion at mid-latitudes in the northern hemisphere and 70% depletion at mid-latitudes in the southern hemisphere, about 10 times larger than today; and (2) at least a doubling of surface UV radiation at mid-latitudes in the northern hemisphere and a quadrupling at mid-latitudes in the southern hemisphere compared with an unperturbed atmosphere. This compares with current increases of about 5 and 8% in the northern and southern hemispheres, respectively, since 1980.

The implications of ozone depletion are horrendous with nearly 19 million additional cases of non-melanoma skin cancer up to the year 2060 and 3 million more cases up to 2030. The number of eye cataracts could increase by about 130 million cases by the year 2060, about 50% in developing countries. There are a number of unquantifiable effects, such as damage to the immune system, adverse impact on animals, lower productivity of crops, damage to aquatic ecosystems, and so on. It must be pointed out, however, that a number of scientists (Ellsaesser, 1992, 1994; Maduro and Schauerhammer, 1992) have warned against overemphasizing such dangers, because of ozone depletion.

Ozone problems have become essential parts of such global scale international programmes as the WCRP with its Stratospheric Processes and their Role in Climate (SPARC) Project (Chanin, 2001) and the IGBP, including the International Global Atmospheric Chemistry (IGAC) Core Project, as well as many other ozone-related projects (Kondratyev, 1998).

Marked progress has been achieved in the development of satellite remote sensing aimed at obtaining information on global scale spatio–temporal ozone and ODS variability and trends (Fichmann *et al.*, 1999). A number of comprehensive observational programmes have been carried out in the high latitudes of the northern hemisphere and in the Antarctic, to study the stratosphere and ozone dynamics. As a result, not only has the well-known ozone hole in the Antarctic been discovered, but also ozone "mini-holes" and substantial depletion of ozone over vast territories of Siberia (Harris *et al.*, 1998). Additional reliable assessments of the atmospheric effects of aircraft emissions on both stratospheric and tropospheric ozone have been obtained (Kondratyev and Varotsos, 2000). Important progress has been made using interactive modelling of general atmospheric circulation and ozone dynamics (Takigawa *et al.*, 1999), in assessing the impact of volcanic aerosols on stratospheric processes (Timmreck *et al.*, 1999). A new analysis of ozonesonde data for the troposphere and the lower stratosphere has been carried out (Logan, 1999) as well as further studies about the nature of the Antarctic ozone hole (MacKenzie *et al.*, 1999). New studies have been conducted to assess the influence of ozone depletion on surface UV radiation and the subsequent impact on human health and the biosphere (Zerefos *et al.*, 1998).

Problems associated with tropospheric ozone change have recently attracted much more attention than before, especially in the field of numerical modelling (observations still remain inadequate). Important progress has been made in interactive simulation of the troposphere dynamics and ozone change. The impact of both stratospheric and tropospheric ozone on climate has been studied (Kondratyev and Varotsos, 2000). Special attention has been paid to heterogeneous chemical reactions on the surfaces of aerosol particles in the stratosphere (especially in the case of polar stratospheric clouds) and troposphere (Kondratyev, 1998b, 1999b).

As has been pointed out in the *Scientific Assessment of Ozone Depletion* (1999):

- Enhancing anthropogenic emissions of ODS (nitrogen oxides, carbon monoxide, and hydrocarbons) lead to the large-scale production of ozone, which, through long-range transport, influences the ozone concentration in large regions of the troposphere in both hemispheres. Such changes are characterized by strong spatio–temporal variability.
- The increase of ozone in the troposphere since pre-industrial times is estimated to have augmented the average RF by $0.35 \pm 0.15\,W/m^2$, which is thought to have contributed 10–20% of the warming owing to the increase of long-lived GHGs during the same period.

Undoubtedly, marked progress has been achieved in studying atmospheric ozone and anticipating the consequences of its changes, however, quite a number of related problems remain unsolved thus far. Many aspects of ozone research have been discussed in a number of recent publications (Varotsos, 1997; Bojkov and Visconti, 1998).

5.4.1 Stratospheric ozone

5.4.1.1 Ob′servations and interpretation of observational data

Although TOZ observations started about 150 years ago, with the discovery of the existence of the ozone layer in the stratosphere, the global archive of TOZ data has been accumulated mainly during the past few decades due to satellite observations. Various remote sensing techniques have been used for this purpose using satellite measurements of backscattered solar UV radiation, attenuation of UV radiation by the atmosphere under conditions of "occultation geometry", and thermal emission in the 9.6-μm ozone absorption band (Kondratyev, 1989).

The most important results from observations are detection of the mean global TOZ decrease and the appearance of the ozone hole in the Antarctic as well as a short-term but sharp TOZ decrease in the northern hemisphere high latitudes.

Over the past few years, substantial data on ozone depletion in the Arctic has been obtained. Hansen and Chipperfield (1999) studied, for instance, ozone depletion at the edge of the Arctic polar vortex in the winter of 1996/1997, when the Arctic polar stratospheric vortex was extremely long-lived. The ozone lidar at the Arctic Lidar Ob′servatory for Middle Atmosphere Research (ALOMAR) operated from mid-December 1996 until mid-May 1997, when the vortex was centered at the pole, and its edge was almost permanently located over northern Scandinavia. Ozone depletion of up to 40% was observed at the 475 and 550-K levels. Maximum depletion occurred around 5 May at levels up to 550 K and around 20 April at 675 K. Model analysis performed by Hansen and Chipperfield (1999) showed that while much of the early spring ozone depletion was due to halogen chemistry, associated with chlorine activation on polar stratospheric clouds, the ongoing depletion in late April and early May was due to "summertime" NO_x ($NO + NO_2$) chemistry. The unusual persistence of the vortex, with the isolation of high-latitude air masses until early May, allowed this depletion to occur.

Fichmann *et al.* (1999) have discovered, through the analysis of ERS-2 satellite data for the Arctic spring periods 1997 and 1998, a substantial stratospheric ozone depletion. Extensive regions of low ozone total column were observed, and it has been shown that the major decrease is dominating in the lower and middle stratosphere inside the polar vortex. In the spring of 1998 an ozone mini-hole was observed (ozone profiles under mini-hole conditions were derived for the first time).

In order to explore the mechanisms of Arctic ozone destruction and relevant chemical–radiative interaction, MacKenzie *et al.* (1999) have undertaken a numerical simulation of three northern hemisphere winter/spring seasons with contrasting thermal and dynamic characteristics, using a mechanistic model of the middle atmosphere with an interactive chemistry scheme. The calculations indicate that cooling of the lower stratosphere induced by the enhanced ozone loss ranges from virtually zero in the warmest winter to ~5 K in the coldest one. An analysis of the role of polar stratospheric clouds (PSC) has shown that chemical–radiative interaction via PSC processes has had only a small impact on the ozone distribution in recent Arctic winters, but it has the potential to become more important, should an unprecedentedly cold winter occur while atmospheric halogen concentrations remain artificially high.

Guirlet *et al.* (2000) have used a 3-D stratospheric model to simulate the Arctic ozone depletion in the three winters from 1995/1996 to 1997/1998, winters with quite different dynamical conditions. Despite the very different meteorological conditions for these three winters, by mid-February the model-calculated mean vortex loss was similar in each year, at around 20% at the 480-K level. By late March, relevant total ozone losses were 53 DU (Dobson Units) in 1998, 71 DU in 1996, and 80 DU in 1997.

Rinsland *et al.* (1999) discussed observations made inside the November 1994 Antarctic stratospheric vortex and inside the April 1993 remnant Arctic stratospheric vortex, using the Atmospheric Trace Molecule Spectroscopy (ATMOS) fourier transform spectrometer. By comparing vortex and extravortex observations of NO_y (total reactive nitrogen) obtained at the same N_2O volume mixing ratios, the Arctic vortex denitrification of 5 ± 2 ppbv at 470 K (~18 km) has been inferred (these results are robust for a wide range of winter conditions).

Since stratospheric ozone is an intensive absorber of solar radiation and tropospheric ozone is a GHG, there is no doubt that changes in the content of ozone in the atmosphere must affect the climate.

Many mentioned publications categorize the following as the most important aspects of the problem:

- From 1979, the rate of mean annual TOZ decline in mid-latitudes of both hemispheres was within 2.6–3.7% over a period of 10 years. Analysing ozone

Table 5.5. Total ozone trends (% per decade since January 1979) calculated as averages of individual station trends in the indicated regions and groups of months (ground-based observations). All values are negative and statistically significant at 2σ. Every column corresponds to a time period of three months (e.g., DJF represents the December–January–February period)

Region	DJF	MAM	JJA	SON	Year
Arctic	7.9	7.7	2.5	3.6	5.7
35°N–60°N	4.1	5.7	2.9	1.6	3.7
35°S–60°S	2.9	2.2	3.4	2.0	2.6
Antarctic	6.3	2.4	6.5	20.0	8.9

decline in the northern polar and mid-latitudes during winter and spring, Bojkov *et al.* (1998) concluded that the ozone mass deficiency O_3 (MSD) (defined from the pre-1976 base average and areal extent with negative deviations greater than ~2 and ~3σ), integrated for the first 105 days of each year, has increased dramatically from ~2,800 Mt in the winter–spring periods of 1993 and 1995. Bojkov (1999) pointed out that total ozone levels in the latitudes 60°N–60°S were at their lowest in 1993 because of the large increase in stratospheric aerosol caused by the eruption of Mount Pinatubo in 1991 (Kondratyev and Galindo, 1997; Kondratyev, 1999b). Since 1992/1993, TOZ values over this part of the globe have been variable around a fairly constant level. Table 5.5 illustrates the TOZ trends (Bojkov, 1999).

- The amplitude of the annual cycle of ozone at mid to high latitudes has decreased by 15% in the last two decades, because larger declines of maximum ozone values have occurred during winter and spring.
- Ozone depletion increases with latitude, particularly in the southern hemisphere. Little or no downward trends are observed in the tropics (20°N–20°S). Analysis of global TOZ data through early 1998 shows substantial decreases in ozone in all seasons at mid-latitudes (35°–60°) of both hemispheres. There were downward trends in the northern hemisphere of about 6% per decade between 1979 and 1998, observed in winter and spring and about 1.5–3% per decade were observed in summer and autumn. In the southern hemisphere the difference between seasons was smaller and the mid-latitude trends averaged within 2.0–3.4% per decade. Satellite and ozonesonde data show that much of the downward trend in ozone levels occurs below 25 km.
- According to *Scientific*... (1999), TOZ decreased significantly at mid-latitudes (25°–60°) between 1979 and 1991, with estimated linear downward trends of 4.0%, 1.8%, and 3.8% per decade, respectively, for northern mid-latitudes in winter/spring, northern mid-latitudes in summer/fall, and southern mid-latitudes all year round. However, since 1991 the linear trend observed during the 1980s has not continued, but rather the TOZ was almost constant at all mid-latitudes in both hemispheres after recovery from the 1991 Mount Pinatubo eruption. Present-day understanding of how changes in halogen and aerosol loading affect

ozone explains why linear extrapolation of the pre-1991 ozone trend to the present is not suitable.

- A complex 3D-spatial structure of ozone distribution was recently discovered taking the form of lamination (Varotsos, 1997) and formation of low-ozone air pockets during the northern winter in the middle stratosphere outside the polar vortex.
- Nair *et al.* (1998) pointed out that the rapid ozone loss localized in pockets is due to the isolation of air at high latitudes (and high solar zenith angles). Thus low ozone levels are due to a decrease of oxygen production rates and not an increase in the loss rate by reaction with halogen species, as in the classical ozone hole. For an explanation of low ozone pockets see Morris *et al.* (1998) who developed a simulation model that takes account of photochemical processes and stratospheric dynamics.
- Russel and Lastovicka (1998) concluded, on the basis of ozonesonde data, that the number of thin ozone layers per one ozone vertical profile decreased by about 1.9 to 1.0 from 1970 to 1993 (with depletion of the ozone content in such layers of about 50%).
- The lowest values of TOZ were obtained between 1992–1993 at the ozone hole in the Antarctic and the lowest TOZ values in the whole observational period were obtained over the densely populated regions of the northern hemisphere. The conclusion that anthropogenic chlorine and bromine compounds, coupled with surface chemistry on natural polar stratospheric particles, are the cause of polar ozone depletion has been further strengthened. The links to halogen chemistry have also been established with regard to ozone losses detected in the Arctic winter stratosphere.
- According to *Scientific* ... (1999), the large ozone losses during the spring in the southern hemisphere continued unabated with approximately the same magnitude and areal extent as in the early 1990s. In Antarctica, the monthly total ozone in September and October has continued to be 40 to 55% below the pre-ozone hole values of approximately 320 m-atm/cm (DU), with up to a 70% decrease for periods of a week or so. This depletion occurs primarily over the 12–20-km altitude range, with most of the ozone in this layer disappearing during early October.
- Ob′servations made after the eruption of Mount Pinatubo in 1991 resulted in the discovery of the substantial impact such a major volcanic eruption has on stratospheric ozone (Kondratyev and Varotsos, 2000; Timmreck *et al.*, 1999).
- Impacts of subsonic and supersonic aircraft on upper tropospheric and lower stratospheric ozone have been confirmed (Kondratyev and Varotsos, 2000; Penner *et al.*, 1999).
- The earlier conclusion about the substantial role of methyl bromide as an ozone-destroying component has been verified, the main sources of this being crop fumigation, biomass burning, and cars. The most recent data (*Scientific*..., 1999) indicate, however, that the role of methyl bromide as an ozone-depleting compound is now considered to be less than was estimated in

UNEP...(1994), although significant uncertainties remain (the science of atmospheric methyl bromide is complex and still not well understood).

- The planned countermeasures to control emissions of ozone-depleting compounds will not greatly change the maximum level of TOZ decline forecast to be reached during the next 10 years.
- According to model simulations much of the observed downward trend in lower stratospheric temperatures (about 0.6°C per decade over 1979–1994) is attributed to the ozone loss in the lower stratosphere (*Scientific*..., 1999). The decrease in stratospheric ozone since 1980 may have offset about 30% of greenhouse forcing, which favoured the global climate warming. However, the increase in tropospheric ozone in the same period has intensified the atmospheric greenhouse effect in the northern hemisphere by ~20%. The problem of ozone trends with opposite signs in the stratosphere and in the troposphere deserves special attention with respect to TOZ variation impact on UV radiation.
- From some published data (Michaels *et al.*, 1994), the TOZ decline in 1992–1993 was not followed by any increase in UV radiation. The necessity of a thorough analysis of data on the trends of TOZ, UV, and related biological consequences was expressed by Ellsaesser (1992, 1994). However, Feister and Grewe (1995) observed higher UV radiation from low ozone values at northern mid-latitudes in 1992 and 1993 and pointed out that this could lead to adverse biological effects. Numerous further studies have confirmed this conclusion.
- *Scientific*... (1999) concludes that the understanding of the relation between increasing surface UV radiation and decreasing TOZ has been further strengthened by ground-based observations, and newly developed satellite methods show promise for establishing global trends in UV radiation. The satellite estimates for 1979–1992 indicate that the largest UV increases occur during spring at high latitudes in both hemispheres.

Logan (1999) presented an analysis of ozonesonde data with the aim of using them to evaluate models of transport and chemistry. These data show that between 10 and 30% of the ozone column are located between 100 hPa and the tropopause at mid and high latitudes, and this region drives much of the seasonal TOZ variation. The sonde data quantify the build-up of ozone in the lowermost stratosphere of the northern hemisphere in winter and its loss in late spring and summer. The amount of ozone between the tropopause and 100 hPa in the northern hemisphere decreases from 175 to 75 Tg from March to September, with the maximum rate of decrease in May to June, about half of the decrease being caused by the increase in the height of the tropopause. There is a lag of about a month, when ozone starts accumulating in the lowermost stratosphere from September near 40°N to October or November near 80°N, and a similar time lag, when ozone starts decreasing from February or March near 40°N to March or April near 80°N.

Despite caution against overemphasis of the ozone impact on the environment and ecosystems, this problem undoubtedly deserves very serious attention.

Prather *et al.* (1996) posed a number of important questions in this context:

- What would have happened if CFC use had followed a free-market growth into all sectors and countries before the observation of stratospheric ozone depletion?
- What if the ozone hole had been discovered in 1985 (as indeed it was) without the prior suspicion about chlorine from CFCs?
- When should we expect to see recovery of the ozone layer as provisions of the Montreal Protocol begin to show their effect on the global atmosphere?
- How much stratospheric chlorine and associated ozone depletion could the global atmosphere have been committed to in a worst case scenario?

Answering these questions, Prather *et al.* (1996) emphasized that:

- whether we had ignored the scientific evidence or nor, ozone depletion would be dramatically worse than what we are experiencing today;
- if CFCs had followed free-market growth until 2002, the Antarctic ozone hole would be a permanent fixture through the 21st century, instead of disappearing by 2050 as predicted in the Copenhagen 1992 scenario; and
- ozone depletion is expected to reverse and recover measurably in the first decades of the 21st century, with unambiguous detection of the recovery taking a decade or more.

Kaye (1998) and McCormick (1998) have summarized discussions of satellite TOZ observations using the Total Ozone Mapping Spectrometer (TOMS) and Stratospheric Aerosol and Gas Experiment (SAGE) instrumentation, respectively. Up to the present time, a total of four TOMS instruments have flown in space: Nimbus-7 TOMS (1978–1993), Meteor-3 TOMS (1991–1994), ADEOS TOMS (1996–1997), and Earth Probe TOMS (1996–present-day). TOMS instruments have provided excellent data for TOZ, although there are some limitations that must be considered in certain applications, especially for studies of tropospheric ozone (currently no data are released for zenith angles greater than 84°, although data are in principle available for up to 88°).

Assimilation of TOZ satellite measurements in a 3-D tracer transport model has been discussed by Jeuken *et al.* (1999). Fioletov *et al.* (1999) assessed the world ground-based TOZ network performance by comparison with satellite data (TOMS and solar backscatter ultraviolet (SBUV) instrumentation on board the Nimbus-7 and Meteor-3 satellites). The results indicate that about 80% of all Dobson, Brewer, and filter ozonometer stations have standard deviations of monthly mean differences with TOMS that are less than 2.5%. Typically, results of the comparisons between ground-based stations and SBUV are similar to those for ground-based and TOMS comparisons.

Bhatt *et al.* (1999) evaluated the quality of the satellite Halogen Occultation Experiment (HALOE) ozone profiles in the lower stratosphere by comparison with coincident ozonesonde profiles. It was found that the HALOE ozone agrees, on average, to within 10% of their coincident ozonesonde measurements down to

100 hPa at tropical/subtropical latitudes and to 200 hPa at extratropical latitudes. Bacmeister *et al.* (1999) conducted an intercomparison of satellite and aircraft observations of ozone, CFC-11, and NO_y using trajectory mapping. Correlation coefficients for coincident pairs of satellite and aircraft data are between 0.68 and 0.80.

Apart from TOZ data, processing of TOMS information allows retrieval of tropospheric ozone content and surface flux of ultraviolet radiation, especially if ratios such as UVL/UVS (UV radiation longward and shortward of 320 nm) are considered. TOMS instrumentation made it possible to accumulate a data set covering the period from late 1978 to February 1998 for zonal mean TOZ. One of the applications of the data obtained was the study of TOZ long-term trends and interannual variability in mid-latitudes (particular emphasis was placed on dynamically driven contributions). The results suggest that changes in lower stratospheric temperature and/or tropopause height could contribute significantly to the mid-latitude decreases observed by TOMS (Schubert and Munteanu, 1998). It was seen that short-term meteorological events could significantly affect the zonal mean distribution of TOZ, especially in February, when there can be enormous variation from one winter to the next.

To test retrieval algorithms, observational data from the Satellite Ozone Limb Sensing Experiment (SOLSE) and the Limb Ozone Retrieval Experiment (LORE) instruments that flew on the space shuttle in the autumn of 1997, are available. Measurements of a number of species, with UV absorption in the ultraviolet, using the Global Ozone Monitoring Experiment (GOME) instrument aboard the ERS-2 satellite have been conducted (column distributions of BrO, ClO, and SO_2 have been studied).

In continuation of earlier SAGE observations, three new ozone-measuring instruments for SAGE-III have been developed (McCormick, 1998). The first flight took place in 1999 (Russian Meteor-3M satellite in a sun-synchronous orbit) while the second flight will be an opportunity to fly into a mid- to high-inclination orbit. The SAGE-III instrument will be flown aboard the International Space Station (ISS), scheduled for mid to late 2002. Using solar occultation, SAGE-III data will allow retrieval of 1-km vertically resolved mid-atmosphere profiles of ozone, aerosols, and a number of aerosol properties, water vapour, temperature, nitrogen dioxide, and cloud presence. In addition, using lunar occultation, it will be possible to retrieve vertical profiles of nitrogen trioxide and chlorine dioxide. For validation purposes, the SAGE-III Ozone Loss and Validation Experiment (SOLVE) has been planned, whose aim is to use a joint balloon and airborne measurement campaign in the Arctic designed to examine the processes that control polar to mid-latitude stratospheric ozone levels and also aid in SAGE-III validation. An important role in solving validation problems belongs to the ozonesonde network. Guirlet *et al.* (1998, 2000) described the Third European Stratospheric Experiment on Ozone (THESEO) conducted in 1998–1999 as a continuation of the earlier European Arctic Stratospheric Ozone Experiment (EASOE, 1991–1992) and the Second Stratospheric Arctic and Mid-latitude Experiment (SESAME) with the principal purpose of better understanding the processes that determine stratospheric ozone trends in mid-latitudes on the basis of studying interactions between relevant chemical and dynamical processes. The aim is to investigate ozone formation both within and outside of the circumpolar vortex. The observations covered a broad range of latitudes from the tropics to the Arctic using ozonesondes, lidars, small balloons, and aircraft. On 19 December, 1999, the NASA Terra satellite was launched, opening a new perspective for ozone observation.

Discussing ozone problems, it is appropriate to quote the Policy Statement on Atmospheric Ozone accepted by the American Meteorological Society (AMS) Council, 28 January, 1996 (Hales, 1996):

- The AMS recognizes ozone as an atmospheric constituent that has a number of important beneficial as well as detrimental effects on the atmosphere, on surface ecosystems, and on humankind. These include effects on both UV and longwave radiation, resulting effects on atmospheric wind systems and direct impacts on plants and animals. Moreover, ozone is recognized as the dominant progenitor of much of the trace-gas chemistry that occurs in both the troposphere and the stratosphere. The AMS also notes that extensive couplings exist among these chemical, radiative, and dynamical components of ozone's behaviour and add substantial uncertainty to many of the currently available assessments of ozone's impact.
- Despite many uncertainties, ample evidence exists to substantiate that atmospheric ozone has been affected in important and even critical ways by human activity. In the case of the stratosphere, the existence of the Antarctic ozone hole is unquestionable and our evidence that it results from human-produced halocarbons is overwhelming. Ozone depletion in the mid-latitude stratosphere is less dramatic; however, the general agreement of the large numbers of available column and profile measurements make it reasonable to speculate that such depletion is indeed occurring and that it is largely human-induced. Anthropogenic activities also significantly influence tropospheric ozone. Humankind is directly responsible for the excessive ozone levels that often occur near the surface in and downwind from populated areas. These effects are also felt throughout significant regions of the free troposphere. The effects in both the stratosphere and troposphere are sufficiently profound to mandate substantial concern, on both a local and a global basis.
- While the above manifestations of human impact on ozone are clear, there remain important gaps in our understanding of ozone's complex behaviour. For example, currently we cannot quantitatively explain the mid-latitude observations of insignificant surface UV increases. Impacts of tropospheric ozone build-up on the free troposphere's chemistry are also in a highly speculative state. These issues and others must be resolved in order to forecast potential future manifestations satisfactorily and provide a firm basis for policy analysis and associated policy actions.
- Many of the noted uncertainties arise because of the strong couplings between chemistry, radiation, and atmospheric dynamics. Because of this, their

resolution will require substantial combined efforts from scientists having chemical and meteorological backgrounds. The AMS intends to actively provide one component of a continuing forum for this scientific interaction and act in conjunction with other like-minded organizations for this purpose.

- Current ozone-control legislation (such as the US Clean Air Act, with 1990 amendments) and international agreements (such as the Montreal Protocol) tend to reflect the uncertainties noted above. The State Implementation Plan approach of the current US Clean Air Act is burdened by older and largely invalid concepts of chemistry and atmospheric transport, which make ozone-standard attainment difficult if not impossible at many locations. Future amended versions of this act should incorporate more manageable and scientifically valid approaches to ozone control and substantial current meteorological and chemical research should be directed to determining effective strategies for this purpose. With regard to the Montreal Protocol, it is encouraging to note that halocarbon limitations under this agreement appear to have resulted, recently, in decreases of some of the shorter lived halogen-containing species. Owing to the noted uncertainties and complexities associated with stratospheric ozone depletion, however, the effects of this and other international agreements must be monitored continuously and carefully, to ensure their effectiveness and to establish the basic understanding required for more effective maintenance during future years.

5.4.1.2 Biological implications of total ozone depletion

The biological consequences of the TOZ decreasing trend are connected mainly with the enhancement of biologically active UV solar radiation at the Earth's surface. It is proper in this context to quote basic conclusions made up in the UNEP Scientific Assessment (*UNEP*..., 1994):

- Large increases in UV radiation have been observed in association with the ozone hole at high southern latitudes. The measured UV enhancements agree well with model calculations.
- Clear sky UV measurements at mid-latitude locations in the southern hemisphere are significantly larger than at corresponding sites in the northern hemisphere, in agreement with expected differences due to TOZ and Sun–Earth separation.
- Local increases in UVS were measured in 1992/1993 at middle and high latitudes in the northern hemisphere. The spectral signatures of the enhancement clearly implicate the anomalously low ozone observed in those years, rather than variability of cloud cover or tropospheric pollution. Such correlations add confidence to the ability to link ozone changes to UVS changes over relatively long timescales.
- Increases in clear sky UV over the period 1979 to 1993 due to observed ozone changes are calculated to be greatest at short wavelengths and at high latitudes. Poleward of 45°, the increases are greatest in the southern hemisphere.

- Uncertainties in calibration, influence of tropospheric pollution, and difficulties of interpreting the data from broadband instruments continue to preclude the unequivocal identification of long-term UV trends.
- Scattering of UV radiation by stratospheric aerosols from the Mount Pinatubo eruption did not alter total surface-UV levels appreciably.

Recent information on the biological implications of total ozone depletion have been discussed at the European Conference on Atmospheric UV Radiation in Helsinki (*Abstracts*..., 1998) as well as by Martens (1998) and in numerous journal publications.

As far as impacts of variable UV radiation on humans and ecosystems are concerned, the most important aspects of the problem include (Kondratyev and Varotsos, 2000) human health (one of the most dangerous consequences is skin cancer) and health of terrestrial and marine ecosystems.

5.4.2 Tropospheric ozone

Increasing pollution of the troposphere has stimulated studies of anthropogenic impacts on tropospheric chemical composition (Kondratyev and Varotsos, 2001). In this context, tropospheric ozone occupies a special place in view of its importance for human health and ecosystem functioning as well as its role as a GHG. Pszenny and Brasseur (1997) pointed out that surface ozone is a human respiratory irritant. Relatively small ozone amounts can cause chest pain, coughing, nausea, throat irritation and congestion in healthy people. It may also worsen bronchitis, heart disease, emphysema, and asthma. Surface ozone is also phytotoxic. It can produce foliar injuries, reductions in crop yield and biomass production, and shifts in competitive advantages of vegetation species in mixed populations. Photolysis of tropospheric ozone by UV radiation in the presence of water vapour, the primary source of hydroxyl radicals (OH), is very important. Hydroxyl radicals are responsible for the removal of many trace gases (such as CH_4, HFCs, and HCFCs) through oxidation.

A number of fundamental publications (*UNEP*..., 1994; Høv, 1997; Isaksen, 1988; *Scientific*..., 1999) contain basic information on various features of tropospheric ozone dynamics and relevant problems.

Basic achievements up to 1994 are shown in *UNEP*... (1994):

- Recent measurements of the NO_y/O_3 ratio have basically confirmed earlier estimates of the flux of ozone from the stratosphere to be in the range 240–280 Tg (O_3)/yr, which is in reasonable agreement with results from general circulation models.
- The observed correlation between ozone and alkyl nitrates suggests a natural concentration of 20–30 ppb in the upper planetary boundary layer (at about 1-km altitude), which agrees well with the estimate from the few reliable historic data.

- Measurements of the gross ozone-production rate yielded values as high as several tens of ppb per hour in the polluted troposphere over populated regions, in good agreement with theoretical predictions. Likewise, the efficiency of NO_x in ozone formation in moderately polluted air masses was found to be in reasonable agreement with theory.
- Direct measurements of hydroxyl and peroxy radicals have become available. While they do not survey to establish a global climatology of OH, they do provide a test of our understanding of the fast photochemistry. Today, theoretical predictions of OH concentrations (from measured trace gas concentrations and photolysis rates) tend to be higher than the measurements by up to a factor of two.
- Measurements of peroxy radical concentrations in the remote free troposphere are in reasonable agreement with theory, however, significant misunderstanding exists with regard to the partitioning of active nitrogen and the budget of formaldehyde.
- Measurements have shown that export of ozone produced from anthropogenic precursors over North America is a significant source of ozone in the tropics during the dry season. These findings show the influence of human activities on the global tropospheric ozone balance.
- Photochemical net ozone destruction in the remote atmosphere has been identified in several experiments. It is likely to occur over large parts of the troposphere with rates up to several ppb per day. Consequently, an increase in UVB radiation (e.g., from stratospheric ozone loss) is expected to decrease tropospheric ozone in the remote atmosphere, but in some cases it will increase production of ozone in and transport from the more polluted regions. The integrated effect on hydroxyl concentrations and climate is uncertain.

It was emphasized in *UNEP*...(1994) that uncertainties in the global tropospheric ozone budget, particularly in the free troposphere, are mainly associated with uncertainties in the global distribution of ozone itself and its photochemical precursors, especially CO and NO_x. The role of heterogeneous processes including multiphase chemistry in the troposphere is not well characterized, and the catalytic efficiency of NO_x in catalyzing ozone formation in the free troposphere has not been confirmed by measurements.

A new phenomenon discovered during the last decade in the Arctic were episodes of tropospheric ozone depletion (ODE) in the inversional layer (a few hundred metres thick). A very severe drop of ozone concentration (up to zero level) was observed in a few hours. Röckmann *et al.* (1999) have pointed out that there is a high correlation of ODE with filtered bromine concentration.

Foster *et al.* (2001) have undertaken observations of Br_2, BrCl, and Cl_2 at Alert (Canada) to find out the adequacy of the hypothesis that bromine atoms play a central role in the depletion of surface-level ozone in the Arctic at polar sunrise. The measurements indicate that in addition to Br_2 at mixing ratios up to ~25 ppt, BrCl was found at levels as high as ~35 ppt, but molecular chlorine was not observed, implying that BrCl represents the dominant source of chlorine atoms during polar sunrise (this is consistent with recent modelling studies).

Recent results and problems of tropospheric ozone research have been discussed by Høv *et al.* (1997), in connection with the completion of the first phase of the Tropospheric Ozone Research (TOR) Project, as well as by Pszenny and Brasseur (1997) in the context of the development of tropospheric ozone studies within the IGAC Core Project of the IGBP (Bates *et al.*, 1998; Kondratyev, 1999a).

Pszenny and Brasseur (1997) pointed out that the most important tropospheric ozone precursors are nitrogen oxides ($NO_x = NO + NO_2$), methane and other "non-methane" hydrocarbons (NMHCs), and carbon monoxide. All of these precursors are products of fossil fuel and biomass burning, but each also has significant sources from the biosphere. The only ozone sinks being deposition to surfaces (vegetation, soil, oceans), chemical destruction *in situ*, and some export back to the stratosphere in the tropics. It is believed that the increase of ozone observed in near-surface air, which may also be occurring in the free troposphere, is the result of increasing NO_x emission.

Cox (1999) developed a 1-D box model to describe the ozone budget and NO_x chemistry in the marine boundary layer. Results using small prescribed NO_x concentrations gave compensation points, where ozone loss by photolysis and physical removal is balanced by its production, via NO_x chemistry, of ~30 ppt to ~15 ppt at the two sites considered (Mace Head, Ireland and Cape Grim, Tasmania), in line with conclusions from observational data.

According to *Scientific*... (1999), trends in tropospheric ozone since 1970 in the northern hemisphere show large regional differences, with increases in Europe and Japan, decreases in Canada, and only small changes in the USA. The trend in Europe since the mid-1980s has reduced to virtually zero (at two recording stations). In the southern hemisphere, small increases have been observed in surface ozone.

A number of field observation programmes carried out recently obtained new important information on ozone dynamics. One of the programmes was the Pacific Exploratory Mission–West A (PEM–West A) which is a major component of IGAC's East Asia/North Pacific Regional Experiment (APARE) Activity (Pszenny and Brasseur, 1997). The principal aim of PEM–West A has been investigations of the tropospheric ozone distribution over the North Pacific and of the growing emissions of ozone precursors from eastern Asia, which will be a major contributor to expected ozone increase on a hemispheric scale in the coming decades. Ob'served ozone trends were best described in terms of two geographical domains: the western North Pacific rim and the western tropical North Pacific. For both regions, ozone photochemical destruction decreased more rapidly with altitude than did photochemical formation. The ozone tendency was typically found to be negative below 6 km and positive from 6 to 8 km. On the basis of lidar sounding data from a DC-8 aircraft, Newell *et al.* (1997) studied the influence of Asian continental pollution on the western Pacific in September–October 1991 and February–March 1994.

Schultz *et al.* (1999) studied the budget of ozone and nitrogen oxides ($NO_x = NO + NO_2$) in the tropical South Pacific troposphere by photochemical point modelling of aircraft observations at 0–12-km altitudes from PEM-Tropics A flights in September–October 1996. It was found, in particular, that chemical production of ozone equals one-half of the chemical loss in the tropospheric column over the tropical South Pacific. The net loss is 1.8×10^{11} molecules/cm^2/s. The missing source of ozone is matched by westerly transport of continental pollution into the region. Application of a global 3-D model corroborates the results from the point model and reveals the importance of biomass burning emissions in South America and Africa for the ozone budget over the tropical South Pacific (biomass burning increases average ozone concentrations by 7–8 ppbv throughout the troposphere). The NO_x responsible for ozone production within the South Pacific troposphere below 4 km can largely be explained by decomposition of peroxyacetylnitrate (PAN) transported into the region with biomass burning pollution at higher altitudes.

Fenn *et al.* (1999) discussed *in situ* and laser remote measurements of gases and aerosols, made with airborne instrumentation to establish a baseline chemical signature of the atmosphere above the South Pacific Ocean during PEM-Tropics A observations. Between 8°S and 52°S, biomass burning plumes containing elevated levels of O_3 over 100 ppbv, were frequently encountered by the aircraft at altitudes ranging from 2 to 9 km. Air with elevated O_3 was also remotely observed up to the tropopause, and these air masses were observed to have no enhanced aerosol loading. Frequently, these air masses had some enhanced potential vorticity (PV) associated with them, but not enough to explain the observed O_3 levels. A relationship between O_3 and PV was developed from cases of clearly defined O_3 from stratospheric origin, and this relationship was used to estimate the stratospheric contribution to the air masses containing elevated O_3 in the troposphere.

The aim of the summer 1993 intensive North Atlantic Regional Experiment (NARE) was to investigate how major sources of gases in the surrounding industrial regions affect ozone distribution over the northern Atlantic Ocean (Lin *et al.*, 1998). The results strongly suggest that NO_x emissions from human activities have substantially increased ozone concentrations in much of the troposphere over the North Atlantic. In the boundary layer and lower free troposphere the largest increases are seen in the industrial regions of the USA and Europe.

Lee *et al.* (1998) discussed 8-year ozone sensing data obtained between 1989 and 1997 at Naha (Okinawa Island). These data demonstrate the continuous increase in tropospheric ozone in north-eastern Asia in the 1990s. It was suggested that this tropospheric ozone trend was due to the increasing emission of NO_x from north-eastern Asia (China, Japan, South Korea, and Taiwan) during this period, with a rate of 3.9% per year.

Logan (1999) conducted an analysis of ozonesonde data, synthesizing what is known about the distribution of tropospheric ozone (it was pointed out that a minimum of 20 sensings are required for 95% confidence intervals of ozone monthly mean levels reliable to better than ±30% near the extratropic tropopause; 20 sensings also ensure mean levels are reliable to better than ±15% for 500–500 hPa for the extratropics and 800–1,000 hPa for the tropics). The results indicate that ozone variability is higher in the upper troposphere for subtropical locations than for tropical locations, and 35 sensings are required for 400–1,000-hPa mean levels to be defined to better than ±15%.

For the northern middle and high latitudes, the broad summer maximum in ozone in the mid-troposphere extends all the way up to the tropopause. Median concentrations at the tropopause are highest in June and July (typically 120–200 ppb), and are smaller by a factor of 2 in the winter. Highest values of ozone occur in the spring 2 km above the tropopause. Mean ozone values in the mid-troposphere are approximately constant from 35° to 50°N in the summer. In the northern subtropics, there is a summer minimum in mid-tropospheric ozone over the Pacific and a summer maximum over the Atlantic, which appears to be related to differences in circulation. Ozone is the most similar in May and June over the Atlantic and the Pacific – the months with minimum burning in the tropics. On the basis of ozonesonde data in combination with surface and satellite data, Logan (1999) derived a climatological grid for tropospheric ozone.

Using a large ozone-profile data set obtained at Réunion Island (21°S, 55°E) since 1992, Taupin *et al.* (1999) showed that tropospheric ozone above the south-western Indian Ocean is strongly linked to dynamical conditions prevailing in the tropics. The average seasonal variation of tropospheric ozone reaches 25 DU and peaks at 55 DU in October. A large ozone enhancement was observed that occurs in parallel with the biomass-burning season in southeast Africa and Madagascar (this season extends from August to December and peaks in October). The ozone maximum concomitantly results from the transport of ozone and its precursors westward from the biomass-burning zones. High ozone values recorded in the upper troposphere during the austral winter are likely to originate from additional stratosphere–troposphere exchanges. Winter anticyclonic conditions further favour the photochemical production of ozone. The low tropospheric ozone-mixing ratios observed across the entire troposphere during the austral summer could result from the heterogeneous destruction of ozone and from less favourable conditions for its photochemical production, because of the development of deep convection.

Since the production of tropospheric ozone is due to oxidation of CO and hydrocarbons in the presence of nitrogen oxide radicals ($NO_x = NO + NO_2$), which may originate from lightning, downwelling of stratospheric air, aircraft emissions and surface emission (including fossil-fuel combustion) transported convectively in the upper stratosphere, as well as hydrogen oxide radicals ($HO_x = OH +$ peroxy radicals), Jaeglé *et al.* (1998a) examined the origin of NO_x in the upper troposphere over the central United States using aircraft observations during the Subsonic Aircraft: Cloud and Contrail Effects Special Study (SUCCESS) campaign in April–May of 1996. Simultaneous measurements of NO, NO_y, N_2O, O_3, OH, HO_2, H_2O, CO, CO_2, and CH_4 concentrations were made up to 12.5-km altitude together with aerosol, cloud, and radiative observations. Observational data on the correlation between NO_y (the sum of NO_x and its oxidation products) and CO at 8–12-km altitudes indicate that NO_x originates primarily from convective transport of polluted boundary layer air. Lightning and aircraft emissions appear to be only minor sources of NO_x. Simulation modelling has led to the conclusion that the NO_x/NO_y ratio is maintained above the chemical steady state by frequent convective injections of fresh NO_x from the polluted boundary layer and by the long lifetime of NO_x in the upper troposphere (5–10 days) (see also Molemaker and Arellano, 1998). In contrast with earlier studies, Jaeglé *et al.* (1998b) found no evidence for fast heterogeneous recycling from HNO_3 to NO_x in the upper troposphere.

The sources of HO_x (OH + peroxy radicals) and the associated production of ozone at 8–12 km over the USA were examined by Jaeglé *et al.* (1998a) by modelling observations of OH, HO_2, NO, and other species during SUCCESS (the 3-D photochemical model was used, developed at Harvard University). Comparison with observations indicates that HO_x concentrations measured in SUCCESS are 3 times higher than can be calculated from oxidation of water vapour photolysis of acetone. The highest discrepancy was seen in the outflow of a convective storm. Jaeglé *et al.* (1998b) showed that convective injection of peroxides (CH_3OOH and H_2O_2) and formaldehyde (CH_2O) from the boundary layer to the upper troposphere could resolve this discrepancy. More generally, the data collected over the central USA during SUCCESS suggest that the local convection was a major source of HO_x and NO_x to the upper troposphere. An average net ozone production of 2 ppbv/day was found between 8 and 12 km over the continental USA in spring. Ozone production was NO_x-limited under essentially all conditions encountered in SUCCESS. The high levels of HO_x present in the upper troposphere stimulate ozone production and increase the sensitivity of ozone to NO_x emissions from aircraft and other sources.

An important component of IGAC is the Biomass Burning Experiment (BIBEX): studying impacts on the atmosphere and biosphere. This was set up to monitor the prominent role of biomass burning as a source of atmospheric pollutants and of the important ecological impact of vegetation fires in the tropics. The Southern Africa Fire-Atmosphere Research Initiative (SAFARI-92) field experiment was conducted in the 1992 dry season in southern Africa under the auspices of BIBEX, together with the joint Brazilian and US Transport and Atmospheric Chemistry near the Equator – Atlantic (TRACE-A) Project, field studies of which were also conducted in the 1992 BIBEX IGAC's Southern Atlantic Region Experiment (STARE) campaign.

The results of SAFARI confirmed that it is justified to consider biomass burning as a significant contributor to the overall increase in GHGs that has occurred over the last 150 years, accounting for some 10–25% of current emissions. A major finding of TRACE-A was that widespread biomass burning in both America and southern Africa is the dominant source of precursor gases necessary for the formation of the huge amounts of ozone observed over the southern Atlantic Ocean. The generation of ozone is usually enhanced in the upper troposphere, where relatively high concentrations of NO_x prevail.

It has been found that ozone in the tropical troposphere plays a key role in determining the global oxidizing power of the atmosphere. Most of the oxidation of long-lived gases by hydroxyl radicals takes place in the tropics, where intense sunlight and high humidity promote the formation of OH from the photolysis of ozone.

The release into the atmosphere of biogenic gases produced by the African rain forest and savanna, as well as from biomass burning, and their role in the formation of tropospheric ozone and aerosol is the focus of the IGAC Experiment for Regional Sources and Sinks of Oxidants (EXPRESSO). Under IGAC, the Global Tropospheric Ozone Project (GTOP) is now being formulated so that all forthcoming data can be used intelligently to provide an understanding of the global tropospheric ozone cycle (Pszenny and Brasseur, 1997). A major aim of IGAC is to understand the role of biological processes in producing and consuming atmospheric trace gases.

An important contribution to studying tropospheric ozone dynamics was made by the TOR Project (Borrell *et al.*, 1998). As pointed out by Høv *et al.* (1997), the main achievement of TOR was the development of suitable instrumentation for high-quality measurements of ozone precursors (NO_x, NO_y, VOC), intermediates (carbonyl compounds, and RO_2), and photolysis rates as well as the implementation of quality assurance procedures for these measurements, the establishment of a large high-quality network, and accumulation of a significant database. The high-quality data sets provide detailed mechanistic information on the chemical and physical processes that control the budget of ozone and its precursors in the polluted boundary layer and in the free troposphere. The vertical sensings were analysed for distribution and seasonal variation in ozone in the free troposphere over Europe and provide insight into various processes – such as the exchange between the boundary layer and the free troposphere and between the troposphere and the stratosphere.

Examples of the results obtained include the concept of O_x for studying chemical ozone formation, the O_x/NO_x relation for studying the catalytic efficiency of NO_x, the use of alkyl nitrates to measure peroxy radicals, the role of individual hydrocarbons in ozone formation, and the use of the H_2O/O_3 correlation for the ozone budget in the upper troposphere.

A hierarchy of models was developed and/or applied in TOR for data analysis. 3-D transport models with coupled chemistry modules were developed in TOR and have been used to study important transport processes – such as convection – and for budget studies.

Of course, after the completion of the first phase of TOR many problems still remain unsolved because of the complexity of the processes that govern tropospheric ozone dynamics. A different aspect of the problem is the unseparable influence of transport and chemistry on the concentrations observed at a given location. It is also important that the net chemical balance of ozone depends, in non-linear fashion, on the concentrations of NO_x, VOCs, H_2O, and O_3 itself, and on the UV radiation flux, and thus it is closely coupled with the atmospheric lifecycles of other trace gases. At continental surface sites, dry deposition plays an important role in the net balance of ozone, in addition to advection and vertical exchange. It is obvious that future success may be achieved only on the basis of combined analysis of sufficiently complete observational data and results of simulation modelling, with the use of fully coupled models of chemistry and transport.

In this context, very important progress was recently made in simulations of tropospheric ozone distribution using a chemistry–climate model. An illustration of this success is the analysis of tropospheric ozone over the Indian Ocean carried out by De Laat *et al.* (1999) using the chemistry–general circulation model of the European Centre Hamburg (ECHAM). Such an approach allowed interactive dynamics of the atmospheric circulation and ozone to be studied. It was shown, in particular, that large-scale upper tropospheric ozone minima were caused by convective transport of ozone-depleted boundary layer air in the intertropical convergence zone (ITCZ). Similarly, an upper tropospheric ozone minimum was caused by cyclone Marlene, south of the ITCZ. The mid-tropospheric ozone maxima were caused by transport of polluted African air.

As pointed out earlier, the satellite remote sensing data are an important source of information on tropospheric ozone distribution (Borrell *et al.*, 1998; *Scientific ...*, 1999). There are many techniques capable of retrieving tropospheric ozone from the TOMS data (Kaye, 1998). One of them is the use of TOMS-SBUV residuals (in which tropospheric ozone is taken as the difference between the TOMS total ozone and integrated stratospheric ozone from one of the SBUV instruments) or the difference in ozone columns over mountains and nearby sea level areas (as has been carried out over the Andes Mountains and nearby eastern Pacific). Another residual technique suggested by Chandra (1996) is based on the difference between the TOMS total ozone and stratospheric ozone determined by combining the Microwave Limb Sounder (MLS) and the HALOE instruments aboard the UARS satellite. Another similar method, known as the Convective Cloud Differential (CCD) technique, derives tropospheric ozone from the average difference between TOZ over clear regions and over regions of high clouds (in the latter case reliable cloud top information becomes very important). Initial studies used the TOMS data and the TOMS/SAGE residuals only; recently work has been based on a residual approach using the TOMS and the SBUV instrument data (in this case precise determination of the tropopause height is very important).

The Tropospheric Emission Spectrometer (TES) planned for the EOS CHEM spacecraft (Kaye and Rind, 1998) is a new method of obtaining the tropospheric ozone data. TES is a fourier transform spectrometer measuring in the 650–3000 cm^{-1} wavelength region with the use of both nadir and limb-viewing geometry, spectral resolution being 0.025 cm^{-1} in the limb mode and 0.1 cm^{-1} in the nadir mode. The primary focus of the TES instrument is on measurements of ozone and its precursors in the troposphere, although TES has an excellent capability for detecting a broad range of species because of its high resolution.

Munro *et al.* (1998) discussed a retrieval of vertical profiles of ozone in the troposphere from the data of the European Space Agency's GOME on board the ERS-2 satellite launched in April 1995. GOME is a UV/visible spectrometer that measures solar radiation backscattered from the Earth's atmosphere in four contiguous wavelength bands between 137 and 790 nm at moderate resolution (0.2–0.4 nm). An important advantage of GOME in comparison with the SBUV instrument is the availability of data for the ozone Huggins bands (310–340 nm). With knowledge of the temperature profile, the temperature-dependent spectral structures in the Huggins bands yield additional ozone profile information below the ozone peak (in the troposphere and lower stratosphere). The comparison between retrieval results obtained and ozonesonde data indicates that, in cloud-free scenarios and above low clouds, ozone profiles extending down through the troposphere can be retrieved from GOME measurements. Munro *et al.* (1998) demonstrated the GOME's ability to detect tropospheric ozone produced by biomass burning over Africa.

Future tropospheric ozone research faces a number of still unsolved problems, including:

- Accumulation of adequate global information on observed variability of tropospheric ozone for sufficiently long time periods (the lack of data is especially severe in the tropics and subtropics); observations of atmospheric ozone precursors are equally important.
- Investigations about the chemistry and photochemistry of tropospheric ozone under cloudy conditions.

As emphasized by Crutzen (1998), the role of clouds as transporters of chemical constituents (such as reactive hydrocarbons, CO, and NO and their oxidation products) from the boundary layer to the middle and upper troposphere (and possibly into the lower stratosphere) needs to be better understood and quantified, so that they can be parameterized for inclusion in large-scale photochemical models of the atmosphere. Similarly, the production of NO by lightning and its vertical redistribution by convective storms also needs to be better quantified.

The interaction of chemical constituents emanating from the boundary layer with liquid and solid hydrometeors in the clouds, are of special importance. The influence of clouds on the photochemically active UV radiation field is a potentially important research topic. An issue of key significance is the interactions between gases and atmospheric aerosols. Since the continental biosphere is a large source of hydrocarbons, quantification of relevant sources in terms of physical (e.g., temperature, humidity, light levels) and biogeochemical (solid physical and chemical properties and land use) parameters are urgently needed for inclusion in atmospheric models. The formation of ozone, carbon monoxide, partially oxidized gaseous hydrocarbons, and organic aerosol may be better quantified and parameterized for inclusion in a chemical transport model, if the hydrocarbon oxidation mechanisms in the atmosphere were better understood.

Monod and Carlier (1999) have conducted a box model daytime study of the multiphase photochemistry (both gas and aqueous phases) of Cl organic compounds within a non-precipitating cloud on a local scale. According to their results, when ozone accumulates in clear sky conditions, as soon as a cloud is formed, the tropospheric ozone changes drastically: the net production decreases by a factor of 2 or more and, depending on NO_x concentrations and pH values, can actually lead to a net chemical destruction. Monod and Carlier (1999) pointed out that both indirect and direct impacts, caused by the presence of the liquid phase, explain this. The indirect impact is the result of the much higher solubility of HO_2 and RO_2 radicals than NO and NO_2, resulting in lower gas phase efficiency of the NO_x cycles in producing ozone. The direct impact is caused by a very fast reaction of ozone towards O_2^- radicals within the liquid phase. The relative importance of these two impacts has been determined for different NO_x concentrations and pH values. When $NO_x = 1$ ppbv, the direct impact is of negligible importance at pH = 3 (for a total decrease in ozone production by a factor of 2), but it accounts for 6% at pH = 4.17 (for a total decrease in ozone production by a factor of 6.6), 16% at pH = 5.2 (for a total net chemical destruction of ozone), and 28% at pH = 6 (for a total net chemical destruction of ozone). At this NO_x level and at pH = 5.2 and 6, despite its small contribution, the effect of the direct impact is the net chemical destruction of ozone. When $NO_x = 0.1$ ppbv, both direct and indirect impacts contribute to the net chemical destruction at all pH values. However, the direct impact now contributes to a larger extent, accounting for 15% at pH = 3, and up to 77% at pH = 6. The direct impact involves the $O_3 + O_2^-$ reaction which leads to the production of aqueous-phase OH radicals. The latter species in turn react with dissolved organic compounds to produce additional HO_2 and O_2^- radicals. Therefore, the aqueous-phase sink of ozone is auto-catalytic in the presence of soluble organic compounds.

Warnings about potential surprises in the photochemistry of tropospheric ozone were made by Ravishankara *et al.* (1998), who discussed the role of the hydroxyl radical OH as a natural atmospheric detergent responsible for cleansing the atmosphere of pollutants by means of their oxidation. Ravishankara *et al.* (1998) pointed out that, in general, the reactions of electronically excited species are of negligible importance in the chemistry of the lower atmosphere, but the case of $O(^1D)$ is a notable exception. Its role is pivotal. Even though most of the $O(^1D)$ is deactivated to the ground state ($O(^33P)$) – the small fraction that survives to react with H_2O and CH_4 turns out to be the major source of OH. Knowledge of how $O(^1D)$ is formed in the atmosphere is therefore critical in understanding the reaction of OH. Recent surprising findings are beginning to reveal the importance of the longer wavelength "tail" in the chemistry of $O(^1D)$ formation. The longer wavelengths are important because stratospheric ozone screens most of the short wave UV from the lower atmosphere.

Bojkov (1999) emphasized the significant uncertainties that remain in the budget of tropospheric ozone.

5.4.3 Concluding remarks

Investigations of changes in total ozone (especially in high latitudes) highlight that the TOZ variability problem as a very important aim for the study of global-scale environmental dynamics, having great significance for humans and ecosystems. The recognition of such an environmental danger led to the signing of the Montreal Protocol and further amendments to avoid dangerous environmental consequences of ozone depletion. There are many lessons to be learned from the Montreal Protocol which can be applied to solving other global environmental issues (*Scientific* ..., 1999).

The primary lesson is an application of the "precautionary principle": taking the necessary actions in time to prevent damage, rather than waiting for proof of the damage, by which time the damage could be irreversible. Another important lesson of the Protocol is how to react to issues when there is no scientific certainty (in 1982, when the Protocol was signed, many questions remained open). The solution has been to undertake successive steps to phase out ozone-depleting substances and get the scientific community to advise the governments periodically on the further steps needed to protect the ozone layer and recommend alternative technologies. Over the last 10 years governments have changed the Protocol four times in accordance with relevant scientific advice. Another lesson of the Protocol is in promoting universal participation, including developing countries, in the Protocol by recognizing "common and differential responsibility". One more important lesson is the integration of science, economics, and technology both in developing control measures and implementing them.

Sarma (1998) emphasized that one measure of success of the Montreal Protocol and its subsequent amendments and adjustments was the forecast of "the world that was avoided" by the Protocol:

- The abundance of ozone-depleting substances in 2050, the approximate time when the ozone layer is now projected to recover to pre-1980 levels, would be at least 17 ppb of equivalent effective chlorine (this is based on the conservative assumption of a 3% annual growth in ozone-depleting gases), which is about five times larger than today's value.
- Ozone depletion would be at least 50% at mid-latitudes in the northern hemisphere and 70% at mid-latitudes in the southern hemisphere, about 10 times larger than today.
- Surface UV radiation would at least double at mid-latitudes in the northern hemisphere and quadruple at mid-latitudes in the southern hemisphere compared with an unperturbed atmosphere. This compares with current increases of 5% and 8% in the northern and southern hemispheres, respectively.

As far as the Arctic ozone is concerned, Fergusson and Wardle (1998) have pointed out that over the next few decades its total column changes will depend on climatic changes. Not only anthropogenic impact due to ODS but various natural phenomena will lead to ozone changes, including weather systems, the quasi-biannual oscillation (a periodic reversal of the direction of stratospheric winds over the equator), El Niños, slight variations in solar radiation associated with the sunspot cycle, and volcanic eruptions.

A set of thematic issues has been devoted to the Photochemistry of Ozone Loss in the Arctic Region In Summer (POLARIS) Mission conducted in 1997 with special emphasis on photochemical investigation and detailed comparison of radical and reservoir measurements with models. In such a way, Müller *et al.* (2001) used N_2O as a long-lived tracer to identify chemical ozone depletion in the Arctic vortex in the presence of ozone variations caused by dynamical effects. Ob′servations conducted during the 1991/1992 winter and early spring indicated (consistent with the dynamical development of the polar vortex and with observed chlorine activation), that the major fraction of the ozone decline had occurred before February 1992. The reduced ozone levels persisted over the lifetime of the polar vortex until late March

1992. Hitchman *et al.* (1999) studied, in particular, the influence of synoptic waves on total ozone during the Arctic summer of 1997.

Shindell *et al.* (1998) have investigated polar stratospheric ozone losses and delayed eventual recovery owing to increasing GHG concentrations. Temperature and wind changes induced by the increasing GHG concentrations alter the planetary wave propagation, reducing the frequency of sudden stratospheric warmings in the northern hemisphere. This results in a more stable Arctic polar vortex, with significantly colder temperatures in the lower stratosphere and concomitantly increased ozone depletion. Increased concentrations of GHGs might therefore be at least partly responsible for the very large Arctic ozone losses observed in recent winters. Due to model calculations made by Shindell *et al.* (1998), the Arctic ozone losses will reach a maximum between 2010 and 2019, roughly a decade after the maximum in stratospheric chlorine abundance is reached.

Of serious importance may be the development of supersonic transport aircraft that would fly in the lower stratosphere. A fleet of 500 to 1000 of these aircraft would release large quantities of nitrogen oxides, water vapour, and sulphates, all of which have the potential to increase ozone depletion. Because these substances contribute to PSC formation as well, their impact on the Arctic ozone levels could be particularly harmful (Fergusson and Wardle, 1998).

Special attention has been paid to problems of polar ozone changes during the Eighth Session of the SPARC Scientific Steering Group and SPARC 2000 General Assembly (Chanin, 2001). Important subjects were the internal variations in the stratosphere–troposphere coupled system, and the issue of the AO. The importance of the modulation due to large internal dynamical variations was repeatedly pointed out. Model calculations made by Tabazadeh *et al.* (2000) show that an important factor of the Arctic ozone changes may be widespread severe denitrification, which can enhance future Arctic ozone loss by up to 30%.

One of the main causes of the current focus on tropospheric ozone is its role as a GHG and toxic component. While reviewing the sources of change in radiative forcing and their influence on climate, Hansen *et al.* (1997) emphasized the need for information about the exact nature of the change in the vertical distribution of ozone (especially to see if increases in tropospheric ozone may have countered decreases in stratospheric ozone), the nature of aerosol particles (especially single scattering albedo), as well as their height. Derwent *et al.* (2001) have emphasized the importance of considering indirect climatic impact due to tropospheric ozone, which dictates the necessity to take into account ozone precursors as radiatively active trace gases. In this context further development of numerical climate modelling, taking into account various minor gaseous components, is urgently needed. Brasseur *et al.* (1998) described simulations of ozone and related chemical tracers obtained by the Model for Ozone and Related Chemical Tracers (MOZART) developed at the National Center for Atmospheric Research (NCAR). One of the results was the simulation of changes in surface ozone since pre-industrial time, which shows changes in ozone amounts of some 30–40 ppb in July over the USA and much of Europe along with changes of 20–30 ppb over most of Eurasia. The average RF coming from these increases should be approximately 0.45 W/m^2 (higher in the northern hemisphere, lower in the southern hemisphere). Predictions for the future (up to 2050) show that the largest changes of surface ozone are likely to come in the tropics.

Jacob *et al.* (1995) described the Chemistry, Aerosols and Climate Tropospheric Unified Simulation (CACTUS) at Harvard University, the present targets of which include the climatic effects of changes in emissions of precursors on tropospheric ozone formation, the feedback of climate change on tropospheric ozone, the climatic effects of changes in emissions of sulphate precursors, and the response of sulphate aerosols to climate change.

Scenario calculations for the year 2025 made by Lelieveld and Dentener (2000) indicate that the man-made emissions at low northern latitudes, in particular, in southern and eastern Asia, will become a very strong tropospheric O_3 source in the next few decades.

5.5 POLAR AEROSOLS

5.5.1 Introduction

Studies of climate-forming processes in the Arctic have been of particular concern for a long time. The basic features of these processes are very specific because they are determined by the presence of an ocean isolated from the atmosphere by ice cover, and because of the high sensitivity of the Arctic climate to various external impacts, including atmospheric pollution. Of special interest is the phenomenon of the Arctic haze which results from the transport of aerosol to the high latitudes from the industrial regions of Europe, Asia, and North America. Another kind of pollutant in the Arctic atmosphere is formed by CFCs which reach the stratosphere and destroy the ozone layer. The processes of wet and dry deposition are responsible for the surface pollution in the Arctic by heavy metals and other toxic components. This leads to surface albedo changes. General problems of atmospheric aerosol studies have been recently discussed by Houghton *et al.* (2001) and Kondratyev (1999b, 2002a,b). It is of interest that in discussing perspectives of the IGAC Project, Brasseur *et al.* (2002) has particularly emphasized the importance of aerosol studies.

Specific features of the pollution of the Arctic atmosphere will be discussed below. The emphasis has been placed on studying three phenomena of principal importance: (1) formation and development of the Arctic haze; (2) deposition of heavy metals in the Arctic; and (3) extended cloudiness and radiation. Investigations of pathways and physical properties of aerosols transported to the Arctic have been reviewed in the context of a possible impact of the Arctic haze on climate.

For intercomparison purposes, in Tables 5.6 and 5.7 we reproduce general information about atmospheric aerosols from the Intergovernmental Panel on Climate Change (IPCC) Report-2001 (Houghton *et al.*, 2001).

The high transparency of the summertime Arctic atmosphere is well known. However, in winter a strong haze is often observed in high latitudes

Table 5.6. Annual source strength for present-day emissions of aerosol precursors. (Tg N, S, or C per yr).

	NH	SH	Global
NO_x			
Fossil fuel	19.9	1.1	21.0
Aircraft	0.43	0.03	0.46
Biomass burning	3.3	3.1	6.4
Soils	3.5	2.0	5.5
Agricultural soils			2.2
Natural soils			3.2
Lightning	3.7	2.4	7.0
NH_3 (as Tg N/yr)			
Domestic animals	17.5	4.1	21.6
Agriculture	11.5	1.1	12.6
Human	2.34	0.31	2.6
Biomass burning	2.5	2.2	5.7
Fossil fuel and industry	0.29	0.01	0.3
Natural soils	1.36	1.07	2.4
Wild animals	0.10	0.02	0.1
Oceans	3.65	4.50	8.2
SO_2 (as Tg S/yr)			
Fossil fuel and industry	68	8	76
Aircraft (1992)	0.04	0.003	0.04
Biomass burning	1.2	1.0	2.2
Volcanoes	6.3	3.0	9.3
DMS or H_2S (as Tg S/yr)			
Oceans	11	13	24
Land biota and soils	0.6	0.4	1.0
Volatile organic emissions (as Tg C/yr)			
Anthropogenic			109
Terpenes	67	60	127

Note: NH = northern hemisphere; SH = southern hemisphere.

(Barrie, 1986, 1996; Kondratyev and Binenko, 1981; Kondratyev *et al.*, 1996 and many others).

Ob′servations performed in the early 1970s at several Arctic stations, revealed much higher turbidity than was expected, and the annual change of transparency was opposite to that recorded in mid-latitudes. The concentration of aerosols in winter turned out to be 10 to 40 times higher than in summer in the North American sector of the Arctic, whereas in the Russian part of the Arctic (Cape Schmidt) this value varied from 5 to 20 times, depending on the concrete meteorological situation.

The airborne measurements near Point Barrow (Alaska) and at Cape Schmidt revealed an increase in the haze concentration with altitude, reaching a maximum at a height of 2–3 km (sometimes higher). Hence, the sources of the haze aerosol had not been local. Studies, started in 1976, revealed an increased concentration of

Table 5.7. Primary particle emissions for the year 2000 (Tg/yr).
Bobylev *et al.* (2003)

	NH	SH	Global	Min.	Max.
Organic matter (0–2 μm)					
Biomass burning	28.3	26.0	54.3	45	80
Fossil fuel	28.4	0.4	28.8	10	30
Black carbon (0–2 μm)					
Biomass burning	2.9	2.7	5.6	5	9
Fossil fuel	6.5	0.1	6.6	6	8
Aircraft	0.005	0.0004	0.006		
Industrial dust, etc.			100		
Biogenic (0–2 μm)			50	10	100
Sea salt					
$d < 1$ μm	23	31	54	18	100
$d = 1$–16 μm	1,420	1,870	3,290	1,000	6,000
Total	*1,443*	*1,901*	*3,344*	*1,018*	*6,100*
Mineral (soil) dust					
<1 μm	90	17	110	180	300
1–2 μm	240	50	290	340	900
2–20 μm	1,470	282	1,750	480	1,800
Total	*1,800*	*349*	*2,150*	*1,000*	*3,000*

Note: NH = northern hemisphere; SH = southern hemisphere.

vanadium and manganese due to industrial pollution; this was a first indication that the Arctic haze had not been of natural origin but anthropogenic. This was also confirmed by the colouration of dust samples (whereas in summer they had been colourless and slightly brown) due to the presence of soot carbon, which could have been the product of incomplete combustion. Organic aerosols constitute about 1 μg/m^3, with surprisingly large amounts of black carbon, 0.3–0.5 μg/m^3. The total concentration of the submicrometre fraction ($r < 1$ μm), including SO_4, V, Mn, Pb, Cd, Fe, As, Sb, and Se, was 4 μg/m^3.

The most convincing indicator of the anthropogenic origin of haze is a high concentration of sulphates. From the data of observations at Barrow, the sulphates constitute most of the mass of haze particles and, probably, result from oxidation from sulphur dioxide, especially in summer.

The seasonal variation of sulphates and vanadium sampled at Barrow showed that maximum concentrations were observed in March (2 μg/m^3 and 1 ng/m^3, respectively); in summer, minimum concentrations were in August and constituted 0.1 μg/m^3 (SO_4) and 0.02 ng/m^3 (V). On some days in March–April at Barrow and Spitsbergen, concentrations of sulphates of 5 and 10 μg/m^3, respectively, were observed. For comparison, the concentrations of sulphates in the background regions of the northern hemisphere constitute 0.1–0.2 μg/m^3. The high acidity of precipitation also confirms a great amount of sulphates in the Arctic.

Halsall *et al.* (2001) have studied the size distribution of the chemical com-

Table 5.8. Parameters for lognormal size distribution.

Mode	r_j (μm)	σ	N (cm^{-3})	M ($\mu g/m^3$)
Fine	0.02	1.8	400	0.05
Accumulation	0.15	1.4	340	5.0
Coarse	2.5	2.0	0.41	5.0

position of atmospheric aerosols at high Arctic latitudes. The results indicate that two principal mechanisms of aerosol formation are functioning: (1) formation of sea salt particles; and (2) gas-to-particle transformation of biogenic DMS.

Shaw (1993) has substantiated a model of the polluted Arctic troposphere to estimate the magnitude and seasonal variation of the climate forcing function of Arctic haze. For such an assessment they have used three lognormal mass models with properties consistent with the observations in Alaska (Table 5.8) where r_j is the geometric mean radius of the jth mode, σ is the geometric mode standard deviation, and N and M are the particle number and mass concentrations, respectively. The estimates show that the major perturbation to the radiation budget is the lowering of the albedo (heating) of the Earth–atmosphere systems around the vernal equinox and is due to a trace amount (about 5% by mass) of black carbon associated with the removal-resistant submicrometre mode of aerosols. The decrease of the radiation budget at the troposphere level is about $5\,W/m^2$. The black carbon over the reflecting polar ice/snow introduces a heating of about 1.5°C per day into the haze layer.

5.5.2 Aerosol observations

From 1973 onwards, the scientists of the Main Geophysical Observatory and the Institute of Arctic and Antarctic (Russia) have been carrying out ground-based and airborne-based investigations of aerosols in the polar regions. Measurements have been made of the aerosol vertical profiles (up to 8 km) and the size distribution and chemical composition of aerosols (Kondratyev and Zhvalev, 1981). Simultaneously with aerosol measurements, radiative properties of the atmosphere were measured, including SW and LW radiation fluxes. Observations were made in the area of the Chukotka Peninsula, in the regions of Anadyr, Cape Schmidt, and the Stations Chokurdakh and Tixi.

Morphological analysis of aerosol samples has shown that the lower layers of the troposphere are characterized by the presence of oval particles whose shape is close to a sphere. They are sufficiently dense and look homogeneous. Particles in the samples taken over open waters are especially typical. The particles in samples taken in the middle troposphere were less spherical than those from the lower troposphere. Roughly, 70% of the total number of particles were irregular in shape.

The upper troposphere is characterized by the presence of loose irregular coagulants, particles of the "chain" type, small particles of dense matter. Observations

Table 5.9. Elemental composition of polar aerosols (ng/m³).

Region	Period of observations	Na	Md	V	Mn	Cu	Pb	Ca	Br	SO_4
Greenland	February–April 1978	130	23	0	0.7	0.3	8	–	9.4	–
Novaya Zemlya	April–May 1984	440	300	1.0	6	–	1.3	0.05	12	–
Wrangel Island	May 1984	1,180	60	1.0	10	12	1.8	0.06	3.5	520
Alaska	February 1978	310	–	1.5	–	–	–	–	–	2,200

yielded valuable information about the chemical composition of Arctic aerosols in regions where measurements have not been made earlier. Table 5.9 presents these data together with the results for other polar regions (Kondratyev and Zhvalev, 1981).

The amount of soot particles can be considerable from the viewpoint of their effect on the albedo of the snow–ice surface and cloud cover properties. A thorough analysis of the results available is required from the point of view of the Arctic haze impact on the regional and global climate. The effect of the aerosol deposition on the ecologically sensitive biomes of the tundra can also be considerable. This has been demonstrated by Adamenko *et al.* (1991a,b).

Later on, balloon impactor measurements of the vertical structure of atmospheric aerosols have been accomplished by scientists of the St. Petersburg University Institute of Physics and the Voyeykov Main Geophysical Ob′servatory. The impactor used to sample atmospheric aerosols is a continuously moving substratum (formavar film) located under a nozzle, through which an air flow containing aerosol particles is pumped at high speed. Starting from 1987, balloon measurements were based on using a two-cascade impactor. The first cascade caught gigantic particles, among which there were particles that had possibly been emitted to the atmosphere by different sources such as fuel combustion, cement production, the ceramic industry, etc. The first cascade protected the substratum of the second one from possible contamination during the impactor's landing.

Basic information concerning the structure and size distribution of atmospheric aerosols has been obtained from electron microscope analysis of the impactor samples. The density of particles whose size exceeded the minimum, d_{min}, has been estimated. The formavar films made it possible to identify particles with $d_{min} > 0.005\,\mu m$ with the help of the microscope. The minimum size of particles was restricted to $0.01\,\mu m$. Within the range of particle sizes up to $d = 0.2\,\mu m$, the uncertainty in estimating the coefficient of efficient trapping of the particle on the substratum is very large, particularly for altitudes of $H < 10\,km$. The data for $d < 0.2\,\mu m$ can be considerably underestimated.

Morphological analysis makes it possible to identify the following types of particles: particles of mineral origin, particles of organic substance, H_2SO_4 and sulphate particles, conglomerates of smaller particles, particles of the soot fractal type, particles of mixed origin, homogeneous spherical particles of dense matter (high temperature condensate, e.g., micrometeorites), and crystals resulting from reactions in the atmosphere (*in situ*). Chemical and elemental analyses were made using the methods of mass spectroscopy, infrared (IR) spectroscopy, neutron activation, and X-ray fluorescence. For this purpose, samples taken from different altitudes within a broad range were used in combination (usually three layers are considered during a flight: 4–10 km, 10–15 km, and over 15 km). In some cases, the neutron activation analysis makes it possible to identify elements with a sufficiently high vertical resolution.

In the cases of observations performed in 1989, in Appatity, Kola Peninsula, all the measured concentrations of chemical compounds and elements turned out to be at the noise level. It is only when IR spectral analysis is employed for altitudes above 15 km that an absorption band appears, determined by the presence of H_2SO_4 in the samples. Its concentration has been estimated at 100–250 $\mu g/m^3$. A neutron activation analysis has revealed the presence of trace amounts of Fe 10–15 $\mu g/m^3$ in the lower atmospheric layers.

One can speak with confidence about two layers of increased concentration of aerosol particles in the polar stratosphere: at altitudes 14–16 km and 20–21 km, respectively. The data from the electron microscope analysis reveal certain regularities of altitude-dependent variations in both the size distribution and morphological structure of aerosol particles. To increase the reliability of size distribution data, the results obtained for different altitudes were classified into three groups: 5–10 km, 15–20 km, and 24–30 km, respectively.

Examination of these data shows that the amount of particles with $d < 0.1\,\mu m$, has been greatly underestimated in the Arctic stratosphere, in comparison with the lower atmosphere. The particle size distribution is characteristic of a slow decrease in the concentration of particles with the growth of their size. In the lower stratosphere, there were a large amount of particles in every size range. At altitudes above 24 km, the number of the smallest particles ($d < 0.03\,\mu m$) decreases with altitude. Apparently, there are no sources there to generate particles of these sizes. The maximum particle size distribution is in the range of $d = 0.05$–$0.15\,\mu m$. This can be accounted for by the large lifetime of such particles in the stratosphere; hence, their growth due to condensation and coagulation.

The regular features of variations in the morphological structure of aerosol particles as a function of altitude reveal themselves very clearly. The general trend is as follows: the troposphere contains sufficiently large particles of mineral origin, conglomerates of smaller particles of organic substance, and sometimes H_2SO_4 particles. About 10% of the total amount may be classified as soot particles. As altitude grows, the proportion of H_2SO_4 particles also increases markedly, reaching 50–60% of the total amount at an the altitude of 9–10 km. On the whole, these are spherical liquid particles with individual small solid inclusions (the electron microscope shows traces of dried-up droplets). Even the observed conglomerates of solid particles apparently contain H_2SO_4 (among individual microparticles).

In the lower stratosphere, the proportion of spherical H_2SO_4 particles continues to increase. There are hardly any particles containing organic substance. Particles of mineral (probably volcanic) origin can be identified. In higher atmospheric layers, particles of irregular shape can hardly be found. Actually, all particles are either

H_2SO_4 or sulphate. There are structures representing irregular conglomerates that look like array webs of fine particles. It is possible that an array of such small particles was formed out of material sedimented on the film. Data available are insufficient to model optical characteristics of stratospheric aerosols: first of all, there is no information about the soot component. One can only assume that there are soot particles in polar aerosols and that these exist mostly as independent aggregates of small nuclei of the size $d \cong 0.01\,\mu m$. This circumstance reduces the share of radiation absorbed by aerosols, in comparison with the role of soot as a shell on larger particles of a different origin.

Data on the shape and size distribution of irregular particles (plates, chains, cylinders, etc.) should be specified. An assumption can be made that at night these particles have a certain predominant orientation, and if this is so, their substitution by equivalent spheres is impossible.

Pirjola *et al.* (1998) have conducted aerosol size distribution measurements throughout the whole year (June 1993–August 1994) at the Värriö environmental measurement station located in northern Finland near the Russian border (67°46′N, 29°35′E) 400 m above sea level. This area is practically free of local sources of pollutants. Therefore, the air quality at Värriö is normally very good, and the signs of long-range transport can be clearly detected, especially because sources of strong pollution from Nikel (69°25′N, 30°14′E) and Montchegorsk (67°N, 32°40′E) smelters, responsible for occasional pollution events in Värriö, are located less than 200 km away.

During the time period of observations, numerous nucleation events, associated both with marine air masses and with air pollution by smelters located on the Kola Peninsula, were recorded. To compare observations with calculations, a combined aerosol and chemistry model was run along calculated back trajectories with the special purpose of finding out whether homogeneous nucleation of sulphuric acid and water explains the observed nucleation mode concentration. Pirjola *et al.* (1998) have shown that with the Kola Peninsula events the simulated number densities were almost an order of magnitude higher than those observed, and it thus appears likely that H_2SO_4/H_2O nucleation produced the observed particles. The simulated nucleation mode number densities in the southern case were lower than those observed, although it is impossible to rule out the possibility that the measured particles were actually produced by H_2SO_4/H_2O nucleation. In the two marine cases the simulated nucleation mode concentrations were essentially zero. It may be suggested therefore, that the marine particles were formed via a different nucleation route than the pollution-associated particles.

Kawamura *et al.* (1998) have pointed out that organic compounds in polar air and snow play a prominent role as tracers of distant sources, as well as in physical and chemical reactions that influence the composition of atmosphere and snow. Table 5.10 summarizes relevant observational data on chemical composition of Arctic aerosols. According to these data, aerosol carbon comprises 2.4% to 11% of the total aerosol mass (detailed information on biogenic aerosols have been discussed by Kondratyev (2001b)).

Considering observational data on concentrations and the seasonal cycle of

Table 5.10. Concentrations of aerosol total carbon (TC), total nitrogen (TN), water soluble organic carbon (WSOC), and dicarboxylic acids, ketoacids, α-dicarbonyls as well as lipid class compounds in Arctic aerosols collected from February to June, 1991 at Alert Station.

Components	Range	Average
Aerosols	2500–9100 ng/m^3	5200 ng/m^3
TC	88–639 ng/m^3	359 ng/m^3
TN	16–154 ng/m^3	86 ng/m^3
C/N weight ratio	2.4–7.1	4.5
TC/aerosol	2.4–11.1%	6.8%
TN/aerosol	0.48–2.4 %	1.6%
WSOC	40.7–300 ng/m^3	186 ng/m^3
WSOC/TC	30–72%	53%
WSOC/aerosol	1.2–5.5%	3.4%
Dicarboxylic acids ($C_2 - C_{11}$)	7.4–84.5 ng/m^3	36.6 ng/m^3
Ketoacids ($C_2 - C_6$)	0.76–8.9 ng/m^3	3.7 ng/m^3
α-Dicarbonyls ($C_2 - C_3$)	0.05–2.8 ng/m^3	0.88 ng/m^3
Diacid-C/TC	1.5–9.1%	3.8%
Ketoacid-C/TC	0.18–0.78%	0.34%
α-Dicarbonyl-C/TC	0.019–0.17%	0.073%
Lipid class compounds:		
n-alkanes ($C_{18} - C_{35}$)	0.15–2.7 ng/m^3	0.85 ng/m^3
PAHs	0.0002–0.85 ng/m^3	0.11 ng/m^3
n-alcohols ($C_{13} - C_{30}$)	0.24–0.95 ng/m^3	0.50 ng/m^3
Fatty acids ($C_7 - C_{32}$)	1.3–6.5 ng/m^3	3.2 ng/m^3
Long-chain diacids ($C_{22} - C_{26}$)	0.074–0.56 ng/m^3	0.27 ng/m^3

black carbon in aerosol at a coastal Antarctic station, Wolff and Cachier (1998) have identified three sources of carbon aerosol (CA) in the atmosphere: (1) natural primary and secondary emissions of organic gases by vegetation; (2) fossil fuel burning; and (3) biomass burning. The observational data available indicate that the two latter sources (which are mostly of anthropogenic origin) dominate in the global CA budget with the two principal regions of CA formation in the tropics (due to biomass burning) and mid-latitudes (due to fossil fuel burning). Basically, two types of CA are distinguished: organic carbon and black carbon (BC). Wolff and Cachier (1998) have examined BC observational data at Halley station, which are characterized by pronounced annual variations (with maximum concentrations in summer). An interesting phenomenon of long-range transport of biomass burning products from the tropics to Antarctica has been discovered. Shaw (1988) has published a review of Antarctic aerosol properties.

An unusual phenomenon of mass transfer in Arctic plumes and the implications for global climate have been analysed by Crane *et al.* (2001), who emphasized the potential importance of plumes observed in the Russian Arctic by NOAA satellites. Such plumes provide evidence that gas hydrates may be destabilizing under the ocean floor and transferring into the atmosphere large quantities of light hydrocarbons including methane, a known GHG.

Polissar *et al.* (1998a,b) have studied atmospheric aerosol over Alaska between 1986 and 1995 analysing its spatial and seasonal variability as well as elemental composition and sources. Due to this observational data, it was noticed that sulphate mass concentration varied within 0.01–3.9 μg/m^3, while BC concentration changed between 0.01 and 5 μg/m^3. Almost everywhere annual variations of sulphate concentration were observed with a maximum concentration in winter–spring and a minimum in summer. Similar annual variations were absent, however, in cases of BC and submicron aerosols. Forest fires in the central part of Alaska in the summer resulted in a BC maximum during this season. In all cases an impact of long-range transport on aerosol mass concentration and chemical composition was observed, especially during the winter and spring seasons. Four principal components of aerosol chemical composition were identified: anthropogenic aerosols (due to long-range transport), sea salt aerosol, local soil aerosol, and BC aerosol (generated by forest fires).

Ob′servation data on the dependence of aerosol properties on particle size for the central part of the Greenland ice sheet in 1993–1995 were obtained by Kerminen *et al.* (1998). The Aitken mode (particle diameter <0.1 μm), one or two accumulation modes (0.1–1 μm), and supermicrone particles were investigated. Aitken particles are characterized by the highest number density but relatively small (<5%) total mass. The prevailing components of the accumulation mode particles were sulphate, ammonium, MSA, and bicarboxylic acid. Aerosol particles were, probably, internal mixtures of the components mentioned. Concentration ratio for MSA and sulphate depends on particle size (within accumulation mode). Supermicron particles usually contained less than 20% sulphate and more than 95% nitrate.

Nagel *et al.* (1998) have considered spectral aerosol optical depth (AOD) data in the Arctic from the summer of 1993 to the spring of 1996. A specific feature of this time period was that it was influenced by volcanic aerosols after the Pinatubo eruption with the removal of volcanic material from the stratosphere into the troposphere. Mean values of AOD changed during the period from 1993 to 1996 from 0.09 to 0.02 at 403 nm and from 0.065 to 0.01 at 1041 nm/1057 nm, respectively. A bimodal distribution was found with main radius modes at 0.1 μm to 0.3 μm and 0.75 μm to 0.9 μm. The annual change of the spectral AOD was characterized by a minimum in summer and high values in spring. The distinct layering of Arctic haze was identified as well as aerosol size spectra change with altitude.

Many authors (Crane *et al.*, 2001; Rahn, 1981; Brasseur *et al.*, 2002; Kashulina *et al.*, 2003) analysed observations of size resolved aerosol mass and gas concentrations during the European Arctic Aerosol Study (EAAS), the principal aims of which were to get information on aerosol chemical and optical properties and their annual variations over the region of northern Finland during the time period July 1997–July 1999. An important task was also an investigation of gas–aerosol interaction to examine processes of gas-to-particle conversion (also with the use of numerical modelling results). It has been shown, on the basis of model results, that all submicrometre accumulation-mode aerosols appeared to be in near equilibrium with NH_3. On the other hand, supermicrometre, coarse-mode aerosols of recent marine origin appeared to be out of equilibrium with HNO_3 and closer to equilibrium with HCl, while continentally influenced aerosols sometimes appeared to be near equilibrium with both HNO_3 and HCl.

5.5.3 Arctic haze

The phenomenon of Arctic haze has been discussed in detail by Kondratyev *et al.* (1996). Its characteristic features are connected with the fact that the entire Arctic is exposed to air pollution, exhibiting a strong seasonal variation with a maximum of pollution in winter and spring and a minimum in summer. Jaeschke *et al.* (1999) have pointed out that the haze is composed of fine aerosol, predominantly very tiny suspended particles. These pollutants are mainly of anthropogenic origin, but they also contain sea salt and soil components from wind erosion. In relation to chemical mechanisms, which lead to gas-to-particle conversion, gases emitted from anthropogenic sources can act as precursors of the particulate matter of the haze.

SO_4^{2-} is the dominant inorganic constituent of aerosol particles. SO_4^{2-} is transported into the Arctic from mid-latitudes, local formation occurs from its precursor substance SO_2. The latter exists in polluted air masses with a banded structure originating from a plume of pollutants. Chemical conversion to sulphate leads to a haze layer.

An appearance after 1976 of a network in the Arctic to sample aerosols has made it possible in 1981 to make the following conclusions:

(i) a considerable part of the aerosol haze observed in the wintertime Arctic consists of anthropogenic aerosols which come, apparently, from Eurasia;
(ii) the formation of haze starts in late autumn and reaches its maximum in March–April;
(iii) the horizontal visibility in the presence of haze decreases to 3–8 km;
(iv) the anthropogenic component of haze consists mainly of sulphates (with an average concentration of 2 $\mu g/m^3$), organic carbon (1 $\mu g/m^3$) and soot carbon (0.3–0.5 $\mu g/m^3$); and
(v) the aerosol that reaches the Arctic overcomes great distances and during its motion the aerosol survives significant ageing.

In March–April 1982, complex aircraft studies were made of aerosol atmospheric chemistry and radiative characteristics of haze, using the flying laboratory WP-3D "Orion" on extended routes (up to transpolar). Visual observations revealed not only the presence of a multiyear haze but also the brown–orange colour of the surface of the central Arctic ice cover. The haze covered the whole polar cap and propagated up to the flight ceiling (8.4 km). Actinometric observations revealed considerable (not less than 10%) absorption of short-wave radiation by haze. The bottom layer of the overcast haze was at a height of 3 km, and above it there were observed broken layers at different altitudes, which looked like bands. The concentration of soot carbon in the haze particles was only a factor of 3–4 times less than in the region of Denver (USA), only 1 order of magnitude less than in New

York, and several orders of magnitude greater than levels observed over the Pacific Ocean.

Becker *et al.* (1999) have accomplished Fourier spectrometer observations of the IR atmospheric emission spectra in Ny-Ålesund (79°N, 11°E) from 1995–1997 to retrieve column amounts of trace gases present in the atmosphere as well as aerosols. The retrieval problem has been solved through comparison between measured and calculated emission spectra. In order to measure the atmospheric optical depth, a star photometer (during polar night) in the UV/visible range was used. The results obtained indicate a strong decay in the optical depth of the tropospheric aerosols over the two spectral regions considered. Aircraft observations made by Jaeschke *et al.* (1999) indicate that sulphur oxides are important constituents of haze particles.

Atmospheric aerosol can contribute substantially as a cause of climatic variations in polar latitudes. Particles, strongly absorbing radiation, can cause the heating of respective layers of the atmosphere, comparable with its heating due to molecular absorption. The submicrometre particles of atmospheric aerosol, which are cloud condensation nuclei, affect the optical properties and the size distribution of particles and, hence, the radiative characteristics of clouds. The optically active aerosol, falling onto snow or ice, reduces the albedo, which promotes a more rapid thawing of the snow and ice cover in polar regions.

As has been mentioned above, it was discovered in the late 1970s that considerable concentrations of aerosol particles were observed in the Arctic. If the background concentrations are considered to be 0.1–0.2 $\mu g/m^3$ then in the Arctic wintertime they reach 1–2 $\mu g/m^3$ (i.e., the concentration turns out to be comparable with that observed in some industrial regions (3–6 $\mu g/m^3$)). The seasonal change of the aerosol concentration is clearly expressed: in winter it is an order of magnitude higher than in summer. A maximum is observed in February–April, a minimum in June–July.

Based on numerous measurements, carried out at the network of stations, the conclusion has been drawn that the whole Arctic is subject to gas and aerosol pollution of anthropogenic origin, with local sources contributing least. The aerosol transport to the Arctic takes place mainly from three regions (the northwest coast of the USA, northern Europe and south-east Asia). Among the polluting components, the following have been discovered: graphite carbon, sulphates, vanadium, magnesium, organic matter (including polycyclic aromatic hydrocarbons), silicon-containing substances, and condensation nuclei.

The annual course of pollution is explained by the effect of a number of factors:

(i) an increased amount of pollutant outbreaks in winter;
(ii) an intensified meridional circulation in winter;
(iii) a retarded sedimentation of aerosol particles in conditions of the wintertime inversion; and
(iv) a reduced washing out of aerosol by clouds and precipitation.

The characteristic time for the transport of polluted air masses from the mid-latitudes to the Arctic at a distance of about 5–10,000 km constitutes about 20 days. Apparently, a large-scale turbulent diffusion and not a regular advection is the prevailing factor in transport. If this holds, then the spatial distribution of aerosol in the Arctic should be homogeneous with only characteristic scales of more than 2000 km, which exceeds the characteristic size of the Arctic cyclones.

In the chemical analysis of precipitation and aerosol in the regions of its sources and sinks, a consideration of the ratio of non-soil elements Mn/V (as a product of oil burning), compared to soil aerosols of the Al type (whose average composition was Mn (950 ppm) V (135 ppm) and Al (81 ppm), and the concentration of the principal circulation processes in the northern hemisphere atmosphere shows that it is impossible to explain an increase of the Mn/V ratio in the Arctic through only the influence of the North American regions. Vast regions of Siberia also contribute substantially through long-distance transport to high latitudes (Crane *et al.*, 2001).

Many authors have discussed the results of the three-year study (1992–1994) of atmospheric aerosol properties (aerosol mass and number concentrations, light scattering, size distribution, and ion composition) in the Novosibirsk (Western Siberia) region. It was shown that submicron aerosol concentration is higher in winter than in summer. In winter, the main contribution (by mass) to aerosol particle composition belongs to soluble inorganic compounds of condensation origin, such as ash. In summer, the major part of aerosols comprise particles of organic and biogenic origin, as well as insoluble erosion particles. In summer, an anthropogenic impact in Siberia on the formation of global-scale aerosols is limited by regional-scale influence of local pollution sources. In wintertime, the scale of these local sources may be larger. The techniques of synchrotron radiation X-ray fluorescence analysis (SRXFA) and virtual impactor were applied to study the elemental composition of atmospheric aerosols in Siberia. It has been shown that the composition differs for different size fractions. The most dramatic changes occur between submicron and coarse aerosol size fractions. The observation results have revealed that some areas of western Siberia are being polluted from industrial sources in northern Kazakhstan. Basically, aerosols were generated by processes of soil erosion and mineral fuel combustion. In a recent publication Kutsenogii K. and Kutsenogii P. (2000) have summarized the results of a 7-year study of atmospheric aerosols in Siberia with the use of the Siberian network of stations aerosol data within the "Siberian Aerosols" programme, started in 1991 as a component of the international "Arctic Haze" project, Raputa *et al.* (2000) have analysed aerosol composition measurement data for the northern part of western Siberia (the Tyumen region) with the principal purpose of studying the impact of local pollution sources.

5.5.4 Antarctic aerosols

For intercomparison purposes it is of interest to summarize the present-day knowledge of Antarctic aerosols. Systematic studies of atmospheric aerosols in the Antarctic, started in 1968 by A. I. Voskresensky at Mirny Station, have been continued between 1975 and 1978 and later on at various locations, including the South Pole, with the summertime values exceeding 10 times those in winter, which suggests a substantial impact of meteorological conditions on the concentration of aerosols.

An analysis of aerological sensings and aerosol measurements at the South Pole plateau has made it possible to draw the following conclusions:

- the mean total concentration of aerosols varies from 15 particles per cm^3 (in winter) to 100–150 particles per cm^3 (in summer), with the particle size of 0.02–0.2 μm prevailing;
- in the periods of downward atmospheric motion from the middle and upper troposphere, concentrations of 500–1500 particles per cm^3 are observed, due to increasing concentrations of particles smaller than 0.01 μm (in many cases the particles smaller than 0.005 μm prevail);
- with an advection of moist air masses from the Weddell Sea to the South Polar Plateau and strong winds near the surface, the concentration of aerosol at the South Pole was about 200–600 particles per cm^3; and
- the aerosol concentration increases usually on the summertime polar day.

The aircraft measurements of aerosol concentrations were performed in November 1977 over the South Pole plateau using a portable counter of condensation nuclei. A maximum particle concentration was observed in lower, wetter layers of the atmosphere, located over the near-surface inversion as well as in moist air masses over the Ross Ice Shelf and in the moist air–surface layer of the Ross Sea.

In all cases (even with crossing the tropopause) a comparatively stable decrease of aerosol concentration with altitude was observed. The number density (number of particles per 1 g of air) at the South Pole was higher in the 650–450-hPa layer than near the surface. A homogeneous distribution of the aerosol concentration is observed in the 5000–450-hPa layer in the 75°S–90°S belt. However, above the level of 450 hPa, the concentration decreases weakly with altitude, which can be explained by the effect of a relatively calm and very cold atmosphere above this level.

Such conditions determine the possibility of the formation of ice crystals through homogeneous nucleation and "washing-out" of aerosols through the Brownian coagulation of slowly falling particles. The data obtained show that the aerosol is transported with the moist air masses from the seas. As these air masses move over the continental surface, ice crystals form, which fall out onto the surface, removing water and aerosols from the lower layers of the atmosphere. Therefore, the air flow from the lower atmosphere to the South Pole turns out to be dry and almost devoid of aerosols.

However, with strong winds ensuring atmospheric mixing, a relatively high concentration of aerosols can be observed near the surface. In the case of strong downward motions, the air masses enriched with submicrometre aerosols can reach the surface from the 550–540-hPa layer. In each case, when the tropopause was crossed by the aircraft, lower values of aerosol concentrations were recorded in the stratosphere than near the surface. It follows that only the middle and upper troposphere can be the source of aerosols in the southern polar zone.

The observations considered never revealed concentrations exceeding 10^3 particles per cm^3 near the surface, as has been recorded by earlier observations. Hence, an important problem of the origin of rare aerosol "storms" remains unsolved.

The balloon measurements of aerosol have made it possible to determine the vertical profile of particle concentration with radius more than 0.15 μm and to reveal the presence of a stratospheric aerosol layer 7 km thick localized at a level of 10 km, on average. The calculated aerosol optical thickness of the atmosphere at a wavelength of 500 nm was 0.0035 ± 0.002, but despite its small value, it constitutes one-quarter of the total aerosol optical thickness of the atmosphere over the Antarctic.

As has been mentioned above, a specific feature of the Antarctic atmosphere, located far from other continents, is the presence of a considerable concentration of aerosol particles. This testifies to the existence of the long-range transport of particles under the influence of the processes of diffusion and large-scale circulation.

The Antarctic is a unique natural laboratory for studying the transport of particles at long distances and the transformation of minor atmospheric components.

In the troposphere and lower stratosphere over the elevated Antarctic ice sheet plateau, submicrometre particles of aerosol are observed whose concentration varies from several particles to thousands of particles per cm^3. Shaw (1988) summarized the results of studies of the origin and properties of the Antarctic aerosol and its sources and interaction with the ice cover of the Antarctic Plateau. In this connection, a model of the diffusional transport of particles (as a problem with boundary conditions) was considered, which assumes the existence of two modes of aerosol (Table 5.11).

The smallest particles (Aitken nuclei) consist, apparently, of the transformed products of the nucleation of gas components. The age of these particles does not exceed 2–3 days, they form locally over the ice caps at a rate of about

Table 5.11. Characteristic values of the parameters of the two-mode model of the Antarctic aerosol.

Parameter	Large particles	Aitken nuclei
Average radius (μm)	0.14	0.005
Average concentration (cm^{-3})	40	1,000
Density of particles (g/cm^3)	2.16 (NaCl)–1.18 $(NH_4)_2SO_4$	5.23×10^{-19}
Mass of particles (g)	5.78×10^{-13}	
Average mass concentration (ng/m^3)	300	0.65
Mixing ratio with respect to air (ppm)	0.38	0.001
Layer thickness (km)	10	10
Total content in the atmosphere (g/cm^2)	1.88×10^{-7}	3.9×10^{-10}
Total number (cm-2)	5.0×10^5	1.5×10^9
Total optical thickness (at 0.5 μm/s)	0.013	10^{-8}
Stokse rate of sedimentation (cm/s)	5.0×10^{-4}	2.8×10^{-7}
Coefficient of diffusion (cm^2/s)	2.0×10^{-6}	2.0×10^{-3}

4×10^{21} g/cm^3/s. The concentration of the Aitken nuclei in the troposphere varies within the range 10^2–10^3 cm^{-3}, except for the lower turbulent boundary layer several hundred metres thick, where the concentration of the smallest particles decreases strongly. Apparently, the Aitken particles consist of the transformed gaseous sulphates and hydrocarbons.

The concentration of larger particles of the second mode, with a radius of about 0.14 μm, constitutes approximately 0.5 cm^{-3}. These particles contribute most to the mass of aerosol and are of great interest, since they, no doubt, come from remote regions and sediment onto the surface, forming the aerosol inclusions of the polar ice. Long-term variations in their concentrations in ice cores can serve as indicators of variations of palaeoclimate.

The sources of the particles of the second mode, by their priority, are as follows:

1. unidentified sulphate sources (the share of sulphate aerosol is 83%);
2. oceanic sources encircling the Antarctic (about 7%);
3. arid zones of the southern hemisphere (Australia, and the Kalahari, and Atacama deserts), from which the aerosol is transported above the clouds (4%);
4. extraterrestrial sources (6%); and
5. bases on the Antarctic continent.

The total mass concentration of particles in Antarctic air constitutes 170–300 ng/m^3, comprising various elements: Na, 3–70 ng/m^3; Al, 0.6–0.8 ng/m^3; and S, 60 ng/m^3 (soil aerosol, 7–10 ng/m^3; sea salts, 10–24 ng/m^3; sulphates, 150–250 ng/m^3; and chondrites, 0–9 ng/m^3). In snow the total mixing ratio is 150–250 ppb, comprising various elements: Na, 3–20 ppb; and Al, 0.6–1.2 ppb (soil aerosol, 7–15 ppb; sea salts, 10–69 ppb; sulphates, 60–115 ppb; and chondrites, 0.12 ppb).

Turbulent diffusion can be considered as a major mechanism for the transport of particles. Estimates of turbulent diffusion confirm the assumption concerning the sources of the particles of the second mode and make it possible to estimate their contributions. The transport of particles from continental sources through turbulent diffusion prevails, no doubt, in the middle and upper troposphere.

The particles are removed from the free atmosphere over the Antarctic through diffusion to the turbulent boundary layer or as a result of coagulation with cloud particles or precipitation. The major mechanisms for removal are sedimentation onto the plateau surface (2.0×10^{-14} g/cm^2/s), snow flakes (1.6×10^{-14} g/cm^2/s), ice crystals (0.2×10^{-14} g/cm^2/s), and nucleation (0.6×10^{-14} g/cm^2/s).

The Aitken nuclei are removed mainly through diffusion to ice crystals (0.006×10^{-14} g/cm^2/s), as well as diffusion across the laminar layer to the plateau surface (0.015×10^{-14} g/cm^2/s). Numerous mechanisms for the aerosol removal, functioning near the surface, lead to a decrease of the mass concentration of particles in the turbulent boundary layer by a factor of 5–10, which determines the non-representativity of the aerosol samples taken near the surface, from the viewpoint of the properties of aerosols. The aerosol samples taken at altitudes of 2–3 km should be most representative.

Variations in the atmospheric circulation during the last third of the Wisconsin glaciation could have promoted an intensified desertification in mid-latitudes of the southern hemisphere, which had favoured an increased dust-loading of the Antarctic atmosphere and a substantial sedimentation of dust onto the surface of the polar ice cap. Since the characteristic timescale of the air exchange between the southern and northern hemispheres constitutes 1–2 years, note should be taken that the composition of aerosols in both hemispheres is different. This concerns mainly the polar regions.

A specific feature of the Antarctic atmosphere is its episodical anthropogenic and volcanic pollution. A comparison of the measured optical thickness of the atmosphere (τ_m) along the horizontal and slant paths for the spectral interval 8–10 μm with the results of calculations of radiation attenuation (τ_c) for different models of aerosol consisting of spherical water particles, and for varying meteorological conditions, revealed a sharp decrease of an "excess" optical thickness $\Delta\tau = \tau_m - \tau_c$ with the increasing wavelength. In the interval 8–10 μm, the aerosol attenuation decreases with increasing wavelength. However, the relative contribution of water aerosol to the total optical thickness in the Antarctic condition was estimated at 3%. The estimates of the aerosol optical thickness for the atmospheric transparency window 8–10 μm were much below $\Delta\tau$.

Thus, a consideration of attenuation by water aerosol cannot explain the observational data obtained in different conditions. In the spectral interval 8–10 μm with meteorological visibility exceeding 20–25 km, the contribution of water aerosol to the total attenuation does not exceed 10–20%.

The contribution of aerosol with absorption bands near 9 μm was estimated at 5–13%. Hence, an account of attenuation by both types of aerosol is insufficient to explain the radiation attenuation in the real atmosphere. The absorption by MGCs which is not taken into account may be one of the possible additional factors.

Kondratyev *et al.* (1990) have pointed out that in the case of undisturbed conditions (in the absence of volcanic eruptions) aerosol optical thickness of the atmosphere in the Arctic is higher than in the Antarctic by 42% and 26% in May and July, respectively.

In a comparison of mean aerosol optical thickness in the Arctic (0.09) and Antarctic (0.013), data on the concentration of aerosols in these regions show that the level of pollution over the North Polar region is 7–10 times higher than over the southern one. This is connected with a relative proximity of industrial sources, special features of circulation processes in the atmosphere, and geographic and demographic differences between the hemispheres. This has been confirmed by atmosphere (ozone and carbon monoxide) and meteorological parameters. The flights took place mainly in the 238–162-hPa layer. Measurements of the aerosol concentration were made using an automatic condensation nuclei counter.

At the same time, the US Air Force aircraft LC 130R set its course for the Antarctic at a lower height, carrying a photoelectric counter, which made it possible to obtain data of aerosol measurements taken at two levels. Measurements were also made on board this aircraft of the number density of large particles greater than 0.45 μm in diameter. The concentration of these particles constitutes less than

$1\,cm^{-3}$ in the northern hemisphere stratosphere and in the eastern sector of the northern hemisphere troposphere (the flight was made from San Francisco along 120°E to the North Pole, and then along 5°E, landing in London; the flight was continued along 3°E to the equator with the subsequent turning eastward to land in Cape Town; then the aircraft flew along 180°E to the South Pole and along 173°E to Oakland, New Zealand; the last part of the route was along 156°E across Hawaii to San Francisco).

In the southern hemisphere stratosphere the aerosol concentration grew gradually, reaching a maximum of $1.6\,cm^{-3}$ in the 50–60°S belt. Close to maximum values of concentration were observed in the upper troposphere over the tropical Pacific Ocean and Africa in the regions of the ITCZ.

The latitudinal distribution of the total concentration of the Aitken nuclei is governed by other laws: in the stratosphere of the Arctic and mid-latitude northern hemisphere the concentration varies within the range $10\text{–}40\,cm^{-3}$ and in the respective zones of the southern hemisphere it decreases to $10\text{–}25\,cm^{-3}$. In both polar regions, the concentration of the Aitken nuclei decreases with latitude. The tropical troposphere is characterized by the symmetrical distribution of regions with larger concentrations over land and ocean.

The share of large particles in the South Pole stratosphere constitutes 6–20%, but in the North Pole stratosphere it decreases to 5%, and in the maritime near-surface air it falls to 0.1–1%.

A comparison of measurement data obtained from the two aircraft flights reveals, as a rule, a decrease in the aerosol concentration with altitude, but directly over the tropopause (the 270-hPa level), in the 50–55°S belt, concentrations of about $100\,cm^{-3}$ were recorded near the zone of the strongest wind.

Analysis of the measurement results suggests the conclusion that near the jet streams the tropospheric aerosol penetrates the stratosphere, and the near-surface aerosol gets to the free atmosphere mainly in the ITCZ regions and in the regions of the Antarctic polar front. This is confirmed, for example, by increasing aerosol concentrations in the 4–11°N belt of more than $250\,cm^{-3}$ and sometimes $1500\,cm^{-3}$. However, large particles are distributed over the troposphere and the stratosphere comparatively homogeneously at all latitudes, with a concentration of about $1.0\text{–}0.5\,cm^{-3}$.

Thus, the data presented suggest the conclusion about the presence of the aerosol haze in polar regions, especially in the northern hemisphere, that there is a still increasing proportion of anthropogenic and natural aerosols, which can affect the formation of clouds, the radiation regime, and climate changes in the Arctic.

Wolff and Martin (1998) have pointed out that in recent years, attention has focused on aspects of sea salt aerosol behaviour such as reactions with acid (leading to loss of gaseous HCl) and heterogeneous reactions on sea salt surfaces. A suggestion has been made that for at least the coastal polar areas, sea ice surfaces are a more important source of aerosols than sea water. In this context the data from many years of year-round aerosol and snow analysis at three coastal Antarctic sites (Dumont d'Urville, Halley, and Neumayer) have been brought together and examined. The results obtained clearly indicate the likelihood that sea ice surfaces are actually the source of sea salt aerosol, at least, for coastal Antarctica. Such a conclusion has a very important palaeoclimatic implication, questioning the traditional interpretation of sea salt concentrations in ice cores. Blanchet (1983, 1989) has discussed earlier radiative effects of anthropogenic aerosol components in Arctic haze and snow under conditions of snow surface contamination. Barbante *et al.* (1998) considered in this context the sources and origins of aerosols reaching Antarctica as revealed by lead concentration profiles in shallow snow.

The samples of aerosol particles during austral summer 1998 at the Finnish station Aboa in continental Antarctica was collected to examine components of chemical composition, such as major inorganic ions, methane sulphonate, and dicarboxylates. Particulate sulphate, ammonium, and methane sulphonate were internally mixed and found predominantly in the submicron size range. The submicron mass size distribution had a dominant mode peaking at $\sim$0.3 μm of particle aerodynamic diameter, another mode peaking at 0.6 μm, and an identifiable Aitken mode below 0.2 μm. The mass size distribution of sodium, a tracer for sea salt particles, peaked in most samples slightly below 2 μm; in addition, it had a submicron mode between about 0.5 and 1.0 μm and a mode with great variability between 3 and 10 μm. The overall concentration of sea salt was 1–2 orders of magnitude lower than concentrations measured in coastal Antarctica. Particulate nitrate was found almost entirely in the supermicron size range, the most likely formation pathway being the interaction of nitric acid on some other reactive nitrogen compound with sea salt particles in the Antarctic atmosphere. The only dicarboxylate above the analytical limit of detection was oxalate.

5.5.5 Deposition of heavy metals

Global-scale, long-range transport of aerosols results in the deposition of atmospheric aerosols all over the surface of the globe. Based on observational studies performed at the North Pole, on lakes Ladoga and Onega, in the suburbs of St. Petersburg, watersheds of the Onega–Ladoga system, and Neva Bay, a quantitative estimate has been made of the fallout of dust and some metals from the atmosphere, most of which are either heavy metals or trace elements (Adamenko *et al.*, 1991a,b). These assessments have been made using X-ray-fluorescence analysis of the filters (Vladipore 0.45-μm pore diameter membrane filters) on which a solid deposit from snow cover had been filtered.

Table 5.12 shows data on relative concentrations of conditionally insoluble forms of heavy metals, calcium, and potassium in snow cover on the ice cover of the Ladoga and Onega lakes, at the station North Pole-28 during the annual drift in the north-eastern Arctic north of 84°N, as well as relationships between pre-industrial and present-day concentrations of some chemical elements.

Analysis of these data suggests the following:

1. In the remote regions of the Arctic and north-western Europe a combination of almost the same chemical elements falls out from the atmosphere – the same elements can be identified whose concentrations in snow cover are from

Table 5.12. The ratios of concentrations of chemical elements and dust in snow cover on lakes Ladoga (C_L) and Onega (C_O), at the North Pole in winter (C_{mw}) and summer (C_{ns}), and pre-industrial (C_p) and current (C_c) concentrations in the north-western part of Russia.

Element	C_L/C_O	C_L/C_{ns}	C_L/C_{nw}	C_O/C_{nc}	C_O/C_{nw}	C_c/C_p
K	2.9	12.1	–	4.2	–	–
Ca	1.1	5.6	–	5.0	–	9.8
Ti	3.4	19.6	9.2	5.8	2.8	–
V	3.0	6.0	2.0	2.0	0.7	–
C	1.1	6.2	1.9	5.4	1.6	–
Mn	1.6	25.0	8.5	15.0	5.2	–
Fe	1.8	27.0	15.8	17.7	8.6	10.6
Ni	0.9	5.2	3.0	6.1	3.5	–
Cu	1.3	9.1	7.8	7.1	6.1	18.3
Rb	1.2	2.7	3.3	2.1	2.6	15.4
Pb	2.7	21.8	–	8.2	–	–
Sr	1.6	26.3	10.2	16.8	6.5	–
Zr	1.5	45.5	14.3	29.7	9.3	–
Zn	1.3	10.7	8.2	8.1	6.3	20.4
Dust	1.9	12.6	6.8	6.5	3.5	–

hundredths of a microgram to tens and even hundreds of micrograms per litre of water solution.

2. These elements are located, by order of priority, in the following successions:
   - Ladoga (concentrations in μg/l are given in parentheses): Fe (155), K (82), Ti (17), Ca (13), Mn (2), Zn (2), V (1), Cr (0.8), Zr (0.7), Cu (0.6), Pb (0.6), Sr (0.4), Ni (0.2), Rb (0.2), Br (0.1), and dust (2.6 mg/l).
   - Onega (in μg/l): Fe (84), K (29), Ca (11), Ti (5), Mn (1), Zn (1), Cr (0.7), Pb (0.5), V (0.4), Zr (0.4), Ni (0.3), Sr (0.3), Rb (0.1), Br (0.1), and dust (1.36 mg/l).
   - North Pole (summer, μg/l): K (7), Fe (6), Ca (2), Ti (0.9), Pb (0.2), V (0.2), Zn (0.1), Cr (0.1), Mn (0.99), Cu (0.07), Ni (0.05), Br (0.020), Sr (0.017), Zr (0.015), Rb (0.010), and dust (0.21 mg/l).
3. The wintertime deposition in the Arctic exceeds 2–3-fold that in the summer, which is explained both by the winter duration and by differences in atmospheric stratification, which is more stable in winter than in summer.
4. For almost all elements, the deposition on Ladoga is 20–40% greater than on Onega, which is explained by the proximity of Ladoga to relatively large sources of atmospheric pollution, compared to Onega, as well as by prevailing winds with southern or western components in north-eastern Russia in cold seasons.
5. Analysis of the available data on the fallout of chemical elements from the atmosphere in cities with multimillion populations and with diversified industry suggests that in such cities the deposition of some chemical elements exceeds, by one to two orders of magnitude, the fallout for the remote regions in the north-west (Ladoga and Onega) and exceeds, by three to four orders of

magnitude, the input of metals at the North Pole. The latter data can be considered a measure of the heavy metal deposition for the global background conditions of the northern hemisphere.

6. The heavy metal deposition on Ladoga is 3–45 times stronger than at the North Pole in the summer and 2–16 times more intensive than in the central Arctic in winter.
7. The background deposition of heavy metal on relatively pure Onega is 2–30 times more intensive than at the summertime North Pole and 2–9 times stronger than in the central wintertime Arctic.
8. The present deposition of such elements as Ca, Fe, Cu, Pb, and Zn exceeds, by 10–20 times, that of 150–200 years ago.
9. Differences between the background hemispheric (North Pole) values of dust concentrations and the background regional (Ladoga) values reach one order of magnitude, and for Onega are half as much as for Ladoga.
10. The concentration differences of the deposited Hg are, on average, the same as in the case of total dust. However, the concentrations of lithophyll metals (metals with low enrichment coefficients in particles of atmospheric aerosols – manganese, strontium, and iron) are tens of times smaller in the Arctic dust, whereas the concentration of atmophyll metals (metals with high enrichment coefficients in particles of atmospheric aerosols), for which the industrial contributions are significant, differ less.
11. The atmospheric flux of lithophyll elements (calcium, magnesium, and iron) during the industrial epoch has increased by a factor of 5–11, and that of atmophyll elements (lead, zinc, and copper) by a factor of 15–20. This points to the fact that the present background level of atmospheric deposition assessed for the hemisphere from snow samples at the North Pole are of about the same order of magnitude as in the mid-latitude north-west in the pre-industrial epoch.

5.6 CLOUDINESS DYNAMICS

A very important, climatically significant, process in the Arctic is interaction between cloudiness and radiation. Under the initiative of the World Meteorological Organization, the Global Meteorological Experiment (GME) was carried out in 1978/1979, aimed at obtaining a global database for one year, as well as focused field observational programmes during two Special Ob′serving Periods (SOPs), from 15 February until 15 March and from 15 May until 15 June, 1979.

Recently, new efforts to study Arctic cloud cover dynamics have been undertaken (Randall *et al.*, 2000). Clark *et al.* (1996) have examined cloud cover, precipitation frequency, and temperature using the Comprehensive Ocean–Atmosphere Data Set (COADS) records for the period 1980–1993. An application of principal component analysis revealed two climatic regions, the "North Atlantic" and "Central Arctic" as well as cold (November–April) and warm (June–September) seasons. For the central Arctic, precipitation frequency and cloud cover peak during the warm season. For the North Atlantic, precipitation frequency and the frequency of mode rate/heavy precipitation peak during the cold season, but with relatively constant cloud cover throughout the year.

Curry *et al.* (2000) have emphasized the existence of unusual cloud types that form in the polar regions, and substantiated relevant key scientific issues as well as unsolved problems. We quote below what they have written in this context.

Clouds

Substantial uncertainties exist in our present understanding of Arctic clouds. These uncertainties arise from difficulties in observing these clouds, both from the ground and from satellite, and from the unusual cloud types that form in the polar regions. These unusual cloud types include ice crystal clouds that reach the surface, commonly referred to as "diamond dust"; convective plumes associated with leads or polynyas (openings in the sea ice); persistent mixed-phase clouds; and multiple layers of thin cloud decks that occur in the statically stable Arctic environment. Some key scientific issues relating to Arctic clouds are as follows:

- What is the influence of leads and other open water on cloud properties when large surface–air temperature differences exist?
- How does the extreme static stability and low atmospheric water vapor content of the lower troposphere, especially during winter, affect the flow of energy across the air–sea interface?
- What is the mechanism that leads to the spectacular multi-layering of summer-time cloud systems over the Arctic Ocean?
- How does the transition of low clouds from liquid to crystalline depend on temperature and aerosol characteristics, and how does the springtime transition differ from the autumnal transition?

Radiation

Clouds are the dominant atmospheric modulators of the Arctic radiation climate. Cloud radiative properties depend on the amount of condensed water, the size and shape of the cloud particles, and the phase of the particles (liquid or ice). Recent studies of Arctic clouds suggest that over the course of the year clouds have a net warming effect on the surface, while the top-of-the-atmosphere cloud RF is dominated by the short-wave flux. The radiation environment of the Arctic is complicated considerably by the highly reflective and inhomogeneous snow/ice surface, the complex vertical structure associated with temperature and humidity inversions, low temperatures and humidity, and high solar zenith angles. Key scientific issues related to Arctic radiation are as follows:

- What is the spectral distribution of long-wave radiation? In particular, what is the role of the 20-μm rotation-band "window" region in regulating the surface and atmospheric temperature in the Arctic?
- What are the effects of springtime "Arctic haze" on the absorption of solar radiation in polar clouds?

- What is the role of diamond dust in determining the radiation fluxes?
- What are the short-wave radiative effects of the horizontally inhomogeneous stratocumulus clouds over the inhomogeneous, highly reflective snow/ice surface?
- How do the optical properties of the Arctic surface vary in response to changes in snow and ice characteristics (including meltponds)?

Aerosols

It has been hypothesized that the microphysical and optical properties of Arctic clouds are particularly susceptible to influence by atmospheric aerosol. The concentration and size distributions of cloud droplets depend in part on the aerosol in the atmosphere, specifically on the CCN. The interactions between clouds and aerosol are not simply one way; scavenging by clouds depletes CCN, but aerosol may also be produced and enhanced within the clouds through chemical and physical processes. Ice particle concentrations in clouds may also be affected by aerosol, although this connection is complex and not well understood. The concentration and composition of the ice-forming nuclei (IFN) are hypothesized to be important for determining the phase of Arctic clouds, and therefore, their impact on the radiation balance. Key questions regarding the aerosols found in the Arctic atmosphere are as follows:

- What are the activity spectra of CCN and how do they vary?
- What are the sizes and compositions of CCN, and how and why do they vary?
- How do the CCN distributions interact with the cloud droplet distributions?
- How do processes within Arctic clouds modify CCN?
- What are the nature and source of IFN?
- Is it possible for ice particles to form at relative humidities below water saturation, by deposition nucleation?

The scientific programme substantiated by Curry *et al.* (2000) is not quite complete, because a number of problems also requires discussion, including: (a) interaction between extended cloudiness and radiation (as well as polar cloud radiative forcing (PCRF) in general); and (b) PSC.

As far as the problem of PSCs is concerned, it has been recently discussed in detail by Kondratyev and Varotsos (2000). An extensive survey of cirrus cloud properties and problems is contained in the IPCC Report-2001 (Houghton *et al.*, 2001). Koch (1996) has pointed out in this context that although mean global cloud radiative forcing (CRF) = $-43.6\,W/m^2$, a relevant impact of cirrus clouds is positive and equal to $+23.5\,W/m^2$. This was the principal motivation to accomplish the programme of the European Cloud-Radiation Experiment (EUCREX) in March 1993 in order to undertake measurements of size distribution and optical properties of Ci in the Arctic. It has been discovered that there is a strong disagreement (up to $25–30\,W/m^2$) between measured and calculated (for a 1-D case) radiative fluxes, which signifies a necessity to apply a 3-D radiative transfer parameterization.

Similar results have been recently obtained by Barker and Stephens (2001), who have undertaken an intercomparison of 1-D solar radiative transfer codes for various cases: from simple cloudless atmospheres to complex cloudy atmospheres. The range of estimated fluxes often exceeded 20% (25 1-D solar codes and four 3-D Monte Carlo codes were considered). The principal conclusion made by Barker and Stephens (2001) is that different representations of cloud-related processes may account for much of the uncertainty associated with estimates of climate sensitivity and climate change.

One more unsolved problem (for polar clouds, in particular) is the so-called "excess absorption" of solar radiation by the atmosphere. This problem has been discussed in detail by Kondratyev (1996, 1999b).

Curry *et al.* (2000) have discussed the results of the FIRE (First ISCCP (International Satellite Cloud Climatology Project) Regional Experiment) Arctic Cloud Experiment, which was conducted during April–July 1998 to study Arctic cloud systems under spring and summer conditions. The main goal of the experiment was to examine the effects of clouds on radiation exchange between the surface, atmosphere, and space, and to study how the surface influences the evolution of boundary layer clouds. Curry *et al.* (2000) have summarized the results obtained, including the following:

- A persistent humidity inversion above boundary layer cloud tops, associated with the static stability of the Arctic environment, contributes to the homogeneity and persistence of the cloud by inhibiting evaporative cooling associated with entrainment mixing at the cloud top.
- Evidence was found of direct association of low-level clouds with open water in polynyas.
- Large variations in the relationship between cloud temperature and phase were found, influenced by the presence of ice nuclei, seeding of the cloud by ice particles falling from above, and the size of the liquid drops.
- Aerosol composition is commonly highest aloft, especially during spring, associated with long-range transport and scavenging in the lower atmosphere by boundary layer clouds. Local production of aerosols was observed in dissipating cloud layers, the humidity inversion, and was associated with local production of DMS.
- There is some evidence that the open water in polynias may provide a source of ice forming nuclei.
- Direct evidence was given for the susceptibility of Arctic cloud microphysical and optical properties to modification by anthropogenic aerosol.
- The spatial inhomogeneity of the surface physical and optical characteristics is very large, especially during the summer melt season. Temporal variations in surface albedo were documented to occur in response to atmospheric forcing: the first seasonal rainfall markedly decreased the snow surface albedo and heralded the onset of snowmelt; substantial variations (up to 20%) in surface albedo arose from variations in cloud optical depth that were not associated with any variations in surface characteristics; and late summer surface albedo

changes occurred in response to freezing and thawing of the surface in response to storms.

- Measurements of the asymmetry parameter show the substantial effect that habit and size of the ice crystals have on the radiative transfer.
- The combination of visible, near-IR, submillimetre, and microwave wavelengths shows considerable promise in discriminating the characteristics of the surface and clouds under many conditions that have hitherto been difficult to identify using only a single sensor.
- The European Centre for Medium-Range Weather Forecasts (ECMWF), the Charles Sturt University (CSU), and other models tend to underpredict low cloud amount and the column liquid water path, resulting in simulations of surface net short-wave radiative fluxes that are too large, and downwelling long-wave radiative fluxes that are too small.

Within the Russian national programme of the GME, during the SOP-2, aircraft measurements over the station North Pole (NP-22, the East Arctic) were made, as well as a subsatellite experiment flights in the regions of the volcanoes of Kamchatka and Kuril Islands. The results obtained will be summarized below in detail since they are still unique with regard to completeness of observational data, including both cloud and radiation characteristics.

The choice of the region of the East Arctic has been determined by the fact that it belongs to the very important weather and climate-forming regions of the globe. The presence in this region of the drifting station NP-22, operating within the POLEX-North programme, has made it possible to carry out the complex observations using a flying laboratory in this part of the Arctic. Studies near Kuril Islands were the first attempt to investigate the radiation regime of the atmosphere in the region of active volcanism.

The aquatory of the eastern Arctic in June is characterized by the oceanic polar climate due to a special radiative regime of the atmosphere, by the presence of snow and ice covers, open water surfaces, specific cloud fields, and the effect of air masses from the continent and the Pacific Ocean. The extended cloudiness prevailing in late spring–early summer, was the principal object of studies.

5.6.1 The climatological and synoptic characteristics

The Arctic Basin gets heat from the southern latitudes, being a region representing a powerful heat sink as a result of the negative radiation budget of the surface–atmosphere system. Therefore, the Arctic Basin strongly affects the atmospheric circulation over the adjacent continent and oceans.

In the warm seasons the Arctic is influenced by a weakly expressed low-pressure area, whereas weak anticyclones are observed over the adjacent regions of the North Pole Ocean (the eastern Arctic, in particular, also embraces the sector of warm surface waters of the Pacific Ocean). Most often, the cyclones move to the eastern Arctic from the Bering Strait. The map of the mean surface pressure for July shows a decrease in the average wind speed and the prevailing air flow with a clear-cut eastern component over the Eurasian coastline.

The monthly mean pressure in the NP-22–Wrangel Island region is about 1014 ± 2 hPa, and the surface wind speed is small, averaging about 3–4 m/s. With the inversions typical of this region, the surface layer is quite isolated from the air masses rapidly moving over it. This fact, together with lacking topographic effects, determines a lower repeatability of strong winds near the surface. The air and surface temperatures in the NP-22–Wrangel Island region in June were 0–1°C and −1°C, respectively.

The special features of the formation of heat balance are of primary importance in understanding the weather and climate variability. The surface energy balance constituents in June in the Arctic are characterized by the prevailing effect of the radiative component. So, for example, the short-wave radiation absorbed by the surface constitutes 13 W/m^2, and the downward and upward long-wave radiation is 28 and 32 W/m^2, respectively. The latent heat fluxes due to the phase transformations of water and turbulent heat exchange with the surface and the heat exchange with deeper layers of the ocean are 10, 11, and 65 W/m^2, respectively.

Thus, diabatic processes are very important for the formation of the climatic conditions of the Arctic. The diabatic impact favours cyclogenesis in air masses moving over warm surfaces and anticyclogenesis over cold surfaces.

The summertime melting of pack ice leads to the formation of stable fog and low-level cloudiness observed for 100 days and longer. Usually, they are caused by an advection of relatively warm and moist air masses over the melting ice or cold water. Fog is not observed when the wind speed exceeds 10 m/s.

The repeatability of extended stratus clouds and fogs in the summertime Arctic varies from 50% to 80% and, on average, they cover up to 70% of the area of the ocean outside the polar circle (i.e., the lower cloudiness and fogs are phenomena of not only regional but also planetary scale significance).

Two types can be mentioned among the variety of fogs: advective and radiative. The radiative fogs are formed, as a rule, due to the radiative cooling of the surface and the adjacent air layers in the presence of either moist air and (or) aerosols in the subinversion layer, 0–300 m. Most often, radiative fogs are formed at a wind speed of about 1–3 m/s and temperatures from −20 to +20°C. They are characterized by the existence of the zone of wind intensification (mesojet) near the top of the fog, by typical vertical profiles of the turbulent mixing coefficients, and water content.

The advective fog appears as a result of warm air masses moving over a cooler surface (water–ice and land–sea).

The formation of stratified clouds in the boundary layer takes place in three stages: (i) water evaporation; (ii) the upper transport of water vapour and its accumulation beneath the inversion; and (iii) cooling and condensation. Near the upper boundary a temperature inversion begins and a mesojet is observed; the average radius of droplets and water content increase with height. Here a powerful long-wave cooling and increasing radiative heating due to short-wave radiation are observed, too.

The weather in the test area during the whole period of observations (10 days) was determined by the impact (from the north) of the high-pressure area in the Arctic water basin, with cyclones moving along the eastern coastline of Russia. The synoptic situation considered was typical of this season in the Arctic.

5.6.2 Results of aircraft measurements of the characteristics of radiation fields and cloudiness

During the expedition 15 flights were made: from the 21–28 May at Kamchatka and in the region of Kuril Islands, and from the 1–7 June, 1979, in the eastern sector of the Arctic in cloudy and cloudless conditions over the water, ice, and NP-22.

Based on radiation measurements of the short-wave radiation fluxes, the radiative characteristics of the surface–atmosphere system albedo A over the weakly reflecting water surface of the Okhotsk Sea increased with altitude H from 0.2 km to 8.4 km with $\Delta A/\Delta H \approx 0.004\,\mathrm{km}^{-1}$ for the solar height $h = 44°$. Over the Chukchi Sea, for $h = 35°$ and a thin haze $\Delta A/\Delta H \approx 0.006\,\mathrm{km}^{-1}$. Earlier data from the programme over the Azov Sea for $h = 80°$ gave $\Delta A/\Delta H \approx 0.009\,\mathrm{km}^{-1}$ and in the case of the dust outbreak from Sahara $0.037\,\mathrm{km}^{-1}$. The albedo of the Earth–atmosphere system over the ice decreases with altitude in the 0.2–8.4 km range: $\Delta A/\Delta H \approx 0.01\,\mathrm{km}^{-1}$, this decrease being most pronounced in the lower 200-m layer. For example, the albedo of ice from the data of NP-22 was 0.77–0.79, and at a height of 200 m it was 0.70–0.72, especially in the presence of the cloud layer located above. The albedo of the system in a cloudy atmosphere over the water surface decreases with a gradient of about $0.003\,\mathrm{km}^{-1}$, and over the ice by about $0.01\,\mathrm{km}^{-1}$. In the presence of an aerosol layer or haze the vertical profile of the Earth–atmosphere system albedo can be more complicated. The ice albedo as a function of ice concentration and type varies from 0.42 to 0.74, from measurements at an altitude of 200 m.

An intercomparison of the cloud albedo values at an altitude of 1 km over the water and ice has shown that the increase of the cloud albedo due to the effect of the surface located below the clouds can constitute 0.2 to 0.3. The major difference in the albedos of stratified clouds over the water and ice is connected with different albedos of the surface and, to a lesser extent, it is determined by the effect of the phase state of clouds and by the different relationship between the liquid and solid phases of cloud particles.

The phase state of clouds is more substantial for clouds of middle, and especially upper, levels. The presence of large crystals in the cloud leads to a stronger forward scattering, which determines a lower albedo for such one-layer clouds when compared to lower clouds (Ac, As $\sim$ 0.55–0.65 and Cs $\sim$ 0.15–1.30).

The albedo of altocumulus and altostratus, as well as cumulostratus clouds over the ice and underlying layer of stratus clouds is characterized by the values 0.60–0.76 and 0.52. The albedo of fog over the water (for a thickness of the fog layer $\Delta H = 200\,\mathrm{m}$) is 0.39. The presence of the fumarole plume over it (i.e., the presence of hydroscopic compounds: water vapour, halogens, sulphur dioxide, and carbon dioxide, ammonia, hydrogen sulphide, and chlorine compounds) leads to an

Table 5.13. Average values of the size of cloud particles $\bar{r}$, liquid water content $\bar{w}$, attenuation coefficient $\bar{\gamma}$, and inhomogeneity scale $\bar{x}$.

Types of clouds		Characteristics			
		$\bar{r}$ (μm)	$\bar{w}$ (g m^3)	$\bar{\gamma}$ (km^{-1})	$\bar{x}$ (m)
Si	Arctic	4.4	0.15	41.7	1.57
	ETR	4.5	0.31	42.6	1.62
Sc	Arctic	4.0	0.1	24.9	0.49
	ETR	4.7	0.23	47.9	1.47
As	Arctic	–	0.07	4.4	–
	ETR	–	0.28	25.5	1.15
Ac	Arctic	4.7	0.20	20.5	0.82
	ETR	50–100	–	22.5	–
Cs	Arctic	80–150	–	–	–

Note: ETR is the European Territory of Russia.

increase of albedo by 0.03–0.04, but at an altitude of 4 km this difference is held within the error of albedo measurements.

Clouds in the eastern part of the Arctic were characterized by values of water content of 0.05–0.15 g/m^3, with average values of the attenuation coefficient γ (km^{-1}) differing from the respective values over the European territory of Russia (Table 5.13).

Based on the technique of integral parameters, with the average water content and attenuation coefficient of stratified clouds known, the average diameter of cloud droplets and their number density was estimated at 3.9 μm and 30 cm^{-3}, respectively, which was below the respective data for the European Territory of Russia (ETR) (i.e., the clouds in the Arctic, especially the stratocumulus ones, are thinner and particles are smaller than over the ETR).

Knowing the geometric thickness and the average attenuation coefficient, one can determine the optical thickness of clouds, using the formula:

$$\tau = \sum_{(i)} \gamma_i \Delta H_i$$

and draw the dependence of the albedo of stratified clouds over the ice on their optical thickness.

It follows from the results obtained that the albedo of stratus clouds over the ice at a lower Sun elevation is higher than over land, water, or towns (the albedo of the surface of water, towns, suburban zones, and ice was, respectively, 0.04, 0.20, 0.13, and 0.70; these are the earlier Canadian Experiment (CANEX) results).

The absorptance of stratified clouds is at a maximum over towns, which is connected with the presence in the cloud of an optically active (absorbing) aerosol as well as with the changing size distribution of cloud particles over towns under the influence of hydroscopic condensation nuclei. The absorptance of clouds over the water and ice are nearly the same (within the error of measurements) and constitute

2–7% for stratified clouds with ΔH not more than 500–600 m. The rate of the radiative heating of lower clouds varies from 0.1 to 0.8°C/h and that of longwave cooling from 0.1 to 0.5°C/h.

An analysis of the vertical profiles of the radiative heat flux divergence in cloudless and cloudy conditions over the NP-22 for 4 June, 1979, and 7 June, 1979, showed that the atmosphere in the clear sky weather was characterized by a sufficiently high transparency 0.80 ± 0.02, aerosol optical thickness 0.082 at a wavelength of 0.4 μm, and water content 0.5–0.6 cm. The aerosol number density at the surface level from the data of NP-22 was 0.4 cm^{-3}, at an altitude of 200 m was 0.2 cm^{-3}, and at altitudes above 4 km it was negligibly small for the particle sizes from 0.2 to 20 μm. In the presence of clouds the total equivalent water content of an air column from the aerological data was 1.2 cm, the relative humidity at the surface level was 96%, and within the cloud 70%, with average liquid water content of 0.10 g/m^3. The vertical profile of the radiative heat flux divergence due to the short wave, long wave, and total radiation is characterized by a maximum in the lower 500-m layer and in the presence of clouds in the lower 200-m layer. A small maximum in the 2.8–4.2-km layer is connected with the effect of a humid haze at this altitude. The total radiative heat flux divergence decreases with altitude, and above the 2.85-km level radiative cooling prevails.

The simultaneous measurements of brightness temperature T_B in the atmospheric window 10–12 μm and temperature of clouds, as well as long-wave radiation fluxes at the top and bottom levels of clouds made it possible to estimate their emissivity using the formulas:

$$\varepsilon = \frac{B(T_B)}{B(T)} \qquad \varepsilon = \frac{E_{e1} - E_{e2}}{E_{b1} - E_{b2}}$$

where $B(T_B)$ and $B(T)$ are blackbody emissions for the respective temperatures; E_{e1} and E_{e2} are equivalent long-wave radiation fluxes at the boundaries of the cloud layer; and E_{b1} and E_{b2} are blackbody emission fluxes at the respective values.

Calculations based on the observational data for the Arctic clouds have shown that the emissivity of lower clouds in most cases is below unity and varies from 0.65 to 0.85. For mid-level altocumulus and altostratus clouds it constitutes 0.35–0.50, and for cirrostratus clouds 0.10–0.25.

Table 5.14 shows data for the spectral albedo and coefficient of reflection anisotropy of ice covered with snow, at the surface level and at altitudes 0.2 and 8.0 km.

The ice–snow albedo in the visible is higher than the total albedo and markedly decreases with altitude and in the IR. The snow reflection anisotropy grows with wavelength and altitude. The reflection anisotropy for clouds and water surfaces as compared with the snow-covered ice grows markedly (approximately by factors of 2 and 5, respectively) at the same Sun elevation. Thus, the Arctic is characterized by surfaces, which reflect radiation non-isotropically, and this fact should be taken into account, especially when determining the short-wave outgoing radiation fluxes and the radiation budget of the Earth–atmosphere system.

Table 5.14. The spectral albedo (numerator) and the coefficient of reflection anisotropy (denominator) of the snow-covered ice at different altitudes, at a Sun elevation of 26 ± 4°.

	λ (μm)					
H (km)	0.51	0.99	1.13	1.24	1.38	0.3–3.0
0	0.90/0.90	0.65/0.65	–	–	–	0.77/0.77
0.2	0.82/1.04	0.70/1.10	0.68/1.10	0.52/1.19	0.40/1.16	0.68/1.14
8.0	0.74/1.05	0.63/1.08	0.51/1.05	0.42/1.08	0.18/1.09	0.56/1.10

Table 5.15. The real n_1 and imaginary n_2 parts of the ice–snow refraction index at $t = -7$°C from 0.3 to 28 μm

λ (μm)	n_1	n_2	λ (μm)	n_1	n_2
0.3	1.380	3.95×10^{-9}	1.3	1.291	1.25×10^{-5}
0.4	1.319	1.30×10^{-9}	1.6	1.263	3.17×10^{-4}
0.5	1.312	3.20×10^{-9}	1.9	1.275	4.25×10^{-4}
0.6	1.309	9.75×10^{-9}	2.2	1.265	3.88×10^{-4}
0.7	1.306	3.35×10^{-8}	2.5	1.225	8.50×10^{-4}
0.8	1.304	1.13×10^{-7}	2.8	1.084	2.80×10^{-2}
0.9	1.301	1.10×10^{-7}			
1.0	1.299	1.93×10^{-6}			

Note should be taken that in some cases, under certain respective synoptic situations in the Arctic, an increase of the aerosol content can be observed as well as a decrease of the atmospheric transparency coefficient down to 0.6 in the presence of large-scale forest fires in Siberia, eruption of volcanoes at Kamchatka, outbreaks of anthropogenic aerosol from the lower latitudes, and dust–sand outbreaks from the Gobi Desert.

Deposition of small (~0.1 μm), strongly absorbing particles with a mixing ratio of only 1 ppm, can reduce the albedo of snow and ice in the visible by 5–15% as compared to the albedo of pure snow (~96%), especially in the visible (Table 5.15).

In the IR the ice (or snow) is a strong absorber by itself, which determines a sharp decrease of albedo in this spectral region. The values of the real and imaginary parts of the refraction index for ice at some wavelengths are given in Table 5.15, and they are much less than the respective values of the imaginary part of the refraction index for minerals ($n_2 = 10^{-2}$–10^{-4}), products of volcanic eruptions ($n_2 = 10^{-1}$–10^{-4}), and soot carbon ($n_2 = 0.5$–1). Therefore, the presence of soot in concentrations of only 0.3 ppm is sufficient to explain the decrease in the ice and snow albedo in the visible as well as a more rapid melting of ice in the Arctic.

Based on the accomplished observations in the Arctic, the following conclusions can be drawn:

- The albedo of the surface–atmosphere system decreases over the weakly reflecting (water) surface and significantly increases over the strongly reflecting (snow/ice) surface and depends on the presence of aerosol haze.
- The albedo of the sea ice of different concentrations and type can vary from 0.4 to 0.8 as observed from the 0.2-km altitude and is at a maximum in the visible region of the spectrum.
- The albedo of clouds in the Arctic is markedly higher than the albedo of clouds of the same optical thickness over the city, land, and water surface over the ETR and depends on the surface albedo as well as on a number of micro and macro-parameters.
- The presence of hygroscopic particles as a result of fumarole activity of volcanoes in the cloudy atmosphere raises the albedo of clouds, but their effect on the radiative properties is small compared with the impact of anthropogenic hydrophobic aerosols (chiefly soot).
- The net radiative heat flux divergence is at a maximum in the lower 500-m layer, especially in the presence of lower clouds; it decreases with altitude and is characterized by cooling in the atmospheric layers located above clouds.
- The emissivity of clouds (even at a low altitude) in the Arctic is below unity. This is connected with the prevailing crystal phase in clouds and is confirmed by a considerably different backscattering coefficient as compared with the attenuation coefficient in clouds.
- The Arctic is characterized by a surface with a strong reflection anisotropy.
- The Arctic together with the Antarctic, of all regions are the least subject to the effect of anthropogenic factors, but in the case of an outbreak of optically active aerosols, its effect can lead to a faster snow and ice melt and, hence, a more substantial warming.

Hobbs *et al.* (2001) have conducted airborne studies of cloud structures over the Arctic Ocean and comparisons with retrievals from ship-based remote sensing measurements. Shupe *et al.* (2001) have discussed some of the results of the NASA/FIRE Arctic Clouds Experiments which took place during April–July 1998 with the primary goal of investigating cloud microphysical, geometrical, and radiative properties with aircraft and surface-based measurements. An overview of the microphysical properties of Arctic clouds observed in May and July, 1998, during FIRE Advanced Composition Explorer (ACE) has been made by Rangno and Hobbs (2001). They studied ice particles in stratiform clouds in the Arctic and possible mechanisms for the production of high ice concentrations.

5.6.3 Numerical modelling of the dynamics of extended cloudiness and its interaction with radiation

One of the problems of studying clouds in the Arctic is an analysis of the dynamics of the formation and development of fog and clouds over the ice during the polar day over the NP-22. The periodicity of sensings was 6 and 12 hours. The complex of surface and airborne instruments was not completely adequate to meet the requirements for studying the evolution of clouds. Therefore, the results of numerical modelling of the development of clouds were compared, mainly, with the data on radiation, optical, and micro and macro-characteristics of clouds in the course of their temporal evolution with the purpose of assessing the contribution of radiation to the energetics of the cloud atmosphere (Kondratyev and Zhvalev, 1981).

Different heat flux divergences in the cloud layer can be considered, in general, to be based on the heat balance equation. So, for example, the rate of change of the humid entropy of the cloud mass per unit time can be expressed through the rate of change in the pseudo-equivalent potential temperature Θ_e, which is:

$$\frac{\partial \Theta_e}{\partial t} = -v\Delta\Theta_e - \bar{w}\frac{\partial \Theta_e}{\partial z} - \frac{\partial}{\partial z}\overline{(w'\Theta_e')} - \frac{\Theta_e}{c_p T}\varepsilon_r$$

where v corresponds to the horizontal wind speed; w is the vertical wind component; ε_r is the radiative temperature variation; and $\overline{w'\Theta_e'}$ is the vertical turbulent heat flux. An advective term $v\Delta\Theta_e$ and the contribution of large-scale downward motion $\bar{w}(\partial\Theta_e/\partial z)$ (whose value can apparently explain the difference in clouds over the continent and the ocean) are very difficult to estimate precisely, both based on the observational data and using the modelling results. Therefore, usually a 1-D time-dependent model is considered on the basis that it approximately corresponds to a 2-D stationary model of cloud dynamics.

Three components correspond to the sensible heat flux $\overline{w'\Theta_e'}$: at the expense of the change of potential temperature $\overline{w'\Theta'}$; water vapour $\overline{w'r_v'}$; and droplet water $\overline{w'r_d'}$, which together with the radiative heat flux divergence, determine the energetics of clouds. Therefore, it is important, based on calculations and observational data, to assess the role of these components in the evolution of extended clouds in the Arctic and to compare, where possible, the results of calculations and observations.

In the numerical modelling of the process of formation and evolution of clouds in the atmospheric boundary layer (ABL) with air masses moving from the continent to the ocean covered with ice, the same numerical model as used for the modelling of low clouds and fogs over the land was used (Kondratyev *et al.*, 1992; Khvorostyanov, 1995; Marchuk *et al.*, 1986).

The model includes an equation of the ABL dynamics, the transfer equation for heat, water vapour, and long-wave and solar radiation, the latter being calculated in a two-stream approximation. In this model the microphysical technique has been used to calculate the characteristics of cloudiness: oversaturation of water vapour and spectrum of droplets, content, droplet concentration, average size of droplets, and visibility.

Figure 5.1 shows a block-diagram of the model. Each group of processes in the ABL is simulated with an equation system of its own. In the numerical modelling a splitting technique has been used. The blocks in Figure 5.1 correspond to the breaking down of the model into partial problems: arrows show the calculation succession and the wording in the panels beneath show the exchange between them.

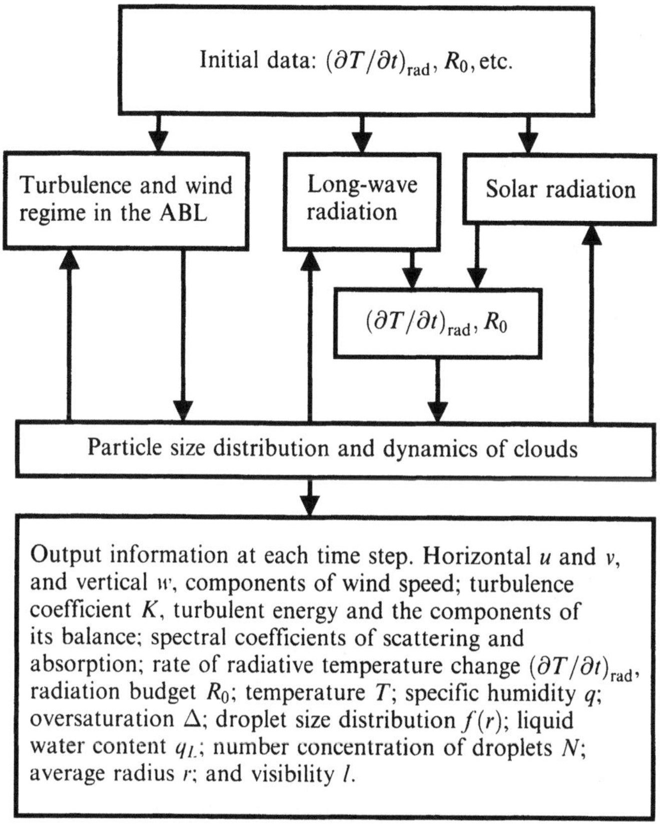

Figure 5.1. The model block-diagram.

The synoptic situation during the experiment in the course of aircraft and aerological sensings was characterized by warm air masses covering a cooler ice surface, which led to the clear sky weather change and formation of an advective fog, and its subsequent transformation into a stratus cloud.

Analysis of the vertical profiles of temperature t and relative humidity U in clear skies at the moment of formation of fog (which took 48 hr) followed by stratus cloud (after another 24 hr), showed that the process of cloud formation was followed by a decrease in air temperature and a raising of the upper boundary of the temperature inversion and relative humidity, so that it coincided with the cloud layer top, along with the growth of the geometrical thickness of the cloud from 200 to 300 m. The wind speed, according to data from airborne navigation instruments, was 3 m/s. Turbulent exchange took place when clouds appeared, and at this time the water content of fog averaged 0.06 g/m^3 and that of cloud 0.11 g/m^3. With a droplet radius of about 6 μm the content of precipitable water in the surface layer in conditions of clear skies, fog and stratus clouds, was 0.29 cm and 0.6 cm, respectively. The coefficient of turbulent mixing and vertical speed pulsation constituted about 15 m^2/s and

1 m^2/s, respectively, with a relative error of 50%. The surface temperature was −2°C and north-easterly winds prevalied.

Thus, a situation was modelled in which a warmer air mass moved over a cold surface, using the prescribed initial data. The melting ice at a temperature of −2°C was taken as the underlying surface, initial relative humidity was 80%, ice albedo 40%, geostrophic wind speed 10 m/s, and Sun elevations 20–35° – all characteristic of the polar day. The temperature was considered to decrease with altitude with a lapse rate of 6°/km, with constant relative humidity.

Numerical modelling has shown that when air masses move over ice, they cool and sublimational sinking of water vapour onto the ice takes place, hence the temperature and specific humidity in the lower 300–400-m layer decreases in time and inversions form. However, the cooling rate exceeds that of water decrease and 4 h 40 min later a cloud formed in the 150–180-m layer. The mechanism for its creation is the same as for the formation of clouds over land covered with snow. This is followed by downwind water vapour transport: its boundaries are 70 m and 360 m, and its maximum water content (0.17 g/kg) is reached in the upper third of the cloud. The average radius of droplets is 5–6 mm. This picture coincides with observational data, but does not correspond to the boundaries of the observed developing cloud, which were 80 m and 280 m.

The maximum rate of the long-wave radiative cooling (−2.9°/h) is observed 60 m below the upper cloud boundary and a maximum of solar heating (0.65°/h) is located 15 m lower at a Sun elevation of 23°. Since the cloud is not thick and its bottom is sufficiently low, there is no long-wave radiative warming in its lower part. Owing to a shift of maximum radiative cooling to the upper part of the cloud, the air–surface temperature inversion broke and was replaced by the elevated inversion with its bottom at the 300-m altitude. A maximum of the turbulence coefficient (10.2 m^2/s) is reached at a height of 100 m.

An interesting effect is observed in the radiation budget of cloud. Before the cloud forms at a Sun elevation of $\sim 20°$, the long-wave fluxes are $F_{l\uparrow} = 276.41$ W/m^2 and $F_{l\downarrow} = 196.8$ W/m^2; the long-wave balance at this moment is $R_l = -79.57$ W/m^2 and the short-wave balance at this moment is $R_s = 92.13$ (i.e., the net budget is positive but small, which is explained by the high ice albedo and agrees with the data of radiation measurements).

After cloudiness forms, $R_l = -7.1$ W/m^2, and $R_s = 83.76$ W/m^2, the net positive budget increases (i.e., the cloudiness is opaque for long-wave and almost transparent for solar radiation), which coincides with the conclusion drawn by Herman and Goody (1976) and leads to a strong greenhouse effect: in the presence of clouds the ice should melt more rapidly. Thus, the low cloudiness can markedly affect the ice regime of the Arctic and is an important climate-forming factor.

Further downward development of the cloud is limited by undersaturation with respect to water near the ice surface, and the upward development is slowed down with the cloud top approaching the top of the boundary layer (i.e., a relative stabilization sets in). During 8 hours the top rose 200 m (in reality, by 100 m), with the bottom sinking 20 m.

The maximum of liquid water content increased and shifted to the top. The temperature lapse rate also increased, especially in the upper part of the cloud, which led to the increase of the turbulence coefficient and appearance of the second maximum near the cloud top.

Most substantial is that the oversaturation of ice was positive only in the upper 100-m cloud layer, where radiative cooling determines the condensation. In the lower 400-m layer, due to solar heating, the oversaturation is negative, and the droplets evaporate. As a result of evaporation the average radius of particles in the bottom part of the cloud increases.

If, as is often done in hydrodynamic models, the bottom of the developing cloudiness is determined from the condition of saturation, its height will be strongly overestimated (in this case the humidity becomes saturated 400 m below the bottom), and the radiative heating, both long-wave and solar, will happen to be under the cloud. This may be one of the reasons that calculations by Herman and Goody (1976) have given a stratified cloudiness (evaporation of the middle layer). The numerical modelling considered (Marchuk *et al.*, 1986) has shown that at an altitude of 200–300 m evaporation is most intensive, and with the use of the hydrodynamic approach, the water can completely evaporate here, leading to a stratification. This experiment has not revealed any stratification of clouds, however.

The existence of a two-layer cloudiness, described by Herman and Goody (1976), could have been connected with the proximity of the frontal surface. During the further development of the cloud no qualitative changes take place. Even over 24 hours its boundaries still stay at altitudes of 60 m and 750 m.

The principal difference in the dynamics of macro-parameters of both calculated and observed clouds is that, in fact, the rate of the transfer from clear sky weather to cloudy weather is lower than in the numerical experiment, with a slower increase in the geometrical thickness of the cloud, whose lifetime in the Arctic can be 3 to 5 days. Apparently, this is connected with the fact that in the calculations the initial relative humidity U has been assumed to be constant with altitude, whereas the observed vertical profile of humidity before the cloud formation had been characterized by a decrease with height. In the numerical experiment, with the initial humidity decreasing with altitude, the cloud developed upward much more slowly, and the whole process became closer to the quasi-stationary one. Thus, a rapid decrease of relative humidity with altitude (due to constant low temperatures and attenuation of the turbulent exchange with the surface) is an important feature of the Arctic regions, substantially determining the cloud regime in the Arctic.

During the transformation of the clear weather and the formation of fog and stratus clouds the rate of radiative temperature change due to both short-wave (solar) and long-wave radiation increases, but in the net rate of the radiative temperature change the cooling in the cloud layer of the atmosphere prevails. In this regard, it is of interest to consider the contribution of various heat flux divergences to the dynamics of the development of low clouds. At initial stages of cloud formation the long-wave cooling and negative turbulent influx due to the temperature inversion is compensated for (by 30–40%) through latent heat and solar heating, so that the net rate of cooling can reach 1–3°C h. With the development of the cloud and increase of the elevation of the Sun the rearrangement of the temperature stratification leads to a change in the sign of the sensible heat flux divergence, which, together with positive phase transformations and increasing solar heating, compensate substantially the long-wave cooling, so that the net rate of temperature decrease reduces by an order of magnitude and constitutes 0.1–0.3°C/h. The changes of temperature due to advection and vertical motions are of the same order of magnitude. Thus, the principal conclusion drawn from calculations – and agreeing with observations that reveal the quasi-stationary existence of low clouds – is that a mutual compensation takes place of long-wave cooling, on the one hand, and positive heat flux divergence due to latent and sensible heat exchange, solar heating, on the other.

A very important effort in the numerical modelling (1-D thermodynamic model) of the impact of clouds on the surface radiation budget (SRB) of the Arctic Ocean has been undertaken by Curry *et al.* (1993). The most important results of this study have been summarized as follows:

- clouds have a net warming effect on the surface, in contrast to clouds at lower latitudes;
- the annual cycle of CRF shows warming at all times except for cooling during a few weeks in mid-summer;
- for annually averaged cloud fraction less than 0.4, the CRF is positive (warming) over the entire annual cycle;
- variations in cloud fraction result in variations in the melting period, which may account for observed interannual variability in the melt season;
- low clouds have the greatest impact on the surface radiative flux;
- mid-level clouds have a smaller impact on the surface radiative flux than low-level clouds, but show a larger period and amplitude of summertime net cooling;
- high clouds have relatively small net effect on the surface energy balance, showing a small warming effect over the entire annual cycle;
- cloud optical depth is parameterized to increase with increasing atmospheric temperature, resulting in a net surface warming;
- during summertime, the net flux decreases both for an increase and decrease in drop size;
- the impact of changing ice crystal size is much smaller than that for changing cloud water drop size;
- the modelled equilibrium sea ice thickness decreases with increasing cloud fraction; and
- surface radiative fluxes vary non-linearly with cloud fraction, due to changes in surface characteristics.

The model experiments by Curry *et al.* (1993, 1996) have provided insight into the nature of two cloud feedback processes in the Arctic. Clouds have been shown to have a net warming effect on the Arctic Ocean surface. This is in contrast to the global average condition, which indicated that clouds have a net cooling effect on the surface. Increasing global temperature is hypothesized to increase the amount of condensed water and thus the cloud optical depth. The results of the calculations for the Arctic showed that increasing atmospheric temperature and thus cloud optical depth results in an increased net surface radiation flux (due to observational results obtained by Dong *et al.* (2001) the Arctic stratus clouds produce a net warming of 20 Wm^2 in the surface layer during the transition season suggesting that these clouds may accelerate the springtime melting of the ice pack). The so-called "cloud optical depth feedback" is thus a positive climatic feedback mechanism in the Arctic in contrast to the global average assessment obtained earlier. Increasing amounts of pollution aerosol have been hypothesized to decrease water drop size and thus increase cloud optical depth. The results of calculations for the Arctic show that the response of surface radiation fluxes to a decrease in drop size is complex, showing a decrease in the summertime net surface radiation flux and an increase in the wintertime net radiation flux. Overall, Arctic clouds do not provide a consistent negative feedback that can be expected to counteract the effects of greenhouse warming. Recently the simulation of Arctic low-level clouds observed during the FIRE using a new bulk microphysics scheme has been accomplished. Similar efforts of simulations of ice cloud evolution during FIRE using the Canadian Centre for Climate Modeling and Analysis (CCCMA) single-column model (SCM) have been undertaken by Lohmann *et al.* (2001), while Khvorostyanov *et al.* (2001) have accomplished modelling with explicit spectral water and ice microphysics of a two-layer cloud system of altostratus and cirrus observed during the FIRE ACE.

Wang and Zwiers (2001) have accomplished a study of the detailed interactions among the long-wave radiation, clouds, turbulent structure, and snow surface temperature. The key process of such interactions is turbulent transport, which directly links the long-wave radiative cooling at the cloud top to the SAT.

5.7 MODELLING THE PROCESSES OF PROPAGATION OF POLLUTANTS AT HIGH LATITUDES

5.7.1 The Arctic Basin and its role in global changes of the environment

As has been mentioned, the Arctic basin is quite special in the formation of the global processes in the environment, determining numerous feedbacks in the Earth's climate system. Lowered temperatures, a considerable level of atmospheric circulation and available large ice-covered water basins – all this distinguishes high latitudes from other latitudes of the globe.

An intensive development of the northern territories in Russia, Canada, USA, and in the Scandinavian countries has led to a considerable change of natural conditions in these regions. The development of the oil- and gas-extracting industry on Yamal, Taimyr, northern West Siberia and the coal- and gold-extracting industry in Yakutia and Chukotka, as well as the functioning mining industry on the Kola Peninsula make the northern territories of Russia most dangerous for the Arctic environment. The geological services of Finland, Norway and the Russian Central

Expedition of the Kola Peninsula have drawn maps of the distribution of land lichen, lakes and five horizons of soils (Kashulina and Reimann, 2002; Kashulina *et al.*, 2003; Urquizo *et al.*, 2003). This will mark the beginning of documenting the state of the northern territories.

In some territories of the Arctic the vegetation cover has been broken, and the areas and productivity of the pastures for deer have decreased. The hydrological regime of the Arctic Basin rivers has changed greatly, too. River runoffs take pollutants to the coastal seas of the northern coast of Russia, which damages the functioning of the Arctic Basin ecosystems. Although the population of the Arctic areas are less industrially occupied compared with the low-latitude population, for the purposes of global modelling it is necessary to take into account the prospects for broadening this activity.

Further affecting the frail Arctic ecosystems will lead to negative consequences on a global scale. Therefore, the problem of development of the northern territories, especially in Russia, needs a thorough analysis of the dynamics of all types of ecosystems, formation of databases on their state, and development of efficient ways to ensure the cooperative development of natural and anthropogenic processes. Here are some pointers for further development of the northern territories:

1. Intensive development and differentiation by the territorial characteristics (explored and perspective) of deposits of the mining industry and energetics.
2. Formation of preserves, national parks, reservations, reserves and other forms of protection of the ecosystems of the northern territories.

These two directions should be realized using the renewable database under the control of the changes in the spatial structure of ecosystems. The monitoring system should monitor violations of the balance of ecosystems and anthropogenically affected landscapes and estimate the state of the habitats of animals and population of the northern territories. In this regard the following studies are of first priority:

- Complex monitoring of land areas and sea basins to create a cadastre of land resources and databases on the parameters of biogeocenoses and ecosystems.
- Study of the social problems of peoples living in northern latitudes and an assessment of damage made to their habitat.
- Division of the northern territories into districts, based on the combination of landscape and administrative principles and taking into account migration processes.
- Revealing and ranging the functional problems of the systems of the nature-protective profile.

First of all, violations of land cover should be revealed and included in the database, as well as places of various dumps, contaminated territories, routes of oil and gas pipes, sources (functioning and hypothetic) of pollutants of soil, water,

Table 5.16. Estimates of some parameters of the Arctic Basin.

Parameter	Published estimate of the parameter
Area of the Arctic Basin (10^6 km^2)	14.09
Fluxes of water masses through the straits	
Faroe–Scotland	+135; −45
Dutch	+30; −130
Bering	±1.8
Faroe–Iceland	+40
Salinity of ice (‰)	
One-year ice (~150 cm)	5
Perennial ice (~3 m)	1
River runoff (km^3/year)	
Yenisey	603
Ob′	530
Lena	520
Mackenzie	340
Pechora	130
Kolyma	132
Northern Dvina	110
Outflow of freshwater with ice from the basin (km^3/year)	1,500

Table 5.17. Characteristics of the Arctic basin water bodies.

Water body	Area (10^3 km^2)	Volume (10^3 km^2)	Depth (m)	
			Average	Maximum
Central Basin	4,977	12,442	2,500	4,000
Barents Sea	1,424	316	222	600
Greenland Sea	1,195	1,961	1,641	5,527
Norwegian Sea	1,340	2,325	1,735	3,970
White Sea	90	6	67	350
Buffin Bay			200	700
Kara Sea	893.4	101	113	600
Chukchi Sea	582	23	40	60
East-Siberian Sea	944.6	18	20	30

and atmosphere, and zones of anthropogenic floods and marshes. Estimates of some parameters of the Arctic basin systems are given in Tables 5.16 through to 5.18.

There is an impact on the high-latitude environment independent of the territory of the Arctic Basin. The global structure of interactions between the global regions includes, for instance, the trans-boundary transfer of gaseous and solid chemicals. Therefore, the sources of pollutants located in middle and low latitudes can be dangerous for the Arctic. Numerous observations of the Arctic atmosphere

Table 5.18. Characteristics of the freshwater balance of the Arctic Basin water bodies.

Water body	Area 10^3 km^2	Supply of fresh water in ice (km^3)	Volume of melted freshwater (km^3/year)
Central Basin	4,977	13,000	1,990
North-European Basin	4,065	2,470	1,170
Seas of the Siberian shelf	3,025	5,330	2,260
Canadian Basin	2,632	4,700	1,800

pollution reveal the presence in it of heavy metals, POPs, polychlorinated biphenyls (PCBs), SO_2, and other long-lived compounds carried by atmospheric currents from Europe, Asia, and North America (Bard, 1999).

Various aspects of these problems are considered in many international and national programmes on the environmental studies. So, in 1991 the programme ARCSS was launched (McCauley and Meier, 1991) which is a part of the international "Global Change" programme. The main goal of this programme was to develop methods, technologies, algorithms and software, which would permit assessment of the sensitivity of global changes in the nature–society system (NSS) to changes in its Arctic area. In fact, the ARCSS programme should provide an understanding of the role of physical, geological, chemical, biological, and social processes, taking place in the Arctic region, on the global changes of the environment and create thereby the scientific basis for solving the basic problem of prediction of such changes on different timescales, from years to centuries. The US national "Biocomplexity" programme launched in 2000 supplements the goals enumerated above, broadening them to global scales. Within the framework of this programme, plans are made to study and understand the interaction between the complexity of biological, physical, and social systems as well as trends in changes of the present environment. The complexity of the system interacting somehow with the environment implies a phenomenon happening when the living system contacts its environment in global conditions.

Biocomplexity is a derivative of biological, physical, chemical, social, and behavioural interactions of the environmental subsystems, including living organisms and the global population. The notion of biocomplexity in the surrounding world is closely connected with the "laws of functioning of the biosphere as a unity of its ecosystems and natural–economic systems of different scales, from local to global" (Kondratyev, 1999). Therefore, to determine the biocomplexity and to estimate it, a combined formalized description is needed of biological, geochemical, geophysical, and anthropogenic factors and processes taking place at a given level of the spatial–temporal hierarchy of scales.

Manifestation of biocomplexity is a characteristic feature of all systems of the environment connected with life. The elements of this manifestation are studied within the theory of stability and vitality of ecosystems. It should be noted that formation of biocomplexity includes indicators of the degree of mutual modification of interacting systems, and it means that the biocomplexity should be studied considering both spatial and biological levels of organization. The difficulty of this problem is determined by the complicated behaviour of the object under study, especially with the human factor taken into account. Within this study, the Arctic systems are considered to be the NSS subsystems.

The problem of regular monitoring of the northern territories and the Arctic Basin is far from its solution, stimulating many scientiests to develop new methods for the synthesis of geoinformation monitoring systems in Arctic conditions. During the last few years, attempts have been made in Russia and the USA to consolidate efforts of specialists to solve these problems (Krapivin *et al.*, 1997, 1998; Phillips *et al.*, 1997; Rochon *et al.*, 1996; Krapivin and Phillips, 2001).

To understand and predict correlations between the processes taking place in the Arctic environment and other global regions is only possible within the complex scientific–technical approach to studies and analyses of these processes, which foresees both expeditions and theoretical studies with satellite, aircraft, ground mobile and stationary measurements, as well as new information technologies, such as GIS and GIMS-technologies (Arsky *et al.*, 1998; Kondratyev *et al.*, 2002, 2003).

Interactions between the atmosphere, land, and marine ecosystems in conditions of the Arctic climate can be on numerous spatial–temporal scales. Understanding the internal connections at each level is a key problem of organization of monitoring. Each scale is characterized by a certain type of landscape, vegetation cover, topography, by the character of the hydrophysical and synoptic structures, as well as by wild animals. Revealing the internal and external cause-and-effect feedbacks between these elements and other elements of the global NSS will make it possible to form the base of knowledge of the future system of geoinformation monitoring of the Arctic. Figure 5.2 explains methods for studying these feedbacks. Methods utilizing the combination of remote and *in situ* monitoring with numerical modelling shown in the right-hand part of Figure 5.2 have been suggested in Kondratyev *et al.* (2002). Their practical use will enable one to obtain more accurate estimates of the role of the Arctic latitudes in the global processes taking place in the NSS. The present division of land and marine processes should not interfere with complex studies. The developed numerical models of the NSS functioning will make it possible to overcome this division. Such models will enable one to synthesize a combined pattern of migration of chemicals in the Arctic latitudes and to assess the consequences of large-scale anthropogenic processes in the northern territories. Determination of critical situations and processes will become a partial problem of the complex system of monitoring the Arctic region.

Really, the formation of a software package is the basic problem. The spatial distribution of land and marine ecosystems as well as climate scenarios and anthropogenic trends are to be the input information for the application programs. A supplemented database and renewable set of partial models will ensure reliable predictions of the dynamics of these hypothetical scenarios of managing the Arctic environment. A 3-D model to calculate the dynamics of the radionuclide pollution of the Arctic Basin (Preller and Cheng, 1999) and a 2-D model of the ice conditions of

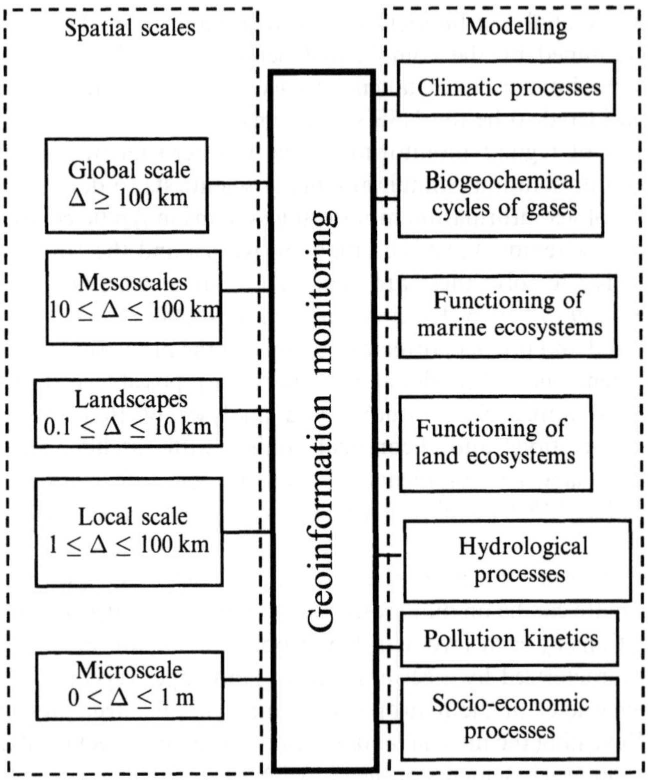

Figure 5.2. Conceptual scheme of organization of the monitoring of the northern-latitude environment reflecting the relationship between spatial and problematic levels which should be studied to understand the role of the Arctic region in global changes of the NSS.

the Arctic Ocean (Riedlinger and Preller, 1991) can serve as examples of successful solutions to some of these problems.

5.7.2 The atmospheric transport of pollutants to the Arctic

As has been mentioned, the cold Arctic Ocean becomes a sink for numerous pollutants generated by humans. Only during the last century the content of sulphates in the Arctic ice increased by 70%. At present, the input of sulphur (75% as SO_x and 25% as sulphates) is estimated at 2.3 Mt/yr. The contribution of the Euro-Asiatic and North American sources constitutes 94% and 6%, respectively. Europe contributes 300,000 t SO_2 annually to the atmosphere. The concentration of sulphate aerosols in the Arctic atmosphere has two maxima – in winter (56 $\mu g/m^3$) and in summer (14 $\mu g/m^3$). In some regions the concentration of sulphur compounds reaches even higher levels: Norilsk (140–160 $\mu g/m^3$), Kola Peninsula (>1000 $\mu g/m^3$), and northern Norway and Finland (400 $\mu g/m^3$). As a result, high latitudes reveal a reduction of biodiversity, shortening of nutrient chains, changes of albedo, and violation of radiation budget in some territories.

Bard (1999) studied the problem of the global transport of anthropogenic pollutants of the environment to high latitudes and analysed possible consequences for marine ecosystems and human health in the Arctic Basin. As a result of dry and moist deposition, these pollutants reach the seawater and soil, and due to the processes of biological accumulation get to the nutrient chains of the ecosystems. It is especially manifested through marine ecosystem produce. Consumption of fish and other sea products results in the pollutants infiltrating people living both in the northern territories and in middle and low latitudes.

The following programnmes are dedicated to studying the problems of the Arctic ecosystems pollution: the Northern Contaminants Programme, and the multinational Arctic Monitoring and Assessment Programme. An atmospheric pollutant can be transported from the zone of its generation to the regions tens or even thousands of kilometers apart. The Arctic atmosphere turned out to contain POPs (HCHs – 480 Pg/m^3; PCB – 17 Pg/m^3; DDT – 1.5 Pg/m^3; chlordan – Pg/m^3; toxaphene – 40 Pg/m^3), whose sources are located in middle and low latitudes. As Bard (1999) points out, the presence of outside pollutants in the Arctic atmosphere was discovered in the mid-20th century. At that time, the Arctic haze contained SO_x, NO_x, PAHs, heavy metals, and large particles of soot. As a result, mechanisms of chemical transport from low to higher latitudes were discovered and explained.

Numerous observations in distant freshwater basins of Alaska and Canada revealed pesticides and other chemicals whose sources were not located there. Another indicator of the growing concentration of pollutants in the Arctic is the content of chemicals in the ice of Greenland. So, during the last 500 years the content of mercury increased from 0.6 μg/g to 2.6 μg/g, and the input of mercury to the Arctic and sub-Arctic lakes changed from 0.7 μg/m^2/yr to 54.35 μg/m^2/yr. Another source of information about the dynamics of mercury concentration in the Arctic ecosystems is the study of the level of mercury in human hair. From pre-industrial times, it changed from 3.1 μg/g to 9.8 μg/g.

Many industrial countries, which still use pesticides, are located in southern latitudes with increased temperatures and intensive precipitation. As a result, in these latitudes, due to evaporation, the atmosphere obtains various chemicals which, being transported to high latitudes, precipitate onto land or sea surfaces. The mechanism of such precipitation are connected with more powerful processes of condensation of organic vapour on particles in conditions of low temperatures. Thus there is an effect of accumulation of pollutants in the Arctic ecosystems not generated in these latitudes. Of course, the atmospheric transport of pollutants between low and high latitudes is rather complicated. So, for instance, mercury, more volatile organochlorines and PAHs can return from northern latitudes to southern ones. Therefore modelling of such a transport should be oriented towards a consideration of the discovered effects. In particular, the Arctic haze is noted to concentrate mainly in the lower troposphere at altitudes up to 3 km, and it is manifested most in cold periods of the year, between December and April. There are also seasonal oscillations in the fluxes of pollutants caused by a seasonal cycle of anthropogenic ejections characterized by the summertime maxima between April and June due to the use of pesticides in agriculture. However, this cycle correlates weakly with oscillations of pollutant concentrations in the northern latitude environment. This discrepancy is explained by a combination of anticyclones with various atmospheric stagnant phenomena, which lead to an accumulation of pollutants in the zones of their sources with their subsequent transport to distant regions. This results in a strong decline from the stationary process of large-scale oscillations of the concentration of pollutants in the Arctic atmosphere.

The rate of atmospheric transport of pollutants from their sources to high latitudes is estimated at a period of 2–10 days. Such a range results from differences between the rates of the transport of pollutants to the Arctic from the Euro-Asiatic and North American sources. More than 50% of the chemicals of the Euro-Asiatic origin reach the Arctic more rapidly than substances coming to the atmosphere from the sources in the territory of North America. Moreover, during winter months, a strong Siberian anticyclone drives air masses from central Euro-Asia to the Arctic, so that pollutants have time to move onto the territory of North America and southern Greenland before precipitating onto the Arctic surface. In spring, an increased atmospheric pressure in Siberia enlarges the contribution of western Euro-Asia to the Arctic atmosphere pollution. In summer, there is a weak north–south transport of air masses alternating with their input from the northern sectors of the Pacific and Atlantic Oceans. Here, of importance, is the pollutants lifetime, which depends on numerous factors determined mainly by the rates of photodegradation, washing-out, and exchanges on the "atmosphere–water" and "atmosphere–vegetation" boundaries.

The Arctic Basin can be polluted both through the atmosphere and directly due to water exchange with the Atlantic and Pacific Oceans across narrow straits. It is known that 76% of HCH gets to the Arctic due to the oceanic currents, 22% – due to dry and moist deposition from the atmosphere, and only 2% – via river runoff. Moreover, 75% of HCH are transported from the Arctic due to runoffs through the Canada Archipelago and East-Greenland Current.

In connection with the problem of the greenhouse effect, there is an urgent question of the role of the Arctic ecosystems in CO_2 assimilation from the atmosphere as well as an urgent problem of assessing the consequences of pollution of the Arctic Basin from the viewpoint of global ecodynamics and climate changes. The effect of acid rains on rather vulnerable high-latitude water and land ecosystems is observed in the lakes of northern Norway, in the Russian territory of the Arctic, and sub-Arctic regions of Sweden, where the values $pH \in [4.9, 6.0]$ are recorded. The vulnerability of the Arctic ecosystems is connected with their small diversity and simple nutrient chains.

The Arctic environment is considerably threatened by heavy metals. According to estimates, 2–6% of total emissions of heavy metals in the industrial regions reach northern latitudes. Countries of West and East Europe contribute most to this flux. Heavy metals are observed everywhere, in plants, birds, fish and other marine animals, both growing and living in high latitudes. From available estimates, the total annual flux of heavy metals to the Arctic is characterized by the following values (Bard, 1999): fish (0.005–0.569 μg/g), maritime birds (0.046–2.67 μg/g), seals, whales and polar bears (<21.6 μg/g). The concentration of cadmium in the liver of the northern latitude animals constitutes: fish (0.034–2.11 μg/g), *Crustacea* and polar bears (0.477–7.79 μg/g), birds, whales and seals (0.853–36.6 μg/g). During the last 15–20 years the content of mercury in the seal population has increased 3 times. Mercury accumulates in the insoluble biologically inert form of inorganic Hg-Se crystals. But in the human stomach the meat of such seals due to oxidation frees the mercury and acts as a poison. So, the content of mercury in the blood of the population of one of the settlements in Quebec Province reached 14.1 μg/l, whereas for the population of Quebec City this level does not exceed 2 μg/l. On the whole, studies of the Canadian Arctic have shown that the concentration of mercury in the blood of newborn children averages 20 μg/l, varying from 0.6 to 214 μg/l. From the estimates of the World Health Organization, the content of mercury in blood at a level of 1200 μg/l can cause neurological changes in adults. A risk for the foetus is represented by a concentration of mercury in the mother's blood of 40–80 μg/l. Moreover, about 10 μg/g of mercury in the mother's hair can lead to a 2-month delay in the development of her newborn child's intelligence. In general, the human central nervous system starts malfunctioning with concentrations of mercury in the body of about 1280 μg/kg. However, all problems arising here should be thoroughly studied, since there are no single conclusions about the consequences of the environmental pollution for living beings. So, for instance, there is an opinion that the high-level presence of selenium in food can neutralize mercury due to the formation of insoluble mercury–selenide crystals, which are not dangerous to life.

Data on studies of the effect of pollutants on animals and humans are important for specification of the parameters of the demographic unit of the global model. Knowledge of correlations between distant ecosystems enables one to simplify the functional presentations of many processes and to increase the adaptive indicators of GSM. One of the elements of such correlations is the transport of pollutants by migrating birds not taken into account in global models.

Thus the control of the content of pollutants in the high-latitude environment needs the development of technologies, which will make it possible to promptly assess the state of the elements of the environment, which acts as a regulator of the NSS. The climatic conditions and weakly developed monitoring systems in high latitudes complicate the solution of this problem. Apparently, the increasing role of the methods of ecoinformatics permits the reduction in the Arctic database requirements due to the technology of modelling.

5.7.3 Model of the transport of pollutants to the Arctic Basin from Angara and Yenisey runoffs

An intensive development of industry in the northern territories of Russia has led to considerable changes of nature in these regions. Extraction of minerals and creation of a network of enterprises to process materials for nuclear energy have brought forth the problem of estimation of the levels of pollution of the environment with radionuclides, heavy metals, and oil products. The presence of a powerful river network of the Arctic Basin has also necessitated studies and assessment of the global-scale propagation of pollutants due to their transport by rivers from the territory of Russia and, in particular, central Siberia. The Angara–Yenisey (AY) system is one of the important systems in this process of river systems.

The Yenisey River, with its tributaries covers central Siberia over the watershed area of 2,580,000 km^2, reaching 4,102 km from south to north. The average annual runoff in the Yenisey mouth is estimated at 19,800 m^3/s. Angara as one of the large tributaries of the Yenisey, constitutes 24% of its runoff. The Lower Tunguska constitutes 20% of the Yenisey runoff. An important characteristic of the Yenisey is its shallow waters and small temperature gradients both vertically and over the surface. From hydrochemical indicators, the waters of the Yenisey are characterized by low turbidity, low content of minerals, and they are rich in dissolved oxygen. So, the concentration of salts averages 54 μg/l, the average annual runoff of ions is 43.2 Mt, the solid runoff is estimated at 14.5 Mt. According to available estimates, the average annual concentrations (μg/l) of chemicals in the Yenisey mouth constitute: for oil products – 0.13; NH_4 – 0.06; SO_4 – 15.1; P_{total} – 0.026; Cu – 0.01. From ecological indicators, the Yenisey is oligotrophic with prevailing diatomic algae and benthic animals. The anthropogenic aspects of the Yenisey are connected with the impact on its ecosystem of industrial (chemical, metallurgic, timber processing), agricultural, and urban municipal enterprises. The Krasnoyarskaya, Sayano-Shushenskaya, and Irkutskaya hydroelectric power stations are located on the Yenisey and the Angara. Regulating of the AY system by dams has reduced the flow and turbulence of waters over large areas, water exchange has reduced, stagnant zones appeared, and the effect of the adjacent territories increased through the processes of eutrophication of shallow-water areas.

One of the anthropogenic aspects of the AY system functioning is a possible input into it of radionuclides and heavy metals. Measurements carried out by the Russian–American ecological expeditions in the summer of 1995 and 1996 revealed their traces in the waters of the Angara and the Yenisey. Five cities are located along the Angara: Irkutsk, Angarsk, Svirsk, Bratsk, and Uzolye-Sibirskoye. These cities generate a broad spectrum of chemicals, an input of which into the AY system was recorded by the expeditions mentioned above (Krapivin *et al.*, 1998). The scheme in Figure 5.3 shows points of measurements, and Tables 5.19 and 5.20 give fragments of the obtained data.

To estimate the AY system in the process of the spatial propagation of pollutants, it is necessary to determine the dependence of their flux in different sections up to the Yenisey delta, on hydrological and ecological characteristics, which, to a great extent, are determined by the structure of the AY system. Solution of this problem requires a topologically isomorphic graph of the AY system, and for each of its elements it is necessary to construct a model of pollutant transition (AY-model). Here, of importance, is the problem of an optimal combination of the model of the AY system structure with the real metric of elements, as well as a weighted combination of gross indicators of discrete sections of the AY system with its general efficiency.

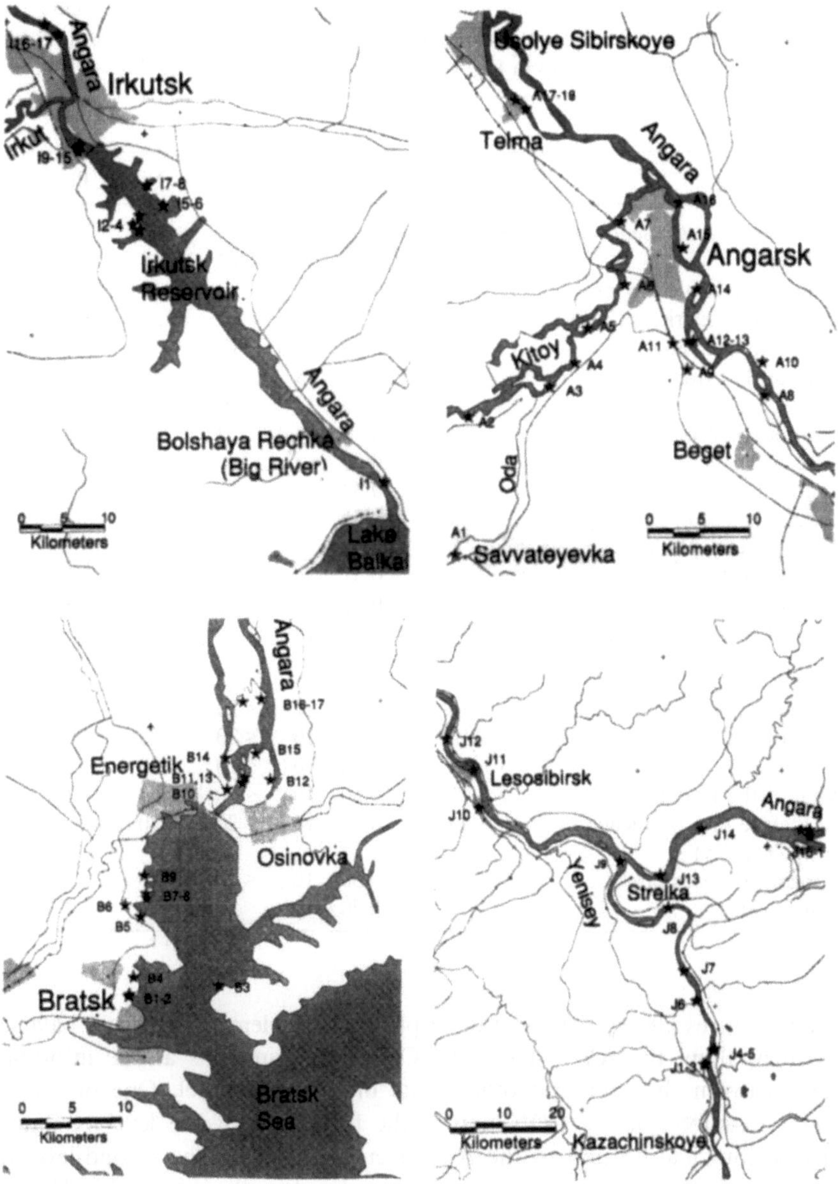

Figure 5.3. Maps of sample locations (marked by stars) during the American–Russian ecological expedition of 1996.
From Krapivin *et al.* (1998).

According to the scheme in Figure 5.4, the AY-model has units of three levels of cosubordination. The first-level units are the hydrological (CHD) unit and control unit (CON). The CHD unit simulates a river runoff with consideration of its correlation with topographic and synoptic parameters of the territory under

Table 5.19. Laboratory analysis of the concentrations of heavy metals in sediments and in water measured in July 1996 during the American–Russian hydrophysical expedition. Points of sampling are marked on the map in Figure 5.3. The measurement error did not exceed one sigma.

From Krapivin *et al.* (1998).

Point of sampling	As	Cd	Cr	Cu	Ni	Pb	Zn
Samples of river bottom sediments (ppm)							
Grand River	7.2	<0.52	26	19	25	15	64
Upstream of the Irkutsk dam	5.0	<0.51	41	22	44	11	55
Downstream of the Irkutsk dam	4.5	<0.51	20	11	21	9	17
Angarsk (technical canal)	0.9	<0.50	7.0	11	9.3	7.9	31
Upstream of the Angarsk	4.3	<0.51	48	38	40	14	86
Kitoy River	2.4	<0.50	27	20	37	7.8	40
Upstream of the Bratsk dam	<0.5	<0.50	6.4	9	12	2.3	22
Downstream of the Bratsk dam	3.4	<0.50	31	210	39	6.1	50
Angara (up the flow into Yenisey)	3.2	<0.50	14	13	18	4.5	31
Place of confluence of the Angara and the Yenisey (the Strelka settlement)	5.1	<0.51	47	35	37	14	100
Downstream of Lesosibirsk	6.9	<0.51	34	31	36	17	100
Kazachinskoye	5.8	<0.51	54	43	390	18	150
Upstream of Kazachinskoye	2.2	<0.50	18	8.2	20	3.9	47
Water samples (ppm)							
Downstream of the Irkutsk dam	12	<0.50	<10	<20	<20	<5.7	<5.7
Upstream of the Irkutsk dam	10	<0.50	<10	<20	<20	<5.7	<5.7
Angarsk	12	<0.50	<10	<20	<20	<5.7	<5.7
Upstream of the Bratsk dam	16	<0.50	<10	<20	<20	<5.7	<5.7
Angara (before the confluence with the Yenisey)	12	<0.50	<10	<20	<20	<5.7	<5.7
Downstream of Lesosibirsk	13	<0.50	<10	<20	<20	<5.7	<5.7
Kazachinskoye	8.6	<0.50	<10	<20	<20	<5.7	<5.7

study, and realizes all algorithms of description of the elements of water balance of the territory along the AY-system. The CON unit functions as the information interface between the user and other units of the AY-model. One of the most important functions of the CON unit is the formation of the level of database responsible for identification of the AY-system's elements. The second-level units realize partial models of various processes considered by the AY-model. The third-level units realize various versions of models. The units of all levels are described in Table 5.21.

Let the watershed area of the AY-system cover the territory Ω. We digitize Ω by latitude φ and longitude λ with steps $\Delta\varphi$ and $\Delta\lambda$, respectively, so that $\Omega = \cup\Omega_{ij}$, where Ω_{ij} is part of the territory Ω with an area $\sigma_{ij} = \Delta\varphi\Delta\lambda$. From the set Ω we select the cells $\Omega_{i^*j^*} = \Omega_k$, which contains fragments of the riverbed of the AY-system. We number these cells from $k = 1$ near the Angara mouth to $k = N$ at the Yenisey mouth. The process of this digitizing is provided by the IAS

Table 5.20. Results of measurements of the content of radionuclides in river bottom sediments made in July 1996 (averaged data of isotopic activity, Bq/kg of dry weight). Places of sampling are marked on the map in Figure 5.3.

Place of sampling	^{60}Co	^{137}Cs	^{152}Eu	^{235}U	^{238}U
Angara					
Upstream of the Irkutsk dam	<2.0	2.2 ± 1.0	<3.1	<1.9	<280
Downstream of the Irkutsk dam	<3.5	<3.6	<5.3	27.2 ± 4.6	720 ± 300
Angarsk (technical canal)	<2.0	<1.8	<3.1	<1.7	<320
Upstream of Angarsk	<4.1	25.2 ± 3.1	<6.0	<4.2	<630
Upstream of the Bratsk dam	<2.3	<2.3	<3.9	<2.2	<420
Downstream of the Bratsk dam	<2.6	3.4 ± 1.2	<4.0	<2.5	<440
Before the flow of the Angara into the Yenisey	<1.9	2.2 ± 1.0	<2.8	<1.7	<290
Yenisey					
Upstream of Kazachinskoye	8.6 ± 1.9	22.9 ± 2.3	6.7 ± 3.0	<2.7	<500
Kazachinskoye	241 ± 11	392 ± 12	151 ± 27	<8.2	<1,340
Downstream of Kazachinskoye	30.1 ± 3.0	203 ± 5	42.3 ± 9.5	<3.8	<680
Place of confluence of the Angara and the Yenisey (the Strelka settlement)				<4.5	<840
Downstream of Lesosibirsk	96.9 ± 5.1	211 ± 6	55.7 ± 13.5	<6.4	<860
	<5.7	27.1 ± 3.8	<8.2		

unit via the structure of identifiers $A_k = \|a_{k,ij}\|(k = 1, \ldots, 7)$ included into the database of the AY-model. Apart from them, the database contains information about the coefficient of the model and some sets of draft scenarios. The CON unit provides a direct input to the database both to obtain reference information and to introduce changes. The user can change the base estimates of all parameters or, after forming the scenario, change them within this scenario. Identifiers forming an image of the set Ω and being the basic attribute of the function of the formation of scenarios are the most important level of the database. Identifiers are symbolic matrix structures whose elements connect the AY-model with an object under study.

Identifier $A_1 = \|a_{1.ij}\|$ relates the AY-model to the system under study:

$$a_{1.ij} = \begin{cases} 0 & \text{when } (\varphi_i, \lambda_j) \notin \Omega \\ 1 & \text{when } (\varphi_i, \lambda_j) \in \Omega \end{cases}$$

With the help of A_1 the user determines the spatial configuration of topography of the AY-system with an accuracy of up to $\Delta\varphi\Delta\lambda$ over the area.

Identifier $A_2 = \|a_{2.ij}\|$ gives the subject configuration of the AY-system:

$$a_{2.ij} = \begin{cases} 0 & \text{for water surface} \\ 1 & \text{for land} \\ 2 & \text{for the combined territory} \end{cases}$$

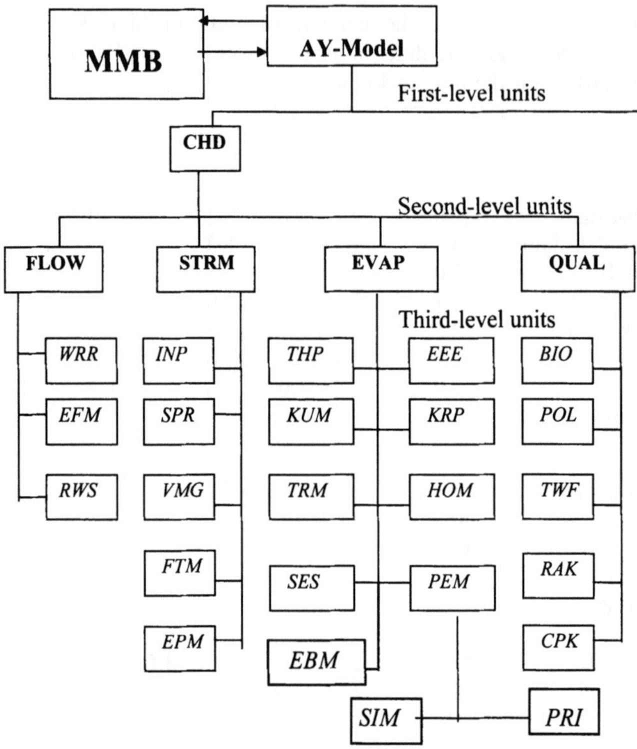

Figure 5.4. The AY-model structure. See Table 5.21 for an explanation of the abbreviated units.

Identifier $A_3 = \|a_{3,ij}\|$ determines the types of water elements of the AY-system:

$$a_{3,ij} = \begin{cases} 0 & \text{river has no tributaries} \\ 1 & \text{river has tributaries} \\ 2 & \text{reservoir without tributaries} \\ 3 & \text{reservoir with tributaries} \\ 4 & \text{natural river broadening with tributaries} \\ 5 & \text{natural river broadening without tributaries} \end{cases}$$

Identifier $A_4 = \|a_{4,ij}\|$ determines the types of soil–plant formations:

$$a_{4,ij} = \begin{cases} 0 & \text{arctic desert} \\ 1 & \text{tundra} \\ 2 & \text{northern taiga forest} \\ 3 & \text{mid-taiga forest} \\ 4 & \text{southern taiga forest} \\ 5 & \text{mountain tundra} \\ 6 & \text{mountain thinly planted northern taiga forest} \\ 7 & \text{mountain mid- and southern taiga forest} \\ 8 & \text{anthropogenic landscape} \end{cases}$$

Table 5.21. List of units of the AY-model whose scheme is presented in Figure 5.4. From Bras (1990), Krapivin and Kondratyev (2002), and Kondratyev *et al.* (2002, 2003).

Identifier of the unit	Description of the unit
CHD	Control of the parameterization of hydrophysical and hydrological processes in the AY-system
CON	Controlling interface with functions of service of the man–machine dialogue when realizing simulation experiments
FLOW	Model of the river runoff
STRM	Simulation procedure that reflects the water masses motion in the AY-system
EVAP	The choice of the model to parameterize the process of evaporation
QUAL	The choice of the criterion to assess the water quality
WRR	Model of the water regime in a water body
EFM	An empirical model of the river runoff
RWS	Model of spreading the river runoff beyond the river bed
INP	Model of infiltration
SPR	Model of the sink taking into account the effect of vegetation and soil cover
VMG	Model of the vertical uplifting of groundwater in evaporation, feeding, and ex-filtration
FTM	Model of filtration
EPM	An empirical model of precipitation
THP	A specified model of transpiration
RUM	The Kuzmin model (Kuzmin, 1957)
TRM	A simple model of transpiration
SES	Model of snow melting and evaporation from snow surfaces
EBM	Parameterization of evaporation using the method of energy balance
PEM	The Penman model (Penman, 1948)
PRI	The Penman model as specified by Priestley (Bras, 1990)
KRP	The Penman model as specified by Kohler and Richards (1962)
SIM	A simplified version of the Penman model
EEE	Empirical models of evaporation
KOH	The Kohler model (Bras, 1990)
HOM	The Horton model (Bras, 1990)
BIO	A simulation model of sedimentation and biological assimilation of pollutants
POL	A simulation model of the formation of anthropogenic fluxes of pollutants
TWT	Model of the process of the water temperature formation
RAK	Model of kinetics of the radionuclide pollution of water
CPK	Model of kinetics of the chemical pollution of water
COD	Database correction
SCE	The choice and formation of scenarios for simulation experiments
VIP	Visualization of the simulation experiment results
AAS	Adaptation of the AY-model to the scenario of the simulation experiment
IAS	Referencing the AY-model to the space and timescales

Identifier $A_5 = \|a_{5,ij}\|$ forms the field of the sources of pollution:

$$a_{5,ij} = \begin{cases} 0 & \text{pollution sources do not exist on the territory } \Omega_{ij} \\ 1 & \text{anthropogenic source of pollutants exists on } \Omega_{ij} \\ 2 & \text{natural source of pollutants exists on } \Omega_{ij} \end{cases}$$

By analogy with the previous ones, identifier $A_6 = \|a_{6,ij}\|$ enables one to introduce the types of chemicals in the sources of the AY-system's pollution.

Since the AY-model permits various modifications of the applied methods of modelling the hydrophysical, biogeochemical, and hydrological processes, the user can chose among them with the help of identifier A_7. Use of the concrete method is referenced to geographic coordinates, which permits planning of the structure of the numerical experiment in order to combine various approaches to the description of the pollution dynamics at various sites of the territory Ω. Various modifications are important both for an adequate use of the database and for combination of various methods of modelling individual elements of the AY-system. The internal structure of all identifiers is formed in the dialogue regime and its examination is one of the functions of the CON unit.

The computer image of the AY-system formed by a set of identifiers $\{A_i\}$ is a set of images of the geographical cells Ω_{ij}. Thus the functioning of the hydrological network in the territory Ω and formation of distribution of pollutants in water can be presented as an interaction of the cells Ω_{ij}. Each cell Ω_{ij} is described by a standard block-diagram of the regional water balance shown in Figure 6.14.

Each territory Ω_{ij} is characterized by the presence of part of the river network, water basins, and land sites. So that in general, the territory Ω_{ij} of an area σ_{ij} is divided into a water part of area $p_{ij}\sigma_{ij}$ and a land part of area $(1 - p_{ij})\sigma_{ij}$. With this in mind, the closed system of the balanced equations of the AY-model is as follows:

$$\sigma_{ij}\partial W_{Aij}/\partial t + \xi_\varphi \partial W_{Aij}/\partial\varphi + \xi_\lambda \partial W_{Aij}/\partial\lambda = V_{ij} - B_{ij}\sigma_{ij} + D_{ij} + L_{ij} + T_{ij} \tag{5.1}$$

$$\begin{aligned}\partial C_{ij}/\partial t + \mu_{ij}\partial C_{ij}/\partial x = p_{ij}B_{ij} + [K_{ij} - V_{ij} - U_{ij} - F_{ij} \\ - M_{ij} - H_{ij} + J_{ij}]/(p_{ij}\sigma_{ij}) + \Gamma_{ij}\end{aligned} \tag{5.2}$$

$$\begin{aligned}(1 - p_{ij})\sigma_{ij} d\Phi_{ij}/dt = FUM_{ij} - K^*_{ij} - P_{ij} + N_{ij} - T_{ij} \\ - L_{ij} + W_{Aij}(1 - p_{ij})\sigma_{ij}\end{aligned} \tag{5.3}$$

$$\sigma_{ij}\partial G_{ij}/\partial t + \rho_{ij\varphi}\partial G_{ij}/\partial\varphi + \rho_{ij\lambda}\partial W_{Aij}/\partial\lambda = H_{ij} + P_{ij} - N_{ij} - J_{ij} - D_{ij} \tag{5.4}$$

where:

$$FUM_{ij} = \begin{cases} 0 & \text{for } a_{2,ij} \neq 2 \\ F_{ij} + U_{ij} + M_{ij} & \text{for } a_{2,ij} = 2 \end{cases}$$

$$K^*_{ij} = \begin{cases} 0 & \text{for } a_{2,ij} = 1 \\ K_{ij} & \text{for } a_{2,ij} = 2 \\ [K_{ij-1} + K_{ij+1}]/2 & \text{for } a_{2,ij} = 0 \text{ and } a_{2,ij-1} = a_{2,ij+1} = 1 \\ [K_{i-1j} + K_{i+1j}]/2 & \text{for } a_{2,ij} = 0 \text{ and } a_{2,i-1j} = a_{2,i+1j} = 1 \end{cases}$$

p_{ij} is the share of the area σ_{ij} covered with water; x is the axis of coordinates along the river bed with the stream; Γ_{ij} is the contribution of the springtime melting of snow accumulated on the surface; $\rho(\rho_\varphi, \rho_\lambda)$ is the velocity of movement of subsurface waters; ξ_φ and ξ_λ are projections of wind speed onto latitude and longitude, respectively; and μ_{ij} is the current velocity in Ω_{ij}.

In the form shown above the synthesis of the AY-model of partial models is realized through boundary conditions for variables W_A, C, and G. The transport of soil moisture Φ between the cells Ω_{ij} is considered to be absent. The functional presentations for all currents of the scheme of the water balance of the territory are written based on the laws of hydrodynamics and with available experimental information taken into account. The empirical units EPM and EFM of the AY-model, based on data series $B(\varphi, \lambda, t)$ and $K(\varphi, \lambda, t)$, calculate the functions $B(\varphi, \lambda, i)$ and $K(\varphi, \lambda, t)$. The calculation algorithms are based on methods of interpolation. In general, the model of the formation of the runoff should take into account the watershed topography and the spatial distribution of its soil characteristics as well as special features of vegetation cover. As an approximation of the function K_{ij} we take the following dependence:

$$K_{ij}(\varphi, \lambda, t) = \begin{cases} (q_{ij} - k_{ij}[1 - \exp\{-a_{ij}X_{ij} - c_{ij}S_{ij}\}])\sigma_{ij} & \text{for } t_1 \leq t \leq t_2 \\ (t_4 - t_3)^{-1}\int_{t_2}^{t_3} (1 - p_{ij})\sigma_{ij}B_{ij}(\varphi, \lambda, t)\, dt & \text{for } t_3 \leq t \leq t_4 \\ 0 & \text{for } t_2 < t < t_3 \end{cases} \tag{5.5}$$

where X_{ij} and S_{ij} are, respectively, the density of vegetation (m/km^2) and the thickness of soil layer (m) over the area σ_{ij}; q_{ij} is the coefficient of the relief runoff; k_{ij} is the coefficient of penetration of water through vegetation and soil cover; a_{ij} and c_{ij} are coefficients of holding precipitation by vegetation and soil cover; t_1 is the onset of summer period; t_2 and t_3 are the beginning and the end of winter; and t_4 is the end of the period of snow melting.

Model (5.5) needs data on the vegetation cover and characteristics dependent on it. It is the function of the SPR unit, so that during the winter period $[t_2, t_3]$ the snow is accumulated, during the spring period $[t_3, t_4]$ it melts uniformly, and only in the summer $[t_1, t_2]$ are the role of soil–plant formations in the formation of the surface run-off due to precipitation taken into account.

To parameterize the processes of evaporation and transpiration, numerous theoretical and empirical methods are known, and the choice of each method is determined by the structure and content of the available database (Kondratyev *et al.*, 2002). In general, evaporation from the surface of a water basin or land is parameterized with the method chosen by the user. The freedom of choice is determined by the list of units in Table 5.21 responsible for modelling the fluxes L, V, and T. The EBM unit uses the method of energy balance:

$$V = Q_e/(\rho L_e) \qquad Q_e = [Q^* + (Q_v - Q_o)]/(1 + R) \tag{5.6}$$

where $Q^* = Q_S - Q_r + Q_a - Q_{ar} - Q_{bS}$; Q_S is the incoming solar radiation (kcal/m^2/day); q_r is the reflected solar radiation; Q_a is the incoming long-wave radiation; Q_{ar} is the reflected long-wave radiation; Q_{bS} is the long-wave radiation emitted by water masses at their temperature; Q_v is the advective energy; Q_o is the accumulated energy; ρ is water density (kg/m^3); L_e is the latent heat of evaporation (kcal/kg); $R = Q_h/Q_e$; and Q_h is the heat removed from the system to the atmosphere.

To calculate the coefficient R we use two formulas:

$$R = C_{pa}P_aK_h(T_{w,S} - T_{w,z})/\{0.622L_eK_m(e_{w,S} - e_{w,z})\} \tag{5.7}$$

or

$$R = 0.61 \times 10^{-3}P_a(T_{w,S} - T_{w,z})(e_{w,S} - e_{w,z}) \tag{5.8}$$

where $T_{w,S}$ is the average sea surface temperature (SST); $T_{w,h}$ is the average air temperature at height h; $e_{w,S}$ is the saturated vapour pressure at the temperature $T_{w,S}$; $e_{w,h}$ is the average vapour pressure at the height h; C_{pa} is the characteristic heat of the atmosphere at constant pressure; K_h and K_m are coefficients of the vortex diffusion for heat and momentum, respectively ($K_h/K_m = 0.92$); and P_a is the atmospheric pressure on the water surface.

The PEM unit uses the Penman formula (Bras, 1990):

$$V = (\rho L_e)^{-1}[\Delta Q^*/\gamma + L_eB_p(e_{Sh} - e_h)](\Delta/\gamma + 1)^{-1} \tag{5.9}$$

where:

$$B_p = 0.622k^2(K_w/K_m)(\rho_a/P_a)U_h/\ln^2(h/h_o) \tag{5.10}$$

h_o is the surface roughness (m); U_h is the wind speed at height h; k is the constant ($k \approx 0.4$); $K_w/K_m \approx 1.13$; e_{Sh} is the saturated vapour pressure at height h at temperature T_h; Δ/γ is an empirical coefficient (Table 5.22); and $Q^* = Q_e - Q_h$. The parameter h_o is either given in the database as a function of time and space, or is calculated with the formula: $h_o = \alpha(U^*)^2/g$, where $a \approx 0.009$, g is the gravity multiplier, and U^* is the friction rate ($h_o = 0.001$ for $U_h < 2.25$ m/s and $h_o = 0.005$ for $U_h \geq 2.25$ m/s (Table 5.23)).

Model (5.9) can be simplified to the Priestley model:

$$V = 1.63 \times 10^{-9}(0.93 + U_2)(e_S - e_2) \tag{5.11}$$

where U_2 is the wind speed at height $h = 2$ m above the water surface (m/s); and e_S and e_2 are measured in mbar.

Similar to Equation (5.11), the SIM unit uses a simplified model:

$$V = \kappa U - U_2(e_S - e_2) \tag{5.12}$$

where the coefficient κ is estimated in Table 5.24.

Thus the PEM unit has internal modifications, the choice of each of them being made by the user of the AY-model through the formation of a scenario. Via the information interface the AY-model asks the user about the type of the Penman model in accordance with the mnemonics given in Table 5.24.

Other parameterizations of the process of evaporation, such as formulas by Horton (the HOM unit) and Power (the ROW unit) have been described by Bras

Table 5.22. Estimates of the temperature coefficient Δ/γ.
From Bras (1990).

T (°C)	Δ/γ	T (°C)	Δ/γ	T (°C)	Δ/γ	T (°C)	Δ/γ
0.0	0.67	11.0	1.30	22.0	2.38	33.0	4.12
0.5	0.69	11.5	1.34	22.5	2.45	33.5	4.22
1.0	0.72	12.0	1.38	23.0	2.51	34.0	4.32
1.5	0.74	12.5	1.42	23.5	2.58	34.5	4.43
2.0	0.76	13.0	1.46	24.0	2.64	35.0	4.53
2.5	0.79	13.5	1.50	24.5	2.71	35.5	4.64
3.0	0.81	14.0	1.55	25.0	2.78	36.0	4.75
3.5	0.84	14.5	1.59	25.5	2.85	36.5	4.86
4.0	0.86	15.0	1.64	26.0	2.92	37.0	4.97
4.5	0.89	15.5	1.68	26.5	3.00	37.5	5.09
5.0	0.92	16.0	1.73	27.0	3.08	38.0	5.20
5.5	0.94	16.5	1.78	27.5	3.15	38.5	5.32
6.0	0.97	17.0	1.82	28.0	3.23	39.0	5.45
6.5	1.00	17.5	1.88	28.5	3.31	39.5	5.57
7.0	1.03	18.0	1.93	29.0	3.40	40.0	5.70
7.5	1.06	18.5	1.98	29.5	3.48	40.5	5.83
8.0	1.10	19.0	2.03	30.0	3.57	41.0	5.96
8.5	1.13	19.5	2.09	30.5	3.66	41.5	6.09
9.0	1.16	20.0	2.14	31.0	3.75	42.0	6.23
9.5	1.20	20.5	2.20	31.5	3.84	42.5	6.37
10.0	1.23	21.0	2.26	32.0	3.93	43.0	6.51
10.5	1.27	21.5	2.32	32.5	4.03	43.5	6.65

(1990). There is a Kohler model, which is a modification of the Dalton law (the KOH unit):

$$L = (e_S - e_a)^{0.88}(0.42 + 0.0029U_p) \quad \text{(mm/day)} \tag{5.13}$$

where U_p is the wind speed at a height 1.5 m above the evaporating surface (km/day).

The model by Kohler and Richards specifies the PEM model to calculate L (the KRP unit):

$$\Delta = (0.00815T_a + 0.8912)^7 \qquad \text{for } T_a \geq -25°\text{C} \tag{5.14}$$

$$\Delta/(\Delta + \gamma) = [1 + 0.66/(0.00815T_a + 0.8912)^7]^{-1}$$

$$e_S - e_a = 33.86[(0.00738T_a + 0.8072)^8 - (0.00738T_d + 0.8072)^8] \qquad \text{for } T_d \geq -27°\text{C}$$

$$Q^* = 7.14 \times 10^{-3}Q_S + 5.26 \times 10^{-6}Q_S(T_a + 17)^{1.87} + 3.94 \times 10^{-6}Q_S - 2.39 \times 10^{-9}(Q_S)^2(T_a - 7.2)^2 - 1.02$$

where T_{is} is the average daytime atmospheric temperature (°C); and T_d is the average temperature of condensation (°C).

Table 5.23. Sea surface roughness z_o as a function of the type of surface and wind speed.
From Bras (1990).

Surface	Wind speed U_z at $z = 2\,\mathrm{m}$ (m/s)	Z_o (cm)
Open water	2.1	0.001
Calm dirty surface	2.1	0.001
Calm snow-covered low grass	2.1	0.005
Bare soil	1.8	0.02
Snow-covered steppe	1.8	0.1
Bushes (135 cm)	6.2	14.0
Coniferous forest		
5 m	6.2	65.0
27 m	6.2	300.0
Mowed grass		
1.5 cm	1.8	0.2
3.0 cm	1.8	0.7
4.5 cm	2.0	2.4
4.5 cm	6-8	1.7
High grass		
60–70 cm	1.5	9.0
60–70 cm	3.5	6.1
60–70 cm	6.2	3.7
Broadleaved forest		
17 m	6.2	270.0

Table 5.24. Empirical estimates of the coefficient κ in Equation (5.12).
From Bras (1990).

Mnemonics of the PEM unit	Estimate of κ
$M1$	1.523×10^{-9}
$M2$	1.16×10^{-9}
$M3$	1.572×10^{-9}
T	2.37×10^{-9}
E	1.326×10^{-9}
MD	1.59×10^{-9}

The AY-system functions for a long time at low temperatures, when land is covered with snow, and the water surface is covered with ice. Therefore, to complete the AY-model, we include the KUM unit, which functions as the model of evaporation from snow, proposed by Kuzmin (1957):

$$V = L = (0.18 + 0.098 U_{10})(e_o - e_2) \quad (\mathrm{mm/day})$$

The T flux is parameterized in the TRM and THP units. The TRM unit does not take into account dependences of transpiration on the type of vegetation cover. A more detailed description of the process of transpiration is connected with the consideration of the scheme of water movement in plants. For trees, this process starts with water assimilation by roots from the soil, which then moves via the vessels of the xylem of the stem of tree to leaves and via leaves, as vapour, it enters the atmosphere. Part of the water is spent on the growth of the plant biomass. A similar part of the pollutant is considered to remain in the formed wood (the THP unit):

$$T = \begin{cases} 0 & \text{for } t \in [t_2, t_3] \\ \min\{fP_r, (1 - p_{ij}\Phi)\} & \text{for } t \notin [t_2, t_3] \end{cases}$$

where P_r is the vegetation produce; and f is the rate of water consumption per unit produce.

The model of infiltration (the INP unit) is described by the Darcy simplified equation (Bras, 1990):

$$P = \begin{cases} \Phi(1 - p_{ij})\sigma_{ij}/\psi & \text{for } t \notin [t_2, t_3] \\ 0 & \text{for } t \in [t_2, t_3] \end{cases}$$

where ψ is the time of complete water infiltration into soil at fixed porosity and temperature.

To parameterize the fluxes J, N, and D, we use simplified models (the VMG unit):

$$J_{ij} = \begin{cases} 0 & \text{when } G_{ij} \le G_{ij,\max} \\ \theta_j(G_{ij} - G_{ij,\max}) & \text{when } G_{ij} > G_{ij,\max} \end{cases}$$

$$N_{ij} = \begin{cases} 0 & \text{for } t \in [t_2, t_3] \\ 0 & \text{when } G_{ij} \le G_{ij,\max} \\ \theta_N(G_{ij} - G_{ij,\max}) & \text{when } G_{ij} > G_{ij,\max} \text{ and } t \notin [t_2, t_3] \end{cases}$$

$$D_{ij} = \begin{cases} 0 & \text{for } t \in [t_2, t_3] \\ 0 & \text{when } G_{ij} \le G_{ij,\max} \\ \theta_D(G_{ij} - G_{ij,\max}) & \text{when } G_{ij} > G_{ij,\max} \text{ and } t \notin [t_2, t_3] \end{cases}$$

where θ_J, θ_N, and θ_D are empirical coefficients; and $G_{ij,\max}$ is the critical level of groundwaters.

Other elements of the water balance of the AY-system are described by the model of the hydrological regime of the territory presented in Kondratyev *et al.* (2002, 2003).

The territory Ω_{ij} may contain a sector of a river or a part of a water basin. We assume that the discrete grid $\Delta\varphi \times \Delta\lambda$ excludes a possibility of combinations of various elements of the AY-system in one cell Ω_{ij}. Considering in a first approximation the AY river waterway as the base one, we assume that at each site Ω_{ij} the water medium is vertically homogeneous. Then any sector of the river or water reservoir can be attributed to one class of elements of the AY-system, differing only in current velocity. All tributaries Y_{ij} are considered as the prescribed functions of time, and in Equation (5.2) they play the role of free parameters.

As seen from Equation (5.2), the level of water in Ω_{ij} is distributed along the x-axis. With:

$$Y_{ij} = \begin{cases} Y_{ij}^* & \text{when } a_{3,ij} = 1, 2, 5 \\ 0 & \text{when } a_{3,ij} = 0, 3, 4 \end{cases}$$

where the value Y_{ij}^* is kept in the database.

Calculations begin with Ω_1 under the initial condition $C_1(t, 0) = C_{1,0}$. For Ω the initial condition is determined by the value $C_s = C_{s-1}(t, x_{s-1})$, where x_{s-1} is the maximum distance with the stream, from entrance to exit from Ω_{s-1}.

In a general case, at a site $\Omega_k (k = 1–N)$ the vertical structure of the river basin is made up of snow, ice, and water of thicknesses g_k, r_k, and C_k. An accumulation g_k is connected with the flux B:

$$dg_k/dt = \begin{cases} 0 & \text{for } t \notin [t_2, t_3] \\ B & \text{for } t \in [t_2, t_3] \end{cases}$$

so that the term Γ_k in Equation (5.2) plays the role of a buffer:

$$\Gamma_k = \begin{cases} 0 & \text{for } t \in [t_1, t_3] \\ p_k \sigma_k g_k & \text{for } t_3 < t \leq t_3 + \Delta t \end{cases}$$

where Δt is the period of the springtime melting of snow (days).

Finally, describe the units of the AY-model simulating the fluxes of pollutants in the territory Ω. Consider concrete types of pollutants: radionuclides, heavy metals, and oil hydrocarbons. Each of them behaves specifically in water. The dynamics of radionuclides in the AY system is determined by its hydrological regime (i.e., by their transport by the current, accumulation in bottom deposits, in ice, living biomass, by washing-out and input with water flows, interaction with the atmosphere, and anthropogenic activity). A variety of the types of radionuclides are described with the parameter ν. Each type of ν is characterized by the half-life period τ^ν. Let us denote the concentration of radionuclide ν in water $Q_\nu (nCi/\text{m}^3)$, in a soil $E_\nu (nCi/\text{m}^2)$, in bottom sediments $\Xi_\nu (nCi/\text{m}^2)$, and in groundwater $S_\nu (nCi/\text{m}^2)$. With these notations and the structure of the hydrological network of the AY-system, equations of the dynamics of radionuclide ν are written as follows:

$$\begin{aligned} \partial Q_{\nu,k}/\partial t + \mu_k Q_{\nu,k}/\partial x = {} & [H_{\nu,1} + H_{\nu,2} + H_{\nu,3} + (1 - b_1)H_{\nu,8} - H_{\nu,4} - H_{\nu,5} \\ & + H_{\nu,6} - H_{\nu,7} + H_{\nu,12}]/\Delta - Q_{\nu,k} \ln 2/\tau^\nu \end{aligned} \tag{5.15}$$

$$\begin{aligned} \partial E_{\nu,k}/\partial t = {} & [b_1 H_{\nu,8} + H_{\nu,4} + H_{\nu,9} - H_{\nu,10}]/[(1 - \rho_k)\sigma_k] \\ & - E_{\nu,k} \ln 2/\tau^\nu + \mu_\Phi H_{\nu,12}/\sigma_k \end{aligned} \tag{5.16}$$

$$\partial \Xi_{\nu,k}/\partial t = (H_{\nu,5} - H_{\nu,6})/(\rho_k \sigma_k) - \Xi_{\nu,k} \ln 2/\tau^\nu \tag{5.17}$$

$$\partial S_{\nu,k}/\partial t + \rho_k \partial S_{\nu,k}/\partial x = [H_{\nu,10}/(1-p_k) + H_{\nu,7}/p_k - H_{\nu,2} - H_{\nu,9} - H_{\nu,11}]/\sigma_k - S_{\nu,k} \ln 2/\tau^{\nu}$$

where b_1 is the share of the anthropogenic ejection of radionuclides getting to soils; $H_{\nu,m}$ is the transport of radionuclide ν by the flux m from one medium to another (nCi/day), $m = 1$ – washing out from soil by the run-off K, $m = 2$ – input from groundwater with the flux J, $m = 3$ – inflow with tributaries, $m = 4$ – runoff with fluxes U, F and M, $m = 5$ – deposition due to gravity and with dead organisms of the ecosystem, $m = 6$ – washing-out from bottom deposits, $m = 7$ – outflow with the flux H, $m = 8$ – anthropogenic source, $m = 9$ – input to the soil from groundwater with the flux N, $m = 10$ – washing-out into groundwater by the flux P, $m = 11$ – deposition in groundwater and $m = 12$ – deposition with precipitation; $\Delta = C_{\cdot,k}\, p_k\, \sigma_k$ is the water volume in the river sector Ω_k (m^3); and $C_{\cdot,k}$ is the average depth of the site Ω_k (m).

The fluxes $H_{\nu,m}$ in Equations (5.15)–(5.18) are parameterized by the following models:

$$H_{\nu,1} = a_1 E_{\nu,k} K \qquad H_{\nu,2} = J\Xi_{\nu,k} \qquad H_{\nu,3} = (1-b_1)H_{\nu,8}$$

$$H_{\nu,4} = Q_k(U+F+M) \qquad H_{\nu,2} = m_1 Q_{\nu,k} + m_2 m_3 d_0 Q_{\nu,k}$$

$$H_{\nu,6} = p_k \sigma_k \mu_\xi \Xi_k \qquad H_{\nu,7} = HQ_{\nu,k} \qquad H_{\nu,9} = G^{-1} NS_{\nu,k}$$

$$H_{\nu,10} = A^{-1} PE_{\nu,k} \qquad H_{\nu,11} = \mu_0 S_{\nu,k} \qquad H_{\nu,12} = BC_{\nu,B}$$

where μ_0 is the coefficient of radionuclide deposition in groundwater (m^2/day); a_1 is the coefficient of the capture of radionuclides by water flow (m^{-1}); m_1 is the coefficient of gravitational deposition (m^3/day); m_2 is the coefficient of dying-off of a living organisms biomass (m^3/day); m_3 is the coefficient of the capture of radionuclides by biomass (m^3/day); d_0 is the concentration of the biomass of living organisms (μg/m^3); μ_ξ is the coefficient of washing-out of radionuclides from bottom deposits (day^{-1}); and $C_{\nu,B}$ is the content of radionuclide ν in precipitation (nCi/m^3).

Equations (5.15)–(5.18) form the basis of the RAK unit, which realizes calculations of the dynamics of radionuclides in the whole AY-system with compartment structure of the territory $\Omega = \{\Omega_{ij}\}$ taken into account. To simplify the control of this process and to introduce into the AY-model the required modifications, the structure of the QUAL unit has subunits to simulate the anthropogenic flux $H_{\nu,8}$ (POL), the flux $H_{\nu,5}$ (BIO), and the process of the formation of water temperature (TWF).

The fluxes of heavy metals and oil hydrocarbons in the AY-system are described by similar relationships. Denote concentrations of dissolved and weighted fractions of heavy metals in river waters (bottom deposits) as ψ_w (ψ^*) and e_w (e^*), respectively. Then the dynamic equations of the movement of heavy metals in the AY-system can be described as follows:

$$\partial e_{w,k}/\partial t + \mu_k \partial e_{w,k}/\partial x = [H_{e,1} + H_{e,2} + H_{e,3} + (1 - b_2)H_{e,8} - H_{e,4} - H_{e,5} + H_{e,6} - H_{e,7} + H_{e,12}]/\Delta$$

$$\partial e_{S,k}/\partial t = [b_2 H_{e,8} + H_{e,4} + H_{e,9} - H_{e,10} + H_{e,12}]/[(1 - p_k)\sigma_k]$$

$$\partial \psi_{w,k}/\partial t + \mu_k \partial \psi_{w,k}/\partial x = [H_{\psi,1} + H_{\psi,2} + H_{\psi,3} + (1 - b_2)H_{\psi,8} - H_{\psi,4} + H_{\psi,6} - H_{\psi,7} + H_{y,12}]/\Delta$$

$$\partial \psi_{S,k}/\partial t = [b_2 H_{\psi,8} + H_{\psi,4} + H_{\psi,9} - H_{\psi,10} + H_{\psi,12}]/[(1 - p_k)\sigma_k]$$

$$\partial e_{*,k}/\partial t = (H_{e,5} - H_{e,6})/(p_k \sigma_k)$$

$$\partial \psi_{*,k}/\partial t = (H_{\psi,5} - H_{\psi,6})/(p_k \sigma_k)$$

$$\partial e_{g,k}/\partial t + \rho_k \partial e_{g,k}/\partial x = [H_{e,10}/(1 - p_k) + H_{e,7}/p_k - H_{e,2} - H_{e,9} - H_{e,11}]/\sigma_k$$

$$\partial \psi_{g,k}/\partial t + \rho_k \partial \psi_{g,k}/\partial x = [H_{\psi,10}/(1 - p_k) + H_{\psi,7}/p_k - H_{\psi,2} - H_{\psi,9} - H_{\psi,11}]/\sigma_k$$

where ψ_S (e_S) and ψ_S (e_S) are concentrations of dissolved (weighted) fractions of heavy metals in soils and groundwater, respectively; $H_{\kappa,m} (\kappa = S, g)$ are fluxes of heavy metals similar to the fluxes $H_{\nu,m} (m = 1\text{–}12)$; and b_2 is the share of the anthropogenic heavy metals getting into the soil.

Oil hydrocarbons O can be observed in a soil (O_S), river water (O_w), bottom deposits (O_*) and in groundwater (O_g). Processes of their transformation and movement between various media include dissolving $(H_{O,\kappa,1})$, evaporation $(H_{O,\kappa,2})$, deposition as droplets (clusters) $(H_{O,\kappa,3})$, oxidation $(H_{O,\kappa,4})$, adsorption by weighted organic substance $(H_{O,\kappa,5})$, biosedimentation $(H_{O,\kappa,6})$, and bacterial decay $(H_{O,\kappa,7})$:

$$\partial O_{w,k}/\partial t + \mu_k \partial O_{w,k}/\partial x = \left[H_{O,w,8} + H_{O,w,9} - \sum_{i=1}^{7} H_{O,w,i} \right] \Big/ \Delta$$

$$\partial O_{S,k}/\partial t = [H_{O,S,8} - H_{O,S,9} + H_{O,S,12} - H_{O,S,15}]/[(1 - p_k)\sigma_k]$$

$$\partial O_{*,k}/\partial t = [H_{O,w,3} - H_{O,w,4} - H_{O,w,7} - H_{O,w,13}]/(p_k \sigma_k)$$

$$\partial O_{g,k}/\partial t + \rho_k \partial O_{g,k}/\partial x = [H_{O,S,15} + H_{O,w,14} - H_{O,g,16} - H_{O,g,1} - H_{O,g,3} - H_{O,g,4} - H_{O,g,7}]/\sigma_k$$

where $H_{O,w,8}$ and $H_{O,S,8}$ are anthropogenic fluxes of oil hydrocarbons into river waters and soils, respectively; $H_{O,S,9}$ is washing out of oil hydrocarbons from soil by runoff K; $H_{O,g,10}$ is the inflow of oil pollution from groundwater with the flux J; $H_{O,w,11}$ is the inflow with the tributaries of the AY-system; $H_{O,w,12}$ is the runoff with the fluxes U, F, and M; $H_{O,*,13}$ is the washing out from bottom sediments; $H_{O,w,14}$

is the outflow with the flux H; $H_{O,S,15}$ is the washing out into groundwater with the flux P; $H_{O,g,16}$ is the return of oil pollution into the soil from groundwater due to the flux N.

To verify the AY-model, the Russian–American ecological expedition was carried out in the summer of 1996 (Phillips *et al.*, 1997). Some results of the expedition are given in Tables 5.19, 5.20, and 5.25 (Krapivin *et al.*, 1998). The results of analysis of the radionuclide pollution are given in Table 5.20. The results of the laboratory elemental study of water samples are given in Tables 5.19 and 5.25.

The anthropogenic radioisotope ^{137}Cs was detected in all samples taken upstream of the dam of the Irkutsk water reservoir. The concentration of radioisotope ^{137}Cs varied within 2 to 12 Bq/kg (dry weight). These values agree with background levels, which can be expected from global processes. Down stream of the Irkutsk dam, near the cities of Angarsk and Irkutsk, the measured concentrations of ^{137}Cs vary from <4 to 30 Bq/kg, which points to exceeded background global levels of ^{137}Cs. Near Bratsk, only the background level of ^{137}Cs in Angara was observed. Up stream of Angara, from its inflow into the Yenisey, the levels of ^{137}Cs constituted about 2 Bq/kg of dry deposit. Downstream of the confluence of the Angara and Yenisey the concentration of ^{137}Cs varies from 3 to 27 Bq/kg. A 10–15-fold increase of the background level is observed in the region of the Strelka settlement. Concentrations of ^{137}Cs, ^{152}Eu, ^{60}Co, substantially exceeding the background levels, were observed in the Yenisey downstream of its confluence with the Angara: from 9 to 240 Bq/kg for ^{60}Co, from 14 to 400 Bq/kg for ^{137}Cs, and from 7 to 150 Bq/kg for ^{152}Eu. Apparently, such variations in the radionuclide pollution of the Yenisey are caused by anthropogenic sources in Krasnoyarsk, which is located about 270 km upstream of Strelka.

With respect to the levels of pollution of the AY-system with heavy metals, the expedition obtained the following estimates. Five heavy metals (Cd, Cr, Cu, Ni, Pb) have concentrations at background levels, two metals (As, Zn) exceed the background concentrations by 4–5 times. Table 5.25 compares concentrations of heavy metals. It shows that the pollution of the AY-system with heavy metals is increasing.

The results of the expedition and the archived data of the Irkutsk Scientific Centre of the Russian Academy of Sciences enabled the use of the AY-model to carry out several simulation experiments, to calculate the fluxes of radionuclides, heavy metals, and oil hydrocarbons from AY-system into the Arctic Basin. The water balance of the AY-system was calculated with the data given in Figure 5.5. It is assumed that the vertical distribution of pollutants in river water is uniform, pollutants in soil and plants are absent, and the processes of atmospheric cleaning follow an exponential law with the characteristic time τ_a ($\tau_a = 10$ days for $t \in [t_1, t_2]$, $t_a = 20$ days for $t \in [t_2, t_4]$). It is also assumed that the sources of pollution located in the territory Ω, do not change their regime of generation, the function of distribution of these regimes being homogeneous at time intervals shown in Table 5.26. The values μ, ρ, and ξ have been determined as averages over a 3-year period from the data in Rovinskiy *et al.* (1995). The slope of the river bed of the AY-system is assumed to be 0.21 m/km. The ecological elements of the river medium have the biomass $\beta(t)$ constant over each time interval $[t_i, t_{i+1}](i = 1, 2, 3)$. The equation for

Table 5.25. Comparison of results (ppm) of the laboratory analysis of the 1996 expedition materials on the Angara water quality.

From Analytical Services Center of Ecology and Environment, Inc., New York.

Type of Data	Data of the Irkutsk ecological service on the Angara water quality control						Results of water samples analysis	
	1978		1979		1980			
	*T*1	*T*2	*T*1	*T*2	*T*1	*T*2	August 1996	
The Ershov water collector (upstream of the Irkutsk dam)							Map ID#-I15	Map ID#-I11
Mg	3.2	3.0	3.3	3.1	3.4	3.3	3.5	4.9
Fe	na	na	0.01	nm	na	na	<0.05	<0.05
As	na	na	na	na	na	na	0.0063	0.012
Zn	na	na	na	na	na	na	0.15	0.12
Cu	0.002	0.002	0.003	0.002	0.001	0.003	<0.02	<0.02
Mn	0.001	0.001	0.003	0.005	0.003	0.006	<0.01	<0.01
Al	0.005	0.005	0.018	0.008	0.004	0.007	<0.1	<0.1
Co	nm	na	nm	na	na	na	<0.02	<0.02
Ni	na	na	nm	0.003	0.001	0.002	<0.02	<0.02
V	na	na	na	na	na	na	<0.02	<0.02
Pb	na	na	na	na	na	na	<0.0057	<0.057
The water collector near the Sukhov station upstream of Angarsk							Map ID#-A14	
Mg	3.4	3.4	3.7	3.5	3.3	3.4	3.2	
Fe	0.02	0.3	0.07	0.001	na	na	<0.05	
As	na	na	na	na	na	na	0.012	
Zn	na	na	0.0053	na	na	na	0.24	
Cu	0.001	0.002	0.004	0.007	0.004	0.008	<0.02	
Mn	0.011	nm	0.018	0.007	0.012	0.011	0.028	
Al	na	0.015	0.02	0.028	0.013	0.012	<0.01	
Co	nm	na	0.004	0.002	na	na	<0.02	
Ni	nm	0.001	0.003	0.002	0.002	0.0005	<0.02	
V	na	na	nq	na	0.001	na	<0.02	
Pb	na	na	na	na	na	na	<0.0057	
The water collector 0.5 km downstream of the Bratsk dam							Map ID#-B2	
Mn	4.4	3.8	4.4	3.9	4.8	5.2	4.9	
Fe	na	na	0.02	0.13	na	na	<0.05	
As	na	na	na	na	na	na	0.016	
Zn	na	na	na	na	na	na	0.087	
Cu	na	na	na	na	na	na	<0.02	
Mn	na	na	0.004	0.001	0.002	0.001	<0.01	
Al	nm	0.002	na	0.002	0.003	nm	<0.1	
Co	na	na	na	na	na	na	<0.02	
Ni	na	na	na	na	na	na	<0.02	
V	nm	na	na	0.001	0.001	0.15	<0.02	
Pb	na	na	na	na	na	na	<0.0057	

Note: nm = no measurements; na = beyond measurement accuracy (sensitivity of devices); $T1 = [t_1, t_2]$; $T2 = [t_2, t_4]$.

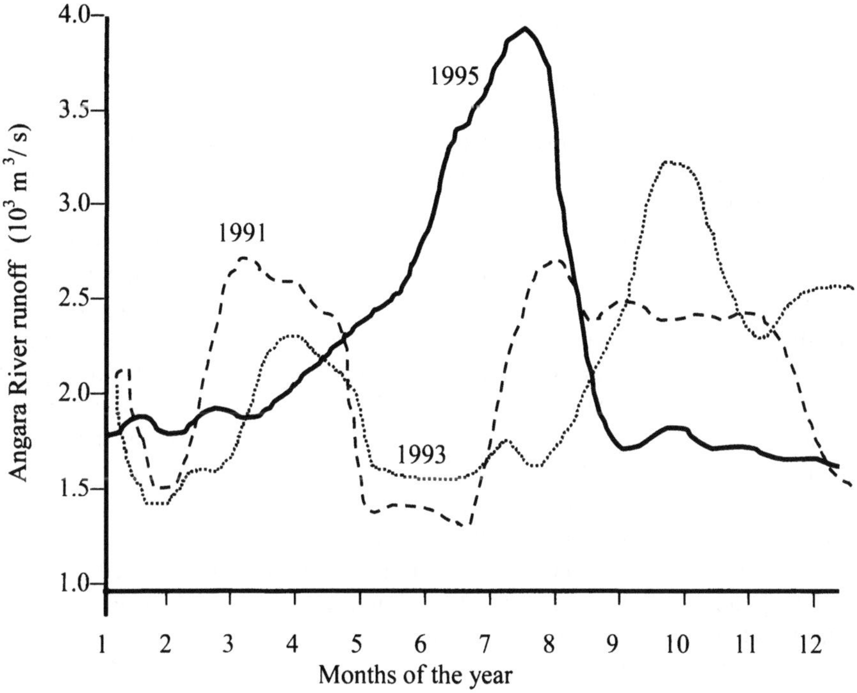

Figure 5.5. Estimates of the Angara River runoff through the Irkutsk dam (m3/s). Years are marked on the curves.

Table 5.26. Relative concentrations of ^{137}Cs in water (γ_w) and in bottom sediments (γ_d) normalized for values at $x = 0$.

Distance from Lake Baikal (km)	Time from the beginning of the simulation experiment (days)							
	30		60		90		120	
x	γ_w	γ_d	γ_w	γ_d	γ_w	γ_d	γ_w	γ_d
0	1.0	1.0	1.0	1.0	1.0	1.0	1.0	1.0
250	8.2	20.4	6.7	18.7	7.4	17.8	8.3	17.1
500	9.5	14.4	9.4	12.3	9.3	15.3	8.4	14.9
1,000	10.3	4.1	8.8	9.8	7.8	11.6	7.9	10.3
1,500	9.7	3.0	10.2	5.2	6.3	8.5	4.8	7.6
2,000	8.9	2.9	6.9	2.9	4.9	4.1	4.5	2.8
2,500	6.5	2.3	5.9	2.2	3.1	2.4	3.1	1.8
3,000	3.2	1.9	2.7	1.7	2.9	1.9	3.0	1.4
3,500	2.5	1.6	1.8	1.1	1.8	1.2	2.4	1.1
4,000	1.7	1.1	1.6	0.9	1.8	1.2	1.4	1.1

the flux $H_{v,5}(v = e, \psi)$ is written as:

$$H_{v,5} = [g_1 + m_1 m_2 \beta] Q_{v,k} \Delta_k$$

where g_1 is the coefficient of gravity deposition ($\approx 0.05\,\text{day}^{-1}$), m_1 is the indicator of mortality ($0.01\,\text{day}^{-1}$); and m_2 is the indicator of the impact of pollution on mortality ($\approx 0.03\,\text{kg}^{-1}$).

Table 5.26 exemplifies the use of the AY-model to estimate the distribution of ^{137}Cs in Ω, where the cells Ω_i are located along the river bed with a digitizing step of $\Delta x = 10\,\text{km}$. The distribution of ^{137}Cs has one maximum with changing amplitudes and locations. A similar pattern is observed for other radionuclides from Table 5.20. This result cannot be obtained with simple consideration of the location of the sources of the radionuclide pollution. The observed variations of the amplitude and location of the maximum of ^{137}Cs distribution are explained by the intensive water mixing in the AY-system and by the presence of reservoirs and vortices. As a result of a combination of the effects of non-uniform turbulence, conditions are created for accumulation of pollutants in local bottom sediments and their spontaneous washing out when the pollutant's concentration exceeds a critical level. Calculations took into account the sailing directions on the Angara.

The simulation experiment shows that the inflow of radionuclides into the Kara Sea is persistent with the dispersion $\pm 32\%$. The role of biota of the AY-system in the transformation of the radionuclide pollution is negligible and constitutes below 3%. The contribution of biota into the process of radionuclides sedimentation varies depending on seasons from 0.1 to 0.7%.

Figure 5.6 shows the distribution of concentrations of heavy metals calculated with the use of the AY-model along the AY-river system, from Lake Baikal to the

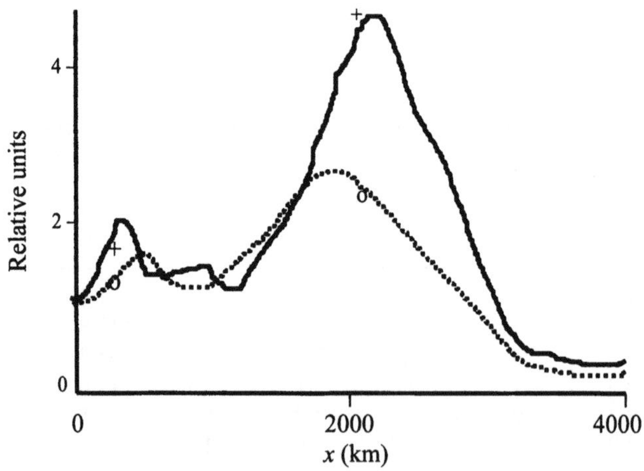

Figure 5.6. Distribution of the concentration of heavy metals in water (dashed line) and bottom sediments (solid line) as a function of the distance x from Lake Baikal. The signs O and + correspond to concentrations of heavy metals in river water and bottom sediments, respectively, measured in the American–Russian expedition in 1996.

Kara Sea. It is seen that there are three maxima in this distribution, located at distances 200, 1200, and 2000 km from Baikal. A comparison of this distribution with that of the sources of heavy metals suggests the conclusion that the AY-system neutralizes about 70–75% of its pollution with heavy metals at a distance of 600–1000 km from their source. The location of maxima is a function of the current velocity. For the data in Figure 5.5 the shifting of the maxima can extend over 150 km. The river system protects the Kara Sea waters from pollution with heavy metals so that their concentration at the Yenisey mouth does not exceed 2% of the nearest maximum (see Figure 5.6).

The AY-model makes it possible to assess the dependence of the level of pollution in the estuary of Yenisey as a function of anthropogenic activity. Suppose the intensity of the sources of heavy metals is such that their concentration in the water near the cities Angarsk, Irkutsk, Krasnoyarsk, Bratsk, and Ust'-Ilimsk is described by a stationary function, maintaining the concentration of heavy metals at a level h in each city mentioned above. The computer experiment shows that there is a persistent correlation between h, concentration of heavy metals in the estuary of the AY-system, and the velocity μ of its water flow. A 10% increase of h leads to a 2.5% increase of the inflow of pollutants to the Kara Sea. A 1% increase of μ causes a 0.7% increase of pollutants in the Kara Sea. These results are valid only with variations of h and μ around their averages. In critical situations, these results become unstable and, apparently, the AY-system requires more accurate parameterizations.

5.7.4 Simulation model of pollutant propagation in the Arctic Basin

An attempt has been made here to synthesize the accumulated knowledge in the form of a spatial simulation model describing the dynamics of the Arctic ecosystem pollution (SSMAE) to estimate the distributions of pollutants over the Arctic seas. The boundary of this basin is determined from its maximum configuration, including the Norwegian and Bering Seas. The model can be used to study the dynamics of any pollutant in the ocean medium. To take into account the physical–chemical characteristics of the pollutant, the model should be supplemented with the respective parametric unit. As the base units, taking into account the specific features of pollutants, the models of interaction of heavy metals, oil hydrocarbons, and radionuclides with the components of the ecosystems of the Arctic waters have been considered.

5.7.4.1 The block-scheme of the model

A mathematical model of the biosphere (MMB) with the geographical grid of land surface digitizing $5^\circ \times 4^\circ$ by latitude and longitude has been described in Kondratyev *et al.* (2002). The unit MMB, which describes the functioning of the World Ocean, foresees a set of point models of the four oceans. Therefore, (SSMAE) is a broadened ocean unit MMB for the Arctic Ocean. This correlation is shown in Figure 5.7 and in Table 5.27.

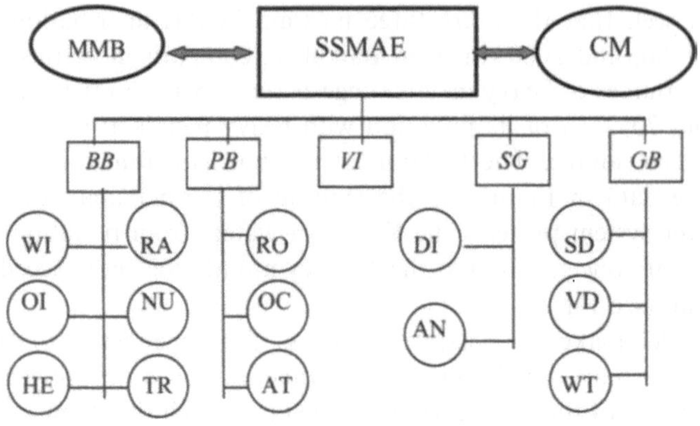

Figure 5.7. The SSMAE block-scheme. Notations are given in Table 5.27.
From Krapivin (1993).

Table 5.27. List of the SSMAE units.

Unit identifier	Description of the unit
AT	Model of the atmospheric transport of pollutants
AN	A set of anthropogenic scenarios
BB	Model of biota functioning in conditions of energy exchange in the trophic chain of the Arctic Basin's ecosystem
DI	Simulator of model scenarios describing the ice fields, synoptic situation, and change of hydrological regimes
VD	Model of the Arctic Basin's water dynamics
GB	Hydrological unit
HE	Model of the input of heavy metals into nutrient chains
TR	Model of the illumination field formation
NU	Model of the biogenic background formation
OC	Model of the pollutants transport through water exchange between the Atlantic and Pacific Oceans
OI	Model of the input of oil hydrocarbons into nutrient chains
RA	Model of the input of radionuclides into nutrient chains
RO	Model of the river runoff into the Arctic Basin
SG	Scenarios enerator (DI and II)
SD	Model of the dynamics of seawater salinity
PB	Model of the input of pollutants into the Arctic waters
WI	Model of energy fluxes in the trophic pyramide of the Arctic Basin's ecosystem
WT	Formation of the water temperature field
VI	Programme units providing a visualization of the results of simulation experiments

The functioning of SSMAE is maintained by the MMB and any climate model (CM), which gives the temperature and probably of other characteristics of the climate system. There are several versions of the use of these data. The user chooses, based on his or her experience, whether all the output data of the CM should be used to substitute the data calculated in the MMB or only some of them. In the case of the absence of a CM, the user forms a scenario of the climatic situation and puts it into the SSMAE. The SSMAE input data represent information about the land sources of pollutants in the coastal zone of the Arctic Basin, about climate, the ice fields, and mapped currents. The sources of this information are prognostic estimates based on multiyear series of *in situ* and remote measurements. The most concentrated and complete source of data about the Arctic Basin is held on a laser disk (SEDAAR, 1996) "Sea Ice in the Polar Regions" issued in 1996 by the CIESIN (`http://www.ciesin.org`).

Divide the territory of the Arctic Basin Ω into cells Ω_{ij} of size $\Delta\varphi$ by latitude and $\Delta\lambda$ by longitude. These cells are the base grid in the numerical algorithms of calculations of the ecological and hydrophysical fields in the territory Ω. In a general case, the cells Ω_{ij} are heterogeneous by parameters and functional characteristics. There are cells among them which adjoin the mouths of rivers Ω_R and ports Ω_P, bordering the land Ω_Γ, located in the Bering Strait Ω_B and at the southern boundary of the Norwegian Sea Ω_N. In depth z, the water basin Ω is stratified into layers with a step Δz. The distribution of depths over Ω is prescribed in the form of the matrix $\|h_{ij}\|$ where $h_{ij} = h(\varphi_i, \lambda_j)$. As a result, the whole volume of water in the basin $V^* = \sum \Omega_{ij} h_{ij}$ is divided into the volume compartments $\Omega_{ijk} = \Delta\varphi_i \times \Delta\lambda_j \times \Delta z_k$. Within Ω_{ijk} the water masses are considered to be homogeneous by temperature, salinity, density, and biomass of all the trophic levels of the Arctic ecosystem.

The SSMAE can vary the input information. When it is impossible to use the MMB or the CM, the user of SSMAE can include the unit SG, which makes it possible to consider the hypothetic situations of interactions between the atmosphere, land, and ocean.

Anthropogenic impacts on the territory Ω have a seasonal character. Therefore, for simplicity purposes we introduce notations for time intervals: τ_w – winter, τ_f – spring, τ_u – summer, and τ_a – autumn.

5.7.4.2 *The biotic unit*

The ice–water ergocline and the community of algae getting into snow and ice play an important role in the formation of the biological productivity of the northern seas. The primary produce in the nutrient chains of the Arctic Basin's ecosystems is determined by phytoplankton productivity, whose formation is connected with the composition of complex variations of synoptic, hydrodynamic, geochemical, and energetic parameters of water. This problem has been studied in detail in Legendre and Legendre (1998). Table 5.28 demonstrates the structure of the possible seasonal composition of the conditions of primary produce formation in Ω. This scheme is realized in each cell Ω_{ijk}.

Nutrient salts – $n_A(\varphi, \lambda, z, t)$. Formation of the biogenic background at a depth z is determined by the rate of photosynthesis, advection from deeper layers and input due to decay of the weighted dead organic substance. The role of hydrodynamic conditions manifests itself through maintaining the concentration of biogenic salts,

Table 5.28. The vertical structure of the Arctic Basin's water bodies.

	Parameters of the layer					
Layer (A)	Δz	T_A	E_A	k_A	α_A	β_A
Surface		T_0	E_0			
Snow	g	T_g	E_g		α_g	β_g
Floating ice	r	T_r	E_r		α_r	β_r
Drowned ice	f	T_f	E_f		α_f	β_f
Water	$z-f$	T_W	E_W	k_W	α_W	β_W

Note: Δz = layer thickness, T_A = temperature, E_A = illumination, k_A = turbulence coefficient, α_A = coefficient of illumination attenuation, β_A = coefficient of light reflection.

needed for photosynthesis, by their transport from other layers, where the concentration of chemicals is sufficiently high.

As the basis of the NU unit of the BB model (see Table 5.27) we take the scheme of circulation shown in Figure 5.8.

The dynamic equation for the nutrient salts n_A in the medium $A = \{s$ – snow, i – ice, w – water$\}$ will be as follows:

$$\partial n_A/\partial t = k_{w,2}\partial^2 n_A/\partial z^2 - V_{A,\varphi}\partial n_A/\partial\varphi - V_{A,\lambda}\partial n_A/\partial\lambda - V_{A,z}\partial n_A/\partial z + Q_A - \rho_A R_A + \varepsilon_{A,1}H_{h,1}$$

where R_A is the rate of photosynthesis in the medium A; Q_A is the inflow of nutrient salts into the medium A due to detritus lysis ($Q_A = \delta_n R_{A,D}$); ρ_A is the rate of consumption of nutrient salts by unit biomass of pure production of phytoplankton; δ_n is the content of nutrient salts in dead organic substance; $R_{A,D}$ is the rate of detritus decomposition ($R_{A,D} = \mu_A D_A$); μ_A is the intensity of detritus decomposition; $\varepsilon_{A,1}$ is the share of radionuclide h, which is a chemical analogue to the nutrient salts n_A; and $V = (V_{A,\varphi}, V_{A,\lambda}, V_{A,z})$ is the vector of the current velocity at the point with coordinates (φ, λ, z).

Illumination – $E(\varphi, \lambda z, t)$. Solar radiation E_0 (W/m^2/day) through the atmosphere reaches the surface Ω. This value can be estimated from monitoring data or use of a model. The flux E_0 is attenuated in snow, ice, and water, according to the scheme in Table 5.28. In separate cells Ω_{ij}, depending on season, the structure of these layers changes. Within each layer, attenuation of illumination with depth is described using the exponential law:

$$E(\varphi, \lambda, z, t) = \begin{cases} E_0 & \text{for } z = -r-g \\ E_0 \exp(-\alpha_s z) & \text{for } -r-g < z \leq -r \\ E_s(\varphi, \lambda, -R, t)\exp(-\alpha_{i1} z) & \text{for } -r < z \leq 0 \\ E_{i1}(\varphi, \lambda, 0, t)\exp(-\alpha_{i2} z) & \text{for } 0 < z \leq r \\ E_{i2}(\varphi, \lambda, r, t)\exp(-\alpha_w z) & \text{for } z > r \end{cases} \qquad (5.19)$$

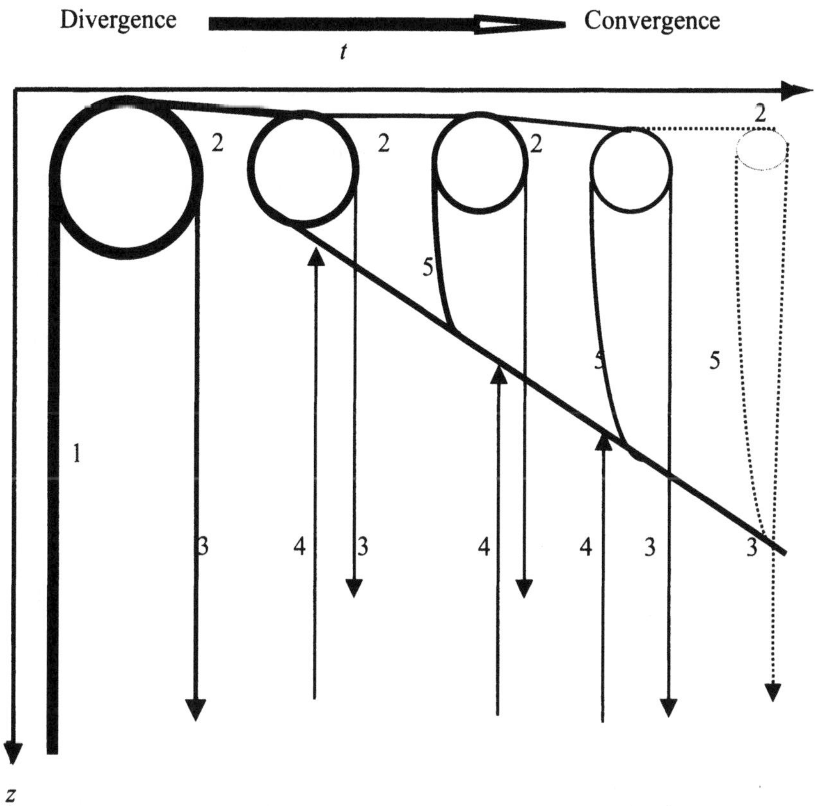

Figure 5.8. Scheme of the cycle of nutrient elements and organic matter in the succession of the ecosystem in the zone of upwelling after Vinogradov *et al.* (1975). Notations of the characteristic zones of development of the upwelling ecosystem: 1 – uplifting of nutrient elements and dissolved organic matter (DOM) with water from deep layers in the zone of upwelling; 2 – their multiple use in the production/destruction cycles of the surface layer communities; 3 – loss of nutrient elements with descending organic remains and migrating organisms; 4 – turbulent ascent of nutrients and DOM from deep layers and their delay in the zone of the lower maxmum; and 5 – inclusion of nutrients ascending from the layer of the lower maximum with migrating organisms into the production cycles of the surface community.

The parameters α_N $(N = s, i1, i2, w)$ depend on salinity, turbidity, temperature, and concentration of biogenic elements. The form of this dependence can be prescribed as a scenario, and these parameters are considered to be constant. The function E is calculated in the TR unit (see Table 5.27).

Photosynthesis – R_A. The rate of photosynthesis in the medium A is the function of illumination E_A, concentration of nutrient salts n_A, temperature T_A, phytoplankton biomass p_A, and concentration of pollutants γ_A. To describe this function, we use the Michaelis–Menten equation (Svirezhev, 1987, 2002):

$$R_{A,p} = ap_{A,\max}E_A/(E_A + k_{A,E})$$

where $k_{A,E}$ is the constant of semi-saturation of photosynthesis by illumination ($\mu g/m^3$); $p_{A,\max}$ is the maximum quantum output; and a_A is the coefficient of the proportion which can be calculated with an assumption that $\partial p_A/(p_A \partial z) \to 0$ for $n_A \to 0$ and $\partial p_A/(p_A \partial z) \to$ const. with increasing n_A. The function $a_A = a_1 k_0(T,t)/[1 + p_A/(a_2 n_A)]$, where a_1 is the maximum rate of capture of biogenic elements by phytoplankton (day^{-1}) and a_2 is the indicator of the rate of photosynthesis saturation, satisfies this condition:

$$k_0(T,t) = a_3 \max\{0, (T_c - T)/(T_c - T_{opt}) exp[1 - (T_c - T)/(T_c - T_{\text{opt}})]\}$$

where a_3 is the weighted coefficient; T_c and T_{opt} are temperatures (°C), critical and optimal for photosynthesis. The effect of pollutant on photosynthesis is manifested through changes of the coefficients in these equations.

The dependence of the rate of photosynthesis on the concentration of biogenic elements is described with the relationship:

$$R_{A,n} = a_{A,n} p_{A,\max} n_A/(n_A + k_{A,n})$$

where $k_{A,n}$ is the constant of semi-saturation of photosynthesis by biogens ($\mu g/m^3$); and $a_{A,n}$ is the coefficient of proportion.

Considering these formulas, we obtain the final expression for the rate of photosynthesis:

$$R_A = \min\{R_{A,p}, R_{A,n}\}$$

The kinetic equation for phytoplankton biomass in the medium A will be:

$$\begin{aligned}\partial p_A/\partial t &= k_{A,2}\partial^2 p_A/\partial z^2 - V_{A,\varphi}\partial p_A/\partial\varphi - V_{A,\lambda}\partial p_A/\partial\lambda - V_{A,z}\partial p_A/\partial z \\ &+ R_A - \theta_{A,p} - M_{A,p} - [k_{A,Zp}R_{A,Z}/\xi_{A,Z} + k_{A,Fp}R_{A,F}/\xi_{A,F}]p_A\end{aligned}$$

where $\theta_{A,p} = \kappa A p_A$; $M_{A,p} = y_A p_A$; κ_A is the rate of energy exchange between the phytoplankton cells and the medium A; y_A is the rate of the phytoplankton cells dying off; $\xi_{A,Z}$ and $\xi_{A,F}$ are rations of zooplankton Z and nekton F in the medium A, respectively ($\xi_{A,Z} = \xi_{A,F} = 0$ for $A = s, i1, i2$); $k_{A,Zp}$ and $k_{A,Fp}$ are shares of phytoplankton in the ration of zooplankton and nekton, respectively; and $R_{A,Z}$ and $R_{A,F}$ are productivities of trophic levels Z and F in the medium A, respectively ($R_{A,Z} = R_{A,F} = 0$ for $A = s, i1, i2$).

Trophic pyramid. Despite the severe climatic conditions, biology of the Arctic Basin is diverse and spatially heterogeneous. However, within the SSMAE it is impossible to take into account all aspects of the spatial distribution of the species of plankton algae and elements of the trophic chains. It can be realized after the formation of the respective databases – the goal of numerous scientific programmes. Therefore, a simple version of the parametric description of biotic elements of the Arctic water basins was chosen. It is assumed that the trophic pyramid in Ω is homogeneous and described by the scheme in Figure 4.1. Parameters of trophic levels are heterogeneous. Considering the assumed notations, the WI model (see Table 5.27) is described by the following system of equations:

$$\partial b_A/\partial t = k_{A,2}\partial^2 b_A/\partial z^2 - V_{A,\varphi}\partial b_A/\partial\varphi - V_{A,\lambda}\partial b_A/\partial\lambda - V_{A,z}\partial b_A/\partial z + R_{A,b} - M_{A,b} - k_{A,Zb}R_{A,Z}b/\xi_{A,Z}$$

$$\partial Z/\partial t + V_{w,\varphi}\partial Z/\partial\varphi + V_{w,\lambda}\partial Z/\partial\lambda + V_{w,z}\partial Z/\partial z = R_Z - M_Z - [k_{w,rZ}R_{w,F}/\xi_r + k_{ZZ}R_{w,Z}/\xi_{w,Z}] - \theta_Z$$

$$\partial r/\partial t = R_r - M_r - \theta_r - k_{rr}R_r r/\xi_r$$

$$\partial D/\partial t = R_D - M_D - \theta_D - [k_{rD}R_r/\xi_r + k_{DD}R_D/\xi_D]D$$

$$\partial d_A/\partial t + V_{A,\varphi}\partial d_A/\partial\varphi + V_{A,\lambda}\partial d_A/\partial\lambda + V_{A,z}\partial d_A/\partial z = M_Z + M_{A,b} + M_{A,p} + M_D + M_r - R_{A,d} - [k_{Dd}R_D/\xi_D + k_{rd}R_r/\xi_r]d_A$$

where θ_ω is the expenditure of the element $\omega = (Z, r, D)$ on energy exchange with the environment; M_ω is an amount of dead biomass of the element ω; and k_{ij} is the indicator of the share of the element j in the ration ξ_i of the element i. The presence of the index A represents the possibility of the element's functioning in medium A, otherwise it is assumed that $A = w$. For instance, if we assume that bacteria are not preserved in the snow cover, then $b_A = 0$ for $A = s$. Similarly, with the movement in snow and ice neglected, $V_s = V_{i1} = V_{i2} = 0$. In the latter equation, at $A = s, i1$ or $i2$ all terms without the index A are equal to zero.

The process of biomass limiting by the environment is parameterized using the method described in Krapivin and Kondratyev (2002). Inhibiting the element's biomass with pollutants is described by the threshold function.

5.7.4.3 The hydrologic unit

Circulation of water masses in the Arctic Basin is a complex system of cycles and currents of different scales and signs in the totality of the Arctic sea basins. The unit HU describes circulations and dynamics of salinity and temperature of waters over the whole water basin Ω.

Subunit VD. Dynamics of water masses in Ω is presented with fluxes between the compartments Ω_{ijk}. Directions of the water flux exchanges are prescribed at each horizon z_k. Fluxes of water masses and salts between Ω_{ijk} are described by balance relationships. The external boundary is determined by the shoreline, bottom, Bering Strait, southern boundary of the Norwegian Sea, and the water surface–atmosphere border. The structure of the movement of water masses can be simulated with the help of scenarios generated in the SG unit (see Table 5.27). Data on the hydrological regimes of water basins are synthesized into a 4-level time structure according to the seasonal division of the timescale into intervals τ_w, τ_f, τ_u and τ_a. For the Bering Strait, the velocity of the current is estimated by the function:

$$V(t) = \begin{cases} V_1 & \text{for } t \in \tau_u \cup \tau_a \\ V_2 & \text{for } t \in \tau_w \cup \tau_f \end{cases}$$

The transport of water masses across the southern boundary of the Norwegian Sea is assumed to be equal to V_3.

Subunit WT (see Table 5.27). The temperature of the medium $A(T_A)$ in the territory Ω is formed due to evaporation, precipitation, ice motion, water inflows from the Atlantic and Pacific Oceans, river runoffs, and other factors. Its changes in the compartment Ω_{ijk} are described by the heat balance equation:

$$\zeta C\Omega_{ijk} dT_{w,ijk}/dt = \sum_{s,l,m} (W_{ijk,slm} + f_{ijk,slm}) - W_{ijk}$$

where $T_{w,ijk}$ is the water temperature in the compartment Ω_{ijk}, (°K); ζ is the seawater density (g/cm^3); C is the water heat capacity (cal/g/deg); $W_{ijk,slm}$ is the input of heat with water flow from the compartment Ω_{slm} into the compartment Ω_{ijk}; $f_{ijk,slm}$ is the heat exchange between the compartments Ω_{slm} and Ω_{ijk} as a result of the turbulent mixing of water masses; and W_{ijk} is the net output of heat from the compartment Ω_{ijk}.

Dissipation of the kinetic energy of motion, the geothermal flux at the bottom of the ocean, the thermal effects of the chemical processes in seawater, as well as water freezing and ice melting are considered as not being globally determining factors in the formation of the field of temperatures. However, for the upper-layer compartments a seasonal correction ΔT_a is introduced:

$$T_s = \begin{cases} T_a - \Delta T_a & \text{for } t \in \tau_u \cup \tau_a \\ T_a + \Delta T_a & \text{for } t \in \tau_w \cup \tau_f \end{cases}$$

where T_a is the air temperature.

The heat flux H_{wA} across the snow–ice cover is calculated with a simplified formula:

$$H_{wA} = \zeta_i[T_a - T_w(\varphi, \lambda, r, t)]/[R + r + \zeta_i G/\zeta_S]$$

where ζ_i and ζ_S are the heat conductivities of ice and snow, respectively.

Unit SD (see Table 5.27). The dynamics of salinity of the Arctic water has been well studied. The Arctic Basin is known to have four types of basic water masses with characteristic formation of the field of salinity. This fact is used to verify the SD unit. The process of the formation of the field of salinity we describe without any a priori information about its structure, based only on the balanced relationship:

$$\Omega_{ijk} dS_{ijk}/dt = \sum_{s,l,m} [\alpha_{ijk,slm} q_{slm} S_{slm} + E_{ijk,slm}(S_{slm} - S_{ijk})] - q_{ijk} S_{ijk} + F_{ijk}$$

where S_{ijk} is the concentration of salts in Ω_{ijk}; $E_{ijk,slm}$ is the volume coefficient of diffusion; q_{slm} is the total flux of water from Ω_{slm}; $\alpha_{ijk,slm}(\sum_{s,l,m} \alpha_{ijk,slm} \leq 1)$ is the share q_{slm} directed into Ω_{ijk}; and F_{ijk} is the function of the source that describes the processes of the input of salts into Ω_{ijk} due to ice melting, precipitation, uplifting of bottom sediments, and transformation across the boundary $\Omega_R \cup \Omega_\Gamma \cup \Omega_B \cup \Omega_P \cup \Omega_N$. The ice salinity in Ω_{ij} is determined by the two-stage scale: multiyear s_1 and young s_2.

5.7.5.4 *The Arctic Basin pollution*

The pollution unit (PU) of the territory Ω simulates the processes of the input of pollutants due to atmospheric transport, river and surface coastal runoffs, navigation, and other economic activity. A variety of pollutants are confined to three components: oil hydrocarbons, heavy metals, and radionuclides. A variety of fluxes of pollutants is presented by three units described in Table 5.27. Denote the type of pollutant as d (e – weighted functions of heavy metals, g – dissolved fraction of heavy metals, O – oil hydrocarbons, h – radionuclides). Pollutants are assumed to get into living organisms only through nutrient chains.

Unit RO. The river runoff contributes significantly to the formation of the field of the Arctic waters pollution. It is assumed that the content $\gamma_{\Xi,d}$ of the pollutant d in the river waters Ξ is prescribed by the scenario in the AN unit. The pollutant gets to the compartments $\Omega_{slm} \in \Omega_R$ at a rate c_d. Further propagation of the pollutant over Ω is described by the units BU and VD.

Unit OC. The effect of water exchange of the Arctic Basin with the Pacific and Atlantic Oceans on the level of pollution manifests itself through the direct transport of pollutants from one basin to another across Ω_N and Ω_B.

Unit AT. The atmospheric transport of heavy metals, oil hydrocarbons, and radionuclides can be described by many models. Use of these models to retrieve the fields of pollution over Ω makes it possible to estimate the permissible parameters of digitizing Ω and the time step. Taking $\Delta t = 10$ days, $\Delta\varphi = \Delta\lambda = 1°$ and using only data on wind speed and direction, we chose the version of the Euler one-level model.

Unit OI. Oil products $O(\varphi, \lambda, z, t)$ differ by their multicomponent nature, which broadens the spectrum of situations of their behaviour in seawater. Here the following processes of transformation of oil products are postulated: dissolution H_{O1}, evaporation H_{O2}, penetration into deep water layers in the form of droplets H_{O3}, oxidation H_{O4}, adsorption by a weighted organic substance H_{O5}, biosedimentation H_{O6}, and bacterial decomposition H_{O7}:

$$\partial O/\partial t = k_{w,2}\partial^2 O/\partial z^2 - V_{w,\varphi}\partial O/\partial\varphi - V_{w,\lambda}\partial O/\partial\lambda - V_{w,z}\partial O/\partial z + Q_O - \sum_{i=1}^{7} H_{Oi}$$

where Q_O is the anthropogenic source of oil products.

Unit HE. Fluxes of heavy metals in seawater include an assimilation of the dissolved fraction by phytoplankton H_{gZ} and nekton H_{gr}, sedimentation of solid fractions H_{e1}, deposition with detritus H_{gD}, assimilation from bottom deposits by detritus consumers H_{egD}, and release from bottom deposits due to diagenesis, diffusion and human interference H_{ega}. All these fluxes together with the external sources of heavy metals form a complex system of their circulation in the marine medium:

$$\partial e_w/\partial t = k_{w,2}\partial^2 e_w/\partial z^2 - V_{w,\varphi}\partial e_w/\partial\varphi - V_{w,\lambda}\partial e_w/\partial\lambda - V_{w,z}\partial e_w/\partial z - H_{el}$$

$$+ \alpha_1 H_{ega} + \sum_{i=1}^{3} \alpha_{i2} Q_{egi}$$

$$\partial g_w/\partial t = k_{w,2}\partial^2 g_w/\partial z^2 - V_{w,\varphi}\partial g_w/\partial\varphi - V_{w,\lambda}\partial g_w/\partial\lambda$$

$$- V_{w,z}\partial g_w/\partial z - H_{gZ} + (1-\alpha_1)H_{ega} - H_{gr} - H_{gd} - H_{ga}$$

$$\partial e^*/\partial t = H_{el} - \alpha_1(H_{egD} + H_{ega})$$

$$\partial g^*/\partial t = H_{gD} - (1 - \alpha 1)(H_{egD} + H_{ega})$$

where e^* and g^* are concentrations of heavy metals in bottom deposits in solid and dissolved fractions, respectively; H_{ga} is the outflow of heavy metals from the sea to the atmosphere with evaporation and spray; Q_{egi} is the input of heavy metals to seawater from river runoff ($i = 1$), with precipitation ($i = 2$), and with wastes discharged by ships ($i = 3$); α_{i2} is the share of weighted particles in the ith flux of heavy metals; and α_1 is the share of the solid fraction of heavy metals in bottom sediments.

Unit RA. A variety of the types of radionuclides may be described by the parameters h. Each type is characterized by the period of half-life τ^h, velocity H_{hl} of the inflow to the water basin Ω, velocity H_{ha} of accumulation in living organisms $a(p, b_A, Z_A, r, D)$, and velocity H_{hd} of sedimentation with dead components of the ecosystem. As a result, the concentration Q_h of radionuclide h in Ω_{ijk} is described by a system of equations:

$$\partial Qh/\partial t = k_{w,2}\partial^2 Q_h/\partial z^2 - V_{w,\varphi}\partial Q_h/\partial\varphi - V_{w,\lambda}\partial Q_h/\partial\lambda$$

$$- V_{w,z}\partial Q_h/\partial z + \sigma_{ij}H_{hl}/\sigma - H_{ha} - H_{hD} + H_{h*} - Q_h \ln 2/\tau^h$$

$$\partial Q_{h*}/\partial t = H_{hD} - H_{h*} - Q_{h*} \ln 2/\tau^h$$

where Q_{h*} is the concentration of radionuclide h in bottom sediments; and H_{h^*} is the rate of outflow of radionuclide h from bottom sediments. The transport of radionuclides between the water layers due to migration of living organisms is considered to be negligibly small compared to the flux H_{hd}.

5.7.5.5 *Scenarios*

The anthropogenic impacts on the environment of the Arctic Basin is described by the unit SG consisting of two subunits DI (direct impact) and II (indirect impact). Direct impacts include the inflow of pollutants d in the Arctic waters with the northern river runoffs Q_{d1}, the springtime runoff from the shore Q_{d2}, the oil products discharge from the ships Q_{d3}, the transport of pollutants through the atmosphere Q_{d4}, changing hydrological regimes, etc. Indirect impacts are connected with changing climatic parameters. The indicators Q_i are parameterized in the units PB and SG depending on their nature. A consideration of direct impacts on the ecosystem of the Arctic Basin is realized via the models in the unit PB and using scenarios in SG. Indirect impacts are taken into account through parametric interfaces (MMB/SSMAE/CM). To demonstrate the simulation procedures we formulate the following three scenarios.

Scenario 1. One of the important factors of the functioning of the Arctic Basin's biogeochemical system is the sea ice area σ_I, whose interannual variability is described by the Markov chain with a several months attenuation. Therefore, the intraannual anomalies σ_I are presented in the form of the prescribed function $\sigma_I = G(t, t_B, t_F, t_z, \sigma_B, \sigma_F)$ uniform with respect to φ and λ. Here t_F (t_B) is the beginning of the period of the summertime (wintertime) stabilization of σ_I at the level $\sigma_F(\sigma_B)$ and t_z is the beginning of the springtime instability.

The parameters G and $R + r$ are either calculated from the data of the unit CM or postulated from mean statistical estimates with reference to seasonal intervals τ_V, τ_f, τ_u, and τ_a. In particular, a threshold level of salinity S_p is introduced, with respect of which the ice formation regime changes. In the demonstrated calculations it is assumed that $S_p = 34.75‰$. With a 0.1‰ decrease (increase) of the upper water layer salinity Ω_{ij1} with respect to S_p the rate of increase (decrease) of σ_I is estimated at 0.05 million km^3.

Scenario 2. Modelling the subwater illumination $E(\varphi, \lambda, z, t)$ in the Arctic conditions requires the solution of the complicated problem of energy exchange on the atmosphere–ocean border. In a general case, instead of Equation (5.19) the following equation should be considered:

$$\partial E(\varphi, \lambda, z, t)/\partial z = -(b_1 + b_2 b_A + b_3 Z + b_4 r + b_5 d + b_6 n_A)E$$

where b_1 and $b_i (i = 2\text{–}6)$ are indicators of light attenuation by water and by the ecosystem's components (m^2/μg). With the dependence of $b_i (i = 1\text{–}6)$ on the spatial coordinates taken into account, complicated demands are made to databases, which in the conditions of the inadequately studied Arctic Basin leads to an insoluble problem. As preliminary calculations have shown, when modelling the global processes in the ocean it is sufficient to prescribe $E(\varphi, \lambda, z, t)$ in the form of Equation (5.19) with E and $\partial E/\partial z$ being continuous at the media interface. The coefficients k_{A2}, β_A, and α_A (Table 5.28) are constant within the layer A and can be broken at the media's boundaries.

Scenario 3. The input of a pollutant of the type ξ into the basin Ω is described by a set of determinate functions:

$$Q_{\xi 1}(\varphi, \lambda, z, t) = \begin{cases} q_1 & \text{when } (\varphi, \lambda, z) \in \Omega_R \text{ and } t \in \tau_f \text{ or } (\varphi, \lambda, z) \in \Omega_B \cup \Omega_N \\ c_{\xi 1} & \text{when } (\varphi, \lambda, z) \in \Omega_R \text{ and } t \in \tau_V \cup \tau_u \cup \tau_a \\ 0 & \text{in other cases} \end{cases}$$

$$Q_{\xi 2} = \begin{cases} p_1 & \text{when } (\varphi, \lambda, z) \in \Omega_\Gamma \text{ and } t \in \tau_f \\ 0 & \text{in other cases} \end{cases}$$

$$Q_{\xi 2} = \begin{cases} f_1 & \text{when } (\varphi, \lambda, z) \in \Omega_P \text{ and } t \in \tau_f \cup \tau_a \cup \tau_u \\ 0 & \text{in other cases} \end{cases}$$

$$Q_{\xi 2} = \begin{cases} \delta_1 & \text{when } (\varphi, \lambda, z) \in \Omega_P \cup \Omega_R \cap \Omega_N \cup \Omega_\Gamma \cup \Omega_B \\ 0 & \text{in other cases} \end{cases}$$

5.7.4.6 Simulation experiments

SSMAE enables one to estimate the dynamics of the Arctic Basin pollution under various a priori suppositions about intensities of the fluxes of pollutants and other anthropogenic impacts on the ecosystems of this region. Here some of the possible situations will be considered. The thermal regime of the Arctic Basin is prescribed as a normal distribution with mean temperatures and dispersions over water reservoirs (Skiles, 1998). The scheme of the pollutant transport in the atmosphere is taken from Crane *et al.* (2001). The parameters for the SSMAE units are estimated using the published works and personal recommendations. The difference schemes for calculations are described in Kondratyev *et al.* (2002). The vertical distribution of pollutants at an initial moment t_0 is assumed to be uniform. Quantitative estimates of some parameters of SSMAE are given in Table 5.29. The average diameter of solid particles ranged between 0.12 and 1000 μm, and the vertical rate of sedimentation was assumed to be 0.003 m/s. The concentration of biogenic elements in ice and snow was equal to zero. It was also assumed that in the boundary conditions for the Stephan problem (Krapivin and Kondratyev, 2002) $P(\varphi, \lambda, t) = 0°C$, $f(\varphi, \lambda, t) = 3°C$ at $(\varphi, \lambda) \in \Omega$. The ratio of solid to dissolved fractions of heavy metals at the moment t_0 is assumed to be 1 : 2 (i.e., $e(\varphi, \lambda, z, t_0)/g(\varphi, \lambda, z, t_0) = 0.5$. The fluxes H_{gZ}, H_{gr}, H_{gd}, and H_{gD} are described by linear models, $H_{e1} = 0.01e_w$, $H_{ega} = 0$. The boundaries of the Norwegian and Bering Seas are approximated by the lines $\varphi_N = 62°N$ and $\varphi_B = 51°N$, respectively. The values of other parameters of SSMAE are determined from the literature (Legendre and Legendre, 1998; Rudels *et al.*, 1991). Initial data are given in Tables 5.30 and 5.31.

The dynamics of the radionuclide pollution of the Arctic Basin. Intensities of external fluxes across the boundaries of the Arctic Basin and the fluxes $H_{hD}, H_{h^*}, H_{h\alpha}$ are described by linear models. Some results of simulation experiments are given in Figures 5.9–5.12. Figure 5.9 characterizes the average trend in the content of radionuclide pollutant over the whole Arctic water basin. The curves describe the dynamics of the vertical distribution of the concentration of radionuclides in two layers of the water thickness and in bottom sediments. Radionuclides move from the upper layers to deeper waters at a rate providing a 43% reduction in the level of radionuclide pollution over a period of 20 years. These distributions for each Arctic sea are shown in Figure 5.10. Local variations in the vertical profiles of distributions of radionuclides are determined both by hydrological and ecological conditions. The correlation between these conditions is a seasonal function. Table 5.32 estimates the role of ecological processes in the formation of the vertical distribution of

Table 5.29. Estimates of some parameters used in the SSMAE simulation experiments.

Parameter	Estimate of the parameter
Step of space digitizing by	
Latitude ($\Delta\varphi$)	1°
Longitude ($\Delta\lambda$)	1°
depth (Δz)	
$z \leq 100$ m	1 m
$z > 100$ m	h-100 m
Coefficient of the ice heat conductivity (λ_1)	2.21 W/m/K
Coefficient of the water heat conductivity (λ_2)	0.551 W/m/K
Coefficient of turbulent mixing (k_2^W)	
For open waters	0.5×10^{-4} m^2/s
For ice-covered water bodies	5×10^{-6} m^2/s
Characteristic heat of ice melting (q)	
Content of biogenic elements in dead organic matter (δ_n)	334 kJ/kg
Intensity of detritus decomposition (μ_A)	0.1
$A = g, r, f$	0
$A = W$	0.01
Velocity of current in the Bering Strait (V)	
$i = 1$	0.2 m/s
$i = 2$	0.05 m/s
Water heat capacity (C)	4.18 kJ/kg/K
Ice salinity (s_i)	
$i = 1$	5‰
$i = 2$	1‰
Water salinity at $z > 100$ m (s_0)	34.95‰
Area of the Arctic Basin (σ)	16,795,000 km^2
Half-life period of radionuclides (τ_ε)	
$\varepsilon = {}^{60}$Co	5.271 years
$\varepsilon = {}^{137}$Cs	30.17 years
Critical temperature for photosynthesis (T_c)	−0.5°C

radionuclides over all the Arctic seas. These estimates show that the biological community plays an insignificant role in the transport of radionuclides from the upper layers into the deep ocean. As is seen, the content of radionuclides in the waters of the East Siberian, Chukchi, and White Seas as well as in the Laptev Sea varies markedly. A marked accumulation of radionuclides is observed in the Central region of the Arctic Basin. The aqua-geosystems of the Greenland and Kara Seas are characterized by certain conservativeness, and in the Norwegian Sea the level of pollution is even reduced. A sufficiently persistent pattern is observed in the vertical distribution of radionuclides. With respect to the initial time moment t_0, the vertical distribution becomes stabilized after 5–7 years. The Kara Sea, Laptev Sea, and East Siberian Sea are characterized by special regimes of stabilization. Here the stabilization of the vertical distribution sets in later compared to other water bodies (by 10–12 years).

Table 5.30. Initial data for SSMAE on the distribution of pollutants over the Arctic water bodies at the moment t_0.

		Concentration			
		Radionuclides (Bq/l)		Heavy metals	Oil hydrocarbons
Water body	Symbol	^{137}Cs	^{60}Co	(mg/l)	(mg/l)
Greenland Sea	Γ	0.05	0.05	0.5	0.2
Norwegian Sea	N	0.05	0.05	0.7	0.4
Barents Sea	B	0.07	0.07	0.8	0.6
Kara Sea	K	0.10	0.10	1.0	0.4
White Sea	∇	0.10	0.10	1.1	0.4
Laptev Sea	Λ	0.05	0.05	0.9	0.5
East Siberian Sea	E	0.01	0.01	0.9	0.5
Bering Sea	S	0.02	0.02	0.8	0.7
Chukchi Sea	X	0.01	0.01	0.8	0.6
Beaufort Sea	Φ	0.05	0.05	0.7	0.2
Central Basin	U	0.00	0.00	0.1	0.1

Table 5.31. Input fluxes of radionuclides, heavy metals (suspended particles e and dissolved fraction (Ψ)), and oil hydrocarbons O in the Arctic Basin due to river runoff and other sources considered in SSMAE.

		Concentration of pollutant				
Source	Flow into the basin km^3/ year	^{137}Cs (Bq/l)	^{60}Co (Bq/l)	e (mg/l)	Ψ (mg/l)	O (mg/l)
Rivers						
Yenisey	600	0.5	0.5	0.3	5.1	2.3
Ob′	400	0.1	0.1	0.4	6.9	4.7
Lena	500	0.0	0.0	1.1	8.8	6.9
Pechora	130	0.1	0.1	0.3	1.5	3.0
Northern Dvina	100	0.0	0.0	0.2	1.1	4.0
Other Siberian rivers	200	0.1	0.1	0.1	0.5	2.3
Rivers of North America	600	0.2	0.2	0.1	1.0	1.0
Evaporation	3,500	0.0	0.0	0.0	0.0	0.0
Precipitation	5,300	0.0	0.0	0.1	0.1	0.0
Southern boundary of the Norwegian Sea	12,000	0.1	0.1	0.6	2.2	2.4
Bering Strait	10,560	0.0	0.0	0.5	1.9	1.9

As the results of modelling show, variations of the data in Table 5.30 within ±100% change the time of stabilization of the level of the radionuclide pollution by not more that 30%. So, in 4–8 years with respect to t_0, we have distributions presented in Figure 5.11. The quantitative estimates vary widely depending on initial conditions and half-lives.

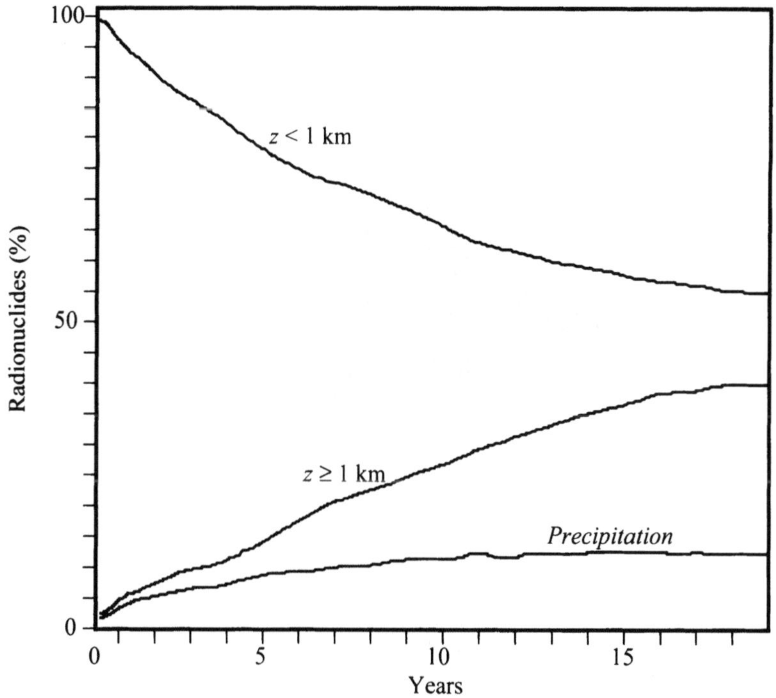

Figure 5.9. Dynamics of distribution of radionuclides in the Arctic Basin. It is assumed that at the moment $t_0 = 0$ the radionuclide pollution (^{137}Cs, ^{60}Co, see Table 5.31) is in the layer $z < 1$ km. The curves show the dynamics of the radionuclide pollution in two layers of water thickness and in bottom sediments. The results are obtained by averaging over all Arctic water basins.

One of the unstable parameters is the river runoff to the Arctic Basin. Figure 5.12 shows changes in the results of modelling with varied volumes of river runoff into the Arctic Basin. The radionuclide pollution is reduced by 80% if the river runoff decreases by 50%. With a doubled river runoff the level of radionuclide pollution increases by 5.8%. However, an error in estimation of the volume of the river runoff by ±50% can cause a deviation of the modelling results up to 100% when calculating the level of radionuclide pollution. As follows from other curves in Figure 5.12, such deviations are less for heavy metals and oil hydrocarbons.

SSMAE permits an estimation of numerous parameters of the radionuclide pollution. So, for instance, one can estimate the dependence of mechanisms of biological transformation of the radionuclide pollution on initial supposition. The biological outflow of radionuclides from the layer of mixing varies over wide timescales (from months to several years). The percentage of the vertical transport of radionuclides by living elements of the Arctic ecosystem varies between 11% for the process of migration and 89% for dead organisms. There is a stable result in that the lower trophic levels of the Arctic ecosystem have a greater concentration of radionuclides than the higher trophic levels. It is clear that for a higher accuracy of the

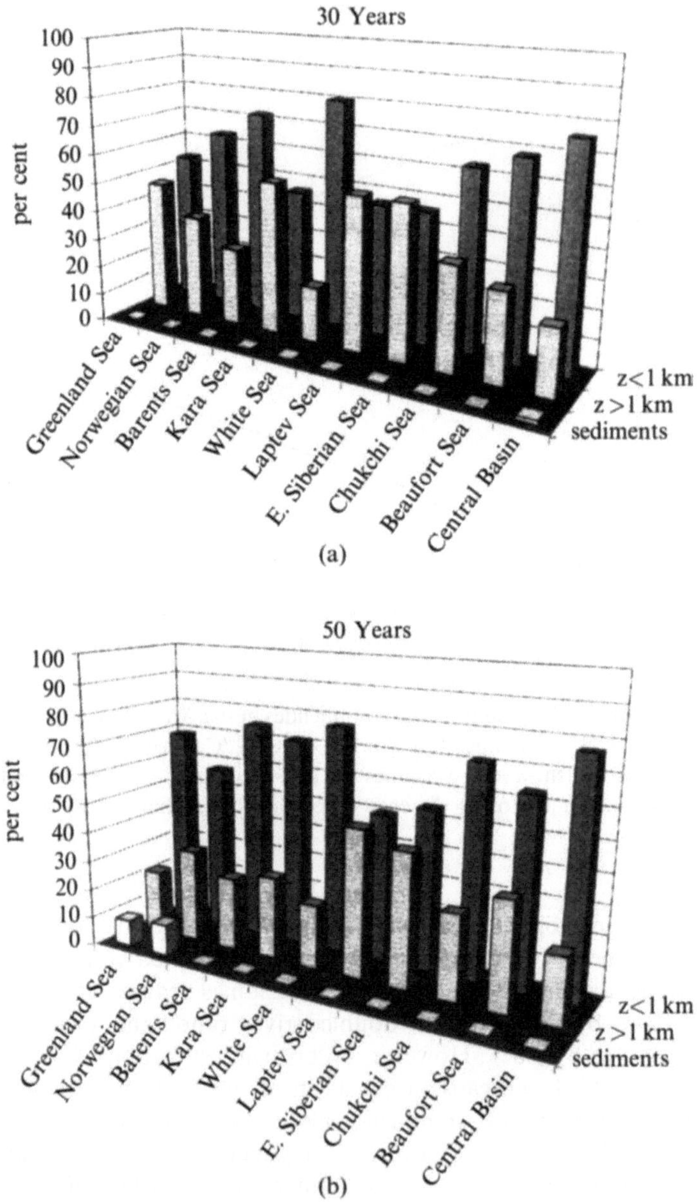

Figure 5.10. Spatial distribution of oil hydrocarbons in the Arctic Basin 5 years after t_0. Results are obtained using the following limitations: $O(\varphi, \lambda, z, t_0) = 0$, $H_{O1} = 0.1\,\mathrm{mg/m^3/day}$; $H_{O2} = 0$ for $g(t) > 0$ and $H_{O2} = 0.01\,\mathrm{mg/m^3/day}$ for $g(t) = 0$ and $z \geq \Delta z_1$, $H_{O3} = 0.01\,\mathrm{mg/m^3/day}$; $H_{O4} = 0.02\,\mathrm{mg/m^3/day}$; $H_{O5} = k_D(D_A)^{1/3}$ (k_D is the coefficient of adsorption equal to zero for media $A = s, i_1, i_2$ and $0.005\,\mathrm{day}^{-1}$ in the medium $A = w$); $H_{O6} = k_Z Z^{1/4}$ (k_Z is the coefficient of biosedimentation equal to zero for media $A = s, i_1, i_2$ and $0.004\,\mathrm{day}^{-1}$ in the medium $A = w$); and $H_{O7} = k_B B_A$ (k_B is the coefficient of bacterial decomposition equal to 0.01 for media $A = s, i_1, i_2$ and $0.05\,\mathrm{day}^{-1}$ in the medium $A = w$).

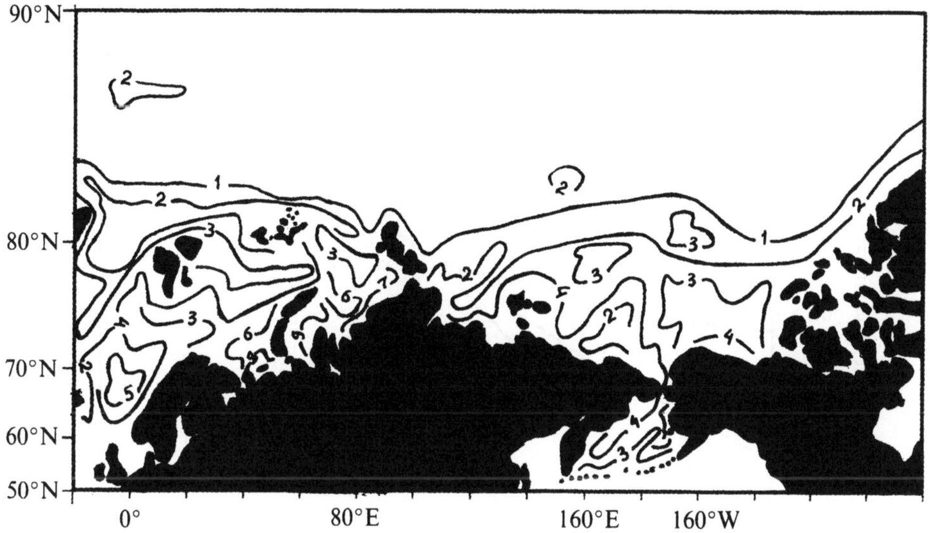

Figure 5.11. Spatial distribution of the forecast of the oil hydrocarbons in the Arctic aquatories 5 years after t_0. The scale step is 0.003 mg/l. The results were obtained under the following restrictions: $O(t_0, \varphi, \lambda, z) = 0$, $H_{O1} = 0.1\,\mathrm{mg/m^3/day}$; $H_{O2} = 0$ for $g(t) > 0$ and $H_{O2} = 0.01\,\mathrm{mg/m^3/day}$ for $g(t) = 0$ and $z \geq \Delta z_1$; $H_{O3} = 0.01\,\mathrm{mg/m^3/day}$; $H_{O4} = 0.02\,\mathrm{mg/m^3/day}$; $H_{O5} = k_D D_A^{1/3}$ (k_D is the adsorption coefficient which is equal to zero for $A = g, r, f$ and $0.005\,\mathrm{day}^{-1}$ when $A = w$); $H_{O6} = k_Z = Z^{1/4}$ (k_Z is the biosedimentation coefficient which is equal to zero when $A = g, r, f$ and $0.004\,\mathrm{day}^{-1}$ for $A = w$); $H_{O7} = k_B B_A$ (k_B is the bacterial destruction coefficient which is equals to $0.01\,\mathrm{day}^{-1}$ for $A = g, r, f$ and 0.05 when $A = w$). Identifiers g, r, f, and w correspond to the different layers of the Arctic Basin (see Table 5.28).

unit RA it is necessary to take into account in the model of the biological processes the role of each radionuclide in the dynamics of the plankton community. The version of the unit RA realized in SSMAE takes into account mainly the physical processes of capturing radionuclides in nutrient chains.

The dynamics of the Arctic Basin pollution with heavy metals. Results of simulation experiments are given in Table 5.33. It is seen that the functional stabilization over the whole Arctic Basin takes 32 months. The concentration of heavy metals in the compartments $\Omega_R \cup \Omega_P$ (river mouths and ports) exceeds 6 times their level in the central basin and twice the level in $\Omega_B \cup \Omega_\Gamma \cup \Omega_N$ (near the coastline, in the Bering Strait, and at the southern boundary of the Norwegian Sea). The concentration of heavy metals in phytoplankton is less (by 18%) than in zooplankton and is higher (by 26%) than in nekton. There is a process of accumulation of heavy metals at the upper trophic levels. Besides, the ratio of the concentration of heavy metals in phytoplankton to that at other trophic levels varies from 0.3 in $\Omega_R \cup \Omega_P$ to 0.5 in open waters.

The spatial features of the distribution of heavy metals in seawater are characterized by a faster accumulation in the water reservoirs bordering on the western coastline of Novaya Zemlia and located between the Yan-Mayen and

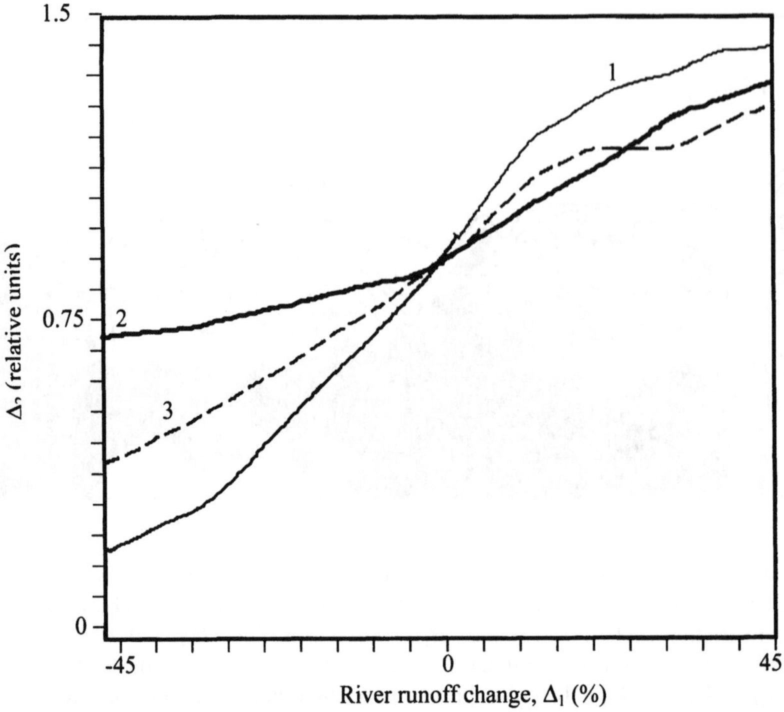

Figure 5.12. Effect of variations of the river runoff on the level of the Arctic Basin pollution. Here $\Delta_2 = (O, e, \varepsilon, \psi)$ is an average content of pollutant normalized for initial data; Δ_1 is the percentage change of the river runoff in Ω with respect to an average over three years 1991, 1993, and 1995. Note: 1 – radionuclides, 2 – heavy metals, 3 – oil hydrocarbons.

East Iceland Currents. In the central water basin, concentrations of heavy metals are distributed quasi-uniformly. The vertical gradients of the concentration of heavy metals vary within 0.11–1.54 $\mu g/m^2/day$. The forms of the vertical distributions of the dissolved fraction (g_w) and suspended particles (e_w) of heavy metals over large water basins are not clearly manifested. An average relationship of the concentrations of heavy metals both in bottom deposits and in water ($[g_* + e_*]/[g_w + e_w]$) varies over the Arctic Basin between 1.9 and 5.7. For instance, this relationship for the Bering Sea is equal to 3.3. The contribution of the process of biosedimentation into the formation of the vertical distribution of heavy metals is determined by the coefficient whose value varies between 0.23 and 1.24 $\mu g/m^2/day$.

SSMAE provides the possibility to assess the characteristics of various types of heavy metals. Such estimates are shown in Table 5.34.

The dynamics of the Arctic Basin pollution with oil hydrocarbons. Figure 5.11 shows a map of the distribution of oil hydrocarbons in the Arctic Basin drawn with the help of SSMAE. Calculations show that the oil pollution Q_O estimated from literature has no time to spread over large water reservoirs due to high decomposability. After the oil discharge into the water, processes of self-cleaning provide a transformation

Table 5.32. A portion of the simulation experiment results using the SSMAE for the estimation of the vertical distribution of radionuclides in the Arctic Basin. The contribution of ecological processes to the formation of the vertical distribution in the radionuclide content of the water is represented by the parameter ξ (%). The average content of the phytoplankton biomass is represented by the parameter p_w (g/m^2).

	Seasons							
	Winter		Spring		Summer		Autumn	
	τ_w		τ_s		τ_u		τ_a	
Aquatory	p_w	ξ	p_w	ξ	p_w	ξ	p_w	ξ
Greenland Sea	3.2	2	8.4	10	5.7	5	6.3	5
Norwegian Sea	2.9	2	7.8	9	5.9	5	6.7	6
Barents Sea	2.1	1	8.9	11	6.8	6	7.1	6
Kara Sea	2.4	1	9.2	12	5.3	5	6.0	5
White Sea	2.2	1	7.6	9	6.3	6	6.4	5
Laptev Sea	0.9	1	2.4	4	1.3	2	1.4	2
East Siberian Sea	1.3	1	2.7	4	1.9	3	2.1	3
Bering Sea	2.5	2	7.1	9	3.9	4	5.3	4
Chukchi Sea	2.3	2	6.9	8	4.1	4	5.1	4
Beaufort Sea	1.9	2	5.7	7	4.8	4	4.9	4
Central Basin	1.0	1	1.7	2	1.5	2	1.6	2
Average value	*2.1*	*1.5*	*6.2*	*7.7*	*4.3*	*4.2*	*4.8*	*4.0*

of oil pollution into other forms (50% in the media s, i_1, and i_2 and 72% in the medium w). The distribution of oil hydrocarbons stabilizes in three years. As is seen in Figure 5.11, average levels of oil pollution of the water body Ω are taken to be 0.005 mg/l, which is well below the background pollution of the World Ocean. In the zones effected by the waters of Gulf Stream and Pacific Ocean this level is exceeded by a small amount. The Barents and Kara Seas where the concentration of oil hydrocarbons reaches 0.03 mg/l, turn out to be the most polluted. The total content of oil hydrocarbons averages 65,331 t, and the dispersion constitutes 32%. The contribution of the processes $H_{Oi}(i = 1–7)$ is estimated by the relationship $H_{O2} > H_{O4} > H_{O1} > H_{O5} > H_{O6} > H_{O7}$. This row changes for each sea. The prevalence of one process of oil hydrocarbon destruction over others is determined by seasonal conditions. The process of cleaning due to oil evaporation from the water surface (H_{O2}) prevails over other processes in summer. In fact, the evaporated oil hydrocarbons return all or partially to the Arctic Basin with precipitation, but they also get into other water bodies. These processes are simulated in the units EPM, EEE, KOH, KRP, and EBM (Table 5.21). The maximum rate of dissimilation of oil hydrocarbons constitutes 0.0028 g/m^2/day. The flux H_{O7} averages 27 t/year. For various seas this flux is as follows: the Bering Sea – 3.7; the Greenland Sea – 11; the Norwegian Sea – 2.2; the Barents Sea – 3.4; the Kara Sea – 2.3; the White Sea – 2.3; the Laptev Sea – 2.8; the East Siberian Sea – 2.8; the Chukchi Sea – 3.4; the Beaufort Sea – 2.5, and the central basin – 0.5.

Table 5.33. Results of the simulation experiment on the estimates of the dynamical characteristic parameters for pollution of the Arctic waters by heavy metals.
From Kondratyev *et al.* (2002).

	Estimate of the parameter after Δt years					
Parameter	$\Delta t = 1$	$\Delta t = 3$	$\Delta t = 5$	$\Delta t = 10$	$\Delta t = 15$	$\Delta t = 20$
Average concentration of heavy metals in the biomass (ppm)						
Phytoplankton	0.011	0.012	0.016	0.024	0.036	0.037
Zooplankton	0.013	0.014	0.019	0.028	0.041	0.043
Nekton	0.015	0.017	0.022	0.04	0.07	0.07
Detritophages	0.033	0.037	0.048	0.088	0.015	0.016
Average content of heavy metals in the Arctic waters (ppm)	0.022	0.027	0.036	0.037	0.038	0.038
Flow of heavy metals from the upper layer to the deep waters (mg/m²/d)						
Norwegian Sea	0.71	1.07	1.14	1.17	1.19	1.22
Barents Sea	0.72	1.08	1.25	1.19	1.24	1.16
Greenland Sea	0.26	0.62	0.71	0.82	0.76	0.89
White Sea	0.11	0.23	0.24	0.21	0.19	0.2
Kara Sea	0.34	0.47	0.57	0.61	0.63	0.64
Laptev Sea	0.55	0.78	0.81	0.89	0.74	0.77
East Siberian Sea	0.59	0.79	0.95	0.97	1.02	1.07
Chukchi Sea	0.88	0.83	1.54	1.49	1.31	1.44
Beaufort Sea	0.34	0.67	0.66	0.81	0.74	0.69

Table 5.34. Estimates of heavy metal flows to and from the atmosphere.
From Kondratyev *et al.* (2002).

	Parameter	
Heavy metal	Atmospheric deposition $Q^2_{e\psi}$ (mg/m²/h)	Evaporation and spray H^{ψ}_{α} (t/yr)
Ag	0.7	7
Cd	1.1	58
Co	0.3	5
Cr	1.8	188
Cu	15	169
Fe	599	894
Hg	0.6	3
Mn	4.2	283
Ni	5.5	60
Pb	48	5
Sb	0.05	123
Zn	109	4,471

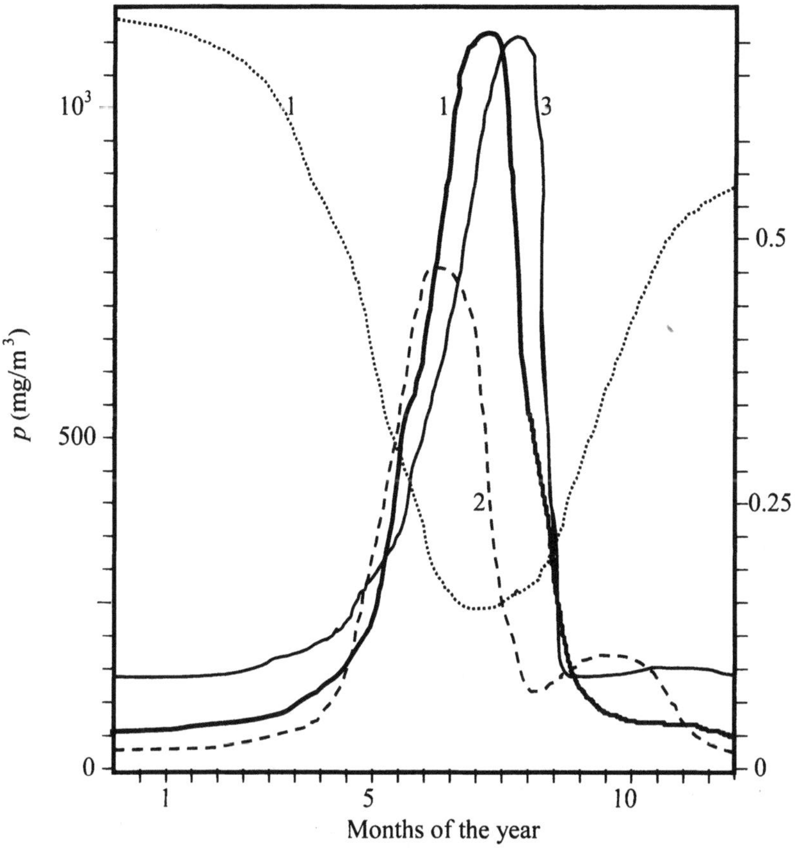

Figure 5.13. Effect of the Barents Sea ecosystem on the dynamics of oil hydrocarbons (dashed curve) in the seawater. Dynamics of phytoplankton are presented by solid curves: 1 – modelling results; 3 – distribution of phytoplankton in the south-western (northern and north-eastern) sectors of the sea.

An adequate estimate of the role of the Arctic ecosystem in the dynamics of oil hydrocarbons has characteristic features for each sea. As an example, Figure 5.13 gives such an estimate for the Barents Sea. The difference between the results of modelling and curves 2 and 3 is explained by an assumption that the trophic structures of various Arctic water bodies are similar and described by a single scheme, with the available differences in the trophic structures and their parameters being neglected.

Figure 5.13 shows that the vegetation period equal to 4.9 months for the Barents Sea affects the contribution of the ecosystem into the process of its self-cleaning of oil pollution. In the case considered, the ecosystem of the Barents Sea neutralizes about 25% of oil pollution during the vegetation period. In another season, this value oscillates round 3%. A dispersion of these estimates, with latitudinal variations taken into account, exceeds 53%. For instance, in the

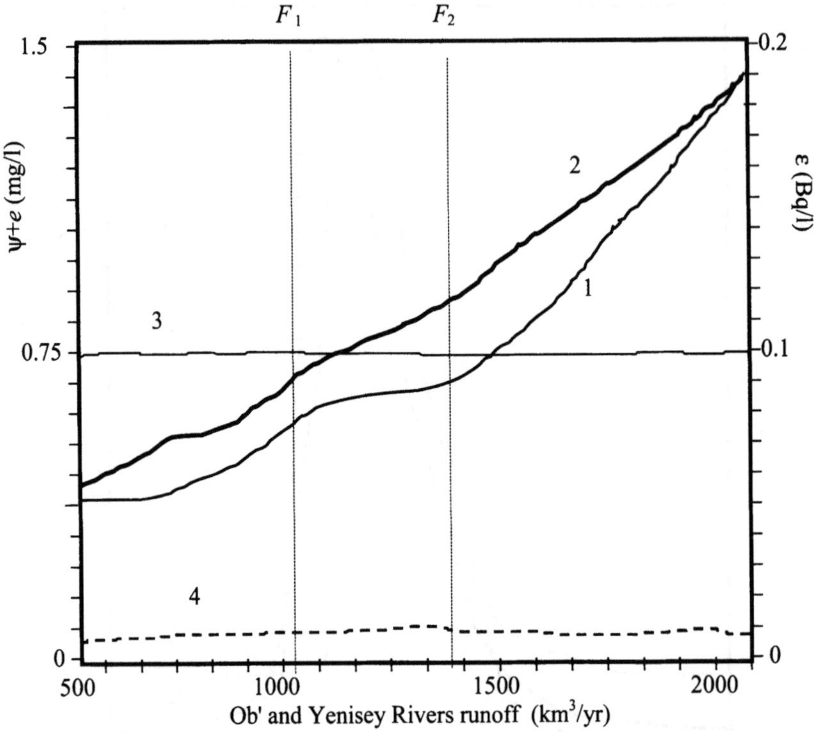

Figure 5.14. Dependence of concentrations of heavy metals $(e + \psi)$ and radionuclides $(\varepsilon = ^{137}\text{Cs} + ^{60}\text{Co})$ at different geographical points on the river runoffs of the Ob′ and Yenisey Rivers (F). Integral F_1, F_2 corresponds to real variations of the volume of the river runoff. Note: 1(2) – heavy metals (radionuclides) at the point with coordinates $\varphi = 75°$N and $\lambda = 65°$E; 3(4) – heavy metals (radionuclides) at the point with coordinates $\varphi = 72°$N and $\lambda = 150°$W.

northern sector of the Barents Sea the vegetation period continues from 2.3 to 3.1 months, whereas far-southern latitudes are characterized by a period from 5.2 to 5.8 months. Therefore, the role of the ecosystem in the sea's self-cleaning of oil is estimated at 8% for northern latitudes and 36% for southern latitudes of the Barents Sea, respectively. SSMAE provides such estimates for each cell $\Delta\varphi \times \Delta\lambda \times \Delta z$ of the Arctic Basin.

The dynamics of pollution in the Arctic Basin. There is a problem of determination and understanding of the role of various sources of pollution in the formation of water quality in different water bodies of the Arctic Basin (McCauley and Meier, 1991). One of the base functions of SSMAE is an estimation of dynamic characteristics of pollution in any cell $\Omega_{ij} \subseteq \Omega$ as a function of time. The influence of pollutant sources on the Arctic ecosystem occurs through the boundary area $\Omega_R \cup \Omega_P \cup \Omega_B \cup \Omega_\Gamma \cup \Omega_N$ and through the atmosphere. The total pattern of the spatial distribution of a pollutant is formed via the local processes of its kinetics in the waters of the Arctic Basin. Incompleteness of the database for simulation experiments with the use of SSMAE makes one consider some scenarios, within the framework of which suppositions are made about concrete values of uncertain parameters of the Arctic natural system. Clearly, for a detailed formation of databases on the fluxes of pollutants into the Arctic Basin, these fluxes including the river runoff should be studied. One efficient attempt has been made by Preller and Cheng (1999). In particular, the rivers Ob′ and Yenisey flowing into the Kara Sea contribute to the pollution of the central basin by up to 2.1% with the volume of runoff varying from 500 to 1500 km^3/year, and with the runoff increasing to 2000 km^3/year (this contribution grows by linear law up to 7.6%). Hence, there is a critical level of pollution of the Kara Sea, at which the aqua-geosystem has no time to neutralize the flux of pollutants.

SSMAE permits assessment of the fluxes of pollutants between various water bodies of the Arctic Basin. For instance, the transport of heavy metals and oil hydrocarbons from the Barents Sea to the Kara Sea constitutes 631 and 474 kg/year, respectively. The total flux of pollutants from the Russian territory to the region of Alaska, within the framework of suppositions in Table 5.31, varies from 0.3 to 0.9% of the initial flux. As follows from the curves in Figure 5.14, the Ob′ and Yenisey Rivers do not practically affect the level of pollution of the Arctic waters near Alaska.

6

Biogeochemical cycles of pollutants in the environment

6.1 SPECIAL FEATURES OF GLOBAL BIOGEOCHEMICAL CYCLES

A discussion of the problems of global ecodynamics cannot be constructive without a complex consideration of the feedbacks of the cycles of chemicals in the environment with the biogeochemical processes in the presence of numerous elements, such as nitrogen, sulphur, phosphorus, methane, ozone, and others (Hadley and Toumi, 2002; Fan *et al.*, 1992, 1999; Faluvegi *et al.*, 1999; Demirchian and Kondratyev, 1998). Only this complex approach makes it possible to obtain reliable estimates of the greenhouse effect. The processes of CO_2 assimilation from the atmosphere are affected by numerous natural and anthropogenic factors manifested through a long chain of cause-and-effect bonds. For instance, the acid rains affect the state of the vegetation cover and the latter affects the CO_2 exchange on the atmosphere–land border. The use of nitrogen and phosphorus fertilizers in agriculture changes the role of cultural plants in the assimilation of atmospheric CO_2 and affects the rates of decomposition of the soil organic matter. In totality, many chemicals, especially greenhouse gases (GHGs), after getting to the environment from anthropogenic sources become an object of not only biogeochemical analysis but also economic consideration. Such a diverse analysis was carried out, for instance, in connection with the CH_4 dynamics problem at the Second International Conference in Novosibirsk in 2000 (Bazhin, 2000; Byakola, 2000). Relevant connections should be thoroughly systematized and parameterized. Without analysis, the reliability of the estimate of the role of the biosphere in the assimilation of excess CO_2 from the atmosphere would be in doubt. Complex studies in this direction are being carried out, for instance, in several laboratories in the USA. Measurements of the spatial and temporal distributions of gases involved in the global CO_2 cycle are made at surface observatories, with the use of flying laboratories, and by satellite remote sensing. An accumulation of such information will make it possible to reveal dependences needed for global or regional modelling (Gillon and Yakir, 2001; Hakola *et al.*, 2003; Hidy, 2003; Kennedy *et al.*, 2002; Krupa, 2003; Nikolo and Zeller, 2003; Ericksen *et al.*, 2003; Sun *et al.*, 2001; Stein and Lamb, 2000; Mabuchi *et al.*, 2000; Kennedy, 2001; Craig *et al.*, 1998).

All this confirms the apparent fact that conclusions drawn from fragmentary studies of the global carbon cycle not based on a constantly complicating global model of the type described in Krapivin and Kondratyev (2002), will always be in doubt. For the global conclusion like the KP recommendations, these doubts become dangerous with unpredicted global consequences. Nevertheless, such conclusions and assessments are necessary. Unfortunately, most of the international programmes on the subject considered are not aimed at the development of the global modelling technology and do not concentrate the efforts of specialists on deriving numerical models of the nature–society system (NSS).

The existing global models are simple and inadequately provided with sufficient databases. Three directions in global modelling have been described in the works by Kondratyev *et al.* (2002a,b), Sellers *et al.* (1996a–c), Boysen (2000), and Krupchatnikov (1998). In each of them one or several components are absent, but on the whole, conceptually they are mutually additive. This makes it possible to combine them and, hence, to derive a global model which would take into account the most important processes in the NSS.

The photochemical system of the atmosphere is a poorly studied one and is practically neglected in global models. Knowledge is required of the laws of changes in the incoming solar radiation intensity in connection with the effects of its absorption by gases and aerosols. First of all, of importance here is the role of molecular nitrogen, ozone, water vapour, nitrogen oxide, sulphur dioxide, nitrogen dioxide, CO_2, and other gases. Therefore, in this chapter some models are suggested which make it possible to parameterize part of these elements.

An interaction of abiotic factors of the environment and living organisms of the biosphere is followed by continuous cycles of matter in nature. Various species of living organisms assimilate substances needed for their growth and life support, emitting into the environment the products of metabolism and other complex mineral and organic compounds of chemical elements in the form of unassimilated food or dead biomass. The biospheric evolution has resulted in a stable chain of global biogeochemical cycles, whose breaking in the second-half of the 20th century has posed many principal problems, such as unpredictable climate changes due to the possible greenhouse effect, a decrease in biodiversity, progressive desertification, and many others. The questions of what happens with the Earth's climate and what are the prospects for the ozone layer depletion remain unanswered despite huge economic expenditures on their study. Now it is clear that these and other nature-protection questions cannot be answered without developing an efficient global system of monitoring based on the global model of the NSS, one of the basic units of which simulates the biogeochemical cycles of the main chemicals in the biosphere. This approach will enable one to assess the anthropogenic fluxes of pollutants and to estimate the permissible emissions of hydrogen, chlorine, fluorine, methane, and other chemicals into the environment.

The global biogeochemical cycle of CO_2 has been the centre of attention for scientists and specialists from many countries try to answer the following questions. (1) What are the concentrations of CO_2 that can be expected in the future with the existing or predicted rates of organic fuel burning? (2) What climatic changes can result from the increasing concentrations of CO_2? (3) What are the consequences of such climatic changes for the biosphere? (4) What procedures can mankind undertake in order to either reduce the negative consequences of climate change or prevent them?

Clearly, now, according to rough model estimates, industry should search for new sources of energy, which reduce the rates of organic fuel burning and, hence, reduce external impacts on natural biogeochemical cycles.

During the last decade the notion of the "greenhouse effect" has been used in numerous publications on the problems of global climate change on the Earth. This notion refers to an explanation of changes in the thermal regime of the atmosphere caused by the effect of some gases on the process of solar radiation absorption as well as long-wave radiation transfer. Many gases are characterized by a high stability and long residence times in the atmosphere. Carbon dioxide is one of them. Its time of residence in the atmosphere is estimated at 2–3 years.

Numerous long-term observations in various latitudinal belts show a high level of correlation between temperature and CO_2 content (Connors *et al.*, 1999). The atmosphere–ocean interaction contributes most towards this dependence. Although the atmosphere and the ocean are in equilibrium with respect to CO_2 exchange, it still regularly gets broken. The most serious causes of this break include: (1) sea surface temperature (SST) variations; (2) changes in the ocean volume; and (3) changes in the regime of the vertical circulation of the ocean. In general, an efficiency of these causes can be characterized by the following ratio of the forcing of CO_2 concentrations in the atmosphere. The first cause contributes about 65% towards the change of the CO_2 partial pressure in the atmosphere (p_a). The remaining 35% is accounted for by the second and third causes. Quantitatively, this relationship is characterized by a 6% increase of the atmospheric CO_2 partial pressure per 1°C increase of the temperature of the upper layer of the ocean. Also, a 1% decrease of the ocean volume increases p_a by 3%.

An assessment of the greenhouse effect requires complex consideration of the interaction of all processes of energy transformation on the Earth. However, in the diversity of the processes (from astronomical to biological) that affect the climatic system on various timescales, there exists a hierarchy of significance. This hierarchy cannot be constant, since the role of some processes can vary in their significance to climatic variations. In fact, the impact of the greenhouse effect is determined by an exceess of surface temperature T_L over the efficient temperature T_e. The Earth's surface temperature T_L is a function of surface emissivity κ. The effective temperature T_e is a function of emissivity α of the atmosphere–land–ocean system. In a general case, the parameters κ and α depend on many factors, in particular, on the CO_2 concentration in the atmosphere. There are a lot of simple and complicated numerical models where attempts have been made to parameterize these dependences. Unfortunately, there is not a single model that would meet the requirements of adequacy and reliably describe the pre-history of the climatic trends on the Earth. Nevertheless, one can state that the greenhouse effect depends non-linearly on the difference $T_L - T_e$ (i.e., on atmospheric turbidity, especially in the long-wave region). The more CO_2 in the atmosphere, the stronger is the atmospheric turbidity. The strongest effect of CO_2 on the atmospheric turbidity is in the long-wave region (12–18 μm). This effect is weaker in the wavelength intervals 7–8, 9–10, 2.0, 2.7, and 4.3 μm. It is clear that with the increasing partial pressure of CO_2 in the atmosphere the role of various bands of CO_2 will grow, and this means that with the intensified CO_2 absorption bands the upward long-wave radiation flux will decrease. At the same time, the downward long-wave radiation flux on the Earth's surface will increase. From the available estimates, a reduction of the global mean upward and increase of the downward fluxes are estimated at 2.5 and 1.3 W/m^2, respectively.

Thus to estimate the level of the greenhouse effect due to CO_2 and other gases, it is necessary to know how to predict their concentration in the atmosphere with account of all the feedbacks in their global biogeochemical cycles (Watson *et al.*, 2000; Grant *et al.*, 2001; Grogan *et al.*, 2001; Guaduong and Masao, 2001; Grossman, 2001). This problem touches upon several spheres of science – biogeochemistry, geochemistry, soil science, ecology, agrichemistry, geology, oceanography, physiology, and radiochemistry. The present methods of global ecoinformatics enable one to combine the knowledge accumulated in these fields (Krapivin and Kondratyev, 2002).

Of course, the global cycle of chemicals should be studied not only to be able to assess the climatic consequences of the anthropogenic activity but also to understand the prospects of environmental dynamics from the viewpoint of its quality and possibility for life. Since the cycles of chemicals in nature are closely connected with living substance activity, one can single out the geological, biogenic, and biological cycles of this rotation. The biogenic cycle includes subcycles, such as biogeochemical, biogeocenotic, and geochemical (Haan *et al.*, 2001; Heans, 2001; Hensen *et al.*, 1998; Korner, 1998; Nepstad *et al.*, 2002).

6.2 PHYSICAL PROCESSES OF PROPAGATION OF CHEMICALS IN THE ATMOSPHERE AND THEIR MODELLING

6.2.1 Ratio of the scales of physical processes of mixing in the atmosphere and the choice of model type

An experience of numerous studies on modelling the atmospheric processes of the transition of pollutants suggests a necessity to classify these processes according to relationships between space and timescales. The necessity of this classification is based on many international programmes – Global Change, Programme on Global Atmospheric Chemistry, and Modelling the Global Biogeochemical Cycles (Marchuk and Kondratyev, 1992). The need to classify the physical

Table 6.1. List of scales of cartographic information presentation characteristics of the developed systems of monitoring.

Spatial resolution (km)	Scale	Spatial resolution (km)	Scale
0.5 × 0.5	1:1,250	40 × 40	1:50,000
1 × 1	1:2,500	250 × 250	1:250,000
5 × 5	1:10,000	500 × 500	1:625,000
10 × 20	1:25,000		

processes of atmospheric mixing is dictated by parameters of the systems of monitoring the atmospheric characteristics, requirements for simplicity of models of aerosol and gas transport in the atmosphere, as well as by limited databases. Of course, this leads to a natural gradation of space scales, which ensures an efficient parameterization of the processes of propagation of atmospheric pollutants and agrees with international standards. At present, the most widely used is the standard Geophysical Data System (GEODAS) which provides nine levels of space resolution from one degree to a one half-second by latitude and longitude. In this case, seven scales of cartographic reproduction of data are provided (Table 6.1).

Data for these scales can be obtained by synthesis of satellite data and data of national monitoring systems. The latter are important since they concretely define the subject area and establish priorities characteristic of this region. For instance, for developed countries, of importance are prompt estimates of the quality of atmospheric air in the zones of megalopolises and large industrial enterprises. For developing countries, of principal importance are control of trans-boundary transport of pollutants and estimations of possible changes of atmospheric air in connection with the construction of industrial enterprises. The ratios of scales given in Table 6.1 meet most situations of atmospheric monitoring.

Along with the problem of the choice of scales of spatial resolution, there is a problem with timescales. This problem is important for the formation of the structure of numerical models, which would provide a description of the dynamics of pollutants in the atmosphere. According to preliminary estimates within the International Geosphere–Biosphere Programme (IGBP), there is a spatial–temporal scale of transitions between complexity and depth of the hierarchical structure of feedbacks considered in the model. Earlier models did not take into account this fact and often could not be applied to the natural object under study. The coordinated scale of digitizing natural phenomena to be used in construction of the models proposed by Tiller (1986) enables one to classify natural phenomena reflecting their subordination in space and timescales. This classification is based on a fundamental understanding of the hierarchical order in the general theory of systems. According to this theory, the behaviour of any complex system is determined by a triad of frequencies of its variability, which provides an agreement between connectedness and stability of the system. Such a view makes it possible to cut off excess details in the model structure by an a priori taken timescale or to establish a minimum time step from the data on the space scale. For instance, if the time step in the model is chosen to be equal to one year, there is no sense in taking into account processes such as atmospheric turbidity. In other words, in this case a point model can reliably describe the atmosphere, and all attempts to construct complex models, which would describe the processes of atmospheric motion cannot raise the reliability of the model, but will only increase its complexity.

A more strict theoretical substantiation of this approach to the choice of the model structure is provided by the Tikhonov (1952) theorem. A combined description of dynamic processes of different characteristic timescales brings forth parameters, which refer to different processes and strongly (by orders of magnitude) differ from each other. This makes it possible to divide the processes under consideration into three groups:

- Processes of the chosen timescales.
- Processes which, with respect to the chosen timescale, can be considered to be in dynamic equilibrium (quasi-stationary) and for which, on these grounds, some parameterization can be introduced (fast processes).
- Processes which can be considered, with respect to the chosen timescale, as unchanging (i.e., static (slow processes)).

As an example, we take a numerical model of the biogeochemical carbon cycle, which includes both processes of the geological timescale and very fast processes (photosynthesis and respiration). If the goal of the model is to study the dynamics of the CO_2 content in the atmosphere during several decades, one should exclude from consideration the geological processes with characteristic timescales of millions of years and parameterize the processes with characteristic timescales of days, for instance, the processes of the living biomass formation via photosynthesis (Krapivin and Kondratyev, 2002). In general, these problems have not been solved yet, and there is no constructive mechanism of agreement between space and timescales. Each specialist has a concept of his own for choosing the elements of the model for their subsequent realization. Unfortunately, this stage is characterized by model deviations from reality that are not eliminated in the future.

The characteristic time intervals of variability for most of the natural processes are well known. Here are some of them:

- Processes of sedimentation – minutes, hours;
- Transpiration of plants – hours, days;
- Formation of plant biomass – days, months;
- Changes of plants and animals in communities – months, years;
- Formation of soil – years, centuries;
- Geomorphological processes – centuries, millennia.

Therefore, before synthesizing a system of models, timescales of the considered processes should be systematized. As a result, the priorities of models and their units

Table 6.2. Characteristics of the most developed models of atmospheric motion.

Model	Characteristic of the model	
	Spatial resolution	Temporal resolution
IKE (Institut für Kernenergetik und Energiesysteme)	1 × 1 km	Hours
LRTAP (Long-Range Transport of Airborne Pollutants)	500 × 50 km	Days
ADOM (Acidic Deposition and Oxidants Model)	20 × 20 km	Days
TADAP (Transport and Deposition of Acidifying Pollutants)	5 × 5 km	Hours
ECMWF (European Centre for Medium-Range Weather Forecasts)	150 × 150 km	6 hours
ICLIPS (Integrierte Abschatzung von Klimaschutzstrate-gien)	500 × 500 km	Year
THD (Technishe Hochschule Darmstand)	100 × 100 km	Hours

can be established and, thereby, the structure of the programme provision can be determined.

Table 6.2 synthesizes some information about available models of atmospheric dynamics. As is seen, the type of the model correlates strongly with the space scale.

Consideration of the model of the atmospheric physical mixing is determined by its accuracy. Table 6.2 characterizes a set of models of atmospheric pollution having different spatial and temporal resolutions. These models, depending on their space scale, take into account the processes of physical transformation of pollutants, from microprocesses in clouds to large-scale motions of the atmosphere. Depending on this, models are divided into dispersible, Gaussian, Eulerian, and Lagrangian. Within this gradation, there is a hierarchy of models including or excluding an account of the vertical structure of the atmosphere, the interaction of atmosphere and the Earth's surface (land and water), exchange processes between clouds as well as the vertical air fluxes, and depending on relationships between the synoptic and the physical parameters of the atmosphere.

It is impossible to determine a unique bond between the scale of the model and its internal infrastructure without considering other characteristics of the model. Therefore, the estimates given in Table 6.3 should be considered as recommended, aimed at orientation of the model's constructor, enabling him to assess the speed of inclusion of these components into the model.

6.2.2 Classification of atmospheric pollutants

Formation of the fields of atmospheric pollution from natural and anthropogenic sources depends strongly on physical characteristics of pollutants. Sections of the physics of clouds and precipitation, where mechanisms of the microphysical and macrophysical processes of formation and growth of polluting clouds are studied in detail, are dedicated to the analysis of this problem. Clearly, to understand the

Table 6.3. A fragment of the Tiller (1986) scale used to model atmospheric pollution dynamics.

Spatial resolution of the model	Processes to be taken into account in the model
Industrial region, landscape, megalopolis, city (up to 50 km)	Use of the Gauss-type model. Burning of wastes, deforestation and reconstruction of land cover, pollution of drinking water and water bodies, industrial emission of pollutants, soil pollution, washing out of pollutants by precipitation, production processes, medico-biological estimate of territory, and division of the atmosphere into many levels.
Large region, oblast, country (up to 1000 km)	Use of the Lagrange and Euler-type models. Large-scale atmospheric circulation with the selection of the upper and lower atmosphere, irrigation and other water systems, integral areal sources of biospheric pollution, biogeochemical cycles, erosion, large-scale fires, desertification and swamping, succession of land cover, river runoff, and interactions on shelfs.
Continent, globe (>1000 km)	Use of unit models. Averaged characteristics of the atmosphere and climate, oceanic circulation, interactions in the "atmosphere–land–ocean" system, biogeochemical cycles, and succession of large forest areas.

processes of cloud formation and precipitation, it is necessary to consider the dynamic microphysical processes altogether. These processes are determined by a combination of the physical parameters of the atmosphere itself and pollutants with widely changing characteristics, such as weight, size, form, phase state, and electrical charge. For instance, the classification of atmospheric pollutants assumed by the US National Agency on the Environment includes three base classes and eight subclasses (Table 6.4). This classification can be used in the models which do not take into account the size of aerosols and particles and do not include the processes at ionic-level. The existing classification of particles by size covers the diameters of particles from 0.0001 μm to 1 cm. Within this range, solid particles with diameters from 0.0001 to 1 μm are considered as smoke components, and liquid particles are elements of fog. Particles exceeding 1 μm in diameter are interpreted as dust or spray. Depending on the size, the role of particles in the dynamic processes of atmospheric pollution changes. Particles less than 1 μm form smog, tens of μm form haze and drizzle, and thousands of μm form rain. Table 6.5 summarizes some typical aerosols and their characteristics. This classification simplifies the choice of the model's structure, if the character of atmospheric pollution has been determined. In a more complicated situation, when the spectrum of pollutions is sufficiently broad, division of pollutants by their physical characteristics enables one to synthesize a complex model as a set of hierarchically subordinated partial models and to thereby simplify the procedure of calculations of the dynamic characteristics of the polluted atmosphere.

Table 6.4. Classification of atmospheric pollutants.
From Straub (1989).

Basic classes	Subclasses	Characteristic elements
Inorganic gases	Nitrogen oxides (NO_x)	Nitrogen dioxide, nitric acid
	Sulphur oxides (SO_x)	Sulphur dioxide, sulphuric acid
	Other inorganic compounds	Ammonia, carbon dioxide, chlorine, hydrofluorine, sulphide, ozone
Organic gases	Hydrocarbons	Petrol, butadiene, butane, ethylene, isooctane, methane
	Aldehydes, ketones	Acetone, formaldehyde
	Other organic compounds	Acids, spirits, chlorinated hydrocarbons, peroxiacyl nitrate, polycyclic aromatic substances
Aerosols	Solid particles	Dust, smoke
	Liquid particles	Vapour, oil fog, products of reactions with polymers

Table 6.5. Properties of some types of aerosols.
From Straub (1989).

Type of dispersal system	Size range of particle diameters (μm)
Raindrops	5000–500
Natural mist	500–40
Natural fog	40–1.0
Foundry sand	2000–200
Limestones, fertilizers	800–30
Sand tailings from floatation	400–20
Pulverized coal	400–10
Sulphide ore for floatation	200–4
Dust from foundries	200–1
Cement	150–10
Fly ash	80–3
Silica dust in silicosis	10–0.5
Pigments	8–1.0
Pollens	60–20
Plant spores	30–10
Bacteria	15–1.0
Metallurgic smoke	100–0.1
Fog of the sulphuric acid thickener	10–0.1
Spirit evaporation	2–0.1
SO_3 fog	3–0.5
NH_4Cl vapour	2–0.1
Zinc oxide evaporation	0.3–0.03
Oil smoke	1–0.03

The physical characteristics of atmospheric pollutants also include the rate of gravitational sedimentation, lifetime in the atmosphere, and phase state. Some of the gaseous air components, such as N_2, O_2, He, Ne, Ar, Kr, Xe, and H_2, have very long lifetimes. The lifetimes of others, such as CO_2, O_3, N_2O, and CH_4, are between several and tens of years. Such gases as H_2O, NO_2, NO, NH_3, SO_2, H_2S, CO, HCl, and I_2 live in the atmosphere for only days or weeks. Depending on only this characteristic to describe the dynamics of various gases in the atmosphere one can choose an adequate model with a minimum requirement for a database.

The size of the polluting particles is important for the choice of the model type. Gradations given in Table 6.5 only partially cover the classification of aerosols. Here it is necessary to add information about the source of pollutant, which helps to specify the parametric space of the model. Knowledge of the cause of pollution makes the choice of the model type easier. Figure 6.1 demonstrates an attempt to formalize one of the main physical processes affecting the level of concentration of pollutants in the atmosphere – sedimentation (Zhang *et al.*, 2002). Here are shown the processes which should be studied and modelled in order to raise the objectivity of monitoring.

Of course, aerosols and gases can be classified and typified in more detail. There are tens of types of smoke. The sizes of smoke particles can be 1–0.01 μm for resin smoke, 0.15–0.01 μm for tobacco smoke, etc. Here the model must reflect micro-processes connected with the motion of these particles. For instance, the average run of a particle of carbon smoke for *t* seconds is equal to 0.00068 t/D/cm (D is the particle diameter (μm)). Knowledge of such dependencies in the system of monitoring is a priority problem for ecoinformatics.

When modelling the scattering of gases and particles in the atmosphere, it is important to know the difference between the polluted and clean atmosphere. Also, one should always bear in mind the vertical heterogeneity of the atmosphere. On a global scale, the formation of the atmospheric air quality is dependant on all levels of the atmosphere: troposphere, stratosphere, chemosphere, and ionosphere. If, for instance, the problem of the effect of aviation on the atmosphere is considered, then it is necessary to take into account the interaction between the troposphere and the stratosphere. When studying the fluxes of pollutants from land sources, first of all one should consider the motion of the lower atmosphere. Certainly, spatial scales and lifetimes of the pollutants in the atmosphere are important for the investigation of their fluxes. Table 6.6 gives data for a pure atmosphere, which should be used in the control of numerical experiments. The formula to recalculate the volume characteristics of gaseous components into weight characteristics can be useful here:

$$1\,\mathrm{ppm(vol)} = \frac{1\ \text{litre of pollutant}}{10^6\ \text{litres of air}} = 40{,}9\,\mathrm{M}\,\mu\mathrm{g/m}^3$$

where M is the molecular weight. Table 6.7 lists the coefficients of re-calculation between weight and volume units for some pollutants.

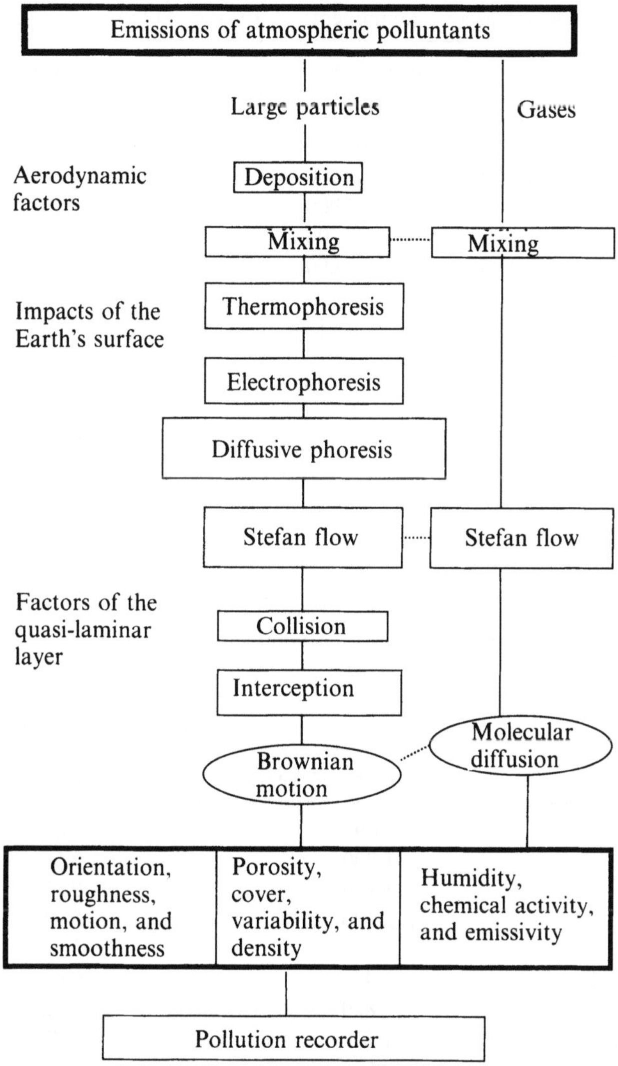

Figure 6.1. Diagram of the processes affecting the rate of the motion of gases and large particles in the atmosphere.

6.2.3 Passive and active propagation of pollutants in the atmosphere

A pollutant ejected into the atmosphere suffers gravitational sedimentation, turbulent mixing, wind-driven transport, and washing out by rain. All these forcings determine the behavioural characteristics of the polluting cloud, the form and appearance of the flux of the pollutant, as well as the spatial distribution of the pollutant density over a given territory.

Smoke and other atmospheric aerosols are subject to gravity and interact with solar radiation, gases, and ions. In the surface air layer this interaction is

Table 6.6. Gaseous components of a non-polluted atmosphere.

	Basis (ppmv)		Basis (μg/m^3)	
Gas	Dry	Wet	Dry	Wet
Nitrogen	780,900	756,500	8.95×10^8	8.67×10^8
Oxygen	209,400	202,900	2.74×10^8	2.65×10^8
Water	–	31,200	–	2.30×10^7
Argon	9,300	9,000	1.52×10^7	1.47×10^7
CO_2	315	305	5.67×10^5	5.49×10^5
Neon	18	17.4	1.49×10^4	1.44×10^4
Helium	5.2	5.0	8.50×10^2	825
Methane	1.0–1.2	0.97–1.16	656–787	635–763
Krypton	1.0	0.97	3,430	3,320
Nitrogen oxide	0.5	0.49	900	873
Hydrogen	0.5	0.49	41.3	40
Xenon	0.08	0.08	429	417
Organic vapour	0.02	0.02	–	–

Table 6.7. Coefficients of transition between volume and weight units at 25°C and 760 mm Hg.

	Transition coefficient	
Pollutant	ppm → μkg/m^3	μkg/m^3 → ppm ($\times 10^{-3}$)
Ammonia (NH_3)	695	1.44
CO_2	1,800	0.56
CO	1,150	0.87
Chlorine	2,900	0.34
Ethylene	1,150	0.87
Hydrogen chloride	1,490	0.67
Hydrofluorine	820	1.22
Sulphide chloride	1,390	0.72
Methane (carbon)	655	1.53
NO_2	1,880	0.53
Nitrogen oxide	1,230	0.81
Ozone	1,960	0.51
Peroxyacetilnitrate	4,950	0.20
SO_2	2,620	0.38

supplemented by various effects of the Earth's surface (vegetation cover, soil, land surface roughness, and sea surface roughness). On the whole, the mechanism of dry sedimentation of gases and aerosols in the absence of precipitation is an important reason for reducing the volume of pollutants in the atmosphere. The rate of activity of this mechanism is different over water and solid surfaces. In each case there are specific processes, which determine the rate of sedimentation of the pollutant. In the first case these are water surface roughness, humidity, and spray. In the second case – the type of land surface and vegetation cover.

The role of sedimentation manifests itself most in cases of large particles with diameters much greater than submicron sizes. Small particles themselves sediment much slower than they are transported by the moving atmosphere, and therefore many models neglect this vertical component. Note that for the process of sedimentation the diameter of the particle is not as important as its density. For instance, soot structures with low efficient density and large aerodynamic sections are easily transported by wind and sediment much slower than compact particles with the same mass. According to Zhang *et al.* (2002), the rate of sedimentation of particles with diameters 0.1–1 μm averages 0.001 m/s, which is negligibly small compared to the rate of atmospheric transport.

In heavily polluted places one can observe the process involved in the uplifting of particles due to non-uniform heating by sunlight. Both the possibility of this phenomenon and its characteristics have been poorly studied. Therefore, in a first approximation many authors do not take this into account. In this regard, during time intervals longer than one day, due to the Brownian motions, the irregularity of particle heating decreases. Finally, note such a physical process as coagulation of particles – capturing of some particles by others due to different rates of motion. In this case, particles can either stick together or repell each other. Their interaction determines the form of the cloud of pollutant and can prolong their residence times in the atmosphere.

The washing out of particles is an important process of removing pollutants from the atmosphere. Here two situations are possible. One is connected with a simple capture of particles by rain droplets, the other with the so-called process of nucleation. This process is connected with condensation of the oversaturated water vapour on the surface of aerosol particles, which leads to the formation of water droplets or ice crystals with subsequent sedimentation onto the Earth's surface.

The final estimate of the time of residence of a given pollutant in the atmosphere is obtained using the respective models. This estimation has been discussed in detail in Zhang *et al.* (2002). The authors described the meteorological features of the pollutants' propagation in the atmosphere, analysed the scales of the transport and scattering of pollutants, constructed the models which predict the concentration of pollutants, described the algorithms of parameterization of the processes of formation of clouds and jets of pollutants, and gave formulas to describe the vertical structure of the atmosphere. Components of the Earth's radiation budget were analysed and the simplest characteristics of correlation between pressure, wind, temperature, and humidity were given. The states of the atmosphere were classified as neutral, unstable, and stable using the scale of the vertical temperature lapse rate, which simplifies the parameterization of the vertical gradients and velocities. The scale of atmospheric phenomena is determined in the time range from 1 second to 1 month and with spatial scales varying between 20 km and 10,000 km. Within this scale, an analysis was made of the processes of transport and scattering of pollutants in the atmosphere from the point sources, moving, and distributed sources. The Meteorological Service of Canada (MSC) developed A Unified Regional Air-Quality Modelling System (AURAMS) aimed at studying particulate matter and other atmospheric pollutants in the regions of North America. The models developed earlier by MSC serve as the base units for AURAMS:

- the Chemical Tracer Model/Mesoscale Compressible Community Model (CTM/MC2);
- ADOM;
- the Canadian Aerosol Module (CAM); and
- the Canadian Emissions Processing System (CEPS).

CTM/MC2 provides a simulation of photochemical oxidations in the regional-scale atmosphere, based on the prognostic meteorological model. ADOM covers most substantial episodes of acid sedimentation, which can happen in the climatic zones of North America. The CAM describes the processes of aerosols sedimentation with their size taken into account and makes it possible to combine AURAMS with the global or regional climate model. Finally, the CEPS forms the input files with data needed for other units of AURAMS.

Zhang *et al.* (2002) determine the process of dry sedimentation of pollutants from the atmosphere by the velocity v_1 of its vertical transport calculated as an inverse value of total air resistance $R_t = R_a + R_b + R_c$, where R_a is the aerodynamic resistance, R_b is the quasi-laminar sublayer resistance, and R_c is the surface or canopy resistance. The constituent R_a depends on meteorological conditions and surface roughness characteristics and is independent of the chemical properties of a pollutant. The value R_b is a function of the rate of friction and characteristics of molecular scattering of chemicals. The surface resistance R_c is calculated by the formula:

$$\frac{1}{R_c} = \frac{1 - W_{st}}{R_{st} - R_m} + \frac{1 - W_c}{R_{cutd}} + \frac{W_c}{R_{cutw}} + \frac{1 - W_g}{R_{gd}} + \frac{W_g}{R_{gw}}$$

where W_{st} is the fraction of stomatal closure when the leaves are wet; W_c is the fraction of canopy that is wet (if there is a canopy); W_g is the fraction of the ground surface that is wet; and R_{st}, R_m, R_{cukd}, R_{cutw}, R_{gd}, and R_{gw} are the canopy stomatal, mesophyll, dry cuticle, wet cuticle, dry ground, and wet ground resistance, respectively. For bare soil, the first three terms in this formula are excluded from consideration.

In conditions of increased humidity, canopy and ground can be either totally or partially covered with water. Therefore, the parameters W_{st}, W_c and W_g have a wide range of values. Zhang *et al.* (2002) proposed a simplified presentation of these parameters: $W_{st} = 0.5$, $W_c(W_g) = 0.9(0.9)$ for rainy conditions, 0.7(0.5) for dewy conditions, and 0.2(0.2) for humid conditions. A more detailed parameterization of W_c and W_g is given in Sellers *et al.* (1996c).

The calculation of the indicators of surface resistance requires additional specifications of the structure of the elements of vegetation cover. In particular, to calculate R_{st}, the canopy is divided into illuminated and shadowed parts:

$$R_{st} = 1/[G_{st}(E)f_1(T_a)f_2(D)f_3(\psi)D_i/D_\nu]$$

where G_{st} is the unstressed canopy stomatal conductance; E is the photosynthetically active radiation (PAR); $f_i(i = 1, 2, 3)$ are the dimensionless functions representing the conductance-reducing effects of air temperatute T_a, water vapour pressure deficit D, and water stress (leaf water potential) ψ, respectively, on leaf stomatal conductance; and D_v and D_i are the molecular diffusivities for water vapour and the pollutant gas, respectively.

The parameter G_{st} depends on leaf illumination:

$$G_{st}(E) = F_{\text{sun}}/r_{st}(E_{\text{sun}}) + F_{\text{shade}}/r_{st}(E_{\text{shade}})$$

where F_{sun} and F_{shade} are the total sunlit and shaded leaf area indexes (LAIs), respectively; E_{sun} and E_{shade} are PAR received by sunlit and shaded leaves, respectively; and r_{st} is the unstressed leaf stomatal resistance (LSR), which is calculated by the formula:

$$r_{st}(E) = r_{st,\min}(1 + b_{rs}/E)$$

Here $r_{st,\min}$ corresponds to a minimum value of LSR, and the coefficient b_{rs} is determined empirically. Parameters F_{sun} and F_{shade} are estimated by the formula:

$$F_{\text{sun}} = 2\cos\theta[1 - \exp\{1 - 0.5\text{LAI}\cos\theta\}] \qquad F_{\text{shade}} = \text{LAI} - F_{\text{sun}}$$

Parameters E_{sun} and E_{shade} are calculated in accordance with approximations:

$$E_{\text{sun}} = \begin{cases} R_2\cos\alpha/\cos\theta + E_{\text{shade}} & \text{for LAI} < 2.5 \text{ or } E < 200\,\text{Wm}^{-2} \\ R_2^{0.8}\cos\alpha/\cos\theta + E_{\text{shade}} & \text{for all other conditions} \end{cases}$$

$$E_{\text{shade}} = \begin{cases} R_1\psi_1 - 0.07R_2\psi_2 & \text{for LAI} < 2.5 \text{ or } E < 200\,\text{Wm}^{-2} \\ R_1\psi_3 - 0.07R_2\psi_2 & \text{for all other conditions} \end{cases}$$

where $\psi_1 = \exp\{-0.5\text{LAI}^{0.7}\}$; $\psi_2 = (1.1 - 0.1\text{LAI})\exp\{-\cos\theta\}$; $\psi_3 = \exp\{-0.5\text{LAI}^{0.8}\}$; θ is the solar zenith angle; α is the angle between the leaf and sun; and R_1 and R_2 are the downward visible radiation fluxes above the canopy from diffuse and direct-beam radiation, respectively. The angle α averages 60° for a spherical canopy.

Estimates of R_m are rather uncertain. Zhang *et al.* (2002) use the following approximation: $R_m = 100$ s/m for species with both relatively limited solubility and very small oxidizing capacity, and $R_m = 0$ for all other species. Finally, the values R_{cutd}, R_{cutw}, R_{gd}, and R_{gw} are estimated depending on the properties of various chemicals and soil–plant formations. Tables 6.8 and 6.9 give some estimates of R_{cutd} and R_{gd} for O_3 and SO_2. A variety of such estimates as functions of the types of surfaces and seasonally varying characteristics of the Earth's cover complicates the parameterization of the rates of sedimentation of air pollutants and increases, thereby, the uncertainty of the results of modelling the processes of

Table 6.8. Examples of R_{cutd} (s/m) estimations for SO_2 and O_3 during two seasonal categories. See for details Zhang *et al.* (2002).

	Midsummer with lush vegetation				Late autumn after harvest, no snow			
Surface category	LAI	z_0	$10^3\ SO_2$	$10^3\ O_3$	LAI	z_0	$10^3\ SO_2$	$10^3\ O_3$
Evergreen needleleaf trees	5.2	0.8	1.0	1.0	4.2	0.9	2.0	2.0
Evergreen broadleaf trees	4.5	2.65	1.0	1.0	4.5	2.65	2.0	2.0
Deciduous needleleaf trees	1.1	0.85	2.0	2.0	0.3	0.8	8.0	8.0
Deciduous broadleaf trees	3.4	1.05	1.2	1.2	0.1	0.95	9.0	9.0
Mixed broadleaf and needleleaf trees	4.5	1.15	1.0	1.0	2.3	1.15	2.0	2.0
Grass	2.0	0.1	1.5	1.5	1.0	0.05	3.0	3.0
Crops, mixed farming	2.0	0.1	1.5	1.5	1.0	0.02	3.0	3.0
Shrubs and interrupted woodlands	0.5	0.1	4.0	4.0	1.5	0.1	2.0	2.0
Wetlands with plants	0.2	0.03	6.0	6.0	0.1	0.02	9.0	9.0
Urban	0.3	1.0	6.0	6.0	0.1	1.0	9.0	9.0

Note: z_0 = surface roughness (m).

Table 6.9. Examples of R_{gd} (s/m) estimations for SO_2 and O_3 during two seasonal categories. See for details Zhang *et al.* (2002).

	Autumn with cropland before harvest				Winter, with snow, subfreezing			
Surface category	LAI	z_0	$10^3\ SO_2$	$10^3\ O_3$	LAI	z_0	$10^3\ SO_2$	$10^3\ O_3$
Evergreen needleleaf trees	5.2	0.9	0.4	0.5	5.5	0.9	0.2	3.0
Evergreen broadleaf trees	4.5	2.65	0.3	0.6	4.5	2.65	0.2	32.0
Deciduous needleleaf trees	0.8	0.9	0.4	0.45	0.0	0.55	0.2	3.0
Deciduous broadleaf trees	1.9	1.05	0.4	0.5	0.0	0.55	0.2	3.0
Mixed broadleaf and needleleaf trees	3.5	1.15	0.4	0.5	2.3	1.15	0.2	3.0
Grass	1.5	0.1	0.35	0.4	0.5	0.02	0.2	3.0
Crops, mixed farming	1.5	0.1	0.35	0.4	0.0	0.02	0.2	3.0
Deserts	0.0	0.04	0.7	0.5	0.0	0.04	0.2	3.0
Tundra	0.0	0.03	0.4	0.8	0.0	0.03	0.2	3.0
Shrubs and interrupted woodlands	2.5	0.1	0.4	0.4	1.2	0.1	0.2	3.0
Wetlands with plants	0.2	0.03	0.05	1.0	0.0	0.02	0.2	3.0
Ice caps and glaciers	0.0	0.01	0.1	2.0	0.0	0.01	0.2	3.0
Inland water, ocean	0.0	$Z_0(u)$	0.01	2.0	0.0	$Z_0(u)$	0.01	2.0
Urban	0.2	1.0	0.5	0.5	0.0	1.0	0.4	2.0

Note: z_0 = surface roughness (m); u = wind speed (m/s).

atmospheric self-cleaning. According to the estimates in Zhang *et al.* (2002), some suppositions can be made on the character of these dependences. In particular, canopies with larger LAIs and younger leaves have smaller cuticle resistance. But taller canopies have relatively larger and in-canopy aerodynamic resistances. Canopies with constant wet soil have smaller ground resistances for SO_2 and slightly larger resistance for O_3.

For the parameter R_{gw} the following approximation is valid (Zhang *et al.*, 2002):

$$R_{gw} = R^*_{gw}(\mathrm{LAI} + 0.1)^{-1.3}$$

where R^*_{gw} (s/m) is an empirical constant estimated at:

$$R^*_{gw}(SO_2) = \begin{cases} 100 & \text{under rainy and high-humidity conditions} \\ 200 & \text{under dewy conditions} \end{cases}$$

$$\mathrm{R}^*_{gw}(O_3) = \begin{cases} 400 & \text{under rainy and high-humidity conditions} \\ 800 & \text{under dewy conditions} \end{cases}$$

The AURAMS considers 48 chemicals, whose surface resistance to sedimentation is estimated from the data for SO_2 and O_3:

$$\frac{1}{R_{cutd}(i)} = \frac{\alpha(i)}{R_{cutd}(SO_2)} + \frac{\beta(i)}{R_{cutd}(O_3)}$$

$$\frac{1}{R_{cutw}(i)} = \frac{\alpha(i)}{R_{cutw}(SO_2)} + \frac{\beta(i)}{R_{cutw}(O_3)}$$

$$\frac{1}{R_{gd}(i)} = \frac{\alpha(i)}{R_{gd}(SO_2)} + \frac{\beta(i)}{R_{gd}(O_3)}$$

$$\frac{1}{R_{gw}(i)} = \frac{\alpha(i)}{R_{gw}(SO_2)} + \frac{\beta(i)}{R_{gw}(O_3)}$$

where i is the identifier of the element; and α and β are some constants characteristic of each chemical whose calculation is based on Henry's Law constant $H^*$and value for the negative log of electron activity for half-redox reaction in neutral aqueous solution $pe^0(w)$. Based on these two input characteristics, a binary classification of v_1 is usually considered:

$$v_1 = \begin{cases} 0 & \text{for } pe^0(w) < -2.0 \text{ and } H^* < 100 \\ >0 & \text{for all of other conditions} \end{cases}$$

Some estimates of these characteristics are given in Table 6.10. Parameters α and β for other chemicals can be calculated on the basis of their estimates for SO_2 and O_3 as basic components of the gas-phase processes of dry and wet sedimentation of atmospheric pollutants. Here some additional remarks should be made. Estimates of α and β given in Table 6.10 are not always based on direct measurements. For example, sulphuric acid vapour (H_2SO_4) has a high soluability and an ability for oxidation. That is why sulphuric acid is rapidly removed from the atmosphere. On

Table 6.10. Chemical and physical parameters of some chemical species.
From Zhang *et al.* (2002).

Chemical species	σ	H^*	$pe^0(w)$	α	β
Sulphur dioxide	64	2,650	75.5–−7.6	1	0
Sulphuric acid	98	$>2.1 \times 10^5$	4.9–−4.3	1	1
Nitrogen dioxide	46	$(0.3–4.1) \times 10^{-2}$	28.4–−8.2	0	0.8
Ozone	48	$(0.9–1.3) \times 10^{-2}$	28.1–−18.6	0	1
Hydrogen peroxide	34	$(0.7–1.4) \times 10^5$	24.8–−9.7	1	1
Nitric acid	63	3,200	14.1–−8.9	10	10
Nitrous acid	47		17.5–−14.8	2	2
Peroxyacetylnitrate	121	2.8–5.0	30.2–−1.5	0	0.6
Peroxypropylnitrate	135	2.9	37.8–−2.3	0	0.6
Aromatic acylnitrate	183	5	46.9–−11.2	0	0.8
Peroxymethacrylic nitric anhydride	147	1.7	3.1	0	0.3
Formaldehyde	30	4,900	3.0–−0.1	0.8	0.2
Acetaldehyde	44	≥15	1.0–−3.9	0	0.05
C3 carbonyls	58	2.4–3.7	−1.3–−1.8	0	0.05
C4-C5 carbonyls	72	9–18	1.3–−1.8	0	0.05
C6-C8 carbonyls	128	4–110	1.5	0	0.05
Aromatic carbonyls	106	35–42	−1.0–−2.3	0	0.05
Organic nitrates	77	2.0	10.5–−5.0	0	0.5
Organic peroxides	48	10–310	4.2–−3.6	0.1	0.8
Acetic acid	60	9.6×10^5	3.1–−9.6	1.5	0
Formic acid	46	9.8×10^6	1.9–−6.4	2	0
Creosol	104	830	−2.5	0.01	0
C3 alcohol	60	≥90–170	0.3	0.4	0
Erhyl alcohol	46	≥120–230	−1.3–−2.9	0.6	0
Methyl-vinyl-ketone	70	21–44	0.2	0	0.05
Methacrolein	70	4.3–6.5	−1.2	0	0.05
Methylgloxal	72	4–32	−0.7	0.01	0
Methyl alcohol	32	≥140–230	3.0	0.6	0.1

Note: H^* is the effective Henry's Law constant; $pe^0(w)$ is the negative log of electron activity for the half-redox reaction in a neutral aqueous solution; σ is the molecular weight.

the contrary, NO_2 is characterized by low soluability and a very high ability for oxidation. Dry sedimentation of NO_2 has the same speed as O_3.

There are direct measurements of v_1 for NO_2 which show that in this case there are a certain variety of deviations from this supposition. In particular, the rate of dry sedimentation of NO_2 depends strongly on its concentration. When its concentration is small, NO_2 sediments with the same rate as O_3. With high concentrations of NO_2, chemical and photochemical reactions with other components of the atmosphere begin to take effect. Therefore, in the daytime, the rate of NO_2 dry sedimentation is higher than at night. Also, NO_2 precipitates more rapidly over the vegetation covered soil than over a bare surface (bare soil, water). Here the effect of stomatal uptake is manifested. As Zhang *et al.* (2002) point out, in the daytime $v_l(NO_2) = (0.2\text{–}0.7)v_l(O_3)$, and at night $v_l(NO_2) = (0.3\text{–}1.0)v_l(O_3)$. According to Pilegaard *et al.* (1998), the greatest daytime rate of dry sedimentation of NO_2 and O_3 over harvested wheat is 0.35 and 0.5 cm/s, respectively. There is a certain set of such estimates for other substances. For instance, for nitrous acid over grass cover $v_l \in [0.2, 1.7]$ cm/s, and for ozone and formaldehyde over broadleaved forest v_l is 0.2 and 0.5 cm/s at night and 1.1 and 1.3 cm/s in the daytime, respectively. The dry sedimentation rate for formic acid and acetic acid over savannah during the rainy season changes from 0.3 to 1.0 cm/s, and in the dry seasons it is 0.2–0.64 cm/s. At the same time, with the transition to a forested territory in humid conditions, the values of v_l become higher: $v_l \in [0.6, 1.1]$ for formic acid and $v_l \in [0.42, 0.68]$ for acetic acid. The rate of dry sedimentation for NH_3 is of the same order (0.8–1.0 cm/s). Experimental observations show that v_l is highly dispersive and, hence, measurements are required in every climatic zone and over various surfaces. It is enough to say that there are experimental estimates of v_l, which show that $v_l(NH_3)$ can reach 20 cm/s, and for SO_2 it can even reach about 50 cm/s.

In view of the importance of the process of dry sedimentation of atmospheric pollutants, many authors develop models of various degrees of complexity, so that when synthesizing the respective units of a Global Simulation Model (GSM) there is a possibility to duplicate them and to create a flexible structure in order to adapt the global model to the available database. Among these models is the model developed by Zufall *et al.* (1999) based on the Reynolds-averaged Navier–Stokes equations with a modified version of the Renormalization Group (RNG) model for turbulence closure. The Reynolds stresses are calculated by the equation:

$$\rho u_i' u_j' = \frac{2}{3}\left(\rho k + \mu_i \frac{\partial u_i}{\partial x_j}\right)\delta_{ij} - \mu_i\left(\frac{\partial u_i}{\partial x_j} + \frac{\partial u_j}{\partial x_i}\right)$$

where ρ is the density of air; u' is the fluctuating component of the velocity; k is the turbulent energy per unit mass; δ is the Kronecker delta; $\mu_i = 0.0845\,\mathrm{k}^2/c$ is the turbulent viscosity; $c = k^{3/2}/l_z$ is the rate of energy dissipation; and l_z is the length scale. The characteristics k and l_z are calculated by solving equations of the pollutant transition.

There are two zones near the surface: a viscosity-affected region and a fully turbulent region. The viscosity-affected wall region is defined as the region with turbulent Reynolds number:

$$R_{y'} = \rho\sqrt{k}y/ \leq 200$$

where y is the normal distance to the wall and μ^* is the dynamic viscosity of air.

Modelling of the rough surface is usually based on the use of 2-D sinusoidal waves, though, of course, real roughness is characterized by a spectrum of wavelengths and other parameters, differing from sinusoidal law.

The motion of the pollutant in the atmosphere is subject to not only gravitational sedimentation. Therefore, in modelling dry sedimentation the so-called trajectory models are used based on relationships for the velocities of both horizontal and vertical motion, changes of kinetic energy, rate of energy dissipation, and shear stress. Ultimately, at each time moment, the movement of particle is determined by solving the equation of the balance of gravitational, retarding and uplifting forcings on particles. The uplifting force can be neglected. Therefore, the equation of particle motion is written in the form:

$$\rho_p dU_p/dt = -0.75 d_p^{-1} \rho_a C_D (U_p - U_a)\sqrt{(U_p - U_a)^2 + (W_p - W_a)^2}$$

$$\rho_p dW_p dt = -0.75 d_p^{-1} \rho_a C_D (W_p - W_a)\sqrt{(U_p - U_a)^2 + (W_p - W_a)^2} + g(\rho_p - \rho_a)$$

where U is the streamwise velocity; W is the normal velocity; d_p is the particle diameter; g is gravitational acceleration; and C_D is the dimensionless drag coefficient calculated as a function of particle Reynolds number. Subscripts "p" and "a" refer to the particle and air, respectively.

Analysis of solutions for equations of particle motion in the atmosphere suggests the following most significant conclusions:

- the rate of particle sedimentation approaching the surface decreases as a function of the type of surface;
- intensity of vertical motion of the particle increases with decreasing height over the surface;
- rough or uneven surfaces increase the turbulence and lead to an intensive collision of particles; and
- uneven surfaces due to hollows and slopes slow down the sedimentation of particles.

In a general form, a change of the concentration of any pollutant C is described by the equation:

$$\partial C(t, \varphi, \lambda, h)\partial t + \nabla \vec{V} C = \nabla D \nabla C + R \tag{6.1}$$

where $\vec{V}(V_\varphi, V_\lambda, V_h)$ is the wind speed; φ is latitude; λ is longitude; h is the height; t is time; D is the coefficient of molecular diffusion, and R represents the changes due to atmospheric turbulence, emission, and mixing.

Detailing of the components of Equation (6.1) requires an analysis of the partial processes of the atmospheric propagation of pollutants and the construction of the respective units of the general model (dynamic, correlative, probabilistic, system, evolutionary, etc.). As examples of such units we consider parameterizations used successfully in the models given in Table 6.2.

Problems of chemical interaction of atmospheric pollutants are also important, and their consideration in modelling further complicates the study. Therefore, most of the models of pollutant propagation in the atmosphere assume a priori that all components are mutually neutral. However, in some cases a parameterization of the processes of chemical transformation of pollutants becomes possible due to the use of statistical characteristics of chemical reactions or through descriptions of the laws of phase transitions. In particular, the transformation of SO_2 into H_2SO_4 can be described by the following model:

$$d[H_2SO_4]/dt = d[SO_4]/dt = W[SO_2]$$

where $W = 0.1\%/\text{day}$ in the daytime and $0.01\%/\text{day}$ at night.

Many different models have been created to describe the process of sedimentation of pollutants. So, Sinik *et al.*(1985) proposed several parameterizations for the coefficient of washing out:

$$r = 10^{-4} I^{1/2} \qquad r = \theta I^a \qquad r = -C^{-1} dC/dt \qquad r = 3.3 \times 10^{-4} I^{0.9}$$

where $I = RR/(24N)$ is the precipitation rate (mm/hr); RR is the monthly amount of precipitation (mm); N is the number of rainy days; and θ and a are parameters.

The diffuse equation is widely used:

$$\frac{\partial C}{\partial t} + V_\varphi \frac{\partial C}{\partial \varphi} + V_\lambda \frac{\partial C}{\partial \lambda} + V_h \frac{\partial C}{\partial h} = -\frac{\partial}{\partial \varphi}\left(\overline{V_\varphi C}\right) - \frac{\partial}{\partial \lambda}\left(\overline{V_\lambda C}\right) - \frac{\partial}{\partial h}\left(\overline{V_h C}\right) \tag{6.2}$$

If we suppose that in Equation (6.2) an advection prevails over diffusion in the direction h (i.e., $\partial\left(\overline{V_h C}\right)/\partial h \ll V_h \partial C/\partial h$), then an integration of Equation (6.2) with respect to λ gives:

$$\frac{\partial C_\varphi}{\partial t} + V_h \frac{\partial C_\varphi}{\partial h} + V_\lambda \frac{\partial C_\varphi}{\partial \lambda} = -\frac{\partial}{\partial \lambda} \int_{-\infty}^{\infty} \overline{V_\lambda C} d\varphi$$

where

$$C_\varphi = \int_{-\infty}^{\infty} C d\varphi$$

This transformation results in a 2-D problem.

Chobadian *et al.* (1985) proposed two formulas to estimate the depth of the mixed layer of the atmosphere, which is important when determining the vertical structure of the model:

$$H = 8.8 x_1 U_a^{-1} \Delta\theta \qquad H = \frac{U_*}{U_a} \left\{ \frac{x_1 |\theta_1 - \theta_2|}{|\Delta\theta/\Delta x_3|} \right\}^{1/2}$$

where x_1 is the rate of shifting relative to land surface (m/s); $\Delta\theta$ is the vertical lapse rate of the potential temperature in the inversion layer (°C); U^* is the rate of friction over the leeward surface; θ_1 is the lower level of the potential temperature over the source of atmospheric pollution (°K); and $|\Delta\theta/\Delta x_3|$ is an absolute value of the rate of motion over the source.

The desire to simplify the parametric descriptions of some subprocesses of the dynamics of the atmospheric pollution leads many authors to the development of simple and efficient models, which require a small database. Numerous works have been dedicated to the classification of situations taking place in the real atmosphere when pollutants are emitted by point sources (chimney stacks). A sufficiently complete classification can be found in Scorer (1990), where ten gradations are given of the behaviour of the pollutant jet (smoke) in the vicinity of an isolated point source (stack): in the form of a loop, spiral, flag, as well as precipitating, spreading, swelling, breaking, bifurcational, fumigating, and rising. In each of these cases the use of the Gaussian model with minimum input information is possible. The monitoring system should only distinguish between these situations, based on measurements of the meteorological parameters. For instance, the loop form of the jet of the pollutant can appear because of surface roughness (high buildings, hilly landscapes) in clear sunny weather. Smoke particles are wind-driven in zig-zags within the swelling and gradually precipitating jet. The breaking jet is typical of the second-half of a summers day, when the atmosphere is well heated and convective air motions prevail.

One of the units of the model of atmospheric pollution is a description of the process of emission of the pollutant into the environment. In most cases the Gauss law is used:

$$C(r, h) = f(h)\frac{M}{2\pi\sigma_r^2}\exp(-r^2/(2\sigma_r^2)) \tag{6.3}$$

where r is the distance to the centre of the pollutant jet; σ_r is the horizontal size of the jet; $f(h)$ is the vertical distribution of the mass of the pollutant; $M = Q\Delta t$ is the mass of the pollutant in the jet; Q is the rate of emission from the source; and Δt is the time step. The function f can be approximated by the following dependence: $f(h) = 1/H$ at $0.5H < h < 1.5H$ and 0 otherwise (M is an efficient height of the jet). For instance in Despres *et al.* (1986) the following formula is proposed to calculate M:

$$H = \begin{cases} Z + 0.29V^{-1}Q_h^{1/2} & \text{for normal conditions of stability} \\ Z + 0.49V^{-1}Q_h^{0.29} & \text{for weak conditions of stability} \end{cases}$$

where Q_h is the rate of the thermal flux from the source (stack) (cal/s); and Z is the height of the source (m).

To simplify the process of simulation of the field of pollutant in the zone of the source, the flux is digitized into individual formations (clouds of smoke, dust), each being considered individually as a homogeneous cloud. In particular, to digitize the flux into n parts during the time Δt, the following formula can be used: $n = 2V\Delta t/H$. For instance, at the wind speed $V = 5\,\mathrm{m/s}$ and $Z = 100\,\mathrm{m}$ we obtain that clusters will be emitted from the stack every 10 s. As a result, the process of modelling the dynamics of the whole emitted jet is substantially simplified. So, for instance, in modelling the propagation of sulphur, the sulphur-forming components can be additionally separated. Figure 6.2 is a schematic illustration of the process of sulphur transport in a smoke jet.

6.2.4 Types of models and their information bases

By methods of parameterization of the described processes of pollutant scattering, models of the atmospheric pollution dynamics are divided into Gaussian, Eulerian, and Lagrangian. Within each type there are static, box, correlative, and determinate models. The most important section of the model space is its section according to the applied methods since it is in this case that a strong differentiation of models by their databases is observed.

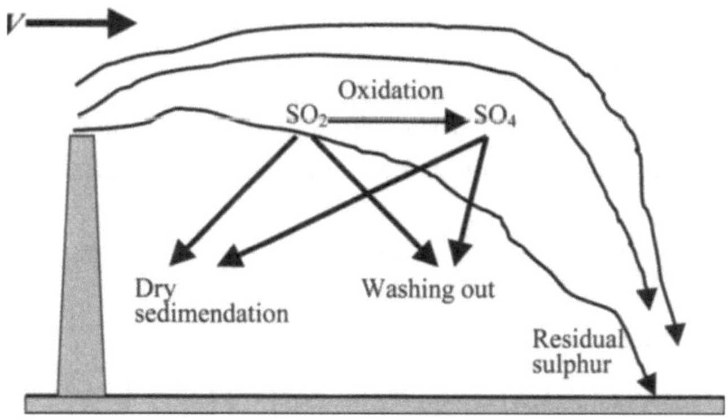

Figure 6.2. A simplified scheme of the model of propagation of sulphur from a point source.

The Gauss-type models need data on the height of the pollution source and dispersion characteristics within the polluting cloud. Various approximations of the Gaussian jet in the vicinity of a point source depend on the parameters of the surface air stability. The configuration of the model is affected by geophysical characteristics. For instance, when the source is located in the zone of transition from water surface to land, the clusters of pollutants can be emitted with the stable and unstable behaviour of the pollutant jet. In this case, with respect to the horizontal coordinate x on land, with the wind blowing along this coordinate, three zones with characteristic indicators can be identified: the zone of undisturbed dispersion, the zone of sedimentation, and the zone of lagging.

In the zone of undisturbed dispersion the distribution of the flux of pollutant is formed under the influence of the homogeneous and stable layer:

$$C(\varphi,\lambda,h)=\frac{Q}{2\pi V\sigma_\lambda\sigma_h}\exp\left\{-\frac{1}{2}\left(\frac{\lambda}{\sigma_\lambda}\right)^2\right\}\left\{\exp\left[-\frac{1}{2}\left(\frac{h-Z}{\sigma_h}\right)^2\right]+\exp\left[-\frac{1}{2}\left(\frac{h+Z}{\sigma_h}\right)\right]\right\}$$

where φ is wind direction; λ is the coordinate in the Cartesian system; h is the vertical coordinate; and σ_λ and σ_h are the horizontal and vertical dispersions (scattering).

In the second zone the behaviour of the polluting cloud is formed in conditions of instability and therefore calculations of the concentration of the pollutant require an extended database:

$$C(\varphi,\lambda,h)=\frac{Q}{(2\pi)^{3/2}}\int_0^{\varphi}\frac{1}{\sigma'\sigma_h(V,\varphi,\varphi')V}\exp\left\{-\frac{(p')^2}{2}-\frac{1}{2}\left(\frac{\lambda}{\sigma'}\right)^2\right\}$$
$$\times\left\{\exp\left[\frac{-(h-Z)^2}{2\sigma_h^2(V,\varphi,\varphi')}\right]+\exp\left[-\frac{(h+Z)^2}{2\sigma_h^2(V,\varphi,\varphi')}\right]\right\}\frac{\partial p'}{\partial\varphi'}d\varphi'$$

where φ' is the leeward distance in the intersection of the cloud and the upper part of the boundary layer; σ' is the average value of scattering on the stable and unstable zones border; $p' = [L(\varphi) - Z]/\sigma_h$; and $L(\varphi)$ is the height of the boundary layer in the leeward direction φ.

The third zone is special due to conditions of the formation of the polluting cloud. This cloud is held totally within the atmospheric boundary layer, and the concentration of the pollutant in it can be calculated by the formula:

$$C(\varphi, \lambda, h) = \frac{Q}{2\pi V \sigma_\lambda \sigma_h} \exp\left[-\frac{1}{2}\left(\frac{\lambda}{\sigma_h}\right)^2\right]\left\{\exp\left[-\frac{1}{2}\left(\frac{h-Z}{\sigma_h}\right)^2\right] + \exp\left[-\frac{1}{2}\left(\frac{h+Z}{\sigma_h}\right)^2\right]\right\}$$

for $\sigma_h < 0.47L(\varphi)$; and

$$C(\varphi, \lambda, h) = \frac{Q}{\pi V \sigma_\lambda L(\varphi)} \exp\left[-\frac{1}{2}\left(\frac{\lambda}{\sigma_\lambda}\right)^2\right]$$

for $\sigma_h > 0.8L(\varphi)$. Within the interval $0.47 \leq L(\varphi) \leq 0.8L(\varphi)$ the concentration C is calculated between the points $\sigma_h = 0.47L(\varphi)$ and $\sigma_h = 0.8L(\varphi)$ by the interpolation formula.

This model describes the fields of the pollutant over the territories of scales up to 100 km. For larger territories, other types of the models are used.

The information base of the Eulerian and Lagrangian models is formed according to the problem with initial conditions for Equation (6.1). Depending on various real situations, the content of the database that provides a realization of the model of the pollutant transition becomes hierarchical. This hierarchy determines the correlation of space and timescales, it governs the accuracy of the model and its other characteristics. There are numerous global databases. As a rule, they consist of information levels with a subject orientation.

6.3 MODELLING THE ATMOSPHERIC PROCESSES OF POLLUTANT PROPAGATION

6.3.1 Modelling the wind field

Knowledge of the wind field components is a necessary element of modelling the process of the atmospheric pollutant propagation over the territory of the region independent of the spatial scale. With numerous sources of atmospheric pollution in the region, information is needed about the 3-D wind field, and this is only possible through a combination of measurements and modelling. For large regions or in studying the conditions for propagation of pollutants on the land–water border, the Eulerian classic grid model is usually used:

$$\partial u/\partial x + \partial v/\partial y + \partial w/\partial z = 0$$

$$\partial u/\partial t + u\partial u/\partial x + v\partial u/\partial y + w\partial u/\partial z = f(v - v_g) + \partial(K_M \partial u/\partial z)/\partial z$$

$$\partial v/\partial t + u\partial v/\partial x + v\partial v/\partial y + w\partial v/\partial z = f(u_g - u) + \partial(K_M \partial v/\partial z)/\partial z$$

$$\partial T/\partial t + u\partial T/\partial x + v\partial T/\partial y + w\partial u/\partial z = \partial(K_H(\Gamma + \partial T/\partial z)/\partial z$$

where f is the Coriolis parameter; Γ is the rate of dry adiabatic sedimentation; T is the air temperature; K_M and K_H are coefficients of diffusion; and u_g and v_g are the thermal wind constituents:

$$u_g = -\frac{gT}{f}\int_0^z \frac{1}{T^2}\left(\frac{\partial T(t,x,y,\xi)}{\partial y}\right)d\xi$$

$$v_g = \frac{gT}{f}\int_0^z \frac{1}{T^2}\left(\frac{\partial T(t,x,y,\eta)}{\partial x}\right)d\eta$$

For practical use, the following should be borne in mind: deviation and uplifting of the flux over hills, thermal friction, tunnel effect in valleys, thermal effects of islands, lakes, etc. Additions to this system of equations are made proceeding from the real configuration and topology of the region. In particular, the equation of mass conservation can be written as the Poisson equation.

6.3.2 The Gauss-type models

The Gauss-type models are used, as a rule, to parameterize the processes of pollutant propagation near high sources. A series of modifications of the Pasquill–Gifford–Turner (PGT) disperse model, adequately describing the near-surface fluxes of pollutant and less accurately the fluxes from high-mounted sources is well known. There are numerous modifications of the PGT model, which provide it with more universal functions for an arbitrary point source. Modifications are based on the base Gauss-type model, which describes concentrations of the pollutant in the lower atmosphere:

$$C(\varphi,\lambda,0) = \frac{Q}{\pi\bar{u}\sigma_z\sigma_\lambda}\exp\left[-0.5\left(\frac{h_{ef}}{\sigma_z}\right)^2\right]\exp\left[-0.5\left(\frac{\lambda}{\sigma_\lambda}\right)^2\right] + \text{terms of reflection} \qquad (6.4)$$

where Q is the source power; h_{ef} is the average height of the polluting cloud; and $\bar{u}$ is the efficient rate of transport.

Parameters of the pollutant jet scattering are important parameters of the model (6.4). They are complex functions of the meteorological situation. Theoretical and experimental studies of many scientists have shown that turbidity and dispersibility of the convective boundary layer are controlled by two important parameters: the height of mixing z_i and the scale of convection rate w_*:

$$w_* = \left(\frac{g}{T\rho c_p}Yz_i\right)^{1/3}$$

where Y is the flux of the surface sensible heat; g is the acceleration due to terrestrial gravity; T is the air temperature; and c_p is the specific air heat capacity at a constant pressure.

The vertical σ_w and horizontal σ_v constituents of the rate of deviation of the turbulent flux from the centre of the convective cloud are proportional to the value

w_* and, hence, depend on z_i. The total energy of turbulence is a composition of two energies generated by mechanical (σ_{wm}) and convective (σ_{wc}) forces:

$$\sigma_w^2 = \sigma_{wm}^2 + \sigma_{wc}^2$$

Here it is assumed that mechanical and convective forces do not correlate. Then, by analogy, we can write:

$$\sigma_z^2 = \sigma_{zm}^2 + \sigma_{zc}^2 \tag{6.5}$$

The parameter σ_{zc}^2 is calculated on the assumption that:

$$\frac{d}{dt}\sigma_{zc} = \sigma_{wc}(z') \qquad \text{where } t = \varphi/\bar{u} \tag{6.6}$$

Here z' is the efficient height at which the characteristic of the vertical turbulence is calculated. We parameterize dependences of σ_{wc} on the height by the following relationships:

$$\sigma_{wc}^2 = \begin{cases} 1.54w_*^2(z/z_i)^{2/3} & \text{for } z < 0.1z_i \\ 0.33w_*^2 & \text{for } z \geq 0,1z_i \end{cases}$$

where the level $z = 0.1z_i$ corresponds to similar values of the vertical constituents of turbulence.

Let h_s be the height of the pollutant jet and t have the scale $x/\bar{u}$. Then from Equation (6.6) at $h_s \geq 0.1z_i$ we obtain the following approximation: $\sigma_{zc}^2 = 0.33w_*^2t^2$. For $h_s < 0.1z_i$ we have:

$$\sigma_{zc}^2 = \begin{cases} 1.54w_*^2(h_s/z_i)^{2/3}t^2 & \text{for } \sigma_{zc} < h_s \\ (0.83w_*z_i^{-1/3}t + 0.33h_s^{2/3})^3 & \text{for } h_s \leq \sigma_{zc} < 0.1z_i \\ (0.581w_*t + 0.231h_s^{2/3}z_i^{1/3} - 0.05z_i)^2 & \text{for } \sigma_{zc} \geq 0.1z_i \end{cases}$$

The mechanical constituent in Equation (6.5) is calculated on the assumption that the variation of the vertical rate due to mechanical mixing is constant in the boundary layer and is determined by the relationship: $\sigma_{wm}^2 = 1.2u_*^2$, where u_* is the rate of friction.

The relationship between the parameter of mechanical scattering σ_{zm} and the rate of mechanical turbulence σ_{wm} is not as simple as in the case of convective turbulence. This is connected with the fact that the size of scale characterizing the mechanical scattering, in contrast to the convective case, varies with height, and therefore it is less than the convective scale. Hence, for the unstable state the following dependencies are used:

$$\sigma_{zm}^2 = \begin{cases} \sigma_{zmu}^2 = 1.2u_*^2t^2\exp(-0.6tu_*/h_s) & \text{for } tu_*/h_s < 1 \\ \sigma_{zmu}^2 = 1.2u_*^2t^2\exp(-0.6) & \text{for } tu_*/h_s \geq 1 \end{cases}$$

For stable conditions the following approximation is valid:

$$\sigma_{zm}^2 = \sigma_{zmu}^2/(1 + 1.11tu_*/L)$$

where L is the Monin–Obukhov (1954) length.

Thus both the right-hand terms in Equation (6.5) are estimated. Now we determine σ_λ in Equation (6.4). According to Berkowicz *et al.* (1986), the horizontal scattering of the pollutant is accurately approximated by the formula:

$$\sigma_\lambda = \left(\frac{0.25 w_*^2}{1 + 0.9 \varphi w_* / u z_i} + u_*^2 \right)^{1/2} \frac{\varphi}{\bar{u}} \tag{6.7}$$

In this expression the first and second terms in brackets denote the contribution of convective and mechanical turbulence, respectively. For nocturnal conditions, the first term is equal to zero. For a stable atmosphere and with a weak wind, the ratio $u_*/\bar{u}$ decreases rapidly with increasing stability. Nevertheless, numerous observations have shown that the horizontal scattering based on an hourly averaging can exceed the respective values when the atmosphere is unstable. In this case the horizontal rate fluctuates with a constant amplitude of about 0.5 m/s. While the horizontal scattering remains proportional to fluctuations of the horizontal wind, for the case of stable stratification the parameter u_* in Equation (6.7) can be substituted for 0.5 m/s. It means that σ_λ will never be less than $0,5\varphi/\bar{u}$.

Let us determine the parameter of horizontal scattering:

$$\sigma_{\lambda f} = \left(\Delta d / \sqrt{2\pi} \right) \varphi$$

where Δd is the change of wind direction in radians.

According to Equation (6.4), the accuracy of calculations of the pollutant concentration at a point with the coordinates (φ, λ) depends substantially on wind speed. The estimate of the pollutant density in a given space volume depends on the accuracy of the wind speed estimate at a given height. For instance, in the case of unstable stratification, the wind speed is approximated by its value at the height of the source, and in stable conditions the wind speed is calculated by its averaging over that height. According to Monin and Obukhov (1954), we have:

$$u(z) = \frac{u_*}{k} \left[\ln\left(\frac{z + z_0}{z_0} \right) - \psi_m\left(\frac{z}{L} \right) + \psi_m\left(\frac{z_0}{L} \right) \right] \tag{6.8}$$

where z_0 is the linear size of surface roughness and ψ_m are universal functions of similarity calculated, for instance, in Paulsen (1970). Proceeding from these calculations, the wind speed is estimated by Equation (6.8) at $z \leq z_B$ and is assumed to be equal to $u(z) = u(z_B)$ for all heights $z > z_B$, where $z_B = \max\{0.1 z_i, |L|\}$.

Practical use of the Gauss-type models has many aspects, which appear depending on numerous factors of the natural–anthropogenic environment. These are the effect of buoyancy of the pollutant in the atmosphere, intermittent layers of the atmosphere, and many others. Berkowicz *et al.* (1986) proposed to calculate the parameter h_{ef} in Equation (6.4) using the following expression: $H_{ef} = h_s + \Delta_h$, where:

$$\Delta h = \min\{\Delta h_{in}, \Delta h_f\} \qquad \Delta h_{in} = 1.6 \left(\frac{F}{u} \right)^{1/3} \varphi^{2/3} \qquad F = \frac{V}{\pi} \frac{g}{T_e} (T_e - T_a)$$

Here F is the buoyant flux; h_s is the height of the stack; V is the flux volume; T_e is the temperature of the flux emitted from the stack and T_a is the environmental temperature.

The height at which the jet of emitted pollutant rises due to the effect of its buoyancy is calculated by the formula:

$$\Delta h_f = \begin{cases} 1.3\dfrac{F}{uu_*^2}\left(1+\dfrac{h_s}{\Delta h_f}\right) & \text{neutral conditions} \\ 4.3\left(\dfrac{F}{u}\right)^{3/5}\left(\dfrac{Yg}{T_a\rho c_p}\right)^{-2/5} & \text{conditions of convection} \\ 2.6\left(\dfrac{F}{us}\right)^{1/3} & \text{stable conditions} \end{cases}$$

where $s = (g/T_a)\partial\theta_a/\partial z$. Here $\partial\theta_a/\partial z$ is the potential air temperature lapse rate at the level of the stack.

A multitude of possible situations when the Gauss-type model is used is determined by landscape topology, the presence of boundaries of the type "sea–land", mountains, etc. A consideration of the concrete situation makes it possible to simplify the formulas and improve their reliability. In particular, on the water–land border the air flux gets constantly adjusted to the configuration of the wind field, forming into jets that correspond to mechanical and temperature conditions. In conditions of the stable near-shore thermal layer, a stable gradient of the pollutant concentration is observed, which depends on the distance of the source from the shore. Here the following formula is valid:

$$\frac{3X}{\beta'^2 Usz_b^3} = \frac{3}{2}\left[\frac{z_{eq}}{z_b}\right]^2\left[\frac{z_{eq}}{z_b}-1\right]+\frac{2}{9} \tag{6.9}$$

where z_{eq} is the final stable height of the jet over the stack; z_b is the height of the bottom of the stable atmospheric layer over the stack; U is the wind speed at the height of the stack; β' is the effective coefficient of the polluting particle capture ($\beta' \cong 0.4$); s is the stability parameter; and X is the parameters of the jet buoyancy. The parameters s depends on the potential temperature lapse rate θ and the environmental temperature T_a: $s = (g/T_a)\partial\theta/\partial z$ (g is the gravity coefficient). The parameter F can be calculated as a function of the rate v of the jet emission from the stack, the inner radius R of the stack and the temperature T of the emitted jet:

$$F = gvR_2(T - T_a)/T$$

With the value z_d known from Equation (6.9) the value z_{eq} is easily found, and the effective height Y of the jet is found as the sum: $Y = z_{eq} + h_s$, where h_s is the height of the stack. The whole zone of propagation of the polluting cloud with the wind blowing from the water to land is divided, as a rule, into three characteristic zones. Digitizing of space in accordance with these zones simplifies the model, reducing the required database and raising its reliability. In each zone the use of the Gauss-type model leads to a simplified parameterization of the process of scattering of the pollutant and makes easier an interpretation of the modelling results. In the undisturbed disperse zone nearest to the water surface the distribution of the pollutant jet that gets to a homogeneous stable layer of the atmosphere is described by the base Gaussian equation of scattering:

$$C(\varphi, \lambda, z) = \frac{Q}{2\pi U \sigma_\lambda \sigma_z} \exp\left[-\frac{1}{2}\left(\frac{\lambda}{\sigma_\lambda}\right)^2\right] \times \left\{\exp\left[-\frac{1}{2}\left(\frac{z-Y}{\sigma_z}\right)^2\right] + \exp\left[-\frac{1}{2}\left(\frac{z+Y}{\sigma_z}\right)^2\right]\right\}$$

where U is the wind speed; Y is the effective height of the jet; and σ_λ and σ_z are parameters of the horizontal and vertical dispersion.

In the next zone of fumigating, forces are acting which distort the pollutant jets due to instability of heat fluxes in the surface air layer. In this zone the following approximation is valid:

$$C(\varphi, \lambda, z) = \frac{Q}{(2\pi)^{3/2}} \int_0^\varphi \frac{1}{\sigma' \sigma_z(u, \varphi, \varphi') U} \exp\left[-\frac{1}{2} p'^2 + \left(\frac{\lambda}{\sigma'}\right)^2\right] \times \left\{\exp\left[-\frac{(z-Y)^2}{2\sigma_z^2(u, \varphi, \varphi')}\right] + \exp\left[-\frac{(z+Y)^2}{2\sigma_z^2(u, \varphi, \varphi')}\right]\right\} \frac{dp'}{d\varphi'} d\varphi'$$

where φ' is the leeward distance of the intersection of the pollutant jet with the upper boundary of the internal heat boundary layer (IHBL); $p' = [L(\varphi) - Y]/\sigma_z(s, \varphi)$; $L(\varphi)$ is the IHBL height at the distance φ; $\sigma_\lambda(s, \varphi)$ is the σ_λ value for the stable atmospheric layer at the distance φ; $\sigma_z(s, \varphi)$ is the σ_z value for the stable atmospheric layer at the distance φ; $\sigma_\lambda(u, \varphi, \varphi')$ is the σ_λ value for the unstable atmospheric layer at the distance φ after passing the leeward distance φ'; and $\sigma_z(u, \varphi, \varphi')$ is the σ_z value for the unstable atmospheric layer at the distance φ after passing the leeward distance φ':

$$\sigma' = \sqrt{\sigma_\lambda^2(s, \varphi) + \sigma_\lambda^2(u, \varphi, \varphi')}$$

In the zone where the upper boundary of the polluting cloud is close to $L(\varphi)$ and is totally located within the IHBL, the following approximations are valid:

$$C(\varphi, \lambda, z) = \begin{cases} xEG/\sigma_z(u, \varphi) & \text{for } \sigma_z < 0.47L(\varphi) \\ xE/L(\varphi) & \text{for } \sigma_z > 0,8L(\varphi) \end{cases} \tag{6.10}$$

where

$$x = Q/[\pi U \sigma_\lambda(u, \varphi)] \qquad E = \exp\left[-\frac{1}{2}\left(\frac{\lambda}{\sigma_\lambda(u, \varphi)}\right)^2\right]$$

$$G = \exp\left[-\frac{1}{2}\left(\frac{z-Y}{\sigma_z(u, \varphi)}\right)^2\right] + \exp\left[-\frac{1}{2}\left(\frac{z+Y}{\sigma_z(u, \varphi)}\right)^2\right]$$

For the leeward distance φ, where $0.47L(\varphi) \leq \sigma_z(\varphi) \leq 0.8L(\varphi)$, the concentration of the pollutant is calculated by interpolation between its estimates in the bordering zones according to Equation (6.10).

6.3.3 Modelling the planetary boundary layer

Modelling the processes of pollutants propagation in the surface air layer is connected with the description of the time dependencies of wind speed, temperature, and humidity of the atmosphere as 3-D functions of time and space coordinates. One such model of the planetary boundary layer (PBL) has been described in Bornstein *et al.* (1986). The model contains two units. The first unit analytically describes the surface boundary layer (SBL), in which the field of the pollutant is averaged and represented by a 1-D structure. Another transition layer of the atmosphere is described by hydrodynamic and thermodynamic equations solved numerically with the use of finite-difference schemes. Equations of the transition layer of the atmosphere are derived on the assumption that the atmosphere is incompressible and hydrostatic, water vapour does not change its properties, the potential temperature is calculated by the formula $\theta = T + \Gamma z$, where T is the air temperature, Γ is the rate of dry adiabatic cooling, and z is the vertical coordinate. Equations of the atmospheric transition layer are written as:

$$\frac{\partial u}{\partial t} + \frac{\partial(uu)}{\partial x} + \frac{\partial(vu)}{\partial y} + \frac{\partial(wu)}{\partial z} = -\frac{1}{\rho_a}\frac{\partial p_M}{\partial x} + f(v - v_g) + \frac{\partial}{\partial z}\left(K_m \frac{\partial u}{\partial z}\right) + K_H\left(\frac{\partial^2 u}{\partial x^2} + \frac{\partial^2 u}{\partial y^2}\right)$$

$$\frac{\partial v}{\partial t} + \frac{\partial(uv)}{\partial x} + \frac{\partial(vv)}{\partial y} + \frac{\partial(wv)}{\partial z} = -\frac{1}{\rho_a}\frac{\partial p_M}{\partial y} + f(u - u_g) + \frac{\partial}{\partial z}\left(K_m \frac{\partial v}{\partial z}\right) + K_H\left(\frac{\partial^2 v}{\partial x^2} + \frac{\partial^2 v}{\partial y^2}\right)$$

$$\frac{\partial u}{\partial x} + \frac{\partial v}{\partial y} + \frac{\partial w}{\partial z} = 0$$

$$g\frac{(T_M + T_n)}{\theta_a} - \frac{1}{\rho_a}\frac{\partial p_M}{\partial z} = 0$$

$$\frac{\partial \theta}{\partial t} + \frac{\partial(u\theta)}{\partial x} + \frac{\partial(v\theta)}{\partial y} + \frac{\partial(w\theta)}{\partial z} = \frac{\partial}{\partial z}\left(K_h \frac{\partial \theta}{\partial z}\right) + K_H\left(\frac{\partial^2 \theta}{\partial x^2} + \frac{\partial^2 \theta}{\partial y^2}\right) + \frac{1}{\rho_a c_p}\frac{\partial Q_N}{\partial z}$$

$$\frac{\partial q}{\partial t} + \frac{\partial(uq)}{\partial x} + \frac{\partial(vq)}{\partial y} + \frac{\partial(wq)}{\partial z} = \frac{\partial}{\partial z}\left(K_q \frac{\partial q}{\partial z}\right) + K_H\left(\frac{\partial^2 q}{\partial x^2} + \frac{\partial^2 q}{\partial y^2}\right)$$

where

$$u_g = -\frac{1}{f\rho_a}\frac{\partial p_n}{\partial y} \qquad v_g = \frac{1}{f\rho_a}\frac{\partial p_n}{\partial x}$$

(u, v, w) are projections of wind speed onto the rectangular coordinates x, y and z, respectively; K_H is the coefficient of the horizontal turbulent motion; ρ_a is the air density; and q is the specific humidity. In a general case, a strict solution of these equations is impossible and therefore a suitable discrete grid is needed, over which these equations are substituted for finite-difference equations. As shown in Bornstein *et al.* (1986), the time step of digitizing and the size of the discrete space grid should meet the requirements:

$$\Delta t \leq \min\{\Delta x/u, \Delta y/v, \Delta z/w\}$$

In practice, instead of this sufficiently cumbersome scheme, simulation schemes are often used based on the acceptable digitization of space. In this case the territory of the region or of the globe is divided, as a rule, into homogenous sites with linear sizes $\Delta\varphi$ by latitude and $\Delta\lambda$ by longitude: $\Omega = \{\Omega_{ij}; i = 1, \ldots, n; j = 1, \ldots, m\}$. Giving the vertical structure of the atmosphere and taking into account physical and chemical processes of the transformation of pollutants, relationships are derived recurrent by the time step Δt to recalculate their concentrations. In the presence of control measurements at several points of space $\Xi = \{(\varphi, \lambda, z), (\varphi, \lambda) \in \Omega, 0 \leq z \leq H\}$ the steps of digitization of the geographical grid and time are optimized on the basis of one or several criteria of accuracy. As a rule, such an approach provides the needed stability of the modelling results (Krapivin and Kondratyev, 2002).

6.3.4 The Euler-type model

Denote $C_s(t, \varphi, \lambda, h)$ as the concentration of the pollutant of the sth type at the height h over the point with the geographical coordinates (φ, λ) at the time moment t. For the Euler-type model, general equations of the pollutant transport in the environment according to Equation (3.1) are as follows:

$$\frac{\partial C_s}{\partial t} + V_\varphi\frac{\partial C_s}{\partial \varphi} + V_\lambda\frac{\partial C_s}{\partial \lambda} + V_h\frac{\partial C_s}{\partial h} = \frac{\partial}{\partial \varphi}\left(K_\varphi\frac{\partial C_s}{\partial \varphi}\right) + \frac{\partial}{\partial \lambda}\left(K_\lambda\frac{\partial C_s}{\partial \lambda}\right) + \frac{\partial}{\partial h}\left(K_h\frac{\partial C_s}{\partial h}\right)$$
$$+ E_s(t, \varphi, \lambda, h) + P_s(C_1, \ldots, C_q) - \nu_{1s} - \nu_{2s}$$

where K_φ, K_λ, and K_h are coefficients of turbulent diffusion; E_s is the characteristic function of the sources of emission of the pollutant of the sth type; P_s is the operator describing the physical and chemical transformation of the pollutant; ν_{1s} is the rate of dry sedimentation; ν_{2s} is the rate of washing out with precipitation; and $V = \{V_\varphi, V_\lambda, V_h\}$ are the components of wind speed.

As a rule, models of this type are used to calculate the concentration of pollutants in the atmosphere for territories with an areal extent of more than 50 km. The simplified schemes of calculation of the function C are derived by dividing the space into units $\Delta\varphi \times \Delta\lambda \times \Delta h$, and at each height h_k a step-by-step calculation of the concentration $C_s(t, \varphi_i, \lambda_j, h_k)$ is realized. The calculation scheme can be simplified further by diving the whole procedure into two stages. First, distributions of $C_s(t, \varphi_i, \lambda_j, h_k)$ are found for each level of the vertically digitized space, and then the processes of the vertical transition of pollutants is considered. This scheme makes it possible to easily move on to vertically averaged levels depending on the available information about the parameters of the vertical stratification of the atmosphere. The convergence of this procedure depends on the relationship between the parameters $\Delta t, \Delta\varphi, \Delta\lambda$, and Δh.

The Euler-type model contains many degrees of freedom up to a consideration of scenarios (Gyalistras, 2002). The base knowledge of the modelling system contains sets of parameterizations of partial processes of transformation of pollutants, with the concrete choice being made by the user. All information in the bases of data and knowledge are structured according to a multitude of spatial and subject identifiers – matrix structures $\{A_m\}$.

Emission of the pollutant

In accordance with the structure of the identifiers describing the sources of pollution, the passport information about the rates of emission of the sth pollutant can be prescribed in each compartment $\Delta\varphi \times \Delta\lambda$: minimum and maximum rates $E_{s,\min}$ and $E_{s,\max}$, respectively. Without additional information about any source, the value E_s is calculated according to the procedure of the uniform distribution in the interval $[E_{s,\min}, E_{s,\max}]$ or another law of distribution is assumed (e.g., Gaussian).

Dry sedimentation

The functional description for v_{1s} and v_{2s} is important, and an adequacy of the model depends on it. Therefore, many studies have been dedicated to this problem. Many authors prove that a linear approximation of the functions v_{1s} and v_{2s} is most acceptable:

$$V_{is} = K_{is}C_s \qquad i = 1, 2$$

Let the coefficient of proportionality K_{ican} be the function of time and space coordinates and depend on the physical state of the Earth's surface. Using the identifier $\|\mu_{ij}\|$ one can broaden this dependence. For instance, the version $\mu_{ij} = 0$ can mean that dry sedimentation in the model is not considered at all. For $\mu_{ij} = 1$ the scheme proposed in Zlatev *et al.* (1992) is considered:

$$K_{2s} = K_{2s}^* v_{ds} d / H_{\text{mix}}(t, \varphi, \lambda)$$

where K_{2s}^* is the coefficient of proportionality; v_{ds} is the rate of sedimentation of the pollutant of the sth type (e.g., $v_{ds} = 0.008$ m/s for SO_2, 0.002 m/s for SO_4, and 0.001 m/s for NO_2 and HNO_3); H_{mix} is the height of the atmospheric mixed layer (m); and d is the parameter for consideration of the physical state of the Earth's surface ($d = 1$ for dry surface, $d = 2$ for wet surface, etc.).

With the help of the parameter μ_{ij} the modelling system can control the wide choice of various dependencies of the rate of dry sedimentation on synoptic parameters and other characteristics of the environment. For instance, at $\mu_{ij} = 2$ the following model is chosen:

$$\nu_{2s} = \begin{cases} 0.002U_* & \text{for } L \geq 0 \\ 0.002U_*[1 + (-300/L)^{2/3}] & \text{for } L < 0 \end{cases}$$

where L is the Monin–Obukhov length (m); and U_* is the rate of friction.

Washing-out

Similarly to the case above, it is expedient in modelling to consider the rate of wet sedimentation of pollutants from the atmosphere from different points of view, since in a general case the value ν_{1s} is the function of many factors, some of them not always being estimated reliably. For the convenience of the computer modelling, the identifier $\|\nu_{ij}\|$ is introduced, and, for instance at $\nu_{ij} = 0$ it is automatically assumed that $\nu_{1s} = 0$ (no washing out). For $\nu_{ij} = 1$, the version is identified when:

$$\nu_{1s} = K_{2s}^* C_s r$$

where r is the rain rate.

Letting $\nu_{ij} = 2$ corresponds to the model of washing out:

$$\partial C_s / \partial t = -\phi r(t, \varphi, \lambda) C_s(t, \varphi, \lambda, h)$$

For $\phi =$ constant we obtain a simple explicit scheme of recalculation of C_s in each cell:

$$C_s(t + \Delta t, \varphi_i, \lambda_j, h_k) = C_s(t, \varphi_i, \lambda_j, h_k) \exp\{-\phi r(t, \varphi_i, \lambda_j)\Delta t\}$$

6.4 SIMULATION SYSTEM TO MODEL THE PHYSICAL PROCESSES OF THE TRANSPORT OF POLLUTANTS IN THE ATMOSPHERE

Figure 6.3 demonstrates a set of units of the Simulation System for the Atmospheric Pollution Physics (SSAPP). The software for its functioning is tabulated in Table 6.11. It is assumed that all input information in SSAPP is concentrated in thematic files, which can be formed based on various sources, such as distant sensors, lines of information communication, portable standard and non-standard accumulators of observation data, direct input of data from the keyboard or other compatible input device. The base version of SSAPP is meant for the last case. Switching of SSAPP onto other regimes requires additional software elements, conforming the information carrier and providing its transformation into the file accumulator. In general, the interaction of all SSAPP elements is described by the scheme shown in Figure 6.4.

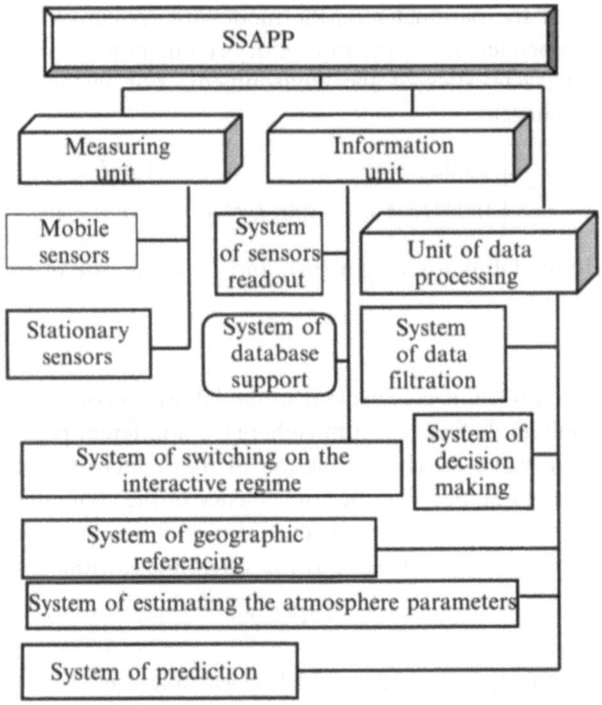

Figure 6.3. The SSAPP structure with the functions of complex estimation of atmospheric pollution over a given territory.

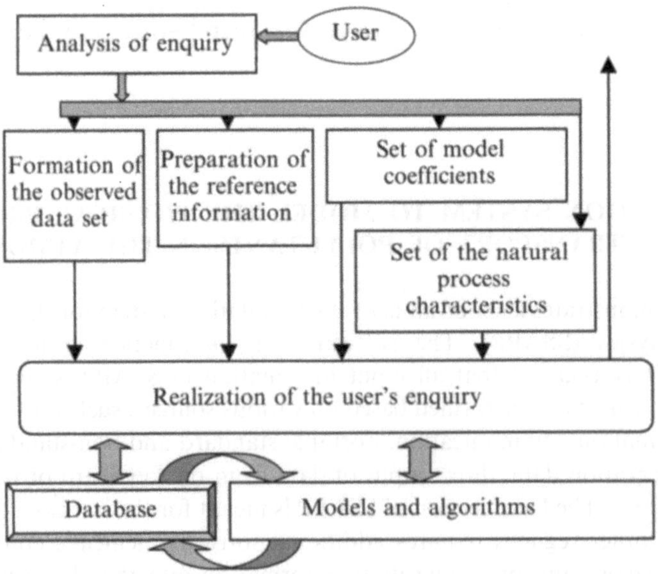

Figure 6.4. Exemplary scheme of the SSAPP functioning in the regime of the dialogue with the user.

Table 6.11. The SSAPP software.

Identifier of the software	Characteristic of the software
STANDARD	Calibration and scaling of input information
FILTER	Filtration of input data
ASSENT	Spatial–temporal coordination of different types of data
APPROX	Interpolation of measurement data over space and formation of space samples
RESTOR	Data restoration
GAUSS	The Gauss-type model
EULER	The Euler-type model
DEVIDE	The choice of the spatial digitations grid and transition to the unit modelling
CHOICE	Taking the information needed for modelling from the database
ATMOSEA	Modelling the processes of pollutant transport due to the atmosphere–sea interaction.
ATMOLAND	Modelling the processes of the pollutant transport with the atmosphere–plant–soil systems functioning taken into account.
SULPHU	Model of sulphur fluxes in the environment
MAP	Mapping the modelling results
SIGNAL	Analysis of the modelling results and their visualization with the standard levels taken into account
IBMMEN	Man–machine interface
NITROG	Model of nitrogen fluxes in the environment
OXYGEN	Model of oxygen budget in the natural–anthropogenic environment.
CARBON	Model of the carbon dioxide cycle in the environment
MAIN	Control and correlation of the SSAPP units.

6.4.1 Formation of database fragments

The algorithms for the formation and processing of the bases of space data have been approved by geographical information systems (GIS)-technology. Choose the matrix hierarchical topological structure (space identifier) reflecting the space and subject structure of the territory Ω in order to refer the SSAPP to a concrete subject, so that:

$$\Omega = \bigcup_{(i,j)} \Omega_{ij} \qquad \Omega_{ij} = \bigcup_{(s,l)} \Omega_{i_s j_l} \text{ etc.}$$

Schematically this is shown in Figure 6.5. According to this algorithm, the territory Ω in the SSAPP database is presented by a set of matrices:

$$A_k = \|a_{ij}^k\| \qquad k = \overline{1, N}$$

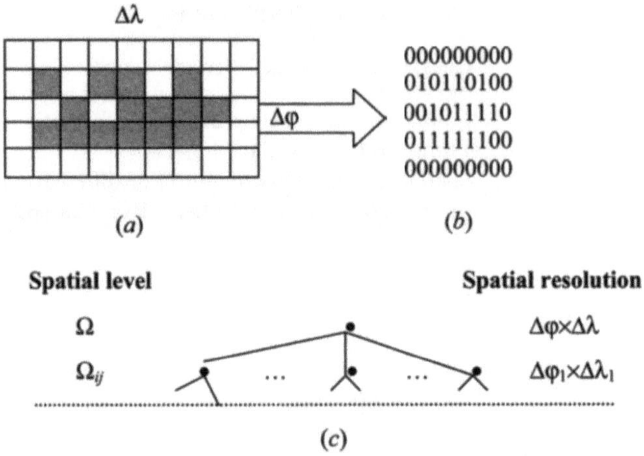

Figure 6.5. Procedure of topological reference of the territory to fragments of SSAPP: (*a*) outlines of the territory Ω; (*b*) the form of identifier A_1 in database; and (*c*) hierarchy of the database levels corresponding to the spatial and thematic structure of Ω.

where

$$a_{ij}^1 = \begin{cases} 0 & \text{for } (\varphi, \lambda) \notin \Omega \\ a_1 & \text{for } (\varphi, \lambda) \in \Omega \end{cases}$$

$$a_{ij}^2 = \begin{cases} A - \text{forest vegetation} \\ B - \text{urban territory} \\ C - \text{other types of surface} \end{cases}$$

$$a_{ij}^3 = \begin{cases} A - \text{no pollution sources} \\ B - \text{pollution source of type ``}B\text{''} \\ C - \text{pollution source of type ``}C\text{''} \\ \quad\ \text{other types of pollution sources} \end{cases}$$

Matrices A_4 and A_8 contain geographical information about the level of the territory with respect to sea level, and data on rain rate, respectively. The identifier A_5 defined the distribution of population density in the region. The structures A_6 and A_7 determine average statistical data on wind speed and direction over the territory Ω. All information is prescribed for a certain time period and, hence, the SSAPP database contains sets of matrices $\{A_i\}$ marked as belonging to some time interval. Each element of these matrix structures are hierarchically decoded by levels, but not necessarily in a uniform way. Matrix A_3 considers the types of pollution sources given in Table 6.12. For a better use of SSAPP, the database foresees the possibilities to identify the sources of atmospheric pollution with a set of pollutants. The

Table 6.12. Classification of the sources of atmospheric pollution.

Category of the source of pollution	Examples of enterprises fitting this categorization	Atmospheric pollutants
Chemical plants	Oil processing, woodworking, superphosphate extraction, cement production	Sulphur oxides, sulphides, fluorine compounds, organic vapour, particles, aromatic substances
Spraying and pulverizing in agriculture	Control of pesticides and weeds	Organic phosphates, chlorinated hydrocarbons, arsenic, lead
Crushing, grind, bolting	Road building plants	Minerals and organic substances
Destruction	Rebuilding of cities	Minerals and organic particles
Burning of the fields	Burning of stubble and clearings	Smoke, ashes, soot
Fuel burning	Power stations	Sulphur and nitrogen oxides, carbon oxide, smoke and aromatic substances
Fuel production	Gas dissipation	Fluorine compounds
Production of printing ink	Photo production and printing	Fluorine compounds
Metallurgic plants	Fusing, production of steel and aluminium	Lead, arsenic, zinc, fluorine compounds, sulphur oxides
Mills, car engines	Elevators, carriers	Sulphur and nitrogen oxides, aromatic substances, carbon oxide, smoke.
Use of nuclear fuel	Nuclear reactors	Argon-41
Ore extraction	Grind, crushing, bolting	Dust of uranium and beryl
Burning of refuse and waste	Dumps, refuse recycling plants	Sulphur oxides, smoke, ashes, organic vapour, aromatic substances
Waste recycling	Dumps of scrap metal, car dumps, plants processing second-hand materials	Smoke, organic vapour, aromatic substances, soot
Dry-cleaning, fuel evaporation, paint pulverizing	Dry cleaning, degreasing, chemical separators, painting of cars and furniture	Iodine-131, hydrocarbons, organic vapour

structure of elements of the identifier A_9 identifies each space cell Ω_{ij} with the type of the polluting enterprise.

6.4.2 Subsystems to make statistical decisions

A combination of the SSAPP units SIGNAL and IBMMEN provides a realization of the function of decision making about the signal for the user in accordance with an a priori prescribed elemental structure: "the air standard" (AS). To preserve generality, it is assumed that the AS can be established in each cell Ω_{ij} or in some of them by an independent set of elements. This information is contained in the identifier A_{10}.

In the process of measurements or calculations, for each pollutant of the type ξ the function of concentration distribution $P(\varphi, \lambda, h, t)$ is formed. In due course, a data series is formed at each point (φ, λ, h):

$$P_{\xi 1}, P_{\xi}^2, \ldots, P_{\xi}^n, \ldots$$

Because of errors in measurements and calculations based on inaccurate data, the choice of situations $P_\xi < \mathrm{AS}_\xi$ or $P_\xi \geq \mathrm{AS}_\xi$ is realized at each time moment t on the basis of the respective statistical estimate with division into two versions: (1) decision is made by a totality of indicators; and (2) decision is made by each indicator separately. In the latter case, the stability of the inequality $P_\xi \geq \mathrm{AS}_\xi$ is tested during a given time period Δt. In the case of meeting the condition $P_\xi \geq \mathrm{AS}_\xi$ during this time period, the display shows the signal with the respective information. The time period Δt is determined at an expert level during system exploitation. The real choice of Δt depends on the multitude of local industrial and other factors and refers to the user's competence. To simplify an expert estimation or decision making in real time mode, especially when an integral criterion is used, the user obtains estimates of all indicators on a homogeneous normalized scale.

Let $q(l, i, j, k)$ be an indicator of the atmospheric quality in the compartment Ω_{ijk} by the lth parameter. To reduce all indicators to a single scale, we normalize them:

$$Q_l = Q(l, i, j, k) = q(l, i, j, k) / \max_{i,j,k} q(l, i, j, k)$$

Then the results of normalization by height (k), latitude (i), longitude (j), and the whole territory Ω are smoothed over. The smoothed results are used to estimate the air quality of territories of the user's choice. The final decision is made on the basis of comparison of the estimates of Q_l with a set of threshold values contained in the database. The complex estimation of the state of the atmosphere over a given part of the territory Ω is realized by the vector parameter $Q = (Q_1, \ldots, Q_n)$. This n-dimensional vector belongs to some indicating space X, each point of which corresponds to a concrete combination AS_ξ (prescribed by the user). The whole space X is divided into two parts: X_0 – the pollution does not exceed permissible levels, and X_1 – pollution reaches dangerous levels. The concrete realization of Q is estimated by attribution into one of these subspaces.

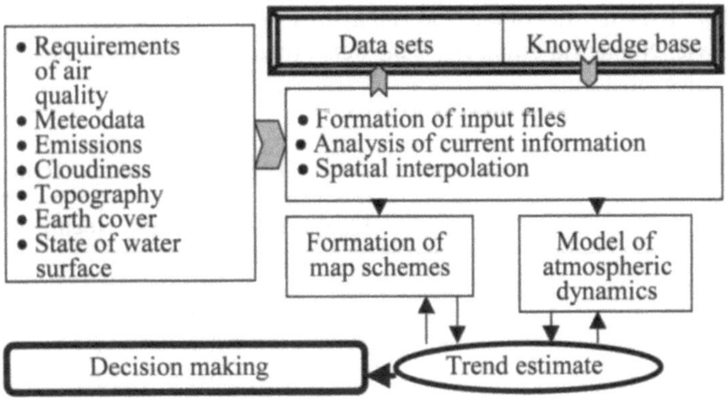

Figure 6.6. Main components of the SSAPP functioning.

Thus the basic stages of the SSAPP functioning can be presented by the scheme shown in Figure 6.6. Of course, the quality of the decisions made depends on the adequacy of the database. The control of the database fragments and of various modifications of parametric descriptions of the pollutant transformation is accomplished by a set of identifiers available to the user through a user-oriented interface. Therefore, a set of identifiers is used at the system manager level observing the compatibility of its functions.

6.5 NATURAL–ANTHROPOGENIC SULPHUR CYCLE

The global model of the biosphere can be improved by increasing the number of biogeochemical cycles taken into account. The necessity to include a unit describing the sulphur fluxes in natural systems is dictated by the dependence of biotic processes on the content of sulphur in the compartments of the biosphere. The available data on supplies and fluxes of sulphur compounds in the atmosphere, soils, vegetation cover, and hydrosphere enable one to formulate numerical relationships to simulate the global cycle of sulphur (the Model of the Global Sulphur Cycle (MGSC) unit).

The global sulphur cycle consists of the mosaic structure of local fluxes of its compounds with other elements formed due to water migration and atmospheric processes. The conceptual schemes of the global and regional cycles of sulphur have been described in detail by many authors (Nitu *et al.*, 2000b; Kashulina *et al.*, 2003; Hadley and Toumi, 2003; Austin *et al.*, 2002). However, the existing models have been developed for autonomous functioning and usage, which makes it difficult to include them into the global model without substantial changes of their parametric space. The model of the global sulphur cycle proposed here has been derived in the form of a unit with inputs and outputs, which enables one to match it with other units of the global model via their inputs and outputs.

Table 6.13. The characteristics of the land and hydrospheric fluxes of sulphur in the structure of its global cycle. Numerical estimates of the fluxes (mg/m^3/day) are obtained by averaging over the respective territories.

From Krapivin and Nazarian (1997).

	Land		Hydrosphere	
Sulfur flux	Identifier	Estimate	Identifier	Estimate
Volcanic eruptions				
H_2S	C_1	0.018	H_3	0.0068
SO_2	C_5	0.036	H_5	0.0073
SO_4^{2-}	C_{20}	0.035	H_9	0.0074
Anthropogenic emissions				
H_2S	C_2	0.072	H_1	0.00076
SO_2	C_6	0.92	H_6	0.038
SO_4^{2-}	C_{10}	0.47		
Oxidation of H_2S to SO_2	C_4	1.13	H_2	0.3
Oxidation of SO_2 to SO_4^{2-}	C_9	1.35	H_8	0.16
Dry sedimentation of SO_4^{2-}	C_{12}	0.37	H_{11}	0.11
Fallout of SO_4^{2-} with rain	C_{11}	1.26	H_{10}	0.38
Biological decomposition and emission of H_2S into the atmosphere	C_3	1.03	H_4	0.31
Assimilation of SO_4^{2-} by biota	C_{15}	0.41	H_{13}	1.09
Biological decomposition and formation of SO_4^{2-}	C_{16}	1.13	H_{17}	0.43
			H_{23}	0.12
Sedimentation and deposits	C_{18}	0.22	H_{15}	0.98
	C_{19}	0.11	H_{16}	0.55
			H_{19}	0.0076
			H_{25}	0.036
Wind driven return to the atmosphere	C_{13}	0.25	H_{12}	0.33
Replenishing sulphur supplies due to dead biomass	C_{17}	0.86	H_{14}	1.1
Assimilation of atmospheric SO_2	C_7	0.46	H_7	0.18
Washing out of SO_2 from the atmosphere	C_8	0.27	H_{24}	0.061
River runoff of SO_4^{2-} to the ocean	C_{14}	1.17		
Transition of gas phase H_2SO_4 to H_2S	C_{21}	0.018	H_{26}	0.0076
Assimilation of the washed out part of atmospheric SO_2 by biota	C_{22}	0.036	H_{27}	0.015
Oxidation of H_2S to SO_2 in the water medium			H_{18}	0.045
			H_{22}	0.19
Advection of SO_2			H_{20}	0.38
Advection of H_2S			H_{21}	0.37

In contrast to hydrogen, sulphur compounds cannot be attributed to the long-lived elements of the biosphere. Therefore, in the unit of sulphur the spatial digitization of its natural and anthropogenic reservoirs should be planned to reflect the local distributions of sulphur in the vicinity of its sources and to enable one to estimate the intensities of the inter-regional fluxes of sulphur compounds. The version of the sulphur unit proposed here, in contrast to the known hydrodynamic models of the long-distance transport, takes into account the fluxes of sulphur compounds between the hydrosphere, atmosphere, soil, and biota. The model does not consider the vertical stratification of the atmosphere. The characteristics of sulphur fluxes over land and oceans averaged vertically are calculated. The spatial digitization of the biosphere and the World Ocean corresponds to the criterion inherent to the global model. A description of the fluxes of sulphur compounds is given in Table 6.13. This scheme is realized in every cell Ω_{ij} of the Earth's surface and in every compartment Ω_{ijk} of the World Ocean. The interaction between the cells and the compartments is organized through the climate unit of the global model. Therefore, the equations of the sulphur unit lack the terms reflecting the dynamic pattern of the spatial transformation of the sulphur reservoirs. With account of notations assumed in Figure 6.7 and in Tables 6.13 and 6.14, the equations

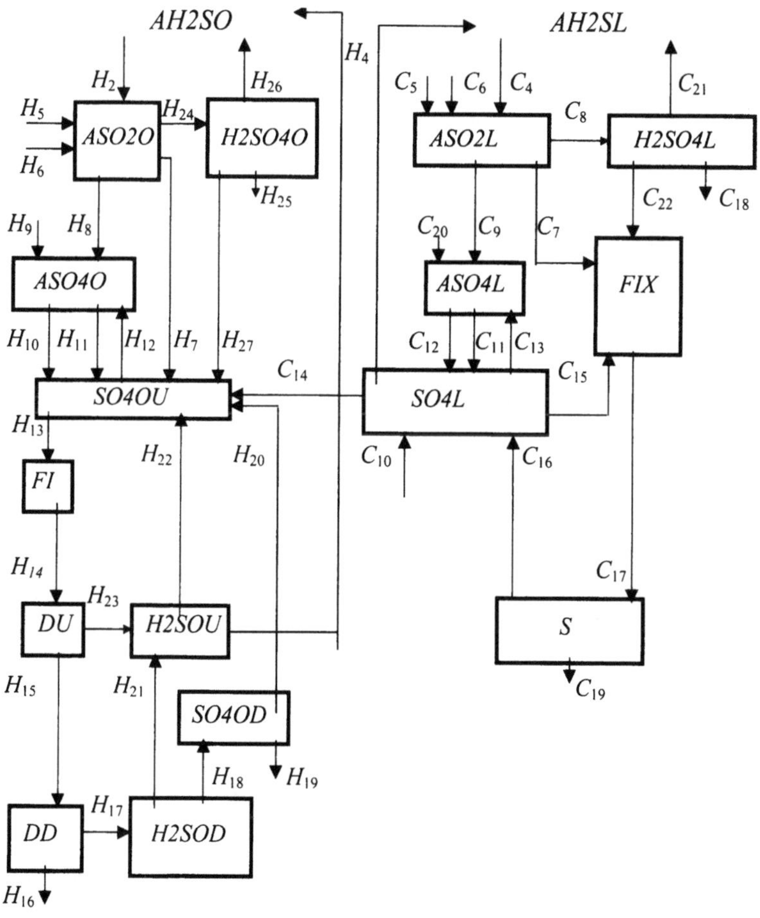

Figure 6.7. The scheme of sulphur fluxes in the environment. Notation is given in Tables 6.13 and 6.14.

Table 6.14. Some estimates of the sulphur reservoirs that can be used as initial data.

Reservoir	Identifier in equations of the MGSC unit	Quantitative estimate of the sulphur reservoir (mg/m^2)
The atmosphere over the oceans		
H_2S	AH_2SO	10
SO_2	ASO_2O	5.3
SO_4^{2-}	ASO_40	2
The atmosphere over land		
H_2S	AH_2SL	36.9
SO_2	ASO_2L	17.9
SO_4^{2-}	ASO_4L	12.9
Land		
SO_4^{2-}	SO_4L	11.2
Biomass	FIX	6,00
Soil	S	5,000
Photic layer of the World Ocean		
H_2S	H_2SOU	1.9
SO_4^{2-}	SO_4OU	19×10^7
Biomass	FI	66.5
DOM	DU	730
Deep layers of the World Ocean		
H_2S	H_2SOD	2×10^6
SO_4^{2-}	SO_4OD	3.4×10^9
DOM	DD	13,120

describing the balanced relationships between the reservoirs of sulphur compounds will be written in the following form:

$$\frac{d\mathrm{AH_2SL}}{dt} = C_1 + C_2 + C_3 - C_4 + C_{21}$$

$$\frac{d\mathrm{ASO_2L}}{dt} = C_4 + C_5 + C_6 - C_7 - C_8 - C_9$$

$$\frac{d\mathrm{ASO_4L}}{dt} = C_9 + C_3 + C_{20} - C_{11} - C_{12}$$

$$\frac{dS}{dt} = C_{17} - C_{16} - C_{19}$$

$$\frac{dSO_4L}{dt} = C_{10} + C_{11} + C_{12} + C_{16} - C_3 - C_{13} - C_{14}$$

$$\frac{d\mathrm{FIX}}{dt} = C_7 + C_{15} - C_{17} + C_{22}$$

$$\frac{d\mathrm{H2SO4L}}{dt} = C_8 - C_{18} - C_{21} - C_{22}$$

$$\frac{d\mathrm{AH2SO}}{dt} = H_1 + H_3 + H_4 + H_{26} - H_2$$

$$\frac{dASO2O}{dt} = H_2 + H_5 + H_6 - H_7 - H_8 - H_{24}$$

$$\frac{d\mathrm{ASO4O}}{dt} = H_8 + H_9 + H_{12} - H_{10} - H_{11}$$

$$\frac{\partial \mathrm{SO4OU}}{\partial t} + v_z \frac{\partial \mathrm{SO4OU}}{\partial z} + k_z \frac{\partial^2 \mathrm{SO4OU}}{\partial z^2} = H_7 + H_{10} + H_{11} + H_{20} + H_{22} + H_{27} + C_{14} - H_{12} - H_{13}$$

$$\frac{\partial \mathrm{H2SOU}}{\partial t} + v_z \frac{\partial \mathrm{H2SOU}}{\partial z} + k_z \frac{\partial^2 \mathrm{H2SOU}}{\partial z^2} = H_{21} + H_{23} - H_4 - H_{22}$$

$$\frac{\partial \mathrm{H2SOD}}{\partial t} + v_z \frac{\partial \mathrm{H2SOD}}{\partial z} + k_z \frac{\partial^2 \mathrm{H2SOD}}{\partial z^2} = H_{17} - H_{18} - H_{21}$$

$$\frac{\partial \mathrm{SO4OD}}{\partial t} + v_z \frac{\partial \mathrm{SO4OD}}{\partial z} + k_z \frac{\partial^2 \mathrm{SO4OD}}{\partial z^2} = H_{18} - H_{19} - H_{20}$$

$$\frac{\partial \mathrm{DU}}{\partial t} + v_z \frac{\partial \mathrm{DU}}{\partial z} + k_z \frac{\partial^2 \mathrm{DU}}{\partial z^2} = H_{14} - H_{15} - H_{23}$$

$$\frac{\partial \mathrm{DD}}{\partial t} + v_z \frac{\partial \mathrm{DD}}{\partial z} + k_z \frac{\partial^2 \mathrm{DD}}{\partial z^2} = H_{15} - H_{16} - H_{17}$$

$$\frac{\partial \mathrm{FI}}{\partial t} + v_z \frac{\partial \mathrm{FI}}{\partial z} + k_z \frac{\partial^2 \mathrm{FI}}{\partial z^2} = H_{13} - H_{14}$$

$$\frac{\partial \mathrm{BOT}}{\partial t} = H_{16} + H_{19}$$

where v_z is the velocity of the vertical water motion in the ocean (m/day) and k_z is the coefficient of turbulent mixing (m^2/day).

The equations of the MGSC unit in each cell of the spatial division of the ocean surface are supplemented with initial conditions (Table 6.14). The calculation procedure to estimate the sulphur concentration consists of two stages. At each time moment t_i, first for all cells Ω_{ij} the equations of the MGSC unit are solved by the method of quasi-linearization, and all reservoirs of sulphur are estimated for $t_{i+1} = t_i + \Delta t$, where a time step Δt is chosen from the condition of convergence of the calculation procedure. Then at the moment t_{i+1} with the use of the climate unit of the global model these estimates are specified with account of the atmospheric transport and ocean currents over the time Δt. The sulphur supplies in the reservoirs are measured in mg S/m^3, and the sulphur fluxes have the dimensions mg S/m^3/day. The sulphur supplies in the water medium are calculated with the volumes of the compartments Ω_{ijk} taken into account. To estimate the sulphur supplies in the atmosphere, it is assumed that an effective thickness of the atmosphere h is an input parameter either introduced into the model by the user, prescribed as the constants from Table 6.14, or received from the climate unit of the global model. The quantitative estimates of the fluxes in the right-hand parts of the equations of the MGSC unit are obtained in different units of the global model. The anthropogenic fluxes of sulphur H_1, H_6, C_2, C_6, and C_{10} are simulated in the unit of scenarios. The fluxes H_3, H_5, H_9, C_1, C_5, and C_{20} are prescribed either by the climate unit or formed in the unit of scenarios.

The accuracy of different functional presentations of the fluxes in the equations of the MGSC unit corresponds to the accuracy of similar fluxes of the biogeochemical cycles of hydrogen, phosphorus, and nitrogen. The rate of emission of H_2S into the atmosphere through humus decomposition is described by the linear function $C_3 = \mu_1(\text{pH})\text{SO4L}T_L$, where μ_1 is the proportion coefficient depending on soil acidity (day^{-1}/K; and T_L is the soil temperature (°K). The initial value of SO4L is estimated from the humus supply considering that the content of sulphur in humus is prescribed by the parameter a_g (%). According to the available observations of the input of H_2S into the atmosphere from the ocean, the flux H_4 varies widely from low values to high values at the transition from stagnant waters to zones of upwelling. The flux H_4 is assumed to be the function of the ratio of the rates of H_2S oxidation in the photic layer to the rate of the vertical uplifting of water. Therefore, to describe the H_4 flux, use the parameter t_{H2SU}, which reflects the lifetime of H_2S in the water medium: $H_4 = \text{H2SU}/t_{\text{H2SU}}$. Determine the value of t_{H2SU} as a function of the rate of the vertical advection v_z and concentration of oxygen O_2 in the upper layer Z_{H2S} thick: $t_{\text{H2SU}} = \text{H2SOU}v_z(\theta_2 + O_2)/[O_2(\theta_1 + v_z)]$, where the constants θ_1 and θ_2 are determined empirically, the value of O_2 is either calculated by the oxygen unit of the global model or prescribed from the global database (Keeling *et al.*, 1998; Lueker *et al.*, 2001).

The reaction of oxidation of H_2S to SO_2 in the atmosphere, on land, and over the water surface is characterized by a rapid process of the reaction of hydrogen sulphide with atomic and molecular oxygen. At the same time, the reaction of H_2S with O_3 in the gas phase is slow. It is impossible to describe within the global model the diversity of the situations appearing here. However, an inclusion of the fluxes H_2 and C_4 into the unit of sulphur enabled one to take into account the correlation between the cycles of sulphur and oxygen. These fluxes are parameterized with the use of the indicator t_{H2SA} of the lifetime of H_2S in the atmosphere: $C_4 = \text{AH2SL}/t_{H2SA}$, $H_2 = \text{AH2SO}/t_{\text{H2SA}}$. The mechanism to remove SO_2 from the atmosphere is described by the fluxes H_7, H_8, H_{27}, C_7, C_8, and C_9. Schematically this mechanism consists of a set of interconnected reactions of SO_2 with atomic oxygen under the influence of various catalysts. A study of the succession of the reaction enables one to estimate the lifetime of SO_2 for oxidation over land t_{SO2L} and water surface t_{SO2A1}. This makes it possible to assume the following parameterizations of the fluxes H_8 and C_9: $H_8 = \text{ASO2O}/t_{\text{SO2A1}}$, $C_9 = \text{ASO2}L/t_{\text{SO2L}}$.

Sulphur dioxide is assimilated from the atmosphere by rocks, vegetation, and other Earth coverings. Over the water surface this assimilation is connected with the intensity of turbulent gas fluxes and surface roughness. We describe a dry deposition of SO_2 over the vegetation by the model $C_7 = q_2RX$, where

$q_2 = q_2'\text{ASO2L}/(r_{tl} + r_s)$, where r_{tl} is the atmospheric resistance to the SO_2 transport over the vegetation of l-type (day/m); r_s is the surface resistance of the s-type to the SO_2 transport (day/m); RX is the produce of vegetation of the X-type (mg/m^2/day) (calculated by the biogeocenotic unit of the global model); and q_2' is the proportion coefficient. The parameters r_{tl} and r_s are functions of the types of the soil–vegetation formations and estimated, respectively, at 0.05 and 4.5 for the forests, 0.9 and 3 for grass cover, 0.5 and 2.6 for bushes, 0.8 and 1 for bare soils, 1.9 and 0 for water surfaces, and 2 and 10 for snow cover.

The process of washing out of SO_2 from the atmosphere with changing phase to H_2SO_4 and a subsequent neutralization on the surface of the l-type is described by the function $C_8 = q_{1l}W\text{ASO2L}$, with the Langmuire coefficient q_{1l} and precipitation intensity $W(\varphi, \lambda, t)$. An interaction of acid rain with the Earth's surface elements is reflected in the scheme in Figure 5.1 by the fluxes C_{18}, C_{21}, and C_{22} for land and H_{25}, H_{26}, and H_{27} for water surfaces. To parameterize these fluxes, assume the hypothesis that the reservoirs of H2SO4L and H2SO4O are spent in proportion to the outflows, and the coefficients of this proportion are the controlling parameters of the numerical experiments: $C_{18} = h_1\text{H2SO4L}$, $C_{22} = h_2RX\text{H2SO4L}$, $C_{21} = h_3T_a\text{H2SO4L}$, $H_{25} = h_6\text{H2SO4O}$, $H_{26} = h_4T_a\text{H2SO4O}$, $H_{27} = h_5RFI\text{H2SO4O}$, $h_1 + h_2RX + h_3T_a = 1$, $h_4T_a + h_5RFI + h_6 = 1$, where $T_a(\varphi, \lambda, t)$ is the air surface temperature. Parameterize the fluxes H_7 and H_{24} by the relationships: $H_7 = \text{ASO2O}/t_{\text{SO2A2}}$, $H_{24} = q_{1l}W\text{ASO2O}$, where t_{SO2A2} is the lifetime of SO_2 over the water surface.

Sulphates interacting with the ecosystems and establishing an interaction of the sulphur cycle with other biogeochemical processes are one of the most important elements in the global cycle of sulphur. Numerous complicated transformations of sulphates in the environment are described by a set of fluxes H_7, H_8, H_{10}, H_{11}, H_{12}, C_9, C_{11}, C_{12}, and C_{13} for the atmospheric reservoir and the fluxes H_{13}, H_{18}, H_{19}, H_{20}, H_{22}, C_3, C_{14}, C_{15}, and C_{16} for land and the World Ocean.

Physical mechanisms for the transport of sulphates from the atmosphere to the soil and water medium are connected with dry and wet sedimentation. An efficient model of the wet removal of particles and gases from the atmosphere was proposed by Langmann (2000): where a substitution of the mechanism of the aerosols and gases by a simplified binary model enables one to match it with other units of the global model: $H_{10} = \mu W\text{ASO4O}$, $H_{11} = \rho v_O\text{ASO4O}$, $C_{11} = b_3W\text{ASO4L}$, and $C_{12} = d_1v_a\text{ASO4L}$, where v_O and v_a are the rates of the aerosols dry deposition over the water surface and land, respectively; and μ, b_3, ρ, and d_1 are the constants.

The return of sulphates from the soil and water medium to the atmosphere is connected with rock weathering and spray above the rough water surface: $C_{13} = d_2RATE\text{SO4L}$, $H_{12} = \theta RATE\text{SO4U}$, where $RATE(\varphi, \lambda, t)$ is the wind speed over the surface (m/s); and d_2 and θ are the empirical coefficients.

The flux C_{14} relates the surface and water reservoirs of sulphur. Let σ be the share of the river system area on land and d_3 – the proportion coefficient, then $C_{14} = d_3W\text{SO4L} + (C_{11} + C_{12})\sigma$.

The surface part of the sulphur cycle is connected with the functioning of the atmosphere–vegetation–soil system. Plants adsorb sulphur from the atmosphere in the form of SO_2 (fluxes C_7 and C_{22}) and assimilate sulphur from the soil in the form of SO_4^{2-} (flux C_{15}). In the hierarchy of the soil processes two levels can be selected defining the sulphur reservoirs as "dead organics" and "SO_4^{2-} in soil". The transitions between them are described by the flux $C_{16} = b_2 ST_L$, where the coefficient $b_2 = b_{2.1} b_{2.2}$ reflects the rate $b_{2.1}$ of transition of sulphur contained in dead organics into the form assimilated by vegetation The coefficient $b_{2.2}$ indicated the content of sulphur in dead plants.

The fluxes of sulphur in the water medium according to studies by Bodenbender *et al.* (1999), depend on the biological processes in the water bodies and constitute an isolated part of the global cycle of sulphur that contains only the fluxes that connect it with the atmospheric and surface cycles. Rough estimates show that the rates of the sulphur cycle in the water of the seas and oceans do not play a substantial role for the remaining parts of its global cycle. And though this specific feature does exist, for the purity of the numerical experiment, in the proposed model the internal hydrospheric fluxes of sulphur compounds are separated in space and parameterized with the same details like other fluxes of sulphur in the atmosphere and on land. This excessiveness is important for other units of the global model as well. In particular, it is important for the parameterization of photosynthesis whose rate *RFI* affects the closure of other biogeochemical cycles. Finally, assume: $H_{13} = \gamma RFI$, $H_{14} = bMFI$, $H_{15} = f\mathrm{DU}$, $H_{16} = p\mathrm{DD}$, $H_{17} = q\mathrm{DD}$, $H_{18} = \mathrm{H2SOD}/t_{\mathrm{H2SOD}}$, $H_{19} = u\mathrm{SO4D}$, $H_{20} = a_1 v_D \mathrm{SO4D}$, $H_{21} = b_1 v_D \mathrm{H2SOD}$, $H_{22} = \mathrm{H2SOU}/t_{\mathrm{H2SOU}}$, $H_{23} = g\mathrm{DU}$, where MFI is the mass of dead phytoplankton, t_{H2SOU} and t_{H2SOD} is the time of complete oxidation of H_2S in the seawater in the photic and deep layers, respectively; γ, b, f, p, q, u, a_1, b_1, and g are the constants.

The anthropogenic input to the sulphur unit is realized through the prescribed fluxes C_2, C_6, C_{10}, C_1, and C_6 in the form of the functions of the spatial coordinates and time.

6.6 PHOSPHORUS CYCLE

In contrast to nitrogen, the main reservoir of phosphorus in the biosphere is not the atmosphere but the rocks and other deposits formed in the past geological epochs, which, being subject to erosion, emit phosphates. Besides, there are other mechanisms of the return of phosphorus to the biospheric cycle, but, as a rule, they are not that efficient. One of these mechanisms is fishing, returning to land from the hydrosphere about 60×10^3 t P/year, as well as an extraction of phosphorus-containing rocks estimated at 1–2×10^6 t P/year. The present cycle of phosphorus is closed by its fluxes to the bottom deposits in the World Ocean.

The flux diagram of the model of the global phosphorus cycle (MGPC) is presented in Figure 6.8, according to which the balance system of equations will be:

$$\frac{\partial P_A}{\partial t} + V_\varphi \frac{\partial P_A}{\partial \varphi} + V_\lambda \frac{\partial P_A}{\partial \lambda} = H_1^P + H_{19}^P + \begin{cases} -H_{16}^P & (\varphi, \lambda) \in \Omega_O \\ H_7^P - H_8^P & (\varphi, \lambda) \in \Omega/\Omega_O \end{cases}$$

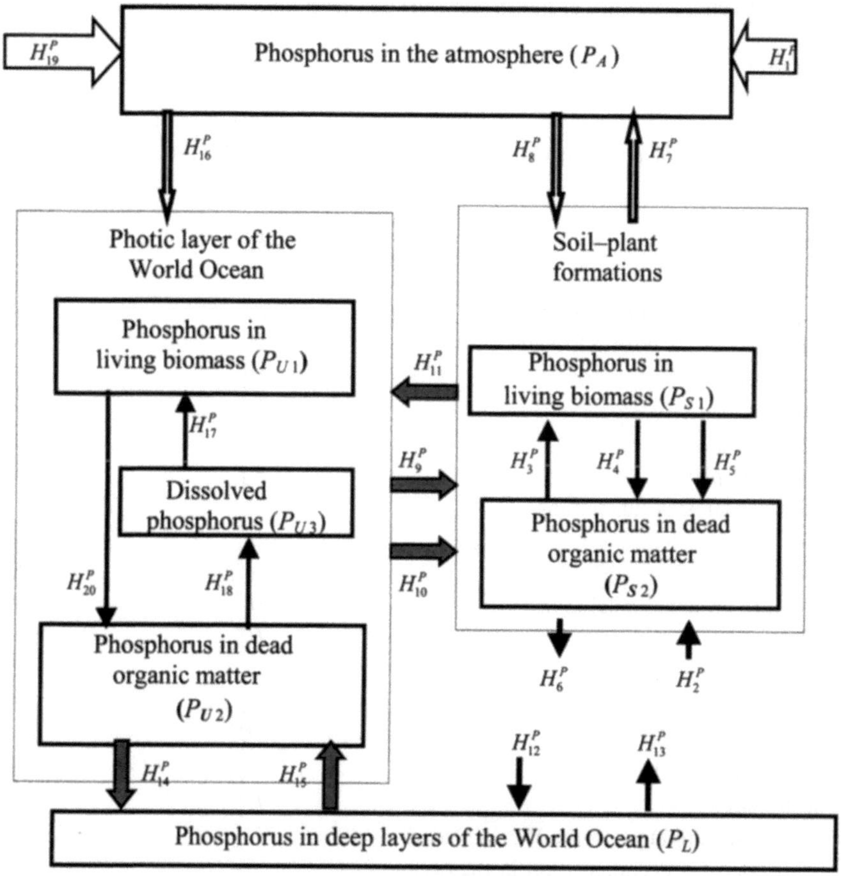

Figure 6.8. Scheme of phosphorus fluxes in the biosphere. Notations are given in Table 6.15.

$$\frac{\partial P_U}{\partial t} + v_\varphi \frac{\partial P_U}{\partial \varphi} + v_\lambda \frac{\partial P_U}{\partial \lambda} + v_z \frac{\partial P_U}{\partial z} = H^P_{11} + H^P_{15} + H^P_{16} - H^P_9 - H^P_{10}$$

$$\frac{\partial P_L}{\partial t} + v_\varphi \frac{\partial P_L}{\partial \varphi} + v_\lambda \frac{\partial P_L}{\partial \lambda} + v_z \frac{\partial P_L}{\partial z} = H^P_{12} + H^P_{14} - H^P_{13} - H^P_{15}$$

$$\frac{\partial P_S}{\partial t} = H^P_2 + H^P_8 + H^P_9 + H^P_{10} - H^P_6 - H^P_7 - H^P_{11}$$

where $P_U = P_{U1} + P_{U2} + P_{U3}$, and $P_S = P_{S1} + P_{S2}$. With this detailing we have:

$$\frac{\partial P_{U1}}{\partial t} + v_\varphi \frac{\partial P_{U1}}{\partial \varphi} + v_\lambda \frac{\partial P_{U1}}{\partial \lambda} + v_z \frac{\partial P_{U1}}{\partial z} = H^P_{17} - H^P_9 - H^P_{10} - H^P_{20}$$

$$\frac{\partial P_{U2}}{\partial t} + v_\varphi \frac{\partial P_{U2}}{\partial \varphi} + v_\lambda \frac{\partial P_{U2}}{\partial \lambda} + v_z \frac{\partial P_{U2}}{\partial z} = H^P_{20} - H^P_{18}$$

Table 6.15. The characteristics of fluxes (10^6 t/yr) and reservoirs (10^6 t) of phosphorus in the biosphere.
Krapivin (2000b).

Reservoirs and fluxes of phosphorus	Identifier of the MGPC unit	Evaluation
Phosphorus supplies		
In the atmosphere	P_A	3
On land	P_S	1,546
In the photic layer of the World Ocean	P_U	2×10^4
In deep layers of the World Ocean	P_L	12×10^4
Volcanic emissions	H_1^P	0–2
Fertilization	H_2^P	19
Assimilation by plants	H_3^P	45.34
Input with dead plants	H_4^P	39.34
Input due to organisms functioning		
On land	H_5^P	5
In the World Ocean	H_{20}^P	81.5
Transition into an unassimilable form	H_6^P	2.9
Weathering	H_7^P	5
Falling out with precipitation		
On land	H_8^P	1.8
On the oceans	H_{16}^P	2
Removal with fish catch	H_9^P	0.06
Removal by birds	H_{10}^P	0.04
Washing out and sinking into the World Ocean	H_{11}^P	4–14
Input due to detritus decomposition in the oceans		
Photic layer	H_{18}^P	550
Deep layers	H_{12}^P	159
Exchange between photic and deep layers of the ocean		
Lifting	H_{15}^P	96.1
Descending	H_{14}^P	22
Precipitation	H_{13}^P	13–83.9
Rock weathering	H_{19}^P	1
Photosynthesis	H_{17}^P	630–1,300

$$\frac{\partial P_{U3}}{\partial t} + v_\varphi \frac{\partial P_{U3}}{\partial \varphi} + v_\lambda \frac{\partial P_{U3}}{\partial \lambda} + v_z \frac{\partial P_{U3}}{\partial z} = H_{11}^P + H_{18}^P - H_{17}^P$$

$$\frac{\partial P_{S1}}{\partial t} = H_3^P - H_4^P - H_5^P$$

$$\frac{\partial P_{S2}}{\partial t} = H_2^P + H_4^P + H_5^P + H_8^P - H_3^P - H_6^P - H_7^P$$

Determine now the functional and dynamic characteristics of the fluxes of phosphorus (Table 6.15) based on analysis of the existing ideas about their nature. The atmospheric cycle is governed by rock weathering, volcanic eruptions, and by washing out of phosphorus by precipitation. From the available estimates, the phosphorus content in the lithosphere constitutes 0.093%, and the processes of weathering annually deliver into the atmosphere about 0.2×10^6 t P, on average. Since these processes are of a complicated stochastic nature and their models are absent, in the first approximation the fluxes H_1^P and H_{19}^P will be considered constant.

The continental cycle of phosphorus is determined by ten fluxes (Figure 6.8) closed by a single component P_S indicating the phosphorus supplies on land in the soil–vegetation formations and animals. The supplies of phosphorus in soils are replenished due to the fluxes H_l^P $(l = 2, 4, 5, 8, 9, 10)$. The loss of phosphorus from the soil is determined by the fluxes H_j^P $(j = 3, 6, 7, 11)$. With a more complicated detailing of the surface reservoirs of phosphorus and consideration of more ingenious effects in the interaction of these reservoirs, the classification of the surface fluxes of phosphorus becomes more complicated. Within the assumed detailing of the surface reservoirs of phosphorus the following functional presentations of the fluxes $\{H_i^P\}$ can be considered:

$$H_2^P = p_1 G M G_0^{-1} M_0^{-1} \qquad H_3^P = p_V R_V \qquad H_4^P = p_4 M_L \qquad H_5^P = p_5 H_F$$

$$H_6^P = p_6 P_{S2}/P_{S2,0} \qquad H_9^P = p_3 I \qquad H_7^P = p_7 \theta^{\Delta T} P_{S2}/P_{S2,0}$$

$$H_{11}^P = p_2 P_{S2}[1 - \exp(-k_{su} W_{so})]/P_{S2,0}$$

$$P_{S2,0} = P_{S2}(\varphi, \lambda, 0)$$

where p_V is the content of phosphorus in the living biomass of plants; p_4 and p_5 are the content of phosphorus in organic matter of vegetable and animal origin, respectively; I is the amount of products extracted from the ocean; G is the population; M is mineral resources; R_V is the product of vegetation photosynthesis; H_F is the biomass of the unassimilated food of animals; $M_L = \mu_V L$ where μ_V is the rate of vegetation dying off and L is the vegetation biomass; θ is the temperature coefficient of the rate of the dead organic matter decomposition on land; ΔT is the surface air temperature (SAT) variation with respect to the control value; W_{so} is the volume of the river run-off into the oceans; and p_i $(i = 1, \ldots, 7)$ are the constants. The index "0" in G_0, M_0, and $P_{S2,0}$ attributes these parameters to some control time moment t_0, when all parameters of the MGPC are known.

Describe the hydrochemical cycle of phosphorus by a totality of fluxes $H_k^P (k = 9\text{–}18, 20)$: $H_{12}^P = p_{14} R_{DL}$, $H_{13}^P = p_8 P_L/P_{L,0}$, $H_{14}^P = p_9 P_U/P_{U,0}$, $H_{15}^P = p_{10} P_L/P_{L,0}$, $H_{16}^P = p_{12} R_{WO} P_A/P_{A,0}$, $H_{17}^P = p_{13} R_\Phi$, $H_{18}^P = p_{15} R_{DU}$, and $H_{20}^P = p_{16} M_\Phi$, where R_D is the rate of the dead organic matter decomposition, R_{WO} is precipitation over the ocean, R_Φ is the product of phytoplankton and other living organisms in the ocean, M_Φ is the rate of living biomass dying off, and p_i $(i = 8\text{–}16)$ are the proportion constants.

6.7 NITROGEN CYCLE

The MGNC unit simulating the fluxes of nitrogen in the environment is necessary in the global biospheric model for several indisputable reasons: nitrogen compounds can affect the environmental conditions, change food quality, affect the climate, and transform hydrospheric parameters. An abundant use of nitrates leads to water pollution and deterioration of the quality of food products. It is well known that an intensive exploitation of soils without taking into account the consequences of the misuse of nitrogen fertilizers breaks the stability of the agro-ecosystems and human health. Besides, nitrogen protoxide (N_2O), nitrogen dioxide (NO_2), and nitrogen oxide (NO), being minor gas components (MGCs) of the atmosphere, substantially affect the formation of the processes of absorption of optical radiation in the atmosphere. Small deviations in their concentrations can cause significant climatic variations near the Earth's surface (Kondratyev, 1999b; Gorshkov and Makaryeva, 1999; Sellers *et al.*, 1995; Stockwell *et al.*, 1999; Valiela and Bowen, 2002).

The nitrogen cycle is closely connected with the fluxes of hydrogen, sulphur, and other chemicals (Smith *et al.*, 1998; Dimitroulopoulou and Marsh, 1997; Hellebrand *et al.*, 2003).

6.7.1 Conceptual schemes of the nitrogen cycle in the biosphere

The global cycle of nitrogen as one of the nutrient elements is a mosaic structure of local processes of its compounds formed due to water migration and atmospheric processes. The present-day nitrogen cycle is strongly subject to anthropogenic forcings manifested through interference with the nitrogen cycle both directly and via the influence of related processes. Therefore, the construction of an adequate model of the nitrogen cycle should be based on the description of the whole complex of natural processes and those initiated by humans.

The natural sources of nitrogen oxides are connected with the vital functions of bacteria, volcanic eruptions, as well as several atmospheric phenomena (e.g., lightening discharges). The biogeochemical cycle of nitrogen includes processes such as fixation, mineralization, nitrification, assimilation, and dissimilation. The structural schemes of these processes have been described in detail by many authors. Their complexity level is determined by the goal of the study, availability of data on the rates of transformation of nitrogen-containing compounds and by many other circumstances.

Nitrogen moves in the biosphere by a complicated meandering structure of fluxes consisting of a hierarchy of cycles at various levels of life organization. From the atmosphere, nitrogen enters the cells of micro-organisms, from which it enters the soil and then higher plants, animals, and humans. Dying off of living organisms results in the return of nitrogen to the soil, from which it is either consumed by plants and living organisms or is emitted to the atmosphere. A similar scheme of nitrogen oxide cycling is inherent in the hydrosphere. The characteristic feature of these cycles is their openness connected with the available processes of removal of nitrogen from the biospheric balance into rocks, from

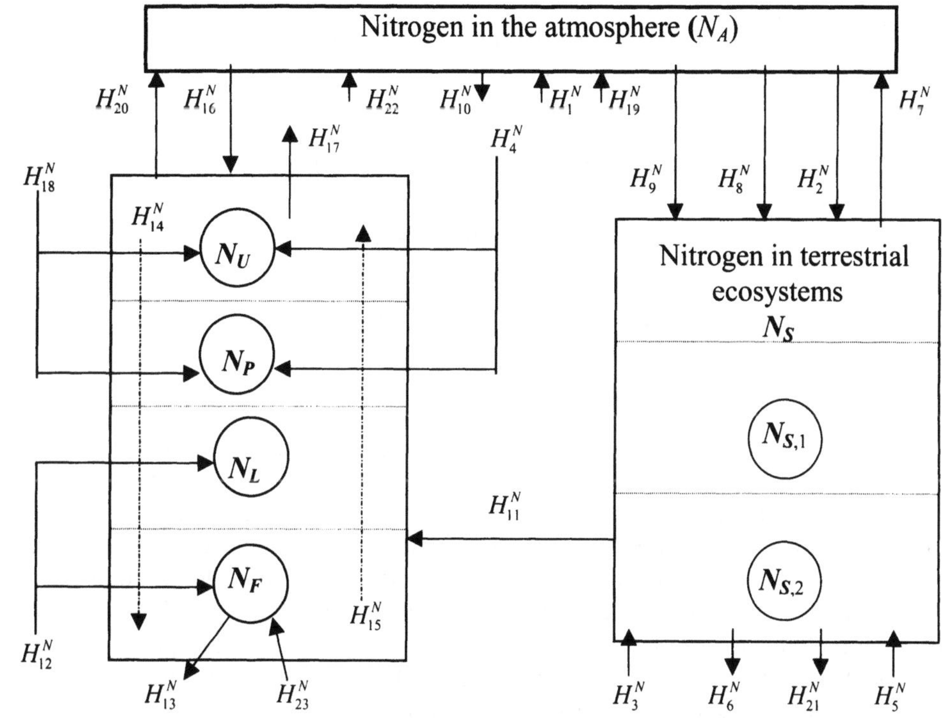

Figure 6.9. The scheme of nitrogen fluxes in nature (see Table 6.16 for explanatory notation).

where it returns much slower. Taking into account the nature of the nitrogen cycle in the biosphere and its reservoir structure enables one to formulate a global scheme of nitrogen fluxes.

A comparative analysis of the model schemes of the flux diagrams of nitrogen compounds in nature proposed by various experts makes it possible to construct the block-scheme presented in Figure 6.9. Here the atmosphere, soil, lithosphere, and hydrosphere are considered as nitrogen reservoirs. The first three reservoirs are described by 2-D models, and the hydrosphere is described by a 3-D multilayer model. The characteristics of nitrogen fluxes between these reservoirs are given in Table 6.16. The equations of the model are written as

$$\frac{\partial N_A}{\partial t} + V_\varphi \frac{\partial N_A}{\partial \varphi} + V_\lambda \frac{\partial N_A}{\partial \lambda} = H_1^N + \begin{cases} H_{20}^N - H_{16}^N & (\varphi, \lambda) \in \Omega_O \\ H_7^N + H_{19}^N - H_8^N - H_9^N & \\ \quad + H_{22}^N - H_2^N - H_{10}^N & (\varphi, \lambda) \in \Omega / \Omega_O \end{cases}$$

$$\frac{\partial N_{S1}}{\partial t} = H_8^N + H_6^N - H_3^H$$

$$\frac{\partial N_{S2}}{\partial t} = H_2^N + H_3^N + H_5^N + H_9^N - H_6^N - H_7^N - H_{11}^N - H_{21}^N$$

$$\frac{\partial N_U}{\partial t} + v_\varphi \frac{\partial N_U}{\partial \varphi} + v_\lambda \frac{\partial N_U}{\partial \lambda} = H_{16}^N + H_{4,U}^N + H_{18,U}^N + H_{11}^N - H_{17,U}^N - H_{20}^N - H_{14,UP}^N - H_{15,UP}^N$$

$$\frac{\partial N_P}{\partial t} + v_\varphi \frac{\partial N_P}{\partial \varphi} + v_\lambda \frac{\partial N_P}{\partial \lambda} = H_{18,P}^N + H_{4,P}^N + H_{14,UP}^N + H_{15,PL}^N - H_{17,P}^N - H_{14,PL}^N - H_{15,UP}^N$$

$$\frac{\partial N_L}{\partial t} = Q_L + H_{12,L}^N + H_{14,PL}^N + H_{15,LF}^N - H_{14,LF}^N - H_{15,PL}^N$$

$$\frac{\partial N_F}{\partial t} = Q_F + H_{12,F}^N + H_{23}^N + H_{14,LF}^N - H_{13}^N - H_{15,LF}^N$$

where $V(V_\varphi, V_\lambda)$ is the wind speed; $v(v_\varphi, v_\lambda)$ is the current velocity in the ocean; and Q_L and Q_F are functions describing the mixing of the deep waters of the ocean.

To simplify the calculation scheme presented in Figure 6.9, advection processes in the equations of the MGSC unit can be described by a superposition of the fluxes H_{14}^N and H_{15}^N. The computer realization of the equations of the MGNC unit introduces into the equations of the MGSC unit some corrections for the agreement between the dimensionalities of the variables in conformity with the spatial digitization of Ω. Therefore, the estimates of the fluxes H_i^N given below for their consideration in the MGNC should be corrected following this criterion.

6.7.2 Atmospheric components of the nitrogen cycle

The atmospheric part of the nitrogen cycle is an example of a complicated mechanism of transformation of gas substances characterized by an intricate set of fluxes on the borders between the basic reservoirs of nitrogen. Nevertheless, the results obtained by many experts make it possible to formulate clear ideas about the flux diagram of nitrogen compounds in the atmosphere.

Nitrogen resides in the atmosphere both in a free state (N_2) and in the form of various compounds – ammonia (NH_3), nitrous oxide, nitrogen oxide, nitrogen dioxide, and other nitrogen oxides (NO_3, N_2O_3, N_2O_4, N_2O_5), which play an intermediate role in chemical reactions. From the available estimates, the active part of atmospheric nitrogen constitutes 3.92×10^{12} t, (i.e., $N_A = 0.77 \times 10^4$ t/km^2). Detailing of the atmospheric reactions of nitrogen is incomplete because of insufficiently studied sources and behaviours of various forms of ammonia. The most important reactions in the atmosphere are the following:

$$NO_2 \rightarrow NO + O \qquad \varphi K_a = 0\text{–}25\ 1/\text{hr}$$

$$O_3 + NO \rightarrow NO_2 + O_2 \qquad K_1 = 1320\ 1/\text{ppm/hr}$$

The photochemical equilibrium is described by the relationship:

$$(NO)(O_3)/(NO_2) = \varphi K_a / K_1$$

Table 6.16. The characteristics of the reservoirs and fluxes of nitrogen in the biosphere (Figure 6.8).

Reservoirs (Gt) and fluxes (10^6 t/yr)	Identifier	Evaluation
Nitrogen supplies		
Atmosphere	N_A	39×10^5
Soil	N_S	280
Photic and intermediate layer of the ocean	$N_U + N_P$	2,800
Deep and bottom layer of the ocean	$N_L + N_F$	36,400
Natural sources of the hydrosphere	H_1^N	0.392
Technogenic accumulation		
In fuel burning	H_2^N	22.8
In fertilizer production	H_9^N	41.8
Input due to dead organisms		
On land	H_3^N	42.2
In the upper layers of the World Ocean	H_{18}^N	5
In deep layers of the World Ocean	H_{12}^N	7.8
Input due to organism functioning		
On land	H_5^N	0.1
In the World Ocean	H_4^N	0.3
Biological fixation		
On land	H_6^N	20.3
In the World Ocean	H_{17}^N	10
In the atmosphere	H_{10}^N	40
Denitrification		
On land	H_7^N	49.8
In the World Ocean	H_{20}^N	
Atmospheric fixation		
Over land	H_8^N	4
Over the World Ocean	H_{16}^N	3.6
Runoff from the land into the World Ocean	H_{11}^N	38.6
Precipitation	H_{13}^N	0.5
Vertical exchange processes in the oceans		
Descending	H_{14}^N	0.2
Ascending	H_{15}^N	7.5
Anthropogenic emissions to the atmosphere	H_{19}^N	15
Removal of nitrogen from the cycle due to sedimentation	H_{21}^N	0.2
Input of nitrogen to the atmosphere in the process of rock weathering	H_{22}^N	0.217
Input of nitrogen to the water medium with dissolving sediments	H_{23}^N	0.091

The time of relaxation in this case constitutes 16 seconds and therefore the equilibrium between NO, NO_2, and O_3 in the atmosphere can be considered stable. However, the equilibrium $N_2 + O_2 \leftrightarrow 2NO$ under anthropogenic conditions is connected with the transformation of NO into NO_2 taking several hours. Therefore, here also, from the viewpoint of global modelling, a separate consideration of the components NO and NO_2 is inexpedient. In other words, we shall consider nitrogen of the atmosphere as a generalized component of the global model.

Atmospheric fixation

As a result of various physico-chemical processes, taking place in the atmosphere, free nitrogen can move from the atmosphere to soil and water bodies. Fixation of atmospheric nitrogen due to electrical charges and photochemical processes constitutes annually not more than 0.035 t/km^2 (from more accurate estimates, it may constitute 0.027 t/km^2 for land and 0.01 t/km^2 for the oceans). Since the nitrogen fluxes due to atmospheric fixation are mainly determined by meteorological conditions, it is quite natural to consider them independently for each region of land and each water body of the World Ocean in the form of the functions of temperature and precipitation.

The flux H_{16}^N of nitrogen fixed in the atmosphere over any ocean basin is described by the relationship:

$$H_{16}^N = [\lambda_1(\theta_1)^{\Delta T} + \lambda_2 R_W]N_A$$

where ΔT is the atmospheric temperature variation; θ_1 is the indicator of the temperature dependence of the rate of atmospheric fixation of nitrogen; R_W is precipitation; and λ_1 and λ_2 are the coefficients.

The equation of atmospheric fixation over a land site Ω_{ij} is written by analogy to H_{16}^N:

$$H_{8,ij}^N = [\lambda_3(\theta_1)^{\Delta T} + \lambda_4 R_{W,ij}]N_A$$

where λ_3 and λ_4 are coefficients.

To estimate the coefficients λ_i $(i = 1, \ldots, 4)$, as a first approximation one can use average data of nitrogen fluxes and precipitation. If we assume $H_{16}^N = 9.96 \times 10^{-3}$ t/km^2/year and $H_8^N = 0.027$ t/km^2/year, and estimate the local precipitation over the ocean and land at 1.01 m/year and 0.24 m/year, respectively, and the convective precipitation over the ocean and land at 0.19 m/year and 0.116 m/year, respectively, then we obtain $\lambda_1 = 0.00498$, $\lambda_2 = 0.00458$, $\lambda_3 = 0.0135$, and $\lambda_4 = 0.0285$. These estimates are easily specified with an account of local data at a fixed time moment for smaller regions and water bodies.

Geospheric sources of nitrogen

The flux of nitrogen H_1^N is determined by the geothermal activity of the Earth. Its estimates testify to the necessity for considering of this constituent in the global model. For instance, in the nitrogen fumaroles of Vesuvius the content of nitrogen by weight constitutes 98%, the gases of the lavas of the Hawaiian volcanoes constitute only 5.7% of nitrogen, and over the globe an input of juvenile nitrogen averages 0.4×10^6 t/year. Let H_1^N be a function of time approximating a statistical series of observations. A more strict account of this flux of nitrogen in the model can be realized by using the algorithms of parameterization of random processes, for instance, with the use of evolutionary modelling. However, within the global model, orientated towards describing the processes with a time step of decades, it is enough to use a consecutive approximation with an average annual step-by-step discretization.

The flux H_1^N can be, to some extent, interpreted as a compensation for the fluxes H_{13}^N and H_{21}^N.

6.7.3 The surface part of the nitrogen cycle in the biosphere

The nitrogen supplies on land consist of the assimilable nitrogen in soil $N_{S2} \approx 0.19 \times 10^4\ \mathrm{t/km^2}$, in plants ($12 \times 10^9$ t), and living organisms (0.2×10^9 t). A diversity of nitrogen fluxes is formed here of the processes of nitrifucation, denitrification, ammonification, fixation, and river runoff. The intensities of these fluxes depend on climatic conditions, temperature regime, moistening, as well as chemical and physical properties of soil. Many qualitative and quantitative characteristics of these dependences have been described in the literature (Hellebrandt *et al.*, 2003). Let us consider some of them.

Nitrification

The return of nitrogen to the cycle due to living micro-organisms is one of the stabilizing natural processes. Simplifying the whole process of the transformation of ammonia salts into nitrates, let us present the activity of heterotrophic micro-organisms and saprophagans as a generalized process of organic matter decomposition. The rate of the organic matter decomposition and nitrification increases with increasing temperature, reaching its optimal value at $T_a = 34.5$°C. Therefore, for the flux H_3^N an approximation $H_3^N = \lambda_{N\kappa} M_\kappa$ can be assumed, where M_κ is the rate of dying off of the component κ and $\lambda_{N\kappa}$ is the content of nitrogen in the component κ.

Denitrification

The processes of denitrification of H_7^N on land are important channels for the input of nitrogen into the atmosphere. The intensity of these processes depends on temperature, humidity, pollution of soils with poisonous chemicals, and pH. The quantitative and functional characteristics of these dependences have been well studied. Within the global model it is possible to take into account only the factors of temperature and humidity:

$$H_7^N = \lambda_6 \theta_2^{\Delta T} W_S \frac{N_S}{k_1 + N_S}$$

where W_S is soil moisture; θ_2 is the temperature coefficient; and λ_6 and k_1 are empirical parameters. If we assume $H_7^N = 0.318\,t/km^2/year$, then $\lambda_6 = 0.496$ and $k_1 = 0.556$.

Biological fixation

In the biological cycle of nitrogen of importance are the processes of fixation by micro-organisms and plants whose intensity is estimated at 148×10^6 t/year. The rate of fixation, depending on the character of the medium, can vary reaching 3×10^9 t/year in highly productive regions. The nitrogen flux H_{10}^N depends on the distribution of vegetation cover and can be described by the equation $H_{10}^N = \sigma_\kappa \lambda_\kappa R_\kappa / \sigma_{ij}$, where σ_κ is the area under vegetation of the κ-type on the territory Ω_{ij} of the area σ_{ij}, R_κ is the productivity of plants of the κ-type, and λ_κ is the coefficient.

Fixation of nitrogen by plants directly from the soil via the root systems (flux H_6^N) occupies a principle place in the cycle of nitrogen, especially on the territories with cultural vegetation. For instance, an increase of the share of legumes in agriculture can raise H_6^N up to 35 t/km²/year. Therefore, a consideration of this flux in the model is necessary and can be realized in the following form:

$$H_6^N = \sigma_\kappa R_\kappa \mu_\kappa / \sigma_{ij} \tag{6.11}$$

where μ_κ is the constant.

The rate of assimilation of nitrogen by the roots of plants is known to depend on the soil temperature regime, decreasing slowly with temperatures lowering to 8–10°C and dropping at temperatures below 5–6°C. The motion of nitrogen from the roots to the upper parts of the plants slows down, too. Equation (5.26) reflects this regularity through the respective reactions of plant productivity with the time lag neglected.

On land, plants annually assimilate about 30×10^6 t N from the atmosphere and more than 5.3×10^6 t N directly from the soil. Approximate estimates of the productivity of various types of vegetation average $R_\kappa = 710\text{–}3,243\,t/km^2/year$. Hence, from Equation (5.26) we have $\mu_\kappa = 0.134 \times 10^{-5} - 0.506 \times 10^{-4}$.

The loss of nitrogen through washing out from soils

On global scales, the means of nitrogen migration include the transport of its compounds between land and oceans due to water runoff. The annual input of nitrogen from land into the World Ocean is estimated at 38.6×10^6 t. If the total sink to the ocean from land is described by the function W_{SO}, then the nitrogen flux H_{11}^N can be approximated by the expression:

$$H_{11}^N = \lambda_N N_{S2}[1 - \exp(-k_N W_{SO})]$$

where λ_N and k_N are the coefficients. This dependence provides for zero nitrogen flow from the land to the ocean if there is no runoff and stabilization at the level ρ, when the volume of water dranage to the ocean drastically increases. To estimate the parameters λ_N and k_N, it is necessary to take into account the spatial heterogeneity of the types of soil–vegetation formations, relief, and other geophysical parameters.

In particular, the content of nitrogen compounds in water differs as a function of the runoff territory. The river waters in the forest regions with temperate climates contain 0.4 mg/l of nitrates, for the arid climate this value is 1.45 mg/l. The concentration of nitrates increases sharply in the drainage waters of the irrigation systems (5.5 mg/l), in the river waters of densely populated regions (25 mg/l), and reaches a maximum in the soil solutions of the salted irrigated soils (200 mg/l). Groundwaters contain from 10 to 100 mg/l of nitrates. With the total runoff of water into the World Ocean reaching 50×10^3 km^3, 30% of which represents underground flow, the total flux of nitrogen per unit area of the ocean will be 0.107 t/year. Assuming $W_{SO} = 0.337$ m and that a 95% level of the sink saturation is reached at a fivefold increase of W_{SO}, we obtain $k_N = 1.367$ and $\lambda_N = 0.708$.

The surface part of the nitrogen cycle contains a constant process of removal of nitrogen from the biosphere into deposits, in particular, as a result of accumulation of saltpetre (potassium nitrate) on the Earth's surface through erosion and alkalizing. From available estimates, $H_{21}^N \approx 3.9 \times 10^{-4}$ t/km^2/year, with $H_{21}^N < H_{22}^N$, but $H_{21}^N + H_{13}^N \cong H_1^N + H_{22}^N + H_{23}^N$. This relationship follows from the fact that during the geological period the loss of nitrogen had been balanced by its input. Of course, in the present biosphere, with the changing intensity of most of the fluxes enumerated in Table 5.4, this balance has started shifting due to increasing fluxes H_9^N and H_{19}^N.

6.7.4 The hydrospheric component of the global nitrogen cycle

In seawater, nitrogen is present in the form of dissolved gas, ions of ammonium NH_4^+, nitrite NO_2^-, nitrate NO_3^-, and in the form of various organic compounds. Inorganic nitrogen compounds are assimilated by algae and phytoplankton and thus transfer into organic forms that serve as food for living organisms. An expenditure of inorganic nitrogen supplies is compensated by atmospheric precipitation, river runoff, and mineralization of organic remains in the process of living organisms functioning and dying off. A diversity of means of nitrogen transformation in water has been studied inadequately, although the available information is sufficient for the global model. The processes, such as replenishing nitrogen supplies in water due to lysis of detritus and functioning of living organisms, the nitrogen exchange between photic and deep layers of the ocean, as well as nitrogen fixation at photosynthesis and denitrification, have been thoroughly studied and described in the literature. Also, there are rough estimates of nitrogen supplies in the ocean, according to which one can assume, on average, that $N_U = N_P = 0.77 \times 10^4$ t/km^2 and $N_L = N_F = 10^5$ t/km^2. More detailed spatial distributions of nitrogen supplies in the hydrosphere can be calculated from the data on biomass, dissolved organic matter, and concentration of dissolved oxygen. The volume relationships of dissolved nitrogen are related to the volume of oxygen as $\mathrm{ml}N_2/\mathrm{l} = 1.06 + 1.63\,\mathrm{mlO_2/l}$.

The nitrogen supplies in water bodies are replenished due to bacterial decomposition of organic sediments and dissolved organic matter. Consider the component D denoting the content of dead organic matter in the water medium. On its basis, one can write the following relationship: $H_{18}^N = \lambda_D D(\varphi, \lambda, z, t)$, where λ_D is the indicator of the nitrogen content and the rate of detritus lysis. The free nitrogen supplies in the water medium are also replenished in the process of the functioning of various organisms. With an account of phytoplankton Φ and nekton r, we have:

$$H_4^N = H_r^N + H_\Phi^N \qquad H_r^N = \lambda_r T_r \qquad H_\Phi^N = \lambda_\Phi T_\Phi$$

where T_r and T_Φ are the characteristics of metabolic processes, respectively, in nekton and phytoplankton; and λ_r and λ_Φ are the coefficients. To determine the averages of these coefficients, assume $T_r = 0.194\,\mathrm{t/km^2/year}$, $T_\Phi = 0.125\,\mathrm{t/km^2/year}$, and $H_r^N = H_\Phi^N = 0.83 \times 10^{-3}\,\mathrm{t/km^2/year}$. Then $\lambda_r = 0.00428$ and $\lambda_\Phi = 0.00664$.

The process of denitrification in the water medium delivers considerable amounts of nitrogen into the atmosphere: $H_{20}^N = \lambda_5(\theta_2)^{\Delta T} N_U(\varphi, \lambda, z, t)$, where λ_5 and θ_2 are the constants.

The biological fixation of nitrogen in the water medium constitutes about 10×10^6 t/year, reaching 20.7×10^6 t/year in the photic layer of the ocean, and $(36\text{–}1800) \times 10^4\,\mathrm{t/km^3/year}$ in small lakes. For the World Ocean, $H_{17}^N = 0.0277$ t/km^2/year, on average. Assuming $H_{17}^N = \lambda_R R_\Phi$, where R_Φ is the phytoplankton produce averaging $168.8\,\mathrm{t/km^2/year}$, we obtain $\lambda_R = 0.164 \times 10^{-3}$.

The characteristic feature of the nitrogen cycle in the water medium is its transport due to gravitational sedimentation, vertical convection, turbulent diffusion, and convergence. The processes of nitrogen transport by migrating animals are almost negligible and can be neglected in the global model. The simplest form of the description of the vertical fluxes of nitrogen is the model: $H_{14}^N = \lambda_\kappa \Delta N_\kappa$ and $H_{15}^N = \lambda_\rho \Delta N_\rho$, where $\kappa = (U, P, L)$ and $\rho = (P, L, F)$.

6.7.5 Anthropogenic factors of the global nitrogen cycle

The present contribution of human activity to the general biospheric cycle of nitrogen has reached a level where the consequences of the introduced changes become unpredictable and probably rather catastrophic. The epidemiological studies testify to the growth in numbers of respiratory diseases in territories with high concentrations of nitrogen and sulphur oxides as well as photochemical oxidizers. The harmful effect of nitrogen oxides on living organisms starts manifesting itself when a level of $940\,\mathrm{mkg/m^3}$ is exceeded. In general, the consequences of nitrogen pollution of the biosphere are more complicated. For instance, on the one hand, the technogenic accumulation of nitrogen from the atmosphere during fertilizer production plays a positive role by raising the productivity of the land and water ecosystems. On the other hand, it causes an undesirable eutrophication of water basins. Removal of nitrogen from the atmosphere for industrial and agricultural needs is compensated for by a technogenic input of nitrogen into the atmosphere via the burning of solid and liquid fuels. A considerable share is attributed to transport emitting nitrogen oxides reaching, for instance, in the USA, 11.7×10^6 t/year. However, even an observance of this physical equilibrium cannot substitute for the chemical–biological balance. Therefore, in this multifunctional hierarchical set of global fluxes of nitrogen, the most vulnerable knots and bonds should be revealed, something only possible within a well-planned numerical experiment.

The quantitative estimate of the main stages of the nitrogen cycle with an account of anthropogenic factors enables one to see an integral effect of the breaking of the global balance of nitrogen. From the available data on the global distributions of violations of the nitrogen cycle it is impossible to reliably estimate the contribution from the industrial synthesis of nitrogen compounds and their scattering over the globe into its biogeochemical cycle. During the last years of the 20th century industry increased the total amount of nitrogen circulating in the biosphere by 50%. As a result, the natural equilibrium between the processes of nitrification and denitrification has turned out to be violated with an excess of nitrates reaching 9×10^6 t.

Preliminary estimates of an increasing anthropogenic pressure on the nitrogen fluxes between the biospheric elements suggest the hypothesis that a strong correlation exists between fertilizer production H_9^N and population density G, technogenic accumulation of nitrogen at fuel burning H_2^N and mineral resources expenditure R_{MG}, anthropogenic input of nitrogen into the atmosphere H_{19}^N and intensity of emissions of general pollutions Z_{VG}. The quantitative characteristics of the factors of these dependencies can be obtained from the known trends. From some estimates, the amount of nitrogen oxides emitted into the atmosphere is proportional to the weight of used fuel with a 4% annual increase in trend. The scales of industrial fixation of nitrogen for the last 40 years increased by a factor of 5, reaching a value that represents what could have been fixed by every ecosystem on the Earth before the use of present-day agricultural technology. In 1968 the global industry produced about 30×10^6 t of fixed nitrogen, and in 2000 this value reached 1 billion.

Let us formalize these correlations as the following models:

$$H_9^N = \min\{U(K)\bar{G}(K,t); N_A\sigma_K/\sigma\} \tag{6.12}$$

$$H_2^N = \lambda_{AG}R_{MG} \qquad H_{19}^N = \lambda_{GA}Z_{VG}$$

where K is the number of an economic region; $\bar{G}$ is the average population density of the K region; and σ_K is the area of the K region. The coefficients U, λ_{AG}, and λ_{GA} are determined from the analysis of available information about the processes in Equation (6.12). If we assume that $H_2^N = 0.154\,\mathrm{t/km^2/year}$, $H_{19}^N = 0.102\,\mathrm{t/km^2/year}$, and $H_9^N = 0.283\,\mathrm{t/km^2/year}$, then with $\bar{G} = 24.4\,\mathrm{men/km^2}$, $R_{MG} = 30.5$ oil units/km^2/year, and $Z_{VG} = 3.39\,\mathrm{t/km^2/year}$, we obtain $U = 0.283$, $\lambda_{AG} = 0.504 \times 10^{-2}$, and $\lambda_{GA} = 0.03$.

The anthropogenic interference into the nitrogen cycle can also have medical and biological consequences expressed, for instance, through increased mortality with the growing levels of NO_2 at 190–320 mkg/m^3. From the data of the World Health Organization the natural background concentration of NO_2 over the continents constitutes 0.4–9.4 mkg/m^3.

Finally, the human impact on the nitrogen cycle can affect the structure and intensity of the energy exchange in the biosphere.

6.8 BIOSPHERIC BUDGET OF OXYGEN AND OZONE

The oxygen cycle in nature is composed of characteristic biogeochemical transitions between the reservoirs of the basic constituents circulating in the biosphere. Therefore, the block-scheme of the oxygen exchange is similar to those of sulphur, nitrogen, carbon, and phosphorus (Figure 6.10). However, oxygen constituents are widely distributed over the globe, making it one of the must substantial components of the biogeochemical cycles. Its amount in the Earth's crust, including the hydrosphere, reaches 49% by mass. The lithosphere (without the ocean and the atmosphere) contains 47.2% of oxygen, water 88.89%. In the ocean water, oxygen constitutes 85.82%, with living organisms representing 65% by mass. These estimates testify to a significance of oxygen for the biosphere. At present, about 39×10^{14} t O_2 circulate in the biosphere.

Oxygen is present in the biosphere in the form of molecular oxygen (O_2), ozone (O_3), atomic oxygen (O), and as a constituent of various oxides. On the one hand, oxygen maintains life on the Earth due to the process of respiration and formation of

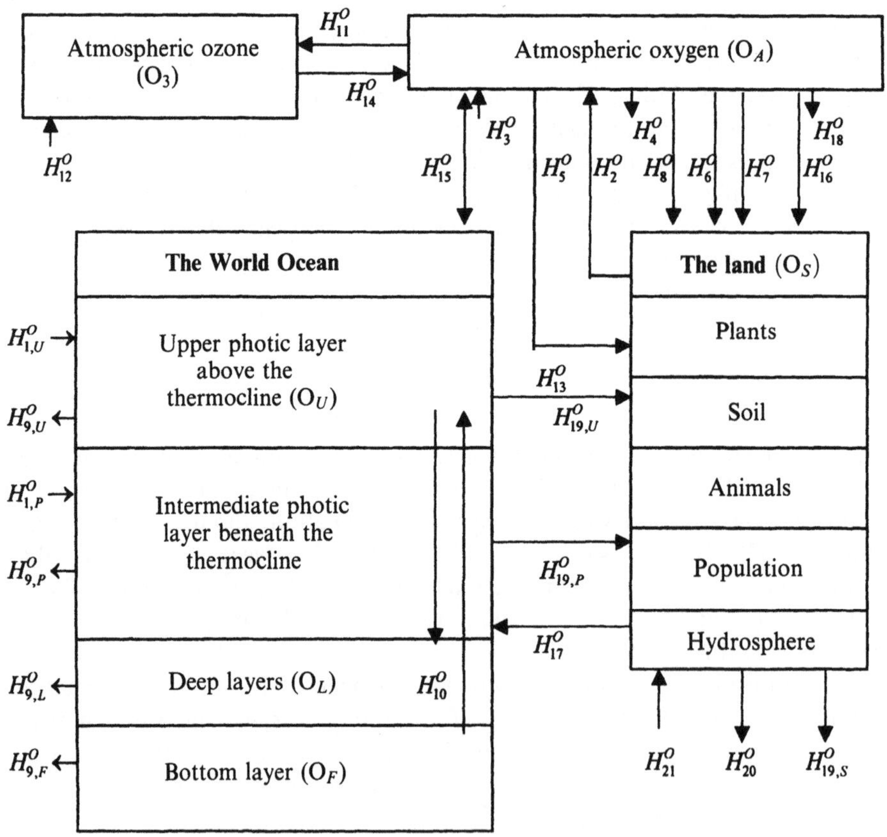

Figure 6.10. Oxygen fluxes in the biosphere.

the ozone layer, and on the other hand, it is the product of organism functioning. This fact hinders a description of its cycle, since it requires a synthesis of the descriptions of various processes. Here an attempt has been made to describe the cycle and to derive a model of the global oxygen cycle (MGOC).

Many authors believe that in the near future nothing threatens the stability of the global biogeochemical cycle of oxygen. This statement is not valid for ozone, whose concentration and spatial distribution suffered serious changes during the last decades. According to Kondratyev and Varotsos (2000), the available observations of the vertical profile of atmospheric ozone show a very complicated spatial and temporal variability that depends on many characteristics of the NSS. The MGOC unit, for the parameterization of ozone fluxes, follows a numerical model by Aloyan and Aruiunian (Kondratyev and Varotsos, 2000), with a required correction taken into account. This correction consists of the substitution of some functional dependencies for scenarios reflecting the dynamics of changes in concentrations of the chemicals not described in the global model of the carbon cycle.

6.8.1 Oxygen sources

Today and in the geological past there are two sources of oxygen – endogenic and photosynthetical. Without dwelling upon the respective scientific discussions and all existing concepts, we shall try to describe the sources of oxygen in the present biosphere, following numerous studies in this field.

The basic source of atomic oxygen is the photosynthesis of plants, whose equation has the form:

$$2H_2O + CO_2 \xrightarrow[\textit{chlorophyll}]{\textit{light}} COH + H_2O + O_2$$

Photosynthesis annually produces more than 50×10^9 t of oxygen (i.e., of the order of $3.3 \times 10^{14}\%$ of its supply in the atmosphere). Hence, we see that it is only due to photosynthesis that oxygen supplies in the atmosphere can be totally renewed during a time period of 300,000 years. About 80% of the total amount of oxygen produced by photosynthesis results from phytoplankton functioning, and land vegetation communities produce only 20%. If we denote, by $R_\Phi(\varphi, \lambda, t)$ and $R_\kappa(\varphi, \lambda, t)$, the produce of phytoplankton Φ and land surface of the κ-type at the Earth surface point (φ, λ) at a time moment t, then the oxygen fluxes to the hydrosphere and from land to the atmosphere can be described by the relationships:

$$H_1^O = a_\Phi R_\Phi \qquad H_2^O = a_\kappa R_\kappa \qquad H_{21}^O = a_S R_\Phi$$

where the coefficients a_Φ, a_κ, and a_S depend on phytoplankton species and the type of vegetation. For their averaging we use the data on the fluxes: $H_1^O = 140\,\mathrm{t\,O_2/km^2/year}$, $H_2^O = 70\,\mathrm{t\,O_2/km^2/year}$, $H_{21}^O = 600\,\mathrm{t\,O_2/km^2/year}$, $R_\Phi = 401.3\,\mathrm{t/km^2/year}$, and $R_\kappa = 102.4\,\mathrm{t/km^2/year}$. Then $a_\Phi = 0.35$, $a_\kappa = 0.68$, and $a_S = 1.49$. Of course, these estimates have a considerable spatial and temporal scatter. In particular, using the data on the productivity of some oceans, we obtain the values of the coefficient a_Φ for the Atlantic Ocean – 0.53, the Indian Ocean – 0.25, the Arctic Ocean – 11.1, and the Pacific Ocean – 0.64.

Apart from photosynthesis, photolysis can be a source of oxygen in the atmosphere (i.e., the decomposition of water vapour under the influence of UV radiation in the upper layers of the atmosphere). However, the intensity of this source in present-day conditions is negligible. Nevertheless, denote this flux by $H_3^O = a_H W_A$, where W_A is the water vapour content in the atmosphere and a_H is the empirical coefficient. If we assume that in the upper layers of the atmosphere a constant share of W_A can reside, then at $H_3^O = 0.0039\,\mathrm{t\,O_2/km^2/year}$ and $W_A = 0.025\,\mathrm{m}$, we have $a_H = 1.56 \times 10^{-7}\,\mathrm{1/year}$.

Vernadsky (1944) considered rock metamorphism, basaltic volcanism, and underground radioactive waters as possible sources of oxygen. However, there are no sufficiently reliable estimates of these fluxes and therefore it is impossible to parameterize them.

6.8.2 Oxygen sinks

The oxidation process both on land and in the water medium is the basic consumer of oxygen on the Earth. The ability of oxygen to react with many elements of the Earth's crust creates the fluxes of oxygen leaving the biospheric reservoirs. A balance between the income and expenditure fluxes of oxygen has been reached in the course of biospheric evolution.

Oxygen is spent on respiration of plants, animals, humans, and on dead organic matter decomposition both in the hydrosphere and on land. To parameterize the income parts of the oxygen balance, we use the following models: $H_5^O = a_1 T_\kappa$, $H_6^O = a_2 T_F$, $H_7^O = a_3 T_G$, $H_{19}^O = a_6 T_R$, $H_8^O = a_Q R_D$, and $H_{20}^O = a_5 R_S$, where T_m is the energy expenditure on respiration ($t = \kappa, F, G, R$), and R_ς is the rate of the dead organic matter decomposition ($\varsigma = Q, D, S$).

6.8.3 Ozone

Atmospheric ozone constitutes 0.64×10^{-6} of the atmospheric mass and belongs to the group of optically active gases. It absorbs UV solar radiation in the range 200–300 nm, strongly affecting thereby the thermal regime of the stratosphere and affecting the vegetation cover (Coyle *et al.*, 2003). Ozone has a number of vibration–rotation bands of absorption in the IR spectral region (9.57 μm) and partially absorbs visible radiation in the Chappuis band (0.6 μm). The formation and destruction of ozone have been described in detail (Kondratyev and Varotsos, 2000; Kondratyev, 1999a).

Ozone forms in the upper stratosphere from molecular oxygen under the influence of UV solar radiation. In the lower stratosphere and troposphere, the source of ozone is the decomposition of nitrogen dioxide under the influence of UV and visible radiation. The formation of the vertical profile of the ozone concentration is connected with its meridional and vertical transport. The general characteristic of this profile is the total amount of ozone measured by the thickness of its layer in Dobson units (1 = 0.001 cm).

The ozone destruction is represented by a complex set of photochemical reactions with participation of the compounds of hydrogen, nitrogen, and chlorine. From the available estimates, 50–70% of ozone is destroyed by nitrogen compounds, 20–30% by oxygen (O), 10–20% by water-containing particles of HO_x, and less than 1% by chlorine compounds. The prevailing role of nitrogen compounds in ozone destruction has been confirmed (Wauben *et al.*, 1997) for all latitudes. The equation of photochemical equilibrium between concentrations of ozone and nitrogen oxides is $[NO][O_3]/[NO_2] = \mu$, where the equilibrium constant μ depends on the solar radiation intensity and can range from 0 to 0.02.

There are various approaches to a parameterization of the process of formation and destruction of the ozone layer. The complexity of derivation of the dynamic models of the ozone cycle in the atmosphere is connected with its participation in more than 75 chemical reactions, a qualitative and quantitative description of which is impossible without deriving detailed models of numerous MGCs of the atmosphere. Nevertheless, there are empirical models of the ozone layer, which make it possible, in the present climatic situation, to obtain adequate spatial distributions of ozone. For instance, Bekoriukov and Fedorov (1987) derived a simple empirical model of the total ozone content confirmed by the observational data for the southern hemisphere:

$$O_3(\varphi, \lambda) = \sum_n \sum_{n \leq m} P_n^m(\varphi)\left[a_{n,m} \cos(m\lambda) + a_{n,-m} \sin(m\lambda)\right] \qquad (6.13)$$

where P_n^m are non-normalized spherical functions of the nth degree, and of the mth order; and $a_{n,m}$ and $a_{n,-m}$ are the empirical coefficients whose values are given in Bekoriukov and Fedorov (1987) and in Krapivin and Kondratyev (2002).

There are also static models to describe the vertical profile of the ozone density distribution. One such model is the Kruger formula:

$$O_3(h) = 51.4 \exp\left[-(h-40)/4.2\right] \qquad (\text{mkg/m}^3)$$

Other ozone models have been reviewed by Kondratyev and Varostos (2000).

The simplest dynamic model of the ozone layer can be written in the form of a balanced equation that reflects its income–expenditure components. The ozone supplies are replenished due to the impact of UV radiation on oxygen ($H_{11}^O = e_3 O_A$) and nitrogen dioxide ($H_{12}^O = e_2 N_A$). The ozone layer is destroyed at a rate $H_{14}^O = O_3/T_3$, where T_3 is the lifetime of ozone molecules depending on atmospheric pollution: $T_3 = T_3^O - e_1 B$. The lifetime T_3^O of the ozone molecules in a perfect atmosphere averages 50–60 days. Nitrogen oxides participating contrary to the H_{12}^O cycle of ozone destruction contribute much to the magnitude of B.

6.8.4 Anthropogenic factor in the biospheric oxygen budget

Studies of the history of biospheric evolution reveal a close correlation between the oxygen production intensity and the development of life on the Earth. And though the expected relative oscillations of the oxygen concentration in the near future do not exceed 10%, the considered impacts on the biosphere do not cover all potential anthropogenic trends, and therefore cannot be considered reliable. Therefore, let us analyse the constituents of possible mechanisms for violation of the natural balance of oxygen. Naturally, our concern is for not only an increase but also a decrease in the oxygen content in the atmosphere.

The oxygen cycle is complicated by its ability to take part in a lot of chemical reactions giving a multitude of epicycles. This fact makes the oxygen cycle sufficiently stable but hinders an assessment of this stability.

Anthropogenic forcing on numerous epicycles of oxygen manifests itself both directly through its involvement in other cycles of substances at fuel burning and production of various materials and indirectly, through the environmental pollution and biospheric destruction. Therefore, a parameterization of the anthropogenic impact on the oxygen balance is realized within other units of the global model. The flux H_{18}^O taken into account in the scheme in Figure 6.10 completely covers direct consumption of oxygen both in industry and in agriculture. Assume $H_{18}^O = y_1 R_{MG}$, where R_{MG} is the rate of natural resources expenditure and y_1 is the coefficient ($\approx$0.084).

The fluxes H_{15}^O and H_{16}^O are strongly affected by anthropogenic forcings. Their variations are caused by discharges of high-temperature industrial sewage containing considerable amounts of oxidizers, as well as by the oil-polluted water bodies. The quantitative characteristics of the change of oxygen dissolved in water as a function of temperature have been well studied. The empirical formula to calculate the concentration of the seawater-dissolved oxygen has the form (Ramad, 1981): $[O_2 \text{ dissolved}] = 80/(0.2T_O - 7.1)$, where $[O_2]$ is expressed in mg/l, and T_O in °C. The estimates of oxygen solubility in water are well known (Krapivin and Kondratyev, 2002).

The fluxes H_9^O and H_{19}^O, balancing in natural conditions the oxygen fluxes into the water medium, at an anthropogenic forcing increase, as a rule, due to more active aerobic bacteria and increasing metabolic needs of animals. For instance, a 10°C increase of water temperature increases the oxygen expenditure through respiration of marine animals by a factor of 2.2.

One of the negative manifestations of the anthropogenic impact on the oxygen cycle is a depletion of the ozone layer, especially marked in the polar regions. There are various hypotheses on the causes of sharply changing concentrations of ozone, as well as discussions on the so-called "ozone hole" over the Antarctic. The main cause of all violations is connected with progressing human activity accompanied by the growing volumes of long-lived components emitted to the atmosphere (e.g., freons). The consequences of these violations are very serious, and the real scales of danger threatening life on the Earth can be estimated only with the use of the global model of the NSS (GMNSS).

A diversity of the anthropogenic impacts on the global biogeochemical cycle of oxygen is determined by the direct and indirect causes of breaking the natural balance of oxygen. According to the equation of photosynthesis, the gramme-molecular amounts of assimilated CO_2 and emitted O_2 are equal. Also equal are the gramme-molecular amounts of assimilated O_2 and emitted CO_2 for dead organic matter decomposition and fuel burning. Hence, for time periods of tens and hundreds of years, a change in the CO_2 amount in the atmosphere is accompanied by the same change of O_2, but in the opposite direction. For instance, a CO_2 doubling in the atmosphere leads to a decrease in the amount of O_2. But, since the volume concentration of CO_2 in the atmosphere is now estimated at 0.031%, and that of O_2 at 20.946%, in this case a decrease of O_2 will constitute only 0.15% of the total O_2 content in the atmosphere.

Imagine the following situation. Let the total biomass of the biosphere ($\sim 9.6 \times 10^{11}$ t C), all organic matter of soil ($\sim 14 \times 10^{11}$ t C), and all fossil chemical fuel, the explored supplies of which constitute 128×10^{11} t of conditional fuel (64×10^{11} t C), be burnt. Then the amount of CO_2 in the atmosphere increases by a factor of 12.5, and that of O_2, respectively, decreases by only 1.75%. Hence, the amount of oxygen over hundreds of years has to be practically constant.

However, it should be borne in mind that the region of excess anthropogenic emissions of CO_2 and, hence, O_2 assimilation is concentrated over a small area of cities and forest fires. Since the concentrations in the atmosphere do not equalize instantly, a gradient of O_2 concentrations can appear around these sites, where the oxygen provision will be insufficient for animals and humans. Therefore, the MGOC unit reflecting the spatial heterogeneities in the distributions of O_2 concentrations enables one to reveal such dangerous territories.

An interaction between the cycles of oxygen, nitrogen, sulphur, phosphorus, and carbon manifests itself through the processes of oxidation and decomposition. The level of detailing the global model units does not permit one to reflect the diversity of these processes. Therefore, in the simplest case, when only averaged characteristics of the oxygen cycle elements are taken into account, the scheme in Figure 6.10 of the global O_2 fluxes can be presented in a simplified form reflecting only the main fluxes. The indicated stability of the O_2 concentration in the atmosphere makes it possible to simplify a description of the MGOC unit, using a single balanced equation:

$$\frac{\partial O}{\partial t} + V_\varphi \frac{\partial O}{\partial \varphi} + V_\lambda \frac{\partial O}{\partial \lambda} = k_0 R_F + k_L R_L - \nu_L T_L - b_G G - \nu_F T_F - \nu_G T_G - \mu_Q R_Q$$

where k_0 and k_L are the indicators of the rate of O_2 emission due to photosynthesis in the ocean and on land, respectively; ν_s is the indicator of the role of respiration of land vegetation ($s = L$), animals ($s = F$), and humans ($s = G$) in the removal of oxygen from the atmosphere; and μ_Q is the rate of O_2 consumption due to the decomposition of the soil dead organic matter.

6.8.5 Model of the oxygen budget in the biosphere

Assuming the scheme of the oxygen fluxes in nature (shown in Figure 6.10) to be balanced, let us write the equations of the model in the following traditional form of

Table 6.17. Estimates of the reservoirs and fluxes of oxygen and ozone in the MGOC unit.

Reservoirs (t/km^2), fluxes (t/km^2/year)	Identifier	Evaluation
Oxygen in the upper photic layer of the World Ocean	O_U	0.8×10^8
Oxygen in the transition layer of the ocean	O_P	0.7×10^9
Oxygen in deep ocean	O_L	3×10^4
Oxygen in the bottom layer of the ocean	O_F	9×10^3
Oxygen in the atmosphere	O_A	0.24×10^7
Oxygen in the surface part of the hydrosphere	O_S	0.6×10^8
Ozone	O_3	0.23
Photosynthesis in the ocean	H_1^0	108–388
Photosynthesis on land	H_2^0	70–100
Photodecomposition of water in the atmosphere	H_3^0	0.008
Oxidation processes in the atmosphere	H_4^0	0.009
Respiration of plants	H_5^0	0.07–0.1
Respiration of animals	H_6^0	50–60
Respiration of humans	H_7^0	70–80
Oxidation and restoration processes in soil	H_8^0	1
Oxidation processes in the World Ocean	H_9^0	164
Descending of the oxygen saturated waters	H_{10}^0	190
Decomposition and destruction of O_2 in the atmosphere	H_{11}^0	
Formation of O_3 from NO_2	H_{12}^0	0.23–22.2
Lifting of dissolved oxygen in the upwelling zones	H_{13}^0	36
Decomposition and destruction of ozone in the atmosphere	H_{14}^0	1.48–1.66
Exchange on the atmosphere–ocean border	H_{15}^0	18–140
Exchange on the atmosphere–inland water bodies border	H_{16}^0	18–140
Transport of oxygen into the ocean by river runoff	H_{17}^0	50
Anthropogenic consumption of oxygen	H_{18}^0	60–90
Expenditures of O_2 on metabolism of aquatic animals	H_{19}^0	0.2
Oxidation processes in the continental water bodies	H_{20}^0	90–200
Photosynthesis in continental water bodies	H_{21}^0	100–400

the balanced relationships (Table 6.17):

$$\frac{\partial O_A}{\partial t} + V_\varphi \frac{\partial O_A}{\partial \varphi} + V_\lambda \frac{\partial O_A}{\partial \lambda} = H_2^O + H_3^O + H_{14}^O + H_{15}^O + H_{16}^O - \sum_{i=4}^{8} H_i^O - H_{11}^O - H_{18}^O$$

$$\frac{\partial O_3}{\partial t} + V_\varphi \frac{\partial O_3}{\partial \varphi} + V_\lambda \frac{\partial O_3}{\partial \lambda} = H_{11}^O + H_{12}^O - H_{14}^O$$

$$\frac{\partial O_S}{\partial t} = H_{21}^O - H_{16}^O - H_{17}^O - H_{19,S}^O - H_{20}^O$$

$$\frac{\partial O_U}{\partial t} + V_\varphi \frac{\partial O_U}{\partial \varphi} + V_\lambda \frac{\partial O_U}{\partial \lambda} = H^O_{1,U} + H^O_{13,PU} + H^O_{17} - H^O_{9,U} - H^O_{10,PU} - H^O_{15} - H^O_{19,U}$$

$$\frac{\partial O_P}{\partial t} + V_\varphi \frac{\partial O_P}{\partial \varphi} + V_\lambda \frac{\partial O_P}{\partial \lambda} = H^O_{1,P} + H^O_{10,PU} + H^O_{13,PL} - H^O_{9,P} - H^O_{10,PL} - H^O_{13,PU} - H^O_{19,P}$$

$$\frac{\partial O_L}{\partial t} = Q_L + H^O_{10,PL} + H^O_{13,LF} - H^O_{9,L} - H^O_{10,LF} - H^O_{13,PL}$$

$$\frac{\partial O_F}{\partial t} = Q_F + H^O_{10,LF} - H^O_{9,F} - H^O_{13,LF}$$

Here Q_L and Q_F denote the oxygen fluxes resulting from mixing of the deep and bottom layers of the ocean. The oxygen exchange between the hydrosphere and the atmosphere (fluxes H^O_{15} and H^O_{16}) depend on its partial pressures at the water–air border. The direction of the fluxes H^O_{15} and H^O_{16} depends on the relationship between temperatures T_a, T_U, and T_S. Due to high concentrations of oxygen in the atmosphere, the partial pressure varies negligibly, and therefore the fluxes H^O_{15} and H^O_{16} can be considered to depend only on oscillations in the concentrations of O_U and O_S: $H^O_{15} = k_{OU}(T_U - T_a)O_U$; and $H^O_{16} = k_{OS}(T_S - T_a)O_S$. If we assume $O_U = 5.5\,\text{ml/l}$ and $O_S = 2.1\,\text{ml/l}$, then at $T_U - T_a = T_S - T_a = 2°\text{C}$ and $H^O_{15} = 18\,\text{t/km}^2\text{/year}$, and $H^O_{16} = 140\,\text{t/km}^2\text{/year}$ we obtain $k_{OU} = 0.5 \times 10^{-4}\,\text{km/°C/year}$ and $k_{OS} = 0.1 \times 10^{-2}\,\text{km/°C/year}$.

The ocean layers exchange oxygen by circulation processes, and as a result, depending on latitude, longitude, and season, the intensity of the fluxes H^O_{10} and H^O_{13} can sharply change. In any case, this intensity mainly depends on the velocities v_A of the vertical water upwelling and v_H of its downwelling. In the zones of upwelling the flux H^O_{13} prevails with the flux H^O_{10} prevailing in the zones of convergence. The velocities v_A and v_H range from 0 to 0.1 m/s. The most characteristic values of these velocities range from 10^{-2} to 10^{-4} m/s. For instance, near California $v_A \approx 0.77 \times 10^{-5}$ m/s, and in the region of the Bengal upwelling $v_A = 0.25$ m/s. Thus for the fluxes H^O_{10} and H^O_{13} we use the following approximations: $H^O_{10} = \lambda_{\kappa\gamma}(O_\kappa - O_\gamma)$; and $H^O_{13} = \mu_{\gamma\kappa}(O_\gamma - O_\kappa)$, where $\lambda_{\kappa\gamma}$ and $\mu_{\gamma\kappa}$ are the coefficients of local mixing, $(\kappa = U, P, L; \gamma = P, L, F)$.

In the shelf zone of the ocean the oxygen balance formation is much affected by the runoff from the continents. Despite the complexity of this process, the following simple parameterization can be accepted for this case:

$$H^O_{17} = n_{SU}(O_S - O_U)$$

6.8.6 The role of aviation in changes of the ozonosphere

The problem of monitoring and predicting the ozone layer dynamics in the atmosphere is as equally important as the problem of the atmospheric greenhouse effect (Varotsos and Kondratyev, 1998; Popovicheva *et al.*, 2000). In both cases there are contradictory estimates of the causes of ecological danger at the observed levels of the ozone concentration, ozone destruction, or greenhouse gases. Despite such contradictory estimates, often not without a political nuance, these problems attract the attention of the experts in the fields of natural sciences, who try to create some information technologies which would ensure a high level of objectivity and reliability of the estimates of consequences of the anthropogenic interference into the global biochemical cycles of ozone, carbon dioxide, methane, water vapour, and other MGCs. Here we consider a narrow but important problem of the ozone layer changing, over a limited territory, caused by transport aviation routes. This problem has recently attracted growing attention. The impact of flights of subsonic (altitudes of 9–13 km) and supersonic (16–20 km) aircraft on the ozonosphere has become substantial, at least, on a regional scale. Volumes of global air transport increase by almost 5% annually, and the amount of emitted nitrogen oxides, sulphur compounds, and other MGC increases by about 4% annually. According to average global estimates, the NO_x emissions ($NO + NO_2$) constitute about 500 kt N/year with a predicted increase up to 1100 kt N/year by 2015.

Substances ejected by aircraft engines to the atmosphere include the components, H_2SO_4, HNO_3, HNO_2, HNO, Cl, NO_3, ClO_2, CO, CO_2, CH_4, N_2O, H_2O, SO_2, SO_3, N_2O_5, CH_3Cl, Cl_2, CH_3NO_2, CH_3NO_3, $BrONO_2$, HNO_4, and $ClONO_2$, many of which cause the formation of polar stratospheric clouds, affect markedly the aerosol composition of the atmosphere, and intensify the greenhouse effect.

An analysis of the distribution of these components in the atmosphere requires an understanding of many factors in the field of photochemistry and atmospheric dynamics. Unfortunately, the available ideas about the rates of reactions with participation of the enumerated substances as well as about the coefficients of micro–macro-turbidity and local synoptic characteristics are limited by data averaged in time and space. In this regard, many authors apply various simplifications, which make it possible to overcome information uncertainties.

Atmospheric ozone chemistry has been well studied (Kondratyev and Varotsos, 2000). Nevertheless, this knowledge is insufficient to derive a model of the biogeochemical cycle of ozone whose adequacy would not raise doubts. The main cause is the time-dependent nature of the environmental processes. Unfortunately, the use of both simple and complicated climate models that take into account in detail the compounds of the atmospheric chemistry of ozone, do not give acceptable results. Therefore, one of the possible approaches to raise the reliability of the estimates of the state of the ozone layer over a given territory is an application of technology which combines measurements and modelling and takes into account expert estimates. In this case, to assess the vertical profile of ozone, all available information can be used on its formation and destruction, and the additional background information about anthropogenic and natural processes can be obtained from established correlations or scenarios.

One of the difficulties in the synthesis of the model of ozone dynamics and observational data is a necessity to adequately describe the location of the tropopause. An uncertainty of the aircraft flight route of subsonic aircraft relative to the tropopause appears because of inaccurate determination of the height of the tropopause. This is very important, since depending on whether the flight route is below the tropopause or in the stratosphere, the photochemical reactions with participation of ozone differ. With supersonic aviation, this problem is solved uniquely: all routes lie in the stratosphere. Therefore, to exclude from the model the elements of instability, a hypothesis is assumed on the seasonal change of the tropopause altitude (in spring and summer Z_1, and in autumn and winter Z_2).

This approach excludes the instability of the obtained estimates. Although, of course, there are many successful results of modelling the ozone photochemistry. Among them a number of models of the Lagrange type are rather efficient, which take into account up to 75 chemical elements and compounds. Also efficient is the 3-D Model for Ozone and Related Chemical Tracers (MOZART) (Kondratyev and Varotsos, 2000).

In this section the problem of dynamic estimation of the vertical profile of the ozone concentration in the atmosphere over a limited territory is considered, taking into account only one anthropogenic source – aviation. Since the set of substances ejected by aircraft engines is rather large, the consequences of the aircraft flight over this part of the Earth's surface include numerous changes of both gas and aerosol composition of the atmosphere. Of course, the scale of these changes is determined by the intensity of the aircraft flights and density of the flight corridors. Here a method of integration of these causes is proposed and an information system of expert level is developed, which can be used in the systems of regional and national monitoring. Besides, the presentation of the input information and the spectrum of chemical reactions have been limited. An operator can specify a list of substances contained in exhaust gases.

Units of the simulation system to control the regional ozonosphere (SSCRO) are enumerated in Table 6.18. The controlled territory Ω is covered with the geographical grid with steps $\Delta\varphi$ and $\Delta\lambda$ by latitude and longitude, respectively. Over each cell $\Omega_{ij} = \{(\varphi, \lambda) : \varphi_i \leq \varphi \leq \varphi_{i+1} : \lambda_j \leq \lambda \leq \lambda_{j+1}; i = 1, \ldots, N; j = 1, \ldots, M\}$ and a possible location (altitude and time) of the flight corridor is determined. For this purpose, in the database for SSCRO an indicator is formed of the flight load on Ω (type of engine, fuel, velocity, and flight altitude). The background concentration of ozone and the meteorological situation are assumed to be taken from the data of regional, national, and global systems of the environmental monitoring.

The ozone concentration as a function of the spatial coordinates and time is calculated by the formula:

$$\frac{\partial O_3(\varphi,\lambda,z,t)}{\partial t} = Q + S + U - P - R \tag{6.14}$$

where z is the altitude; and the functionals on the right-hand part of the equation describe the following processes of ozone formation: Q is a change of the ozone concentration due to atmospheric motion and gravitational sedimentation; P and U are photochemical destruction and formation of ozone outside the passageway, respectively; and R and S are photochemical destruction and formation of ozone within the flight corridor.

Table 6.18. Characteristic of the SSCRO units.

Unit	Functions of the unit
AADB	Algorithmic adaptation of the database to the structure of the controlled territory. A matrix structure is developed with elements reliably attaching the environmental elements to geographical coordinates, configuration of the territory of the region, location on it of airports as well as division of the territory into land and water surfaces.
IRDB	Information renewal of the database. The possibility is provided of an operational change of configuration and appearance of the territory of the region described in AADB.
CIF	Management of information fluxes between the SSCRO units. Dimensions of the model parameters are coordinated; dimensions of input data are coordinated with the scales assumed in SSCRO. For instance, the formula 1 ppmv $= 10^{-3}[M/(M_i\rho)]$ mkg/m^{-3}, where M_i is the molecular weight of the ith chemical element. The formulas of the type of $1\,mkgO_3/m^2 \Rightarrow$ 0.467×10^{-7} atm/cm are also re-calculated.
CSBF	Control of the functions of the system units. Depending on the availability of the required information in the database about correlations between various processes, a version is selected from the alternative versions, that does not contradict the database.
PCMD	Parametric agreement of the models and database. Signals of the user's interface are analysed for an efficient removal from the database of the coefficients of models or in the case of disagreements the model is substituted for the scenario.
MPR	Modelling the photochemical reactions in the flight corridor selecting three stages: (1) in the nearest zone after an ejection of the products of fuel combustion from the nozzle of the aircraft engine; (2) scattering the jet of combustion products; and (3) complete mixing with the surrounding atmosphere.
MOST	Modelling the processing of propagation and transformation of ozone in the interaction of the flight passageway (vortex trail) with the surrounding atmosphere.
MOFD	Model of the formation and destruction of ozone with an account of all flight corridors over the controlled territory.
CCAB	Calculation of corrections for the atmospheric balance of ozone due to an account of the effects of land cover and sea surface.
FBOL	Formation of the background level of ozone either from the data of regional and global monitoring system or with the use of the respective model.
SF	Formation of scenarios of location of the flight corridors and of the timetables of their load.
CIUI	Management of the user's information interface. Provision of computer experiments.

The Equation (6.14), with initial data for a time moment $t = t_0$, is solved with account of the mosaic of the flight corridors. The functionals R and S are calculated for $\Omega_{i,j,k} = \{(\varphi, \lambda, z): (\varphi, \lambda) \in \Omega_{ij}; z_k \leq z \leq z_{k+1}\}$ at a time moment t only in the presence of aircraft. Three zones are considered for interaction of the products of fuel combustion in the trail of the working aircraft engine: (1) immediately behind all engines (time duration Δt_1); (2) at the stage of mixing of the exhaust gases with the atmosphere (Δt_2); and (3) on the means of penetration of the mixture into large-scale reservoirs (Δt_3). Hence, after the aircraft transit the passageway exists for a time period $\Delta_t = \Delta_{t1} + \Delta_{t2} + \Delta_{t3}$, after which $R(\varphi_i, \lambda_j, z, k, t) = S(\varphi_i, \lambda_j, z, k, t) \equiv 0$ and in Equation (6.14) the functionals Q, U, and P start working.

During a time period Δt a lot of various processes of transformation of substances ejected by the engines occur within the flight corridor. The term "index of transformations in a jet", represents an integral estimate of the concentrations of these substances as a function of time. Let an exhaust emission take place at moment t_0 (the moment of aircraft transit over a given point of the Earth's surface). Then the index of transformation of chemicals in the trail after the flight can be presented by a 3-step function:

$$J_N(t) = \begin{cases} J_{N1} & \text{for } t_0 \leq t < t_0 + \Delta t_1; \\ J_{N2} & \text{for } t_0 + \Delta t_1 \leq t < t_0 + \Delta t_1 + \Delta t_2 \\ J_{N3} & \text{for } t_0 + \Delta t_1 + \Delta t_2 \leq t < t_0 + \Delta t_1 + \Delta t_2 + \Delta t_3 \end{cases} \tag{6.15}$$

The values J_{Ni} $(i = 1, 2, 3)$ depend on the time of day, season, and on many other parameters (temperature, altitude, geographical coordinates, etc.). The empirical estimates of J_{Ni} are introduced to the database of the system and used in calculations of $J_N(t)$. With further improvement of the SSCRO it will be expedient to include a unit, which would give theoretical estimates of J_{Ni}.

Since the aircraft flies at a velocity V_a along the route $x(\varphi, \lambda, z)$, at a time moment t_0 it is at some point x_0, and all its engines eject $V_M(t_0)$ of M-type substances. With account of Equation (6.15) we obtain:

$$V_M(t) = \begin{cases} L_1 & \text{for } t_0 \leq t < t_0 + \Delta t_1 \\ L_2 & \text{for } t_0 + \Delta t_1 \leq t < t_0 + \Delta t_1 + \Delta t_2 \\ L_3 & \text{for } t_0 + \Delta t_1 + \Delta t_2 \leq t < t_0 + \Delta t \end{cases}$$

where

$$L_1 = V_M(t_0) - \frac{t - t_0}{\Delta t_1}[V_M(t_0) - J_{N1} V_M(t_0)]$$

$$L_2 = J_2 V_M(t_0) + \frac{t_0 + \Delta t_1 + \Delta t_2 - t}{\Delta t_2} V_M(t_0)(J_{N1} - J_{N_2})$$

$$L_3 = J_3 V_M(t_0) + \frac{t_0 + \Delta t_1 + \Delta t_2 - t}{\Delta t_3} V_M(t_0)(J_{N_2} - J_{N3})$$

After some time Δt an apparent effect of the aircraft flight is considered to cease. All processes of transformation and destruction of ozone within the flight corridor after the flight again become natural. The zone of the vortex trail behind the aircraft has a circular section of diameter r, and during the time period $\tau = \Delta t_1 + \Delta t_2$ its interaction with the surrounding atmosphere can be considered negligibly small. At the third stage, this interaction begins with the slight contact between the two media. At any rate, the scenario of interaction of the flight corridor with the surrounding atmosphere needs to be specified and developed, in particular, by forming a set of scenarios.

NO_x is the most important component of exhaust gases. During the lifetime of the aircraft trail NO_x gets oxidized by hydroxyl OH, present in the vortex trail, giving HNO_3 and HO_2NO_2. As laboratory studies have shown, the processes of formation and destruction of ozone are also affected markedly by heterogenic mechanisms of the impact on atmospheric chemistry. This impact manifests itself both within the flight corridor and in a free atmosphere. In particular, the reaction $N_2O_5 + H_2O_2HNO_3$ with sulphate aerosols mainly resulting from aircraft flight, reduces the rate of ozone destruction due to the NO_x cycle, but raises the negative role of Cl_x and HO_x in O_3 destruction.

The second important component of exhaust gases is SO_2, which through ejection by the engines doubles the area occupied by sulphate particles in the atmosphere of the flight corridor, which leads to an increase in O_3 losses. In a number of field experiments (Kraabol and Stordal, 2000) on the F-16 fighter and in laboratory experiments with the F-100 engine using several types of aviation fuel with a high (~1150 ppmm S), moderate (~170–300 ppmm S), and low (~10 ppmm S) content of sulphur, the SO_2 emission changed from 2.49 g SO_2/kg for the fuel with a high sulphur content to 0.01 g SO_2/kg for the fuel with a low sulphur content. For these experiments the following relationship was derived:

$$0.02 \leq \frac{[SO_3]}{[SO_2] + [SO_3]} \leq 0.14$$

The results of studies carried out by Weisenstein *et al.* (1998) show that the mechanisms of evolution of the composition of the engines exhaust gases at their interaction with the atmosphere have been poorly studied, and therefore it is important to further develop the kinetic models describing the role of aircraft flights in changing the atmosphere.

The functional Q in Equation (6.14) is written following the traditional scheme:

$$Q = -V_\varphi \frac{\partial O_3}{\partial \varphi} - V_\lambda \frac{\partial O_3}{\partial \lambda} - V_z \frac{\partial O_3}{\partial z} + \frac{\partial}{\partial \varphi}\left(D_\varphi \frac{\partial O_3}{\partial \varphi}\right)$$
$$+ \frac{\partial}{\partial \lambda}\left(D_\lambda \frac{\partial O_3}{\partial \lambda}\right) + \frac{\partial}{\partial z}\left(D_z \frac{\partial O_3}{\partial z}\right)$$

where $V(V_\varphi, V_\lambda, V_z)$ is the wind speed and $D(D_\varphi, D_\lambda, D_z)$ is the coefficient of eddy diffusion.

The units CCAB, MOFD, and MOST (see Table 6.18) divide the functional Q with an account of the output information from the unit AADB. As a result, the air mass mixing is realized at two stages: (1) mixing of the atmospheric compartment of the zone of aircraft flight with the environment; and (2) mixing of the cells $\{\Omega_{i,j,k}\}$ selected by the unit AADB. At the first stage the volumes and location of the zone of the effect of aircraft are calculated:

$$\omega = \{(\varphi, \lambda, z) : \varphi_0 \leq \varphi \leq \varphi_1; \lambda_0 \leq \lambda \leq \lambda_1; z_0 \leq z \leq z_1\}$$

where $\varphi_1 = \varphi_0 + V_{a\varphi}\Delta t/k_\varphi$; $\lambda_1 = \lambda_0 + V_{a\lambda}\Delta t/k_\lambda$; $\Delta z = z_1 - z_0$ is the diameter of zone of the impact of the aircraft ($\Delta z = r$); φ_0 and λ_0 are latitude and longitude of the aircraft location at a time moment t_0; π_φ and κ_λ is the number of kilometers within 1° of latitude and longitude, respectively; and z_1 and z_0 are the lower and upper boundaries of the flight corridor.

The obtained space ω agrees with the adjacent multitude of atmospheric units $\{\Omega_{ijk}\}$. Then the ozone content is averaged over ω and the adjacent compartments $\{\Omega_{ijk}\}$ with their volumes taken into account.

The second stage realizes a 2-step procedure of recalculation of the ozone concentration over the whole space $\Xi = \{(\varphi, \lambda, z):\ (\varphi, \lambda) \in \Omega;\ 0 \leq z \leq z_H\}$, where z_H is the altitude of the atmospheric boundary layer ($z_H \sim 70$ km), whose consideration is important in estimating the state of the regional ozonosphere. These two steps correspond to the vertical and horizontal constituents of the atmospheric motion. This division is made for convenience, in order that the user of the expert system could choose a synoptic scenario. According to the available estimates (Karol, 2000; Kraabol *et al.*, 2000; Meijer and Velthoven, 1997), processes of vertical mixing prevail in the dynamics of the ozone concentration. It is here that due to uncertain estimates of D_z there are serious errors in model calculations. Therefore, the units CCAB, MOFD and MOST (see Table 6.18) provide the user with the principal possibility to chose various approximations of the vertical profile of the eddy diffusion coefficient (D_z). The database of SSCRO contains the versions of estimates of D_z (m^2/s) already used by many authors in the models of the ozone dynamics: $D_z \approx 3$ m^2/s; $D_z \approx 0.5(V_\varphi + V_\lambda)$; and

$$D_z \approx \begin{cases} 10\,\text{m}^2/\text{s} & \text{for } z \leq 10\,\text{km} \\ 10 + 2.57(10 - z)\,\text{m}^2/\text{s} & \text{for } 10 < z \leq 13\,\text{km} \\ 70 - 1.22(70 - z)\,\text{m}^2/\text{s} & \text{for } 13 < z \leq 70\,\text{km} \end{cases}$$

Since the model retrieval of the background situation over a given territory has numerous versions that require information on regional and global processes, the SSCRO foresees a possibility to apply different models of ozone dynamics. But the scenario is considered basic, which is formed by the operator from the data of the regional or global ozonometric network. The operator can prescribe a discrete function $O_{3,i,j,k,s} = Q_3(\varphi_i, \lambda_j, z_k, t_s)$, a priori considering the value of $O_{3,i,j,k,s}$ – an average estimate of the function $Q_3(\varphi_i, \lambda_j, z_k, t_s)$ in space Ω_{ijk} for a time period $t_{s-1} \leq t \leq t_s$.

Possibilities are foreseen to choose various versions of approximation of the O_3 function over the whole territory by a number of latitudinal and meridional distributions.

Table 6.19. Estimates of some parameters in the SSCRO database.

Parameter	Evaluation
Tropopause altitude (km)	
Summer (Z_1)	10
Winter (Z_2)	13
Cruise altitude (km)	
Subsonic aviation	9–12
Supersonic aviation	16–20
Diameter of the vortex trail zone after the aircraft (m)	100–150
Indices of the supersonic transport aircraft emission	
EL(NO_x)	15
El(CO)	1.5
El(CH_4)	0.2
El(H_2O)	1,230

In this case only the ozone–air mixing ratio can be prescribed. From this ratio the O_3 function is reconstructed, provided the SSCRO database contains information on the vertical profile of air density at any point $(\varphi, \lambda) \in \Omega$. The respective computer dialogue procedure makes this choice automatic.

One of the important elements of the SSCRO functioning is the replenishing of the database over the period of its adaptation to conditions of a given region. The database includes information about the flight timetable over the region and other characteristics of the engines, cruising altitude, airspeed, location of airports, air route, etc. An example is given in Table 6.19.

All background information is concentrated in the form of matrix structures, such as $F = \|f_{ij}\|, D = \|d_{ij}\|, C = \|c_{ij}\|, B = \|b_{ij}\|$, where f_{ij} is a vector whose components contain all the required information for the ith arrival flight (f_{i1} – time of landing, f_{i2} – arrival direction, f_{i3} – type of engine, f_{i4} – airspeed, f_{i5} – cruising altitude, f_{i6} and f_{i7} – latitude and longitude of the airport of entry and other possible characteristics), the d_{ij} vector contains similar information about departing aircraft, the c_{ij} vector describes information about air routes and transit aircraft, and finally, the b_{ij} vector decodes the f_{i3} component, giving the volume of the burnt fuel and the type and composition of exhaust gases. The transit flights, with an intermediate flight stop at the airports of the region, are taken into account by the identifiers F and D separately both before landing and after take-off.

Model estimates of the impact of transport aircraft on the atmosphere and climate (which can be obtained with the use of the SSCRO), will make it possible, with available synoptic information, to solve the problem of optimization of the transport flight corridors and flight timetables. Considering various scenarios of the aircraft load on the regional ozonosphere using the SSCRO, it is possible to determine the location of the flight corridors, which, in other similar conditions, will reduce the consequences of this load. A possibility appears to specify the compounds of other biogeochemical processes with the participation of GHGs.

6.9 RELATIONSHIP BETWEEN THE GLOBAL CYCLES OF WATER AND CARBON DIOXIDE

A specification of the global model of the CO_2 cycle in the biosphere includes a consideration of the role of the process of its washing out from the atmosphere due to precipitation. This process is unidirectional, although studied inadequately. Nevertheless, there are direct observational verifications of CO_2 absorption by rain droplets (Egan *et al.*, 1991). In particular, one of the confirmations of the fact that precipitation washes out CO_2 from the atmosphere is the presence in rain of a considerable amount (up to 15 mg/l) of hydrocarbonate ions HCO_3^-. A combined analysis of the precipitation amount and variations of the atmospheric CO_2 concentration over the same territory performed using the data of the global observational network, has made it possible to reveal a persistent correlation between these processes. The dependence between changes of atmospheric CO_2 and precipitation is known to be sufficiently stable. A detailed analysis of this dependence for various latitudinal belts or for other configurations of limited territories reveals similar patterns independent of geophysical coordinates. Here one should point to a high sensitivity of correlation to the duration and type of precipitation. For instance, during a shower the HCO_3^- concentration in precipitation can either double or halve depending on the presence or lack of thunderstorms. Moreover, this ratio depends strongly on the duration of the precipitation period. Observations showed that with an increasing duration of rain the concentration of HCO_3^- decreases. In other words, the interaction between CO_2 concentration and moisture content in the atmosphere is an important component of the global carbon cycle.

A formalization of the role of rain in the global CO_2 cycle requires a model of CO_2 absorption by water droplets falling at some velocity u (Wang *et al.*, 1997; Wacker, 1995; Chate and Kamra, 1997). The most widely used version of such a model is an equation of the gas balance on the surface of the rain droplets:

$$\frac{dC}{dz} = \frac{3\Delta}{ur^2}\left(1 + 0.3\sqrt{Re}\sqrt[3]{Sc}\right)(C_A - C^*)$$

where Δ is the coefficient of CO_2 diffusion in the air; r is the droplet radius; C^* is the balanced concentration of CO_2 in a droplet; C_A is the CO_2 concentration in the atmosphere; z is the altitude; $Sc = \nu/\Delta$ is the Schmidt number, ν is the kinematic viscosity; and $Re = 2ru/\nu$ is the Reynolds number.

A diversity of the forms of precipitation over the globe complicates consideration of their role in the global CO_2 cycle. This problem can be solved in two ways. The first way is the formal numerical description of all of the processes of precipitation formation. The second way is connected with the use of the present-day means of the global observation of precipitation. In both cases the forms of rain should be clearly classified as functions of meteorological situations. The rain rate can range widely from 1 mm/hr to 8 mm/hr and even more. Besides, there is a certain correlation between the precipitation rate and the size of rain droplets. With the low-intensity rain $r \in [0.1; 0.5]$. A shower can be characterized by the formation of droplets up to $r \sim 6$ mm.

Thus the problem of assessment of the role of precipitation in washing out CO_2 from the atmosphere is urgent, and to solve it, the global model should take into account a change of hydrological cycles separately, over the World Ocean and over land, since these regions of the planet differ in their interaction with the atmosphere.

6.9.1 The global water cycle

Water belongs to substances most widespread in nature. It is present in various forms practically all over the globe, and plays an important role in energy and mass-exchange between the continents, oceans, and atmosphere. The problem of assessing the role of water in the global carbon cycle is only a small part of the general global problem of the nature–society (N–S) interaction.

Oceans, polar ice caps, glaciers, lakes, rivers, soils, and the atmosphere contain $1.4–1.5 \times 10^9\ km^3$ of water. This mass is in constant dynamic interaction with other biospheric components and determines thereby the factors of environmental variability. The developed methods of numerical experiments should be used to assess the role of these factors in present conditions and to show the significance of the water balance in the stabilization of numerous climatic and biogeocenotic processes. Here an attempt has been made, through a systematization of the information about the water balance of the planet, to create a version of the model of the biospheric water balance (MBWB) able, within the general approach to modelling the carbon balance, to take into account the role of water fluxes.

An important moment of the MBWB functioning is its conjunction with the methods of determination of various parameters of the water cycle. Such methods are based on the use of surface, satellite, and airborne measurements. The unit MBWB as part of a global model makes it easier to understand the role of the oceans and land in the hydrological cycle, to choose the main factors that control it, as well as to trace the dynamics of its interaction with plants, soil, and topographic characteristics of the Earth's surface. It is based on an account of the interaction between the elements of the water cycle, as well as natural and anthropogenic factors taken into account through information interfaces with other units of the global model (Krapivin and Kondratyev, 2002).

Consider the block-scheme of the global water exchange and write the respective equations. The basic regularity of the global water exchange is the invariability of water supplies on the Earth in the time period of hundreds of years (i.e., we can reliably write the balanced equation $W_E = W_S + W_O$, where W_E, W_S, and W_O are water supplies on the Earth, on land, and in the oceans, respectively). A compartment of the atmosphere is related to the respective region of water basin. The relationship is valid:

$$\frac{dW_E}{dt} = \frac{dW_S}{dt} + \frac{dW_O}{dt} = 0$$

or $dW_S/dt = -dW_O/dt$. Hence, the trend in changes of water supplies on land is contrary to the similar trend in the oceans.

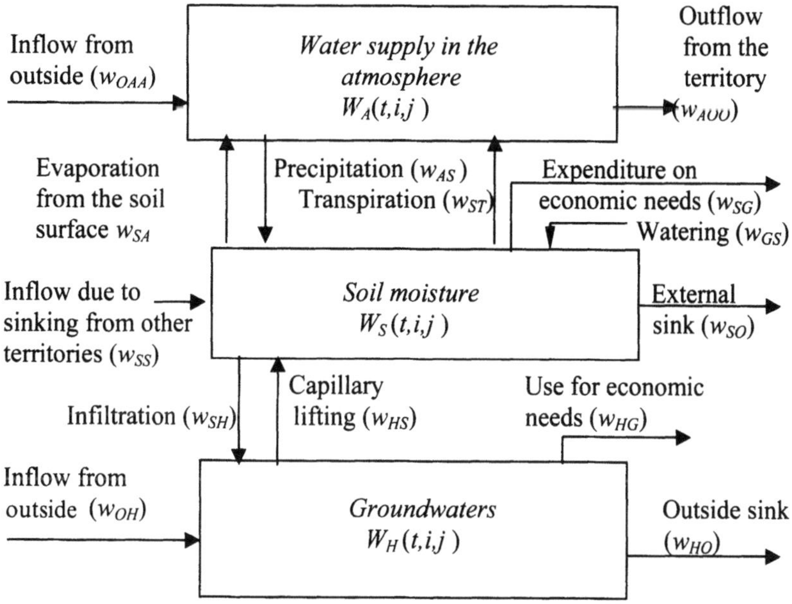

Figure 6.11. Water fluxes across the border of a limited land territory.

With the water supply in the atmosphere $W_A = W_{AO} + W_{AS}$, we obtain $W_E = W_A + W_{S1} + W_{O1}$, where W_{AO} and W_{AS} are water supplies in the atmosphere over the oceans and land, respectively, $W_{S1} = W_S - W_{AS}$, and $W_{O1} = W_O - W_{AO}$. The balanced equation will be:

$$\frac{dW_E}{dt} = \frac{dW_A}{dt} + \frac{dW_{S1}}{dt} + \frac{dW_{O1}}{dt} = 0$$

As is seen, the structure of the trends in the ratios of water supplies is complicated and to analyse it, additional considerations are needed. This complication becomes considerable with further subdivision of the biosphere.

Within the MBWB, small corrections for the water exchange between the Earth and space are not taken into account. The model of the global water cycle can be based on the method of describing the hydrology of comparatively large territories. In this case the basic unit of such a territory is a compartment Ω_{ij} of the Earth's surface of the size $\Delta\varphi_i$ by latitude and $\Delta\lambda_j$ by longitude.

The state of the water component of the compartment Ω_{ij} with the coordinates (φ_i, λ_j) can be characterized by the magnitude of an equivalent liquid water column over unit area. Possible water fluxes across the border of Ω_{ij} are shown in Figures 6.11 and 6.12. The intensities of these fluxes depend on the phase state of the water, temperature, wind speed, and other geophysical and ecological factors. It is difficult to take into account all the small details of these fluxes within the global model because their interactions have been studied inadequately. Therefore, the degree of detailing chosen here has been orientated towards an account of the most important components of their states. Water is considered in liquid, solid, and gas phases.

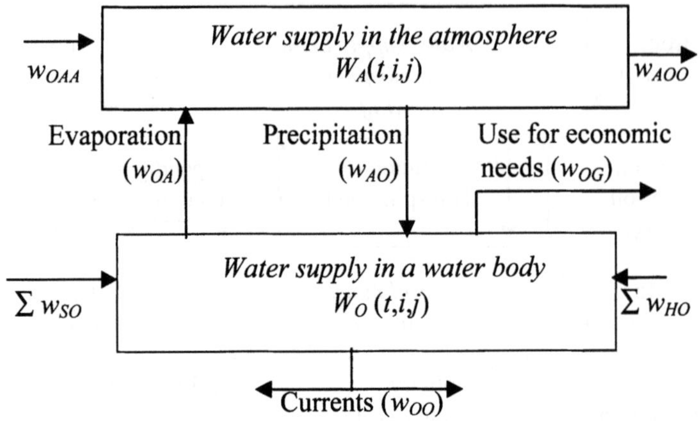

Figure 6.12. Water fluxes across the border between a limited territory and a water body.

Within the compartment Ω_{ij} there is only one state though in the future with the available required information, a vector parameter can be introduced, which determines the share of the precipitation over Ω_{ij} in the form of snow, pellets of snow, granulated snow, pellets of ice, ice rain, hail, rain, drizzle, wet snow, and other forms.

The global water balance consists of the mosaic structure of local balances at the level of Ω_{ij}. The proposed description of water fluxes enables one to trace their balance at any level of spatial digitization – region, water basin, continent, ocean, hemisphere, or biosphere. Clearly, the general balance of evaporation and precipitation at the level of the biosphere is maintained. In other cases, on average, with decreasing spatial sizes of selected units of the biosphere, one should expect an increasing difference between the precipitation amount and evaporation. In this case the water transport through the atmosphere, with river runoff and sea currents will serve as an equalizer. Though the quantitative estimates of all these parameters have been well studied, the water cycle dynamics can be described only with the use of the model. As a first approximation, to assess the role of precipitation in the global CO_2 cycle, one can use only the components W_{AU} and W_{AS}. However, with an account of the spatially heterogeneous distribution of CO_2, the biosphere should be digitized.

With the notations assumed in Figures 6.11 and 6.12, the balanced equations of the water cycle at the level of Ω_{ij} are written as follows:

$$\frac{dW_S(t,i,j)}{dt} = w_{AS} + w_{GS} + w_{SS} + w_{HS} - w_{SO} - w_{SG} - w_{ST} - w_{SA} - w_{SH}$$

$$\frac{dW_H(t,i,j)}{dt} = w_{SH} + w_{OH} - w_{HO} - w_{HG} - w_{HS}$$

$$\frac{dW_O(t,i,j)}{dt} = \sum_{(k,n)\in I_{kn}} [w_{IO}(t,k,n) + w_{HO}(t,k,n)]$$

$$+ w_{AO} + w_O - w_{OA} - w_{OG} - w_{OR} - w_T$$

$$\frac{dW_A(t,i,j)}{dt} = w_{OAA} - w_{AOO} + w_{SA} + \begin{cases} w_V & \text{for water surface} \\ w_{ST} & \text{for land} \end{cases}$$

Detailing of the right-hand parts of these equations with the functional presentations of fluxes in conditions of changing parameters of the environment will determine the level of the qualitative and quantitative reliability of the model. In particular, the model can be simplified by approximating the average value of W_O:

$$\overline{W}_O = \begin{cases} 2{,}500 + 350\sqrt{t} & \text{for } 0 \leq t \leq 70 \\ 6{,}400 - 3{,}200 \exp(-t/62.8) & \text{for } t > 70 \end{cases}$$

where the average depth of the World Ocean is measured in metres and the age of the ocean t is calculated in millions of years. Variations of the ocean volume can also be approximated by the formula: $\Delta V = \Delta W_O A_O + 59.5(\Delta W_O)^2$, where $A_O = 361.06 \times 10^6\,\text{km}^2$.

6.9.2 Moisture cycle in the atmosphere–land system

The land–atmosphere exchange processes include evaporation of the soil moisture, from the leaf surface, stems and trunks of plants, as well as transpiration, precipitation, and evaporation off the surface of unstable water accumulations in low ground. The water flow from the soil through the plant is the least studied link of this chain. The importance of the process of transpiration in the global water cycle can be judged from the available estimates, according to which transpiration takes more water than photosynthesis. For instance, from average estimates, to grow a 20-t yield (wet mass), the plants extract from the soil about 2,000 t of water, with only 3 t of the used water being atomic hydrogen bound with atomic carbon in photosynthesis.

The model description of the process of transportation requires an understanding of the role of the physical and physiological factors in this process. A simplified idea about this role is reduced to the following. If the plant roots are in sufficiently wet soil, then the rate of transpiration is a function of temperature, humidity, wind speed, and insolation. Beyond some threshold of soil moisture, when the water supply in soil ceases, the role of the indicated physical factors sharply diminishes, being inferior to the physiological factors – the type of plants, the construction of the roots, the phase of plant development, the type of soil, and the soil layer thickness. This threshold can vary from 5 to 50 cm of the precipitable water. At any rate, if for a given type of plant the water is not a limiting factor (i.e., water is not limited), then as a first approximation, the total growth of plants can be considered proportional to total potential transpiration for the whole period of growth. The latter is proportional to the amount of the incoming solar radiation.

At present about 12% of the total evaporation from the Earth's surface is used by plants in the process of photosynthesis. In this process about 2250 km^3 of water participate annually with the return coefficient 0.75. Therefore, the simplest description of transpiration will be: $w_{ST}(t,i,j) = \beta_{ij} W_S(t,i,j)$, where β_{ij} depends on the vegetation productivity. The values of β_{ij} are 0.67 for forests, 0.44 for a meadow steppe, and 0.25 for agricultural crops. However, in real situations W_S is a limiting factor in a more complicated dependence of the impact on transpiration rate through the rate of photosynthesis R_p. In other words, $w_{ST} = k_p R_p$, where k_p is the transpiration coefficient for the plants of the p type. As a first approximation, one can use $R_p = \varepsilon_p r_p$, where ε_p is the share of solar energy, assimilated by the pth type of plants in the process of photosynthesis. The value ε_p depends on the presence of water accessible for plants:

$$\varepsilon_p = \varepsilon_{p,0}[1 - \exp(-\varepsilon_{p,1} W_S(t, i, j))]$$

where $\varepsilon_{p,0}$ is the value of ε_p with a sufficient amount of water and $\varepsilon_{p,1}$ is the coefficient reflecting a reduction of solar energy assimilated by plants with a decrease in the accessiblility of water. On average, $\varepsilon_{p,0}$ is reached at $W_S = 10$ mm. Assuming $\varepsilon p/\varepsilon_{p,0} = 0.9$, we have $\varepsilon_{p,1} = 0.23$. In this case $r_p = 9.6\,\mathrm{kg/km^2/day}$ of dry substance (or 37 kg of wet phytomass). The coefficient k_p is estimated for each type of plant and is equal to 368 for maize, 397 for sugar beet, 435 for wheat, 636 for potatoes, and 462 for cotton (kg H_2O/kg of pure substance).

As one of the models of the process of transpiration we can write $w_{ST} = Y_S(24a + b)$, where Y_S is the specific water return, a is the rate of the ground water lifting (cm/hr), and b is the average daily change in the groundwater level (cm).

Within the MBWB unit, to describe the process of the atmosphere–land interaction, the fluxes w_{SA}, w_{ST}, and w_{AS} are used, whose parametric descriptions serve as the basis for this unit. Information on precipitation w_{AS} is usually included in the information bulletins of hydrometeorological services. The pre-history of the distribution of precipitation in the form of a set of matrices $W_{AS}(\theta) = \|w_{AS}(\theta, i, j)\|$, where $(i, j) \in \Psi$ and θ are discrete time moments of the precipitation record by the hydrometeorological service, is used to derive the functional $w_{AS}(t, i, j) = F(W_{AS}(\theta_1), \ldots, W_{AS}(\theta_N), t)$. This is performed with the use of extrapolation, a grouped account of arguments, and evolutionary modelling. Such an approach requires data on precipitation over the discrete geographical grid $\Delta\varphi \times \Delta\lambda$. However, this database can also be modelled by simulating the global cloud field ensuring precipitation. The simplest parameterization of clouds consists in the prescription of the threshold $W_{A,\max}$, beyond which an excess of atmospheric moisture transforms into water and precipitates. To reduce inevitable errors (mainly caused by overestimation of precipitation), it is expedient to introduce the matrix of threshold $W_{\max} = \|W_{A,\max}(i, j)\|$, $(i, j) \in \Psi$. Then:

$$w_{AS}(t, i, j) = \begin{cases} 0 & \text{for } W_A(t, i, j) \leq W_{A,\max} \\ W_{A,\max} - W_A(t, i, j) & \text{for } W_A(t, i, j) > W_{A,\max} \end{cases}$$

If the value $W_{A,\max}(i, j)$ corresponds to a real critical value of the moisture content in the atmosphere over Ω_{ij}, then w_{AS} will be overestimated. It is assumed here that at $W_A(t, i, j) > W_{A,\max}$ the cloud fills the whole cell Ω_{ij}, which does not always correspond to reality. Moreover, the fact that a considerable share of moisture, even with an exceeded critical level, can remain in the cloud and evaporate, is left out of the account. Therefore, to take these special features into account, an adaptive coefficient $\alpha_W < 1$ should be introduced.

Let us divide precipitation into two basic types – solid and liquid. This division can be driven by the thermal principle and by the seasons. The thermal principle is more preferable due to flexibility at sharp climate changes and possible shiftings of the scale of seasonality. The synoptic division of the periods with different precipitation is justified in various regions of the globe. The average daily temperature in the early period of solid precipitation is below zero, ranging from –4°C to –7°C. On the border of this division precipitation of the mixed type is observed. The relationship between the types of precipitation is described by the formula: $x_T = a - bT$, where x_T is the share of solid precipitation; T is temperature; and a and b are empirical coefficients. For the Atlantic climatic zone $a = 50$ and $b = 5$.

To parameterize the process of evaporation from land, numerous formulas are used. Here a simple dependence is assumed:

$$w_{SA}(t,i,j) = \rho_{ij}[1 - \exp(-\delta_{ij}/\rho_{ij})]/\sigma_{ij}$$

where δ_{ij} is a maximum possible rate of evaporation in the region; and ρ_{ij} is the total amount of moisture getting to the soil per unit time.

The choice of the concrete model of the evaporation process is determined by the character of the applied database. Evaporation from the soil surface substantially depends on the type of vegetation cover. The evaporation in the forest and in the field differs by 30–40%. This is connected with the heterogeneous impacts on the water regime, within various vegetation covers, of such factors as soil freezing, snow melting intensity, soil structure, radiation budget, and others. To take into account the dependence of the rate of evaporation from land on temperature T_{ij}, the character of vegetation cover, and soil properties, the following formula is used:

$$w_{SA}(t,i,j) = \delta^*(T_{ij}/\overline{T}_{ij})[1 - \lambda \exp(-\lambda_1 A_{ij}/\overline{A}_{ij})][1 - \lambda_2 \exp(-\lambda_3 X_{ij}/\overline{X}_{ij})]$$

where $\overline{T}_{ij}$ is the surface air temperature in the region Ω_{ij}, averaged over the period considered; $\overline{A}_{ij}$ and $\overline{X}_{ij}$ are the average power of the soil layer and the density of vegetation cover, respectively; and $\delta^*, \lambda, \lambda_1, \lambda_2$ and λ_3 are the empirical coefficients.

A detailed analysis of possible models of evaporation from the land surface with an account of various types of vegetation cover and changes of climatic parameters is given in the works of Bras (1990), Karley *et al.* (1993), and Chock and Winkler (2000). In particular, there are formulas to calculate evaporation as a function of the height and density of vegetation cover, wind speed, and temperature. For instance, the following dependence is proposed for the rate of complete evaporation:

$$E_T = \begin{cases} (\Delta Q_N + m\gamma L_j)/[\Delta + (1+n)\gamma + I(1-C)] & \text{for } T < 0°\text{C} \\ (\Delta Q_N + \gamma L_j)/[\Delta + \gamma] & \text{for } T \geq 0°\text{C} \end{cases}$$

where Δ is the rate of change of the saturated vapour pressure as a function of temperature; Q_N is the amount of energy reaching the evaporating surface; γ is the psychometric coefficient ($\approx$0.66 mb/K); I is the share of complete evaporation due to precipitation caught by foliage; and C is the compensation coefficient due to transpiration. The coefficients m, n, and C are functions of height h and type r_s of vegetation:

$$m = 53\ln^2(20/h + 2.5) \qquad C = (\Delta + \gamma)/[\Delta + (1 + n)\gamma]$$

$$n = r_s[m(1 + U)/100]/250$$

The indicator of the type of vegetation r_s(m/s) for some types is estimated at: 40 (sunflower and alfalfa), 70 (barley and potatoes), 250 (citrus plants), 130 (cotton), 80 (maize and rice), 50 (sugar beet), 60 (wheat), 400 (tundra), 200 (subtropical meadows), 100 (temperate zone meadows),$100 \div 300$ (tropical forests), $200 \div 300$ (coniferous forests), and $100 \div 150$ (deciduous forests in middle latitudes). Typical values of the parameters n, m, and I are: for grass ecosystems – $n \approx 2.5$, $m \approx 3.5$, $I \approx 0.2r$; for woodlands $n \approx 30$, $m \approx 5$, $I \approx 0.3r$ (temperate latitudes) and $I \approx 0.15r$ (tropics) where r is precipitation.

Albedo is an important parameter to calculate the solar radiation energy participating in the process of evaporation. The relationship between the height of plants and albedo, as a first approximation, is described by linear dependence. With the height of plants reaching ~20 m, albedo decreases from 0.25 to 0.1. The albedo values for some types of the Earth cover are known: heather – 0.14; fern – 0.24; natural pastures – 0.25; shrubs – 0.21; savannah – 0.17; deciduous forests in mid-latitudes – 0.1; coniferous forests and orange groves – 0.16; eucalyptus forests – 0.19; wet tropical forests – 0.13; and waterlogged forests – 0.12. The albedo of agricultural fields varies within 0.15 (sugar cane and fruit trees) and 0.26 (sugar beet, barley, cucumber).

The surface part of the land–atmosphere water exchange is connected with the subdivision of the phase space, at least, into two levels – soil and groundwaters. The soil level plays the role of the buffer reservoir between precipitation and groundwaters. A simplest parameterization of fluxes between these levels is reduced to linear dependences: $w_{SH}(t,i,j) = \lambda_{ij} W_S(t,i,j), w_{HS}(t,i,j) = \mu_{ij} W_H(t,i,j)$. However, a more strict description of the soil level is dictated by natural heterogeneity of the structure of Ω_{ij}, where small water bodies and land sites with a certain relief can be located. According to the landscape–hydrological principle, to simulate Ω_{ij}, it is necessary to choose facies and sites of water surfaces, which are typically connected with the floristic background, whose concrete condition is determined by micro-relief, type and properties of soil, surface moisture, depth of groundwater, and other factors. It is possible to choose m_{ij} of facies and n_{ij} of water bodies. In this case the soil moisture forms not only due to the fluxes shown in Figures 6.11 and 6.12, but also due to leakage and filtration of water from the water bodies and aqueducts located in Ω_{ij}.

An important factor of the surface part of the water balance is infiltration of precipitation into the soil both during rains and in runoff. The rate of taking up water into the soil w_{SH} is described by the formula: $w_{SH} = k_S l$, where k_S is the coefficient of filtration and l is the hydraulic slope. If we denote the volume mass of the soil as κ, which on average, varies within 1.4–1.5 g/cm^3, then for k_S it is convenient to use the Azizov formula: $k_S = 256.32\,\kappa^{-7.28} - 1.27\,\kappa^{1.14}$ (cm/day).

The parameter l is calculated using the formula $l = (z_0 + z_1 + z_2)/z_0$, where z_0 is the depth of the column of washingout, z_1 is the capillary pressure, and z_2 is the height of the water layer in the soil surface. At $z_0/z_1 \leq 2$ an approximation $w_{SH} = k_S + t^{-1/2}(0.5k_S z_1 D)^{1/2}$ is valid, where D is the soil moisture deficit and t is time. Other approximations of the function w_{SH} are known: the Horton empirical formula, $w_{SH} = [w_{SH}(t_0, i, j) - k_S] \exp(-\beta t) + k_S$; the Popov formula, $w_{SH} = r \exp(-rt/D) + k_S$, and the Kostiakov formula, $w_{SH} = k_S + \alpha t^{-n}$, where α, n, and β are calibration parameters and r is the rain intensity.

The interaction between the regions of the assumed grid of the biosphere surface division $\{\Omega_{ij}\}$ is realized through fluxes $w_{OAA}, w_{AOO}, w_{SS}, w_{SO}, w_{OH}$, and w_{HO}.

6.9.3 Modelling the moisture cycle in the atmosphere–ocean system

The processes of transport on the atmosphere–water surface border has been well studied. The transport of moisture from the surface of the water body into the atmosphere is one of the fragments of complicated physical processes of mass and energy exchange across the water–air interface. These processes are functions of many climatic parameters and, to a large extent, are regulated by eddy motions in the surface layer of the atmosphere determined by the wind field.

Estimation of the water transport from the water surface into the atmosphere consists of an assessment of the water content of the lower part of the surface layer of the atmosphere, containing spray and water vapour. The eddy flux of water through a unit surface can be described by the relationship:

$$W_V = -\rho K_W(\partial q/\partial z) = -(\rho w)'q' \approx -\rho <w'q'>$$

where W_V is the vertical eddy flux of water vapour (g/cm^2/s); K_W is the coefficient of the eddy transport of water vapour (cm^2/s) (Table 6.20); q is the specific air humidity (g/g); ρ is air density (g/cm^3); z is the vertical coordinate; w is the vertical constituent of wind speed (cm/s); and w' and q are pulsations of w and q values, respectively. Let p be the atmospheric pressure, then we can express q through average water vapour elasticity e: $q = 0.621e/p$.

Evaporation from the water body surface depends on air temperature and can be described by the function $w_{SA} = w^* T^{\omega}$, where w^* and w are the empirical parameters. If measurements are made of wind speed θ (m/s) and the saturated water vapour pressure at a temperature of the evaporating surface E_1 and atmospheric pressure p (mm Hg), then to estimate the rate of evaporation, the

Table 6.20. The coefficient K_W (cm^2/s) of water vapour diffusion in the atmosphere at pressure 1000 mb as a function of temperature T.

T (°C)	−20	−10	0	10	20	30	40
K_W	0.197	0.211	0.226	0.241	0.257	0.273	0.289

Dalton law can be used: $w_{SA} = A(E_1 - e)/p$ and the Shuleikin formula (Kondratyev *et al.*, 2003a,b): $w_{SA} = C\theta(E_1 - e)$, where A and C are the parameters related as $A = C\theta/p$ ($C = 0.45 \times 10^{-6}$ g/cm^3/mb).

6.9.4 Water in the atmosphere

Atmospheric processes of the moisture transport directly connected with the temporal variations of the meteorological elements, play an important role in the global water cycle. The global atmospheric circulation can be described by the Monin model (Monin and Krasitsky, 1985):

$$\partial v_\delta/\partial t + V_z\,\partial v_\delta/\partial z + V_\delta R^{-1}\,\partial v_\delta/\partial\delta + V_\lambda R^{-1}\sin^{-1}\delta\,\partial v_\delta/\partial\lambda$$
$$= R^{-1}(V_\lambda)^2\cot\delta + 2\Omega V_\lambda\cos\delta + (R\rho)^{-1}\,\partial p/\partial\delta + f_\delta$$

$$\partial v_\lambda/\partial t + V_z\,\partial v_\lambda/\partial z + V_\lambda R^{-1}\,\partial v_\lambda/\partial\delta + V_\delta R^{-1}\sin^{-1}\delta\,\partial v_\lambda/\partial\lambda$$
$$= -R^{-1}V_\delta V_\lambda\cot\delta - 2\Omega V_\delta\cos\delta - (R\rho\sin\delta)^{-1}\,\partial p/\partial\lambda + f_\lambda$$

where Ω is the angular rate of the Earth's rotation; $\delta = \pi/2 - \varphi$ is an addition to the latitude; λ is longitude; V_z, V_δ, and V_λ are the components of the velocity of the atmospheric motion; R is the Earth's radius; and f_δ and f_λ are the components of acceleration due to friction expressed through the friction stress tensor u_{sl}:

$$\rho f_\delta = \partial u_{\delta z}/\partial z + (R\sin\delta)^{-1}(\partial u_{\delta\delta}/\partial\delta + \partial u_{\delta\lambda}/\partial\lambda) - R^{-1}u_{\lambda\lambda}\cot\delta$$

$$\rho f_\lambda = \partial u_{\lambda z}/\partial z + (R\sin\delta)^{-1}(\partial u_{\lambda\delta}/\partial\delta + \partial u_{\lambda\lambda}/\partial\lambda) - R^{-1}u_{\lambda\delta}\cot\delta$$

The velocity field equations are closed by prescribed zero boundary conditions on the Earth's surface, determined by the equation of the relief in the form of the prescribed function $z = h(\delta, \lambda)$ and by an added equation of the state of humid air $p = \rho T[r_d + q(r_V - r_d)$, where $r_d = 0.287$ Joule/g/K and $r_V = 0.46$ Joule/g/K are the gas constants of dry air and saturated water vapour and $q \approx 3$–4% is the specific humidity. The distribution of temperature $T(\delta, \lambda)$ and function $q(\delta, \lambda)$ can be described by the respective equations of evolution, and the data from the global archive can be used to substitute them for tabulated values.

A sufficiently complete description of the models of the general atmospheric circulation has been given in Nicolis and Nicolis (1995). A simplified description of the atmospheric part of the hydrological cycle is possible by the equation $\partial W_A/\partial t + \nabla Q = E - P$, where W_A is the vertically integrated specific air humidity in this column and E and P are evapotranspiration and precipitation at the soil level, respectively (Naoum and Tsanis, 2003).

A further simplification of the model of the hydrological cycle is connected with the selection of the following three types of prevailing directions: western, eastern, and meridional. For such an approximation, data are used on the amplitude of the wind speed oscillations and the directions of the moisture transport in the atmosphere. If the mass of water vapour in air column over the area σ_{ij} is $a = W_A\sigma_{ij}$, then, for instance, for the eastern orientation of the atmospheric circulation the water flux between the adjacent cells of the Earth's surface digitizing grid will be $w_{AO} = 2a\theta/d_{ij}$, where θ is the wind speed and d_{ij} is the diameter of Ω_{ij}. Following this scheme, it is easy to recalculate the moisture supplies at each step of time digitization, since it is unnecessary to solve the problems of numerical integration of partial differential equations. The background information about W_A, σ_{ij}, θ, and d_{ij} is accumulated in the database from different sources. The function W_A is calculated by the balance equation or can be prescribed based on other data. In particular, with changing temperature T and partial pressure of water vapour e, then W_A can be estimated from the relationship: $W_A = meh\,(1 + \alpha T)^{-1}$, where h is the height of the effective atmospheric layer and m and α are the proportion coefficients ($m = 0.8$ and $\alpha = 1/273$ when measuring W_A in g/m^2 and T in °C).

6.9.5 Water masses cycle in the World Ocean

First place among all the water reservoirs on the Earth is occupied by the World Ocean whose present volume exceeds 50 times the volume of water in glaciers, second to the World Ocean. This comparison is important for understanding the correlation between the steps of hierarchy of water basins and determining their structure in the model. Within a priori scenarios of anthropogenic activity and possible changes in the biosphere, the correlation between these steps is important. For instance, 1.6% of the global supplies of water are accumulated in the Antarctic. A comparison of these supplies with the volume of the Arctic Ocean where the water content is 20% less than in the Antarctic glacier cover, suggests the conclusion that the global model of the hydrological cycle is inadequate without the role of the Antarctic taken into account (Keeling and Visbeck, 2001).

The hydrology and sea currents of the Southern Ocean with an account of the effect of glacier cover have been described in numerous monographs, and the circulation models of different complexity and degree of detailing have been derived to simulate them. Such models for the World Ocean, on the whole, are based on an account of configurations of the non-penetrating boundaries and topology of the straits. Numerous numerical experiments with such models have made it possible to reveal the principal structure of the global oceanic circulation consisting of a hierarchy of closed ring circulations with the centres of upwelling and downwelling waters and water nuclei, and including the geometry of water basins with the straits between them. To describe the water circulation in the southern basin, it is necessary that the Drake Passage be taken into account. A scheme of the hydrological field circulation in the World Ocean, acceptable for simulation, has been proposed by Seidov (1987) and Chahine (1992). The model is a system of equations and boundary conditions taking into account the outline of the shores, the bottom relief, as well as ice formation and melting. However, on a global scale, to simulate the oceanic circulation, a simplified scheme is necessary reflecting mainly the role of straits. Such a scheme is shown in Figure 6.13. The final unit responsible for the

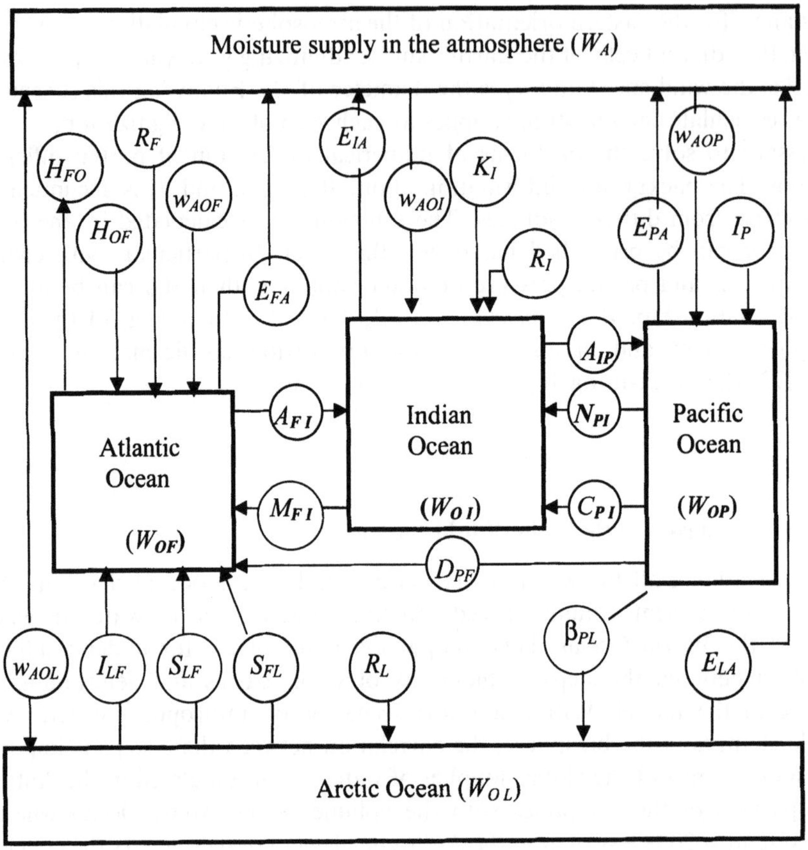

Figure 6.13. Elements of the global water balance with the role of ocean taken into account. Note: w_{AOL}, w_{AOF}, w_{AOI}, w_{AOP} = precipitation; H_{FO}, H_{OF} = Strait of Gibraltar; R_F, R_I, R_L = rivers; E_{FA}, E_{IA}, E_{PA}, E_{LA} = evaporation; A_{FI}, A_{IP} = the Antarctic current; M_{IF} = the Cape Igolny Current; C_{PI} = the East Australian Current; b_{PL} = Bering Strait; I_{LF} = Arctic ice; D_{PF} = Drake Passage; I_P = Antarctic ice; N_{PI} = the Indonesian seas; S_{LF}, S_{FL} = the straits. Some estimates of water fluxes are given in Table 6.21 (page 470).

modelling of the World Ocean circulation has the following form:

$$\sigma_{OF}\frac{dW_{OF}}{dt} = H_{OF} + R_F + I_{LF} + S_{LF} - S_{FL} - H_{FO} - A_{FI} + M_{IF} + D_{PF} + (w_{AOF} - E_{FA})\sigma_{OF} + A_F$$

$$\sigma_{OI}\frac{dW_{OI}}{dt} = A_{FI} + C_{PI} + N_{PI} + K_I + R_I + (w_{AOI} - E_{IA})\sigma_{OI} - A_{IP} - M_{IF}$$

$$\sigma_{OP}\frac{dW_{OP}}{dt} = A_{IP} + R_P + (w_{AOP} - E_{PA})\sigma_{OP} + I_P - B_{PL} - D_{PF} - C_{PI} - N_{PI}$$

$$\sigma_{OL}\frac{dW_{OL}}{dt} = R_L + B_{PL} + (w_{AOL} - E_{LA})\sigma_{OL} + S_{FL} - I_{LF} - S_{LF}$$

$$\sigma \frac{dW_A}{dt} = (E_{PA} - w_{AOP})\sigma_{OP} + (E_{FA} - w_{AOF})\sigma_{OF}$$
$$+ (E_{IA} - w_{AOI})\sigma_{OI} + (E_{LA} - w_{AOL})\sigma_{OL}$$

Within this large-scale approach to the formation of the MBWB ocean unit, the dependencies of the fluxes of water in its different phases on the environmental parameters remain uncertain. Apparently, the mass exchange between the reservoirs s and l can be described by the simplest linear scheme: $w_{sl} = |W_{OS}\sigma_{OS} - W_{OL}\sigma_{OL}|/T_{sl}$, where T_{sl} is the time for equalizing the levels W_{OS} and W_{OL}, σ_{OS} and σ_{OL} are the areas of water basins s and l. For the scheme in Figure 6.13 we have:

$$A_{FI} = \max\{(V_{OF} - V_{OI})/T_{FI}, 0\} \qquad M_{FI} = \max\{(V_{OI} - V_{OF})/T_{IF}, 0\}$$
$$A_{IP} = \max\{(V_{OI} - V_{OP})/T_{IP}, 0\} \qquad N_{PI} = \max\{(V_{OP} - V_{OI})/T_{PI}, 0\}$$
$$C_{PI} = \max\{(V_{OP} - V_{OI})/T^{*}_{PI}, 0\} \qquad D_{PF} = \max\{(V_{OP} - V_{OF})/T_{PF}, 0\}$$
$$S_{LF} = \max\{0, (V_{OL} - V_{OF})/T_{LF}\} \qquad S_{FL} = \max\{0, (V_{OF} - V_{OL})/T_{FL}\}$$
$$b_{PL} = \max\{0, (V_{OP} - V_{OL})/T_{PL}\}$$

where $V_{OS} = W_{OS}\sigma_{OS}(S = F, I, P, L)$.

To estimate the flux K_I, let us take into account the information on the moisture balance in the region of the Red Sea. According to available estimates, the input of water to the Red Sea via the Suez Canal and precipitation can be neglected. Not a single river flows into the Red Sea. The main component of the flux K_I through Bab el Mandeb is rather persistent. Hence, we can assume $K_I = \max\{0, w_{AK}\sigma_{KMP} - E_{KMA}\sigma_{KM}\}$, where w_{AK} and σ_{KMP} are the level and the area of the mainland runoff to the Red Sea, respectively and E_{KMA} is the evaporation from the area σ_{KM} of the Red Sea. The water expenditure through the Strait of Gibraltar H_{FO} (H_{OF}) is determined by the relationship of the levels of W_{OF} and the Mediterranean Sea. In order not to complicate the structure of the model, the level of water in the Mediterranean Sea is determined by its watershed and the difference between precipitation and evaporation. Since the intra-annual distribution of the water inflow into the Atlantic Ocean has a dispersion of 20%, we can reliably assume: $W_{FO} = -W_{OF} = const.$

6.9.6 The regional model of the water budget

Consider the scheme of Figure 6.14 as the basis for modelling the hydrological regime of a limited territory Ω_L, occupied by the aquaecosystem under study. Each territory has the river network, water bodies, and land. According to the landscape–hydrological principle, to derive a simulation model in the zone of the hydrological system functioning, it is necessary to select the facies, which typifies the floristic background, whose concrete appearance is determined by the

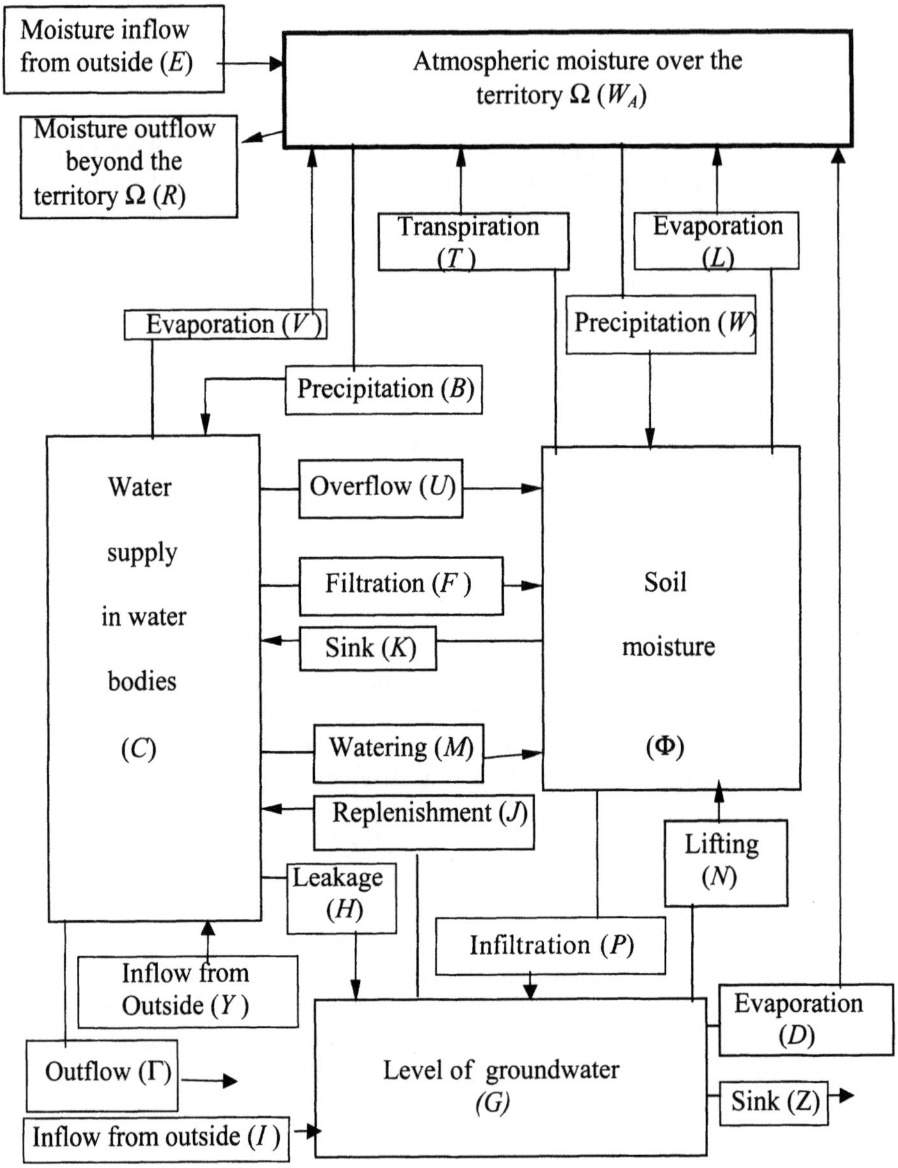

Figure 6.14. The block-scheme of the sample model of water balance in a limited territory.

micro-relief, type, and properties of the soil, surface moisture, depth of groundwater, and other factors. In general, the territory Ω_L is characterized by the presence of *m* facies, and the water network has *n* heterogeneous sites. Bearing this in mind, according to the scheme in Figure 6.14, the closed system of balanced equations has the form:

$$\sigma_{ij} dW_{A,ij}/dt = E_{ij} - R_{ij} + \sum_{k=1}^{n}(V_k - B_k S_k) + D_{ij} + \sum_{l=1}^{m}(L_l + T_l - W_l \sigma_l)$$

$$S_k dG_k/dt = Y_k - V_k + B_k S_k - H_k + J_k + \sum_{l=1}^{m}(K_{lk} - F_{kl} - V_{kl} - M_{kl}) - \Gamma_k + S_k(C_{k-1}V_{k-1}/\Delta_{k-1} - C_k V_k/\Delta_k)$$

$$\sigma_l d\Phi_l/dt = \sum_{k=1}^{n}(F_{kl} + V_{kl} + M_{kl}) + \sum_{k=1}^{m}\psi_l^k \theta_l - L_l - T_l - P_l - \theta_l + N_l + W_l \sigma_l$$

$$\sigma_{ij} dG_{ij}/dt = I_{ij} - Z_{ij} - D_{ij} + \sum_{k=1}^{n}(H_k - J_k) + \sum_{l=1}^{m}(P_l - N_l)$$

In these formulas the following notations are assumed: σ_{ij}, σ_l, and S_k are the areas of the territory Ω_{ij}, of the lth facies and the kth compartment of the river network (km^2), respectively; Δ_k is the linear size of the kth compartment of the river network (km); $W_{A,ij}$, G_k, and Φ_l are, respectively, the levels of water in the atmosphere, in the kth compartment of the river network, and the lth facies on the territory Ω_{ij}; θ_{ij} is the level of groundwater (m); ψ_l^k is the share of the runoff of the kth facies, getting to the territory of the lth facies; the rest of the notations are explained in Figure 6.14.

An application of the model to other regions is realized via the variables E, R, Y_i, Γ_i, I, and Z. Besides, in analysis of the concrete situation, the configuration of the waterway and the level of the water body are taken into account. The required equations are written like those above, proceeding from the condition of the water volume balance. Functionally, all fluxes in the scheme in Figure 6.14 can be described based on the laws of hydrodynamics and with an account of the available observational information. The inflow E_{ij} and outflow R_{ij} of moisture can be determined from the data of remote sensing. Between measurements, the information about wind speed V_{ij} is used, and the functions E_{ij} and R_{ij} are calculated by the formulas: $E_{ij} = E_{H,ij}$; $R_{ij} = W_{A,ij} l^*/(l^* + k_1 V)$, where $l^* = 2\sqrt{\sigma}/H$, E_H is the atmospheric moisture on the windward border of the region, and k_1 is the constant coefficient reflecting the contribution of wind into the circulation of precipitation.

Information on precipitation and runoff is introduced into the information catalogues of the hydrometeorological services (Table 6.22). Based on these data, the respective model units can be derived. Assuming that the distribution of precipitation is proportional to relevant areas, we obtain:

$$B_k = W_{A,ij}\sigma_k/\sigma_{ij} \qquad W_l = W_{A,ij}\sigma_l/\sigma_{ij}$$

The model of the river runoff formation should take into account the watershed topography and the spatial distribution of its soil characteristics, as well as special features of vegetation cover. Let $\theta_l = (g_l + K_l \exp[-a_l X_l - C_l A_l])\sigma_l$, where X_l and A_l are, respectively, the vegetation density (m/km^2) and the soil layer thickness (m) over the area σ_l; g_l is the coefficient of the relief runoff in the lth facies; k_l is the coefficient of water penetration through vegetation and soil cover over the area σ_l; and a_l and

Table 6.21. Quantitative estimates of water fluxes in the scheme in Figure 6.13 (10^3 km^3/year).

Flux	Evaluation	Flux	Evaluation
w_{AOL}	3.6	I_{LF}	0.57
S_{LF}	436	S_{FL}	400
R_L	5.14	A_F	0.3
H_{OF}	23.97	R_F	19.33
w_{AOF}	72.5	E_{FA}	96.6
A_{FI}	6,780.24	M_{IF}	952
D_{PF}	5,771.09	C_{PI}	437
N_{PI}	66.86	I_{OI}	0.765
A_{IP}	6,338.74	E_{IA}	115.4
w_{AOI}	84	R_I	5.386
K_I	0.005	R_P	13.12
E_{PA}	200.4	w_{AOP}	206.7
I_P	0.975	E_{LA}	1.7
β	80.5		

Table 6.22. Estimates of temperature at the beginning and the end of the period with solid precipitation for some global regions.

Region	T_{AF} (°C)
Taimyr Peninsula	−2.4
South-east of the West Siberian Plain	−7.6
Western Siberia	−6.0
South-eastern coastline of Russia	−4.4
Kolyma	−4.0
North of 70°N	−4.0
The Caucasus:	
500–1,000 m above sea level	−6.2
1,500–2,000 m above sea level	−4.8
Finland	−3.0
Japan	−7.0

C_l are the coefficients of precipitation retained by plants or soil in the lth facies, respectively. The parameters of this dependence can be determined from field measurements that establish, for a given type of soil and plants, there is a connection between precipitation intensity, the rate of uptake of water by the soil. So, for saline lands, for instance, the runoff is equal to precipitation. This rather rough approximation can be specified, since the radiometric methods make it possible to classify the soil moisture, at least, into three types: firmly bound, loosely bound, and free water. The bound water is a film moisture adsorbed by the surface of ground particles with the film thickness of 6–8 molecular layers. The content of bound water is estimated at 2–3% in sands and 30–40% in clay and loess. The bound water cannot be assimilated by plants, it does not dissolve salts. In the models considered these specific features are taken into account when determining the respective coefficients of evaporation and transpiration.

The runoff θ_l is distributed between facies and in the form of return water K_{lk} flows into the river. In a general form this is reflected through the coefficients of the runoff distribution $\psi_s^l(\sum_{s=1}^{m+2}\psi_s^l = 1)$, where ψ_{m+1}^l is the share of the runoff from the lth facies beyond the region and ψ_{m+2}^l is the share of the runoff from the lth facies into the river. The coefficients $\omega_{lk}(\sum_{k=1}^{n}\omega_{lk} = 1)$ characterize the runoff distribution from the lth facies by the river compartments and are determined by the landscape relief and the spatial location of the facies and the waterway compartments. Thus $K_{lk} = \omega_{lk}\psi_{m+2}^l\theta_l$.

Evaporation from the soil surface can be described by the formulas of Hitchcock, Horton, Weissman, and others (Bras, 1990). For instance, the formula by Priestly and Taylor (1972) for latent heat of evaporation q_E is $q_E = \alpha S(q^* - q_i)/(S + \gamma)$, where q_i is the soil heat flux (W/m^2); q^* is the net radiation flux (W/m^2); $\gamma = 0.066 \times 10^3$ Pa/K is the psychometric constant; and S is the slope of the curve of the temperature dependence of the saturated vapour pressure (Pa/K);

$$\alpha = \begin{cases} 1.06 & \text{for wet soil} \\ 1.04 & \text{for dry soil} \\ > 1.26 & \text{at a warm air advection over a wet surface.} \end{cases}$$

The Horton formula gives:

$$V = 0.36[(2 - \exp\{-0.44\theta\})l_V - l_a] \qquad \text{(mm/day)}$$

where θ is the wind speed (m/s); l_V is the vapour pressure near the water surface; and l_a is the water vapour elasticity.

The Rower formula is written as:

$$V = 0.771(1.465 - 0.007\rho)(0.44 + 0.26\theta)(l_V - l_a) \qquad \text{(mm/day)}$$

where ρ is atmospheric pressure (mm Hg).

A diversity of the forms for parameterization of the dependence of the rate of evaporation from the soil surface on the environmental parameters enables one to easily adapt the model of water balance to the information base.

The flux T in Figure 6.14 reflects the impact of vegetation cover on the hydrological regime of the territory. One of the simple models of transpiration is the following dependence:

$$T = y(24\alpha^* + \beta^*) \qquad \text{(cm/day)}$$

where y is the specific water return of the soil; α^* is the rate of the groundwater upwelling (cm/hr); and β^* is the daily change of the level of groundwater (cm).

Determine the constituents of the block-scheme in Figure 6.14 characterizing the processes of leakage and filtration of water from the river. Both leakage and filtration depend on the quality of the riverbed and water level. Let:

$$H_i = \begin{cases} \mu_i C_i S_i & \text{for } 0 \leq C_i \leq C_{i,\min} \\ \mu_i C_{i,\min} & \text{for } C_i > C_{i,\min} \end{cases}$$

where μ_i is the coefficient of penetration of water through the river bed. Filtration F_i increases with increasing C_i between two critical values of $C_{i,\min}$, when there is no filtration, and $C_{i,\max}$, when it is at a maximum:

$$F_{i,\max} = \begin{cases} 0 & \text{for } 0 \leq C_i \leq C_{i,\min} \\ \mu_i (C_i - C_{i,\min}) S_i & \text{for } C_{i,\min} < C_i < C_{i,\max} \\ \mu_i (C_{i,\max} - C_{i,\min}) S_i & \text{for } C_i \geq C_{i,\max} \end{cases}$$

The distribution of water filtering from the river between facies depends on the distance r_{ij} between the ith compartment and the jth facies, as well as on the structure of soil and landscape relief. In particular, this dependence can be described by the function $F_{ij} = F_{i,\max}\chi(r_{ij})$, where $\chi(r_{ij})$ is the decreasing function meeting the condition:

$$\sum_{j=1}^{m} \chi(r_{ij}) = 1$$

Evaporation from the river surface depends on the environmental temperature and can be described by the formula $V_i = V_i^* T^\omega$ or by the relationship $V_i = \mu(\theta)(\rho_V - \rho_2)$, where $\mu(\theta)$ is the function reflecting the impact of the wind; ρ_V is the water vapour pressure at the temperature of the evaporating surface (mb); and ρ_2 is the absolute air humidity at a height of 2 m (hPa).

The volume of overflow is determined by the binary regime of the waterway functioning within a maximum possible water level $C_{i,\max}$, so that:

$$V_i^* = \begin{cases} 0 & \text{for } 0 \leq C_i \leq C_{i,\max} \\ C_i - C_{i,\max} & \text{for } C_i > C_{i,\max} \end{cases}$$

The distribution of U_i^* between facies depends on the landscape relief which is characterized by the matrix of the relief runoff $\Psi = \|\Psi_{ij}\|$, and is written as:

$$\sum_{ijj=1}^{m} \Psi_{ij} = 1, \Psi_{ij} \geq 0$$

As a result, $U_{ij} = \Psi_{ij} U_i^*$.

Taking water for irrigation from the ith compartment of the waterway is an anthropogenic factor, and should be considered as a free parameters $M_i^* = \sum_{j=1}^{m} M_{ij}$. To take into account a possible heterogeneity of the distribution of M_i^* between facies, take the matrix of the coefficients of the distribution of watering $\nu = \|\nu_{ij}\| (\nu_{ij} \geq 0, \sum_{j=1}^{m} \nu_{ij} = 1, i = 1, \ldots, n; j = 1, \ldots, m)$, so that $M_{ij} = \nu_{ij} M_i^*$.

The relationship between the surface water fluxes and groundwaters strongly depends on the flux of water infiltrating through the soil layer downwards. This flux, called infiltration, with an account of only the vertical heterogeneity of the soil can be described in a general form by the equation:

$$\frac{\partial P}{\partial t} = \frac{\partial}{\partial z}\left[p(P)\frac{\partial P}{\partial z} + K_z(P)\right]$$

Bras (1990) gave various versions of solutions to this equation. For practical use the following solution is recommended:

$$f = f_c + (f_0 - f_c)\exp(-Pl^2 t) \qquad (6.16)$$

where $f = (P_i - P_0)P/(\pi t)^{-1}$; f_c is the asymptotic value of the rate of filtration; and f_0 is the initial value of the rate of filtration.

The processes of infiltration and evaporation of groundwater depend strongly on the vertical profile of the soil layer. The following soil layers can be selected: saturated and unsaturated. The saturated layer usually covers the depths >1 m. The upper unsaturated layer includes the soil moisture in the zone of the plants roots, the intermediate level, and the level of the capillary water. The water motion through these layers can be described by the Darsy law, and the gravitation term $K_z(P)$ in Equation (6.16) is calculated from the equation:

$$K_z(P) = 256.32\,\delta_s^{-7.28} - 1.27\,\delta_s^{1.14} \qquad \text{(cm/day)}$$

where δ_s is the volume mass of soil (g/cm^3).

Thus the system of equations of the regional water budget with the indicated functional descriptions of water fluxes in the region under study at initial values of $W(t_0), G(t_0), C_i(t_0), \Psi_j(t_0)$ prescribed for a time moment t_0 enables one to calculate the characteristics of the water regime of the whole region for $t \geq t_0$. The initial values are provided by the monitoring system. The regularity of surveys depends on the required accuracy of the prognosis and can be realized by planning the monitoring regime. Based on the synthesis of the model and the remote sensing system, the monitoring can be organized practically for any irrigated agro-ecosystem. In this case problems appear when identifying the airborne measurements with the values of geophysical, ecological, and hydrological parameters. An example of the successful solution of such problems (Vinogradov, 1983) is a determination of the dependence between the coefficient of spectral brightness $\tau_J = \tau_z + (\tau_0 - \tau_z)\exp(-\alpha W^c) + dW^n$, where τ_0 is the coefficient of the dry soil brightness; τ_z is the coefficient of the brightness of soil with the moisture content close to a minimum of the field moisture capacity (when there is no free water in the soil); and α, c, d, and n determine the type of the soil ($\alpha, d, n < 1$ and $c > 1$ – for achromatic loamy soils we have: $\tau_z = 0.09$; $\tau_0 = 0.28$; $\alpha = 0.01$; $c = 2.3$; $n = 0.9$; $d = 0.0001$). Obtaining these estimates is an important problem in the remote sensing of the environment.

Finally, note that the deterministic approach to modelling the water cycle in the zone Ω_L described here cannot be considered as the only possible one. Such an approach gives only average trends in changes of the water cycle components. Their distribution and probabilistic prognosis can be obtained only on the basis of the dynamic–stochastic models of the water balance. In modelling the global carbon cycle this approach enables one to take into account the sink of the atmospheric CO_2 over the region due to washing-out.

6.10 BIOGEOCHEMICAL CYCLE OF METHANE

Methane like carbon dioxide is a GHG. The spectrum of its natural and anthropogenic sources is wide, and its greenhouse effect is substantial (Dementjeva *et al.*, 2000). Before human interference, the natural cycle of methane had been balanced with respect to the climate. Extracting natural combustible gases (90–95% methane), humankind has contributed factors of instability and uncertainty to this cycle. Most authors estimate the level of the global emission of methane into the atmosphere at 535×10^6 t CH_4/year, of which 375×10^6 t CH_4/year is of anthropogenic origin (50×10^6 t CH_4/year being from rice fields). The anthropogenic input of methane is expected to increase within the next 20–30 years, though in some developed regions measures are taken to reduce the anthropogenic emissions of methane into the atmosphere. Nevertheless, the concentration of methane in the present atmosphere increases (by a factor of 7) faster than the growth of the CO_2 concentration, so that its amount increases annually by 2% (i.e., by 2020 the amount of methane in the atmosphere could double compared to 2000, which, from numerous estimates, will lead to a global warming of 0.2–0.4°C). As in the case of CO_2, these estimates will remain rather doubtful and contradictory until the global model mentioned above is synthesized. However, with present knowledge, only the first steps are possible in modelling all these features of the global cycle of CH_4.

Sources of methane are oil, sedimentary and ejected rocks, bottom sediments of lakes, seas, oceans, and other objects of the hydrosphere, as well as soil, peatbogs, and rice plantations. As shown in Table 6.23, numerous processes of methane transformation (70–80%) are of biogenic origin, mostly affected by humans. Of course, the significance of these processes varies depending on many natural and anthropogenic parameters. The relationships of individual elements of correlation between the cycles of CO_2, CH_4, and other chemicals vary, too. At any rate, it is clear that depending on the strategy of the NSS, in due course the composition of the terrestrial atmosphere could change substantially. Burning of only 1 m^3 of methane extracts 2 m^3 O_2 from the atmosphere. From the open dust heaps and municipal and industrial sewage, the atmosphere receives about 2% of the anthropogenic methane ($\sim$270–460 $\times 10^6$ t C) annually. These integral estimates do not permit one to calculate an actual distribution of CH_4 flux in the atmosphere. A certain contribution to a detailing of the spatial distribution of the sources of methane has been made by the Second International Conference on the problems of methane (held in Novosibirsk in 2000), the Proceedings of which contain concrete data on the sources of methane in many regions of the globe. For instance, according to Byakola (2000), within the framework of the international project by the United Nations Environmental Programme/Gross Domestic Product (UNEP/GDP), an inventory of the sources and sinks of CO_2 and CH_4 has been made for the territory of Uganda (236×10^3 km^2). Here, the main anthropogenic sources of methane are agriculture,

Table 6.23. Sources of input of CH_4 into the terrestrial atmosphere.

Source of CH_4	Area of the source (10^6 km^3)	Rate of CO_4 formation (g/m^2/year)	Average rate of CH_4 formation (10^6 t/year)
Rice fields	1.35	206	280
Marshes	2.6	50–100	130–260
Freshwater lakes	2.5	50–100	1.25–25
Arid soils	30	0.44	10
Woodlands	44	0.01–0.09	0.4
Tundra	8	10	0.8–8
Oceans	361	0.012	4–6.7
Shelf zones	1.4	5–10	0.07–1.4
Animals			101–220
Termites			150
Fossil fuels			100
Dust heaps			20–40
Sewage			30–40

municipal sewage, and biomass burning. In 1990, the stockbreeding and rice fields in Uganda contributed 205.45×10^3 t CH_4 and 23.45×10^3 t CH_4, respectively to the atmosphere. The agricultural waste burning added 3.55×10^3 t CH_4.

Naturally, Uganda can be recommended to reduce GHGs emissions, but it is not clear what is the threshold of these emissions. Of course, in the future the stockbreeding and rice production in Uganda will develop, thereby increasing the volumes of CH_4 emitted to the atmosphere. Hence, a balanced correlation should be sought between the economy of the country and the state of the environment. This problem can be solved utilizing new technologies of nature use (Krapivin and Kondratyev, 2002; Kondratyev *et al.*, 2003a,c). In particular, one of the ways of reducing CH_4 emissions is to find a secondary use of organic waste, for instance, in paper production. In Uganda, up to 16% of urban waste is used in paper production.

The gas transport systems are one of the powerful anthropogenic sources of CH_4. The study by Coconea *et al.* (2000) contains information about methane emissions from pipelines in the territory of Romania, a country that signed the Lisbon Protocol in 1994 and now supports the KP. Romania was the first country in Europe where a 50-km pipeline was laid to transport natural gas. At present, natural gas constitutes 37% of the energy resources of the country, the share of oil and coal constituting 32.6% and 15.2%, respectively. Therefore, the problem of anthropogenic input of CH_4 from the territory of Romania into the atmosphere is rather urgent. Here, like in Uganda, technologies play an important role, reducing the leakage of methane from pipelines by 38.9% during the last 20 years, constituting in 1994 55.35% of the leakage in 1987. On the whole, both extraction and distribution of coal, oil, and gas in Romania produce 56% of the entire amount of CH_4 emitted from this territory. Agriculture is in second place (29%).

Table 6.24. Emissions of methane by the coal industry in various countries.
From Gale and Freund (2000).

Country	Coal production (10^6 t/year)	CH_4 emissions (10^6 t/year)	Specific rate of CH_4 emission (kg CH_4/t of coal)
Australia	229	0.8	3.5
England	68	0.5	7.4
Germany	280	1.0	3.6
India	263	0.4	1.5
China	1,141	7.7	6.7
Poland	199	0.6	3.0
Russia	539	4.5	8.3
USA	859	4.3	5.0
Czechoslovakia	88	0.3	3.4
South Africa	182	1.0	0.5
Total	*4,399*	*21.7*	*4.9*

One of the significant sources of CH_4 is Russia, which contributes about 47×10^6 t CH_4/year to the atmosphere, and this flux is expected to reach 78×10^6 t CH_4/year by 2025. This increase will be caused by the developing infrastructure of the gas, oil, and coal industry. On global scales, these trends will be practically observed in all countries. In Table 6.24 the contribution of the coal industry into CH_4 production is estimated for various global regions. These estimates are determined by technologies used in the coal industry. On average, the contributions of various sources into the coal industry itself constitute: 70% – underground ventilation in the coal mines; 20% – underground drainage; 5% – surface loading and unloading operations; 4% – opencast mining of deposits; and 1% – derelict mines.

The global cycle of methane has been studied inadequately, and therefore its modelling faces a lot of unsolved problems. The CH_4 fluxes from the waterlogged territories have been studied best. These fluxes constitute about 20% of the total input of methane to the atmosphere from all sources (Table 6.23). Note, that almost 80% of the sources of methane are biological in nature, so that the anthropogenic interference into this natural cycle is also possible through the violation of various biospheric processes. In particular, on the waterlogged territories, methane forms only due to biological processes. The hydrospheric sources of methane can be presented by a multilayer model (Figure 6.15). This scheme describes the vertical structures of most of the water bodies. Methane forms in the layer of bottom deposits due to bacteria functioning, and in the zone with oxygen methane is partially oxidized giving carbon dioxide, $CH_4 + 2O_2 \rightarrow CO_2 + 2H_2O + E$. Bacteria taking part in methane oxidation use the released energy E for the synthesis of organic matter. The remaining methane gets into the atmosphere and, in contrast to CO_2, does not return to the water medium. This is somehow connected with the

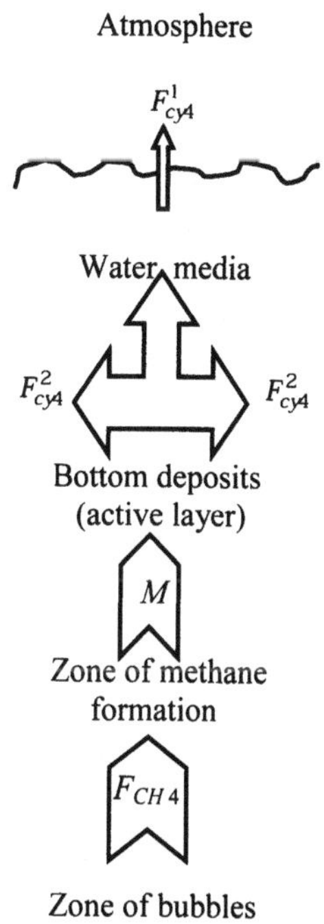

Figure 6.15. The block-scheme of the formation and transport of methane in the waterlogged area.

Note: F^1_{cy4} = methane flux on the atmosphere–water object border; F^2_{cy4} = oxidation of methane in the aerobic zone; F_{CH4} = energy from the source of methane; M = methane concentration.

fact that the CH_4 solubility in water is almost 40 times lower than that of CO_2. The lifetime τ_H of methane in the atmosphere is estimated at about 5 years. Its extraction from the atmosphere takes place due to the participation of methane in photochemical reactions, resulting in methane oxidation first to CO, and then to CO_2. The cycle CO – OH – CP_4 plays an important role in the cycle of methane:

$$OH + CH_4 \rightarrow CH_3 + H_2O$$

$$OH + CO \rightarrow CO_2 + H$$

The participating OH-radicals form in the atmosphere during water vapor photolysis. As a result, the simplest diagram of methane oxidation in the atmosphere is the following:

$$CH_4 \overset{\swarrow}{\longrightarrow} \begin{matrix} OH \\ CO \rightarrow CO_2 \end{matrix}$$

The human interference into the processes described by this diagram breaks the natural stability of the balance $CH_4/CO/CO_2$. In particular, the reclaiming of marshes is one such destabilizing factor. For instance, the drainage of 20% of the global marshes leads to a natural reduction of CH_4 emissions from the marshes by

20%, giving on the whole a reduction in the amount of methane by 4%, which practically does not influence climate, but causes changes in the biogeochemical cycles of ozone and carbon dioxide with as yet unpredicted consequences. These estimates are important for the final conclusion about the level of the integral greenhouse effect. However, the solution of this problem is connected with many factors. For instance, the CH_4 flux on the atmosphere–marsh border depends on the vertical profile of the temperature in the marsh body. In the simplest case, if we denote $T_W(z, t)$ as the temperature at a time moment t at a depth z and write the equation of heat conductivity:

$$\frac{\partial T_W(z,t)}{\partial t} = a^2 \frac{\partial^2 T_W(z,t)}{\partial z^2} \tag{6.17}$$

where $a^2 = Kc^{-1}\rho^{-1}$, K is the coefficient of heat conductivity, C is specific heat capacity, and ρ is the medium density. An estimation of the flux F^1_{CH4} as a time function becomes dependent on the multitude of the poorly assessed characteristics of the environment.

Let the marsh surface temperature vary cyclically with frequency ω and amplitude A, decreasing with depth: $T_W(0, t) = A(z)\cos(\omega t)$, where:

$$A(z) = A(0)\exp\left(-\sqrt{\frac{\omega}{2a}}z\right)$$

The solution of Equation (6.17) enables one to trace the temperature variations $T_W(z, t)$ and suggests the conclusion that in this case these variations weakly depend on $T_W(0, t)$. Even if $T_W(0, t)$ increases by 2°C, then according to Equation (6.17), the amplitude of temperature changes with depth will rapidly decrease to 0.97°C, 0.33°C, and 0.01°C at depths 40 cm, 2 m, and 3 m, respectively. Hence, with a 2°C increase in the average global atmospheric temperature the flux F^1_{CH4} will increase by not more than 1.4%.

Comparing the global significance of the CO_2 and CH_4 cycles in the atmosphere–marshes system, note that the CO_2 cycle promotes a climatic stabilization, whereas the CH_4 cycle intensifies climate changes. With a climate warming the marshes assimilate part of the CO_2 from the atmosphere and reduce thereby the greenhouse effect. On the contrary, when the climate warms due to increasing F^1_{CH4}, the greenhouse effect intensifies.

The western Siberian region of Russia is characterized by numerous intensive natural and anthropogenic sources of methane formation. These are marshes, tundra, permafrost, oil and gas deposits. In this region the flux F^1_{CH4} varies widely over both yearly and shorter time periods. From measurements carried out by Jagovkina *et al.* (2000) at the coastline of Yamal in June 1996, the CH_4 concentration in the atmosphere at a height of 2 m varied from 1.83 ppmv on 18 June to 1.98 ppmv on 23 June, with an average daily value of 0.032 ppmv.

The peatbogs of Siberia are quite special in the global cycle of methane. They play a unique role in the biogeochemical cycles of methane and carbon dioxide. On the one hand they are a non-anthropogenic source of CH_4 and CO_2, but on the other hand they are intensive assimilators of atmospheric carbon, removing it from the natural cycle for long time periods. The marshes of western Siberia, for instance, contain 20–30% of the global carbon supplies. The intensity of CH_4 emissions from the marshes are, on average, almost 2000 times weaker than that of CO_2. 35–50% of all methane emitted from the territory of Russia fall on the marshes. The western Siberian marshes emit to the atmosphere no more than 1.7×10^6 t CH_4/year, which represents less than 1% of the global CH_4 flux. The spatial heterogeneity of the flux F^1_{CH4} is high, determined by the diverse characteristics of the marsh ecosystems. In particular, the upper oligotrophic coniferous–shrubby sphagnous swamps emit 0.9–10 mg C/m^2/hr (Dementjeva, 2000). This estimate is rather approximate, since the scattering of such estimates by various authors constitutes hundreds of percent. For instance, a drained sphagnous swamp of the transitional type can emit 142–204 g C/m^2/hr, and the rush sphagnous bogs 83.5–309 mg C/m^2/hr.

The main mechanism for the formation of methane in a marsh is connected with the functioning of special groups of micro-organisms. Part of the methane is emitted to the atmosphere due to diffusion, but most of the methane remains in the peat layer and is gradually emitted to the atmosphere.

Through interfering with nature, humankind has altered the natural biogeochemical balance of greenhouse gases practically all over the world. One aspect of this has been a reduction of the areas of marshes and their transformation into agricultural fields. Diverse human agricultural activity adds 20% of all the anthropogenic fluxes of GHGs to the atmosphere. In the USA it is 30%. Stockbreeding contributes considerably to this flux. In California and Wisconsin each hectare of pasture annually emits 502 kg CH_4 (or 10,511 kg CO_2) and 134 kg CH_4 (or 2,814 kg CO_2), respectively. In New Zealand such emissions of CH_4 are estimated at 291 kg CH_4 (or 6,110 kg CO_2) (Johnson and Ulyatt, 2000).

Among the Kyoto Protocol signatories, England takes ninth place when considering the volumes of reduced emissions of GHGs. The decreasing trend of methane emissions is part of the general reduction of emissions of six GHGs (CO_2, CH_4, N_2O, hydrofluorocarbons, perfluorocarbons, and sulphur hexafluoride) from 1990. In 2000 the greenhouse emissions decreased by 15% compared to 1990. By 2010, the CH_4 emissions will constitute 20,134 t/year. This reduction will be reached mainly due to new technologies in waste processing and in the coal industry. On the whole, in England, according to the developed scenario, emissions of methane by 2010 will decrease by 14% in agriculture, by 82% in the coal industry, by 29% in the oil and gas industry, and by 73% in waste processing. The possibility of realization of this scenario is confirmed by the decreasing trend in CH_4 in 1998 compared to 1990. For instance, during this period, emissions of methane in the coal industry decreased by 64%, and through waste processing by 29%. In 1990, the share of waste processing in England constituted 32% of all CH_4 emissions, only 3% of these emissions being connected with sewage processing.

In agriculture, emissions of CH_4 in England constituted 1037×10^3 t in 1990 and 998×10^3 t in 1998. The scenario of a reduction in the flux F^1_{CH4} from England due to improved technologies in agriculture foresees emissions of $902–983 \times 10^3$ t CH_4 in 2010. The CH_4 emission from burning agricultural waste is completely excluded, and in stockbreeding, emissions of methane are reduced by 8% compared to 1990.

In the coal industry in England, emissions of methane in 1990 constituted 819×10^3 t with the main contribution to this flux made by underground operations. This constituted 24% of the whole flux of methane from England. In 1998 the flux F^1_{CH4} decreased to 264×10^3 t and by 2010 it should decrease to 218×10^3 t. A similar decreasing trend of methane emissions from England is occurring in the oil and gas industry, too. According to the scenario, the contribution of these sectors of energy production into the flux F^1_{CH4} will decrease from 540×10^3 t in 1990 to 349–464 $\times 10^3$ t in 2010 (Meadows, 2000).

According to Bazhin (2000), the flux F^1_{CH4} in every water basin with a vertical stratiform structure forms in an active layer beneath the water layer. Practically, all aquageosystems have such a structure. The layer, where methane forms, has two areas. In the bottom area located at a depth h, methane takes the form of bubbles. Above this layer, due to diffusion, the concentration of methane decreases, and the bubbles disappear. Denote as $D_{CH4}(z)$ the coefficient of methane diffusion at a depth z, then the stationary behaviour of the whole system shown in Figure 6.15 is described by the equation:

$$\frac{d}{dz}\left[D_{CH4}(z)\frac{d}{dz}M(z)\right] - F_{CH4}(z) + F^1_{CH4} + F^2_{CH4} = 0$$

Model calculations and field measurements performed by Bazhin (2000), show, for instance, that on the rice fields $h_b = 1.3$ m, $F_{CH4} = (1.3\text{–}1.7) \times 10^{-12}$ mol/cm^3/s. According to Khalil *et al.* (2000), the rice fields play a significant role in the gas balance of the atmosphere due to emissions of CH_4, CO, N_2O, H_2, and $CHCl_3$. For instance, in China the rice fields deliver these gases to the atmosphere at the following rates (mg/m^2/hr): CH_4 – 900–50,000; CO – 80–100; H_2 – 5–30; N_2O – 50–1,000; and CHCl – 1–8. A wide scatter in these estimates is explained by the highly unstable nature of the fluxes of these gases due to rice growing technology. For instance, the use of sulphates on the rice fields increases emissions of methane by 12.0–58.9% depending on other characteristics of the fields (Liping *et al.*, 2000).

Thus an estimation of the flux F^1_{CH4} as a function of a given territory with an account of natural and anthropogenic processes taking place there, requires first of all a detailed inventory of methane sources as well as natural and technogenic systems functioning on this territory. Examples of such an inventory given above serve as the basis for development of studies in this area.

The dynamics of the CH_4 content H_A in the atmosphere can be parameterized by a simple balance relationship:

$$\frac{\partial H_A}{\partial t} + V_\varphi \frac{\partial H_A}{\partial \varphi} + V_\lambda \frac{\partial H_A}{\partial \lambda} = F^1_{CH4}(t, \varphi, \lambda, \Xi) - \frac{H_A(t, \varphi, \lambda)}{\tau_H}$$

where Ξ is the identifier of the type of the natural or technogenic system.

7

Modelling the global changes of the environment

7.1 INTRODUCTION

Many scientists have tried to model the global carbon cycle (e.g., Nefedova, 1994; Nefedova and Tarko, 1993; Alexandrov and Oikawa, 2002; Kondratyev *et al.*, 2002a,b; Krapivin and Vilkova, 1990; Nitu *et al.*, 2000a,b; Svirezhev, 2002). Most of the models are based on the data for the pre-industrial period and predict the CO_2 concentration for this century. The number of factors taken into account in the models is constantly growing, with the models becoming more and more adequate. One of the first sufficiently complete models of the global CO_2 cycle was proposed by Bjorkstrom (1979), which took into account the dynamic interaction between the carbon reservoirs in the biosphere and its fluxes within them. For the first time, the unit of the World Ocean has been realistically represented, where the ocean is considered as a multilayer composition of the uniformly mixed reservoirs, with the CO_2 exchange between them being described by linear laws. The models of unit type have been analysed by many experts, who have promoted successful division of the Earth's biogeosystem into heterogeneous compartments and derivation of respective box models (Pervaniuk, 2001). Using the ability of carbon to reside in the atmosphere for a long time, making it possible to describe the atmosphere with a point model, Bacastow (1981) derived a global 4-reservoir model which approximates the dynamics of the CO_2 exchange between the northern and southern hemispheres.

Each of the available models of the global carbon cycle differs by a set of assumptions and therefore is orientated towards selection of some effect. Models allowing a mutual parametric correlation can be used in a more complicated model. For instance, a simple numerical model of gas exchange on the ocean–atmosphere border in the case of wind-driven roughness at wind speeds up to 7 m/s enables one to include within the global model a unit to calculate an established CO_2 flux between the water surface and the atmosphere. The models of the ocean carbonate system described in the literature have the same type. There are also other models of the CO_2 cycle in natural systems (Riedo *et al.*, 2000; Zonneveld, 1998).

All known models of the CO_2 cycle should be modified by detailing the spatial distribution of the soil–plant formations and by specifying the exchange processes both in the ocean and on the atmosphere–ocean border. Also, most accurate parameterizations of all known elements of the biogeochemical cycles of carbon should be combined into a single system. Such an attempt has been made in the block-scheme of the model described in the work by Krapivin and Kondratyev (2002). The basic sources of CO_2 are the functioning of land and marine animals, photochemical reactions, decomposition of dead organic matter, as well as anthropogenic activity. The period of CO_2 migration into the upper atmospheric layers is substantially less than the time step of the model. It is here that problems appear in the adequacy of the obtained estimates of the role of the land and water ecosystems in the assimilation and burying of excess amounts of carbon dioxide in the atmosphere.

A thorough analysis of numerous publications on the role of increasing CO_2 concentrations in the atmosphere leading to an enhanced greenhouse effect and to possible climate warming in the near future shows that predicted respective cataclysms in the environment are only hypotheses. These conclusions are made on the basis of rough models of the global biogeochemical CO_2 cycle that do not take into account the spatial mosaic of most of the effects and feedbacks considered above. Solving such problems is impossible, however, because despite numerous studies of the greenhouse effect, at present there are no complete global databases for greenhouse gases (GHGs) which reflect the seasonal and spatial structure of their fluxes. Besides, even the fluxes enumerated in Table 2.13 have not been modelled and, hence, the global models of CO_2 are only of "educational" character. For example, the results obtained by Alexeev *et al.* (1992) from studies of the CO_2 exchange in the coastal waters of the Caspian Sea show that even for such a small space, the mosaic of CO_2 fluxes is characterized by variations in the Δp parameter within -10 to 234 ppm and of the fluxes H_2 and H_3 from 1.2 to 32.3 mmol/m^2/day. This confirms the necessity for a thorough analysis of local fields p_m and p_a to derive accurate models of gas exchange in the ocean–atmosphere system. Similar conclusions suggest themselves for the soil–plant–atmosphere system. Here it is necessary to consider the effect of a maximum number of types of plants and soils. The available models consider not more than 30 types, whereas there is an atlas of 172 types of plants in this area (Eliasson *et al.*, 1999).

Inaccuracies of the global models of the greenhouse effect due to CO_2 grow, considering the lack of regular reliable data on the biospheric cycle of other GHG, including ozone and water vapour. This is confirmed by a large scatter of the predicted estimates of climate warming by 2075 compared to the pre-industrial period: $1.1°C \leq \Delta T_{\Sigma} \leq 4.2°C$.

The recent detailed analysis of the causes and consequences of the greenhouse effect has been made at the 4th International Conference on greenhouse gases control held from 30 August to 2 September, 1998 in Sweden (Eliasson, 1999). The papers presented at the Conference questioned the reliability of numerous predictions of the possible greenhouse effect dynamics. The authors discussed the biological, energetic, chemical, geological, and social aspects of the greenhouse effect. But the problem of a complex consideration of all these aspects, with an account of additional feedbacks reflecting the correlations between them, remains neglected. A solution to this problem has been suggested by Krapivin and Kondratyev (2002). They proposed to combine in the global model the functioning of several spatially heterogeneous natural–anthropogenic systems and processes. This global model will make it possible, in an adaptive regime, to search for an efficient mechanism to monitor the greenhouse effect and to reliably assess the role of the land and ocean biocoenoses.

The prediction of emissions of anthropogenic CO_2 and other GHGs into the atmosphere is only possible through thorough and regular monitoring of the spatial structure of a huge number of parameters of the nature–society system (NSS). An adaptive mechanism for global model use can be applied in this case (Kondratyev *et al.*, 2000). Only an adaptive–evolutionary approach to the solution of this problem can reduce the uncertainty in the estimates of the NSS parameters and guarantee the harmonic development of the human civilization. A thorough analysis of the publications of numerous experts raises hopes that within the Global Carbon Project (GCP) a constructive step forward will be made in the synthesis of a global model which will adequately reflect the whole spectrum of direct correlations and feedbacks in the environment and will take into account all the developed models of various processes (Yang *et al.*, 2002; Porte and Bartelink, 2002; Yemshanov and Perera, 2002; Sternberg and De Angells, 2002; Peng *et al.*, 2002; Qi *et al.*, 2002; Hadley and Toumi, 2002; Huesemann *et al.*, 2002).

7.2 THE GLOBAL MODEL OF THE NSS

The present stage of scientific and technical progress, with the attendant realization of the global ecological situation that is characterized by insufficiency of energy, geological, biogeocenotic, and other resources, leads us on to the problem of information about global ecology: knowledge of conditions under which humankind and the natural world have co-evolved. For thousands of years this information resource had been weakly correlated to total human activity, which until the beginning of the industrial era had been relatively low. Then, with a historically rapid approach to the situation, when the commercial attitude to the biosphere became the governing strategy of humankind and when an ecological deadlock was realized, the significance of the information resource reached its maximum.

This resource includes not only "academic" knowledge. Every ecological problem is "open", it is part of the system of the global problems of today, the main problem being a necessity to preserve the homeostasis of humankind (Kondratyev, 1998b). It means that the "danger for the biosphere" realized at the end of the 20th century has brought forth the problem of survival for the *Homo sapiens* species, and hence, the problem of attaining a responsible attitude to nature. Both ecological and moral problems have clashed. A rhetorical question – why do we live and work? – has acquired a deep social aspect and is connected with the harmonious relationship between society and nature. The fate of the biosphere will depend on how rapidly humankind solves these problems and finds an optimal balance between the "wise" and the "unwise" attitude to the environment. As the model estimates have shown, 90% of humankind should take part in the solution of these problems. But at present, it is unlikely that many people will deliberately change their position and develop new harmonious relationships with nature. To reach global harmony, it is necessary to concentrate on negative ecological and socio-economic changes, so that our ecological knowledge be put into practice in the form of concrete technologies ensuring that high quality decisions are made in the field of nature-protection activities.

At the present stage of scientific and technical progress, intensive developments have been realized in this field, an analysis of which reveals the characteristic features of ecological knowledge as well as the problems of using applied methods in order to formulate the basic requirements of an efficient information technology. One of the necessary prerequisites for creating the systems to monitor the environmental condition was the availability of data of different quality and, hence, a multitude of numerical models of various types (balance, optimizing, evolutionary, statistical, etc.). Derived on the basis of parameterization and, as a rule, linearization of the laws of natural phenomena, these models include a wide spectrum of determinate and probabilistic descriptions of the geological, ecological, oceanographic, biogeochemical, and biogeocenotic processes of the global, regional, and local nature. Most of them are directed towards a theoretical understanding of the specific features of higher living systems on the basis of available knowledge, and only a small part is aimed at the first steps to an objective assessment of the present global ecological situation. Differing in the goals and mathematical apparatus of description, many models have to be rough due to limitations, inadequacies, and uncertainties in the information base, as well as because of a lack in present-day instrumental systems in the field of numerical experiments. An increase of the number of the biospheric parameters and their relationships (Figure 7.1) in order to raise the adequacy of the models makes them multiparametric (i.e., the problem appears of the "multivariance curse").

As a basic instrument for solving these problems, the method of simulation modelling has been considered, which makes it possible to adjust different data that refer to various mathematical formalisms, and to remove the problem of an excess number of parameters. The required model is constructed on the basis of empirical information not restricted a priori by any mathematical apparatus, which ensures a flexible formalization, unavoidable when the important laws of the phenomena are unknown. In some cases, genetic algorithms are used to model undetermined systems with stochastic elements having no unique solution. One such algorithm is the genetic algorithm for the prediction of species distributions (Anderson *et al.*, 2003). Also, there is a multitude of non-traditional algorithms based on the technology of evolutionary modelling (Bukatova *et al.*, 1991) and self-organizing numerical models (Timoshevskii *et al.*, 2003).

The development of simulation modelling by broadening the information base, combining the formal and informal methods in the stage-by-stage synthesis of the

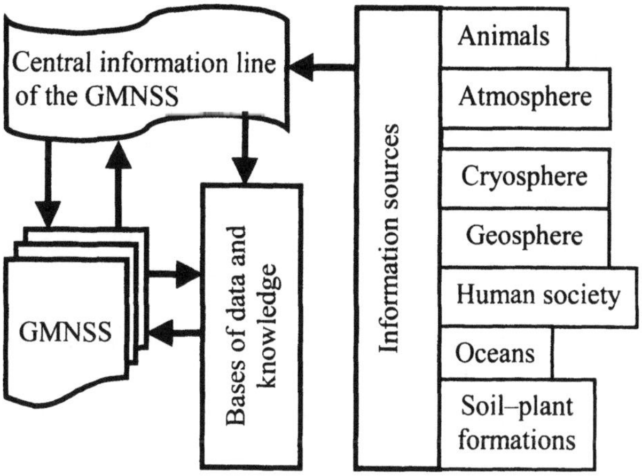

Figure 7.1. Organization of the global model of the nature–society system (GMNSS) functioning.

required model, and finally by producing a means of active man–computer dialogue, will provide an efficient approach to system–ecological modelling. However, it appears now that it is not all that easy. If we compare the information queries in the field of ecological problems and the existing information provision for their solution (different numerical and simulation models, principles of the ecological information processing), we see that not all levels of natural and anthropogenic complexes have a well-developed apparatus for their description as well as for construction of efficient information approaches to obtaining the required estimates of the problematic situations. Difficulties appearing in this regard are not only of technical character. These features manifest themselves especially in global modelling, which reveals a substantial and principally ineradicable inadequacy of knowledge about natural processes, which is shown both by the fragmentary character of empirical data and by the lack of knowledge about the laws that dictate the natural processes of evolution. Now, it is clear that a mechanical set of hierarchies of models and a desire for accumulation of the empirical databases are not adequate. This is just an attempt to reanimate the old school of reasoning that searched for an all-inclusive pattern of development of biospheric processes without hope of success, without the possibility of explaining the permanent self-organizing ability of living systems, and without a substantially improved understanding of the mechanism of the NSS functioning. The situation is such that it is necessary to use computer technologies that combine the methods of both the evolutionary and simulation modelling, as it has been described in Bukatova *et al.* (1991). This makes it possible to take into account the internal dynamics (evolution) of the modelled processes and to synthesize models in conditions of inadequate and partially reliable data.

Traditional approaches to the global model construction face difficulties with the algorithmic description of many socio-economic and natural processes, generating an information uncertainty. The developed approaches to global modelling neglect this uncertainty, and as a result, the models do not adequately consider real processes. A combined use of the evolutionary and simulation modelling makes it possible to remove this problem by synthesizing a combined model, the structure of which is adapted on the basis of pre-history of the biospheric and climatic components. The model can be realized using various classes of models, with the use of traditional computer programmes and special processors of the evolutionary type. The form of such a combination can vary and it depends on the spatial and temporal adequacy of the global databases.

An accumulated experience in global modelling contains a lot of examples of insurmountable difficulties when attempting to find a means of description of scientific and technical progress and of human activity in its various manifestations. Also, difficulties appear in modelling the climate characterized by a superposition of processes with different temporal scales. As for an adequate description in the global model, it is impossible here to clearly define the limits of the information provision and the limits of the required spatial and structural resolution. Therefore, without dwelling upon philosophical analysis of the global problems and without trying to recommend a comprehensive approach to global modelling, we shall discuss only one of the possibilities, showing how the evolutionary modelling in special processor realization enables one to overcome the difficulties mentioned above.

An adjustment of the evolutionary model by the pre-history of natural rhythms makes it possible to obtain a model, which implicitly follows various regularities of the NSS dynamics in the past, and makes it possible to make forecasts in the same temporal rhythm. The special processor version of the model overcomes all difficulties of algorithmic and digital character caused by the multidimensionality of the global model and multitude of parametric uncertainties.

Clearly, it is impossible to completely overcome contradictions between the experts' requirement and NSS models because of insufficient information about the laws of the functioning of the environmental elements and human society. Therefore, a huge number of types of models appeared, each responding to a certain level of required adequacy. Small-sized models mainly based on empirical dependencies are most widespread. Among the successfully used approaches towards overcoming the information uncertainties are neural networks, which, similar to evolutionary modelling, enables one to synthesize the models in conditions when relationships between the characteristics of the environmental systems are poorly known (Aitkenhead *et al.*, 2003). In other words, neural networks permit one to establish the cause-and-effect bonds without equations. Of course, the loss of the accuracy of modelling the process under study with the help of neural networks takes place not because of inadequacy of the parametric description, but due to other input suppositions on the characteristic functions of neural networks (the weights of synoptic connection, node activities, etc.) For instance, Aitkenhead *et al.* (2003) used the activation function y in the form of the curve with saturation: $y = 1/[1 + \exp(-cx)]$, where x is the input variable. The approach based on neural networks, for a small headwater stream Brocky Burn (27 km in length) in north-east Scotland, enabled one to obtain more than 50% of the predictions with an accuracy of up to 10% and above 90% of the predictions within the interval of accuracy up to 30%. This local example confirms an efficiency of neural-based models for synthesis of some global simulation model (GSM) units with a high degree of functional uncertainty, but having a series of observations of the dynamics of their characteristics.

The synthesis of the GMNSS is based on its consideration as a self-organizing and self-structuring system, in which the elements are coordinated in time and space by the process of natural evolution. The anthropogenic constituent in this process breaks this integrity. Attempts to parameterize, on a formal level, the process of co-evolution of nature and humans, as elements of the biosphere, are connected with the search for a single description of all processes in the NSS, which would combine all spheres of knowledge in perceiving the laws of the environment. Such a synergetic approach forms the basis of numerous studies in the field of global modelling (Krapivin *et al.*, 1982; Krapivin and Kondratyev, 2002; Kondratyev *et al.*, 2003a).

Synthesis of the global model requires a choice of spatial grids for climatic, biospheric, and ahthropogenic processes (see Section 1.2.3). Usually they take a uniform geographic grid with steps $\Delta\varphi$ by latitude and $\Delta\lambda$ by longitude. The convenience of this choice is connected with a simplified process of calculations but leads to the loss of data, excessiveness, and associated difficulties in correlation of accumulated knowledge with an available global database. One of the first attempts to reflect the NSS spatial heterogeneity in the model was made in Krapivin (1978), where the space of the NSS was divided into the atmosphere, the World Ocean, and two land regions. It is assumed that the NSS input is the solar radiation energy and irreversible mineral resources. The solar energy is used by photosynthesizing elements of the biosphere (phytoplankton and plants). The flux of this energy is regulated by atmospheric turbidity formed by natural and anthropogenic processes. Atmospheric transparency is determined by the amount of dust particles residing in the atmosphere as a result of dust storms, volcanic eruptions, burning of solid and liquid fuels, emissions from stacks of metallurgic and chemical enterprises, cement plants, and other products as well as due to cloud formation.

Calculations using the first version of the GMNSS have shown that there is a point of balance of the interacting forces between two regions of land. This result has necessitated further digitizing of land territory, which was realized in Krapivin *et al.* (1982). It was found that an assumed scheme of the man–environment interaction (mutual influence of ecological and production processes) is a classical example of the Germeyer's system (Germeyer, 1974). Really, breaking the biota balance or an irreversible climate change will lead humankind from homeostasis, whose condition can be formulated in terms of indicators of the environmental quality. Hence, whatever the contingent of the subjects of the NSS under consideration may be, among the criteria for subjective decisions there is always a criterion general for all subjects (regions, states, or other groups). One such criteria is the survival of humankind.

Since at present the Vernadsky thesis is valid: "none of the living species can exist in the medium consisting of its wastes", we shall not discuss the possibilities of creating a completely artificial habitat for the global population. Nevertheless, it is necessary to postulate the desire to preserve the stability of the environment, homeostasis general for all the NSS subjects, describing the combined functioning of nature and humans. But for the Germeyer's system, a consideration of the problem of the search for persistent collective decisions belonging to the Pareto set is justified. Unfortunately, the latest international forums on the problem of sustainable development do not follow this way. A vivid negative example of this is the Kyoto Protocol discussed above, though the problem of the greenhouse effect due to carbon dioxide becomes more and more urgent every year.

One of the causes of the observed gap between suggestions of scientists and decisions of politicians consists in the absence of an international programme which would combine them and lead to an assumption on some scientifically grounded criteria for estimation of the developing trends of the NSS and its elements. It is for this reason that the global database goes on accumulating without a necessary methodical provision (Field, 2001; 1998; Forest *et al.*, 2002; Lovett, 2002; McCarl and Schneider, 2002; Milford *et al.*, 2002;).

The International Centre for Global Geoinformation Monitoring (ICGGM) described in Section 1.4 (see Figure 1.2), can be an efficient mechanism to solve the problems of the search for stable conditions in the NSS. It is in this way that a single planetary adaptive evolutionary geoecoinformation service can be created, which would have a hierarchical structure of data accumulation and form the multi-level global database (Ivanov-Rostovtsev *et al.*, 2001; Kondratyev, 1999c; Kondratyev *et al.*, 2000; Marchuk and Kondratyev, 1992; Gorshkov *et al.*, 2000).

As has been mentioned above, the existing difficulties of the synthesis of the GMNSS consists mainly in a parameterization of energy fluxes on the atmosphere–land and atmosphere–ocean border. Due to the series of the last works by Sellers *et al.* (1995, 1996a–c, 1997) a possibility appeared of the combined use of the models of the climate and the biosphere. The models SiB1 and SiB2 are orientated towards a description of the energy and heat balance of the Earth's surface. In these models the processes of photosynthesis and evapotranspiration are parameterized with consideration given to plant physiology and the local conditions of vegetation cover formation (see Section 3.9). It is also important that these type of global models are orientated towards an estimation of their parameters within the regime of satellite monitoring. Beginning in 1987, with the use of satellites of the National Oceanic and Atmospheric Administration (NOAA) series and the Advanced Very High Resolution Radiometer (AVHRR), a global series of Normalized Difference Vegetation Index (NDVI) has been created over the geographical grid $1^\circ \times 1^\circ$ with monthly averaging. Also, databases are formed on the types of soils, types of vegetation, absorbed radiation, albedo, cloudiness, soil moisture, etc. But these databases have only an expert justification.

Finally, the Integrated Climate Prediction System (ICLIPS) model can be one possible unit of the global model simulating global socio-economic structure

(Leimbach and Toth, 2003). This model divides the economic section of the NSS into 11 regions:

- Sub-Saharan Africa.
- China, Mongolia, Vietnam, Cambodia, Laos.
- Eastern Europe.
- Former Soviet Union.
- Latin America and the Caribbean.
- Middle East and North Africa.
- North America.
- Pacific OECD (Japan, Australia, New Zealand).
- South Asia (mainly India).
- Western Europe.

7.3 THE ROLE OF VEGETATION IN CLIMATE CHANGE

The land biota is one of the sinks for atmospheric CO_2. Unfortunately, the available data and knowledge of the processes of plant respiration suggest only rough integrated estimates of the CO_2 flux in the vegetation cover. Actually, the role of plants in assimilation of atmospheric CO_2 changes sharply over 24 hours and is a complex function of such factors of the environment as temperature, illumination, and air humidity. Nevertheless, attempts to parameterize the functions of vegetation made in Chapter 3 make it possible to assess the role of all the types of soil–plant formations in CO_2 assimilation (mentioned in Table 2.6).

Figure 7.2 shows the role of the forest vegetation in CO_2 dynamics. In addition to these results, note that experiments with global models make it possible to trace the dependence of the atmospheric gas composition on the structure of the planetary forest cover. From the available estimates, the total area of the forests for $t_0 = 1970$ can be estimated at $\sigma_{L0} = \sigma_L(t_0) \approx 40.3 - 41.84 \times 10^6\,\text{km}^2$ (Watson *et al.*, 2000), with 1% constituting national parks and forest reserves. With the formulated scenario, assume $t_L = 2050$, $\sigma_{X0} = \sigma_X(t_0) = 19.5 \times 10^6\,\text{km}^2$. As seen from Figure 7.2, the increasing rate of deforestation raises considerably the concentration of CO_2 in the atmosphere. Even with a 10% reduction of forest areas by 2050 compared to 1970 (i.e., $\sigma_{L1}/\sigma_{L0} = 0.9$), the atmospheric CO_2 could increase by 44% by the end of the 21st century. On the contrary, a 10% increase in forested areas decreases the concentration of atmospheric CO_2 by 15%. With a 50% increase of the forested areas by 2050, a decrease of atmospheric CO_2 by 2100 will constitute 60% relative to its possible value, with the scale of impacts on the forest ecosystems, observed at the end of the 20th century, preserved. Hence, variations in the forested areas in the biosphere even within $\pm 10\%$ can substantially change the dynamics of numerous components of the global ecosystem.

Table 7.1 exemplifies the calculation of the CO_2 sinks into the vegetation cover in Russia. Such calculations with the use of the GMNSS demonstrate the dynamics of the CO_2 fluxes mosaic in the atmosphere–plant–soil system. Knowledge of this

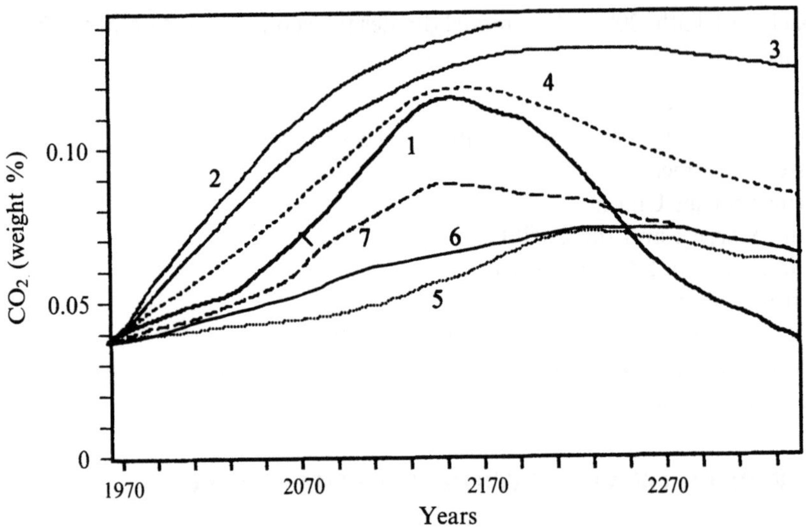

Figure 7.2. The dynamics of the CO_2 concentration for different scenarios of changing forest areas within the framework of the scenario after Demirchian *et al.* (2002): 1 – rate of change in forest areas remains the same as that in 1970; 2 – by 2050 forests have totally disappeared; 3 – by 2050 the area of forest is reduced by 50%; 4 – by 10%; 5 – by 2050 forest areas increase by 50%; 6 – by 30%; 7 – by 10%.

Table 7.1. The dynamics of CO_2 assimilation by plants in the territory of Russia. Emissions of carbon in this territory in 1990 is assumed to be 1.6 Gt C/yr with an annual change after the scenario by Keeling and Bacastow (1977).
From Krapivin and Vilkova (1990).

Soil–plant formation (Table 2.5)	Rate of CO_2 assimilation (10^6 t C/year)				
	Years				
	1990	2000	2050	2100	2150
A	2.6	2.8	6.7	7.1	6.9
C	3.7	4.6	10.9	12.9	12.1
M	4.0	5.1	12.4	14.5	13.8
L	3.2	3.9	9.2	10.3	10.4
F	11.2	14.8	43.6	47.2	44.2
D	31.6	39.9	110.6	121.9	109.3
G	23.3	29.2	72.2	73.4	70.5
R	5.2	6.2	13.1	13.8	10.7
W	4.7	5.1	8.2	8.8	7.9
V	0.7	0.7	0.9	1.1	0.8
@	2.4	2.6	3.7	3.9	2.9
S	0.6	0.7	1.2	1.4	1.0
Q	1.5	1.6	2.2	2.3	1.8
Total	*94.7*	*206.1*	*294.9*	*317.7*	*292.3*

Table 7.2. The dynamics of the ratio of integral rates (H_6^C) of CO_2 assimilation by vegetation cover from the atmosphere with the natural distribution of soil–plant formations (see Table 2.6 and Figure 3.4) and with its transformation according to the scenario in the second column.

Scenario		H_6^C (changed Figure 3.4)/H_6^C (Figure 3.4)		
		Years		
Figure 3.4	Changes in Figure 3.4	2003	2020	2050
A	L	2.70	2.14	1.97
C	L	0.96	0.94	0.95
M	L	1.67	1.15	1.01
F	D	1.68	1.57	1.11
G	D	2.11	1.67	1.45
R	D	3.98	3.70	2.87
P	Z	3.17	2.68	2.43
U	Z	22.52	20.73	17.95
W	Z	23.12	19.44	16.32
E	Z	100.14	77.75	68.54
H	Z	194.56	155.50	138.39
Q	Z	799.14	777.50	751.26
Y	Z	1.43	1.39	1.23
N	Z	69.98	62.20	56.59
J	Z	5.89	5.09	4.67
T	Z	26.54	25.92	23.58
I	Z	17.88	16.37	14.91
#	Z	0.91	1.12	0.97

mosaic makes it possible to assess the role of concrete types of soil–plant formations in the regional balance of carbon, and on this basis, to calculate the global fluxes of carbon dioxide on the atmosphere–land border. Similar calculations are also possible for the atmosphere–ocean system.

Table 7.2 demonstrates the consequences of changing the global structure of the soil–plant formations for the dynamics of CO_2 assimilation by vegetation. As is seen, the anthropogenic change in the vegetation cover substantially changes the balance components in the global carbon cycle. Clearly, such experiments require a thorough analysis of data regarding the possibility of transformation of vegetation cover with an account of climatic zones and biocenological consistency. Nevertheless, such hypothetical experiments are useful for general assessments of possible ranges of anthropogenic impacts on the global carbon cycle. For instance, natural and anthropogenic cataclysms connected with forest fires introduce annually considerable changes into this cycle, because they change numerous fluxes and supplies of carbon over large territories. Tables 7.3–7.5 give estimates of deviations in the content of carbon in the basic biospheric reservoirs with forests fires in different

Table 7.3. Model estimates of the deviation of the carbon content provided all coniferous forests of the northern hemisphere (up to 42°N) are burnt.

Years after impact	Deviations in carbon content (Gt)			
	Atmosphere	Soil	Upper ocean	Deep ocean
0	140.9	−5.4	15.5	0.1
10	104.8	−33.1	29.8	3.1
20	83.1	−44.1	21.5	7.4
30	63.4	−43.5	19.0	8.5
40	47.2	−39.7	14.6	10.4
50	34.2	−34.6	11.6	11.7
60	24.1	−29.4	8.3	12.7
70	16.3	−24.7	6.2	13.5
80	10.2	−20.5	4.5	14.0
90	5.6	−17.1	3.3	14.3
100	2.1	−14.2	2.3	14.5
200	−9.0	−3.4	−0.8	13.5

Table 7.4. Model estimates of the deviation of the carbon content provided all forests of the northern hemisphere (up to 42°N) are burnt.

Years after impact	Deviations in carbon content (Gt)			
	Atmosphere	Soil	Upper ocean	Deep ocean
0	230.8	−7.9	24.9	0.1
10	173.2	−30.9	47.9	4.9
20	138.9	−67.6	39.2	10.0
30	107.9	−89.3	31.1	13.8
40	82.0	−64.3	24.1	16.8
50	60.9	−56.9	18.4	19.1
60	44.2	−49.0	13.9	20.7
70	31.1	−41.6	10.3	21.9
80	20.9	−35.0	7.5	22.8
90	12.9	−29.4	5.4	23.3
100	6.9	−24.7	3.7	23.6
200	−12.7	−5.9	−1.7	21.7

zones. Large-scale impacts on land biota are dampened over 60–100 years after the event. The biosphere turns out to be more resistant to impacts on the forests of southern latitudes and more sensitive to violations of forest areas in temperate latitudes. Hence, the northern hemisphere forests up to 42°N play an important role in stabilizing the carbon cycle in the biosphere.

The scenario of liquidation of forests, as seen from the studies of numerous authors, evokes great interest in studies of the global carbon cycle and the associated

Table 7.5. Model estimates of the deviation of the carbon content provided all tropical forests are burnt.

Years after impact	Deviations in carbon content (Gt)			
	Atmosphere	Soil	Upper ocean	Deep ocean
0	396.0	−20.0	42.2	0.2
10	261.6	−93.7	73.9	8.0
20	162.2	−84.8	48.0	14.9
30	90.6	−38.6	28.2	19.1
40	45.3	−36.5	15.0	21.3
50	18.3	−21.6	7.5	22.4
60	2.9	−12.4	3.0	22.8
70	−5.8	−7.0	0.5	22.8
80	−10.6	−4.2	−0.9	22.6
90	−13.2	−2.6	−1.7	22.5
100	−14.5	−1.9	−2.1	21.8
200	−13.2	−2.3	−1.9	17.7

climate change. A diversity of the possible real situations of the transformation of land cover is so large that it is impossible to assess all the consequences. Note only that, for instance, a liquidation of all northern taiga and mid-taiga forests (types *F*, *D*) over 50 years will lead to a 53% increase of atmospheric CO_2 with subsequent negative consequences for the flux H_6^C. Similar consequences would follow after the liquidation of all wet evergreen and deciduous tropical forests (types *Z*, *Y*). But in this case the indicated increase of atmospheric CO_2 will be reached 20 years later.

The land cover structure changes not only due to human activity. In some regions of the globe, hurricanes introduce considerable changes in the carbon balance of the forest ecosystems. So, in the USA territory every 3 years two hurricanes arrive, which accelerate the transition of living biomass of trees into dead organic matter. If in the USA, 20 Tg C are removed from the forests every year, then 10–15% of this is the result of one hurricane (McNulty, 2002). Hence, the hurricanes accelerate the return of carbon to the atmosphere, and their global inventory is required to more accurately estimate numerous components of the global carbon cycle.

7.4 THE ROLE OF THE WORLD OCEAN IN CLIMATE CHANGE

An estimation of the total contribution of the World Ocean to the CO_2 assimilation from the atmosphere, as in the case of land, is only possible by spatial integration of the difference between the H_3^C and H_2^C fluxes. Table 7.6 gives average estimates of this difference. Even with these rough estimates, we see a mosaic character of the role of various basins of the World Ocean in the atmospheric CO_2 assimilation. Quite special are the water basins in the northern latitudes and in the zones of

Table 7.6. The spatial distribution of $\Delta H_{32} = H_3^C - H_2^C$ (Gt/km^2/year) estimated from averaged values of assimilation and emission of CO_2 at the atmosphere–ocean border during the whole industrial period.

Water basin	σ	ΔH_{32}
South Atlantic Ocean	28.80	0.00138
Equatorial Atlantic Ocean	32.38	−0.00285
North Atlantic Ocean	26.01	0.01025
South Pacific Ocean	49.85	0.00531
Equatorial Pacific Ocean	88.81	−0.00323
North Pacific Ocean	32.45	0.00846
South Indian Ocean	49.63	0.00538
Equatorial Indian Ocean	32.85	−0.00592
Arctic Ocean	7.47	0.00131
The World Ocean	*348.23*	*0.00154*

Note: σ = the area of the World Ocean Basin (10^6 km^2).

upwelling. The cold water basins in the southern latitudes remain practically unknown, where, like in the Arctic Ocean, large territories are covered with ice. From available estimates $\Delta H_{32} = 0.006$ Gt C/year for the ice-covered water basins and 0.022 Gt C/year for ice-free water bodies.

Some ideas about the role of the World Ocean in the CO_2 assimilation from the atmosphere can be obtained from the data in Figures 7.3 and 7.4. Figure 7.5 gives a spatial distribution of CO_2 assimilated from the atmosphere with an account of the combined role of land and ocean ecosystems. We can see that the general role of the environment in stabilizing the CO_2 content in the atmosphere with the considerably increasing share of carbon varies negligibly. It means that the natural medium

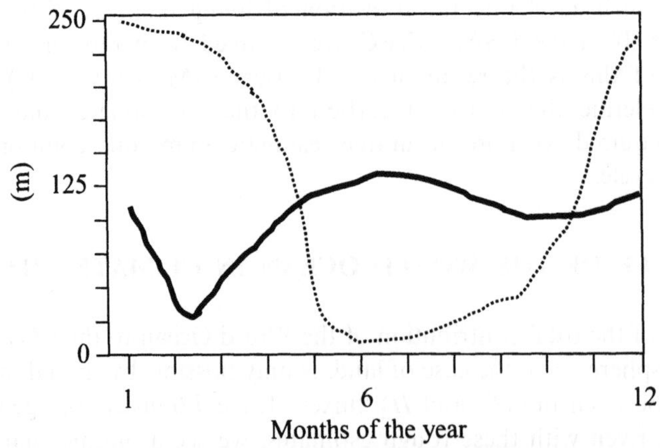

Figure 7.3. Distribution of the depth of the upper quasi-homogeneous layer of the World Ocean in the latitudinal zones 0°–10°N (solid curve) and 60°–70°N (dashed curve).

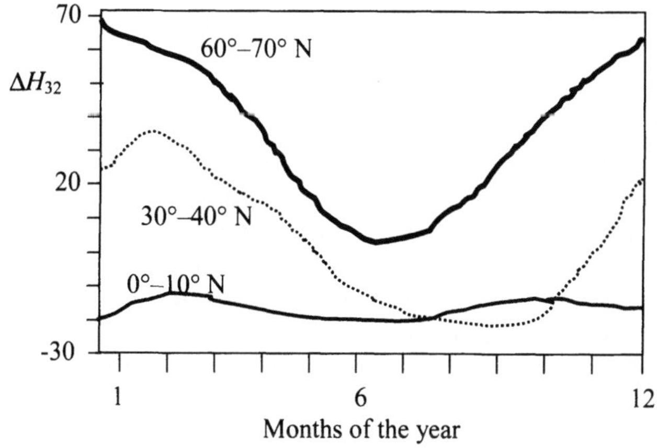

Figure 7.4. The annual distribution of carbon flux across the atmosphere–ocean interface for different latitude zones.

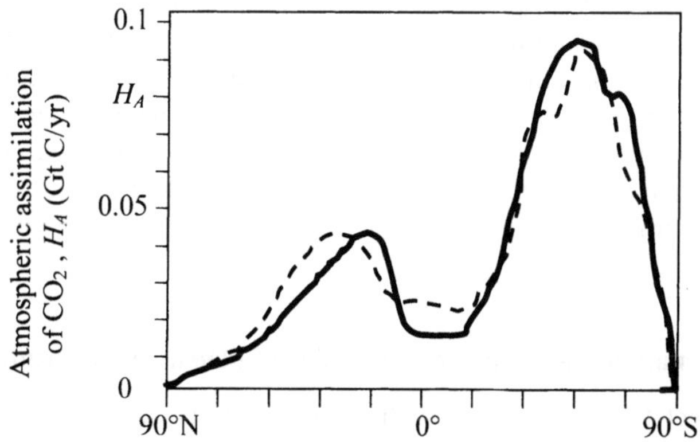

Figure 7.5. The longitude-averaged rates of atmospheric CO_2 assimilation by land and ocean ecosystems with two scenarios of anthropogenic emissions of carbon: 6.26 Gt C/yr (dashed curve, 2000) and 10.6 Gt C/yr (solid curve, predicted for 2020).
Note: $H_A = \Delta H_{32} + H_6^C + H_4^C - H_8^C - H_7^C - H_9^C$.

preserves persistent feedbacks, and the level of variability of reactions of the land and ocean systems to a marked increase of anthropogenic loads on the atmosphere remains stable. A detailed analysis of such reactions shows that almost a 60% increase of the anthropogenic carbon flux during next 20 years will result in a 4.8% increase of CO_2 flux from the tropical water basins of the World Ocean to the atmosphere, but incidentally, the absorptivity of the ecosystems in the northern latitudes will increase by 12.3%. The role of land ecosystems in assimilation of excess carbon from the atmosphere will increase by 11.4%, and the role of the Arctic waters by 13.2%. As a result, in 2020 the following amounts of carbon will be assimilated from the atmosphere: 25.7×10^6 t C by the south-taiga forests; 35.1×10^6 t C by the mid-taiga forests, and 10.7×10^6 t C by the tundra ecosystems.

7.5 DEVELOPMENT OF THE GLOBAL MODEL

As has been mentioned repeatedly, the attention to the problem of carbon dioxide and other minor gas components (MGCs) in the atmosphere was brought about by the concern of scientists, economists, and politicians in connection within the possible consequences of a climate change. Since the fossil fuel combustion and deforestation of the land surface are the most significant sources of anthropogenic increases in concentration of CO_2 and CH_4 in the atmosphere, these gases are being studied the most thoroughly. Both these gases absorb IR radiation from the Earth's surface and therefore influence the planetary climate. The problem of reducing this influence is solved by many politicians without consideration being given to existing knowledge within the environmental sciences. In fact, a real and objective solution for this problem is to thoroughly study the biogeochemical cycles of all gases, somehow related to carbon cycle. The GCP is concentrated on this approach aimed at a complex analysis of the available data on the carbon cycle in the environment and its relevant measurements. The project combines the efforts of the International Geosphere–Biosphere Programme (IGBP), International Human Dimensions Programme of Global Change (IHDP), and the World Climate Research Programme (WCRP). The carbon cycle is considered as a central cycle of the terrestrial system connected with climate, water cycle, cycles of nutrient salts, and biomass production through photosynthesis on land and in the oceans. The human participation in the carbon cycle is determined by food production and habitat formation.

7.5.1 The international programmes on studies of global environmental changes

The GCP project foresees studies in three subjects: (1) models and variability; (2) processes, control, and interactions; and (3) the future dynamics of carbon cycle. Such a structure of studies of the global carbon cycle should create databases and knowledge bases, which would make it possible to specify the elements of the scheme in Figure 2.8 and to increase thereby the accuracy of estimates of the distribution of carbon fluxes between the biospheric reservoirs.

The basic idea of the scheme in Figure 2.8 is that the base components of the global carbon cycle are its reservoirs: the atmosphere (mainly CO_2), oceans (different layers of water masses and bottom deposits), land biosphere (vegetation, litter, and soil), and the carbon of fossil fuels (activated during the last 200 years due to human activity). The required resolution of the description of these reservoirs and fluxes of carbon between them can be found only by modelling the global carbon cycle with a gradual complication of this scheme and respective accumulation of knowledge (Schimel, 1995, 1998; Schimel *et al.*, 1996, 2001).

The present ideas about the carbon flux variability are based on the global-scale space and ground-based observations of the concentrations of atmospheric CO_2 and other gases, on modelling the dynamics of the atmosphere, oceans, and biogeochemical processes, as well as on an application of the principles of mass balance. As a result, the following information of the global balance of carbon in the present biosphere have been formed (IPCC, 2001a):

- About half the CO_2 emitted to the atmosphere from fuel burning is assimilated both by the land and oceanic sinks of carbon.
- Places of fossil fuel consumption are mainly concentrated in the northern hemisphere.
- The concentration of atmospheric CO_2 has increased considerably due to changing land cover in tropical latitudes.
- The amplitude of the intra-annual variability of the atmospheric CO_2 concentration is compatible with carbon emissions due to fossil fuel combustion.
- The inter-annual variability of carbon fluxes between the atmosphere and land biocenoses is greater compared with fluxes between the atmosphere and the oceans.
- A complete pattern of the oceanic sources and sinks of atmospheric CO_2 is known: the tropical waters emit CO_2 and the high-latitude water bodies assimilate CO_2.
- Maximum inter-annual variations of CO_2 flux on the atmosphere–ocean border are observed in the equatorial Pacific (Sabine and Feely, 2001).

As a result, the following questions arise:

- What are the laws following which the sources and sinks of carbon vary in time?
- What are the principles of the inventory of all carbon fluxes in the environment?
- What is the contribution of human activity, including fuel burning and land use, into the distribution of sources and sinks of carbon?
- What are the social aspects of the impact on the carbon cycle?
- What are the mechanisms for the global manifestation of regional and local processes of changing carbon fluxes?

The GCP foresees a search for the answers to these and other questions through choosing the first-priority directions of study:

- To determine the role of the oceans in controlling the seasonal and decadal changes of carbon fluxes in the atmosphere–ocean system.
- To assess the role of land biosphere in the formation of the inter-annual variability of the atmospheric CO_2 concentration.
- To determine the meridional distribution of carbon sinks in the northern hemisphere, marking out the role of North America, Eurasia, Asia, and Europe.
- To assess an importance of carbon fluxes between the oceans and land for the carbon balance, as well as carbon supplies to the inland water bodies.
- To assess the role of sediments and horizontal transport of matter in the coastal zones and open waters in the formation of the structure of the sources and sinks of carbon.

- To understand the spatial pattern and the amplitudes of the sources and sinks of carbon in the southern latitudes of the World Ocean.
- To determine the role of terrestrial tropics in the formation of carbon fluxes in the environment, with an emphasis on changing surface cover.
- To understand the differences in using the fossil fuels in industry and in individual regions within defined countries.
- To explain the leading forces and the consequences of various regional developments of the reservoirs and fluxes of carbon.
- To analyse the social and regional patterns of vulnerability and adaptability to changes in the carbon cycle.
- To determine the role of biological pumps and solubility processes in the global and regional cycles of carbon.
- To understand differences and to remove contradictions in the measurements of velocities and patterns of changes in land use and transformation of land cover.
- To develop methodologies of transition from results of studies of local and regional carbon cycles to global scales.

Clearly, the carbon cycle between the biospheric reservoirs are controlled by a lot of processes still poorly studied. Among them, some of the most significant processes can be selected:

- Physical processes in the atmosphere, oceans, and terrestrial hydrosphere.
- Biological and ecophysiological processes in the oceans and on land.
- Biogeochemical transformations.
- Natural and anthropogenic changes in the terrestrial biosphere.
- Processes of human consumption of the fossil fuel carbon.

To understand these and other processes of the formation of the carbon cycle in the dynamical environment, new paradigms are needed on the total natural and anthropogenic events. Such views have been framed in Kondratyev (1993, 1996, 1999b), Kozoderov (1994, 1998), Gorshkov (1995), Ivanov-Rostovtsev *et al.* (2001), Kondratyev and Krapivin (2002), Krapivin and Kondratyev (2002b), and Schafer and Victor (1999). Of course, a problem-oriented database is needed, which would correspond to the base of knowledge clarifying the cause-and-effect feedbacks in the NSS. There is such a base of knowledge, which explains in detail many processes in this system, revealing mechanisms of their occurrence and recommending a strategy of data accumulation and modelling:

- Direct exchange of CO_2 on the atmosphere–ocean border is regulated by physical processes of the vertical mixing of water layers above the thermocline with deep waters and by additional impacts of biological processes, which pump out carbon from the water and move it down to the sediments.
- The key biological mechanisms of the formation of carbon fluxes in the ocean medium are known. In a general case, these mechanisms are connected with the

functioning of the carbonate system, and in a simpler presentation they are provided with the vital functions of autotrophic and heterotrophic organisms, processes of decomposition of dead organic matter, and respiration of living organisms. This spectrum should be supplemented with a circulation of nutrient elements and photosynthesis (Krapivin and Kondratyev, 2002; Kondratyev *et al.*, 2003c).

- It is established that with an increasing partial pressure of CO_2 in the atmosphere the reaction of the World Ocean is non-linear both in space and in time (Orr *et al.*, 2001).
- The light, water, biogenic elements, and regimes of roughness interact following the non-linear laws described by numerical models (Krapivin, 1996).
- Many significant feedbacks in the system of the global carbon cycle have been well parameterized. Among them are the physiological responses of plants to changes of temperature and humidity of the air and the soil, the aerodynamics of exchange processes between the land surface and the atmosphere, and dynamic and thermodynamic processes within the atmospheric boundary layer (Raupach, 2000).
- At higher concentrations of atmospheric CO_2, without restricting the influence of light, water, and nutrients, the photosynthetical assimilation by land plants becomes physiologically saturated.
- Within regional and continental scales the CO_2 exchange between land ecosystems and the atmosphere is regulated by several factors:

    1. by extreme climatic events such as drought, appreciable changes in seasonal temperatures, and changes in solar radiation flux;
    2. by changes in the frequency of occurrence of fires, by cloud-free skies, and other large-scale fluctuations in the environment, which lead to sensible short-term losses of carbon in numerous biospheric reservoirs;
    3. by changing the boundaries of biomes and by changes of their types (e.g., change of coniferous forests for deciduous forests, of forested areas for grass, and meadows for forested areas); and
    4. by losses of biodiversity and intrusion of exotic species to the territories of sustainable existence of vegetation communities.

All studies foreseen by the plan of the GCP are aimed at obtaining reliable predictions of the consequences of human activity both in the near-term and long-term future. For this purpose, it is necessary to develop an approach to estimate the functions of soil–plant formations and water bodies of the World Ocean in assimilation of CO_2 from the atmosphere. Observations show that during the 500 years up to 1860 the climatic system functioned within a relatively narrow temperature range and CO_2 and CH_4 concentrations (during the last 100,000 years the concentration of CO_2 in the atmosphere varied within 180 to 280 ppmv). The last century has considerably broadened the spectrum of processes of transformation of matter and energy in the environment due to numerous new kinds of impacts on the biospheric components. The human factor in these processes has entailed an uncertainty in the future dynamics of these components, which required scenarios of the development of the economy, science, and politics. The level of uncertainty is so high that the predictions cannot be realistic. According to a report (IPCC, 2001a), the level of carbon emissions by 2100 could reach 30 Gt C/year, which, with the preserved rate of anthropogenic impact on the environment, could lead to CO_2 concentrations in the atmosphere reaching 950 ppm. All these estimates are obtained on the basis of expert opinion and sufficiently primitive models. Therefore, according to the GCP, an accumulation of a large base of knowledge, formation and testing of scenarios, development of models, and analysis of the present trends are first-priority directions for studies of the global changes of the environment. To synthesize the scenarios, the following set of considerations and reasoning is suggested:

- Concentrations of GHGs will be increasing over some time (IPCC, 2001a,b,c).
- The present and expected increase of the concentrations of GHGs, especially of CO_2, will be probably followed by global warming and by other climate changes.
- Apart from physical factors, such as the oceanic circulation, an assimilation of CO_2 by the oceans from the atmosphere will increase in the future with growing concentrations of atmospheric CO_2.
- The response of the oceanic thermohaline and circulation systems to climate changes (especially a temperature increase and change of the hydrological cycle) will probably be non-linear and will be accompanied by an increased stratification of the upper layer of the oceans, which can reduce the ventilation of deep layers.
- An increased stratification of the upper layer of the ocean will change the provision of surface layers with nutrient salts, which will change the role of phytoplankton in carbon assimilation and cause large-scale fluctuations in the atmosphere–ocean interaction, especially concerning the Arctic Basin, the Okhotsk Sea, and other seas in northern latitudes.
- There is a difference in the estimates of direction of changes of the surface sinks of carbon due to a transformation of land biosphere (Cox *et al.*, 2000, 2002; Cramer *et al.*, 2001; Friedlingstein *et al.*, 2001).
- Careless and ill-considered changes to the Earth's surface will lead in the near future to changes of climate and the carbon cycle.
- Instabilities and multiple states of equilibrium can turn out to be bound with biosphysical and biogeochemical systems, containing climatic and hydrological processes and cycles of carbon and nutrient elements. These phenomena can be caused by non-linear feedbacks in redistribution of energy and matter between the atmosphere, land, oceans, and ice.
- Even without dynamic phenomena, such as instabilities and increasing states of equilibrium, an increase of the atmospheric CO_2 concentration during the period 1860–2100 will cause certain shifts in the climate–carbon system, changing the preceding distribution amongst regions.

- The anthropogenic changes in the carbon cycle interact with the biophysical aspects of the formation of this cycle, and hence, have a potential to lead to variations in the global climate system.
- People also react to undesirable climate changes threatening their wellbeing, though these reactions are highly uncertain (e.g., the Kyoto Protocol).
- There is a certain spectrum of potential reactions of human society to undesirable climate changes, including geographical, social, economic, and nature–science directions:
  –changes in CO_2 emissions either due to transition from traditional to non-traditional types of fuel, or caused by changing the technology of its application;
  –improvement of the processes of production and use of energy, changes in life style;
  –transition to renewable energy sources (bioenergy, solar energy, wind);
  –increased role of forests in assimilation of excess carbon;
  –increase of carbon sinks from the biosphere into geological formations and the deep ocean due to engineering surveys;
  –development of new technologies in agriculture, leading to a decrease of GHG emissions.

Eventually, all this will require a universal complex approach to studies of the consequences of realization of alternative scenarios for human civilization development. Clearly, a search for a safe sustainable functioning of the NSS is impossible without modelling alternative scenarios. And for this purpose a global model of the functioning of this system should be derived. Such a model and a scheme of its use have been described in Kondratyev *et al.* (2002a, 2003c).

7.5.2 Improving the structure of global environmental changes

Synthesis of the global NSS is based on the idea of it as a self-organizing and self-structurizing system, whose elements agree in space and time through the process of natural evolution. The anthropogenic impact on this process violates this integrity. Attempts to formally parameterize the process of coevolution of nature and humans as biospheric elements are connected with the search for a single simulation for all processes in the NSS, which would combine the effort spheres of the environmental sciences. Such a synergetic basis is observed in many studies on global modelling.

The global modelling refers to not always appreciated spheres of knowledge, whose intrusion into non-traditional spheres of human life has been actively discussed during the last decades. Despite an existing disagreement on the role of models in studies of the NSS, the fact is that a model makes it possible to predict consequences of an accomplishment of large-scale experiments on the transformation of the environment.

A combination of units within the global model simulating the development of various processes in the NSS requires a flexible technology to coordinate their spatial

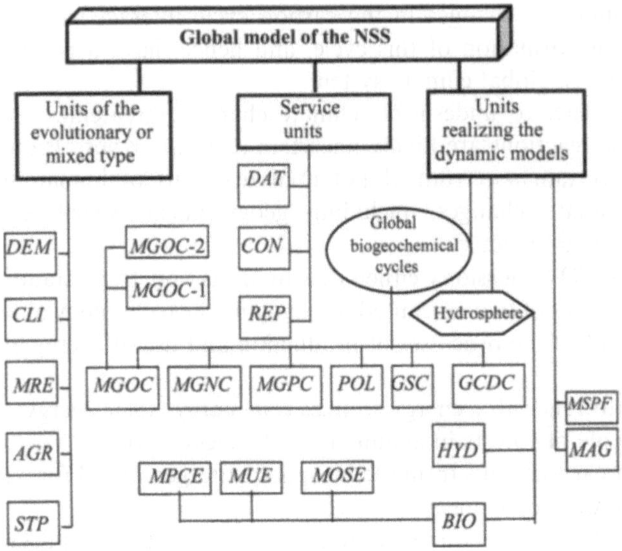

Figure 7.6. The block-scheme for the global model of the NSS (GMNSS). A description of the GMNSS items are given in Table 7.7.

and temporal scales, which in turn, necessitates a synthesis of the object-oriented databases. As a result, the structure of the global model should permit the sustainable development of the established hierarchy of units. This is possible with the use of the semantic mechanisms of adaptation of the model structure to changing simulations of the configuration of the N–S interconnections. Figure 7.1 shows an exemplary scheme of the organization of such a structure where the basic information trunk co-ordinates the inputs and outputs of the model items with the control of their compatibility both among them and with the database.

Observations and studies of global processes in the environment according to the GCP have been aimed at overcoming the difficulties appearing in global modelling and in attempts to adapt an accumulated database and knowledge to the structures of available global models. One of the possible ways to solve these problems is the geographical information monitoring system (GIMS)-technology. This makes it possible to use the global model in the regime of its constant adaptation to the current global monitoring of the environment, providing thereby a combined parameterization of the spatial and temporal variability of biospheric and geospheric processes with the procedure of observations planning.

Figure 7.6 shows a block-scheme of the global model built into the GIMS structure via the DAT unit, which controls the interface between other model units and the database renewed in the monitoring regime. The CON unit adjusts the information base of the global model and its internal connections to the conditions of the simulation experiment and controls the coordination of information fluxes between the model units. The REP unit forms the report documents of the simulation experiment and connects the global model with other information

Table 7.7. A description of the MGNSS items.

Item identifier (see Figure 7.6)	Description
DEM	A set of demographic models that parametrize the population dynamics with a consideration of age structure
CLI	A set of climate models with various spatial resolutions
MRE	Model for the control of mineral resources
AGR	Model of agriculture production
STP	Model of science–technical progress
DAT	Controlling procedure of the interface between the MGNSS items and the database
CON	Informational procedure for the MGNSS items adaptation to the simulation experiment conditions and its control
REP	Reporting and visualization procedure
GCDC	Model of global carbon dioxide cycle
GSC	Model of global sulphur cycle
MGOC	Model of global oxygen (MGOC-1) and ozone (MGOC-2) cycles
MGNC	Model of global nitrogen cycle
MGPC	Model of global phosphorus cycle
POL	A set of models parametrizing the pollutant kinetics within different medias
BIO	A set of models parametrizing the aquatic ecosystems in different climatic zones
HYD	Model of global hydrodynamic processes and the biosphere water balance
MSPF	A set of biocenotic models describing a different classification of soil–plant formations
MAG	Model of the magnetospheric processes related to the global biogeochemical cycles
MUE	Typical model of the upwelling ecosystem of the World Ocean
MOSE	Model of the Okhotsk Sea ecosystem
MPCE	Model of Peruvian Current ecosystem

systems. The remaining units of the global model are divided into two classes with different modelling approaches. The units responsible for the evolutionary modelling are the units parameterizing the demographic (DEM) and climatic (CLI) processes creating the scenarios of managing the mineral resources (MRE) and development of scientific–technical progress (STP) as well as modelling the agricultural production (AGR). The units relevant to the dynamic models are based on a realization of balance relationships. The units parameterizing the global biogeochemical cycles cover the whole phase space of the NSS presented in the scheme in Figure 7.7. This scheme includes the models of the global cycle of carbon dioxide, sulphur, phosphorus, nitrogen, oxygen, ozone, and other chemical contaminants. To simulate the totality of the processes in the hydrosphere, the global model includes the units simulating the hydrodynamic processes and the functioning of water

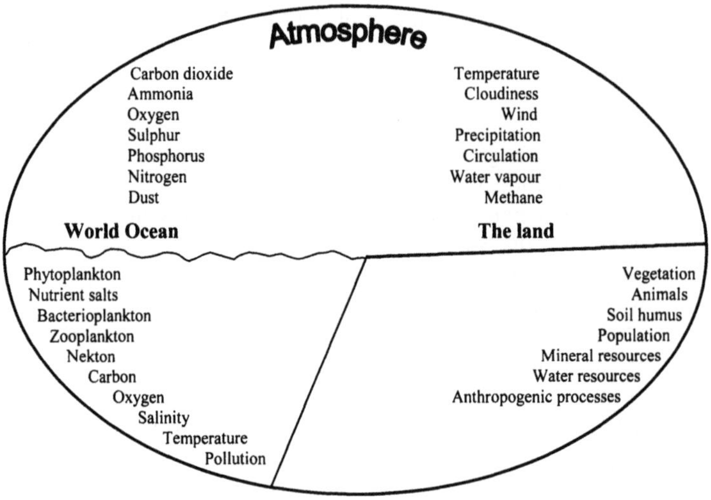

Figure 7.7. The elemental composition of the GSM describing the dynamics of the NSS.

ecosystems in different climatic zones. The global model also includes the ecosystems of the Okhotsk Sea, the Peruvian Current, and the equatorial upwelling. To broaden the adaptive possibilities of the global model, the structure of many units foresees various versions of parametric simulations of the processes in the NSS.

The scheme in Figure 7.7 shows the basic components taken into account in the GSM described here. As seen from Figure 7.6, a concrete realization of each unit of GSM is determined by the level of knowledge of the processes reflected in the unit. So, the units responsible for modelling the biogeochemical and biogeocenotic processes are described by differential equations.

A parameterization of the processes of photosynthesis, dieing, and respiration of plants in land ecosystems is based on knowledge of phytocenology, which includes information about external and internal system connections of the vegetation community, including the temperature dependencies of photosynthesis and evapotranspiration of plants, gas exchange between plants and the atmosphere, impacts of solar radiation on the exchange and growth processes, correlations between vegetation and soil processes, and interaction of vegetation cover with the hydrological cycle (Scanlon and Albertson, 2001).

A parameterization of climate and anthropogenic processes is realized by a combination of equations of motion and balance within the evolutionary model.

7.5.3 Evolution and survivability of the "nature-society" system

The NSS can be presented as a combination of nature (N) and human society (H), which constitutes a single planetary system. Therefore, their separation in deriving global and regional models should be considered conditional. The systems N and H have the hierarchical structures $|N|$ and $|H|$, goals $\underline{N}$ and $\underline{H}$, and modes of behaviour $\overline{N}$ and $\overline{H}$, respectively. From the mathematical point of view, an interaction of the systems N and H can be considered as a random process $\eta(t)$ with an unknown law of distribution, representing the level of intensity in relationships of these systems or estimating the state of one of them. The goals and modes of behaviour of the systems are functions of the index η. The η index can vary in such a way that the modes of behaviour of the systems can be antagonistic, indifferent, and cooperative.

The main goal of the system H is to achieve a high standard of living with guaranteed prolonged survival. The goal of the system N can be determined in terms of survival. The system N behaves itself following the objective laws of coevolution. In this sense the selection of H and N is conditional and can be interpreted as a division of a multitude of natural processes into controlled and uncontrolled. Without dwelling upon the philosophy of this division, consider the systems H and N to be open and symmetrical in their description. The system H includes technologies, science, economic potential, industrial and agricultural production, sociological structure, size of population, etc. The H–N interaction causes changes in η, the level of which affects the structure of vectors $\underline{N}$ and $\underline{H}$. Of course, there exists a threshold of η_{max}, beyond which humankind stops existing, but nature survives. The asymmetric nature of the systems H and N cause changes in the goal and strategy of the system H.

At present, the interaction between these systems $\eta \rightarrow \eta_{max}$ is occurring at a rapid pace, and therefore individual components of vector $\underline{H}$ can be attributed to the class of cooperative behaviour. Since the present socio-economic structure of the world is presented by a totality of states, we shall consider a country as a functional element of the system H. The $\eta(t)$ function reflects the result of interaction of the countries both with each other and with nature. Let us describe the sum of the results of their interaction with the matrix $B = \|b_{ij}\|$, each element of which has a meaning of its own:

$$b_{ij} = \begin{cases} +1 & \text{for cooperative behaviour} \\ -1 & \text{for antagonistic relationships} \\ 0 & \text{for indifferent situations} \end{cases}$$

A country H_i has m_i possible ways to reach the goal $\underline{H}_i$, in other words, it uses a series of strategies $\{\overline{H}_i^1, \ldots, \overline{H}_i^{m_i}\}$. The weight of each strategy $\overline{H}_i^j$ is prescribed by the value $p_{ij}(\sum_{j=1}^{m_i} p_{ij} = 1)$. The resulting value of the η parameter is a function of the indicated characteristics, and at each time moment the situation is described by a theory games model (Krapivin, 1978).

An objective estimate of the dynamics of the natural medium $N = (N_1, N_2)$ is possible with some assumptions using the models of the biosphere N_1 and climate N_2. An accumulated experience of numerous scientists in deriving such models includes the point, regional, box, combined, and spatial models (Marchuk and Kondratyev, 1992; Krapivin, 1993). This experience enables one to move on to synthesis of a new type of global model, covering the key bonds between the hierarchical levels of natural and anthropogenic processes.

In general, the condition of the systems H and N can be described by vectors $x_H(t) = \{x_H^1, \ldots, x_H^n\}$ and $x_N(t) = \{x_N^1, \ldots, x_N^m\}$, respectively. A combined trajectory of these systems in the $n + m$ dimensional space is described by the function $\eta(t) = F(x_H, x_N)$ determined by solutions of the global model equations. The form of F is determined from knowledge of the laws of coevolution, and therefore there is the possibility of investigations in different spheres of science. The available estimates of F (Krapivin, 1996) reveal a correlation between the notions "*survivability*" and "*sustainability*". According to Ashby (1956), the dynamic system is "alive" within the time interval (t_a, t_b), if its determining phase coordinates are within "admissible limits", $x_{H,\min}^i \leq x_H^i \leq x_{H,\max}^i$; $x_{N,\min}^j \leq x_N^j \leq x_{N,\max}^j$. And since the systems H and N have a biological basis and limited resources, one of the indicated boundary conditions turns out to be unnecessary (i.e., for the components of vector $x = \{x_H, x_N\} = \{x_1, \ldots, x_{n+m}\}$ the condition $x_{\min} \leq \eta = \sum_{i=1}^{n+m} x_i$ should be satisfied. This simple scheme requires that the total energy in the system be preserved and its elements be diverse.

Of course, the notion of "the system's survivability" is more capacious and informative. In the system ecology, many authors denote by this term the stability and integrity of the system, meaning the system's ability to resist external forcings. In other words, the survivability is measured by the tendency of the system to suppress large oscillations of its structure and elements, returning the system into its former equilibrium state. Thus the system's survivability should be understood as its ability to actively resist an impact of external factors and to preserve for a long time its characteristics.

7.5.4 Climate unit of the global model

The climatic component of the GMNSS represents the greatest difficulty in the synthesis of the global model, because it is characterized by a lot of feedbacks, most of which are unstable. Among them are ice–albedo, water vapour–radiation, cloudiness–radiation, aerosol–radiation, and many others. The functioning of the Earth's climatic system is determined by the state of the atmosphere, oceans, cryosphere, surface of the continents with land biota, lakes, rivers, groundwaters, and various anthropogenic structures. Therefore, the construction of the model of the climate requires a consideration of numerous factors, whose role in its formation in many cases has been inadequately studied. So far, attempts to derive a numerical model of the Earth's climatic system have not given the results that could be used in GMNSS. Also, it is not safe to say that a realization of the GCP plans will clarify the problem. But without a parameterization of climate the questions concerning the problem of the greenhouse effect will remain unanswered.

There are two approaches to the synthesis of the climatic system. One approach is based on the inclusion of biospheric components into the existing climate models or in those under development. The other approach consists of the development of a unit within the numerical model of the biosphere, which would simulate dependencies of biospheric components on climate parameters. In the former case, problems appear of unstable solutions of the respective systems of differential equations, which makes it difficult to predict global environmental changes. In the latter case there is the possibility of obtaining steady prognostic estimates of environmental changes, but their reliability depends on the accuracy of parameterization of correlations between the elements of the climate and the biosphere. The second approach is preferable, because it permits one to combine the climate models, which can be described on the scenario level, with the numerical model of the biosphere. A detailed analysis of the problems of climate modelling and assessment of the present state of the problems can be found in Marchuk and Kondratyev (1992), Kondratyev (1999b), Kondratyev and Johannessen (1993), and Savinykh and Tshvetkov (2000, 2001, 2002). Some models of individual components of the NSS, corresponding to the second approach, are discussed here. Among them are models of the atmospheric general circulation, atmosphere–ocean interaction, sensitivity of climate parameters to boundary conditions on the Earth's surface, interaction of biogeochemical and climatic processes, and others (Lin *et al.*, 2000; Chen *et al.*, 2000).

The climatic system is a physico–chemico–biological system with unlimited degrees of freedom. Therefore, any attempts to model this complex system are connected with serious difficulties. This explains the diversity of parametric descriptions of some processes in this system. For the global model with a time step of digitization up to one year, the use of two versions can be acceptable. The first version consists of a combined use of correlations between the particular processes of the formation of the climatic situation on a given territory together with the climate scenarios. The second version is based on the use of global monitoring data, which are the basis for the formation of data series about climatic parameters, with territorial and temporal attachment, and are used to retrieve a complete pattern of their spatial distribution. One of the widespread correlation functions is the dependence of variations of average temperature ΔT_g of the atmosphere on the CO_2 content within it:

$$\Delta T_g = \begin{cases} 2.5[1 - \exp\{-0.82(\xi - 1)\}] & \xi \geq 1 \\ -5.25\xi^2 + 12.55\xi - 7.3 & \xi < 1 \end{cases} \tag{7.1}$$

where ξ is the ratio between the present content of CO_2 in the atmosphere $C_a(t)$ and its pre-industrial level $C_a(1850)$.

As seen from Equation (7.1), T_g is an increasing function of the amount of atmospheric CO_2. A 20% increase of CO_2 in the atmosphere raises the temperature by 0.3°C, while a doubling of atmospheric CO_2 causes an increase in T_g of 1.3°C. A detailed analysis of the function (7.1) and a comparison of the observed combined variations of ΔT_g and ξ show that the use of the model (7.1) makes it possible to simplify the climate unit of the N–S model. In particular, if $(\Delta T_g)_{2[CO_2]}$ is calculated by Equation (7.1) with a doubled concentration of atmospheric CO_2, then to assess the present trends in changes of T_g the following formula can be used:

$$\Delta T_g = (\Delta T_g)_{2[CO_2]} \ln \xi / \ln 2 \tag{7.2}$$

where, from the assumed estimates, the pre-industrial value of $C_a(1850) = 270\,\text{ppm}$. The Equation (7.2) satisfactorily approximates the known data with an error of about 50%. It follows from Equation (7.2) with $C_a(1980) = 338$ ppm that $\Delta T_g = 1.3°K$, whereas a real warming is estimated by many authors at 0.6°C.

Of course, the results of the recent discussions on the greenhouse effect in connection with the growth of the CO_2 partial pressure in the terrestrial atmosphere should be reflected in GSM. Equation (7.1) takes into account the impact of CO_2. According to Mintzer (1987), there is the possibility to broaden the consideration of the temperature effect of other GHGs:

$$\Delta T_\Sigma = \Delta T_{CO_2} + \Delta T_{N_2O} + \Delta T_{CH_4} + \Delta T_{O_3} + \Delta T_{CFC11} + \Delta T_{CFC12}$$

where
$$\begin{aligned} \Delta T_{CO_2} &= -0.677 + 3.019 \ln[C_a(t)/C_a(t_o)] \\ \Delta T_{N_2O} &= 0.057[N_2O(t)]^{1/2} - [N_2O(t_o)]^{1/2} \\ \Delta T_{CH_4} &= 0.19[CH_4(t)]^{1/2} - [CH_4(t_o)]^{1/2} \\ \Delta T_{O_3} &= 0.7[O_3(t) - O_3(t_o)]/15 \\ \Delta T_{CFC11} &= 0.14[CFC11(t) - CFC11(t_o)] \\ \Delta T_{CFC12} &= 0.16[CFC12(t) - CFC12(t_o)] \end{aligned}$$

The value of t_o is identified with the year 1980, when the GHG concentrations were known. To calculate ΔT_{CO2}, one can use a more accurate formula by Krapivin (1993).

Among simple formulas of calculations of the latitudinal distribution of average temperature over the globe one can indicate the schemes proposed by Sergin (1974):

$$T(\varphi) = T_g + \gamma(\sin^2 \varphi_T - \sin^2 \varphi) \tag{7.3}$$

where φ is the latitude in radians; γ is the pole–equator temperature difference; and φ_T is the latitude at which $T(\varphi) = T_g$. The latitudinal temperature variations can be described:

$$T_a(\varphi, t) = T_e - 2\varphi(T_e - T_N)/\pi \quad \text{for the northern hemisphere}$$

$$T_a(\varphi, t) = T_e - 2\varphi(T_e - T_S)/\pi \quad \text{for the southern hemisphere}$$

where

$$T_N = \begin{cases} T_{N,\min} + 2t(T_{N,\max} - T_{N,\min})/t_\Delta & t \in [0, 0, 5t_\Delta] \\ T_{N,\min} + 2(t_\Delta - t)(T_{N,\max} - T_{N,\min})/t_\Delta & t \in [0, 5t_\Delta, t_\Delta] \end{cases}$$

$$T_N = \begin{cases} T_{S,\max} + 2t(T_{S,\min} - T_{S,\max})/t_\Delta & t \in [0, 0, 5t_\Delta] \\ T_{S,\max} + 2(t_\Delta - t)(T_{S,\min} - T_{S,\max})/t_\Delta & t \in [0, 5t_\Delta, t_\Delta] \end{cases}$$

$T_{N,\min}(T_{S,\min})$ and $T_{N,\max}(T_{S,\max})$ are minimum and maximum temperatures at the North (South) Pole, respectively (°C); t_Δ is the duration of the year in units of measurements Δ; and T_e is the equatorial temperature of the atmosphere (°C). Many authors use the estimates $T_{N,\min} = -30°C$, $T_{N,\max} = 0°C$, $T_{S,\min} = -50°C$, $T_{S,\max} = -10°C$, and $T_e = 28°C$.

Of course, these zonally averaged temperatures scatter widely, leading to considerable errors. To more accurately reflect the role of various factors in changes of

Table 7.8. Time periods for establishing an equilibrium state for some subsystems of the Earth's climate system.

Domain of influence of the climate system	Period of restoration of the equilibrium state
Atmosphere	
Free	11 days
Boundary layer	24 hours
World Ocean	
Mixed layer	7–8 years
Deep ocean	300 years
Sea ice	from 24 hours to 100 years
Continents	
Lakes and rivers	11 days
Soil–plant formations	11 days
Snow cover and surface ice	24 hours
Mountain glaciers	300 years
Ice sheets	3000 years
Earth's mantle	30 million years

the basic climatic parameter–temperature, it is necessary to calculate the contribution of each factor separately. This can be done assuming additivity of the roles of feedbacks: $\Delta T_{a,\text{final}} = \Delta T_a + \Delta T_{a,\text{feedback}}$. In some models the authors introduce the feedback coefficient β: $\Delta T_{a,\text{final}=\beta\Delta T_a}$. The β parameter is expressed through the amplification index g: $\beta = 1/(1-g)$. The value of the g index is equivalent to albedo α, which on a global scale is a function of T_a. This dependence can be roughly approximated by the formula:

$$\alpha(T_a) = \begin{cases} \alpha_{\text{ice}} & \text{for } T_a \le T_{\text{ice}} \\ \alpha_{\text{free}} & \text{for } T_a \ge T_{\text{free}} \\ \alpha_{\text{free}} + b(T_{\text{free}} - T) & \text{for } T_{\text{ice}} < T_a < T_{\text{free}} \end{cases}$$

Here T_{ice} and T_{free} are average planetary temperatures at which the whole surface of the Earth is ice-covered or ice-free, respectively; and b is the coefficient of transition between critical conditions of the Earth's albedo. Usually it is assumed that $T_{\text{ice}} \in [263, 283]$°K.

Simple and rough climate models can be specified through an account of the characteristic times of feedback functioning. Some estimates of the time of establishing the equilibrium states in the interaction of climate subsystems are given in Table 7.8. The time lag of reactions within NSS varies widely, and its consideration is necessary when assessing the consequences of changes within one or several climate subsystems. In particular, the ice reserves in the Antarctic Ice Sheet are so huge that to raise its temperature to 0°C, the average temperature of the World Ocean would be lowered by 2°C (i.e., to obtain $T_o = 3.7$°C instead of $T_o = 5.7$°C). With an account of the data in Table 7.8, it will take hundreds of years to do this.

Table 7.9. Estimates of the fluxes of particles with radius <29 μm either emitted to the atmosphere or formed *in situ*.

Type of particles	Amount of particles (10^6 t/year)
Particles of natural origin (weathering, erosion, etc.)	100–500
Particles from forest fires and burning of forestry wastes	3–150
Sea salt	300
Volcanic dust	25–150
Particles from gas emissions	
Natural processes	
Sulphates from H_2S	130–200
Ammonium salts from HN_3	80–270
Nitrates from NO_x	60–430
Hydrocarbons from vegetative compounds	75–200
Anthropogenic processes	
Sulphates from SO_2	130–200
Nitrates from NO_x	30–35
Hydrocarbons	15–90
Amounts of particles emitted to the atmosphere	
Natural	773–2,200
Anthropogenic	185–415
Total flux of particles into the atmosphere	958–2615

The observed rate of the anthropogenic climate warming does not dispose of such energy supplies.

The mechanism of the anthropogenic impact on the climate system is manifested through GHG emissions, albedo change due to reconstruction of the Earth's covers, interference with the water cycle, and atmospheric pollution (Austin *et al.*, 2002). Aerosol particles with the radius $10^{-7} \pm 10^{-2}$ cm are observed almost at all altitudes of the atmosphere. Particles of non-anthropogenic origin get into the atmosphere from land or the ocean surface and result from *in situ* chemical reactions. Particles of anthropogenic origin result mainly from fuel burning. Table 7.9 gives an idea about the relationships between these fluxes of particles into the atmosphere.

The mechanism of the effect of particles on atmospheric temperature is explained by the fact that solar radiation coming to the Earth, mainly within the range 0.4–4 μm, is partially reflected and absorbed by them. The global albedo of the surface–atmosphere system varies, too. Besides, the particles affect the processes of moisture condensation in the atmosphere, since clouds, rain, and snow form with their participation. Take the heat balance equation of the surface–atmosphere system:

$$(1-\alpha)E_o^* + E_a - \sigma T_s^4 = 0 \tag{7.4}$$

where T_s is the average efficient temperature of the system's emission, close to the temperature of the average energy level near the surface 400 mb; $E_o^* = 0.487\ \mathrm{cal/cm^2}/$

min is the mean hemispherical intensity of the incoming solar radiation; α is albedo; $\sigma = 8.14 \times 10^{-11}$ cal/cm^2/min is the Stefan–Bolzman constant; and E_a is the total intensity of the anthropogenic sources of energy per unit surface.

Let albedo $\alpha = \alpha_o - \Delta\alpha$, where $\alpha_o = 0.35$ is the albedo at present-day conditions; and $\Delta\alpha$ is a small part of the albedo determined by the impact of anthropogenic aerosols. From Equation (7.4) we obtain an expression for temperature:

$$T_s = [E_o^*(1-\alpha)/\sigma]^{1/4}[1 + \Delta\alpha(1-\alpha_o)^{-1} + (E_a/E_o^*)(1-\alpha_o)^{-1}]^{1/4} \qquad (7.5)$$

Assuming $\Delta\alpha \ll 1$ and $E_a/E_o^* \ll 1$, expand the function of the right-hand part of Equation (7.5) into Taylor series by degrees $\Delta\alpha$ and E_a/E_o^* and write the first terms of the series:

$$T_s \approx [\beta_o^*(1-\alpha_o)/\sigma]^{1/4}\{1 + 0.25\,\Delta\alpha(1-\alpha_o)^{-1}[1 + E_a/E_o^*]\} \qquad (7.6)$$

It follows from Equation (7.6) that the temperature at not so strong anthropogenic impacts is the sum of the terms describing the bonds in the surface–atmosphere system, without an account of anthropogenic factors, and the terms T_1 and T_2, representing the contribution of heat and aerosol emissions, respectively:

$$T_1 = 0.25(1-\alpha_o)^{-1}[E_o^*(1-\alpha_o)/\sigma]^{1/4}E_a/E_o^* \approx 96.046E_a/E_o^*$$

$$T_2 = 0.25(1-\alpha_o)^{-1}[E_o^*(1-\alpha_o)/\sigma]^{1/4}\Delta\alpha \approx 96.046\Delta\alpha$$

Note that the contribution of T_1 in present-day conditions is very small. Assuming that $E_a = 4 \times 10^{-5}$ cal/cm^2/min and, hence, $E_a/E_o^* = 8.21 \times 10^{-4}$, then $T_1 = 0.0079$°C. Thus the direct impact of the global energy on the atmospheric average temperature is now negligible. It follows from the expression for T_1 that to raise the atmospheric temperature by 0.5°C due to thermal emissions, the condition $E_a/E_o^* = 0.0052$ should be satisfied, and this means an increase of anthropogenic heat fluxes to the environment by a factor of 63.4. This is equivalent to the energy release from the annual burning of 570×10^9 t of conditional fuel.

If we assume that the energy production is proportional to the size of the population, then $T_1 = 96.046\,k_{TG}\,G\sigma_S/E_o^*$, where G is the population density, men/km^2; σ_S is the land area (km^2); and k_{TG} is the per capita amount of energy production (cal/min).

If we neglect the impact of aerosol on the atmospheric thermal regime, then the direct radiation E, its change dE, and the change of atmospheric turbidity dB will be related as: $dE/E = -k_B dB$, where $k_B = 0.1154$ km^2/t is the proportion coefficient; and B is the amount of anthropogenic aerosols (t/km^2). After integrating this equation, we obtain: $E = E_o^*(1-\alpha_o)\exp(-k_B B)$. On the other hand, according to the definition of albedo, $E = E_o^*(1-\alpha) = E_o^*(1-\alpha_o+\Delta\alpha)$. Equating these expressions for E, we obtain $\Delta\alpha = -(1-\alpha_o)[1-\exp(-k_B B)]$. Hence, a temperature change due to pollution of the atmosphere by anthropogenic aerosols is equal to:

$$T_2 = -0.25[E_o^*(1-\alpha_o)/\sigma]^{1/4}[1-\exp(-k_B B)] = -62.43[1-\exp(-k_B B)]$$

Since the average emission of anthropogenic aerosols is estimated by many authors at 300×10^6 t/year, and the average period of aerosol residence in the atmosphere is estimated at three weeks, 17.262×10^6 tons of particles, on average, reside in the atmosphere. In this case it follows from the formula for T_2 that the atmospheric temperature should decrease by 0.84°C/year.

Some experts consider the factor of atmospheric turbidity B_T instead of the index β, defining it as a ratio of the coefficient of attenuation of solar energy in the real atmosphere α_r to the coefficient of attenuation in an ideal atmosphere α_I: $B_T = \alpha_r/\alpha_I = (\alpha_I + \alpha_W - \alpha_A)/\alpha_I$, where α_W and α_A are the coefficients of attenuation by water vapour and by aerosols, respectively. The following estimates are assumed in the GMNSS:

$$B_T = \begin{cases} 3.0 \text{ in mid-latitudes} \\ 3.5 \text{ in tropical latitudes} \\ 2.0 \text{ with reduced contents of dust and water vapour} \end{cases}$$

An experience of global climate modelling testifies to the fact that the desire of many experts to accurately and completely take into account all the possible feedbacks and compounds of the climate system leads to complicated mathematical problems, the solution of which a huge amount of data is needed, and in most cases the solutions of the respective equations turn out to be unstable. Therefore, the use of these complicated models as a unit of the global model of the GMNSS leads to a negative result (i.e., the impossibility of constructing an efficient model). The most encouraging approach is to combine the climate models with the data of global monitoring. The scheme of this combination is very simple. The existing surface and satellite systems of monitoring the climate-forming processes cover some part of the cells $\{\Omega_{ij}\}$ of the Earth's surface. Over these cells, measurements are made of temperature, cloudiness, water vapour content, aerosols and gases, albedo, and various parameters of energy fluxes. Use of simple climate models and of methods of spatial and temporal interpolations enables one to reconstruct, based on these measurements, a complete pattern of distribution of climatic parameters over the whole territory Ω.

7.6 THE EXTRAORDINARY EVENT OF THE ANTARCTIC OZONE HOLE IN 2002

7.6.1 Introduction

During the last two decades the problem of the Antarctic ozone hole has attracted special attention because of a sharp drop in the Antarctic column ozone. An important factor involved in the formation of the Antarctic ozone hole is the specific dynamics of the Antarctic atmosphere, which is characteristic of the presence of a circumpolar vortex in winter, which captures the circumpolar air; the air temperature then drops to –90°C, provoking the formation of polar stratospheric clouds (PSCs). Heterogeneous reactions on the surfaces of the ice particles of these clouds transform comparatively inert $ClONO_2$ and HCl into active forms of chlorine compounds, catalyzing ozone depletion, and also binding nitrogen

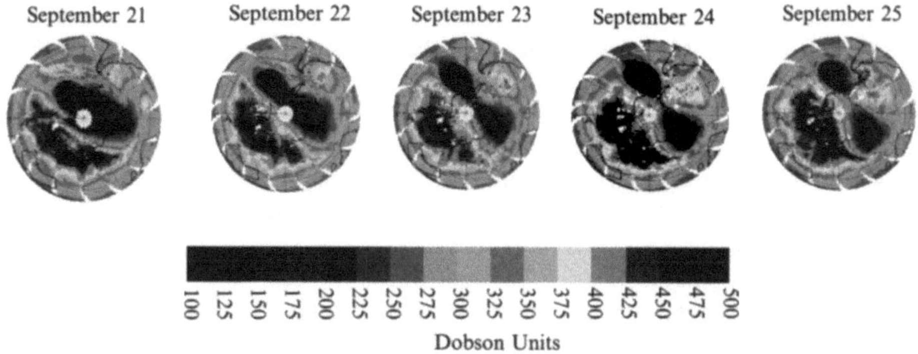

Figure 7.8. The Antarctic ozone hole split in September 2002.
Produced by the Ozone Processing Team at NASA's Goddard Space Flight Center.

compounds. The hole appears during the Antarctic spring (August–September) because these key ozone-destroying reactions are initiated by sunlight, a factor largely absent during winter months (Kondratyev and Varotsos, 2000). According to Solomon (2004) removal of the ozone-destroying chlorine and hence return of the Antarctic ozone layer to its natural state will take decades.

In late September 2002, the Antarctic ozone hole was much smaller than in the previous six years and it had split into two separate holes (NASA/NOAA 2002a, Varotsos 2002a, WMO 2002) (Figure 7.8).

It has been proposed (Varotsos 2002a,b; Varotsos 2004a,b) that a plausible cause for this event was the prevalence of very strong planetary waves, which led to the appearance of a major, sudden stratospheric warming (SSW), an event that had never been observed before in Antarctica. Subsequently, a NASA/NOAA press release of December 6, 2002 (NASA/NOAA 2002b), also reported the prevalence of very strong planetary waves in Antarctica.

Nevertheless, the current understanding of the dynamics of the stratosphere is still limited and most probably this unprecedented episode in Antarctica will further elucidate the causes and the mechanisms of SSWs. The next paragraph illustrates the attempts that are currently being performed worldwide in order to clarify the present understanding of these peculiar events.

7.6.2 The Antarctic Polar Vortex

7.6.2.1 The conventional breakdown of the southern polar vortex

Before starting to examine the unusual behaviour of the Antarctic polar vortex in 2002, it is worthwhile giving a brief description of the relevant past experience, which is based on observational data derived from ground-based and satellite-borne instrumentation.

Using the archived data of the US National Meteorological Center for 7 years (1979–1985), Newman (1986) examined the variability of the southern hemisphere upper stratospheric circulation from winter to summer using information about zonal winds at the 30 hPa level near 60°S, the horizontal temperature gradient at that level, and concentration (at 30 hPa) and total ozone content (satellite information). The transition regime is classified into two extreme categories: (i) early (mid-October) final stage of SSW accompanied by intensive wave activity (1979 and 1982); (ii) late (mid-November) warming stage with weak long waves (1980 and 1981). A change of the sign of horizontal temperature gradient at the 30 hPa level is an indicator of the final stage of SSW.

The onset of the regime of easterly transport near the pole reflects the breakdown of the polar vortex. The formation of long waves in late November and early December, which was observed in all cases considered, leads to a gradual break and finally to the disappearance of the polar vortex. Such "wave events", together with the mean zonal mass transport, weaken the zonal mean gradients of the potential vorticity and ozone content, due to intensified horizontal mixing. This favours the breakdown of the polar vortex and the filling of the ozone hole. In this connection it was shown (Newman 1986) that the final stage of SSW and the breakdown of the polar vortex are events separated by a time interval of 15 to 40 days. The retrospective analysis of an observational data set for the 1960s has demonstrated that the time of the onset of these events has not changed radically.

7.6.2.2 *The exceptional morphology and evolution of the southern hemisphere vortex in 2002*

The investigation of the spatial and temporal features of the Antarctic polar vortex in 2002 is of crucial importance because the relevant findings could provide us with a better understanding on the Antarctic ozone hole split. In this regard, the observational analysis presented by Varotsos (2002a,b) showed that the basic polar vortex in the stratosphere of the southern hemisphere was step-by-step elongated up to the occurrence of the polar vortex split into two centres on 25 September, 2002.

The above mentioned observational result has been verified since by various studies (Allen *et al.*, 2003, Charlton *et al.*, 2003, Konopka *et al.*, 2003, Orsolini *et al.*, 2003). These studies suggest that the middle stratospheric vortex split into two parts; one, rapidly mixed with extra vortex air, while the other returned to the pole as a much weaker and smaller vortex. Looking at this in more detail, the cyclone and the dominant anticyclone moved eastward and poleward, the anticyclone continuing to strengthen while the cyclone decayed. The weaker cyclonic vortex disappeared completely by early October. The remaining cyclonic vortex stayed relatively weak and finally broke down by mid-November. Secondary vortex development by barotropic instability is discussed in Simmons *et al.* (2003).

7.6.2.3 *The plausible cause – an unprecedented dynamical forcing*

Varotsos (2002b) suggested that the amplitude of the zonal wavenumber-2 derived from the Fourier analysis of the isobaric height and temperature fields at 10 hPa) reached a peak when the polar vortex split into two circulations (reversal of the zonal mean wind), a characteristic feature of the occurrence of the unprecedented major SSW in the southern hemisphere. Subsequently, Varotsos and Kondratyev (2003) confirmed the mechanism proposed by Varotsos (2002a,b) and further suggested that this mechanism does not consider the desirable recovery of the ozone layer. The features found using Fourier analysis of the time series of the 100 hPa were similar to the 10 hPa, thus clearly pointing out that the split of the Antarctic polar vortex occurred not only in the stratosphere, but also had extended to the upper troposphere (Varotsos 2003b). In addition, Varotsos (2003c) suggested that a roughly 10-day component in zonal mean temperature and wind at high latitudes of the southern hemisphere, which seemed to originate in the upper stratosphere, might be of crucial importance for the occurrence of this unprecedented event.

Roscoe *et al.* (2003) and Harnik *et al.* (2003) later reached similar conclusions. Specifically, the former study showed a higher wavenumber-2 activity in 2002, since 1979, by analysing the long-term radiosonde record from Halley Bay dating back to 1956. The latter study emphasized that the event, which appeared to set the stage for the SSW, occurred as early as mid-May 2002, comprising a large burst of wave flux into the stratosphere and a strong deceleration of the vortex during its early stage of development, as well as a deceleration of zonal winds at low latitudes. The altered wave geometry of the 2002 vortex allowed internal reflection of travelling wave 2. The actual SSW itself was shown to result from superposition of a stationary wave-1 event with the traveling wave-2.

Thus, the aforementioned analyses reveal that the dynamical forcing from both the troposphere and stratosphere early in the winter of 2002 influenced the morphology of the stratosphere in a significant and unusual manner, setting the stage for the unprecedented major SSW in late September. However, there is a disparity among the considerations of various researchers on whether the primary direct cause of the major SSW is of tropospheric or stratospheric origin.

Manney *et al.* (2003), performing a mechanistic model simulation initialized on 14 September, 2002, reproduced the dynamical features of the 2002 Antarctic major SSW which showed a pattern of strong upward wave propagation throughout the SSW, with a zonal wind deceleration throughout the stratosphere at high latitudes before the vortex split, continuing in the middle and upper stratosphere and spreading to lower latitudes after the split. The latter implies that the anomalously strong tropospheric forcing could be viewed as the primary direct cause of the SSW, which was accompanied by rapid, large-scale transport of air from the subtropics to the polar region, which is also in accordance with Krueger *et al.* (2003).

The tropospheric origin of the major SSW has also been pointed out by Newman and Nash (2003). They noticed that the polar night jet was considerably weaker than normal, and was displaced more poleward than has been observed in previous winters; it resulted from a series of wave events (propagated from the troposphere) that took place over the course of the winter, which preconditioned the polar night jet for the extremely large wave event of 22 September. Furthermore, they also enumerated the morphology of planetary wave energy deposition in the stratosphere, and speculated about a possible role of the tropics in the tropospheric wave forcing. The consideration that the forcing from small-scale gravity waves generated in the troposphere might play a crucial role in initiating the SSW over Antarctica was also suggested by Wang *et al.* (2003).

Charlton *et al.* (2003) argued that the dynamics of the vortex split involved non-linear dynamics in the combined troposphere–stratosphere system, and that the concept of forcing from the lower atmosphere might be inappropriate. They proposed that, as the vortex elongated, it became hydrodynamically unstable, and that this instability could have affected the whole upper troposphere and stratosphere. In this respect, Kushner and Polvani (2003) demonstrated that a simple stratospheric model, which is forced only by tropospheric baroclinic eddies spontaneously, can give rise to a SSW that is similar to the observed southern hemisphere SSW of September 2002. They also suggested (in accordance with Baldwin *et al.*, 2002) that more complicated models to explain the 2002 event were not required, because the extreme events like this one occur over sufficiently long time periods as a result of random fluctuations in the coupled stratosphere–troposphere system (skewed distribution of warming temperatures). In agreement with this, Krueger *et al.* (2003) suggested that the analysis of winter 2002 revealed the dominant role of eastward travelling waves and their interaction with quasi-stationary planetary waves forced in the troposphere which finally led to the sudden break-up of the vortex, in accordance with the results obtained by Varotsos (2002b, 2003b). In addition, it should be emphasized that the break-up of the polar stratospheric vortex in the northern hemisphere is an event that is known to be predictable for up to a week or so ahead. For the split of the southern stratospheric vortex in late September 2002, however, the highly unusual flow conditions nevertheless exposed a computational instability of the forecast model (Simmons *et al.*, 2003). The observations and analyses from 1957 onwards in the ERA-40 archive confirm the extreme nature of the 2002 event.

Finally, it is worth noting that Shindell *et al.* (1998) proposed that temperature and wind changes induced by the increasing concentrations of GHGs alter the planetary-wave propagation, thus reducing the frequency of SSWs in the northern hemisphere. This results in a more stable Arctic and Antarctic polar vortex, with significantly colder temperatures in the lower stratosphere. However, Roscoe *et al.* (2003) have come to the conclusion that the southern hemisphere vortex had probably not split before the major SSW of 2002 since Antarctic records began in the late 1950s, and almost certainly not in August or September. This conclusion appears to contradict the results of Shindell *et al.* (1998) – of a decreasing likelihood for a vortex split given increasing GHGs. It should be emphasized at this point that according to Varotsos (2003b), the SSWs in Antarctica observed before 1980 were considerably more prevalent. In 1968, a year of early SSW, the TOC over the South Pole in October was much higher than that in 1969, a year of late SSW. During 1964–1979, a delay in SSW relative to the time of SSW in 1978 occurred at the South Pole in only 6 out of 16 years. From 1979 through 1985 (and also in 1986 and 1987), SSW delays relative to the 1978 SSW occurred in all years.

It is, therefore, of specific interest to examine the impact of the aforementioned extreme behaviour of the Antarctic polar vortex to the Antarctic ozone hole in 2002. It should be kept in mind that due to the model calculations made by Shindell *et al.*

(1998), the Arctic ozone losses would reach a maximum in the decade 2010 to 2019, roughly a decade after the maximum in the stratospheric chlorine abundance.

7.6.3 The Antarctic ozone hole

7.6.3.1 The diminutive Antarctic ozone hole and its split in 2002

It is known that the ozone hole is not stationary; it rotates and meanders round the Antarctic continent depending on atmospheric circulation conditions. The spatial structure of the TOC field is nearly similar to that of the low stratosphere temperature field (there is a positive correlation between TOC and temperature: a TOC decrease is accompanied by a temperature decrease), but this similarity varies from year to year, reflecting the sensitivity to external factors (Kondratyev and Varotsos, 2000).

This strong correlation in conjunction with the occurrence of the unprecedented major SSW over Antarctica in late September 2002 might explain the fact that the ozone hole in 2002 was much smaller than in the previous six years (Baldwin *et al.*, 2002; Varotsos, 2002a; Charlton *et al.*, 2003; Eskes *et al.*, 2003; Grooss *et al.*, 2003; Hoppel *et al.*, 2003; Krueger *et al.*, 2003; Manney *et al.*, 2003; Scaife *et al.*, 2003; Sinnhuber *et al.*, 2003; Thompson *et al.*, 2003, Wang *et al.*, 2003; Weber *et al.*, 2003). An additional factor for the diminutive ozone hole in 2002 is the fact that after the major SSW, a sudden decline in PSCs activity occurred (Nedoluha *et al.*, 2003).

Apart from the smaller sized ozone hole over Antarctica a second observational finding refers to its splitting into two parts in late September 2002. The main explanation for the latter is based on the arguments discussed in the previous section, notably: due to the major SSW the basic polar vortex in the stratosphere of the southern hemisphere during 21–26 September, 2002 was step-by-step elongated and distorted up to the occurrence of the polar vortex split, thus leading to a corresponding split of the ozone hole. Although there is a close agreement on this explanation in all the existing studies in the field (Baldwin *et al.*, 2002; Varotsos, 2002a,b; Varotsos, 2003a,b,c; Charlton *et al.*, 2003; Eskes *et al.*, 2003; Grooss *et al.*, 2003; Hoppel *et al.*, 2003; Krueger *et al.* 2003; Manney *et al.*, 2003; Newman and Nash, 2003; Scaife *et al.*, 2003; Sinnhuber *et al.* 2003; Thompson *et al.*, 2003; Wang *et al.*, 2003) some of the specific findings on the dynamics of this event differ substantially and are presented below.

First, focusing on the most characteristic features of the variability in the vertical ozone distribution inside and outside the Antarctic polar vortex before, during, and after its break-up, we refer to the study performed by Kondragunta *et al.* (2003), which reported that due to the existence of an ozone depleted region between 0 and 24 km in size, the ozone destruction was present in the elongated but un-split vortex in the lower stratosphere; ozone in the mid to upper stratosphere was not vertically aligned with ozone in the upper troposphere and lower stratosphere. This offset in the vertical distribution of ozone resulted in higher TOC and masked ozone depletion in the lower stratosphere. This finding is in close agreement with the earlier described morphology of the southern polar vortex (Section 7.6.2.3) but it has to be considered against the depicted chemical ozone loss values in other studies.

In this respect, a complementary study by Hoppel *et al.* (2003) showed that the 21–60-km ozone partial column, derived from the Polar Ozone and Aerosol Measurement (POAM) profile measurements, increased by ~150 DU during the major SSW. They also suggested that in the altitude region below the 600-K potential temperature the 2002 ozone loss was similar to previous years up to the time of the major SSW. The ozone loss diminished after this time (about 1 week earlier than usual), resulting in up to 20% less chemical ozone loss within the vortex than in previous winters. This corresponds to partial column ozone values below 600 K inside the polar vortex that were ~25 DU larger in 2002 than in previous years. In addition, an offline 3-D chemical transport model used by Feng *et al.* (2003) found that the ozone chemical loss rates in the polar lower stratosphere in September 2002 were generally less than in 2000, due to smaller amounts of average active chlorine, although around the time of SSW the largest chemical loss rates were similar to those in 2000 (see also, Fonteyn *et al.*, 2003).

Secondly, focusing on the main features of the variability in TOC inside and outside the Antarctic polar vortex before, during, and after its break-up, we refer to the study performed by Eskes *et al.* (2003). According to this study the ozone hole of 2002 as measured by the Total Ozone Mapping Spectrometer (TOMS) showed anomalously high TOC. During the major SSW, the daily minimum of TOMS TOC was about 90 DU higher than in recent winters, and the area contained within the 220-DU contour (nominal ozone hole threshold value) was reduced to 10% of its typical size (see also, Stolarski *et al.*, 2003). In agreement with the later finding, a CTM used by Sinnhuber *et al.* (2003) showed that the Antarctic TOC was significantly higher during the winter and spring of 2002 than during the previous years due to the increased transport during the winter of 2002, while the estimated chemical loss of TOC of about 80 DU south of 60°S is comparable to earlier winters, although the chemical ozone depletion inside the polar vortex was terminated by the major SSW much earlier than in normal ozone hole winters. In the view of chemical simulations, the ozone depletion rates in the polar vortex parts decrease rapidly to zero, while outside the polar vortex high ozone depletion rates were observed, caused mainly by NO_x chemistry (Grooss *et al.*, 2003).

7.6.3.2 The interplay between total ozone and temperature

Nowadays, there is growing evidence that stratospheric variability plays a crucial role in driving climate variability at the troposphere and at the Earth's surface on a range of timescales. Thus, according to the discussion in the previous section, the major SSW of 2002 that provoked large amplitude variation in the strength of the southern hemisphere polar vortex can induce an anomaly in the tropospheric circulation, which can also be followed by coherent changes in Antarctic surface temperatures. In fact, the examination of the temporal evolution of the tropospheric circulation following large amplitude variations in the strength of the southern hemisphere polar vortex in data from 1979–2001, and following the major SSW of 2002, showed that in both cases, anomalies in the strength of the southern hemisphere polar vortex precede similarly signed anomalies in the tropospheric

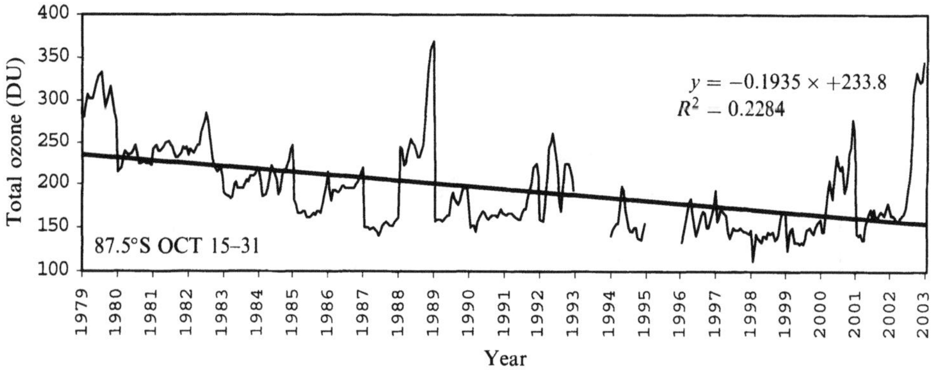

Figure 7.9. Mean TOC measured at South Pole (Antarctica), between the 15–31 October time periods.

circulation, which reflect a bias in the polarity of the southern hemisphere annular mode, and changes in surface temperatures (Thompson *et al.*, 2003).

In this respect, Jadin and Vargin (2003) concentrated on the fact that the relative ozone decrease (increase) in the eastern Antarctic is tightly associated with westerly (easterly) zonal wind anomalies near the southern tip of South America. In particular, the unusual behaviour of the ozone hole in 2002 appears to be caused by great easterlies in this region. For the understanding of long-term changes of the atmospheric circulation and ozone layer, they proposed a simple mechanism of an interference of planetary waves excited by the standing orographic sources (Andes in the southern hemisphere and Rockies in the northern hemisphere) and by the variable thermal sources depending on the sea surface temperature (SST) anomalies with well-defined spatial structure (dipole across Andes in the southern hemisphere and dipole across Rockies in the northern hemisphere).

To investigate further the aforementioned results the TOC observations for the 15–31 October time periods (following the Antarctic polar night) during 1979–2002 over the South Pole made by TOMS flown on Nimbus 7, Meteor 3, and Earth Probe satellites, are shown in Figure 7.9.

The principal characteristics in Figure 7.9 are the well known long-term negative trend in TOC after 1979 (mainly attributed to the stratospheric ozone destruction especially in the presence of PSCs) and TOC fluctuations (mainly induced from the tropical quasi-biennial oscillation (QBO), or the El Niño ENSO effect). Another interesting feature in Figure 7.9 is the unexpected recovery of TOC over the South Pole mainly in 1988 and 2002 (to values comparable with those observed well before the beginning of the TOC depletion effect over Antarctica; Varotsos 2003a,b,c).

To examine further the plausible connection between SST anomalies in the eastern equatorial Pacific and the TOC variability over Antarctica, Figure 7.10 illustrates the difference ($\Delta SST = \max - \min$) in consecutive maxima and minima of seasonal SST anomalies in eastern equatorial Pacific, since 1950.

It can be clearly seen from Figure 7.10 that the two diminutive ozone hole episodes over Antarctica in 1988 and 2002 (to values comparable with those

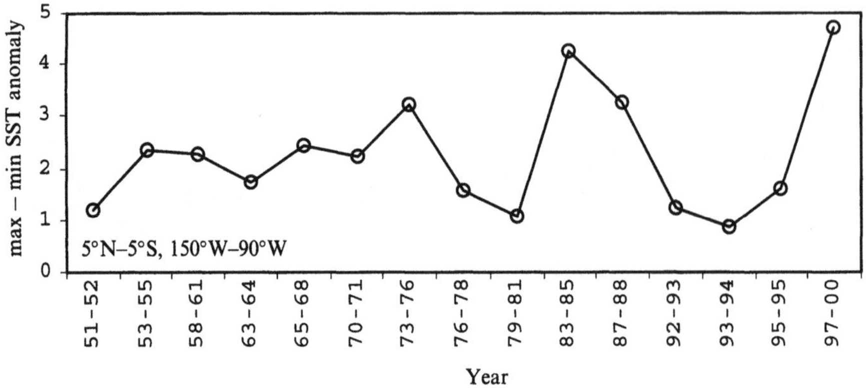

Figure 7.10. The difference (max – min) in consecutive extreme values of the seasonal SST anomalies in eastern equatorial Pacific during the 1950–2002 period.

observed well before the beginning of the TOC depletion effect) followed the two very high differences of the extremes in seasonal SST anomalies in the eastern equatorial Pacific, since 1950. The occurrence of the strongest difference (change) in SST anomalies is an effect that could have strongly increased planetary wave activity in the southern hemisphere leading to an early SSW in Antarctica and the early breakdown of the 2002 Antarctic polar vortex (Varotsos, 2003c). Furthermore, Figure 7.10 implies that one should not only look at the equilibrium atmospheric response to a given SST anomaly, but rather consider the transient problem of how the atmosphere adjusts to fast/slow SST anomaly changes (assuming the atmospheric adjustment time is comparable to the SST anomaly timescale).

Overall, relating TOC anomalies in 1988 and 2002 to the two major El Niño ENSO events in 1982–1983 and 1997–1998 requires proposition of a mechanism to explain the origin of the time delay. Of course, with such a short time record it will be difficult, based solely on observations, to reject the null hypothesis of no relationships between South Pole TOC and ENSO SST. However, Komhyr *et al.* (1991) proposed such a mechanism, according to which a linkage between long-term SST changes in the tropics and global TOC variations exists. This mechanism is that SST changes modulate: the interaction between equatorial QBO winds and extra-tropical planetary waves (by variations in convective activity) that disperse ozone from the tropical stratospheric ozone source region to other parts of the globe; the Haddley Cell circulation; and planetary wave activity at higher latitudes (through teleconnections) affecting ozone and heat transport to polar regions. The latter influences the formation of PSCs and the ozone destruction by heterogeneous photochemical processes (Varotsos, 2003b).

7.6.3.3 The near record size of the 2003 ozone hole over Antarctica

According to recent satellite observations, the Antarctic ozone hole in 2003 became the second largest ever observed. In particular, its size in 2003 (10.9 million square miles on 11 September), was comparable to the largest ever recorded (11.5 million square miles on 10 September, 2000). In both cases (2000 and 2003) the extremely cold temperatures (especially near the edge of Antarctica) allowed the heterogeneous chemical reactions (which are substantially active at very low temperatures) to lead to a considerable chemical ozone loss. In contrast, the occurrence of the "no-ozone-hole" episodes over Antarctica in 1988 and 2002 were accompanied by SSWs (the latter being a major one).

As noted above, the year 2002 "no-ozone-hole" episode does not denote a recovery of the ozone layer. In fact, this was confirmed by the appearance of the near record size 2003 ozone hole. Most probably the year 2002 "no-ozone-hole" episode over Antarctica was dictated by the intrinsic chaos of atmospheric dynamics, but this is currently unknown (Solomon, 2004).

7.6.4 Conclusions

Despite the fact that prior to September 2002 the occasional major SSW in the southern hemisphere were considered a model "defect" (Baldwin *et al.*, 2003), observations showed that the "forbidden" event actually occurred in 2002 and was accompanied by the following characteristic features (Varotsos, 2004b):

1. The dynamic forcing from both thc troposphere and stratosphere early in the winter of 2002 influenced the morphology of the stratosphere in a significant and unusual manner, setting the stage for the unprecedented major SSW in late September. As the vortex elongated, it became hydrodynamically unstable, and this instability affected the upper troposphere and stratosphere. The middle stratospheric vortex split into two parts; one rapidly mixed with extra-vortex air, while the other returned to the pole as a much weaker and smaller vortex.
2. The Antarctic TOC was significantly higher during winter and spring of 2002 than during previous years due to the increased transport during the winter of 2002. The estimated chemical loss of TOC of about 80 DU south of 60°S is comparable to earlier winters, although the chemical ozone depletion inside the polar vortex was terminated by the major SSW much earlier than in normal ozone hole winters.
3. Due to the existence of an ozone-depleted region between 0 and 24 km, the ozone destruction was present in the elongated but unsplit vortex in the lower stratosphere; ozone in the mid to upper stratosphere was not vertically aligned with ozone in the upper troposphere and lower stratosphere. This offset in the vertical distribution of ozone resulted in higher TOC and masked ozone depletion in the lower stratosphere.
4. A relationship between South Pole TOC and ENSO SST is evident. The relative ozone decrease (increase) in the eastern Antarctic is tightly associated with the westerly (easterly) zonal wind anomalies near the southern tip of South America,

and therefore, the unusual behaviour of the ozone hole in 2002 appears to be caused by great easterlies in this region.

5 The appearance of the near record size of the 2003 ozone hole confirmed that the year 2002 "no-ozone-hole" episode does not denote a recovery of the ozone layer.

8

Global climate change and geoinformation monitoring

8.1 REALITY, HYPOTHESES, AND FICTIONS

The unprecedented attention to the problems of the climate during the last several decades (this also refers, in particular, to mass media) has stimulated the calculation of both theoretical and applied developments, provided great advances in understanding the causes of the present climate change, regularities of paleoclimate, and in substantiation of scenarios of possible future climate changes (the question of scenarios and not of predictions are rather doubtful) (Adamenko and Kondratyev, 1999; Lal and Yarasawa, 2001; Lal, 2001). Unfortunately, various speculative exaggerations and apocalyptic prognoses (e.g., complete melting of the Arctic sea ice in the first-half of the 21st century), due to which the problems of climate change formed as a concept of an anthropogenically induced global warming, have become a focus of geopolicy (Kondratyev, 1992b, 1999b; Akovetsky *et al.*, 1991; Savinykh, 1993; Lafreur, 1999; Kukla, 2000). The many questions which have arisen with respect to the Kyoto Protocol have been discussed by the Presidents and Prime Ministers of various countries. There is opinion that the Kyoto Protocol (KP) is not a scientifically substantiated document (Kondratyev, 2000a,b). In particular, the confused character of the situation is determined by the absence of clear and agreed-upon terminology. Digressing from a very complicated situation with the definition of the notion of climate, one should bear in mind, for instance, that so far the notion of "climate change" has been defined as anthropogenically induced, although one of the main unsolved problems arises with the absence of convincing quantitative estimates of the contribution of anthropogenic factors into the formation of the global climate (nobody has doubts, however, that the anthropogenic effects on climate do exist). In the international documents containing an analysis of the present ideas about climate, the notion of "consensus" with respect to the scientific conclusions made in such documents was widely used, as if the development of science was determined not by different views and respective discussions but by a general agreement on concrete problems. Apart from definitions, of importance is the problem of uncertain conceptual estimates of various aspects of climate problems (Kheshgi *et al.*, 1999a,b; Kelley *et al.*, 1999a,b).

These (and other) facts become especially urgent in the context of the World Symposium on Sustainable Development held in September 2002 in Johannesburg. At the meeting of eight countries in Genoa (July 2001) at the suggestion of V. Putin, President of Russia, it was decided to convene in Moscow the World Conference on Climate that was held in September 2003. In this regard it should be mentioned that the success of the UN Second Conference on the Environment and Development (Rio de Janeiro, 1992) and the Special Session of the UN General Assembly "Rio+5" (New York, 1997) only drew to the attention of governments and the public the problems of global changes and sustainable development. Unfortunately, both these World forums were poorly prepared. Perhaps the main indicator of unsatisfactory preparations was the failure of attempts to work out the "Earth's Chart" aimed at the formulation and substantiation of priorities. Instead, an amorphous and declarative document "The Rio Declaration" was approved (Kondratyev, 1998a, 1999c).

Three global ecological problems are not in the focus of the document: (1) climate change ("global warming"), (2) the fate of the ozone layer in the stratosphere, and (3) closed global biogeochemical cycles (the concept of biotic regulation of the environment). A regrettable paradox exists – despite the convincingly substantiated primary importance of the third problem in scientific literature, and the secondary importance of the first two problems, the above-mentioned documents reflect the lack of an adequate understanding of the conceptually important fact that the following succession of events is of the fundamental importance: socio-economic development (stimulated by the growth of population size) → anthropogenic impact on the biosphere → consequences of such impacts for the environment (climate, ozone, etc.).

The lack of understanding has resulted in making the problem of "global warming" of first priority. The UN Framework Convention on Climate Change (UNFCCC), lame, disorientating, and unjust with respect to developing countries, focused without grounds on the anthropogenic origin of the observed global climate warming and on recommendations (first of all, for industrial countries) to reduce greenhouse gases (GHG) emissions into the atmosphere (and first of all, carbon dioxide).

As has been mentioned, in December 1997 more than 160 countries participated in the Third Conference of the UNFCCC signatories held in Kyoto (Japan). Long and sharp discussions were concentrated on whether it is possible to accept the demand to reduce CO_2 emissions by 2008–2012 by about 5% on average (with respect to emissions at the 1990 level). These discussions did not bring about a reduction of CO_2 emissions. Global CO_2 emissions continue to increase not only in developing countries but also in many developed countries. Naturally, the position of the developing countries is that their first-priority goal is to raise the human living standards, not the curtailment of industry for the sake of CO_2 emissions reduction. However, it is the latter that is the condition for ratification of the UNFCCC brought up by the USA and other countries of the "golden billion". The history of the UNFCCC is one illustration of a huge (mainly bureaucratic) activity that absorbs hundreds of millions of dollars annually. From the report of the Global Ecological Foundation (GEF) as of 30 July, 1998, financing of 267 projects by the GEF constituted 1.9 billion dollars (Kondratyev, 2002c). At the World Summit in Johannesburg (South Africa, 2002) an agreement was reached that 2.9 billion dollars will be added to the GEF. Note should be taken that about 10 thousand people participated in the Conference in Kyoto. The recent Conference in the Hague (November 2000) was also large, and representatives of 178 countries (mainly officials and not specialists) participated in the meeting in Bonn (July 2001).

One may think that the situation described is the consequence of an undeveloped scientific basis for the problems of global change. Such a conclusion is only partially valid, since as far back as 1990, for instance, monographs by Gorshkov (1990) and Kondratyev (1990) were published dedicated to a discussion of the key aspects of global ecology. These authors have brought forth and substantiated the fundamental concept of biotic regulation of the environment, and in other works (Krapivin *et al.*, 1982; Kondratyev *et al.*, 2002a, 2003c; Krapivin and Kondratyev, 2002; Svirezhev, 2002) they proved the scientific groundlessness of the "greenhouse" hypothesis of global warming and drew attention to a necessity to study the climate system "atmosphere–ocean–land–ice cover–biosphere" taking into account the complicated feedbacks between its interactive components. A serious analysis has been made of the problem of the global observation system, especially when the developments in the field of remote sensing and the use of respective observation data are concerned. The problem of variability of atmospheric ozone is quite special.

Let us consider the problem of global climate changes as most vividly reflecting the existing delusions in the field of global ecodynamics. The most important circumstances are:

- Observation data (still inadequate from the viewpoint of completeness and reliability) do not contain clear confirmation of the existence of anthropogenic global warming (this especially concerns the data of ground observations in the USA, the Arctic, and the results of microwave satellite remote sensing).
- If an enhancement of the atmospheric greenhouse effect due to a supposed doubling of CO_2 concentration constitutes about $4\,W/m^2$, then uncertainties connected with the consideration of the climate forming role of atmospheric aerosols and clouds as well as with an introduction of the so-called "flux correction" in numerical climate modelling, reaches tens and even hundreds of W/m^2.
- Results of numerical climate modelling that substantiate the hypothesis of the greenhouse global warming and ostensibly agree with observational data, are doing nothing more than fitting the observation data.
- On the basis of these results, recommendations about the levels of GHG emissions reduction are senseless (their realization, however, may have far-reaching negative socio-economic consequences).

From the data of numerical modelling (Wigley, 1998, 1999; Wigley and Raper, 2001) (if its reliability raises no doubts), even the complete realization of the Kyoto Protocol recommendations can ensure only a decrease of mean annual mean global surface air temperature not exceeding several hundredths of a degree.

During the last few years, serious attention has been drawn to the analysis of uncertainties (inadequacy) in the numerical climate modelling. Perhaps, inadequately considered interactive processes in the "aerosol–clouds–radiation" system represent the most serious source of uncertainties (Kondratyev, 2002c). There is no doubt that the most important aspect of numerical climate modelling is connected with the consideration of interactive dynamics of the biosphere, which can be illustrated by two concrete examples that, of course, only to a small degree, reflect the complexity of the problem.

From the observation data for the period 1951–1993 it was found out that the amplitude of the diurnal course of the surface air temperature decreased by about 3°–5°K due to a more rapid increase of the minimum temperature compared to the maximum one. To explain this, it was proposed to take into account the effect of various factors: changes in the amount of clouds, the content of water vapour and tropospheric aerosol, turbulence, and soil moisture. Positive trends of the first three enumerated factors could lead to a decrease in net radiation in the daytime and to the growing counter-radiation of the atmosphere (CRA) at night, whereas changes in the turbulent mixing intensity and soil moisture can result in variations of heat and moisture exchange between the Earth's surface and the atmosphere, which should be much more substantial in the daytime than at night.

The strong interactivity of the processes and insufficient adequacy of their parameterization in climate models complicate the estimation of contributions of various mechanisms for reducing the surface air temperature. In this regard, Collatz *et al.* (2000), using the land biosphere model SiB2 (Sellers *et al.*, 1996b), realized numerical experiments to evaluate the response of the diurnal temperature of the low atmosphere to the external disturbing forcings and to the biophysical state of vegetation cover.

Analysis of the numerical modelling results has shown that an increase of CRA causes an increase in air temperature over the vegetation cover T_m at night, favouring a decrease in the surface air temperature, whereas changes of T_m or increase of T_m + CRA (this can occur in conditions of global warming) favour an increase of both minimum and maximum temperatures, which determines the slight effect of these factors on the surface air temperature. Such a response is determined mainly by the effect of the diurnal course of the aerodynamic stability and radiation balance.

Many numerical experiments on climate modelling are based on the use of the atmospheric general circulation model (GCM) together with land surface models (LSMs). Results of such numerical experiments depend substantially on special interactions between GCMs and LSMs designed for reproducing an exchange of radiation, momentum, and energy between the Earth's surface and the atmosphere.

Desire to take into account diverse ecosystems land has strongly complicated the LSM by inclusion within them of submodels taking into account the processes of photosynthesis, vegetation cover dynamics, and biogeochemical cycles, radically increasing the models' reliability.

Based on the use of a simple model of the biosphere SiB as one of the LSM versions, Kim *et al.* (2001) carried out various numerical experiments on the estimation of its sensitivity. The results have shown that a knowledge of model sensitivity with respect to numerous morphological parameters of vegetation cover is important. The improved SiB2 provided a means of consideration of the biogeochemical processes determining an exchange of water vapour, energy and carbon dioxide between the Earth's surface and the atmosphere. A comparison has been made of the results of numerical modelling of the processes on a test site of rice fields in Thailand (17°03′N, 99°42′E) using models SiB2 and the modified model Sib2-Paddy (rice checks) together with the meso-meteorological model GAME-Tropics (Global Energy and Water Cycle Experiment (GEWEX)-related Asian Monsoon Experiment) with data from meteorological observations in the rainy season (1–6 September, 1999).

The comparison revealed a good agreement between the results obtained with the use of the two considered models with observational data for the diurnal course of the radiation budget and the latent heat flux except for the values of the latter calculated using the Sib2 model. Fluxes of latent heat and soil heat as well as the rates of assimilation of carbon calculated with the SiB2-Paddy model agree satisfactorily with those observed, but the use of the SiB2 model is connected with substantial systematic errors. With some adjustment of the parameters, the SiB2-Paddy model provides reliable estimates of temperature of soil, water, and vegetation cover. Results of calculations of the radiation balance, balances of energy and water, as well as latent heat and the rate of assimilation of carbon dioxide turned out to be quite adequate. Such results create certain prospects for an adequate consideration of the biosphere as an interactive component of the climate system.

De Rosnay *et al.* (2000) estimated the reliability of the schemes of parameterization of land surface processes, used in the models of atmospheric general circulation (MAGC), from the viewpoint of correspondence to observational data of calculated values of mean annual fluxes of energy and moisture depending on a detailed consideration of the vertical structure of the soil. Results of calculations reveal a strong dependence of these fluxes on the vertical resolution. An 11-layer scheme of parameterization of the heat and moisture transport in soils with the upper 1 mm thick layer is quite adequate. Possibilities of realization of the scheme with so thin an upper layer are not clear, however, providing that the horizontal resolution of the MAGC constitutes hundreds of kilometres, the solution of such a problem needs further effort.

A complex of problems concerning the atmospheric chemistry is an important part of the problem of numerical climate modelling. For instance, it is well known that various short-lived minor gaseous components (MGC) substantially affect the formation of the field of concentration of tropospheric ozone (TO) in different conditions (city, regional, and global distribution). These MGCs are precursors of ozone. For instance, nitrogen oxides, methane, numerous organic compounds, hydrogen, and carbon oxides. Each of these MGCs has specific natural (biospheric) and anthropogenic sources.

Since TO is a greenhouse gas, emissions of the enumerated MGCs can indirectly govern the formation of the atmospheric greenhouse effect influencing the field of TO concentration. Moreover, MGCs can change the field of hydroxil concentration and oxidizing capacity of the troposphere. In its turn, the distribution of the hydroxyl concentration in the troposphere controls the lifetime of methane and thus the level of its concentration on global scales.

These circumstances cause a complicated interactivity of processes determining the direct and indirect impacts on the formation of the atmospheric greenhouse effect. Derwent *et al.* (2001) described the global 3-D Lagrangian model that parametrized chemical processes with a consideration of the MGC transport. This model was used to study the fields of TO and methane concentrations when the short-lived TO precursors (CH_4, CO, NO_x and hydrogen) were emitted to the atmosphere. The radiative forcing (RF) due to NO_x depended on the spatial location of NO_x emissions. For each of the short-lived MGCs (TO precursors) the global warming potential (GWP) was calculated for the 100-years period. An introduction of GWP means an estimation of RF, for instance, due to emissions of 1 t of any MGCs calculated (for 100 years) as an equivalent (by RF) emission of carbon dioxide. In the case of a joint effect of methane and TO, the GWP value constituted 23.3.

Analysis of the calculation results has shown that the indirect RF due to changes in the concentrations of methane and TO turned out to be considerable for all MGCs in question. While the RF due to changes of methane concentrations is mainly determined by the effect of emissions of methane itself, in the case of TO it is controlled by all MGC precursors, especially nitrogen oxides. The TO associated indirect RF can be so great that MGCs–TO precursors should be considered as those of MGCs, which should be taken into account when estimating the possible changes of climate and measures to prevent them.

Despite the "anti-Kyoto" statements of President G. W. Bush, the headlines of many American newspapers on January 2001 were dramatic: "Scientists publish frightful prognoses of warming; accelerated climate shifting forecasts a global disaster in this century" (*Washington Post*); "The Earth's warming brings forth a new signal of danger" (*International Herald Tribune*). New scenarios of climate warming in the 21st century, suggesting such changes could be more substantial than earlier supposed, were responsible for this hullabaloo. According to the Third IPCC-2001 Report (IPCC, 2001b), an increase of the mean annual mean global surface air temperature (SAT) could reach 5.8°C by 2100 compared to present-day (Houghton *et al.*, 2001). In 1996 the respective estimate was only 3.5°C.

As Karl and Gleckler (2001) justly pointed out, of far greater importance is the fact that the supposed range of a possible increase of SAT turned out to be wider. However, for many specialists in the field of numerical climate modelling it was unexpected, since this sphere of science is still in the initial stage of development and, besides, the numerical modelling has to be based on the use of a very limited volume of observed data. Even a data series on SAT constitutes only about 100 years.

Though most of the specialists believe that, probably, the observed global warming has been caused mainly by the growing concentrations of GHGs, in some respects the range of estimates of possible climate changes has increased.

Basic uncertainties of the estimates of climate change include three aspects:

1. Detection of global warming from observed data.
2. Attributing the global warming to anthropogenic causes.
3. Prediction of future climate changes.

Karl and Gleckler (2001) believe that new data in the IPCC-2001 report (2001b) narrow the range of uncertainties with respect to the first two aspects of the problem mentioned above, but predictions of future climate became even less convincing. According to IPCC-2001, the observed global warming constituted 0.6°C±0.2°C (at a 95% level of statistical significance), with "most of the warming observed during the last 50 years being probably (with a 66–90% probability) determined by the growing concentration of greenhouse gases". One of the main causes of uncertainties of numerical climate modelling is still connected with an inadequate account of the climate forming role of atmospheric aerosol and clouds. This especially concerns the estimates of the effect of aerosol on clouds and, respectively, on climate, which varies widely (see for details Section 8.2).

Unfortunately, the role of the uncertainties of numerical climate modelling has not been assessed adequately in the IPCC-2001 report (IPCC, 2001c). It is this that prompted many specialists to seriously criticize the report (Kondratyev and Demirchian, 2001a; Essenhigh, 2001; Schrope, 2001).

In the context of the problems of global change, estimates of the present and possible future global climate changes are of primary importance (Kondratyev, 2001a–c). Although until now these estimates demonstrate the prevailing concept of global warming confirmed by the Third IPCC report (2001a), apparently this is only an inertia of the earlier speculative ideas, whose motivation was far from being scientific, which has been convincingly analysed by Boehmer-Christiansen (2000). Radically different opinions in this respect between two presidential contenders in the USA election campaign served as a vivid illustration of the inconsistency of estimates concerning the climate. While A. Gore is known as an eager supporter of the global warming concept and the KP, G.W. Bush's opinion is as follows: "I object to (ecological) policy similar to that corresponding to the

KP, which would lead to a radical increase of prices of petrol, oil products to heat dwellings, natural gas and electricity. Such an agreement would strongly increase the load on the economics of the USA, without any protection from undesirable climate changes. The KP is ineffective, inadequate and unjust with respect to America, since it excludes 80% of the world from participation in realization of the KP recommendations, including the main centers of population concentration, such as China and India" (Kondratyev, 2001c). According to G.W. Bush, of primary importance is the development of new ecologically pure technologies and the use of market mechanisms including freedom from regulation of the markets of electricity and natural gas, taxation as well as "emissions trade" (G.W. Bush believes that natural gas and nuclear energy will play an important role in the reduction of the dangerous dependence of the USA on foreign oil and in providing the country with energy resources in the 21st century). Agreeing with the radical critics of the KP, note however, that the statements of G.W. Bush on market mechanisms are either special for the USA or (in the case of "emissions trade") at least controversial. Nevertheless, the USA were the first country which expressed a desire to free its economics from groundless ecological limitations, bringing forth several deserving suggestions: to consider nuclear energy as an important perspective for future energy and to search and study other ways to maintain the balance of carbon (including measures on reconstruction of forests as sinks for carbon).

It becomes apparent that despite a certain advance in the parameterization of the climate forming processes, the undertaken attempts to compare the results of numerical climate modelling with observed data are rather schematic, controversial and unconvincing. This is connected with extremely complicated climate models and inefficiency of applied modelling technologies. However, an accumulated experience in modelling the climatic processes gives numerous examples of successful parameterizations of some fragments of the global climate system. In particular, several indicative estimates of the role of forests in the formation of regional changes of climatic characteristics are known. So, for instance, Gedney and Valdes (2000) carried out numerical experiments reproducing the present ("control") climate and conditions of complete deforestation of the region of the Amazon Basin. For this purpose, a 190-level spectral model of atmospheric general circulation was used. Deforestation should result in the following changes of the main climate forming parameters: albedo (13.1% → 17.7%); roughness (2.65 m → 0.2 m); share of vegetation cover (0.95 → 0.85); leaf index (4.9 → 1.9); minimum resistance of vegetation cover (150 s/m → 200 s/m); and depth of the root zone (1.5 m → 1.0 m). All this will also lead to changes in the type of soils. A consideration of the numerical modelling results revealed statistically substantial changes of precipitation in winter in the north-eastern sector of the Atlantic Ocean due to deforestation, which propagated further eastward towards western Europe. Such variations are connected with changes in the large-scale atmospheric circulation in middle and high latitudes. Use of a simple model to reproduce such variations confirmed that they are caused by propagation of planetary waves, testifying to the fact that the results under consideration, which reveal connections between processes in the region of deforestation, in the North Atlantic Ocean and in western Europe, are independent of the choice of the model, the level of changes corresponding to estimates of anthropogenic climate changes due to the growth of CO_2 and aerosol concentrations.

Using the model of global climate developed at the National Centre for Atmospheric Research NCAR (USA), Zhang *et al.* (2001) carried out a much more extensive numerical modelling of the climatic consequences of deforestation in the tropics in conditions of growing greenhouse warming due to a doubled atmospheric CO_2 concentration. Calculations suggested a conclusion about a sharp decrease in evapotranspiration (~by 180 mm/year) and precipitation (~by 312 mm/year), as well as an increase of SAT by 3.0°K in the Amazon Basin. Similar but weaker changes take place in South East Asia (a precipitation decrease equal to 172 mm/year and warming by 2.1°K). Still weaker changes occur in Africa (precipitation grows by 25 mm/year). Analysis of the energy balance estimates suggested a conclusion that climate warming takes place not only due to an enhancement of the greenhouse effect but also as a result of reduced evapotranspiration caused by deforestation. Statistically substantial climate changes due to deforestation in the tropics also appear in middle latitudes.

The IPCC-1996 report (IPCC, 1996) contains the sharply debated conclusion: "The balance of available data supposes a marked impact of man on global climate" (Kondratyev, 1998a,b) as well as the statement that, "the anthropogenic signal" manifests itself against a background of natural climatic variability. According to the IPCC-2001 report (IPCC, 2001a), "Studies on detection and attribution regularly reveal the presence of an anthropogenic signal in the data of climate observations for the last 35–50 years... Natural impacts could have played a role in the observed warming during the first-half of the 20th century, but they cannot explain the warming in the second-half of the century." However, the report also contains the following statement: "Reconstruction of climate for the last 1000 years testifies to a small probability that the climate warming observed in the second-half of the 20th century could have been of totally natural origin." Further highly uncertain quantitative estimates of the anthropogenic warming are emphasized, especially from the viewpoint of contributions from various factors of the warming (first of all, it refers to atmospheric aerosols). Inconsistency and weakness of the cited statements and conclusions are so apparent that no comment is needed. Of course, the integral models describing the dynamics of interactions between the socio-economic development and nature, and taking into account the functional multidimensionality of the nature–society systems (NSS), should play the leading role in substantiation of future climate predictions. However, it is still unclear how real are the predictions obtained with the use of such models. One should hope that, at least, in the foreseeable future the integral models may be the only means to obtain rather conditional scenarios, and the expected consequences of anthropogenic activities will be estimated using the global model of the NSS (GMNSS) within these scenarios and give prognoses of climate changes for the periods agreeing with the geographical information monitoring system (GIMS)-technologies of forecasts (Kondratyev *et al.*, 2002b).

8.2 RADIATIVE FORCING STIPULATED BY AEROSOLS

8.2.1 State of investigations

As has been mentioned, the conclusions of the IPCC-2001 report (IPCC, 2001a) are based mainly on the results of numerical climate modelling obtained with the use of 40 scenarios of the possible dynamics of GHG emissions in the future. Such scenarios, which depend on prescribed factors, which cannot be reliably predicted, such as changes of the global population size, socio-economic development, and scientific–technical progress, are very uncertain, which is illustrated in particular, by a wide range of possible cumulative GHG emissions (in carbon equivalent) of 770–2540 Gt C by 2100. The respective estimation of the probable range of CO_2 concentrations in the atmosphere in 2100 gave 540–970 ppm. However, these estimates do not take into account the possible measures for the reduction of GHG emissions as well as the progress of technologies

According to IPCC-2001, an expected increase of SAT by 2100 by 1.4–5.8°C will be caused by a high rate of SAT variability, which will exceed the observed rate of SAT increase in the 20th century by a factor of 2–7. The pre-calculated increase of the level of the World Ocean could reach 9–88 cm, such an increase can (due to gigantic thermal inertia of the climate system) continue over several centuries after stabilization of the level of CO_2 concentration (this also refers to SAT, though to a lesser degree). Such a high uncertainty of the estimates of possible future climate changes is connected not only with the inadequacy of the existing information on the role of GHGs but also, to a high degree, with the role of atmospheric aerosol as a climate forming factor. This especially concerns the estimates of the so-called indirect aerosol radiative forcing (ARF) connected with the effect of aerosols on the microphysical and optical properties of clouds.

The main source of these uncertainties is the deficit of data on extremely variable properties of aerosols. Therefore, during the last years, serious efforts have been undertaken to obtain data on aerosol based on the use of various methods of remote sensing (Savinykh, 1991) (ground, aircraft, satellite) and complex field observations. Bearing in mind the IPCC-2001 report (IPCC, 2001a) and several reviews of the obtained results, we shall discuss only some of them. Among the most significant complex observation experiments are the Advanced Composition Explorer (ACE)-1 (measurements of the main characteristics of aerosols), the Tropospheric Aerosol Radiative Forcing Observational Experiment (TARFOX; for determination of the aerosol radiative forcing), ACE-2 (studies of aerosol in the region of the Pacific Ocean), the Indian Ocean Experiment (INDOEX; studying aerosols in the region of the Indian Ocean), and others. In addition, there was the global network of stations AERONET (Aerosol Robotic Network) equipped with sun photometers to measure the aerosol optical depth (AOD) of the atmosphere, as well as lidar soundings made at various points to obtain information on the vertical profiles of aerosol concentration.

Wagner *et al.* (2001) performed comparative calculations of ARF in the levels of the upper atmospheric layer and surface with the use of observations of the aerosol characteristics for the whole atmospheric thickness with their vertical profiles taken into account. Input data for calculations were results of lidar sounding (at six wavelengths over the range 355–1064 nm) as well as measurements with the sun photometer made on 25 March, 1999 within the INDOEX programme. On this day there was a multilayer haze in the troposphere (at altitudes up to 3.5 km), characterized by the presence of strongly absorbing particles. The AOD at the wavelength 530 nm reached 0.57 (with the Angstrom exponent 0.9).

Calculations of ARF with the prescribed ocean albedo 0.05 and aerosol properties averaged over the atmospheric thickness (the prescribed complex refraction indices are $1.575 - 0.05i$ and $1.65 - 0.035i$) gave ARF values ranging between -5 and -12, and -55 and $-81\,\mathrm{W/m^2}$ at levels of the upper atmospheric boundary and the surface, respectively. With the vertical profile of the aerosol characteristics taken into account (from the data on the coefficients of backscattering obtained from lidar sounding at 6 wavelengths and on the coefficient of extinction at wavelengths 355 and 532 nm, the vertical profiles of the efficient radius of particles, number density, and complex index of refraction were retrieved), the respective values of ARF are equal to -10 and $-60\,\mathrm{W/m^2}$. The calculated values of the ARF efficiency (calculated per unit optical thickness) vary within -11 and $-24\,\mathrm{W/m^2}$ and -101 and $-154\,\mathrm{W/m^2}$, respectively.

Since atmospheric aerosol is one factor having both a direct and indirect effect on the climate, at the initiative of NASA and GEWEX, the Global Aerosol Climatology Project (GACP) was born in 1998. The main goals of GACP are:

- retrieval of aerosol characteristics (including their annual course and interannual variations) on global scales from the data of satellite observations of outgoing short-wave radiation and from routine observations; and
- numerical modelling of the processes of formation, transformation, and transport of aerosol.

The final goal is to obtain representative data on the global climatology of aerosol for the whole period of satellite monitoring, which could be used to study direct and indirect effects of aerosols on climate.

The most important results obtained during the first 3-year phase of GACP are as follows (Mischnenko *et al.*, 2002):

- substantiation of algorithms to retrieve the aerosol characteristics from the data of multipurpose measurements of the outgoing short-wave radiation;
- analysis of the information content of the existing satellite observations from the viewpoint of the aerosol properties retrieval;
- comparative analysis of scenarios of the variability of aerosol properties suggested by various groups of specialists; and
- development of methods to use the data of routine observations to assess the reliability of the aerosol properties retrieved from the data of satellite monitoring.

The results obtained testify to the complex nature of the indirect impacts of aerosol on climate and to the need for a complex approach to the solution of this problem. The most important expected result of realization of the first phase of GACP should be a substantiation of the three versions of the global climatology of aerosols including data on AOD and average size of of particles, with the use of the 1 and 2-channel Advanced Very-High Resolution Radiometer (AVHRR) data as well as the Total Ozone Mapping Spectrometer (TOMS) data. Also plans have been made to apply the numerical models of aerosol transport (with chemical transformations taken into account) to theoretically substantiate aerosol climatology. Bearing in mind the imperfection of the AVHRR calibration, data of improved devices such as the Moderate-Resolution Imaging Spectroradiometer (MODIS) and the Multi-angle Imaging Spectroradiometer (MISR, a multi-angular video-spectroradiometer) are supposed to be used to specify the calibration.

As has been mentioned, most complicated are the estimates of indirect ARF. In agreement with the classical theory, clouds consist of "activated" droplets whose sizes increase spontaneously after reaching a certain critical size (at a certain level of oversaturation of the surrounding water vapour). In this context, clouds are a specific component of the atmosphere with respect to wet non-activated and dry aerosol particles. Sometimes a quite different situation is observed in the atmosphere determined by interactions in the multiphase system consisting of clouds, aerosols, and the gaseous atmosphere. There is an exchange of mass between the components of this system determined by various physical and chemical processes. So, for instance, the sizes of aerosol particles of natural or anthropogenic origin can grow with increasing relative humidity, with such particles finally becoming cloud droplets. Therefore, there is no doubt that clouds should not be considered separately but as a component of the system mentioned above.

Another principally important circumstance consists in an exclusive variety of the spatial–temporal scales (from molecular to global), which are connected with the formation and evolution of cloud cover. The interaction of clouds and aerosols is quite special, substantially affecting both the cloud cover dynamics and aerosols. Analysis of the data from satellite observations has shown that a decrease of the size of cloud droplets caused by aerosol particles functioning as cloud condensation nuclei (CCN) leads to a precipitation reduction, which is manifested over regions of biomass burning, atmospheric pollution, and dust outbreaks in the deserts.

The interaction of clouds with the surrounding atmosphere is also of great importance, especially with water-soluble MGCs such as NH_3, SO_2, HNO_3, organic acids with a small molecular weight, and aldehydes. Although chemical processes with the participation of ice cloud particles remain almost unstudied, the field observations have shown, for instance, that such particles play an important role in removing MGCs from the atmosphere, such as H_2O_2 and NH_3 absorbed by ice crystals. Also it turned out that H_2O_2 reacts with SO_2 much slower in ice particles than in water droplets.

Laboratory measurements have led to the conclusion that soot can be an efficient source of condensation nuclei. An increase of the small droplets number density due to CCN depends not only on the CCN number density but also on a number of other factors, such as available moisture, rate of upward fluxes, air temperature, etc., as well as the microstructure and chemical composition of the aerosol. Over many years it was considered that the chemical composition of CCN is determined by the presence (in them) of soluble inorganic salts – NaCl or NH_4SO_4. Only inorganic aerosol was taken into account in the simplified theory, whereas in the conditions of the real atmosphere organic compounds are of great importance as parts of aerosols. Based on the use of the modified theory, an important role of organic CCN in cloud formation was demonstrated, manifested through a considerable decrease (in the case of atmospheric pollution up to 110%) of the level of critical oversaturation, especially in the presence of small CCN particles. The share of water-soluble organic compounds (WSOC) varies within 20–70% with respect to total content of WSOC.

Numerical modelling has shown that a consideration of the organic component of CCN causes an increase in cloud albedo, which is comparable with the effect of doubling the CCN concentration. Another important factor causing changes of cloud properties are the chemical reactions within them with the participation of MGC and aerosol, depending on the cloud microstructure (since their chemical composition is a function of their size). Two main compounds participating in the reactions in a liquid phase are compounds of sulphur and organic compounds. For instance, oxidation of 80% S(IV) to S(VI) takes place in a liquid phase.

Similar processes with the participation of organic compounds remain, however, almost unstudied. It is one of the causes of the remaining uncertainties of estimates of the contribution of aerosol–cloud interaction as a climate forming factor characterized by a substantial non-linearity, which determines the urgency of laboratory studies of photochemical reactions in clouds with the participation of organic compounds, studies of chemical reactions in ice clouds, chemical composition of aerosol particles and clouds, and consideration of these and other factors in the models of the cloud-cover dynamics.

8.2.2 Direct ARF

An accomplishment of the Complex Energy Experiment (Kondratyev, 1991a, 2003b; Kondratyev *et al.*, 1973, 1983; Kondratyev and Binenko, 2000) has made it possible, based on the use of data from aircraft observations, to obtain for the first time, information on ARF (both integral and spectral) and aerosol properties. However, this complex experiment could not be completed. But, different separate observations were being realized.

As Boucher (2002) noted, an understanding has been reached now that the aerosol-induced RF manifests itself in the form of direct, semidirect, and indirect (of the first and second type) forcing. The direct forcing is the impact of the total attenuation of short-wave radiation by aerosols and the thermal emission of aerosols (this impact manifests itself mainly in conditions of a cloudless atmosphere). The semidirect forcing characterizes the impact of radiation absorption by aerosols in conditions of both clear and cloudy skies on the formation of the vertical profiles of temperature and humidity, which determines the functioning of the feedback that changes the cloud-cover dynamics.

The first type of indirect forcing consists as an increase of the optical thickness and albedo of clouds as a result of a decreasing size and an increasing number density of cloud droplets (with the constant water content of clouds) under the influence of aerosol particles functioning as CCN. The second type of indirect forcing is connected with the growth of the cloud water content, the altitude of the cloud top, or the increase of the clouds' lifetime as a result of decreasing rain rate due to a decrease of the size of cloud droplets.

Direct ARF depends on a set of parameters, which can be classified into three categories:

1. Biochemical parameters (such as intensity of emissions of primary and secondary aerosols, and aerosol lifetime).
2. Characteristics of the atmosphere (relative humidity, cloud amount, and surface albedo).
3. Microphysical parameters (microstructure, optical properties, and hygroscopicity).

Concrete data on the variability of direct ARF and estimates of the respective errors are still limited, which reflects first of all, the need for extending the programmes of field observations of aerosol properties with the use of various observational means including satellites. This will make it possible to carry out a more complete correction of an adequacy of aerosol models and parameterization of the aerosol characteristics in the climate numerical models. The key directions of further developments should include, in particular:

- specification of an available archive of data on aerosol emissions (especially anthropogenic);
- extension of observations of the vertical profiles of aerosol characteristics (methods of ground and satellite lidar sounding should play a special role);
- experimental and theoretical studies of the multicomponent aerosol, including obtaining data on hygroscopic and optical properties (single scattering albedo included);
- analysis of the contributions of natural and anthropogenic aerosols; and
- improving the aerosol models with new data from observations taken into account.

Hansen and Sato (2001) believe that an increase of the mean annual mean global SAT for the last century (by more than 0.5°C), was caused at least partially by anthropogenic forcings on the climate. In this regard, calculations of RF due to the growth of the GHG concentrations were made. The obtained estimates testify to a considerable level of RF due to methane, whose contribution reached about half the RF caused by carbon dioxide: $(1.4 \pm 0.2)\ W/m^2$ (the error is a subjective estimate of dispersion 1σ), including $0.5\ W/m^2$ due to direct and $0.28\ W/m^2$ due to indirect forcing. These results testify to the requirement for special attention to be given to the consideration of CH_4 as a GHG with a separate consideration of RF due to methane.

New estimates of RF due to TO give 0.7–0.8 W/m^2 instead of 0.3–0.4 W/m^2 as is supposed in the IPCC-2001 report (IPCC, 2001a). Apparently, for this RF the interval 0.4–0.8 W/m^2 can be assumed.

Considerably erroneous (due to uncertain input data) calculations of RF due to "black" carbon (soot) gives values of $0.8 \pm 0.4\,W/m^2$. The RF due to "reflecting" aerosols constituting $-1.4 \pm 0.5\,W/m^2$ with an additional contribution from soil aerosols of $-0.1 \pm 0.2\,W/m^2$ plays an important role. A substantial contribution to the total ARF is made by organic aerosols ($-0.3\,W/m^2$) and ammonium sulphate ($-0.2\,Wm^2$).

The sum of all positive components of RF reaches $4.3 \pm 0.6\,W/m^2$ – three times greater than the RF due to CO_2, whereas the sum of the negative component is $-2.7 \pm 0.9\,W/m^2$ and thus the resulting RF is equal to $1.6 \pm 1.1\,W/m^2$ (i.e., close to that due to CO_2 only.

Hansen and Sato (2001) performed an analysis of the trends of different GHGs, from which it follows that the rate of the CO_2 concentration growth increased rapidly during the period from the end of World War II to the mid-1970s, reaching a maximum of about 4%/year, after which it started decreasing to a comparatively stable concentration growth level of 1.5 ppm/year. During the same period the CH_4 concentration grew rapidly (from 5 to 15 ppm/year), slowing down considerably after 1980 (the causes of this are still unknown). The total trend in the rate of RF growth due to 13 chlorofluorocarbon (CFC) compounds was characterized by a rapid intensification until the mid-1980s and a subsequent decrease due to CFC emission reduction in the process of realization of the measures foreseen by the Montreal Protocol. Apparently, before 2010 the RF should start decreasing due to CFC reductions. The maximum rate of the growth of total RF (for the period 100 years) reached about 5 W/m^2. With this rate of growth preserved, the level of the equivalent RF corresponding to a doubled CO_2 concentration should be reached by 2050. However, during the last two decades, the rate of the net RF growth decreased to 3 W/m^2. Taking into account a possible reduction of GHG emissions, the global warming during the several forthcoming decades should constitute $+0.15 \pm 0.05°C$ over ten years under the condition that the level of CO_2 emissions is "frozen". The considerable preserved uncertainty of the estimates is connected, first of all, with the difficulty in calculating the aerosol-induced RF.

Bearing in mind the importance of comparative estimates of RF due to GHGs, let us consider the results of the latest calculations of RF due to GHGs discussed in Hauglustaine (2002). According to the estimates obtained, the mean global total RF due to the growth in the atmosphere of the concentration of well mixed GHGs in the industrial epoch (after 1750) constitutes 2.43 W/m^2, with some of its components having values of (W/m^2): 1.46 (CO_2); 0.48 (CH_4); 0.15 (N_2O); and 0.34 (CFC) with an error of about 10%.

The RF is characterized by a strong spatial variability. So, for instance, in 1980 the mean global total RF reached 5 W/m^2 (per 100 years), and the subsequent decrease was connected with a sharp decrease of emissions into the atmosphere of the ozone destroying CFCs (in conformity with the respective international agreements) as well as with the retarded growth of CO_2 and CH_4 concentrations in the atmosphere (even a decrease of methane concentration continuing for several years was observed). Apart from the GHGs mentioned above, a substantial contribution (up to 8–15%) to the formation of total RF was also made by TO, for which an initial value of RF was $0.35 \pm 0.15\,W/m^2$, however, during the last years this value was specified at $0.7\,W/m^2$.

The errors of the estimates of RF due to various GHGs were mainly caused by a deficit of information about the spatial–temporal variability of the GHG content in the atmosphere, which depends on the power of the surface sources of GHG emissions and GHG sinks in the free atmosphere. This refers, for instance, to natural emissions of compounds, which contribute most to the variability of the long-lived GHGs and the TO-precursors, such as NO_x and non-methane hydrocarbons (NMHC). The effect of indirect factors and feedbacks on RF can be illustrated using carbon monoxide as an example. Calculations have shown that with CO emissions increasing by 40 Tg, the RF due to TO increases, and this increase is 10–20% greater if these emissions originate in the tropics and not in middle latitudes. With the increase of NO_x emissions assumed to be equivalent to 1 Tg, such a difference in the RF variability reaches a factor of 6–8, with the indirect effect of NO_x on RF being negative (i.e., it leads to an ozone-induced decrease of RF). These examples illustrate the critical importance of consideration of the MGC emissions geography.

The height at which the RF takes place plays an important role. In this context, of special importance are processes in the "upper troposphere–lower stratosphere" layer, which necessitates the development of models that take into account chemical processes both in the troposphere and in the stratosphere.

An important factor controlling the lifetime of most of GHGs and other MGCs that pollute the atmosphere is the oxidizing capability of the atmosphere determined mainly by the concentration of OH hydroxyl and other oxidants (O_3, HO_2, H_2O_2, or NO_3). Prognostic estimates of the oxidizing capabilities are extremely uncertain, since they depend on factors such as the content of water vapour and aerosols, the troposphere–stratosphere exchange, intrusion into the troposphere of UV radiation, and the preceding evolution of the oxidizing capability.

A decrease in the level of the MGC-induced uncertainty of RF estimates depends on further progress in the development of interactive models of the tropospheric composition considering the complicated interactions between the atmosphere, ocean, and land biosphere, with respect to physical, chemical, and biological processes taking place in these components of the climate system. Of course, the adequacy of such models requires a thorough test using observational data. The problem of adequacy is connected with the complexity of the models, which takes into account 50–100 chemical compounds and 100–3000 photochemical reactions with a spatial resolution $2^\circ \times 2^\circ$ and consideration of a 20–60-layer atmosphere. One more important aspect of the discussed problems consists of the fact that the RF concept was developed as applied to an analysis of the mean global situation. Therefore, it is necessary to analyse its special features as functions of the scales of spatial–temporal averaging.

In particular, Podgorny and Ramanathan (2001) analysed the results of studies of microphysical, chemical, and optical properties of aerosols (including data from lidar sounding) obtained in accomplishing the field observation experiment INDOEX, and obtained a complete database on atmospheric aerosol, which can be used to control the reliability of the techniques of numerical modelling of radiation transfer in the atmosphere (the main goal of INDOEX was to analyse the effect of atmospheric pollution on the climate forming processes in the tropics of the Indian Ocean). In this regard, calculations have been undertaken of the direct radiative forcing due to aerosols and clouds and of the respective rate of radiative heating for characteristic conditions of the winter monsoon in the tropics of the Indian Ocean. The main goal of the numerical modelling based on the use of the Monte Carlo method was an analysis of the redistribution of short-wave radiation between the ocean mixed layer and the atmosphere above it, caused by natural and anthropogenic aerosols.

According to the observed data, the single scattering albedo $\tilde{\omega}$ at wavelength 500 nm varied between 0.8–0.9, the aerosol optical thickness of the atmosphere τ_A averaged from 0.1 to 0.8, and the cloud amount f (with a characteristic horizontal scale of several kilometres) constituted 25% (this value is typical for lower clouds over the tropical ocean). In many cases the aerosol layer was far above the lower clouds, which substantially affected the formation of ARF.

In conditions of a cloudy atmosphere, the soot aerosol descending onto the ocean surface from Asia and from the Indian subcontinent strongly affected the ARF (even with a 10% contribution into τ_A). For monthly mean conditions, $\tau_A = 0.4$, $\tilde{\omega} = 0.9$, and $f = 25\%$. The diurnal mean ARF at the surface level R_s was equal to $-25\,\mathrm{W/m^2}$ and ARF for the atmosphere (i.e., the absorbed solar radiation R_a) varied between 22–25 $\mathrm{W/m^2}$. The ARF variations at the atmospheric top (R_T) were within 0 and $-3\,\mathrm{W/m^2}$. The presence of aerosol determines an enhancement of the "cloud" RF by 0.5 or 2.5 $\mathrm{W/m^2}$, if the aerosol is located mainly either below or above clouds, respectively. The data in Table 8.1 illustrates the values of the ARF in three typical regions.

The aerosol-induced rate of radiative heating in the trade wind atmospheric boundary layer (ABL) reaches 1–1.5°K/day, which can strongly affect the formation of the trade wind cumulus clouds. Thus the main effect of the atmospheric aerosol over the Pacific Ocean (this aerosol is mainly of anthropogenic origin) is

Table 8.1. Typical estimates of RF in three regions ($\mathrm{W/m^2}$).

Region	f (%)	τ_A	R_T	R_a	R_s
Arabian Sea	15	0.8	−6.3	44.1	−50.4
Maldive Islands	25	0.4	−0.8	24.0	−24
Intertropical convergence zone (ITCZ)	50	0.1	+1.7	6.9	−5.2

manifested through a redistribution of short-wave radiation between the atmosphere and the mixed layer of the ocean.

Remer *et al.* (2002) discussed the possibilities of retrieving the aerosol-induced RF at the top of a cloud-free atmosphere from satellite observations in connection with the beginning of the functioning of new types of space-borne instruments, including MODIS, which opened up potential for obtaining various information about the atmospheric aerosol properties. The numerical modelling of the processes of distant aerosol transport enables one to study the laws of the global distribution of aerosols and in this regard to substantiate requirements for the data of the aerosol satellite remote sensing, including a retrieval of RF.

Remer *et al.* (2002) analysed the results of the numerical modelling obtained with the use of a combined model of the tropospheric general circulation and chemical processes in the troposphere, which determine the formation and evolution of the aerosol properties, including the following types of an accumulative mode of aerosol: sulphates, organic carbon, black carbon, methanosulphonic acid, sea salts, soil dust, and marine aerosol (fractions of large-size dust and sea salt aerosols are considered separately). The total aerosol optical thickness (AOT) is determined by contributions from the accumulative and two rough-disperse modes, although there is also the mode of ultra-small particles, but its contribution into AOT is negligibly small. Since these components of aerosols are considered as a mixed component, this excludes the possibility of evaluating the contribution of individual components to the formation of the total AOT. Naturally, the errors of AOT retrievals from satellite data are at a minimum with high contents of aerosols in the atmosphere.

Remer *et al.* (2002) considered the example of the calculations of the monthly mean spatial distribution of smoke ARF resulting from biomass burning in the southern hemisphere, with the use of two models of distant transport of aerosols. Their results showed that 87–89% of ARF received was due to the smoke exceeded the noise level. This demonstrates that satellite monitoring of ARF is possible. Basic sources of errors are connected with inadequate data about the background aerosol levels, errors in prescribed optical characteristics of the surface, and some characteristics of aerosols. The combined effect of these three sources of errors forms a total uncertainty of ARF of 1.2–2.2 W/m^2. However, it is important that these estimates of errors do not take into account the contribution of the possible (residual) effect of clouds. The level of errors can be reduced with the combined use of the data of satellite and ground observations, as well as results of numerical modelling of the distant aerosol transport.

Estimates of the aerosol-induced direct RF obtained for several types of aerosol (sulphate soot both due to fossil fuel burning and organic) taken into account gave values between −0.25 to −1.0 W/m^2, whereas the indirect RF (caused by changes of the microstructure and optical properties of clouds under the influence of aerosols) varied between 0 to −1.5 W/m^2. The possible interval of the values of RF due to GHGs constituted +2.1 to + 2.8 W/m^2 (i.e., the aerosol-induced RF is close in value but opposite in sign to the greenhouse RF). The problem consists, however, in the considerable uncertainty of the ARF values. In this regard, Christopher and Zhang

(2002) calculated the RF due to aerosols formed from biomass burning in the tropics, which should be important, since more than 114 Tg of smoke observed in the tropics result from biomass burning.

To calculate the short-wave direct RF, the calibrated hourly data of the Geostationary Observing Environment Satellite (GOES)-8 scanner was used for the period 20 June–30 August, 1998, for South American-tropics (4°S, 51°–65°W). Based on the use of the Mie formulas for calculations of optical characteristics of individual aerosol particles as well as methods of discrete ordinates of radiation transport theory, the AOT of smoke formed by biomass burning was calculated at the wavelength 0.67 μm. The results were compared with the AOT from ground observations. Using the AOT values retrieved from the GOES-8 data, fluxes of outgoing short-wave radiation (OSWR) were calculated (applying the 4-flux approximation) and then compared (in some cases) with the OSWR measurements carried out within the framework of the Clouds and Earth's Radiant Energy System (CERES) programme to study cloud cover and the Earth radiation budget (ERB).

The results obtained by Christopher and Zhang (2002) point to the fact that the AOT values retrieved from the GOES-8 data agree well with the data of ground observations from AERONET (with the correlation coefficient 0.97). The correlation coefficient for CERES data and for the numerical modelling of OSWR was 0.94. From the data for August 1998 the daily averaged values of AOT and ARF in the region under study were, respectively, $0.62 \pm 0.39\,W/m^2$ and $-45.8 \pm 18.8\,W/m^2$.

Weaver *et al.* (2002) calculated the short and long-wave RF due to Saharan aerosol using the results of numerical modelling with a 3-D global model of the aerosol transport and the approximated schemes of parameterization of the short and long-wave radiation transport. Calculations enabled the retrieval (with prescribed results of an objective analysis of the fields of meteorological parameters) of the 3-D fields of the concentration of the four sizes of mineral aerosol particles, which served as input information for calculating the radiation fluxes and RF. The results show that the dust aerosol causes a decrease in the outgoing short-wave radiation (this conclusion agrees with the data from the Earth Radiation Budget Experiment (ERBE) satellite measurements). Although a similar agreement is also observed in the case of the outgoing long-wave radiation, in this case the errors of an objective analysis of the temperature field caused by neglect of the effect of aerosol on the temperature field is a complicating circumstance. Depending on the level of the atmospheric dust content, the values of the short-wave RF at the atmospheric top vary within $0 \pm -18\,W/m^2$ (over land).

From observations with space-borne instruments on SeaWiFS over the World Ocean, Chou *et al.* (2002) retrieved values of AOT τ_A which were used to calculate the ARF with a prescribed total atmospheric moisture content (retrieved from Special Sensor Microwave Imager (SSM/I) data), monthly mean total content of ozone (from TOMS data), and the field of air temperature (from data of an objective analysis of the meteorological information carried out by NCAR).

Analysis of the obtained results revealed mainly the zonal distribution of τ_A, whose basic features of spatial distribution are determined by the wind field convergence in the lower troposphere. Maximum values of τ_A are found in the equatorial band, in high latitudes of the southern hemisphere in the Arabian Sea, the Gulf of Bengal, as well as in the coastal regions of eastern and South East Asia. Minimum values of τ_A were recorded in the subtropical convergence zones, especially in the southern hemisphere. During most of the year, great values of τ_A are characteristic of the tropical Atlantic Ocean. The dominating trade winds transport the dust aerosols from Africa across the Atlantic Ocean to the coastlines of Central and South America. Great (small) values of AOT are typical of the zones of convergence (divergence). The mean annual value of τ_A at the wavelength 0.865 μm constitutes 0.105.

The mean annual values of ARF over the World Ocean at the level of the upper atmosphere boundary (LUAB) and of the ocean surface are, respectively, −5.48 and −5.09 W/m^2. The annual course of ARF is negligible, on average, for the whole World Ocean, but it is substantial in the northern and southern hemispheres. The LUAB values of ARF during the period from October to April increase in the northern hemisphere by 33% and in the southern hemisphere (July–January) by 50%. The observed clear sky LUAB value of ARF (−5.48 W/m^2) corresponds to a cooling by about 2.5 W/m^2, if the average cloud conditions are taken into account. Thus the negative ARF is comparable in value with the positive "greenhouse" forcing (about 4.3 W/m^2) which should occur with a doubling of atmospheric CO_2. A strong ARF was observed during the period of intensive fires in Indonesia in 1997. Over large territories the value of ARF in the September–October 1997 period reached −108 W/m^2 at the atmospheric top and exceeded 25 W/m^2 at the surface. The presence of aerosol enhanced the heating of the atmosphere due to solar radiation absorption by about 15 W/m^2. The forest fires in Indonesia coincided with the period of an abnormally intensive El Niño event in 1997–1998, when Indonesia was affected by drought and downward air fluxes. The atmospheric heating and surface cooling caused by aerosols originating from forest fires could increase the atmospheric stability and, hence, reduce the atmospheric circulation thus enhancing drought.

8.2.3 Indirect aerosol forcing

Indirect aerosol forcing is determined by complicated aerosol–clouds interaction processes, which require special investigations (Flossman, 1998; Ten Brink *et al.*, 1999). Therefore, we shall only briefly consider the latest achievements in this field. By now, one can consider as complete the theory of formation of a single cloud droplet by activation and condensation growth of an aerosol particle on a CCN under the condition that CCNs have a simple chemical composition (in particular, they do not contain organic compounds) and they are in equilibrium with chemically active MGCs, such as SO_2, NH_3 and HNO_3. The numerical modelling of the process of cloud droplet formation to reproduce the microstructure of clouds is still a difficult problem, especially in the case of mesoscale processes and, moreover, when the atmospheric general circulation models require an adequate consideration of various properties of cloud cover. The main cause of complexity of such a problem is the necessity for an adequate reproduction of the dynamics of the field of oversaturation ($S_{v,w}$) round the CCN particles during their condensation growth until reaching a maximum of oversaturation ($S_{v,w}^{\max}$) occurring during a short time period. If is known, it raises the possibility of an accurate calculation of a minimum size of aerosol particle, which can be attributed to the category of cloud droplets. Such an approach was realized in detailed microphysical schemes used in the numerical modelling in the case of 1-D models of the pollution tails and 2-D γ-mesoscale models as well as models of large-scale vortices. Another possible (and more economic from the viewpoint of machine time expenditure) approach to a parameterization of the cloud droplet formation as applied to 3-D numerical modelling consists of the use of the concept of the activation spectrum ($N_{CCN} = f(S_{v,w})$), which contains all the necessary information about the ability of the aerosol population to stimulate the formation of cloud droplets.

Cohard *et al.* (2001) proposed an approximate method of parameterization of the relationship between the form of the spectrum of CCN activation and observed properties of the aerosol population (microstructure and solubility) functioning as CCN. In view of considerable differences between the properties of maritime and continental aerosols, these two types of aerosol are considered separately. The numerical modelling of the aerosol particles growth is realized within the interactive scheme describing the process of activation of cloud droplets. With respect to each of the two considered aerosol populations (maritime and continental), when considering the variability of solubility, the chemical composition of particles and the log-normal microstructure are assumed to be homogeneous. The reliability of the parameterization scheme provides the use of a large set of aerosol populations with randomized properties.

Estimates of the indirect climatic impact of aerosols through changing the cloud cover properties proceed from suppositions that either the cloud water content remains constant (the Tuomi effect of the cloud albedo increase due to the growth of the small droplets number density) or the water content increases with the growth of the droplets number density (the Albrecht effect of precipitation suppression in the zone of drizzle). On the other hand, both the results of the numerical modelling of thermodynamics and cloud cover dynamics and the observed data show that when the number density of cloud droplets grows the water content of clouds may decrease.

In this context Han *et al.* (2002) used the data of satellite observations of clouds in order to analyse the dependence of the cloud water content on the droplets number density. With this aim in view, from the data of observations for January, April, July, and October 1987 the sensitivity of cloud water (SCW) was estimated as the ratio of changes of water content to changes of the total number of droplets in clouds. Analysis of the global database for water clouds (the temperature of the clouds upper boundary is above 273°K, optical thickness $1 \leq \tau \leq 15$) revealed the presence of three regimes of changes in the water content depending on the changes of aerosols manifested as an increase, approximate stability, or decrease, with the total number density of cloud droplets increasing.

A consideration of the data of satellite observations suggested the conclusion that in approximately one-third of cases (mainly in conditions of a warm atmosphere) the SCW is negative (i.e., the cloud water content decreases with increasing number of droplets in clouds (the regional variability and the annual negative SCW agree with the data of other observations)). To one-third of these cases (with the constant water content) corresponds the relationship $r_{\mathrm{eff}} \approx N^{-1/3}$ (r_{eff} is an efficient radius of droplets and N is the droplets number density). In other cases, SCW is positive. The results obtained confirm the conclusion that in conditions of increasing cloud droplet number density the size of droplets can decrease and their evaporation directly beneath the cloud bottom can increase, and as a result, the processes in clouds and in the boundary layer are independent in a warm environment. Such a situation is accompanied both by a decrease of water input from the surface and by a decrease of water content in the clouds. The main conclusion drawn from the results discussed is that estimates of the negative indirect RF due to the aerosol–cloud interaction obtained using numerical climate models with the effects of Tuomi and Albrecht taken into account can be overestimated.

One of the manifestations of the indirect climatic impact of aerosols consists in the growth of the number density of small cloud droplets caused by anthropogenic aerosols and by the associated increase in the optical thickness and albedo of clouds, which causes a negative RF (the Tuomi effect) and partial compensation of the positive greenhouse RF. Harshvardan *et al.* (2002) estimated this effect with the combined use of numerical modelling results and data from satellite observations. With this aim in mind, the case was considered of an intrusion of sulphate aerosol polluted air into the North Atlantic Basin (mainly at altitudes of 2–3 km) from the European continent on 2–8 April, 1987, in the region 50°–55°N, 25°–30°W. The numerical modelling was carried out with the use of the global model of transport with chemical processes taken into account and with the prescribed data of the numerical modelling of the weather forecast. The evolution of microphysical characteristics of the lower maritime clouds (temperature (260–275°K) for the period 2–8 April, 1987) was retrieved from the AVHRR data in the visible (0.58–0.68 μm) and near-IR (3.55–3.93 μm) regions.

Analysis of the results obtained on the days with low and high levels of sulphate pollution permitted estimation of the respective changes of the cloud optical thickness at the wavelength 0.64 μm, number density, and efficient radius of particles. These changes testify to the real Tuomi effect. If with a low level of pollution the number density of droplets varied between 65–95 cm^{-3}, at a maximum level of pollution it increased to 300 cm^{-3}, and the efficient radius of droplets decreased simultaneously from 20 μm to less than 12 μm.

The satellite remote sensing data on changes of the efficient radius of cloud droplets (ERCD) can serve as one of the possible indicators of the indirect effect of atmospheric aerosols as CCNs causing an increase of the number density of small cloud droplets and, respectively, an increase of cloud albedo, resulting in a climate cooling. The earlier data revealed the 2 μm difference between the ERCD values over land and over the ocean and the 1 μm difference between the mean hemispherical values, which confirmed the validity of the Tuomi hypothesis. An increase of the concentration of small droplets results in a rain rate decrease, which was confirmed from satellite observations.

To further estimate the aerosol impact on clouds, Breon *et al.* (2002) undertook an analysis of measurements with the polarized radiometer (POLDER) carried by the ADEOS satellite (launched in August 1996) performed over the whole period of its functional life from 3 October, 1996, to 30 June, 1997 (observations could not be continued because of the malfunction of solar batteries). The results obtained demonstrated the considerable effect of aerosols on the microphysical characteristics of clouds on global scales. Maximum values of ERCD (14 μm) were revealed over distant regions of the tropical oceans, with minimum values (6 μm) over heavily polluted continental regions. High concentrations of small cloud droplets are characteristic of clouds located in the direction of the prevailing winds from the continents.

However, some circumstances complicate the final interpretation of the results under consideration:

- In view of specific features of the algorithm for retrieval, the obtained ERCD values are characterized by a certain selectivity, not covering the clouds of all types (they refer mainly to stratified clouds) and complete cycles of their evolution.
- The POLDER data make it possible to retrieve the content of aerosols in the atmosphere, whereas the Tuomi effect is determined by the concentration of CCN in the atmospheric layer where clouds are formed.

These and other circumstances do not permit us to reliably estimate the influence of the Tuomi effect on the RF (Tuomi *et al.*, 1999). Although an analysis of the discussed satellite data cannot be used to distinguish between the contributions of natural and anthropogenic aerosols, one should assume that the anthropogenic aerosol plays the prevailing role.

An active discussion of the indirect impact of aerosol on climate (through the effect of CCN on the processes of formation and various, especially optical, characteristics of clouds) has not led so far to sufficiently satisfactory results. Estimates of the RF varied widely from 0.0 to $-4.8\,W/m^2$, hence, these estimates are very uncertain. Various observed data used to verify the Tuomi effect have led to rather contradictory results. So, for instance, some studies have shown that this effect manifests itself in the presence of stratified clouds but is absent in the presence of cumulus clouds, although according to data in other publications, the Tuomi effect is also observed in the presence of cumulus clouds.

To resolve such contradictions, Feingold *et al.* (2001) carried out an analysis of the satellite remote sensing data obtained within the smoke/sulphate clouds and radiation (SCAR)-B programme on the studies of the interaction of smoke from forest fires, clouds, and radiation in Brazil. Processing of the observed data for two years (in the periods of biomass burning in August–September in 1987 and 1995) covering the region 20°S–5°N; 45°–70°W, suggested the conclusion about an expediency of distinguishing between three types of the cloud response to smoke aerosols:

- An increase of the cloud droplets number density when the content of aerosol increases until reaching saturation with a high concentration of aerosol.
- The same evolution as in the case above, but ultimately, the concentration of droplets grows as the aerosol content still increases. The latter is connected with the fact that high concentrations of large particles determine a suppression of oversaturation and prevent an activation of small particles but open up the possibility of activation of large particles, when the content of smoke aerosols exceeds a certain threshold level.
- The same as in the first case with a subsequent decrease of the droplets number density as the content of smoke aerosol increases, when an intensifying "struggle" for water leads to an evaporation of small droplets. This evaporation results in an unexpected increase in the size of droplets when the content of smoke grows.

A considerable uncertainty in the estimates of indirect (through changing the properties of cloud cover) effects of aerosol on the climate necessitates special efforts to specify the estimates of this effect. An increase in the cloud droplets number density (with constant cloud water content) is accompanied by a decrease in the efficient resource of r_{eff}, which favours an increase of cloud reflectivity. As has been mentioned, this variability was called the first indirect forcing. The second indirect forcing is connected with the fact that with decreasing average size of droplets accompanying the growth of the aerosol number density, the efficiency of rain formation decreases, which leads to the formation of clouds with a greater water content and longer lifetime. Observations confirmed that the anthropogenically induced decrease of r_{eff} can considerably change the water content of clouds and reduce precipitation, but the effect on the lifetime of clouds was not observed.

An experimental verification of the reality of the two types of the aerosol climatic impact is hindered by the following:

- Local character of observations of the properties of aerosol and clouds, which prevents them from being considering as globally representative.
- Difficult filtering out of anthropogenic changes of r_{eff} and water content from the natural ones.

In this regard, Menon *et al.* (2002) performed a numerical modelling of the climatic impact of aerosol based on the combined use of the climate model as well as models of chemical processes in the atmosphere with the participation of sulphur compounds, sources of organic substances, and sea salts determining the three types of aerosols considered in the model. To determine the cloud droplets number density, a diagnostic approach was used with the data of observations over land and the ocean taken into account. Information on the cloud droplets number density thus obtained was used to calculate the variability of r_{eff}, which in turn, permits pre-calculation of the impact of aerosols on the optical thickness of clouds and the rate of microphysical processes.

Estimates of the indirect climatic impact of aerosol were obtained from the data on the difference of values caused by variable characteristics of cloud cover and of the RF at the top level of the atmosphere corresponding to present and pre-industrial conditions. Also, in parameterization of the cloud cover dynamics, calculations were made of the sensitivity of the obtained results to the use of suppositions which determine the vertical distribution of the frequency of occurrence of clouds, rates of auto-conversion, and washing out of aerosol. The content of aerosols in the atmosphere depends on variations of all these parameters.

The numerical modelling results show that the mean global RF due to the indirect impact of aerosol (with the three types of aerosols taken into account) varies from -1.55 to $-4.36\,W/m^2$. These results depend markedly on the content of the pre-industrial background aerosol in the atmosphere (when this content is low, the climatic impact of aerosol is stronger) and depend less on the anthropogenic aerosol, whose increase is followed by an enhancement of the indirect impact of aerosol. A substantial dependence of the content of background aerosol determines an insufficient variability in the factors such as the relationship between the size of cloud particles and cloud albedo and as predictors of the aerosol climatic impact, which can be quantitatively estimated from satellite observations. Such a situation reflects the impossibility of verification of the estimates of the aerosol climatic impact from the observed data.

Analysis of the results of observations of ships' smoke emissions into the atmosphere on the lower clouds reveals a possibility for studying the effect of aerosols on the cloud cover characteristics in real conditions. To estimate the degree of impact of smoke aerosols on cloud properties, Coakley and Walsh (2002) used the satellite AVHRR data on cloud brightness (reflectivity) bearing in mind that when the lower clouds are above the ship, the clouds catch smoke particles which serve as CCNs and favour the formation of new cloud droplets. The growing droplets number density leads to a decrease in their size, which manifests itself, in particular, as an enhancement of reflected radiation at the wavelength 3.7 μm. Therefore, the AVHRR images of cloud cover at this wavelength can be used to reveal the polluted parts of the cloud cover.

An example of successful remote measurements in the visible range (0.64, 3.7, and 11 μm) is the reconstruction of the optical thickness of clouds, and the efficient radius of droplets and brightness temperature of polluted and clean parts of the cloud cover realized in July 1999 in the region of the western coastline of the USA. Analysis of several hundreds of the cloud cover fragments in the sectors of smoke tails about 30 km long has shown that the resulting changes in the optical thickness turn out to be half as much as expected, provided the observed changes in droplet radii and assumed stability of the water content of clouds are taken into account. Theoretical estimates of radiation fluxes have led to the conclusion that such a result cannot be explained only by the effect of radiation absorption by smoke particles. Apparently, the water content of polluted clouds decreases by 15–20% with respect to the original water content. An enhanced absorption of the short-wave radiation by clouds leads to evaporation of a considerable amount of cloud water, which reduces the convection in the ABL. The latter in turn leads to a decrease in the water vapour flux towards the polluted clouds compared to the flux of water vapour to a non-polluted cloud. The processes of absorption and evaporation as well as the contrast between the fluxes of water vapour to polluted and non-polluted parts of the cloud cover result in the loss of water by polluted parts of the clouds.

The conclusion drawn from the numerical modelling (with the observed data taken into account), that the mean global climatic ARF is close to the greenhouse RF, emphasizes the necessity to study the role of atmospheric aerosols not only as climate forming factors but also as an anthropogenic component, which affects the chemical processes in the atmosphere and, hence, its gaseous composition. For instance, a substantial effect of the polar stratospheric clouds particles and sulphate aerosols (though heterogeneous chemical reactions on their surfaces) on the formation of the field of ozone concentration in the stratosphere has been identified.

Since the analysis of the effect of aerosols on the chemical reactions in the troposphere is also important, Tie *et al.* (2001) undertook a numerical modelling of the impact of sulphate and carbon ("black") aerosols on tropospheric chemistry.

With this aim in view, a modified version of the NCAR global interactive model MOZART (Model for Ozone and Related Chemical Tracers) of the chemical processes in the atmosphere and atmospheric circulation (transport of MGCs and aerosol) was used.

A comparison of the results of calculations of the global field of sulphate aerosol concentration and its annual change with the observed data revealed a satisfactory agreement everywhere except the Arctic region. In particular, the vertical profiles of the sulphate aerosol concentration in North America agree well with the observed ones (this refers also to the results of calculations of the carbon aerosol distribution near the Earth's surface). The numerical modelling of the processes of formation of sulphate aerosols and of heterogeneous reactions on the surface of its particles has led to the following conclusions:

- The reaction $SO_2 + H_2O_2$ giving sulphate aerosols in cloud droplets reduces the concentration of hydrogen peroxide. In the gas phase reaction $SO_2 + OH$, OH is transformed into HO_2, but the resulting decrease in OH concentration and increase of HO_2 concentration are negligible ($<3\%$).
- The heterogeneous reaction on the surface of sulphate particles with the participation of HO_2 causes a 10% decrease in the concentration of radical of hydrogen peroxide HO_2 with the coefficient of capture 0.2. However, this value can be overestimated, which determines an expedience of considering this estimate as an upper limit.
- The reaction on the surface of sulphate particles with the participation of N_2O_5 leads to a decrease in concentrations of nitrogen oxides in middle and high latitudes in winter, reaching 80%. Since the intensity of the ozone formation in winter is low, a decrease of the ozone concentration caused by this reaction does not exceed 10%. However, in summer, a decrease of NO_x concentrations can reach 15%, and that of ozone (in middle and high latitudes) 8–9%.

- The heterogeneous chemical reaction on the surface of sulphate particles with the participation of CH_2O (with the upper limit of the coefficient of capture 0.01) leads to a decrease of CH_2O concentration within 80–90% and a decrease of HO_2 concentration in middle and high latitudes in winter by about 8–9%. The obtained estimates should be considered as preliminary.

Developments aimed at numerical modelling of the climate with the participation of an interactive impact of aerosols represented a new stage in the studies of indirect climatic impacts of aerosols and in ARF estimates. This approach is exemplified by the work of Jones *et al.* (2001) in which numerical modelling was undertaken of the impact of anthropogenic sulphate aerosols on cloud albedo and intensity of rain formation (e.g., considering both the first and the second indirect impacts) based on the use of a new version of the global climate model. This model uses a new scheme of parameterization of microphysical processes in clouds with an interactive consideration of the sulphur cycle and parameterization of the effect of sea salt aerosols.

Calculations gave the mean global indirect ARF value of $-1.9\,W/m^2$, with a prevailing contribution of the albedo effect. Estimates of contributions of the first and second indirect ARF gave values of -1.3 and $-0.5\,W/m^2$, respectively. The values of the total indirect ARF can be considered unreliable. Of course, the global geographic distribution of the indirect ARF is characterized by a high uncertainty in the presence of ARF maxima in the regions of the main sources of emissions of sulphur compounds located in the USA, Europe, and China. A comparison of estimates of the ARF due to short-wave and long-wave radiation made by Jones *et al.* (2001) has shown that the contribution of the short-wave ARF (which causes a climate warming) always prevails over that of the long-wave ARF causing a warming. An interactive consideration of the sulphate aerosol climatic impact plays a substantial role.

New ARF estimates with the use of the global climate model developed at the Tokyo University, with the distant aerosol transport taken into account but with a priori prescribed sources and sinks of aerosol, were obtained by Takemura (2002). The following types of aerosol were considered: sulphate, sea salt, carbon ("black carbon"), organic carbon, and soil dust. The mean annual mean global values of direct ARF constituted: $0.32\,W/m^2$ (anthropogenic sulphate aerosol), $+0.19$ and $-0.05\,W/m^2$ (black and organic carbon due to fossil fuel burning), and $+0.15$ and $-0.16\,W/m^2$ (black and organic carbon due to biomass burning). Estimation of the total indirect ARF gave $-1.0\,W/m^2$.

Consideration of the effect of aerosols on the distortion of the remote sensing results is one of the unsolved problems of the satellite monitoring. Therefore, along with the development of the methods of estimation of aerosol parameters it is necessary to develop algorithms of atmospheric correction of remote sensors, especially in the coastal regions, where anthropogenic and maritime aerosols get mixed-up (Smirnov *et al.*, 2003). During the last years (beginning from 1993) due to the AERONET radiometers, an extensive database has been accumulated on maritime aerosols in the zones of islands.

8.2.4 Conclusion

The problem of the assessment of the role of atmospheric aerosol as a climate forming factor is not a new one. An important new stage in understanding and quantitative estimation of this role has become an awareness of its critical importance in the formation of the present climate changes. This conclusion is contained, in particular, in the official fundamental documents, such as IPCC-2001 (IPCC, 2001a–c). The paradox occurs, however, in that with an emphasis on (and, of course, unjustifiably exaggerated) the problem of global warming and the KP (see Chapter 2) at the World Summit on Sustainable Development (Johannesburg, 29 August–6 September, 2002), the problem of the climate forming contribution of atmospheric aerosols was practically ignored. Meanwhile, the remaining uncertainty of the estimates of the contribution of aerosols to the formation of the climate deprives such an important document as the KP of scientific grounds.

The resulting conclusion is apparent: in the very near future it is necessary to place an emphasis on the "aerosol and climate" problem, bearing in mind both an intensification of the respective theoretical developments and, more importantly, obtaining more adequate information which would permit the reliable estimate of the climatic impact of aerosols. In this context, there was the hope that this problem would be thoroughly discussed at the World Conference on Climate Change that was held in Moscow on 29 September–3 October, 2003.

8.3 GLOBAL WARMING ENERGETICS, AND GEOPOLITICS

The problematical multivariance of the dynamics of civilization development is reflected in a set of problems, which have either been solved or are planned to be studied within international programmes. In 1987 the UN Conference on Environment and Development (UNCED) adopted a resolution to organize an International Panel on Climate Change (IPCC). According to this resolution, certain concepts of respective studies have been formulated and some preliminary results have been obtained in the following main directions (IPCC, 2001a–c):

- Systematization of the present ideas about possible anthropogenic climate changes.
- Consideration of the consequences of the impact of these changes on the environment and society.
- Discussion of the required measures to prevent an undesirable development of such impacts.

Publication of the results of the IPCC activity makes it possible to judge the trends of the activity of this organization (Demirchian *et al.*, 2002). Unfortunately, the IPCC emphasizes the catastrophic character of human impact on the global climate, neglecting the opinions of many scientists who show, scientifically, that the role of the anthropogenic factor in the observed changes of climate is more difficult to understand than is suggested by the IPCC reports.

As has been mentioned, the problem of climate change involves several problems not only scientific but also political. Long observations of climate enabled one to select the most important problems, which require a thorough analysis. Nevertheless, other problems should not be ignored either, which at the present stage of civilization development seem secondary. In other words, putting forward some hypotheses on the prospects of the present civilization development should be based on objective laws of nature and data of monitoring, as well as economic and political rules. But the trends of the IPCC activity contradict this concept. According to the IPCC, the rate of the growth of atmospheric temperature over a decade constitutes 0.167°C, whereas according to the data of satellite monitoring, this rate ranges between 0.062 and 0.064°C (Demirchian *et al.*, 2002). This contradiction entails various conclusions about the prognostic raise in temperature when the atmospheric CO_2 concentration doubles: according to the IPCC by 3°C, and from satellite data by 0.53°C. A similar contradiction between the IPCC data and results of independent scientific studies is observed with respect to the estimates of temperature changes in the past, when the temperature trends similar to those nowadays were observed. All this reduces the reliability of the results published by the IPCC.

Climate predictions are of political importance. Therefore, a scientifically grounded estimation of the consequences of anthropogenic activity can principally change public opinion and prevent undesirable social consequences. Many scientists in the USA, Europe, Japan, and Russia, where fundamental studies are being carried out on the development of global climate models capable of reliably predicting the NSS dynamics, understand this.

The role of the mass media is extremely important. It should correctly elucidate the ecological problems, especially global warming. Important decisions have been made at the UN level concerning the prevention of dangerous consequences of global warming for the population. Their realization can markedly hinder the economic development of most countries and further reduce the insufficient level of their life support. The governments interested in making justified decisions in the field of ecological problems should invest considerable material and financial resources into scientific investigations. The global character of ecological and climate problems requires studies of the whole surface of the land and oceans, both in space and deep oceans. To finance these studies, considerable investments from national revenues should be made, which is only possible through the mutual consent of public institutions and firm politicians. Global climate problems should not be used as a tool to frighten the general public. In this context, the IPCC interpreted the results of scientific studies in a most attractive way, distorting thereby the objective information, introducing "frightful" scenarios, and prompting the politicians to made ungrounded decisions.

Since the problems of global warming and, hence, economic development are interconnected, one of the most important principles of scientific work in this field is its independence from the political theatre. On the other hand, the political situation determines the possibility of financing very expensive studies of the climate. At this stage, a situation occurs when political and personal strategies contradict each other. Therefore, the problem has been raised (Demirchian *et al.*, 2002) of obtaining reliable estimates of the role of anthropogenic factors in climate change. The key goal to the solution of this problem is to obtain real estimates of the atmospheric temperature increase (ΔT_{2x}) with a doubled concentration of CO_2. It was shown that this problem, as applied to an improvement of climate models, creates a deadlock situation caused by the impossibility of verifying the modelling results. One of the ways out of this situation is the adaptive–evolutionary technology of geoinformation monitoring (Kondratyev *et al.*, 2000a,b).

The specified value of ΔT_{2x} in the first overview of the IPCC Second Working Group (Chairman Yu. A. Izrael) within 1.6°C–1.7°C enables one to reduce the estimates of the increase of the ocean level and to cease demanding governments to take immediate measures to reduce GHG emissions (Houghton *et al.*, 1992). The observed temperature rise by the year 2000 compared to 1850 by about 0.6°C at $\Delta T_{2x} = 1.66 \pm 0.11$°C constitutes 35–36% of ΔT_{2x}. By 2100 it will reach 67–70% (1.0–1.3)°C of ΔT_{2x}. Thus, there are no grounds for a dangerous increase of temperature in the future due to so-called "delayed" heating (Hansen *et al.*, 2000). These data refer to the case of GHG equivalent concentrations reaching 502–519 ppm and CO_2 concentrations reaching 448–452 ppm by 2100.

Another important direction with numerous discussions and uncertainties is the forecast of possible values of anthropogenic carbon flux (see Tables 2.4, 2.9, and 3.6). There is a sufficient amount of objective data, consideration of which makes it possible to narrow the range of possible levels of anthropogenic emissions of carbon dioxide and its concentration in the atmosphere. Demirchian *et al.* (2002) established a simple link between the size of a population G and the anthropogenic share of CO_2 in the atmosphere, $K_a = 15G(t)$. Such a parameterization substantially simplifies the prediction, since only one assumption on the size of population is needed. Over the period 1750–2000, this parameterization was characterized by a correlation coefficient of 0.994.

Thus the economic aspect of the problem of global warming is reduced to a search for the strategy of reasonable use of natural resources with constantly improving technologies of energy production and its economic use. The geopolitical aspect of the problem becomes urgent in connection with the USSR disintegration and transformation of the USA into a single superpower. From this point of view, the problem of global warming, being a political instrument for various people and groups from the year 1990, acquires a geopolitical importance. The exclusive importance of energy and raw material resources for developed countries determines the necessity for guaranteed extraction and transportation of these resources. Therefore, at the beginning of the 21st century one should expect substantial changes of economic and political levels of interaction between the global population. The world community is prompted to this by the growing intensity of the commodity and finance exchange between the countries. In addition to this, the levels of economic relations has become such that the functioning of an individual country outside these relations becomes impossible. In this context economic and political relations have become global in scale. For this reason, the solution of the problem of regulating the rules of interactions between countries in this globalized world becomes vitally important. Complicating these relationships is these growing size of the global population, bringing forth a multitude of economic and political problems.

8.4 GEOINFORMATION MONITORING CONCEPTION

It is clear that here the climate means a totality of events in the environment connected with the formation of the human habitat and expressed in characteristics such as temperature, precipitation, humidity, composition and atmospheric motion, oceanic dynamics, etc. The climate system is understood in three dimensions: space, time, and human perception. Each of these is historically connected with a certain set of processes in the NSS, and therefore a creation of the global system consisting of models and technical means is urgent for obtaining reliable estimates of the NSS state and especially for predicting its future trends. One complicated and contradictory element is the global climate model, since its synthesis requires a solution of the problem of parameterization of the interaction between the NSS components. The complexity of this problem is determined by the multivariance of the phase space of the climate system. The developed models divide this space into atmospheric, ground, and oceanic parts. Depending on the spatial–temporal division and assumed simplifications and approximations, models of various types appear. This is connected both with the presence of developed conceptual and theoretical approaches and with the available empirical information (see Tables 7.2 and 7.3). Nevertheless, the following four types of most important components of the climate system are present in the models: radiation, dynamics, processes at media interfaces, and the spatial–temporal resolution. Depending on orientation, the models are divided into energy balance, radiative–convective, statistical, dynamic, and models of atmospheric general circulation.

In connection with the global scale of environmental problems, their solution requires the coordinated efforts of the international community, which is now manifested through the appearance of international and national programmes on studies of geoinformation monitoring (Lazarev and Savinykh, 1996; Malinnikov *et al.*, 2000; Savinykh, 1990). The USA pays special attention to these problems. So, for instance, from 1995, the US National Scientific Foundation finances the programme MMIA (Methods and Models for Integrated Assessment) within which the role of large-scale technogenic and socio-political processes in the dynamics of physical, biological and human systems is studied. At the beginning of the 21st century the US National Scientific Foundation supported more than 20 investigation programmes connected with the global problem of environmental change. Among them are programmes such as Greenhouse Gas Dynamics (GGD) the US Global Ocean Ecosystems Dynamics (GLOBEC), the Ecological Rates of

Change (EROC), and others. The emphasis is on studies of the high-latitude environment and the Arctic systems, in particular.

An important aspect of studies of global environmental changes is the technical equipping of the systems of data collection, organization of their functioning (Kondratyev, 1999c; Yakovlev, 2001; Maksudova *et al.*, 2000), and supplying the data in compatible formats. One such format is WEFAX-format (Kramer, 1995) used by the system of geostationary satellites (Meteosat-5 and 7, the Indian National Satellite (INSAT), the FY-2B Weather Faximale (WEFAX), the Geostationary Meteorological Satellite (GMS), GOES-West, and GOES-East), which provide a flux of 400 images of the Earth's surface per day with spatial resolutions from 2.5 to 10 km. Among other widespread formats of image transmission are the Advanced Picture Transmission (APT), the High-Resolution Picture Transmission (HRPT), and the Primary Data User Station (PDUS). The APT satellites in polar orbits transmit on average 12 images of the Earth's surface per day at frequencies 136.5 MHz (NOAA-12 and 15) and 137.62 MHz (NOAA-14 and 17). The HRPT satellites provide resolution of 1.1 km in five spectral regions (two visible and three IR) at a rate of about 12 images per day. The PDUS is used in the METEOSAT series mainly to map the European territories with resolutions of 2.5 km.

These fragments of information indicate that the facilities of remote sensing are sufficient to provide the real time GMNSS with data for the prompt assessment of the state of the environment and prediction of its key characteristics. However, the role of available remote sensing data in studies of the global changes of the environment consists mainly of revealing and recording partial phenomena without analysing their global role (Chen *et al.*, 2003; Del Frate and Ferrazzoli, 2003; Dong *et al.*, 2003; Levesque and King, 2003; Timoshevskii *et al.*, 2003).

Thus, the observed accelerated accumulation of knowledge, the huge growth of the scientific–technical progress, and the global-scale human impact on nature have brought forth the problem of the thorough assessment of the environmental state and analysis of the possibilities of long-range predictions of the present changes. Scientific studies in the field of anthropogenic impacts on the environment suggested the conclusion that the solution of the problem of an objective control of the natural phenomena dynamics, and of the impact of human society on them, is only possible with the development of a single system of the geoecoinformation monitoring based on the GMNSS. Within the international and national programmes on the NSS studies, extensive databases have been accumulated on the environmental parameters and on the dynamics of natural and anthropogenic processes of various scales. In addition, models of biogeochemical, biogeocenotic, climatic, and demographic processes have been developed. The technical base of the global geoecoinformation monitoring includes efficient means of collecting, recording, accumulating, and processing of the data of observations from space, air, ground, and sea platforms.

However, despite considerable progress in many spheres of nature monitoring, the main problem remains unsolved – substantiation of an optimal combination of all technical means, creation of an efficient and economic structure of monitoring, and preparation of reliable methods of prognostic estimation of the environmental dynamics in conditions of anthropogenic forcing. Experience from the latest investigations point to the possibility of developing a global model, which could be used (in the adaptive regime of its use) to substantiate recommendations on the structure of monitoring and to formulate the requirements for databases (Kondratyev, 2002a–c).

The monitoring system should be based on a regional segment, which includes the bank of models of biogeochemical, biogeocenotic, and hydrodynamic processes in the biosphere, with their spatial heterogeneity taken into account, oriented towards a combination of the existing regional databases and inclusion of the functioning systems of environmental observations. In cooperation with specialists who already have climatic and socio-economic models, the model of the NSS can be developed. With the use of this model, one can assess the adequacy and quality of the available databases and give recommendations for the development of their structure.

To further improve the model, it is necessary to study the interaction of the NSS with the processes in the adjacent space and to create a global model, which would consider both the processes in the magnetosphere and the effect of cosmic space.

The final stage is to create a system capable of predicting the development of natural processes and assessing the long-range consequences of the many scale effects on the environment. The sphere of application of such a system covers various problems of environmental protection on regional and global scales with the ecological expertise of the projects of changes of the structure in the Earth's cover, hydrological regimes, and composition of the atmosphere.

The functional structure of the monitoring system is based on the idea of the distributed network structure. The creation of the NSS model requires a realization of purposeful complex studies among which the following problems are important:

- Systematization of data on regional and global changes and formation of complex ideas about the biospheric processes and structure of the biospheric levels. Development of the conceptual model of the biosphere as a component of the global geoecoinformation monitoring system.
- Taking stock and analysis of the existing ecological databases and choosing the structure of the regional and global databases.
- Classification of spatial–temporal characteristics and cause-and-effect feedbacks in the biosphere in order to develop the scale of agreement of spatial and temporal scales of ecological processes.
- Creation of the banks of models of sample ecological systems, biogeochemical, biogeocenotic, hydrological, and climatic processes.
- Study of the processes of the biosphere–climate interaction. The search for laws of the impacts of solar activity on the biospheric systems of various spatial scales.
- Systematization of information about local ecosystems. Description of geophysical and trophic structures with regions, which take into account agreement with spatial–temporal scales.
- Creation of models of regional and global biotechnological fields and development of algorithms for the synthesis of anthropogenic items of global models.

- Formation of the bank of scenarios of coevolutionary development of the biosphere and human society. Development of demographic models. Parameterization of the factors of scientific–technical progress in the context of the use of biospheric resources. Modelling the prospects of the urban construction and of the Earth's resources.
- Search for new information technologies of global modelling which would provide a reduction in requirements for databases and bases of knowledge. Development of the architecture and algorithmic principles of the functioning of calculation systems of neuron-wise elements realizing an evolutionary processing of information with high speed and efficiency.
- Further development of the concept of ecological monitoring and creation of a theoretical base of ecoinformatics. Development of methods and criteria of assessing the stability of regional and global natural processes. Analysis of stability of the biosphere and climate structures.
- Synthesis of the NSS model and development of computer means to carry out numerical experiments within the framework of assessing the consequences of the realization of various scenarios of anthropogenic activity.

Analysis of the recent problem-oriented investigations shows that for an advance in global modelling, new methods should be developed to carry out a system analysis of complicated natural processes, as well as methods of data processing oriented towards the synthesis of balanced criteria for information selection and consideration of the hierarchy of the cause-and-effect feedbacks in the NSS. The global model should be based on a set of regional-scale models in which the region is considered as a natural subsystem interacting through biospheric, climatic, and socio-economic bonds with the global NSS. The respective model describes this interaction and functioning of the whole set of processes in the subsystem at different levels of the spatial–temporal hierarchy. The model covers both natural and anthropogenic processes characteristic of a sample region, and is initially based on the available information. The model structure is oriented towards an adaptive regime for its use.

A combination of the system of collection of information on the environment, the model of the regional geosystem functioning, the system of computer mapping and means of artificial intellect, should all result in the creation of the Geoinformation Monitoring System for a typical region (GIMS-region) capable of the following:

- to assess the impact of global changes on the environment of the region;
- to assess the role of regional changes of the environment in changes of the global climate and biosphere and in adjacent regions;
- to assess the ecological state of the atmosphere, hydrosphere, and soil–plant formations in the region;
- to form and renew the databases of ecological, climatic, demographic, and economic parameters of the region;
- to promptly map the landscape;

- to prevent the ecological consequences of anthropogenic impacts on the environment of the region and its adjacent territories;
- to typify the Earth's cover, natural disasters, settlements, surface pollution of landscapes, hydrological systems, forests; and
- to assess the level of safety of the population in the region.

Substantiation of the GIMS-region is connected with the selection of biospheric components, climate, and social medium characteristics of a given level of spatial hierarchy. The succession of actions on the development of the GIMS-region project and its realization follows on from an experience of the creation of the Geographical Information System (GIS) and derivation of numerical models of the functioning of large natural–economic systems. This experience determines an expediency of the following structure of the GIMS-region:

- subsystem of data collection and express analysis;
- subsystem of primary processing and accumulation of data;
- subsystem of the computer mapping;
- subsystem of the assessment of the atmospheric state;
- subsystem of the assessment of the state of soil–plant cover;
- subsystem of the assessment of the state of the aqueous medium;
- subsystem of the assessment of the level of ecological safety and risk for human health in the region;
- subsystem of the identification of the causes of violation of the ecological and sanitary situation; and
- subsystem of intellectual support.

Based on the simulation–evolutionary technology of modelling, a sample information–calculation complex is formed oriented towards an introduction of methods, algorithms, and procedures of the evolutionary structural synthesis of the discrete–continuous descriptions of objects characterized by information uncertainty and controversial experimental data. Further improvement of this technology is connected with the introduction of the evolutionary neurotechnology as an alternative method for synthesis of future realizations of the NSS model in the monitoring systems and its transfer onto the level of the intellectual computer system of a higher rank compared to expert systems.

Analysis of databases and their supplements shows that the existing monitoring systems lack a closed network of the environment observation means used at present. Based on the experience of the aerospace centres of monitoring used in many countries, it is advised to use a standard system of autonomous multipurpose and specialized ground and airborne centres of assessment for the current state of elements of natural and anthropogenic landscapes, prediction of ecological and sanitary conditions of settlements, environmental diagnostics, abnormal phenomena, natural disasters, and technogenic catastrophes. The network of such centres coordinated with the NSS model can solve the following problems:

- Thematic mapping of the results of ground and aerospace measurements and surveys carried out using *in situ* and remote methods in the optical, IR, and radio super high frequency (SHF) bands.
- The current assessment and prediction of ecological and sanitary conditions of dwellings, industrial and production zones, sources of drinking water and rivers, agriculture, forest and hunting areas, places of storing garbage and industrial waste, and accumulation of mineralized and polluted water as potential sources in emergencies.
- The territorial–seasonal inventory and typification of quasi-homogeneous and abnormal sites as well as the sites of floristic background and soils, water basins and the atmosphere, geophysical fields of various nature; mapping the local parameters of landscape features, identification of the sources of pollution of terrestrial cover, of air and water basins, and their impact on settlements and environmental elements.
- The prompt expertise and high-frequency fixation of the coordinates of small-sized areas and objects with abnormal reflective and radiative characteristics, detection of abnormal phenomena, natural disasters and technogenic catastrophes, and the dynamics of the consequences of such events in settlements and the environment.
- The goal-oriented search and rescue works both on land and sea, collection of expensive metal remains of rockets, river and sea ships, assessment of fire risks in taiga and peat areas, and oil and gas extraction.
- Combination of the results of airborne studies with satellite images and ground measurements to retrieve an objective pattern of the ecological and sanitary situation and to substantiate the rational use of natural resources, agrarian, forest and industrial potential, water bodies and rivers, and transport and irrigation systems.

Autonomous airborne platforms are created on the basis of small, middle-sized, and pilotless vehicles. They are equipped with radars, radiometers, TV, laser and photographic means of observation, as well as with the systems of referencing the observation results to characteristic points on the underlying background and coordinates.

At present it is possible to realize the project of a single system of regional models of the most important natural processes, including the processes in the magnetosphere. Available knowledge and accumulated data make it possible to realize the local and regional levels of the spatial hierarchy. The exploitation of the system in this realization will make it possible to formulate the requirements for databases that provide the transition to a more detailed spatial division of natural systems and processes up to a consideration of the role of some elements of the ecosystems.

The ecological monitoring system includes three main directions:

1. Ecosociology. Study of the impact of human society on nature and the interactions in the NSS.
2. Ecoinformatics. Recording, storage, transmittance, analysis, synthesis, modelling, and presentation of information about the state of the environment.
3. Ecotechnology. Creation of technologies and technical means with the minimum impact on both nature and the human organism.

The problems of ecosociology include:

- analysis of social structures affecting nature, their motives, goals, methods, and technical means;
- analysis of the necessity and adequacy of anthropogenic impacts of the social structure on nature;
- analysis of the optimal allocation of industrial and agrarian infrastructure and dwellings in nature; and
- development of the ecoprotective measures taking into account the socio-regional, socio-national, and socio-global interests.

The problem of ecoinformatics includes obtaining the diagnostic, prognostic, and practical information on the state of nature and its changes due to anthropogenic and natural forcings. The obtained information should reflect the following:

- quantitative estimation of the state of nature and natural resources;
- quantitative estimation of the anthropogenic impact and detection of harmful and surplus production;
- the current state and maintenance of regional (in some cases district) simulation models of interactions in the production–medium, agriculture–medium, and dwelling–medium systems;
- the current state and maintenance of the regional ecomodel;
- the current state used by the ecomodel within national and international co-operation;
- the current state that makes it possible, based on the simulation ecomodel, to construct prognostic schemes of the impact of harmful and surplus production on the nature of the region;
- the current state used in the prognostic modelling of nature changes when extracting natural resources; and
- the current state used, on the basis of simulation and prognostic models, in working out recommendations for the development or curtailment of production and dwellings in the region.

The ecoinformation monitoring system is constructed from a 3-level scheme including the satellite and ground observations as well as airborne observations.

The integrated functional scheme consists of the following main components:

- the complex observing system (data collection) which includes the cosmic, airborne, and ground segments ensuring regular *in situ* and remote measurements of the environmental parameters;
- subsystems of data reception/transmittance on the basis of space and ground-based means using satellite, microwave, wire, and fibre optic communication channels;
- subsystems of primary processing of data from various observing systems;
- subsystems of accumulation and systematization of data (databases); and
- groups of subsystems to solve applied (thematic) problems of monitoring.

By their goal and information–technical characteristics, systems of spaceborne observations refer to the following classes:

(a) Systems of routine detailed observations from spaceborne platforms with the use of data of multi-zonal and topographic surveys.
(b) Systems of prompt observations divided into subclasses:
   - *Systems of the planned survey* of the problems of global operational control of the state of the atmosphere and the Earth's surface with low spatial resolution and with continuous (quasi-continuous) observations using space-borne platforms in geostationary, high-elliptic, and circumpolar orbits;
   - *Systems of detailed observations* operating in continuous (in highly-elliptic or circular orbits) and regular (in mid-altitude circular orbits) regimes, with the high or middle spatial resolution, which provides data for the chosen objects or regions on the Earth's surface.

At present, most of the mid-orbit space vehicles realize both subproblems. In the system of aero-observations, to organize aircraft sensing of the regions with a complicated ecological situation, an aero-scanner is needed, capable of carrying experimental equipment and operating both in pilot and pilotless regimes.

The regional ecomodel and the problems of its operational maintenance require an efficient information–calculation system. The distribution network of modern computers and software are to be the base of this system.

The regional ecological monitoring system should become the focus of the ecological expertise in the territorial management. The system of the ecological monitoring will provide each citizen of the region with an access to information on the state of the environment (generally available sites or pages in any global network). Based on this knowledge, a man will be able to make competent decisions on many vital problems both in the social and in the personal sphere.

The ecotechnological problem includes:

- assessment of real and potential danger of the functioning productions connected with emissions of substances harmful for the environment and humans;
- assessment of technological processes and detection of the cycles of processes emitting these substances;
- correction of technological processes with the parameters of the final product preserved, with reduced harmful impacts on the environment and humans;
- development of the equipment for corrected technological process;
- personnel training in new conditions and with new equipment; and
- qualified expertise in technological processes.

Creation of productions with minimum impact on the environment and humans is connected with the use of high-intellectual computer technologies, which make it possible to control the whole production cycle and to exercise an optimal management of the technological equipment.

All the enumerated directions of the ecological activity can be concentrated in a single organizational structure – regional centre of ecological monitoring, which, according to the scheme in Figure 1.1, is an element of the global geographical information monitoring system (GIMS).

The regional centre for ecological monitoring should consist of three structural units corresponding to the main directions of ecological activity: ecosociology, ecoinformatics, and ecotechnology. Each of them, in accordance with the problems of its own, can be divided into several sections:

1. Ecosociology.
   - Analysis and prediction in the NSS.
   - Nature-protection measures.
   - Decision making and interaction with power structures.
2. Ecoinformatics.
   - Reception and analysis of satellite information.
   - Aero-survey with the mobile avia-group.
   - Ground observations with a network of sensors and ground stations.
   - Control of human health.
   - Modelling and maintenance of the ecomodel of the region.
   - Information–calculation centre.
3. Ecotechnologies.
   - Ecoanalysis of industrial enterprises.
   - Ecoanalysis of the extractive industry.
   - Control technologies.
   - Expertise in technologies and technical decisions.

8.5 GLOBAL SIMULATION EXPERIMENTS

One of the special features of present-day global ecodynamics is an intensified development of science and technology, and for this reason, the environment and human society have changed more intensively than say 50 years ago. The role of interdisciplinary studies of the NSS with the use of accumulated knowledge in mathematics, ecology, sociology, medicine, chemistry, geophysics, etc., has grown. The global simulation experiment based on the use of the latest computer technologies has become a precursor of decision making in the field of nature use. It has principally changed the conception of the experimental study of natural objects. Of course, problems appear in assessing an adequacy of the results of simulation experiments, and ecoinformatics solves these problems (Kondratyev *et al.*, 2002a,d). The late 20th–early 21st centuries are characterized by a new mentality – the transition to experiments with models (not with natural objects themselves). How, for instance, is it possible to carry out a field experiment with such unique systems as oceans, continental ecosystems, and the biosphere as a whole? Such experiments are very dangerous and should be excluded.

The GMNSS described in Chapter 6 makes it possible to realize a wide spectrum of simulation experiments with a multidimensional phase space of the NSS and to predict the development of the NSS for this century. Deeper predictions are likely not possible at the present level of knowledge and global databases. The uncertainties have been analysed, and they are closely connected with climate modelling. Unfortunately, there is no universal model, so far, which would reliably simulate the functioning of the Earth's climate system, reflecting the interaction of the atmosphere, oceans, land, and human society. Therefore, to use the GMNSS, it is necessary to make some a priori assumptions:

- Until the year 2100 the anthropogenic emissions of CO_2 are described by the model of Demirchian *et al.* (2002), and emissions of other GHGs until 2050 follow the pattern of 2000, after which they decrease by 25%.
- Extraction and consumption of non-renewable fossil fuels in the period until 2050 grow by 0.5%/year in developed countries and by 0.7%/year in developing countries, after which they become stabilized.
- The per capita gross home product increases in developed and developing countries at constant rates of 2.5% and 1.5%, respectively.
- The level of medical service in developed and developing countries reduces mortality by 90% and 50%, respectively, due to environmental changes, and in 2020 a remedy for AIDS should be found.
- The atmospheric temperature regime is formed in accordance with the model of Mintzer (1987).
- Investments into agriculture depend on the provision of the population with food (Krapivin *et al.*, 1982).
- Investments of all countries into the struggle against environmental pollution will grow by 0.2%/year until 2050, and then will stop growing.

- Processes of deforestation and afforestation beginning in 2010 are balanced so that the total area of forests as of 2000 is preserved, and the area of tropical forests decreases by not more than 5%.
- The oil pollution of the World Ocean in the 21st century follows the pattern of the last decade of the 20th century.
- The agricultural productivity grows steadily reaching 80% in 2100.

Within these formulated assumptions, the NSS dynamics are characterized by the following indicators:

- The ecodynamic pattern of the NSS functioning is characterized in that derivatives of the functions describing the state of the NSS components are never equal to zero for time periods longer than the time step of the model.
- The size of population by 2100 will reach 12.7 billion people, with the ratio between the age groups being characterized by the following average values: 0–15 years – 32%; 16–65 years – 53%; >65 years – 15%; and the disabled population – 17%.
- The provision of the population with protein food increases by 6.8% by the year 2030 (the World Ocean contributes 12.1%) and then decreases by 3.2% with respect to 2000 (the share of the World Ocean increases up to 22.1%).
- The CO_2 concentration in the atmosphere will reach 486 ppm with a respective increase of average planetary temperature by 0.87°C.
- The dynamics of the spatial distribution of sinks of atmospheric CO_2 is characterized by the growing role of the World Ocean and land ecosystems, whose shares in the mid-21st century will constitute 31% and 19%, respectively, and then by the end of the century the share of the World Ocean will decrease to 26.7% with the role of land ecosystems growing up to 24.4%.
- A change of 5% of tropical forest areas for urban ecosystems will lead to a 1.2% reduction in the total sink of CO_2.
- The complete change of wet tropical forests for grass ecosystem by 2050 will later tell on global climate change due to: a violation of the regional moisture cycle on the atmosphere–land border; an increase of the outgoing long-wave radiation flux from the deforested territory by 4.3 W/m^2; an increase of this territory's albedo by 5.7%; a decrease of the absorbed short-wave radiation flux by 4.1%; an increase in soil temperature by 1.3°C; and a reduction of rains and evaporation by 0.7 and 0.9 mm/day, respectively.

For educational purposes it is of interest to consider hypothetical versions of the man–nature interaction. Consider only some situations, which appear following the hypotheses of the future dynamics of the most important anthropogenic factors. Note that in contrast to many other models. The GMNSS calculates the size of the population as a function of environmental parameters.

Calculations show that the mechanisms of demographic process formation depends strongly on the strategies of nature use. So, a 50% increase of food provision, which is possible with the introduction of more efficient technologies in agriculture and development of aquasystems, results in a 14% decrease in population density. On the contrary, a 10% decrease of the rate of mineral resource consumption causes a 3% increase in population density. The NSS is very sensitive to possible changes in the global structure of soil–plant formations. So, for instance, if by 2050 the forest areas decrease steadily by only 10% with respect to 2000, the atmospheric CO_2 concentration in 2100 will reach 611 ppm (i.e., it will almost double). But if the forest areas increase by 15% by 2050, then in 2100 the partial pressure of CO_2 will constitute 475 ppm.

These results of modelling the NSS response to possible anthropogenic changes in its parameters show that the suggested method enables one to assess the dynamics of various components of this system depending on hypotheses of possible rates of change. An application of this method to reconstruct the integral dynamics of the NSS by some of its parameters for the period 1970–2000 on the basis of data published in Watson *et al.* (2000), Houghton *et al.* (2001), and Kondratyev and Losev (2002), has shown that the error of the prediction for 10 and 20 years does not exceed 15% and 25%, respectively. Hence, GIMS-technology can be recommended as one of the constructive approaches to solve the problems of the global ecodynamics formulated by Kondratyev (1999c). Simulation experiments with the use of the GMNSS with available reliable data on the NSS pre-history enables one to reveal conditions, which confine civilization development, and to find mechanisms for its sustainable development. For this, it is necessary to concentrate the efforts of international and national scientific programmes (see Table 8.2) on the synthesis of the GMNSS database. Koffi Annan, UN Secretary General, emphasized this at his press conference in Johannesburg on the last day of the World Forum (4 September, 2002). He pointed out that the governments came to an agreement to combine efforts

Table 8.2. Some programmes on environmental studies.

Programme	Characteristic of the programme
AAOE	Airborne Antarctic Ozone Experiment
ACC	Anthropogenic Climate Change
ASHOE	Airborne Southern Hemisphere Ozone Experiment
BIBEX	Biomass Burning Experiment
CEPEX	Central Equatorial Pacific Experiment
CLIVAR	Climate Variability and Predictability (WCRP campaign programme)
WOCE	World Ocean Circulation Experiment
WCRP	World Climate Research Programme (since 1980, jointly sponsored by the WMO, ISCU, and IOC)
UNEP	United Nations Environmental Programme
STIB	Stratosphere Troposphere Interactions and the Biosphere
STEP	Stratosphere Troposphere Exchange Project
GOFS	Global Ocean Flux Study
GOCRP	Global Ocean Carbon Research Programme
IHP	International Hydrology Programme (UNESCO)
TOGA	Tropical Ocean and Global Atmosphere Experiment

on biodiversity preservation and finding a solution of the global problems in the spheres of water supply, energetics, medical services, and agriculture. Speaking about the possible ways to solve the global problems of the present civilization, Koffi Annan was fully confident that there is a way "that will reduce poverty while preserving the environment". The GIMS-technology shows us this way. But its realization requires the concentration of accumulated knowledge on the development of the GMNSS by improving its units and, first of all, finding out and formalizing the factors and dynamic strategies of the formation of social evolution.

9

Problems and functions of the environmental monitoring systems

9.1 SYSTEMS OF ENVIRONMENTAL MONITORING

The space-borne means of the Earth observation developed during the last decades cover practically all the significant spheres of the nature–society system (NSS) functioning. Note should be taken here that, unfortunately, these means were created without considering the global modelling results, which in many cases has led to surplus information of satellite systems because most of the global data remained unclaimed. Nevertheless, now the satellite systems give data on the following NSS parameters (Kramer, 1995; Summer *et al.*, 2003; Reichle *et al.*, 1999; Lockwood, 2001; Hotuntsev, 2001):

- interactions in the "Sun–Earth" system (physical mechanisms which control the motion of substance between biospheric and geospheric formations);
- atmospheric dynamics (physical and chemical processes, meteorological parameters, and hydrological fluxes);
- dynamics of the oceans and coastal regions (circulation, wind-driven roughness, water temperature and colour, productivity of surface layers, photosynthesis, transformation of inorganic substance into organic matter, phytoplankton, chlorophyll, marine nutrient chains, fish clusters, and structure of the ocean populations);
- processes in the lithosphere (parameters of the Earth's rotation, tectonic activity, dynamics of continents and glaciers, soil moisture, topographic structures, and resources);
- functioning of the biosphere (characteristics of the atmosphere–biosphere interaction, global structure of soil–plant formations, biomass distributions, agricultural structures, forest areas, snow-covered area, structure of rivers and other aquatic systems, precipitation, food supplies, and pollution); and

- dynamics of the climate system (Earth radiation budget (ERB), minor gaseous components (MGCs) of the atmosphere, greenhouse gases (GHGs), temperature, cloudiness, and long-range climatic trends).

The remote sensing data are the main source of operative information for the systems of control of the global ecological, biogeochemical, hydrophysical, epidemiological, geophysical and even demographic situations on the Earth. By now, the remote geoinformation monitoring of the objects and processes in the environment provide a huge volume of data useful for solving many problems of the NSS control. Unfortunately, the problem of coordination of requirements for global databases and the structure of satellite monitoring has not yet been solved. Development of methods of remote sensing data processing lags behind the progress in the technical provision of space-borne monitoring systems. There are hopes that the use of geographic information monitoring systems (GIMS)-technology will make it possible to solve this problem and to balance the algorithmic and technical aspects of remote sensing.

The remote sensing systems of the Earth in Table 9.1 are only a small part of a broad and diverse spectrum of space-borne systems oriented towards accumulation of data on the state of the land and World Ocean elements and on atmospheric parameters. Among these systems are space-borne observatories "Aqua" and "Terra" as elements of the Earth Observing System (EOS) (Parkinson, 2003). These systems are aimed at control of the planet's health. They measure a great amount of characteristics of land systems, oceans, and the atmosphere. In particular, the satellite "Aqua" transmits data on:

- water vapour supplies in the atmosphere;
- liquid water content in the atmosphere with an assessment of rain rate and concentration of cloud droplets;
- the solid-phase water supply in the form of ice particles in clouds;
- soil moisture;
- water content in snow and ice cover on land and oceans;
- the state of land and ocean vegetation; and
- atmospheric temperature.

Table 9.1 contains information about some space-borne observing systems. These systems are characterized by a certain history of the development of remote sensing and data processing technologies. Of course, the technology of sensing in the optical range was most developed. However, during the last years, technologies of remote monitoring in the microwave spectral region were being intensively developed (Table 9.2). Passive sensing of the environment with the use of microwave radiometers gives efficient results in assessments of the thermal characteristics and determination of the structure of the Earth's cover. Active radiometry determines the physical structure and electric properties of the environmental objects. The combined use of devices in the optical and microwave ranges will make it possible to raise the efficiency of the remote monitoring systems by overcoming the shortcomings of each range. The advantage and good prospects of using

Table 9.1. A brief list of the environmental monitoring systems and their equipment.

System	Characteristic of the system
AARGOS	A340 Atmospheric Research Global Observation System.
ADEOS	Advanced Earth Observation Satellite (NASDA). Equipped with an advanced radiometer in the visible and near-IR regions (AVHRR), and the ocean colour and temperature scanner (OCTS).
ERS-1,2	European Remote Sensing Satellite used within the ESA programme.
EOS	Earth Observing System. Equipped with sensors to record data on clouds and ERB as well as altimeter, acoustical atmospheric lidar, and laser wind gauger.
UARS	Upper Atmosphere Research Satellite launched in 1991 by NASA.
MOS	Marine Observation Satellite (NASDA). MOS-1 launched in 1987, MOS-1b launched in 1989. Equipped with the visible and IR radiometer and the scanning microwave radiometer.
ENVISAT	Environmental Satellite (ESA).
ERBS	Earth Radiation Budget Satellite (NASA).
GMS	Geostationary Meteorological Satellite.
GOES	Geostationary Operational Environmental Satellite.
GOMS	Geostationary Operational Environmental Satellite. Russian geostationary satellite to observe the environment.
JERS	Japanese Earth Resources Satellite.
Aqua (NASA/EOS)	Equipped with six sensors: (1) the Atmospheric Infrared Sounder (AIRS); (2) the Advanced Microwave Sounding Unit (AMSU); (3) the Humidity Sounder for Brazil (HSB); (4) Clouds and the Earth's Radiant Energy System (CERES); (5) the Moderate Resolution Imaging Spectroradiometer (MODIS); and (6) the Advanced Microwave Scanning Radiometer for EOS (AMSR-E). The set of these sensors provides measurements of numerous parameters of the environment to be used in the global model of the NSS (see details in the text).

polarization effects in the measuring systems are also apparent (Krapivin *et al.*, 2002; Krapivin and Shutko, 2002).

Synthesis of the systems of space-borne monitoring of the environment is not only a complicated scientific–technical problem but also an expensive one. Therefore, the quest for efficient solutions to this problem requires a systems approach to analysis of information fluxes needed for a reliable prognostic assessment of the global ecodynamics trends. On the one hand, this requires schematic solutions when planning measurements, and on the other hand, development of adaptive technologies of the data monitoring processing. Among the perspective approaches is the radio occultation method developed by Yakovlev (2001). The concept of this approach consists in the simultaneous use of two satellites: one carrying the signal–transmitting system, the other being a receptor. The propagation of the signal along the "satellite–satellite" track leads to a dynamic change of its characteristics, which enables one to retrieve the atmospheric parameters along the track of the signal propagation. Since this track constantly varies with the changing

Table 9.2. Brief characteristic of some devices for remote sensing (Kramer, 1995).

Device (sensor, radiometer, measuring system)	Used frequencies	Resolution, sensitivity
Advanced Airborne Hyperspectral Imaging Spectrometer (AAHIS)	440–880 nm	3 nm
Airborne Imaging Spectrometer (AIS)	900–2,400 nm	9.3 nm
Compact Airborne Spectrographic Imager (CASI)	418–926 nm	2.9 nm
Compact High Resolution Imaging Spectrograph Sensor (CHRISS)	430–860 nm	11 nm
Airborne Ocean Color Imager Spectrometer (AOCI)	443, 520, 550, and 670 nm	20 nm
	750 nm	100 nm
	11,500 nm	2,000 nm
Wide-angle High-Resolution Line-imager (WhiRL)	595 nm	20 nm
Airborne Imaging Microwave Radiometer (AIMR)	37 and 90 GHz	3 dB
Airborne Multichannel Microwave Radiometer (AMMR)	10, 18.7, 21, 37, and 92 GHz	0.5 K
Advanced Microwave Precipitation Radiometer (AMPR)	10.7, 19.35, 37.1, and 85.5 GHz	0.2–0.5°C
Airborne Water Substance Radiometer (AWSR)	23.87 and 31.65 GHz	0.1 K
CRL Radar/Radiometer	9.86 and 34.21 GHz	0.5 K
Dornier SAR (DO-SAR)	5.3, 9.6, and 35 GHz	1 m
Electromagnetic Imaging Radiometer (EMIRAD)	5, 17, and 34 GHz	1 K
Electromagnetic Imaging SAR (EMISAR)	5.3 GHz	2 × 2 m
Electronically Steered Thinned Array Radiometer (ESTAR)	1.4 GHz	±(3–4)°
Microwave Radiometer RADIUS	0.7, 1.425, 5.475, and 15.2 GHz	0.5–1.5 K

motion of satellites, a situation occurs of constant transmitting of data on the impact of the atmosphere on the propagation of radio waves in a chosen range at different altitudes. On the basis of these data, by solving the inverse problems, it is possible to estimate density, pressure, temperature, water content, and other parameters of the atmosphere. The system is most informative when radio waves propagate along the track "stationary satellite–orbital apparatus". Efficient theoretical models have been developed for such a configuration of the monitoring system. With the use of these models one can determine the dependence of refraction attenuation, angle of refraction, frequency, and absorption of the radio beam by the atmosphere. The transition from the binary structure of the system of the atmosphere microwave sensing to several satellites will make it possible to optimize the global system of environmental monitoring (Krapivin, 2000a; Kozoderov, 1994, 1998).

9.2 THE SEARCH AND IDENTIFICATION OF EXTREME ECOLOGICAL AND TECHNOGENIC SITUATIONS

The catastrophic trends in man–nature interaction during the last decades have necessitated the development of technologies for an optimal combination of anthropogenic and nature-protection activities. According to Moiseev (1990), the 20th century became "a century of warnings", when during a short time period the concept of the environment has drastically changed. This was favoured by studies of Russian scientists on global modelling, which made it possible to assess the possible planetary consequences of the realization of large-scale technogenic projects. One such assessment touched upon the consequences of a nuclear war in the 21st century.

Numerous international and national programmes are dedicated to studies and developments in the methods of resisting natural disasters. One of their main goals is to reduce the consequences of disasters by being able to give warnings about imminent catastrophic events. The spectrum of technological solutions aimed at prompt detection of catastrophic trends in the environment and the reduction of their consequences involves both theoretical and experimental studies. Informing the population about these decisions could have reduced the losses from natural disasters, which cannot be prevented physically. Efficient mechanisms preventing technogenic and natural catastrophes can be created with the use of GIMS-technologies (Kondratyev *et al.*, 2002a,b). The main concept being that observations of a territory of any scale are realized within the framework of the complex system of data collection and analysis.

First experience of the use of GIMS-technology is connected with the creation (in Russia) of the multipurpose flying laboratory IL-18 #75423 (Krapivin and Phillips, 2001; Kondratyev *et al.*, 2001c). The laboratory was meant to record the current state of, and to obtain prognostic estimates of; diverse anthropogenic and natural processes, diagnostics of the dynamics and consequences of abnormal phenomena, natural disasters and technogenic catastrophes in the interests of the safety of the population and the environment, rational use of the Earth's resources, freshwater and glaciers, agrarian and industrial potential, garbage and production waste, and ice reconnaissance and ship steering in polar seas. The problems of the aerospace monitoring are solved by using radar, microwave, TV, laser, and photographic means as a single system of collection, archiving, presentation, and express analysis of the experimental information, its mapping from the data of devices of inertial and satellite navigation, and markers of the spatial–temporal matching.

Experience of using the flying laboratory from 1974, with the annual working load up to 650 flight hours, has demonstrated the possibility of solving the problems of nature monitoring:

- The state and characteristics of soils and vegetation cover.
  –Types of soils, the volume content of water in the layer down to 1 m in depth.
  –The level of the groundwater bedding down to 4 m in depth.

–Violation of the subsurface soil layer and floristic background of tundra, deserts, and semi-deserts, the degree of soil freezing.

- Control of water basins and conduit systems:
  –Zones of filtration and the state of irrigation systems.
  –Hydrological conditions of the flood plains of rivers, typification of floods.
  –The degree of mineralization and chemical pollution of water basins.
  –Places of oil-product spilling, parameters of the polluting films.
- Inventory and typification of forests, marshes, felling areas, and other elements of the anthropogenic landscape.
- Expertise regarding abnormal phenomena, natural and technogenic catastrophes.
  –Fixation of the local anomalies coordinates on land and in water basins.
  –Assessment of the fire risk of forests and peat bogs, oil and gas production and mining enterprises, respective transport networks.
  –Detecting and mapping of the centres of forest and peat bog fires with limited visibility and in extreme meteorological conditions.
  –Localization of the sites with an increased content of methane in the near-surface atmosphere, places of possible accidents at hydrocarbon products pipelines.
  –Dynamics and consequences of volcanic eruptions.
  –Mapping the geothermal fields and lava flows.
  –Indications of metal objects on land and traces of fuel on the sea surface.
- Dynamics of the boundaries of snowfields and glaciers, sea ice solidity.
  –Typification of snow and ice (down to 2 m), and ice aerodromes.
  –Water content in snow.
- Complex subsatellite studies, methodical, technical, and experimental training of the operators for airborne and space-borne laboratories, search and rescue and ecological services.

The fundamental, applied, and practical concepts of the ecological expertise in the state of the Earth's cover, water basins, and the atmosphere and clouds are aimed at development of optimal principles of inventory, preliminary forecasts, rational use, and reproduction of natural resources both within the boundaries of individual regions and countries and on a global scale. The required level of adequacy, reliability, promptness, periodicity, and economy of obtaining information about the physico-geographical, hydrometeorological, and ecological state of vast regions requires a synchronous (in space and time) use of the total arsenal of measuring means in the optical, IR, and microwave regions. This is possible on board the flying laboratory of the type mentioned above.

The problems of mapping the airborne monitoring results are solved with the use of standard systems for the determination of current aircraft location. An operative referencing of the data of airborne experiments to characteristic points at a locality is realized with the use of a video-radiometric complex. These or similar systems are the most important elements for ensuring the meteorological compatibility of the aircraft measurements with results obtained from satellite observations, as well as with data from ground-based support and calibration measurements made at test sites.

Diverse ecological problems occur in detection, outlining, and measurements of geophysical and hydrometeorological characteristics of quasi-homogeneous (in a certain sense) sites on the surface. There are methods of decoding and goal-oriented interpretation of photographic, radar, TV, and other images of a locality. For trace microwave measurements, algorithms and software have been developed which make it possible to perform a real time selection of quasi-homogeneous sites with a priori unfixed boundaries.

The multichannel images of locality (optical, TV, and radar) permit one to typify the floristic background whose concrete image is determined by microrelief, type and salinity of soils, surface moistening, depth of groundwater bedding and degree of their mineralization, etc. Changes of the enumerated factors determine the variations of the reflecting and emitting properties of the surface. Special features of micro- and macro-relief, subsurface anomalies (caverns, groundwater lenses, etc.) affect the energy and polarization parameters of the backscattering of signals from radars and radio-altimeters.

The spatial–temporal variations of the dielectric properties of the near-surface layer of cover determines the spectral and polarization characteristics of the thermal field (and their special features) for water basins, dry and wet soils, peat bogs, mineral and volcanic rocks, lava flows, saline soils, and over-moistened and salted areas.

The multiyear experience of remote monitoring with the use of the IL-18 flying laboratory has shown that such multipurpose systems along with mobility and economic efficiency provide highly reliable expertise of the environment. This was confirmed when determining the topology of conduits and drainage systems in arid regions, outlining the points of leakage in irrigation systems, mapping tree felling areas and cinders, searching for plane wreckages, etc. (Borodin *et al.*, 1997, 1998).

A diversity of developed methods for searching and identifying extreme situations instills hope that in time such functions will be performed by the global monitoring system. For instance, based on the use of empirical orthogonal functions for analysis of data obtained with the TOPEX/Poseidon radar altimeter, Nezlin and McWilliams (2003) created a method to calculate the El Niño characteristics, important for improving the climate unit of the global model. Boschetti *et al.* (2003) demonstrated an efficiency of the combination of remotely sensed data in the visible and thermal-IR bands to detect areas of possible fires. Similarly, based on combined sensors of various ranges, Metternicht and Zinck (2003) proposed a technique for the remote identification and mapping of soils subject to salting.

9.3 EFFICIENCY OF THE DETECTION MONITORING SYSTEMS

In connection with the problem of global studies of the biospheric resources, the detection of anomalies such as polluted spots on a water surface has recently become urgent. An accumulated experience in solving this problem with the use of airborne and space-borne systems of control testifies to the necessity to develop new methods of detection, which would provide a highly reliable independent of the technical characteristics of these systems. In this case it is necessary to accumulate diverse measurements in microwave, optical, and acoustic spectral regions in order to synthesize the multichannel detectors with the hierarchical structure of the search and processing of information.

Also important is a consideration of the uncertainty of the location and the possible dynamic behaviour of the anomaly. Limited parameters of technical means used in environmental studies should be taken into account when making a decision about the presence or absence of the anomaly. This versatile problem of the global detection of anomalies in the biosphere necessitates the use of a systems approach that makes it possible, from united positions, to consider the whole complex of local problems and to approach the formation of a monitoring system of detection capable of self-control.

Limited technical means, memory volume, and fast operating of the on-board computing machinery necessitate a division of the monitoring system into three subsystems: *holder*, *resolver*, and *searcher*. The latter are divided into subsystems-units whose functions are: (1) periodic examination of the Earth's surface elements; (2) memorizing suspicious elements; (3) formation of traces of the moving anomaly from suspicious elements; (4) accumulation in time of data on fixed elements of the Earth's surface to carry out a statistical analysis in order to decide whether the fixed suspicious elements are of noise or signal in character; and (5) stage-by-stage localization of the procedure of the search for an anomaly.

Because of the dynamic variations of anomalies, the monitoring system should perform a periodic survey of the elements of the space and have time to process the flux of statistical information more rapidly. Therefore, there is a problem of assessment of the monitoring system's efficiency as a function of its parameters.

Let the memory volume of the monitoring system be confined to M cells and the information flux go simultaneously from N elements of the space. In each element there is a source of information determined by a random function $\xi_I(t)$. If $M \geq N$ then the solution of the problem of statistical analysis of the information flux is reduced to an accumulation of data in the memory cells and its processing using the methods of mathematical statistics. If $M < N$, it is necessary to develop some algorithm of the accumulation of statistics only about "suspicious" elements in the information flux.

Let, depending on the signal/noise ratio a, the random function $\xi_I(t)$ be determined by the probability density $f_a(x)$. And let the choice of "suspicious" elements for the memory of the monitoring system be carried out for the case $\xi > X$, where X is the prescribed threshold. If $a = a_0$ for the noise character of the flux $\xi_I(t)$, and $a = a_1$ in the presence of both signal and noise, then the probabilities of fixation of the noise and signal elements will be, respectively:

$$p_N = \int_X^\infty f_{a0}(x)\,dx \qquad p_C = \int_X^\infty f_{a1}(x)\,dx$$

Suppose, the shares of the signal and noise elements in the space at the time moment t are equal to $1 - \gamma(t)$ and $\gamma(t)$. Then the random number $\kappa(t)$ of "suspicious" elements at the moment t will be distributed in the following way:

$$P[\kappa(t) = k] = P_{N\gamma,pN}(k_1) \cdot P_{N[1-\gamma],pC}(k_2) \qquad \text{where } P_{n,p}(m) = C_n^m p_m (1-p)^{n-m}$$

Taking the time interval needed for realization of one survey of the whole space under study to be equal to unity, we calculate the probability of overfilling the monitoring system's memory $P[\mu(t) > M]$ at the moment t. Consider the case of discrete time, when the moments of appearance and completion of processing the noise or signal element are identified with the nearest whole number. Let each appearing element be processed in a separate memory cell up to the moment of the final decision about its noise or signal origin, after which the memory cell is cleaned and can receive other information. The number $\mu(t)$ of the memory cells filled at the moment t is a random whole number function.

Let us introduce the generating functions:

$$g_t(x) = \sum_{k=0}^{\infty} P[\kappa(t) = k] x^k \qquad G_t(x) = \sum_{m=0}^{\infty} P[\mu(t) = m] x^m$$

Assume that the time of processing at the moment $t = s$ of one element is a random value $\nu(s)$ with the prescribed distribution $P[\nu(s) < t] = F_s(t)$. Then we have:

$$G_t(x) = \prod_{s=h}^{t} g_s\{1 + [1 - F_s(t-s)](x-1)\}$$

Then we obtain expressions for the mean and dispersion:

$$E\mu(t) = \sum_{s=h}^{t} E\kappa(s)[1 - F_s(t-s)]$$

$$D\mu(t) = E\mu(t) + \sum_{s=h}^{t} [D\kappa(s) - E\kappa(s)][1 - F_s(t-s)]^2$$

If $\kappa(s)$ is characterized by Poisson distribution with the parameter $\lambda(s) = E\kappa(s) = D\kappa(s)$, then $\mu(s)$ is also distributed following the Poisson law with the parameter $\Lambda(t) = E\mu(t)$.

All this reasoning is valid when the value of M is unlimited. If $M = \text{const} > \Lambda(t)$, then $\mu(t)$ has a truncated Poisson distribution:

$$P[\mu(t) = m; M = \text{const}] = P_{\Lambda(t),M}(m) = \left(\frac{\Lambda^m(t)}{m!} \Big/ \sum_{l=0}^{m} \frac{\Lambda^l(t)}{l!} \right) \qquad (m \le M)$$

Hence, the monitoring system's efficiency is determined by the probability of the overfilling of its memory (i.e., the probability of information loss). In a more general case, to assess the efficiency of the monitoring system for detection of anomalies on the Earth's surface, it is necessary to consider the work of all its subsystems with

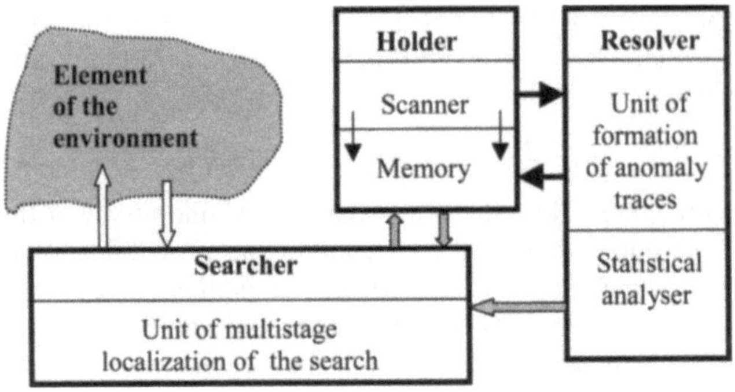

Figure 9.1. Block-scheme of the monitoring system to detect anomalies in the environment.

their individual limitations of memory and rate of data processing taken into account.

Criteria for assessing the monitoring system's efficiency are eventually determined from the probability of them accomplishing their task, which is a complex function of parameters of the systems themselves and the environment. In theory, the model of the environment plays an important role. One of the possible models of such a type is derived on the basis of the use of the notion of "spottiness" of the space under study. The moving anomaly violates the spotty structure, and as a result, its trace is formed. Since the spots can be of various nature, an efficient algorithm of detection should be multichannel capable of analysing the hydrophysical, biological, acoustical, optical, and physico–chemical information.

The general functional scheme of a standard monitoring system can be presented as a set of three subsystems (Figure 9.1): *holder*, *resolver*, and *searcher*. The *fixer* records data from the sensors (radars, radiometers, etc.) and in accordance with the algorithms included in the system, it memorizes "suspicious" elements. The *resolver* analyses the input data from the *holder* and makes a resolve on the signal (anomalous) or noise (background) origin of these elements. The *searcher*, following the *resolver's* command, localizes the anomaly and starts coordinating the functions of the *searcher* and the *resolver* on specifying the data on the observed anomaly.

The probability of overfilling the *resolver's* memory during the working time T_Σ is estimated at:

$$P \geq 1 - T_\Sigma\{1 - \Phi[(M - \Lambda)\Lambda^{-1/2}]\}$$

where Φ is the Gauss integral.

The accuracy of this estimate depends on the algorithms of the monitoring system's units. In particular, with the use of the method of sequential analysis to solve the problem of distinguishing between the hypotheses and evolutionary technology of prediction (Bukatova *et al.*, 1991), the procedure of data processing in the *resolver* is substantially simplified and, hence, the promptness of making a resolve increases, reducing the delay of data between the *holder* and the *resolver*. A consideration of the theoretical assessments of such delays is an independent problem.

9.4 MAKING STATISTICAL DECISIONS IN THE SYSTEMS OF ENVIRONMENTAL MONITORING

A variety of environmental monitoring regimes depend on the data processing and decision making procedures. The statistical analysis of the series of events, which accompany the functioning of the monitoring system, can be realized by numerous methods whose applicability in each case is determined by a set of probabilistic parameters characterizing the phenomenon under study. However, instability and parametric uncertainty in situations when each measurement requires much effort and finance, makes one search new methods of decision making on the basis of measurement data, fragmentary both in space and in time.

With the development of alternative methods of making statistical decisions, the problem of searching an objective estimation of the parameters of the processes taking place in the environment has obtained new substantiation. It is possible to consider and compare two approaches to this problem: classical, based on the procedure of a priori limitation of the number of observations; and a sequential analysis based on the procedure of the step-by-step decision making. The development of computer technologies enables one to realize both approaches in the form of a single system of making statistical decisions.

A procedure of decision making for detection of anomalies or other environmental phenomenon depends on the structure of the monitoring system. The classical approach orients the observing system towards collecting a fixed volume of data. The data are then processed to detect certain effects or properties in the space under study. The method of sequential analysis does not divide these stages but alternates them. In other words, the monitoring data processing is carried out after each measurement. Hence, the algorithmic load in the sequential procedure changes dynamically, whereas in a classical case the data are processed at a final stage. From the viewpoint of the formation of the structure of an adequate automated *resolver*, these approaches should be realized in the form of individual units, the choice between which can vary in monitoring dynamics. However, for this purpose high professionalism is needed. Therefore, in the proposed structure of the *resolver* the operator only controls the parameters of choice between the alternative hypotheses.

The decision making theory based on the sequential analysis procedure to be developed for the problems of the choice between two hypotheses is required for the solution of some problems. The list of problems to be solved before making a resolve include:

- selection of an efficient criterion to estimate the parameters;
- determination of the character of probabilistic characteristics of the process under study; and
- a priori assessment of possible losses on the accuracy of decisions.

One of the features of the automated system of decision making is the presence of a wide spectrum of functions, which should, if possible, provide a solution for most of the intermediate and final problems of assessment of the situation in the regime of monitoring. Therefore, the system should include the following units:

1. visualization of the measured data in the form of direct images, histograms, sum and frequencies, and lables;
2. calculation of statistical characteristics of the observation results (the mean, dispersion, moments of third and higher orders, the excess coefficient, coefficient of asymmetry, entropy, etc.);
3. construction of empirical density and functions of probability distributions;
4. construction of a theoretical distribution of probabilities with an assessment by one of the statistical criteria;
5. calculation of the current volume of samples for the classical procedure by the Neyman–Pearson (Zayezdny *et al.*, 1989);
6. calculation of characteristics of the procedure of sequential analysis and visualization of its state;
7. realization of the functions of the operator's access to the system's units at any step of the system's functioning with the possibilities of decision making on changing the parameters of the procedure or ceasing measurements;
8. visualization of the resolves.

This set of units ensures the formation of the model of the measurement procedure depending on the a priori information about the parameters and character of the process under study. At the same time, the operator can test the accuracy of the input information and promptly change the strategy of monitoring. On the whole, the set of indicated units is an automated system for making statistical decisions.

The sequential procedure of making statistical decisions is based on the Wald distribution. When studying the sum x_n of independent similarly distributed random values, there appear two dual problems of assessment of the distribution function $P(x_n < x) = F_1(x)$ both with fixed n and with its random value. In both cases the x_n value is compared with some threshold C. However, in the second case this problem is transformed into that of the study of the distribution $P(v < n) = F_2(n)$ of a random number v of summands, that leads to a first excess x_v of the threshold $C = C(\alpha, \beta)$: $x_i < C(i = 1, 2, \ldots, v - 1)$; $x_v \geq C$ where α and β are errors of the first and second kind, respectively.

In accordance with the central limit theorem for $n \to 8$ in the first (classical) case, $F(x)$ is described by the normal distribution. In the second case the Wald distribution is realized, the density of which is determined on a half-line $[0, \infty)$ and has one maximum at the point $x = m_c$:

$$w_c(y) = c^{1/2} y^{-3/2} (2\pi)_{-1/2} \exp\left[-0.5c(y + y^{-1} - 2)\right]$$

Let us calculate the derivative of the Wald distribution function and equate it to zero: $dw_c(y)/dy = 0$. The solution for this equation is:

$$y = m_c = [(9 + 4c^2)^{1/2} - 3]/(2c)$$

The location of the maximum of $w_c(y)$ shifts depending only on the value of one parameter c so that we always have $m_c < 1$. For $c \to 0$, $m_c \to 0$, and for $c \to \infty, m_c \to 1$:

$$w_c(m_c) = (3/(2\pi)^{1/2})(1/[m_c(1 - m_c^2)^{1/2}]) \exp\{-3(1 - m_c)/[2(1 + m_c)]\}$$

Hence, for $c \to 0$ or $c \to \infty, w_c(y)$ degenerates into delta-function $\delta(y)$ or $\delta(y - 1)$, respectively.

9.4.1 Distinguishing between hypotheses

The monitoring system is constructed using a standard scheme of man–machine dialogue with the use of a hierarchical menu. The principal scheme of the system has the form of some transformation F whose structure and content are determined by a set of units performing individual functions on analysis and transformation of measurement data. The input data for the system are measurements from n sources (sensors). Resolution of the reliability of this or that hypothesis (H_0 or H_1) can be made for each source or for some set of them $m \leq n$. Various regimes of the operator–system interaction are foreseen. The completion of the resolution procedure is followed by a report on the length of the used succession of data for each acting information channel. In case of addition of two or more channels, one of the hypotheses is finally adopted on the basis of the weighted majority of resolutions made in individual channels. With obstacles appearing in the *resolver* the operator is warned about situations of uncertainty, and can change the weights of channels and introduce corrections into other parameters.

To realize the procedure of adopting complicated hypotheses, when the number of possible decisions exceeds the binary situation, a recurrent procedure is created for realization of the many-criterion problem of decision making in the form of a composition of simple binary situations.

The base procedure realizes the function *resolve* $(\gamma, \xi, \alpha, \beta, G, Y, \theta_0, \theta_1, N, R)$ where γ is the controlling parameter, ξ is the input flux of measured values of the controlled parameters, G is a set of parameters of distribution f_a, Y is the type of distribution f_a, N is the threshold of the number of observations for an automated change of the procedure of sequential analysis of the classical procedure, and R is the indicator of the decision making ($R = 0$, the hypothesis H_0 is valid; $R = 1$, the hypothesis H_1 is valid; $R = 2$, the decision is not made). Introduction of N enables one to avoid a situation when a decision can be not made as long as is needed. When the amount of measurements exceeds the threshold N, the system resolves following the classical procedure. The parameter γ controls the regimes of the use of the *resolver*. Various regimes are possible for realizing the decision making procedures: $\gamma = 0$, the interface of the operator is not planned; $\gamma = 1$, correction of the procedure is foreseen. Depending on γ, the system is adjusted to the respective regime of the units control. At $\gamma = 0$, many intermediate operations are ignored, and the system's operating speed increases. At $\gamma = 1$, the system activates all its functions and practically at each step the operator has a priority right to change the procedure parameters. The system memorizes all actions of the operator and in case of contradictions informs him about it. The parameter γ also controls the regime of data input from information channels of various profiles. There appears the possibility of making the decisions via individual channels or via a group of them. The operator is informed about the state of the final conclusions of individual units. From the channel number the user can ask for statistical characteristics, the type of distribution, and other parameters. The system blinds a senseless query, such as statistical characteristics or probabilistic distribution with small volumes of samples. In the case of uncertainty in the *resolver*, the operator is invited to change the structure of the identifier of the considered channels, to change their weights, or to perform other actions.

9.4.2 Processing the multichannel information

Methods of performing local diagnostics of the environment cannot give a complex assessment of the state of a natural object or process, especially if this element of the environment covers a vast space. In the problems of geoinformation monitoring, situations occur when it is necessary to make a real-time decision using the limited capabilities of the means of data processing (speed, memory, rate of data transmittal, etc.). In this regard, we consider a model of such a situation and calculate its characteristics. The general scheme of sensing the environmental elements assumes that the flux of data $\{T_b\}$ from the jth measuring device can be analysed both separately and together with the data from other channels. In fact, there is a digital processor similar to a personal computer making V operations per second. The task is set to identify an unknown vector parameter $T_b = \{T_{b1}, \dots, T_{bn}\}$ characterizing the state of the controlled object (e.g., for the problems of microwave monitoring these are brightness temperatures). To solve this problem, an algorithm of the parallel-in-time analysis of the vector T_b components is proposed, which makes it possible to reduce the time of data processing. In this case the problem of the data fluxes coordination appears. The procedure consists of the continuous formation of the vector T_b structure by a gradual addition of components whose estimates by local criteria are attributed to T_b: (T_{b1}), $(T_{b2}), \dots, (T_{b1}, T_{b2}, \dots, T_{bi}), \dots, (T_{b1}, \dots, T_{bk})$. Two situations are possible: (1) the system has time to analyse the whole chain of vectors without delay in individual channels; and (2) at some moments, a delay occurs in some channels. In a general case, two types of delay are possible: in time and in volume. Denote constant delays in time $\tau_1, \dots, \tau_k$, and delays in volume $m_1, m_2, \dots, m_k$. In the first case, the process of choosing the vector T_b is realized by a channel-to-channel procedure when the vector components T_{bj} are accumulated with the time delay intervals τ_j and then are transmitted to the chain $\{T_{bj}\}$. The number of delayed estimates of T_{bi} in each channel is random and therefore the buffer memory of each channel should be reserved. In the second case a random value is the time of delay $\tau_i (i = 1, \ldots, k)$, and the volumes of the buffers m_i are fixed. In both cases it is necessary to assess the probability not to miss a real value of the vector T_b and to find optimal values of τ_i and m_i corresponding to the maximum level of this probability.

Let in each channel of the volume $n_i (= 1, \ldots, k)$ and the non-anomalous value T_{bi} be taken as the anomalous value with probability α_i and is rejected with probability $1 - \alpha_i$. Then the probability of appearance of the T_{bi} estimates during the time period $\tau_i = r_i n_i S_i$ claiming to be constituents of the final decision, will be:

$$P\{\mu_i = S_i\} = C_{r_i}^{S_i} \alpha_i^{S_i} (1 - \alpha_i)^{r_i - S_i} = \nu_i(S_i)$$

At $\mu_i \leq r_i \alpha_i$ the output of estimates is slow and they have time to pass from the ith to the $(i+1)$th *resolver* without delay. If $\mu_i > r_i \alpha_i$, then the output of the estimates from the ith to the $(i+1)$th *resolver* is rapid, and between their input to the $(i+1)$th *resolver* the latter has no time to examine all versions of T_{bi+1}. Therefore, the estimates are delayed. The probability that the number of estimates will not exceed an average $r_i \alpha_i$ by more than ε_i, will be:

$$P\{\mu_i \leq r_i \alpha_i + \varepsilon_i\} = \sum_{s=0}^{r_i \alpha_i + \varepsilon_i} \nu_i(s)$$

Denote as M_i the memory volume in each channel for delay. Then the following condition should be met:

$$r_i \alpha_i + \varepsilon_i \leq M_i$$

In this case the probability of not overfilling the memory M_i will be:

$$P\{\mu_i \leq M_i\} = \Phi(u_i)$$

where

$$u_i = \frac{\varepsilon_i}{\sqrt{r_i \alpha_i (1 - \alpha_i)}} = \frac{M_i - r_i \alpha_i}{\sqrt{r_i \alpha_i (1 - \alpha_i)}}$$

$$\Phi(u_i) = \frac{1}{\sqrt{2\pi}} \int_{\infty}^{u_i} \exp\left[-\frac{t^2}{2}\right] dt$$

Utilizing the Boolean formula, let us calculate the probability of non-overflow of memory M_i at the ith channel with the delay τ_i in the process of examination of all values of the ith component. Denote $P_{1,\ldots,N_i}$ as the probability of not overfilling the memory. Then:

$$P_{1.2,\ldots,N_i} \geq 1 - N_i [1 - \Phi(u_i)]$$

Let this probability differ from unity by no more than δ_i. Then we obtain an equation to estimate the delay τ_i:

$$\Phi(u_i) = 1 - \frac{\delta_i}{N_i}$$

or

$$r_i^2 - 2\frac{r_i}{\alpha_i}\left[M_i + (1-\alpha_i)\ln\frac{N_i}{\delta_i}\right] + \left(\frac{M_i}{\alpha_i}\right)^2 = 0$$

In the multichannel monitoring system, along with these special features of decision making, a problem appears of estimating the parameter in the communication channel when it is noise distorted. The problem is especially important in remote monitoring data transmission. In this regard, we study a situation when the communication channel is characterized by the Gaussian additive and Rayleigh multiplicative noises.

9.4.3 Reception of noise-loaded signals

One of the important problems of microwave remote sensing of the environment is the transmission of a useful signal by the noise-loaded channels and its reconstruction at a receiver. This problem is studied by many scientists under the various assumptions about the character of the information channel. Assume a noise-loaded communication channel containing only an additive component v subject to the Gauss law, with $Mv = 0$ and $Dv = \sigma^2$. The input signal $X = f_\lambda(t)$ is the known (usually continuous) function of time t and the transmitted parameter λ (the way to code the parameter). Then the output signal will be: $Y(t) = f_\lambda(t) + v(t)$. With respect to the f_λ function, assume the existence of the first λ derivative. An ideal reception of the transmitted parameter can be realized using optimal discrete statistics of the volumes n of heterogeneous sampled $Y(t)$ values, namely, $y_1, \ldots, y_n$. The sampled values are processes using the method of maximum likelihood, which gives the best estimates of the parameter λ (the Fisher method). For this, it is necessary to derive an equation of likelihood. According to these assumptions, the probability density of a random parameter $Y(t)$ will be:

$$p_\lambda(y) = \left(\sqrt{2\pi\sigma}\right)^{-1}\exp\left[-0.5\sigma^{-2}(y - f_\lambda(t))^2\right] \tag{9.1}$$

The likelihood function is written as:

$$g(y_1, \ldots, y_n; \lambda) = \left(\sqrt{2\pi\sigma}\right)^{-n}\exp\left\{-0.5\sigma^{-2}\sum_{j=1}^{n}\left[y_j - f_\lambda(t_j)\right]^2\right\} \tag{9.2}$$

The likelihood equation will be:

$$L' = \sigma^{-2}\sum_{j=1}^{n}\left[y_j - f_\lambda(t_j)\right]f_\lambda'(t) = 0 \tag{9.3}$$

In a general case Equation (9.3) is transcendental with respect to the estimated parameter $\lambda = \lambda_*$. Therefore, it is expedient to apply an efficient method of successive approximations. As a first approximation, we take some value of the parameter $\lambda = \lambda_1$ and find the next approximation using the formula: $\lambda_2 = \lambda_1 + h_1$, where h_1 is the correction determined by the form of the function (9.2) and an amount of information contained in a heterogeneous sample $y_1, \ldots, y_n$. We obtain:

$$h_1 = \frac{L'(y_1, \ldots, y_n; \lambda_1)}{I(\lambda_1)} \tag{9.4}$$

where

$$I(\lambda_1) = \sum_{k=1}^{n} M_\lambda \left(\frac{p_k'}{p_k}\right)^2$$

where p_k is the probability density of the parameter y_k. We have:

$$I(\lambda_1) = \int_{-\infty}^{\infty} \sum_{k=1}^{n} \left(\frac{p_k'}{p_k}\right)^2 p_k dy = \frac{1}{\sigma^2} \sum_{k=1}^{n} [f'_{\lambda_1}(t_k)]^2$$

Finally we obtain the relationship to estimate the next approximation of the parameter λ:

$$\lambda_2 = \lambda_1 + \frac{\sum_{j=1}^{n} [y_j - f_\lambda(t_j)] f_{\lambda_1}(t_j)}{\sum_{k=1}^{n} [f'_{\lambda_1}(t_k)]^2}$$

By the same reasoning, we move on to the third approximation, etc. In general, if the mth approximation of λ_m has already been found, then the $(m+1)$th approximation is obtained by the formula:

$$\lambda_{m+1} = \lambda_m + \frac{\sum_{j=1}^{n} [y_j - f_{\lambda_m}(t_j)] f'_{\lambda_m}(t_j)}{\sum_{k=1}^{n} [f'_{\lambda_m}(t_k)]^2} \tag{9.5}$$

These successive approximations convert to some limit $\lambda_{**} \approx \lambda_*$, which is assumed as an estimate for λ_*. Of course, the rate of convergence depends on the successful choice of λ_1.

The estimate of λ_* by method of maximum likelihood is distributed asymptotically normal with the mathematical expectation $M\lambda_* \approx \lambda_*$ and dispersion $D\lambda_* = 1/I(\lambda)$, where $I(\lambda)$ is the Fisher information mentioned above. Then the confidence interval λ_* is as follows:

$$P = P\left\{|\lambda - \lambda_*| < u\sqrt{D\lambda_*}\right\} = \frac{1}{\sqrt{2\pi}} \int_{-u}^{u} \exp\left(-\frac{z^2}{2}\right) dz \tag{9.6}$$

Generalizing the problem further, consider the case when the input signal $X = f(\lambda_1, \ldots, \lambda_k; t)$ is the known function of time t and a group of transmitted parameters $\lambda_1, \ldots, \lambda_k$. With respect to the f function, assume the existence and continuity of partial derivatives $\partial f / \partial \lambda_j (j = 1, \ldots, k)$. The output signal will be: $Y(t) = f(\lambda_1, \ldots, \lambda_k; t) + \nu(t)$, with $M\nu = 0$, $D\nu = \sigma^2$. Below we take $\lambda = \{\lambda_1, \ldots, \lambda_k\}$. Then the formulas (9.1) and (9.2) do not change, and the likelihood Equation (9.3) is written as a system of equations:

$$\sum_{j=1}^{n} [y_j - f(\lambda; t_j)] \frac{\partial f(\lambda; t_j)}{\partial \lambda_i} = 0 \qquad (i = 1, \ldots, k) \tag{9.7}$$

In some cases the system (9.7) can be easily solved. But in the general case the system (9.7) is solved approximately. In this case the formula (9.5) is written as:

$$\lambda_i^{(m+1)} = \lambda_i^{(m)} + \sum_{j=1}^{n} \frac{\partial f(\lambda^{(m)}; t_j)}{\partial \lambda_i} [y_j - f(\lambda^{(m)}; t_j)] \Big/ \sum_{j=1}^{n} \left[\frac{\partial f(\lambda^{(m)}; t_j)}{\partial \lambda_i} \right]$$

where some estimates of the parameter λ are chosen as a first approximation $(m = 0)$. At $m \to \infty, \lambda_i^{(m)} \to \tilde{\lambda}_i (i = 1, \ldots, n)$.

The above version of additive noises is an approximation of a real situation of the functioning of radiometric systems. Therefore, we also study the version of multiplicative noises. Take a 1-beam channel with multiplicative noises $\alpha(t)$, whose probability density follows the Rayleigh law (with the respective normalization of the amplitude):

$$p_\alpha(x) = (x/\gamma^2) \exp[-0.5x^2/\gamma^2]$$

and with negligibly small Gauss additive noises $(\sigma^2 \approx 0)$. Then the output signal will be $Y(t) = f_\lambda(t)\alpha(t)$. Let $f_\lambda(t) > 0$. The probability density of the random parameter $Y(t)$ will be:

$$p_Y(y) = \frac{y}{\gamma^2 f_\lambda^2(t)} \exp\left\{ -\frac{1}{2} \left[\frac{y}{f_\lambda(t)\gamma} \right]^2 \right\}$$

Then we derive the likelihood function:

$$g(y_1, \ldots, y_n; \lambda) = \frac{1}{\gamma^{2n}} \prod_{i=1}^{n} \frac{y_i}{f_\lambda^2(t_i)} \exp\left\{ \sum_{j=1}^{n} \left[-\frac{y_j^2}{2\gamma^2 f_\lambda^2(t_j)} \right] \right\}$$

The likelihood equation is:

$$L'(\lambda) = \sum_{j=1}^{n} \left\{ \left[-\frac{y_j^2}{2\gamma^2 f_\lambda^2(t_j)} + \ln \frac{y_j}{\gamma^2 f_\lambda^2(t_j)} \right] \cdot \left(\frac{y_j^2}{\gamma^2 f_\lambda^2(t_j)} - 2 \right) \frac{f_\lambda'(t_j)}{f_\lambda(t_j)} \right\} = 0$$

The denominator in Equation (9.4) is estimated by the formula:

$$I(\lambda) = 4 \sum_{k=1}^{n} \left[\frac{f_\lambda'(t_k)}{f_\lambda(t_k)} \right]^2$$

Hence, the formula for successive approximations of the estimates of the λ parameter is written as:

$$\lambda_{m+1} = \lambda_m + L'(\lambda_m)/I(\lambda_m)$$

9.4.4 Information decoding

In the procedure of making statistical decisions signal decoding is an important stage (Borodin *et al.*, 1998). Consider the set M of equiprobable messages represented in the time interval T by the time functions with similar capacities S. Assume these messages to be transmitted by the channel with continuous time and additive Gauss white noise with spectral density N_0, with carrying capacity $c = S/N_0 \log^2 e$ binary units/s. Then the output signal can be decoded with the error probability described by the integral:

$$P_e = 1 - \int_{-\infty}^{\infty} \Phi'\left(x - \sqrt{2Q}\right)\Phi^{M-1}(x)dx \tag{9.8}$$

where

$$\Phi(x) = (2\pi)^{-1/2}\int_{-\infty}^{\infty} \exp\left(-t^2/2\right)dt \qquad Q = cT$$

For the integral Equation (9.8) we obtain an approximated formula expressing it through a normal distribution. With this aim in view, we obtain the lower and upper estimates of this integral. To obtain the lower estimate, we use the incquality:

$$\int_{-\infty}^{a} \Phi^{M-1}(x)\Phi'\left(x - \sqrt{2Q}\right)dx < \Phi^{M-1}(a)\Phi\left(a - \sqrt{2Q}\right) \tag{9.9}$$

which is valid for any value of a.

As seen from Equation (9.8), the function $\Phi^{M-1}(x)$ tends to unity at $x \to \infty$ and tends to zero at $x \to \infty$. At $M \gg 1$ in some small vicinity of the point $x = x_0$ a sharp transition of the function $\Phi^{M-1}(x)$ takes place from small values to values close to unity. Therefore, using the asymptotic formulas for the probability integral, we can write the following inequality:

$$1 - \frac{M-1}{\sqrt{2\pi}x}\exp\left[-\frac{x^2}{2}\right] < \Phi^{M-1}(x) < 1 - \frac{M-1}{\sqrt{2\pi}x}\left(1 - \frac{1}{x^2}\right)\exp\left[-\frac{x^2}{2}\right]$$

Substituting these relationships into Equations (9.8) and (9.9), we have:

$$\begin{aligned} P_e > P_{e1} = {} & \Phi\left(x_0 - \sqrt{2Q}\right)\left[1 - \Phi^{M-1}(x_0)\right] + \frac{M-1}{\sqrt{2}}\exp\left(-\frac{Q}{2}\right) \\ & \times \left\{\left[1 - \Phi\left(x_0\sqrt{2} - \frac{\sqrt{2Q} - 1}{\sqrt{2}}\right)\right]\exp\left[-\frac{1}{2}\left(\sqrt{2Q} - \frac{1}{2}\right)\right]\right. \\ & \left. - \frac{1}{x_0^3}\left[1 - \Phi\left(x_0\sqrt{2} - \sqrt{Q}\right)\right]\right\} \end{aligned}$$

To obtain the upper estimate, we use an explicit inequality:

$$\Phi^{M-1}(x) > \begin{cases} 1 - (M-1)/(\sqrt{2\pi}x)\exp\left(-x^2/2\right) & x \geq x_0 \\ 2^{1-M} & 0 \leq x < x_0 \\ 0 & x < 0 \end{cases}$$

Introducing this inequality into Equation (9.8), we obtain:

$$P_e < P_{e2} = (1 - 2^{1-M})\Phi\left(x_0 - \sqrt{2Q}\right) + \frac{M-1}{2\sqrt{\pi}x_0}\exp(-Q/2) \times \left[1 - \Phi(x_0\sqrt{2} - \sqrt{Q})\right] \tag{9.10}$$

The formula (9.10) can be specified if we assume:

$$\Phi^{M-1}(x) \approx 2^{2-M}\left(0.5 + (M-1)x/\sqrt{2\pi}\right) \qquad 0 \leq x < x_0$$

Then:

$$P_e \approx P_{e3} = \left(1 - 2^{1-M} - \frac{M-1}{2^{M-2}}\sqrt{\frac{Q}{\pi}}\right)\Phi\left(x_0 - \sqrt{2Q}\right) + \frac{M-1}{\pi 2^{M-1}} \times \left[\sqrt{\pi Q}\Phi\left(-\sqrt{2Q}\right) - e^{-Q} + \exp\left\{-0.5\left(x_0 - \sqrt{2Q}\right)^2\right\}\right] + \frac{M-1}{2\sqrt{\pi}x_0}\exp(-Q/2)\left[1 - \Phi\left(x_0\sqrt{2} - \sqrt{Q}\right)\right]$$

For $M \gg 1$ from Equation (9.10) we have:

$$P_e < P_{e4} = \Phi\left(x_0 - \sqrt{2Q}\right) + \left(2\sqrt{\pi}x_0\right)^{-1}\exp(\ln M - Q/2) \times \left[1 - \Phi\left(x_0\sqrt{2} - \sqrt{Q}\right)\right]$$

In the formulas above for an approximate calculation of the integral (9.8) the parameter x_0 remained non-estimated. Let the estimated Equation (9.10) be at a minimum. Then to estimate x_0 we obtain the following equation:

$$(1 - 2^{1-M})\Phi'\left(x_0 - \sqrt{2Q}\right) - \frac{M-1}{\sqrt{2\pi}x_0}\exp(-Q/2)\Phi'\left(x_0\sqrt{2} - \sqrt{Q}\right) - \frac{M-1}{2\sqrt{\pi}x_0^2}\exp(-Q/2)\left[1 - \Phi\left(x_0\sqrt{2} - \sqrt{Q}\right)\right] = 0 \tag{9.11}$$

For practical use it is expedient to have simpler calculation formulas. With this aim in view, we use the following estimates in Equation (9.10):

$$\Phi\left(x_0 - \sqrt{2Q}\right) < \left[\sqrt{2\pi}\left(\sqrt{2Q} - x_0\right)^{-1}\right] \times \exp\left[-0.5\left(\sqrt{2Q} - x_0\right)^2\right] \qquad Q > x_0^2/2$$

$$1 - \Phi(x_0\sqrt{2} - \sqrt{Q}) < \begin{cases} 1 & x_0^2 < Q/2 \\ \dfrac{\exp\left[-0.5(x_0\sqrt{2} - \sqrt{Q})^2\right]}{\sqrt{2\pi}(x_0\sqrt{2} - \sqrt{Q})} & x_0^2 \geq Q/2 \end{cases}$$

Table 9.3. An example of the calculation of the x_0 parameter by solving Equation (9.11) and its comparison with the estimate from Equation (9.12).

	$M = 32$		$M = 256$		$M = 512$	
Q	x_0	$\tilde{x}_0$	x_0	$\tilde{x}_0$	x_0	$\tilde{x}_0$
2	1.96	2.63	2.71	3.33	2.92	3.53
6	1.98	2.63	2.71	3.33	2.93	3.53
10	2.01	2.63	2.72	3.33	2.93	3.53
14	2.06	2.63	2.73	3.33	2.94	3.53
20	2.15	2.70	2.76	3.33	2.96	3.53

Substituting these expressions into Equation (9.10) and with exponents in the resulting expression of equal value, we obtain the following estimate for x_0:

$$x_0 \approx \tilde{x}_0 = \begin{cases} \sqrt{2\ln(M-1)} & 1/4 < u < 1 \\ \sqrt{2Q}\left[1 - \sqrt{1-2u}/\sqrt{2}\right] & 0 \le u \le 1/4 \end{cases} \tag{9.12}$$

where $u = \ln(M-1)/Q$.

In a similar way we estimate P_e:

$$P_e < \tilde{P}_e = \begin{cases} \dfrac{1}{\sqrt{2\pi Q}}\exp\left[-Q\left(\dfrac{1}{2} - \dfrac{\ln M}{Q}\right)\right] & 0 \le \dfrac{\ln M}{Q} \le \dfrac{1}{4} \\ \dfrac{1}{2\sqrt{\pi Q}}\exp\left[-Q\left(1 - \sqrt{\dfrac{\ln M}{Q}}\right)^2\right] & \dfrac{1}{4} < \dfrac{\ln M}{Q} < 1 \end{cases}$$

Table 9.3 lists the x_0 values obtained as solutions of Equation (9.11) and the $\tilde{x}_0$ values. It is seen that $P_{ei}(x_0) < P_{ei}(\tilde{x}_0)$ and with increasing M the difference between $P_{ei}(x_0)$ and $P_{ei}(\tilde{x}_0)$ decreases. As seen from Table 9.4, the P_{ei} estimates at $M \gg 1$ and $Q > \ln M$ provide a good approximation of P_e. It turns out that the function

$$\bar{P}_e = \frac{1}{4}\sum_{i=1}^{4} P_{ei}$$

Table 9.4. Comparison of different estimates of the probability of signal decoding.

	$M = 16$			$M = 256$		
Q	P_e	P_{ei}	$\bar{P}_e$	P_e	P_{ei}	$\bar{P}_e$
10	0.9×10^{-2}	0.2×10^{-1}	0.2×10^{-1}	0.6×10^{-1}	0.1	0.1
20	0.5×10^{-4}	0.1×10^{-3}	0.1×10^{-3}	0.7×10^{-3}	0.2×10^{-2}	0.1×10^{-2}
30	0.3×10^{-6}	0.8×10^{-6}	0.6×10^{-6}	0.5×10^{-5}	0.1×10^{-4}	0.9×10^{-5}

Table 9.5. Estimates of the Kotelnikov integral.
From Krapivin (1969).

Q	$M = 16$	$M = 32$	$M = 256$
10	0.925×10^{-2}	0.164×10^{-1}	0.635×10^{-1}
20	0.557×10^{-4}	0.111×10^{-3}	0.716×10^{-3}
30	0.301×10^{-6}	0.616×10^{-6}	0.471×10^{-5}
40	0.208×10^{-8}	0.416×10^{-8}	0.333×10^{-7}
50	0.125×10^{-10}	0.251×10^{-10}	0.201×10^{-9}
60	0.771×10^{-13}	0.154×10^{-12}	0.123×10^{-12}
70	0.481×10^{-15}	0.962×10^{-15}	0.770×10^{-14}
80	0.303×10^{-17}	0.606×10^{-17}	0.485×10^{-16}
90	0.193×10^{-19}	0.385×10^{-19}	0.308×10^{-18}

approximates uniformly the P_e parameter at any values of M and Q. Table 9.5 lists the values of integral (9.8).

In conclusion, consider the case of a signal with an unknown phase. In this case the error probability will be:

$$P_e = 1 - \int_0^\infty W_q'(x) W_0^{M-1}(x) dx$$

where

$$W_q(x) = \int_0^x z \exp(-0.5[z^2 + q^2]) I_0(qz) dz \qquad q = \sqrt{2Q}$$

From this we obtain:

$$\begin{aligned} P_e &= 1 - \sum_{k=0}^{M-1} (-1)^k C_{M-1}^k \exp(-q^2/2) \int_0^\infty x I_0(qx) \exp[-x^2(k+1)/2] dx \\ &= 1 - \exp(-q^2/2) \sum_{k=0}^{M-1} (-1)^k C_{M-1}^k \int_0^\infty x \exp\{-x^2(k+1)/2\} \sum_{i=0}^\infty \frac{(qx/2)^{2i}}{(i!)^2} dx \\ &= 1 - \exp(-q^2/2) \sum_{k=0}^{M-1} \sum_{i=0}^\infty (-1)^k \frac{C_{M-1}^k}{i!(k+1)} \left[\frac{q^2}{2(k+1)}\right]^i \end{aligned}$$

Considering an explicit equality:

$$\exp\left[\frac{q^2}{2(k+1)}\right] = \sum_{i=0}^\infty \frac{1}{i!} \left[\frac{q^2}{2(k+1)}\right]^i$$

we finally obtain:

$$P_e = \sum_{k=1}^{M-1} (-1)^{k+1} C_{M-1}^k \frac{1}{k+1} \exp\left[-\frac{q^2}{2} \frac{k}{k+1}\right]$$

9.4.5 Application to geoinformation monitoring problems

To check the *resolver's* efficiency, fragments of the archived remote sensing data obtained with the IL-18 flying laboratory over the Asian territory of the Former Soviet Union have been processed. Measurements were made in the frequency ranges 0.4, 0.8, 2.25, 11, 21, and 27 cm. Averaging was made during one second at an average speed of flight of 600 km/hr. Measured data were chosen for the eastern coastline of the Caspian Sea in the vicinity of the settlement Kyzyk. These are saline lands, dry sands, salt water, saxaul thickets, and anthropogenic landscapes. The characteristic leaps of brightness temperatures correspond to changes of surface types. To be independent of the calibration procedure, measurements were normalized. For this purpose the quasi-homogeneous sets of brightness temperatures $\{T_j\}$ are put in order, the parameters $T_{j,\text{min}}$, $T_{j,\text{max}}$, and $T_{j,\text{middle}}$ are determined for them, and they are transformed into the sets $\{T_j/T_{j,\text{max}}\}$ and $\{\Delta T_j = T_j - T_{j,\text{middle}}\}$. With the help of algorithms of reconstruction of images at each of the quasi-stationary sites we obtain a multitude of distribution functions used as informational characteristics of the Earth's cover. An identification of the type of surface with the type of distribution and its parameters enables one to formulate a learning sample for the system of classification, for instance, from the cluster analysis.

The brightness contrasts used to solve the problem of classification of the Earth's surface elements for the drying and watered parts of the Kara Bogaz Gol Gulf have the form of the Weibull distribution and the exponential distribution, respectively. As a result, the spottiness of landscape is transformed into the spottiness of the clusters of distributions, and the function of the *resolver* is reduced to the division and identification of the latter.

9.5 MANAGING NATURAL RESOURCES

The problem of an optimal management of natural resources appears in regions experiencing the presence of elements of anthropogenic forcing. In this regard, Khudoshina (1996) introduced the term "ecologo-economic system" (EES), which formalizes the problem of managing a totality of nature components and economic activity and enables one to describe numerous contradictions between these components. The efficiency of the monitoring system depends on the correctness of the administrative decisions on the determination of priorities and the distribution and use of resources. It is also important to have a set of efficient indicators of the state of both individual elements of the environment (air, water, soil, flora/fauna, etc.) and the system on the whole. Such indicators enable one to assess the state of the ecological balance of the territory and to simplify thereby the problem of management of the ecosystems located on the territory (Lenz and Beutter, 2003; Yacoubi *et al.*, 2003).

Real problems of managing natural resources are multicriterion and connected with analysis of a great number of factors, dependences between which often cannot be formalized. Khudoshina (1996) proposed a technique to solve such problems for an efficient management of the resources in the coastal regions. This technique is based on the method of logical-information modelling. This method, in turn, is based on the principles of simulation modelling and includes two types of models: dynamic and logical-informational. The dynamic model describes the temporal transformation for all characteristic parameters of the system, combines a group of models of the physical processes, and determines the quantitative changes and relationships in the system. The logical-informational model, with the help of Boolean algebra, assesses the quality of the system for various ranges of the values of its parameters. The logical-informational model is based on the notion of the functioning element characterized by the vector of state $X = (M_i, \{C_{ij}\}, \{P_{ik}\})$ where M_i is the power of the functional element i, C_{ij} is the indicator of consumption by the ith functioning element of the jth resource per its unit power, and P_{ik} is the production of the kth resource per unit power of the functioning element. The state of the functioning element is described by a set of X_{ij} indicators of its provision with the resources C_{ij}:

$$x_{ij} = \begin{cases} 1 & \text{if } C_{ij} \in \left[a_{ij}^H, a_{ij}^B\right] \\ 0 & \text{if } C_{ij} \notin \left[a_{ij}^H, a_{ij}^B\right] \end{cases}$$

In a general case, the estimates a_{ij}^H and a_{ij}^B are functions of many parameters (M_i, in particular).

The functioning elements of the EES include groups of populations in the region, objects of economic activity (transport, agriculture, industrial enterprises, municipal services, parks, beaches, sanatoria, ports and levees, etc.), and natural objects. Of course, in the synthesis of a simulation model it is impossible to reflect the interaction of all the EES elements. Therefore, it is necessary to select a first-priority group of these elements, attributing the rest of them to the background load. In analysis of the EES in the coastal regions, only those objects of economic activity are chosen which directly affect the marine environment. The scheme proposed by Khudoshina (1996) to parameterize the EES functioning in the zone of the sea–land interaction is shown in Figure 9.2.

The functioning of the EES is determined by the volume of consumption and distribution of resources which depends on population density. Hence, the prognostic estimate of the state of the EES is connected with the accuracy of the demographic model and depends on the level of interaction with other EESs. It is

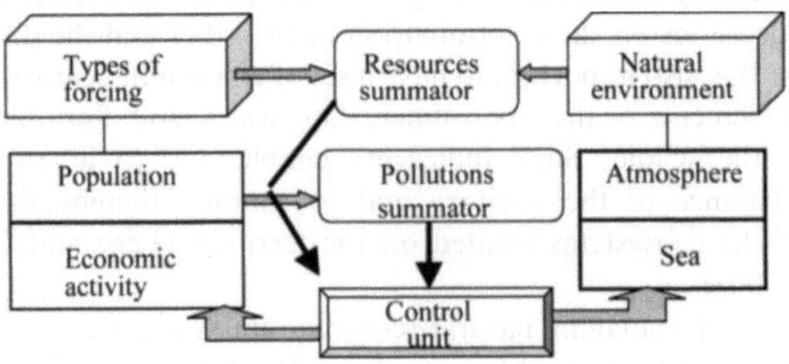

Figure 9.2. Scheme of the interaction of anthropogenic factors of the land and the marine environment.

necessary to bear in mind that all elements of the system consume and produce resources, some playing a progressive role, others (e.g., pollution) being negative. Pollution of the seas violate their main resource – the quality determined by a complex of indicators. The goal of the monitoring system is to find (opportunely) the critical situations when there is a possibility of rapid $x_{ij} \rightarrow 0$.

9.5.1 Model of the functioning element of the EES

The EES complexity is determined by the number ζ of inputs and the number β of outputs: $L(\zeta, \beta)$. Assume the hypothesis of additivity of the quality of the EES complexity:

$$L = \sum_{i=1}^{K} L_i \qquad L_i = \zeta_i^{\xi} \beta_i$$

where $\xi = (\zeta + 1)/\zeta$ (i.e., with an increase of the number of inputs of one subsystem by unity, the subsystem's complexity increases ξ times).

The EES complexity can be decreased by means of its decomposition into two levels of subsystems: EES $= \{E_k = \{\Xi_{kl};\ l = 1, \ldots, L\};\ (k = 1, \ldots, K);\ \Xi_{ki} \cap \Xi_{kj} = 0$ for $i \neq j\}$, where Ξ_{kl} is the subsystem of the lowest level, and L and K are the number of subsystems in first and second levels, respectively. Each subsystem Ξ_{kl} has a specific space of its states $X_{kl} = \{x_{kl1}, \ldots, x_{klm}\}$. As a result, the EES state X can be described by the expression: $X = \{X_k = \{X_{kl} = \{x_{klj} j = 1, \ldots, m\};\ l = 1, \ldots, L\};\ k = 1, \ldots, K\}$. This hierarchical structure makes it possible to simplify the description of the EES structure and to improve the analysis of its dynamics.

9.5.2 Assessment of the state of the EES

The state of the modelled system based on the logical-informational modelling is assessed using Boolean algebra. For economic and natural objects two possible states are considered: normal functioning and the working regime bug (refusal of the system's functioning). The function Y of the state of the EES depends on the function of the state of its elements: $Y = Y_1 \forall \ldots Y_k$. In other words, $Y = 0$ if $\exists Y_i \in \{Y_1, \ldots, Y_k\},\ Y_i = 0$.

Following the method discussed in item 9.5.1 we write:

$$Y = F(Y, \ldots, Y_n)$$

$$Y_i = F_i(Y_{i1}, \ldots, Y_{in_i})$$

$$\ldots$$

$$Y_{ij} = F_{ij}(Y_{ij_1}, \ldots, Y_{in_j})$$

$$\ldots$$

$$Y_{ij\ldots k} = F_{ij\ldots k}(Y_{ij\ldots k_1}, \ldots, Y_{ij\ldots kn_k})$$

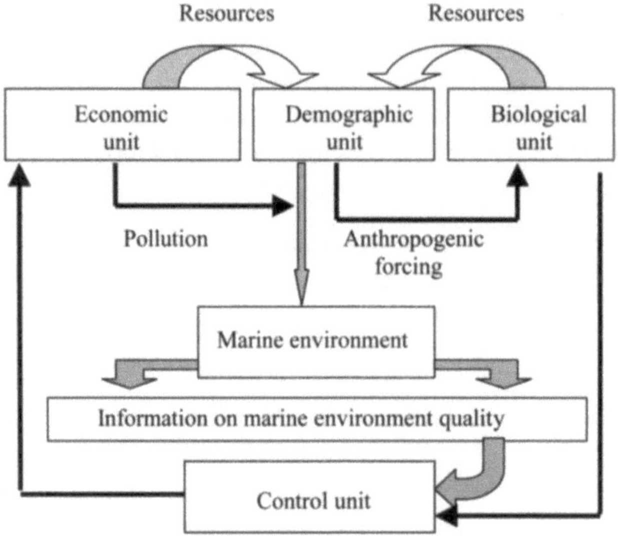

Figure 9.3. Structural scheme of the model of the marine coastal ecotone.
From Khudoshina (1996).

The EES functioning is assessed by the totality of the states of its elements. We have $Y_i = 0$ when there exists a limiting resource of the ith element of the EES. Here the interdependent pairs and chains of the system's elements are taken into account. The concrete structure of the model introduces certain corrections and special features into the logical hierarchy of division of the system into subsystems and elements. This is especially characteristic of the systems with a substantially heterogeneous structure. These systems also include the marine EES with biological, economic, and demographic units (Figure 9.3).

Here pollution plays the role of a negative resource. Information about the quality of the marine environment and the state of the biota, that reflects the impact of economic activity on nature, enters the controlling unit of the model. This unit compares and analyses the information about the state of the EES on the whole. With respect to the biological unit of the marine EES, economic and demographic units simulate an external load in the form of pollutions, extraction of bioresources, and transformation of the habitat. Information about changes of the quality of the marine environment and decreases in the size and diversity of populations of sea animals serve as a feedback in the model, which, of course, reflects on the economic activity of the EES.

The demographic unit models the power and the kinds of external forcings on the natural EES. The input characteristics of the unit reflect the need of a population for economic and biological resources as well as the quality of the environment. The output parameters of this unit are really consumed resources, population size, and the amounts of pollutants emitted to the environment.

The economic unit describes complicated feedbacks between the EES elements, reflecting correlations between input and output parameters of the remaining units as functions of the economic strategy of the system. The biological unit reflects the composition of biotic communities and their trophic structure. The concrete model realization of all elements of the scheme in Figure 9.3 is connected with the use of the models of partial processes described by many scientists (Krapivin and Kondratyev, 2002; Sarancha, 1995; Abakumov, 1993; Polishchuk, 1992; Ougolnitsky, 1999; Ougolnitsky *et al.*, 2000).

9.5.3 Simulation model of the ESS

The most widely used method of the ecologo-economic analysis is simulation modelling which makes it possible to combine, for joint studies, the data of natural and social sciences and to create a computer version of the system of the EES management. This approach and its adequate application in concrete situations has been reflected in many studies accomplished by various scientists. For instance, Gorstko and Ougolnitsky (1990) described two approaches to study the EES using numerical models. The first approach is based on the concepts of the optimization theory and consists in the synthesis of a numerical model of the process. This model relates the internal (endogenic) and external (exogenic) factors of the EES functioning. The problem is to determine the management of the process of the EES functioning, which provides an extremum of some function that characterizes the quality of management. In a general case, the solution to this problem is reduced to mathematical programming or game theory.

The second approach is connected with simulation modelling. Here a set of models of some functions of the EES and expert parameterizations (scenarios) are created (Beese *et al.*, 1995; Nordhaus, 2001), which forms the basis of the simulation experiment for the search of an optimal management of the system. The simulation model is more complicated compared with the numerical model, it has a greater dimensionality and contains a multitude of non-linear direct couplings and feedbacks between the variables. As a rule, the simulation model has more flexible capabilities for assessing the efficiency of the expert choice of the controlling functions to optimize the EES functioning. It requires the development of extensive software for the computer search of optimal or suboptimal management. The structure of the simulation system is shown schematically in Figure 9.4. The most important part of the simulation system is a unit of models of the object under study. This unit can include duplicating elements with different adequacy and, especially important, various requirements for databases. The information retrieval system provides an agreement between the database and the set of partial problem-oriented models and adjusts the simulation model for dialogue with the operator. The existing standard means of many programming languages make it possible to realize these procedures in the forms providing an accomplishment of a simulation experiment by the user without certain skills and knowledge in the field of informatics.

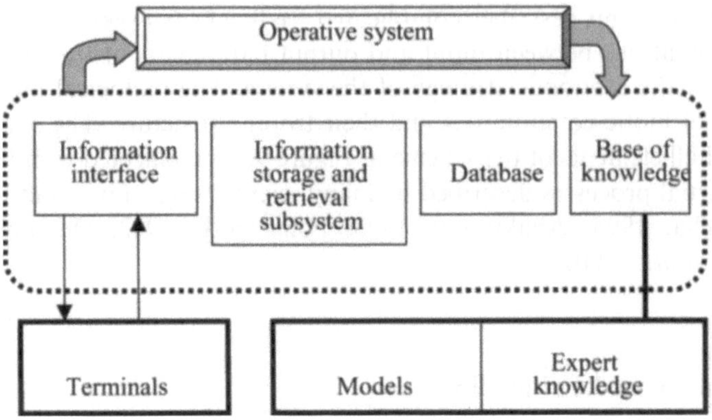

Figure 9.4. The simulation system's structure.
From Gorstko and Ougolnitsky (1990).

The simulation systems of the Okhotsk Sea, the Aral Sea, and the Sea of Azov can exemplify these simulation systems (Krapivin and Kondratyev, 2002; Krapivin and Phillips, 2001). In the first case the anthropogenic factor is sufficiently small compared with the climatic impact on the ecosystem. On the contrary, the Aral Sea is subject to powerful anthropogenic forcing. In each case of the simulation model synthesis it is necessary to take into account the existing knowledge base. Following Gorstko (1990), consider the case of modelling the ecosystem of the Sea of Azov and its related problems. Its catchment basin covers 43°N–54°N and 33°E–46°E with an area of about 0.5 million km^2 and population of 35 million. The Azov Basin undergoes numerous anthropogenic forcings: developed agriculture with irrigation, heavy and light industry, oil and coal extraction, fisheries, network of tarmac roads and railways, food industry, recreation zones. The simulation model of the Sea of Azov created by the Rostov scientists includes 150 components of its state divided into 16 units (Table 9.6). The model database includes extensive information obtained during many decades in the process of natural–scientific observations of the Sea of Azov ecosystem. The components of the vector of the sea state are functions of time and calculated in accordance with dependences included in the base of knowledge. For instance, the equation of the oxygen balance is written in the form:

$$dO_i(t)/dt = f_1 - f_2 - f_3 - f_4 - f_5$$

where O_i is the concentration of oxygen in the water of the ith section of the sea; f_1 is the input of oxygen into the ith section; f_2 is the oxygen consumption by bottom sediments; f_3 is the oxygen expenditure on respiration of benthos animals; f_4 is the oxygen expenditure on oxidation of organic matter; and f_5 is the loss of oxygen on decomposition of pollutants. The simulation procedure, which foresees a digitization of the sea area into non-intersecting zones (sections), provides a correlation of the functions $O_i(t)$ through exchange fluxes between these zones.

Another example of the simulation model is the model of the aquageosystem of

Table 9.6. The main units of the simulation system for the Azov Sea.
From Gorstko (1979); and Gorstko and Ougolnitsky (1990).

Unit	Function of the unit
External factors	Precipitation, change of the wind status, river runoff, motion of soils, outflow of nutrient elements and pollutants, fishery management.
Water dynamics	Change in the water table slope, water mixing, water exchange with the Black Sea, redistribution of ingredients between sea sectors.
Nutrient elements	Outflow of nutrient compounds, their inflow with abrasive materials and precipitation, physico-chemical transformation of the compounds.
Oxygen	Aeration, consumption of oxygen in biochemical processes.
Water quality	Sea self-cleaning, propagation of pollutants over the water body.
Phytoplankton	Photosynthesis, growth, dying off, consumption of nutrients, oxygen release.
Zooplankton	Nutrition, growth, dying off, age transitions.
Benthos	Nutrition, growth, metabolism, dying off, change of the spatial structure.
Nekton (goby, pike, perch, bream, roach, etc.)	Nutrition, growth, dying off, removal by fishing, migration.

the Aral Sea Basin. In contrast to the Sea of Azov, here the scales are greater and the information supply is smaller. There is also a considerable difference in the interaction of natural and anthropogenic factors. During the last decades, the remote monitoring systems have become the main source of observed data (Krapivin and Phillips, 2001).

The ecological and economic situation in many Asian regions of the Former Soviet Union was principally changed due to the development of irrigation systems. On this basis, the development and introduction of simulation technology, as applied to the EES analysis in conditions of arid climates, have become one of the acceptable approaches to developing the management strategy and complex of optimization measures in the NSS (Kondratyev *et al.*, 2000, 2001b,c).

The conceptual models listed in Table 9.7 reflect the basic mechanisms and trends in the change of the natural situation in the Aral Basin between 1960–1980 with a sufficiently negative result – ceasing of inflow of river waters into the Aral Sea and degradation of the natural–economic situation in this region. The situations in other regions are sufficiently problematic (the Tashaus and Khoresm oases, Sarakamysh, the zone of Kara-Kum canal, etc.). It follows from these schemes, in particular, that the water component in the mid-Asian region has been, and remains, the leading object of studies, prediction, and optimization (management).

Table 9.7. Elements of the Aral Sea Basin system's functioning before and after 1960.

Inflow	Outflow		Result
Formation of the Amu-Darya and Syr-Darya runoffs with rivers gravitating towards the Aral Sea Basin in mountains	Evaporation from water bodies, natural geosystems, valleys, and river deltas (lakes included)	Supply of groundwater	Inflow of river waters into the Aral Sea before 1960 varied around 54 m^3/yr
	Irrevocable uptake of water for industry and municipal services	Uptake for irrigation; evaporation from cultural plants, water surface of irrigation systems, etc.	
Decrease of the inflow of river waters to the boundary of the region	Evaporation from the few preserved natural complexes and water basins; evaporation from hydroeconomic and recreational water reservoirs	Replenishment of subsurface waters: irrevocable losses in deserts, water catchment of Uzboi, etc.	Irrevocable uptake for industry and municipal services
Formation of river runoff			Inflow of river waters into the Aral Sea ceased in the 1980s
Filling up of the pores in soils and dead volumes of water reservoirs	Uptake for irrigation, evaporation	Return discharge into the rivers	
		Formation of collector and discharge waters in the irrigation systems	Discharge into lakes preserved in valleys and river deltas

On the whole, the system of units of a general model-simulation of the current natural–economic situation in the region following units: water component, salts, ecological situation, economic load, macro and meso-climatic background, mass and energy-exchange between units direct couplings, and feedbacks (Kojima, 1998).

The scheme of water balance of the aquageosystem "Amu Darya River–Kara Kum canal–Aral Sea–Sarakamysh" includes the following components: precipitation (X), Amu Darya river flow (R_0), the disciplined water flow from the Amu Darya River to the Kara Kum canal (R_1), discharge of seepage waters into Sarakamysh (R_2), watering (R_j), water accumulation in reservoirs, except Sarakamysh (including the dynamics waters, R_d, and interred waters, R_n), outflow from Syr Darya (R_S), water accumulation in the ground lenses (U_0), return water to the reservoirs and rivers (U_v), irrevocable losses of waters in the desert (U_2), infiltration to deep underground horizons (U_g), and evaporation: from water surfaces, E_w, from irrigated areas and oases, E_0, from natural vegetation, E_v, from saline lands, E_1, and from boundaries between land facies, E_2; etc.

Introduced notations make it possible to write the following traditional balance relationships for each subsystem of the Aral Sea Basin:

For the zone of the Kara Kum canal:

$$R_1 + X = R_j + R_d + U_0 - U_v - U_1 + U_2 + E_w + E_0 + E_1 + E_2$$

For the zone of Amu Darya from the canal to Sarakamysh:

$$R_0 + X = R_1 + R_j + U_0 - U_v - U_1 + E_w + E_0 + E_v = R_2 + R_3$$

For Sarakamysh:

$$R_2 + X = E_w$$

For the Aral Sea:

$$X + R_3 + R_S = R_j + U_0 - U_1 - U_v + E_w + E_v + E_1 + E_2 + R_d + R_n - U_w$$

where U_w represents the return subground waters.

For the whole system, many scientists have studied relationships of the water balance, with a consideration of the synoptic situation and of dependences of its components both on each other and on other factors (Krapivin and Kondratyev, 2002; Bortnik *et al.*, 1994). The transition to dynamic equations requires additional information about the character of correlations between the elements of the environment, anthropogenic processes, and hydrophysical constituents of the system. Since there are irrevocable losses in the system ($U_g + U_2$), there should exist the balancing branches of the hydrological cycle. Apparently, these branches exist in the flows of moisture in the region, which, by their character, play a dual role. So, for instance, Krapivin and Kondratyev (2002) considered a situation when a forced irrigation of saline lands and other lowlands was carried out on the eastern coastline of the Caspian Sea at optimally chosen moments in time. As it turned out, the resulting additional evaporation according to a mean-statistical synoptic scenario has improved the hydrological situation in the Turan lowland, and with forced precipitation it was possible to restore the level of the Aral Sea over 10–12 years. Of course, such calculations raise questions, but they show how to study and to search for positive forcings on the natural–anthropogenic system.

9.6 ASSESSING THE ENVIRONMENTAL PARAMETERS

9.6.1 A device to measure the geophysical and hydrophysical parameters

According to algorithms of the multichannel microwave monitoring discussed in Krapivin (2000a), it is possible to promptly measure the wind speed, sea surface roughness, and water temperature as the base characteristics to calculate other parameters of the environment. From these measurements, a mutual control of the accuracy of correction of the estimates of the atmosphere–ocean gas exchange is realized. Solution of the correction equations in a general case requires the use of the criterion of discrepancy between theoretical and empirical estimates of the parameters or correlations. Usually for this purpose, a criterion of mean square deviation is used. The reliability and efficiency of determination of solutions for the system of equations relating the theoretical and empirical estimates can be increased if their dispersive characteristics are taken as a criterion of discrepancy of these solutions.

Let the proper emission of the sea be measured simultaneously with the help of n radiometers, each operating at a fixed wavelength $\lambda_j (j = 1, \ldots, n)$. As a result at the moment t_i at the output of each radiometer the values $Z_{ij} (i = 1, \ldots, M)$ are fixed, so that $Z_{ij} = T_j + \xi_{ij}$, where T_j is the brightness temperature of a sea site at the wavelength λ_j, and ξ_{ij} is the random value (noise) with zero mean and dispersion σ_j^2.

Let the correlation between the brightness temperature and m parameters of the sea be linear:

$$\begin{aligned} A_{11}x_1 + \cdots + A_{1m}x_m &= T_1 + \xi_1 \\ &\vdots \\ A_{n1}x_1 + \cdots + A_{nm}x_m &= T_n + \xi_n \end{aligned} \tag{9.13}$$

where A_{ij} are the coefficients evaluated in the regime of monitoring, and (x_i) are the geophysical, ecological, and biogeochemical parameters. It is necessary to solve Equations (9.13) such that their solution has minimum dispersion. Such a solution is called the σ-solution (Krapivin and Phillips, 2001).

The ith equation of the system in Equations (9.13) is multiplied by the set of parameters $c_{1i}, \ldots, c_{mi}$. An additional condition is given as the following equation:

$$\sum_{i=1}^{n} c_{ji} A_{il} = \delta_{jl} \tag{9.14}$$

where

$$\delta_{jl} = \begin{cases} 1 & \text{for } j = l \\ 0 & \text{for } j \neq l \end{cases}$$

As a result, we have:

$$x_1^0 = \sum_{i=1}^{n} c_{1i} T_i \qquad \tilde{x}_1 = \sum_{i=1}^{n} c_{1i} T_i + \sum_{i=1}^{n} c_{1i} \xi_i \qquad D[\tilde{x}_1] = \sum_{i=1}^{n} c_{1i}^2 \sigma_i^2$$

For $\tilde{x}_i (i \geq 2)$ we write similar relationships. Derive an auxiliary expression:

$$\varphi(c_{11}, \ldots, c_{1n}) - \sum_{i=1}^{n} c_{1i}^2 \sigma_i^2 + \mu_1 \left(\sum_{i=1}^{n} c_{1i} A_{i1} - 1 \right) + \sum_{j=2}^{m} \mu_j \sum_{i=1}^{n} c_{1i} A_{ij}$$

where μ_j are the non-estimated Lagrange multipliers. Putting the first derivative functions φ to zero, we obtain:

$$2c_{1k}\sigma_k^2 + \sum_{j=1}^{m} \mu_j A_{kj} = 0 \qquad (k = 1, \ldots, n)$$

These equalities together with conditions (9.14) form the system $(m+n)$ of equations whose solution enables one to find the desired optimal values of c_{ij}^*. As a result, $D[x_1] = -\mu_1/2$, and the remaining $\mu_j (j \geq 2)$ values meet a system $m \times n$ of equations:

$$\sum_{j=1}^{m} \mu_j \sum_{i=1}^{n} A_{ij} A_{i1} \sigma_i^{-2} = -2$$

$$\sum_{j=1}^{m} \mu_j \sum_{i=1}^{n} A_{ij} A_{i1=} \sigma_i^{-2} = 0 \qquad (l = 2, \ldots, m)$$

For $m = 2$ we have:

$$c_{1,k}^* = \left(A_{k1} \sum A_{i2}^2/\sigma_i^2 - A_{k2} \sum_{i=1}^{n} A_{i1} A_{i2}/\sigma_i^2 \right) \Delta^{-1} \sigma_k^{-2}$$

$$c_{2,k}^* = \left(A_{k2} \sum A_{i1}^2/\sigma_i^2 - A_{k1} \sum_{i=1}^{n} A_{i1} A_{i2}/\sigma_i^2 \right) \Delta^{-1} \sigma_k^{-2}$$

$$\Delta = \sum_{i=1}^{n} A_{i1}^2 \sigma_i^{-2} \sum_{i=1}^{n} A_{i2}^2 \sigma_i^{-2} - \left(\sum_{i=1}^{n} A_{i1} A_{i2} \sigma_i^{-2} \right)^2$$

$$x_j^* = \sum_{i=1}^{n} c_{ji}^* Z_i \qquad (j = 1, 2) \qquad D[x_1^*] = \Delta^{-1} \sum_{i=1}^{n} A_{i2}^2 \sigma_i^{-2} \qquad D[x_2^*] = \Delta^{-1} \sum_{i=1}^{n} A_{i1}^2 \sigma_i^{-2}$$

For the real use of this algorithm in the radiometric monitoring system the following arrangement can be suggested. It consists of radiometers, a microcomputer, multi-channel commutation unit, and units of dispersion evaluation. Its structural and operational schemes are given in Figures 9.5–9.7.

Figure 9.5 shows the structural scheme of an arrangement to measure the parameters of geophysical and hydrophysical objects. The scheme in Figure 9.6 explains the structure of the unit to calculate dispersion. The *resolver* is characterized in Figure 9.7. The apparatus operates as follows. When radiometers 1 are calibrated, the signals Z_{ij} from their outputs through the multichannel switching unit 2 enter the

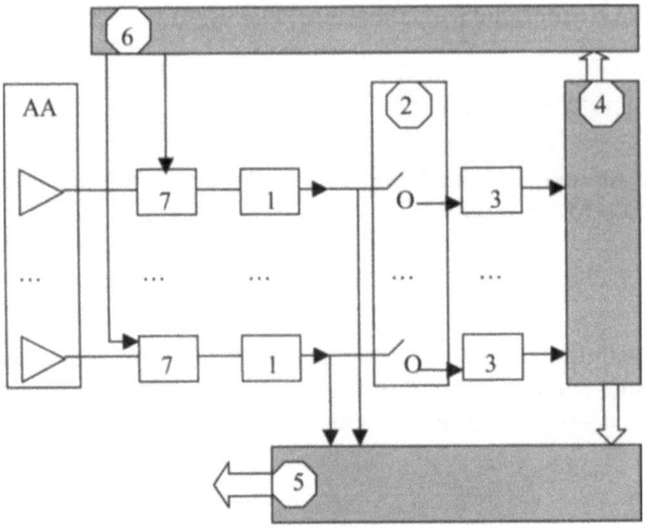

Figure 9.5. The structural scheme of the arrangement to measure the geophysical and hydrophysical parameters in the regime of the microwave monitoring.
Note: 1 = radiometers; 2 = commutation unit; 3 = unit of dispersion calculation; 4 = microprocessor; 5 = resolver; 6 = control unit; 7 = switches; AA = antennas admittance.

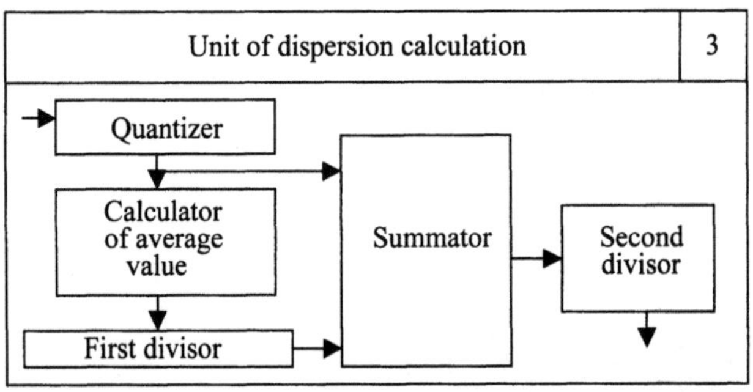

Figure 9.6. Scheme of operations in the unit of dispersion calculation.

unit 3 of the dispersion calculator. The antennas are switched off from radiometers. In unit 3 the dispersion is calculated using the algorithm:

$$\sigma_j^2 = \frac{1}{M-1}\sum_{i=1}^{M}(Z_{ij} - M_j) \qquad M_j = \frac{1}{M}\sum_{i=1}^{M} Z_{ij}$$

The M_j parameters are calculated in the units of the mean calculations. The summator calculates:

$$\sum_{i=1}^{M}(Z_{ij} - M_j)$$

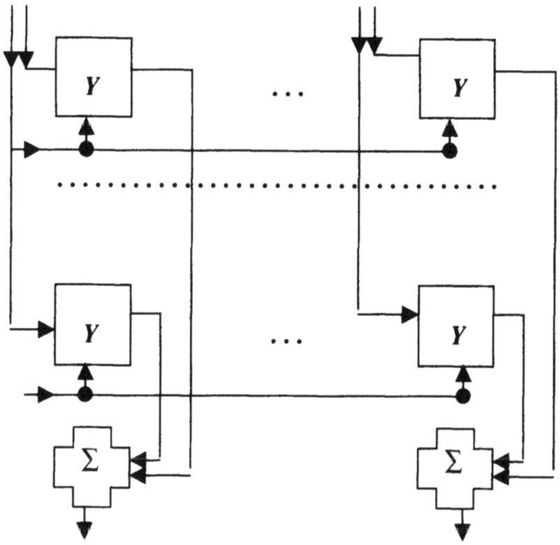

Figure 9.7. Resolver.
Note: Y = multiplier, Σ = accumulator.

The calculated values σ_j^2 enter microprocessor 4 to calculate partial coefficients c_{jl}. From the first group of its outputs they enter the *resolver* 5 where they are used to solve a system of equations, and from the second group of outputs they enter control unit 6 which forms the signal that switches off radiometers 1, for which the partial coefficients are equal or close to zero (i.e., the information channels automatically get excluded). At the same time, the systems of equations to be solved get compressed. This step makes it possible to contract not only the volume of calculations but also to increase the accuracy of estimating the parameters of the sounded objects of the environment.

After determination of dispersions, the commutator 2 blockades the input of data from non-information channels. The data from information channels enter the *resolver* 5. The signal at the multiplier's output corresponds to the value of one of the parameters to be estimated.

9.6.2 Direct and inverse problems of microwave monitoring

An approach to the inverse problem of microwave monitoring considered in Section 9.6.1 is based on the assumption of the linear functional correlation between radiobrightness contrasts and characteristics of the monitored object. The search for analytical relationships relating the radiobrightness temperature of the environment to its thermodynamic and electrophysical characteristics, covers more general classes of the functional models. The most general form of such a dependence is as follows:

$$T_{bi} = \kappa_i \int_0^\infty T(h)\gamma_i(h) \exp\left[-\int_0^h \gamma_i(x)\, dx\right] dh \tag{9.15}$$

where $T(h)$ is the thermodynamic profile of the environment; T_{bi} is the radiobrightness temperature at the frequency ν_i; $\gamma_i(h)$ is the profile of the absorption coefficient of the environment; and κ_i is the emissivity of the environment.

Relationship (9.15) makes it possible using $T(h)$ and $\gamma_i(h)$ to find T_{bi} (direct problem) and from T_{bi} to calculate $T(h)$ and $\gamma_i(h)$ (inverse problem). The proposed method for solving these problems is based on the use of $T(h)$ and $\gamma_i(h)$ and their derivatives at the point $h = 0$ with some additional limitations resulting from the transformations made below. Write the relationship (9.15) in the form:

$$T_{b\lambda} = -\kappa_\lambda \int_0^\infty T(h) dF_\lambda(h) \tag{9.16}$$

where

$$F_\lambda(h) = \exp\left[-\int_0^h \gamma_\lambda(x)dx\right]$$

where λ is the wavelength.

Integrating Equation (9.16) by parts, we obtain:

$$T_{b\lambda} = \kappa_\lambda\left[T_0 + \int_0^\infty T'(h)F_\lambda(h)dh\right] = \kappa_\lambda[T_0 + J_1] \tag{9.17}$$

Let $J_1 = 0$, then $T_{b\lambda} = \kappa_\lambda T_0$. If $dT(h)/dh \equiv 0$, then $J_1 = 0$. Therefore, find conditions under which $J_1 = 0$. Let the following relationships be valid:

$$\gamma_\lambda(h) \equiv c_\lambda > 0 \qquad T(h) = \sum_{k=0}^{n} B_k h^k \tag{9.18}$$

It follows from Equation (9.18) that if the coefficients B_k meet the condition:

$$\sum_{k=1}^{n} k! \frac{B_k}{c_\lambda^k} = 0 \tag{9.19}$$

then $J_1 = 0$.

Integrating Equation (9.17) by parts, we obtain:

$$T_{b\lambda} = \kappa_\lambda\left[T_0 + \frac{T_0'}{\gamma_\lambda(0)} + \int_0^\infty \frac{T''(h)\gamma_\lambda(h) - \gamma_\lambda'(h)T'(h)}{\gamma_\lambda^2(h)} F_\lambda(h)dh\right] \tag{9.20}$$

Equate the integral in Equation (9.20) to zero. It gives:

$$T(h) = c\int \gamma_\lambda(h)dh$$

where c is an arbitrary constant. The integral in Equation (9.20) will be equal to zero if:

$$\sum_{k=2}^{n} B_k k(k-1)k!c_\lambda^{1-k} = 0 \qquad (9.21)$$

Further integrating of Equation (9.20) by parts gives:

$$T_{я\lambda} = \kappa_\lambda \left[T_0 + \frac{T_0'}{\gamma_\lambda(0)} + \left(\frac{T'}{\gamma_\lambda(h)} \right)' \frac{1}{\gamma_\lambda(h)} \bigg|_{h=0} + J_3 \right]$$

where

$$J_3 = \int_0^\infty \frac{T'''(h)\gamma_\lambda(h) - 3T''(h)\gamma_\lambda(h) - T'(h)\gamma_\lambda''(h) + 3T'(h)\left[\gamma_\lambda'\right]^2}{\gamma_\lambda^4} F_\lambda(h)dh$$

This integral will be equal to zero at:

$$T(h) = c \int \gamma_\lambda(h) \int \gamma_\lambda(u)dudh$$

If the condition Equation (9.18) is not met then $J_3 = 0$ at:

$$\sum_{k=3}^{n} B_k k(k-1)(k-2)k!c_\lambda^{2-k} = 0$$

With the process of integration of Equation (9.20) continued N times, we obtain:

$$T_{я\lambda} = \kappa_\lambda \left[\sum_{k=1}^{N} P_k(0) + \int_0^\infty P_N'(h)F_\lambda(h)dh \right] \qquad (9.22)$$

where

$$P_1(h) = T(h) \qquad P_2(h) = T'(h)/\gamma_\lambda(h), \ldots, P_k(h) = P_{k-1}'(h)/\gamma_\lambda(h)$$

But

$$J_N = \int_0^\infty P_N'(h)F_\lambda(h)dh = 0$$

for

$$\sum_{k=N}^{n} B_k k(k-1)\ldots(k-N+1)k!c_\lambda^{N-1-k} = 0 \qquad (n > N)$$

Consider the case when

$$\gamma_\lambda(h) = \exp\{c_\lambda h\}$$

From Equation (9.22) we obtain:

$$T(h) = \sum_{k=0}^{N-1} B_k \exp[kc_\lambda h]$$

Now we move on to the inverse problem. Let measurements be made at wavelengths $\lambda_1, \ldots, \lambda_N$. Solve the equation:

$$P'_{N-1}(h) = c\gamma_\lambda(h)$$

and find

$$P_{k-1}(h) = \int \gamma_\lambda(h) P_k(h) dh \qquad (k = 1, \ldots, N)$$

Let $\gamma_\lambda(h) = \varphi(\lambda)\psi(h)$, that is, the dependence of the absorption coefficient on the frequency and depth of the layer be divided by the product of the known function $\varphi(\lambda)$ and an unknown function $\psi(h)$. As a rule, the function $\varphi(\lambda)$ is set from the empirical table and its analytical approximation can be constructed. Then, excluding from Equation (9.22) the second right-hand term (it is equal to zero), we write the relationship:

$$T_{b\lambda} = \kappa_\lambda \sum_{k=1}^{N} T^{(k-1)}(0) F_k[\varphi(\lambda), \psi^{(N-k-1)}(0), \ldots, \psi'(0), \psi(0)] \tag{9.23}$$

and the solution of the inverse problem is reduced to the solution of a system of algebraic equations with regard to functions:

$$T(0), T'(0), \ldots, T^{(N-1)}(0), \psi(0), \ldots, \psi^{(N-1)}(0)$$

Consider the case $\psi(h) = \text{const.}$ (i.e., $\gamma_\lambda = \varphi(\lambda) > 0$. Equation (9.23) is rewritten:

$$T_{b\lambda} = \kappa_\lambda \sum_{k=1}^{N} T^{(k-1)}(0)/\varphi^{(k-1)}(\lambda) \tag{9.24}$$

In the simplest version, if $T_{b\lambda}$ was measured at N different wavelengths $\{\lambda_i; i = 1, \ldots, N\}$, then the determinant of the system (9.24) will be equal to:

$$\Delta = \prod_{i=1}^{N} \kappa_i \prod_{i>k}^{N} [\varphi(\lambda_i) - \varphi(\lambda_k)]/[\varphi(\lambda_i)\varphi(\lambda_k)]$$

If $\varphi(\lambda_i) \neq \varphi(\lambda_{Jl})$ with none of the combinations $i \neq k$, then $\Delta \neq 0$ and the system of Equation (9.24) has a single solution. Now let:

$$\psi(h) = h^m$$

then it follows from Equation (9.22) that:

$$T(h) = \sum_{k=0}^{N-1} B_k h^{k(m+1)}$$

and we obtain a system of algebraic equations:

$$P_k(0) = (m[k-1])\underset{(m+1)}{!\ldots!} b_{k-1}/\varphi^{k-1}(\lambda)$$

with the following determinant:

$$\Delta = \prod_{i=1}^{N} \kappa_{\lambda_i} \prod_{p=1}^{N} [m(k-1)]^{N+1-p} \prod_{t>k}^{N} [\varphi(\lambda_t) - \varphi(\lambda_k)]/[\varphi(\lambda_t)\varphi(\lambda_k)]$$

The case when the function ψ is approximated by a polynomial is considered in a similar way.

9.7 MONITORING HYDROPHYSICAL AND HYDROCHEMICAL PROCESSES

The control and assessment of the state of the marine environment requires knowledge of numerous characteristics, which can be classified into: (1) factors of the formation of water chemistry; (2) water quality; and (3) dynamic characteristics. The first group includes characteristics of the processes affecting the water quality and determining the dynamics of the parametric space that describes the complex state of the aquageosystem under study. Depending on the conditions of the experiment, these factors are connected with the spatial scale and can be pointwise or areal in character. To describe a set of factors and to determine their spatial significance, we denote the points on the aquageosystem's surface by latitude φ and longitude λ. Choose a rectangular system of coordinates with the axis z directed upward from the surface, so that the value z will denote the depth at the point with the coordinates (φ, λ). Then any measurement will be described by the parameter $\xi(\varphi, \lambda, z, t)$, which in a general case is a non-stationary random value. In a concrete experiment the formation of the set $\Xi = \{\xi\}$ is connected with the factors shown in Figure 9.8. In a

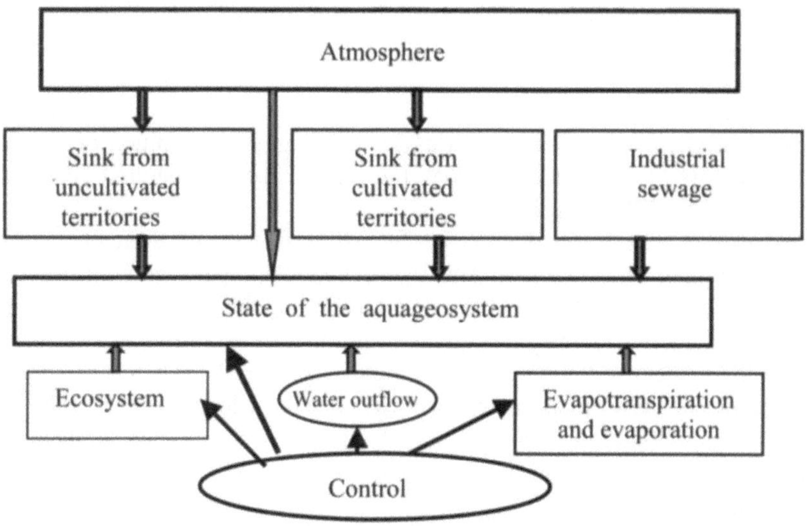

Figure 9.8. Exemplary structure of the factors which contribute to the water quality. These should be taken into account in the hydrophysical experiment.

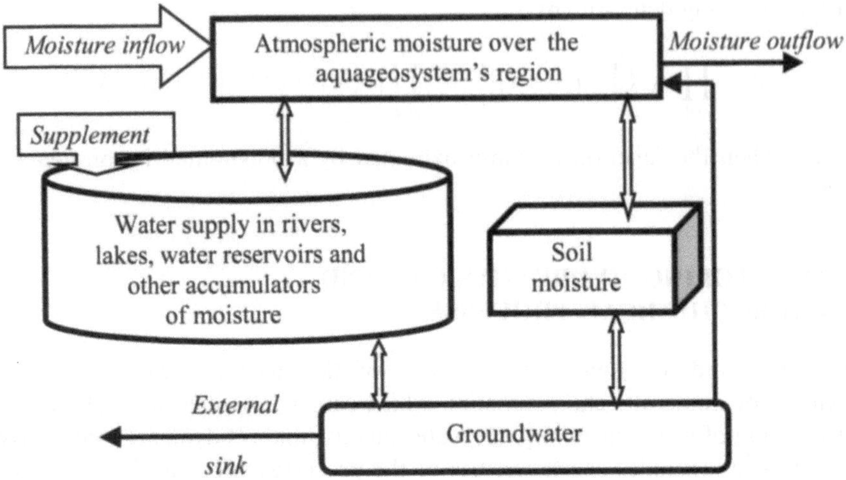

Figure 9.9. Block-scheme of the model of the moisture cycle in the zone of the aquageosystem's functioning with the processes of evaporation, evapotranspiration, leakage, precipitation, and anthropogenic use taken into account.

general form, the scheme of the formation of the set Ξ is shown in Figure 9.9. Depending on the spatial scales of the hydrophysical object under study, a variety of possible schemes of the formation of water quality and other characteristics of the aquageosystem covers a wide spectrum of the physico-chemical interactions of the elements of the water basin ecosystem. A variety of all these schemes is evidence of the complexity of assessing the water quality and the state of the aquageosystem. Therefore, an efficient monitoring of the aquageosystem depends on an adequate description of these interactions and an adequate list of parameters to be measured.

The parameters directly determining the water quality are acidity, turbidity, and the content of chemicals. The water quality criterion is determined depending on numerous factors. Table 9.8 exemplifies the sets of factors required to assess the water quality with different criteria.

These circumstances testify to the fact that it is very difficult to draw a distinction between hydrophysical and hydrochemical experiments. However, it is clear that the quality of water is, to a greater extent, the hydrochemical function. Here the aim is neither to draw this distinction nor to substantiate a set of problems to be solved in the hydrophysical experiment. Apparently, when speaking about data series of the hydrophysical experiment, one should not neglect the hydrochemical data. Ultimately, an assessment of the state of the aquageosystem requires a certain combination of such data, especially when studying the dynamic characteristics.

A change of $\xi(\varphi, \lambda, z, t)$ is connected with a consideration of correlations between all factors affecting the water quality. In the formation of the spatial image of the water object under study, it is necessary to take into account an

Table 9.8. Some criteria and factors – the structure of substances and parameters that determine the water quality.

Factors	
Type of water pollution	**Factors**
Biological	Microorganisms favouring the fermentation of organic substances
Chemical	Toxic substances or changing the composition of water
Physical	Heating and radioactivity
Elements	
Group of characteristics	**Elements**
Basic ions	Chloride Cl^-, sulphate SO_4^{2-}, hydrocarbon HCO_3^-, carbon CO_3^{2-}, natrium Na^+, potassium K^+, magnesium Mg^{2+}, calcium Ca^{2+}.
Dissolved gases	Oxygen O_2, carbon dioxide CO_2, nitrogen N_2, hydrogen sulfide H_2S
Biogenic elements	Nitrogen, phosphorus, and silicon compounds
Microelements	Zn, Cu, Pb, Ni, Co, Li, Rb, Cs, Be, Br, I, F, etc.
Organic substances	Content of carbon, oxidizability, biochemical consumption of oxygen

interaction of climatic, biogeochemical, ecological, and hydrological processes. This is possible by using numerical modelling and, hence, forming new data series in the form of the model coefficients.

The stages of the hydrophysical monitoring are connected with different levels of databases and therefore require the respective algorithms of analysis and interpretation of measurement results. Moreover, an identification of a set of algorithms of data processing with the problems of the experiment cannot guarantee an adequate complex of methods to solve these problems. A combination of the hydrophysical monitoring problems with the system of the automated data processing is only possible in an adaptive regime. Schematically it is shown in Figure 9.10.

The monitoring can be pointwise and areal. In the first case the data processing is connected with the use of a number of algorithms providing an assessment of the needed set of parameters without consideration of the spatial scales of the process or object under study. In the second case the system of data processing requires special methods aimed at reducing their variability and forming a spatial image based on measurements episodic in time and fragmentary in space. In this case it is essential to choose the scales of the cartographic grid and to correlate it with the problems to be solved. Only then can one guarantee a certain level of reliability in the assessment of the state of the aquageosystem.

Algorithms of formation and processing of the spatial databases are approved in the geographical information system (GIS)-technology. However there is a gap between traditional methods of GIS-technology and models as an instrument of prediction. To fill up the gap, let us choose the matrix structures/identifiers with hierarchical topology both in space and in the elemental filling of the aquageosystem

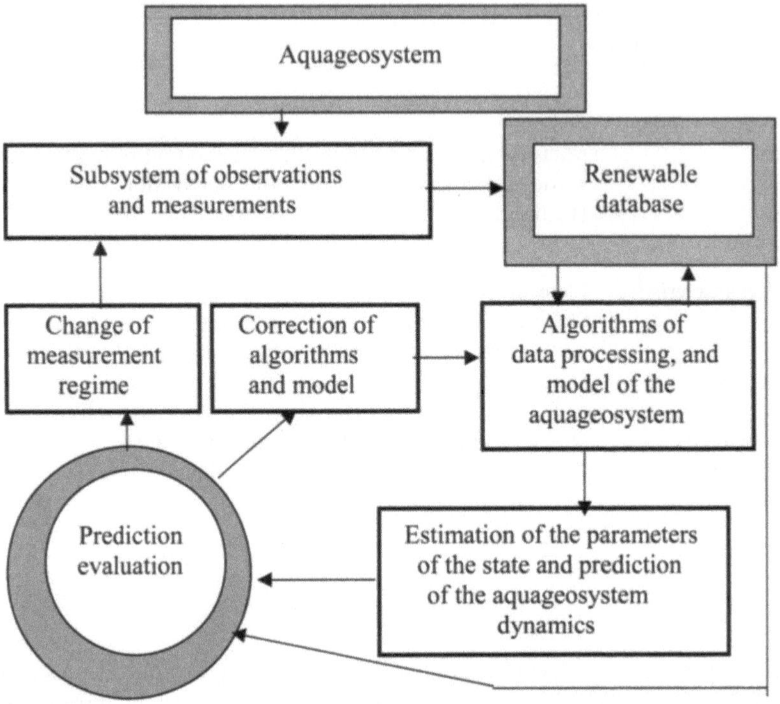

Figure 9.10. Principal block-scheme of the adaptive regime of the hydrophysical experiment.

model. Each element of the identifier describes in the symbol space the image of an element (fragment) of the system under study. In other words, the territory Ω is presented in the form of the hierarchical spatial structure:

$$\Omega = \bigcup_{(i,j)} \Omega_{ij} \qquad \Omega_{ij} = \bigcup_{s,l} \Omega_{i_s j_l}$$

According to this scheme, the territory Ω at the fifth level of the database is represented by a set of matrices $A_k = \|a_{ij}^k\|$ $(k = 1, \ldots, N)$, where algorithms and data from other levels are used to determine the elements a_{ij}^k. An identifier as a simplest binary approximation of the territory Ω can be exemplified as follows:

$$a_{ij} = \begin{cases} 0 & (i,j) \notin \Omega \\ \alpha & (i,j) \in \Omega \end{cases}$$

where the symbol α can identify the sets of elements of the aquageosystem supposed to be taken into account in the system of data processing. This semantics reduces the dimensionality of the fifth level of the database but increases the probability of information distortion in the dialogue mode. In other words, here the problem exists of an optimization of the fifth level of the database.

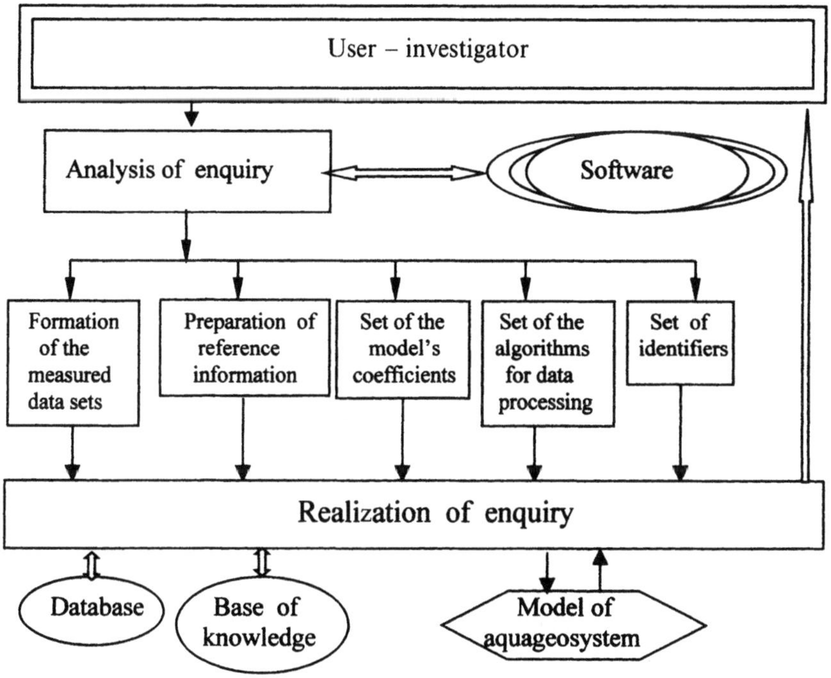

Figure 9.11. Exemplary scheme of the dialogue regime of the use of algorithmic support of the hydrophysical experiment.

On the whole, a totality of the models of the hydrophysical processes over a limited territory or on a global scale permits us to carry out a numerical experiment and to estimate the parameters of a hydrophysical object. Use of the numerical experiment in hydrophysical studies has much in common with other fields of its application. A closed totality of algorithms of analysis of the measured data is constructed and on its basis the fourth and fifth levels of the database are formed. An adaptive procedure of the use of the aquageosystem functioning model (according to the scheme in Figure 9.10) makes it possible to start a dialogue mode of the estimation of many parameters which cannot be reliably estimated from the measured data or which raise doubts.

By applying additional quantitative criteria of stability in the dialogue mode, it is possible to find permissible ranges of changes of some parameters. Schematically this is shown in Figure 9.11.

The most substantial link in the system of automatization of the hydrophysical experiment is the processing of data of the hydrophysical factors, whose knowledge, other conditions being equal, plays a certain role in assessing the complex state of the aquageosystem. These are factors, which introduce into the model some elements of instability and are the source of the aquageosystem's dynamics. Therefore, in all algorithms for the hydrophysical experiment, hydrodynamical equations serve as a reference point for strictness and accuracy. An adequate system of hydrodynamical equations describes changes of the components of the current velocity, pressure, and density and contains five equations: equations of motion for three components, an equation of continuity, and an equation of state. Since we are interested in the problem of pollutant transport in the hydrological network, these equations should be supplemented with equations of water balance within the territory.

9.7.1 Simulation system for hydrophysical and hydrochemical investigation (SSHHI)

The problem of synthesis of the SSHHI in a given territory requires a classification of the elements of the aquageosystem, a consideration of which results from an experience gained in solving such problems as applied to other territories. Analysis of the scheme in Figure 9.12 shows that many fragments of the SSHHI can be realized with the help of GIS-technology. These are formation of the spatial database, modelling the topographic structure, computer cartography, and visualization. Other SSHHI

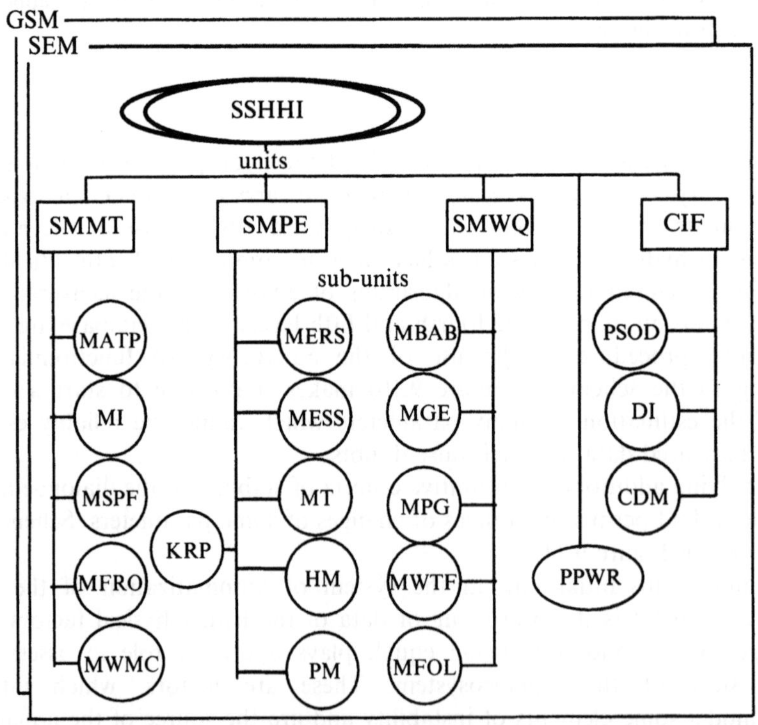

Figure 9.12. Structure of the SSHHI. Notation is given in Table 9.9.

Table 9.9. List of identifiers for Figure 9.12.

Identifier of the SSHHI unit	Description of the unit
GSM	Global simulation model
SEM	System of ecological monitoring
SMMT	Simulation model of the moisture motion over the territory
SMPE	Simulation model of the process of moisture evaporation
SMWQ	Simulation model of water quality
PPWR	Realization of the procedure of predicting the state of the water regime and quality in a given territory
CIF	Control of information fluxes
MERS	Modelling the evaporation from the river surface
MESS	Modelling the evaporation from the soil surface
HM	Horton model (Bras, 1990)
MT	Model of transpiration
MWTF	Model of the water temperature formation
MBAB	Modelling the biotic and abiotic balance of the water body
PM	Penman model (Bras, 1990)
MATP	Model of atmospheric transport of pollutants
MI	Model of infiltration
MSPF	Model of the structure of soil-plant formations
MFRO	Model of the formation of the run-off with the topography of watershed and soil characteristics taken into account
MWMC	Model of water motion in the canal
MGE	Model of gas exchange on the border "water-atmosphere"
MPG	Model of pollutions generation
MFOL	Model of the formation of the oxygen level in water
CDM	Coordination of database with models
DI	Dialogue interface
FSOD	Formation of the structures of output data
KRP	Kohler's and Richard's modification of Penman's evaporation model (Bras, 1990)

fragments are realized in accordance with geographical information monitoring systems (GIMS)-technology.

The GIMS-technology recommends (in the case of the organization of studies in the territories with heterogeneous structures) the formation of a series of maps of different minuteness, as shown in Figure 9.13. For any level of the spatial resolution $\Delta\varphi \times \Delta\lambda$, the database contains an information layer of the sets of identifiers controlled by the control unit and whose data are taken by the respective functions – units of the SSHHI. Here the scheme of cartographic modelling widely used in GIS is realized.

As indicated above, the organization of any nature monitoring requires the solution of the problems of spatial–temporal interpolation. This problem is solved by many methods applied in GIMS-technology. In the SSHHI these methods are

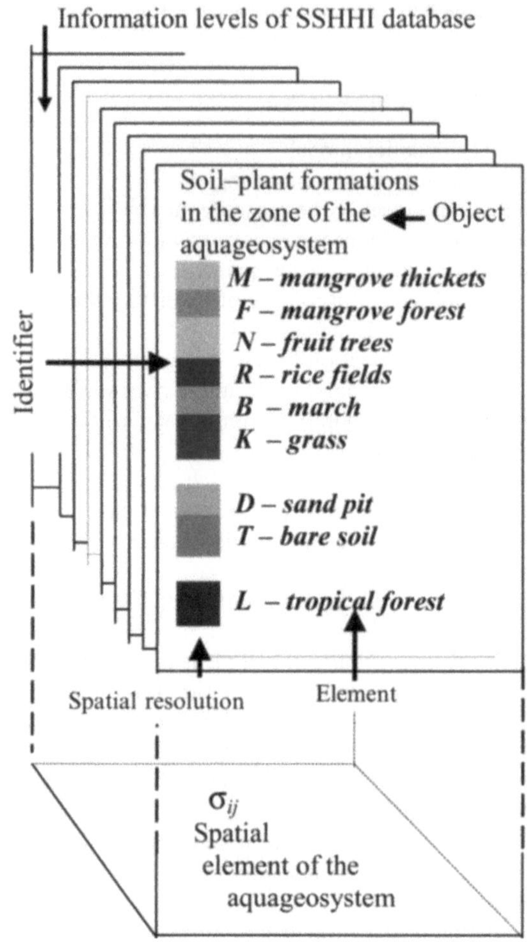

Figure 9.13. Cartographic referencing of data and the scheme of formation of the levels of the SSHHI database in accordance with the structure of the GSM database.

used to integrate information from various sources. This procedure includes the following operations:

- integrating into a single system of spatial data from various sources, such as geographic maps, satellite photographs, remote measurements, ground observations, etc.;
- creation of the geometric description of the Earth's surface within compatible topological structures;
- formation of models and software for transformations between the vector and expanded data (i.e., between maps and images);
- provision of agreement between the series of hydrophysical and hydrochemical characteristics at each level of the spatial scale of the system;

- provision of coordination of information fluxes of spatial, linear, and pointwise data within one unit or for the whole model; and
- synthesis of the scattered spatial information and formation of a representative image of either the aquageosystem or its fragments.

There are elements in the SSHHI database which, according to the scheme in Figure 9.13, cover the whole set of information levels. As a result of the simulation experiment, at each time moment the computer memorizes a set of schematic maps. Depending on the user's query, the FSOD unit forms an output file with the required information structure. Means of computer visualization transform the symbol matrix into a coloured schematic map, and the user obtains information which is easy to analyse.

9.7.1.1 *Method used to form a unit structure of the SSHHI*

To parameterize the hydrospheric part of the investigated site on the Earth's surface and to create a universal schematic envelope of the system, let us cover the territory with a grid with steps $\Delta\varphi$ in latitude and $\Delta\lambda$ in longitude. The area of each cell will be $\sigma_\Delta = \Delta\varphi \times \Delta\lambda \times k_\varphi \times k_\lambda$ where k_φ and k_λ are kilometres in 1° latitude and longitude, respectively. A discrete numbering of the cells gives a totality of the sites $\Xi = \{\Omega_{ij}\}$ where each site Ω_{ij} has an area σ_{ij}. The coordinates of the centre of the site Ω_{ij} are equal to:

$$\varphi_i = \varphi_1 + 0.5\sigma_i + \sum_{s=1}^{i}(s-1)\Delta\varphi_s \qquad i = 1, \dots, N$$

$$\lambda_j = \lambda_1 + 0.5\lambda_j + \sum_{s=1}^{j}(s-1)\Delta\lambda_s \qquad j = 1, \dots, M$$

This grid for digitizing the territory enables one to create a universal procedure of referencing to a given object Ξ under study by using the following symbolic 2-D arrays:

$$\text{TERRITORY}(I, J) = \begin{cases} 0 & \text{for } \Omega_{IJ} \notin \Xi \\ 1 & \text{for } \Omega_{IJ} \in \Xi_S \\ 2 & \text{for } \Omega_{IJ} \in \Xi_W \end{cases}$$

where $\Xi_S \cup \Xi_W = \Xi, \Xi_S$ is part of the land territory; and Ξ_W is part of the territory under a hydrospheric object. The topological structure of the hydrospheric part of the territory is described by the identifier of the HYDROL(I, J) type in which possible configurations of different types of hydrospheric elements are described. The HYDROL array permits selection of a moving part of the water environment in the territory Ω and allows us to more accurately consider thereby the variability of the water chemistry due to heterogeneous catchment areas.

To organize the numerical procedure of calculation of dynamic water flows between the cells Ω_{IJ} let us introduce an array of the heights above sea level:

$$\text{TOPOGR}(I,J) = \begin{cases} 0 & \text{when } \Omega_{IJ} \in \Xi_W \\ \omega_{IJ} & \text{when } \Omega_{IJ} \in \Xi_S \end{cases}$$

where ω_{IJ} is the average height of the cell Ω_{IJ} above sea level in metres.

Each cell Ω_{IJ} is characterized by a certain relationship between water basins, wild and agricultural vegetation, urban constructions, and roads. Let us determine this relationship from a set of the following 2-D arrays: PARTW$(I, J, \text{TIME}) = S_{IJ}^1 \in [0, 1]$ is part Ω_{IJ} under the water basins; PARTA$(I, J, \text{TIME}) = S_{IJ}^2 \in [0, 1]$ is part Ω_{IJ} under cultural vegetation; and PERTS$(I, J, \text{TIME}) = S_{IJ}^3 \in [0, 1]$ is part Ω_{IJ} under wild vegetation. Anthropogenic objects cover the remaining part of the territory: $S_{IJ}^4 = 1 - S_{IJ}^1 - S_{IJ}^2 - S_{IJ}^3$.

Each cell Ω_{IJ} is occupied by one of n types of wild vegetation and by one of m types of cultural vegetation. The schemes of this occupation are presented with 2-D arrays: VEGETA(I, J, TIME) for agricultural vegetation, VEGETS(I, J, TIME) for wild vegetation, and ANTROP(I, J, TIME) for anthropogenic landscapes.

Water can be polluted through the atmosphere, through washing out from soils, and anthropogenic emissions directly into water reservoirs (common and industrial sewage, leakages and discharges from ships). Information about these processes is synthesized in the form of an identification scheme TYPEC(I, J, TIME). Based on the information contained in TYPEC array, the statistical mean situation of pollution in the absence of information is controlled.

The system is filled with data from monitoring or from other information sources that characterize the pollutant emissions into the environment by anthropogenic systems. The system contains the TYPEP(I, J, TIME) array designed for introduction of more accurate information about the polluting productions. The TYPEP array includes productions characteristic of the territory due to which the system provides the information network with average estimates of pollutant emissions in case of the absence of other information. Information about characteristic compositions of sewage in the territory of the productions is used as a database.

9.7.1.2 Formation of the SSHHI information base

Depending on the requirements of the SSHHI efficiency, the corrective information can be obtained from satellites. For instance, information about the arrays TERRITORY, HYDROL, TOPOGR, PARTW, PERTS, VEGETA, VEGETS, and ANTROP can be obtained from satellites. The arrays TYPEC and TYPEP can be replenished and corrected by the regular use of the flying laboratory.

The structure of the SSHHI information base is determined both by the possibilities of its formation and by the need for practical solutions of the problems of water quality control in a given territory. The experience gained in creation and use of such systems suggests a hierarchical structure of the database constructed by the principle of successive digitization of the territory. Following the scheme of hierarchical digitization of the space, one can without damage to the system, constantly increase the minuteness of digitization, with the subject orientation of the territory Ω_{IJ} preserved. Thus the structure and functions of the SSHHI are adjusted to the natural composition of data on various aspects of functioning of natural and anthropogenic landscapes. The SSHHI only synthesizes the available data and suggests what other data should be obtained. Models in the SSHHI structure provide an integration of data in space and by subject areas, due to which an integrated image is formed of the territory with processes taking place in it.

The SSHHI has a set of formalized structures to control the monitoring regime. Use of spatial–temporal and subject identifiers refers the SSHHI to the topology and morphology of the natural or anthropogenic object. The sets of the identifiers enumerated above parameterize the spatial and subject structure of the territory Ω, so that at each level of its spatial digitization a complete description is provided of an image of the system under study. Spatial topology of this parameterization can be heterogeneous and non-uniform as well as excessive.

9.7.1.3 The SSHHI functioning

The user, following the hierarchy of the SSHHI menu, can realize the following operations:

- to ask for data on any identifier (array) and to correct any of its fragments;
- to ask for estimates of all or part of the parameters of simulation units and to correct them;
- to select the sets of parameters and identifiers for a more prompt access to them;
- to synthesize a symbolic schematic map of the distribution of the estimates of the environmental characteristics; and
- to predict the state of the environment down to a given depth or until accomplishing the a priori formulated criterion of assessment of the state of the water environment.

Schematically this is shown in Figure 9.14. The user, through an interface, sends a permission at each step of the command dialogue, which is assessed in the unit of the query analysis, and from its response, the controlling unit realizes a chain of the required actions of the system. Interface message processor supports the roll-out of a progress report in each cycle of service. The final result is presented in the form of a protocol with enumerated characteristics of the water environment by objects and territories as well as in the form of schematic maps or digital information combined with the map.

9.7.2 An expert system to identify the spots of pollutants on the water surface

The problem of detection and identification of pollution spots on the water surface, especially of oil spills, is being solved by many scientists. The recently developed technology of adaptive identification of the environmental elements

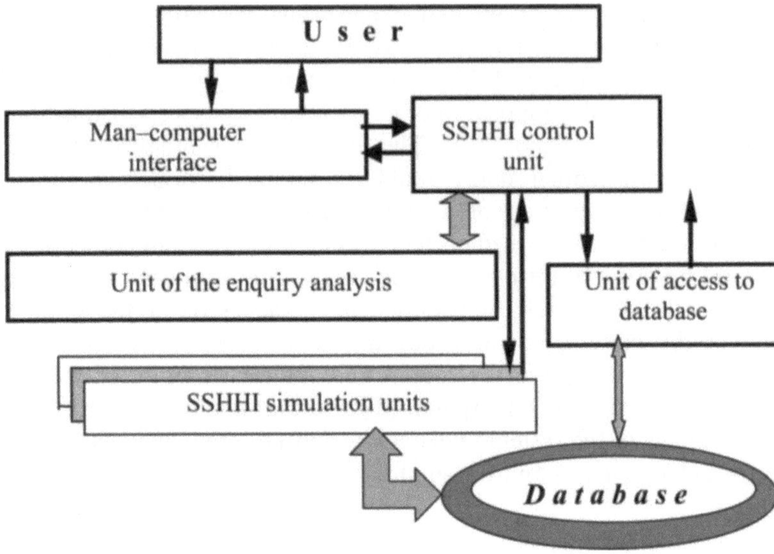

Figure 9.14. SSHHI information units.

from measurements in the visible spectral region permits us to synthesize an expert system for adaptive identification of the environmental parameters (ESAIEP). The system's structure includes a compact multichannel spectropolarimeter (MSP), information interface with computer (IIC), computer software (STW), and extending database (EDB). The STW realizes a number of algorithms in order to process the data fluxes from MSP and provides service functions of visualization and control of the regime of measurements. The EDB consists of the sets of standard spectral images of the spots of pollution represented by points in the multidimensional vector space of indicators, precalculated on the basis of learning samples.

The principle of the ESAIEP functioning is based on fixation of changes of the light flux at the MSP output and their transformation into a digital code. Further processing of these data with respect to their efficiency is determined by the STW composition containing various algorithms of recognition of 2-D objects. The adaptability of the recognition procedure is determined by the level of accumulated knowledge about special features of intensity fluctuations and polarizing properties of the light reflected from the water surface. The STW includes the means that make it possible, and in the case of uncertain identification of the pollution spot, to make an expert decision based on the visual analysis of its spectral image. This procedure is realized in the mode of dialogue with ESAIEP, and if a decision is made, the operator can fix it in the database in the form of a standard for subsequent appearances of similar spots.

The principal scheme of the STW unit providing the procedure of identification is some transformation Φ (Figure 9.15, Table 9.10). The light intensity ξ_j^i recorded at the time moment t_i in the channel λ_j enters the algorithm Φ where two hypotheses H_0 and H_1 are identified (Figure 9.16). The ESAIEP operator determines initial data $\mathbf{v}_i$,

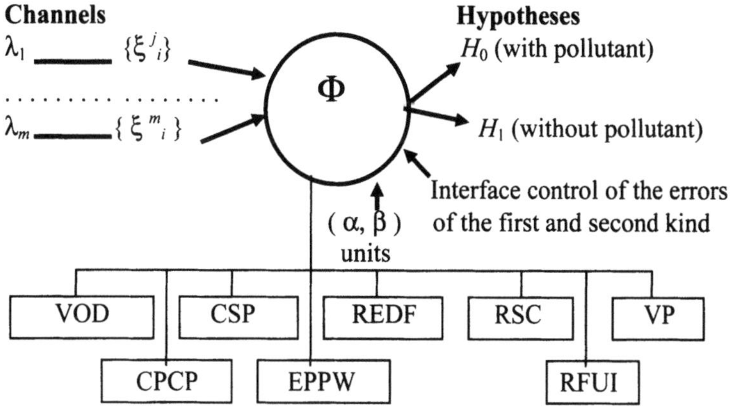

Figure 9.15. The conceptual block-scheme of the procedure of identification of the types of pollution spots on the water surface. Notation is given in Table 9.10.

Table 9.10. Description of the ESAIEP units schematically shown in Figure 9.15.

Unit	Characteristic of the unit
VOD	Visualization of the observed data (spectra, correlations, statistical parameters, empirical and theoretical functions of distribution)
CPCP	Calculation of parameters for the classical procedure of decision making by the Neuman–Pearson criterion
CSP	Calculation of statistical parameters and other characteristics, formation of the vector space $\{\mathbf{u}_i\}$
EPPW	Estimation of the parameters to be used in the procedure of the Wald sequential analysis in decision making
REDF	Reconstruction of the empirical distribution functions
RFUI	Reconstruction of the functions of the user's interface equipped with software to intervene in the working regime of the expert system at any stage of its functioning
RSC	Realization of statistical criterion to estimate the theoretical function of distribution
VP	Visualization procedures

α and β and decides which parameters $\mathbf{u}_i = (u_1, \dots, u_r)$ will be calculated from measurements of $\{\xi_i^j\}$. The service unit IIC makes it possible to form a vector $\mathbf{u}_i$ from statistical characteristics of the series $\{\xi_i^j\}$ or to use direct measurements. A priori information characterizes the type of distribution $f_a(\mathbf{u}_i)$. The function:

$$L_i = \sum_{\hat{i}=1}^{\phi} \psi_j = f_{a1}(\mathbf{u}_i^j)/f_{a0}(\mathbf{u}_i^j)$$

is compared with its ultimate values $L_{i,\min}$ and $L_{i,\max}$. At the first stage, these values are chosen arbitrarily, but then they change until reaching a maximum

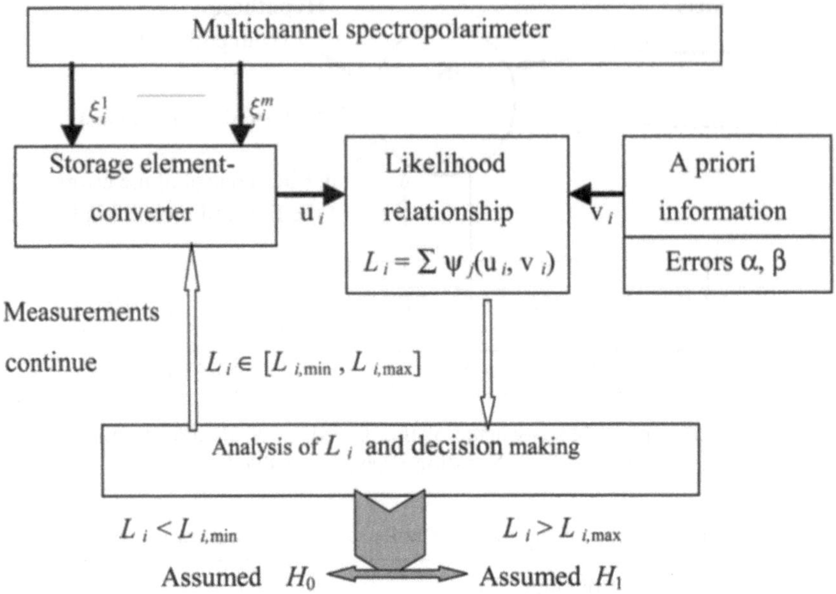

Figure 9.16. Procedure of the sequential analysis to identify hypotheses H_0 and H_1.

accurate recognition of the hypotheses H_0 and H_1. We have $L_{i,\min} \to L^*_{i,\min}$ and $L_{i,\max} \to L^*_{i,\max}$. The values $L^*_{i,\min}$ and $L^*_{i,\max}$ are memorized in the EDB.

After the learning procedure, the functioning of the expert system is limited only by the volume of measurements fixed by the operator, proceeding from statistical reliability and the real time regime. The operator has two possibilities to regulate this regime, establishing the volume of the series $\{\xi_i^j\}$ or fixing the time of their accumulation. Usually the latter characteristic equal to 1 s is chosen. Figure 9.17

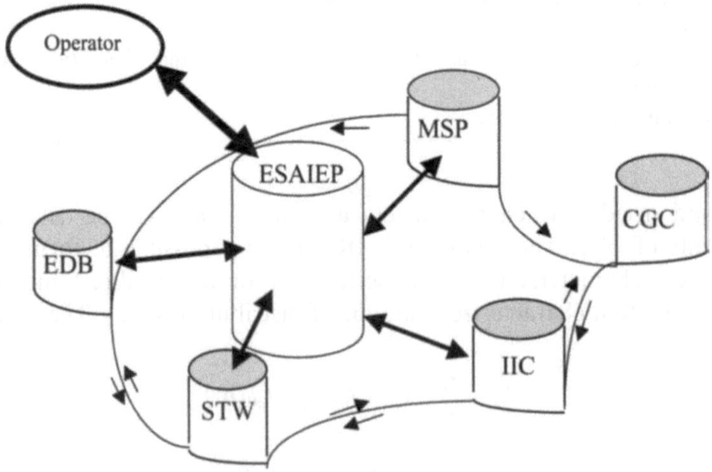

Figure 9.17. The ESAIEP model.

explains this procedure. The operator is combined with the ESAIEP units through the man–machine interface IIC, which provides the selective control of operations in all units.

In the presence of the oil film on the water surface the system analyses its thickness, age, source, and geometry. In this case most informative are the series of measurements at wavelengths 398 nm, 439 nm, and 480 nm. In the case of dissolved or suspended components in the water the system estimates their concentration and from the data in EDB on the hydrodynamic parameters of the water body, it calculates their spatial distribution using the methods of 2-D image processing.

9.7.3 The structural–functional analysis of hydrophysical monitoring data

As an experience of the struggle with large-scale changes of the geochemical and hydrological situations in anthropogenic zones has shown, the problem of holding the process of desertification and ecological degradation cannot be solved without creating a multilevel monitoring system with predictive functions (Krapivin and Phillips, 2001). The GIMS-technology provides the possibility of an adaptive successive analysis of information about the state of the main hydrophysical fields (temperature τ and salinity S) with correction of the simulation model (by principle of feedback) and control of the processes of collection and processing of monitoring data. A simulation model of the hydrophysical fields (SMHF) of the water basin is shown schematically in Figure 9.18. It is an element of the general model that describes the aquageosystem functioning. The SMHF includes a set of items that parameterize different information fluxes (see Table 9.11).

The model of the aquageosystem is the base element of SMHF that provides an adaptation of the monitoring regime due to structural and parametric changes. The functional filling of the model is presented in Table 9.11. The model describes the hydrophysical processes in Ω_A during the ice-free period. Simulations are made of the currents, convective mixing, river runoff, heat and moisture exchange with the atmosphere, as well as turbulent exchange with heat, salt, and momentum. To form the initial conditions for the end of the winter season t_w, a correction is made to the hydrophysical parameters of the aquageosystem obtained in the FIC unit, with multiyear changes of the level and other indicators of the state of the water basin Ω_A taken into account. Corrections are introduced into the calculation grid of depths and initial fields of salinity $S(\varphi, \lambda, z, t)$ as well as seawater temperature $\tau(\varphi, \lambda, z, t)$. When accomplishing these procedures, the characteristic estimates of the indicated parameters obtained from the available fragments of multiyear measurements at the moment t_w should be taken into account. The initial field $\tau(\varphi, \lambda, z, 0)$ is prescribed considering the homothermy of water masses at this moment. The field $S(\varphi, \lambda, z, t_w)$ is formed with the division

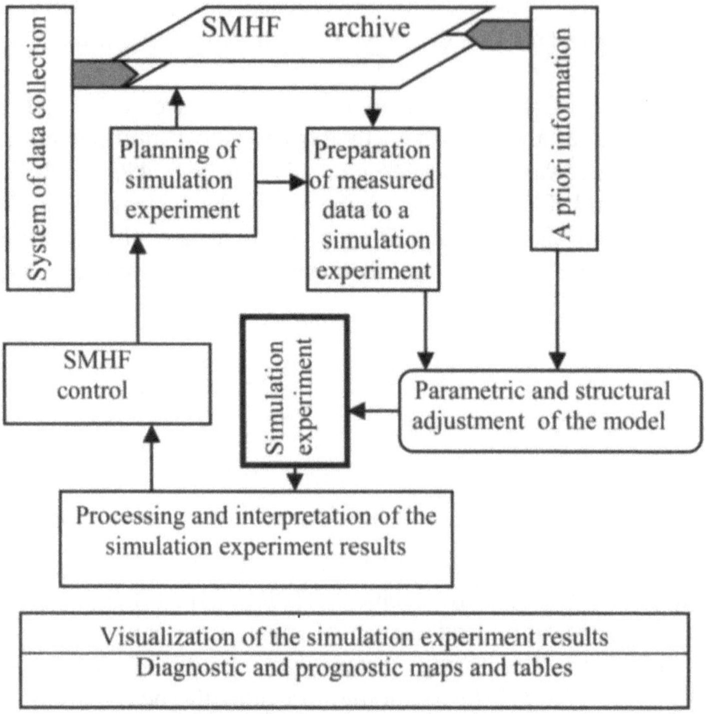

Figure 9.18. Scheme of information fluxes in the SMHF.

Table 9.11. The SMHF units.

Unit	Characteristic of the unit's functions
FIC	Formation of initial conditions for the model considering multiyear changes of the hydrogeophysical characteristics of the aquageosystem. The choice of the steps of spacial digitization in latitude and longitude. Determination of the time step of digitization
MSC	Model of seasonal changes of the aquageosystem level
MFCS	Model of the formation of the current structure
MSTD	Model of the spatial–temporal distribution of the water temperature and salinity
MWD	Model of the formation of the field of water density
SPCM	Simulation of the process of convective mixing of water masses
PCR	Provision of the regime of the structural and parametric correlation of the SMHF units with the modelling results taken into account

of Ω_A into water areas with typical homogeneous hydrophysical characteristics taken into account:

$$S(\varphi, \lambda, z, t_W) = \begin{cases} S_1 & \text{for } (\varphi, \lambda) \in \Omega_D \\ S_2 & \text{for } (\varphi, \lambda) \in \Omega_A \backslash \Omega_D \end{cases}$$

where Ω_D is the deep-water basin; $S_1 = (S_D - S_A)/V_D$, $S_2 = (S_P - S_L)/V_P$; V_D and V_P are water volumes in the deep-water basin and in the shallow-water area; S_D and S_P are water supplies in Ω_D and $\Omega_A \backslash \Omega_D$ respectively; and S_A and S_L are salt supplies in the dried up territories adjacent to Ω_D and $\Omega_A \backslash \Omega_D$ respectively. Such an approximation can be complicated, but in this monograph the binary version is oriented towards the Aral Sea and Nyok Ngot lagoon.

The seasonal or intra-annual variations of the level of the Aral Sea or Nyok Ngot lagoon are described by the prognostic model MSC traditionally considering the balance constituents of the water regime of the aquageosystem: evaporation, precipitation, and river runoff. The MSC uses the mean multiyear values of these constituents of the water balance as well as information about the present hydrometeorological situation in the near-Aral regions and the central regions of central Vietnam. The MSG unit simulates the currents in Ω using the method in Bortnik *et al.* (1994) that makes it possible to estimate the horizontal components of velocity $u(\varphi, \lambda, z, t)$ at any point with the geographical coordinates $(\varphi, \lambda) \in \Omega_A$ and an arbitrary depth z. The vertical constituent of the current velocity is calculated from the equation of continuity. The water motion is assumed to be determined by tangential wind stress over Ω_A and sea surface slopes caused by negative and positive setup as a result of variable winds, with the effect of the coasts and river runoff taken into account. The active forces are balanced by vertical viscosity and friction about the bottom. The coefficient of turbulent viscosity is calculated by the formula $v = 0.25\,aWH/k$, where W is the wind speed, H is the depth, k is the wind coefficient, and a is the coefficient of proportion.

The MSTD unit simulates the spatial–temporal structure of the distribution of salinity S and seawater temperature τ. The water basin Ω_A is divided into compartments $\Omega_1(\cup\Omega_2 = \Omega_A)$ internally homogeneous both in S and in τ. The transport of heat and salt between the compartments Ω_A is realized by the currents and to the difference in gradients. The exchange processes on the border with the atmosphere are described by the linear relationships with multiyear observations taken into account.

To approximate the vertical deviations of water density $\rho(\varphi, \lambda, z, t)$, the linear model MWD was used. The SPCM unit examines the criterion of stability of the water stratification, and on this basis, realizes the procedure of the convective mixing of water masses. Stratification is considered to be stable if $\partial\rho_o/\partial z \geq 0$. In this case there is no convective mixing. When $\partial\rho_o/\partial z < 0$, the stratification is considered to be unstable, and the process of convective mixing is identified with the procedure of S and τ averaging over the adjacent water layers.

An adaptation of SMHF was realized by the PRC unit based on the data of multiyear measurements from the scientific research ship *Otto Schmidt* in the Aral Sea and geophysical expeditions in January 1989 and March–April 2001 on the coast of the South China Sea carried out by the State Oceanographic Institute of the Rosgidromet (Bortnik *et al.*, 1994) and by the Institute of Physics of the Scientific Centre for Natural Sciences and Technology, together with the Institute of Radio Engineering and Electronics of the Russian Academy of Sciences. In the Aral Sea, measurements were made during the ice-free period in the sector Ω_A over 10 days (three times a season) for several horizons with an accuracy of 0.01°C for τ and 0.01‰ for S. In Vietnam, measurements were made with the use of remote technologies of microwave radiometry and ship-borne systems. The obtained measurements were corrected to reduce them to mean diurnal values. An exclusion of random fluctuations in the measured values and their reduction to the form commensurable with the modelling results was realized by averaging and estimating the following characteristics: S value averaged over depth, thermocline depth and the temperature at this depth, and variations of average water temperatures above and beneath the thermocline. The parametric adaptation of SMHF to these values was estimated from a minimum of the mean square deviation. The structural adaptation of SMHF to the spatial distribution of the model components in Q_A was realized in conformity with real data (Bortnik *et al.*, 1994; Bui, 1998; Krapivin and Phillips, 2001). The SMHF adaptation was made using the alternative adaptation method (Aota *et al.*, 1993).

The conditional scheme of the water balance in the zone of the effect of the Caspian and Aral Seas was described in Kondratyev *et al.* (2002a,b). To develop the computer analogue to this scheme we take the geographic grid of digitization of the territory of the region Ξ into quasihomogeneous sites Ξ_{ij} with the sizes $\Delta\varphi$ in latitude and $\Delta\varphi$ in longitude, where $\Delta\varphi$ and $\Delta\lambda$ are free parameters chosen in accordance with the database. Each site Ξ_{ij} is characterized by its type of floristic background α_{ij} and soil β_{ij}, surface moistening Φ_{ij}, and the depth of groundwater M_{ij}. Projecting the grid $\Delta\varphi \times \Delta\lambda$ onto the map of the territory Ξ gives a totality of geophysically heterogeneous objects, the water exchange processes between which follows the balance relationships in accordance with the scheme of the water balance. In each cell Ξ_{ij} of an area σ_{ij} the functioning of the considered scheme is represented by the subset of fluxes $\{H_k\}$ selected by the matrix identifiers $\{A_s\}$ which, according to GIMS-technology, reflect the information structure of the database.

The results of the simulation experiments presented in Figures 9.19 and 9.20 and Tables 9.12 and 9.13 shows the efficiency of SMHF as an element of the system of the Aral Sea monitoring in conditions of the presence or absence of the Amu-Darya and Syr-Darya river runoff. On average, SMHF restores the spatial distribution of S and τ with a relative error up to 10% and gives the prognostic estimates of these distributions for two months with the relative error up to 15%. The *in situ* measurements are recommended to be carried out not less than twice a year in late May and early August. At other time moments the SMHF will provide, with indicated accuracies, estimates of all hydrophysical parameters in Ω_A. Thus the SMHF provides a practically acceptable accuracy and permits its inclusion as an element of the structure of the higher level monitoring system. An adaptation of the SMHF to this function is realized by coordination of the formats of its inputs and outputs with the respective formats of the GIMS.

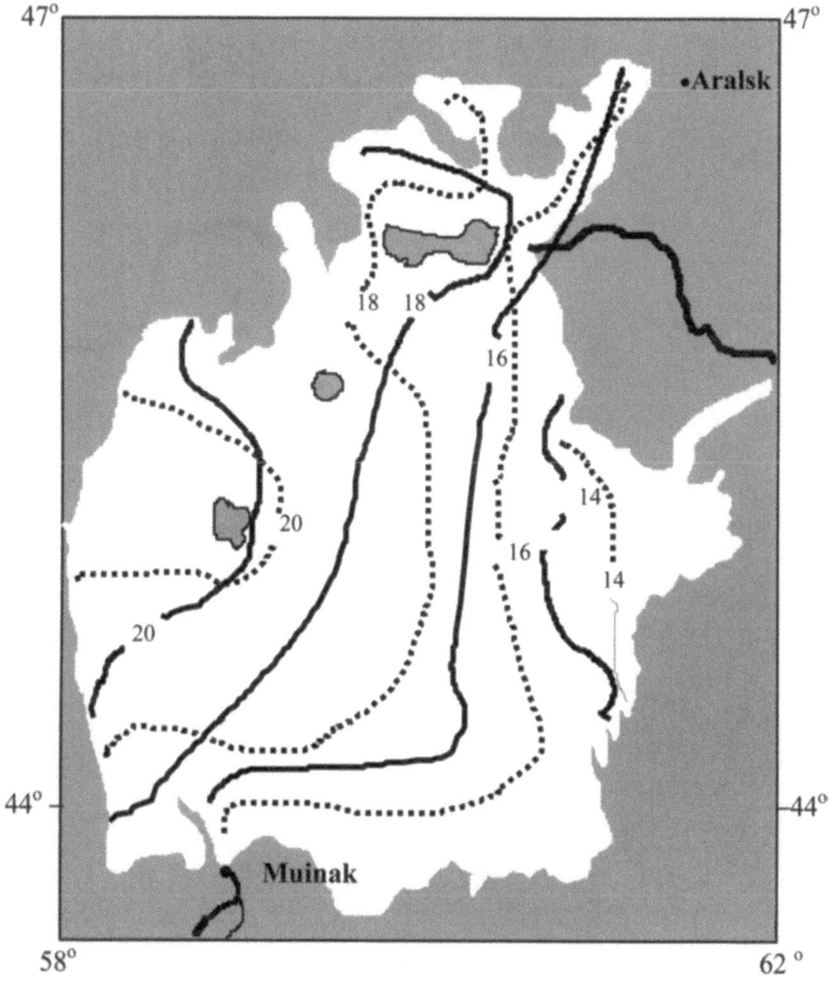

Figure 9.19. Comparison of the predicted (solid curves) temperature (°C) and salinity (‰) of the Aral Sea waters with those measured (dashed curves) in October 1981.

One of the most important functions of SMHF is the prognostic assessment of the hydrophysical object. To analyse this function, we use as initial data the results of measurements in the mouth of Nyok Ngot lagoon made in March 2001. The remaining parameters are taken from mean-statistical estimates for a dry season, except for synoptic parameters and anthropogenic fluxes of chemicals, which were measured during the hydrophysical expedition in March–April 2001. The meteorological data for this period were obtained from the meteorological station Qui Nön. These were estimates of wind speed and direction, atmospheric pressure, temperature and humidity, and precipitation. The volume of the river runoff was calculated proceeding from the condition that 73% refers to the rainy season. The remaining 27% are distributed uniformly over a dry season (January–August). The elements of

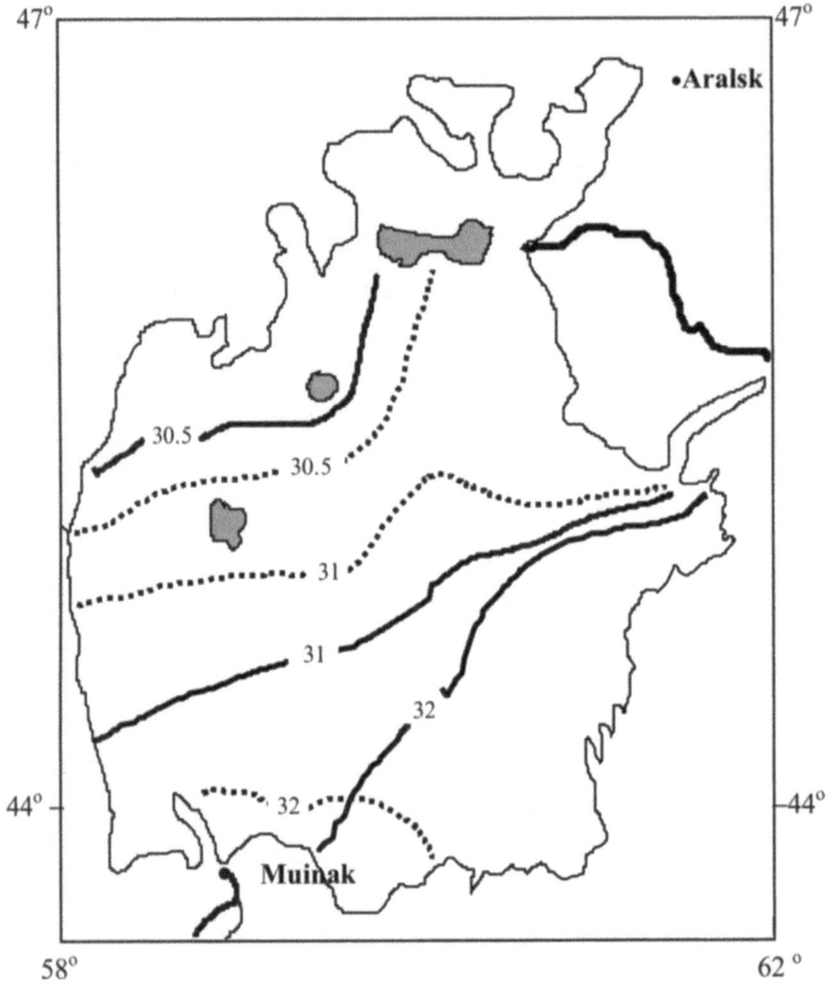

Figure 9.20. Comparison of the predicted (solid curves) temperature (°C) and salinity (‰) of the Aral Sea waters with those measured (dashed curves) in October 1989.

current were calculated from the standard scheme of the MSF unit with correction by measured data. The water expenditure through the hydroalignment near the mouth of the lagoon was also corrected taking into account the measured data.

9.8 REMOTE MONITORING OF THE SOIL–PLANT FORMATIONS

9.8.1 Spectral difference of the types of vegetation cover

Remote sensing technology is an important way to map the soil–plant formations and to solve many applied problems of landscape ecology. Recent achievements in

Table 9.12. Comparison of model and empirical results of estimating the physico-chemical parameters in the Nyok Ngot lagoon (the Vietnam coastline of the South China Sea).

Place of measurement	Salinity (‰)		Turbidity (mg/l)		pH		PO_4^{-3} (mg/l)		NH_4^+ (mg/l)	
	M	*E*	*M*	*E*	*M*	*E*	*M*	*E*	*M*	*E*
1	30.15	33.5	12.40	10	7.04	7.82	0.033	0.03	0.240	0.27
2	23.14	26.0	24.78	21	7.71	7.79	0.034	0.03	0.226	0.20
3	26.48	29.1	29.70	27	7.66	7.58	0.068	0.45	0.115	0.10
4	27.37	32.2	26.88	24	8.01	7.78	0.023	0.02	0.157	0.18
5	28.53	31.7	45.59	47	7.29	7.84	0.042	0.04	0.258	0.30
6	28.40	26.3	29.97	27	7.22	7.52	0.082	0.10	0.177	0.15
7	27.50	25.7	23.94	21	7.45	7.30	0.067	0.35	0.111	0.10
8	26.16	25.4	34.58	38	7.21	7.75	0.023	0.02	0.208	0.25
9	26.54	30.5	48.45	51	7.09	7.71	0.023	0.02	0.179	0.15
10	20.95	26.5	60.48	63	7.63	7.0	0.039	0.05	0.232	0.20
Average error (%)	10.7		11.0		6.0		27.4		14.7	

Note: *M* = model value; *E* = measured value.

this field are closely connected with air and space-borne systems, which can promptly provide information about the spatial structure of the Earth's cover with a high resolution both in space and time (Kramer, 1995; Schmidt and Skidmore, 2003). The resolving characteristics depend on the applied instruments and on measurement principles. The instruments operating in the optical region are developed best. In this case the remotely obtained images of the Earth's cover are in the form of photographs. On their basis the characteristic features of the surface are identified and various versions of the images of the vegetation elements are derived. Here evident progress has been made due to the use of hyperspectral sensors with a high spectral resolution (<10 nm). The hyperspectral sensors cover the spectral region 400–2,500 nm. The spectral image of the vegetation cover depends on the biochemical composition of the vegetation, characteristics of the foliage, concentration of pigments, the canopy structure, and many other parameters. Here of importance is the development of methods to recognize the types of plants from their absorption and reflection of light. The main problems of spectral analysis appears because in many cases the optical images of the geological formations and vegetation cover are alike. Due to their biochemical composition, living plants often demonstrate overlapping features in their electromagnetic wave absorption. There is also a multiple scattering of waves within the canopy, which raises their capability to absorb and reflect. However, the concentration of pigments, the biochemical composition, characteristics of foliage, and canopy structure change from one type of plant to another. This is an area for analysis in order to understand which biophysical and biochemical characteristics are responsible for differences in the spectral images of

Table 9.13. Results of the use of an adaptive system to estimate the characteristics of Nyok Ngot lagoon from the data of control measurements in the lagoon mouth.

Characteristic	Measurement (30.03.2001)	Retrieval	Error (%)	Prediction (30.09.2001)	Measurement (30.09.2001)	Error (%)
Tide duration (hr)	13.2	13.1	1	13.1	13.2	1
Ebb duration (hr)	11.9	12.6	6	12.2	11.8	3
Maximum deviation of depth (m)	0.60	0.63	5	0.58	0.69	19
Minimum deviation of depth (m)	0.50	0.59	18	0.51	0.62	21
Wind speed (m/s)	1.9	2.2	14	5.4	4.2	23
Air temperature (°C)	29.2	29.8	2	28.8	27.9	3
Relative air humidity (mb)	27.9	27.6	1	29.3	31.4	7
Precipitation (mm/day)	2.1	2.2	3	7.1	7.9	11
River runoff (m^3/hr)	20,160	22,176	10	29,007	26,396	9
Salinity (‰)	23.5	25.1	7	28.2	29.9	6
Turbidity (mg/l)	15.9	20.4	28	36.6	28.5	22
DOM (ppm)	34,000	18,020	47	38,423	25,743	33
Suspended substances (ppm)	25.4	28.7	13	34.5	30.4	12
Alkalinity ($CACO_3$) (ppm)	98.1	106.9	9	100.0	85.0	15
pH	7.9	8.8	12	8.3	7.7	7
Fe (mg/100g sed)	23.1	29.8	29	450.3	279.2	38
Cu (mg/100g sed)	0.9	1.2	31	4.5	3.7	18
Pb (mg/100g sed)	1.7	2.0	19	2.8	1.9	29
Zn (mg/100g sed)	3.9	3.0	23	7.2	4.8	34
COD (ppm)	24.6	27.3	11	25.3	23.3	8

the types of plants and soils (Painter *et al.*, 2003; Zarco-Tejada *et al.*, 2003; Magagi and Bernier, 2003).

Wetlands are important elements of the Earth's cover. Their monitoring provides the control of the natural cycle of biogenic elements, which is of principal importance for raising the accuracy of the estimates of the greenhouse effect and protecting the biospheric water resources. The increasing anthropogenic impact on wetlands due to tourism and cattle grazing causes changes in the vegetation cover, and depending on the totality of loads (pollution, rise of sea level, climate change, extraction of minerals, etc.), these territories change their role in the global biogeochemical cycles.

Schmidt and Skidmore (2003) analysed the possibilities of identifying 27 types of plants on the coastal saline lands on the southern side of Schiermonnkoog Island (53°29′N, 6°15′E), which is a part of the Dutch Waddenzee ecosystem. The structure of the Dutch saltmarsh vegetation cover is simple. These are mainly grasses, sedges, rushes, and herbs. Here and there some types of bushes can be seen. The canopy never exceeds 1.5 m being located, on average, at a height of 25 cm. Use of the traditional method of identification of two hypotheses made it possible to demonstrate that of 351 possible pairs of types of cover, the probability of recognition within the 740–1820 nm range constituted 83%, and within 1970–2450 nm – 77%. There were local maxima of the probability of correct recognition of the types of plants: 84% at 404 nm, 81.4% at 501 nm, 80.9% at 579 nm, and 82.9% at 628 nm. These maxima are explained by the absorbing features of most soil–plant formations on coastal saline lands.

Thus the use of hyperspectral sensors enables one, due to increased spectral resolution, to raise the probability of identification of the types of vegetation cover similar in their characteristics. The statistically significant distances between medians of the spectral classes provides a reliable spatial differentiation between these classes, which gives an efficient algorithm of recognition of type of Earth cover from their optical images.

9.8.2 Microwave monitoring of vegetation cover

The remote monitoring of soil–plant formations is aimed at assessing the biological productivity, understanding interactions in the soil–plant–atmosphere system, calculating the biome dynamics, modelling the biogeochemical cycles with consideration of the role of vegetation, and managing the vegetation resources. The radar methods used to solve these problems have been well developed (Krapivin, 2000a).

The Earth's cover is characterized by a great variety of landscapes differing in types of soils, special features of the water–salt and temperature regimes, the character of tilling the surface layer on agricultural lands, the type of vegetation cover, and special features of the local macro-relief. The soil includes solid particles, groundwater (ground solution), and air. Solid particles consist of mineral and organic substances (mineral soils): particles larger than 3 mm – the stony part of the soil; particles from 0.01 to 1 mm – the "physical sand"; and particles smaller that 0.01 mm – the "physical clay". The soil-forming minerals consist mainly of oxides (SiO_2 and Al_2O_3). The content of organic substances (mainly humus) varies in such soils from 1 to 10% by weight. These circumstances, in the synthesis of the system of the Earth cover monitoring, necessitate the choice of information channels for radar monitoring. Numerous and multiyear studies in this field made it possible to determine the most efficient wavelength regions for many types of vegetation cover and soils. The agricultural systems, forests and arid zones have been well studied (Sid'ko and Shevyrnogov, 1998).

Perspective directions in the microwave monitoring of the forest include the control of its hydrological regime, the study of afforestation and deforestation processes, and investigations of the secondary succession in the forest fire areas.

These problems are solved using the decimetric range in which the canopy is transparent for electromagnetic waves in the microwave region and hence, the soil can be sounded down to groundwater.

The possibility of generating an inventory of the tree/bush vegetation from the data of radar sensing is based on the connection of the coefficients of backscattering of microwaves with the structure and dielectric properties of plants. Use of radar in the monitoring of forests, together with remote measurements in the visible and IR regions, is explained by the possibilities of obtaining additional information. The main task of the space-borne radars in monitoring the forest is global mapping of vegetation the scale of which is determined by the ecological and economic needs of a given region.

Recent achievements in the field of remote studies of forest/bush vegetation became possible due to a combination of observations using various wavelength regions. For instance, the radar AN/APQ-96 in the IR region classifies the types of vegetation cover, the synthetic aperture radar "Goodyear" has a resolution of 10–15 m and permits recognition of many types of soil–plant formations, and the multichannel radar SAR-580 with a resolution of about 2 m identifies coniferous and deciduous forests and bushes. A combination of optical range and the SHF region (3 cm, 5 cm, 10 cm, 23 cm) makes it possible to classify the states of regeneration of burnt forests. It is seen in radio images that the dead forest areas have a low reflectivity. This is connected with a strong absorption of radio waves of the 10 cm range in the stems of dead trees. Radio images obtained at cross-polarization make it possible to identify images of restored and mature forest. The dead wood images are usually darker compared to those of clearings.

The fire protection of forests is one of the directions of development of remote monitoring of the state of vegetation cover. The current space and air-borne remote monitoring systems of the type JERS-1 and NOAA/AVHRR permit solution of the following problems:

- prompt assessment of the degree of the forest fire risk;
- mapping the water saturation of the territories in putting out large-scale forest fires;
- mapping the outlines of forest fires through the smoke layer and forest stand canopy;
- estimation of the energetic parameters of the fire; and
- assessment of the post-fire state of forests.

Studies of the thermal emission of natural layers of forest flammable material (lichen, moss, dry grass, dead fir needles, brushwood) enables one to derive a technique to identify the forest fire parameters when measuring the thermal emission in the centimetre range of the electromagnetic waves. Spectral distributions of this range clearly show the front and the rear of the fire as well as the burnt part of the forest. A comparison of brightness temperatures at different wavelengths has shown that radiant emittance of forest fires increases with decreasing wavelength and depends on the character of the fire. Spectra of brightness temperatures for various elements of the forest fire provide a reliable classification of its elements.

The GIMS synthesis to control the forest fire risk over large territories necessitates the development of applied models describing the increasing fire risk of the forests. Of course, these models are based on knowledge of the dependence of thermal emission intensity of an element of the forest landscape on its moisture content. Observations show that an increasing fire risk in the layers of the forest flammable materials causes changes in the statistical characteristics of the fields of SHF emission of forests. In particular, this dependence can be presented as:

$$T_J = \kappa(\lambda, w) T_S(1 - \exp\{-\alpha(\lambda_w)h\} + T_{JS} \exp\{-\alpha(\lambda, w)h\}$$

where κ and α are coefficients of emission and absorption, respectively; w is the moisture content of the environment; h is the thickness of the emitting layer; T_S is the soil temperature; and T_{JS} is the soil brightness temperature. Yakimov (1996) proposed the following rule of solution based on calculation of mathematical expectation M^* and standard deviation σ^* to record T_J of the forest site in the range 2.25 cm. The M^* value is compared with the threshold value $d = 7.2514\sigma^* + 249.876$: if $M^* \geq d$ then the site of the forest is characterized by the middle or higher fire risk; if $M^* < d$ then the fire risk of this site is low. This model should be supplemented with the model of forest fire, for instance, the model that describes a forest fire as a running wave (i.e., as a self-maintaining process of local energy release in the active environment).

The millimetre range (1–10 mm, 30–300 GHz) is also promising for solving the problem of diagnostics of the Earth's cover. As shown by numerous experimental studies, an addition to the monitoring system of the channels of the millimetre range makes it possible to raise the accuracy of classification of the soil–plant formations with roughness and small formations clearly identified. It is especially important in conditions of the anthropogenic landscape.

9.8.3 Microwave monitoring of soil moisture

The soil moisture is divided into solid-bound, loose-bound, and free. The bound moisture is water absorbed by the surface of soil particles in the form of a film up to 6–8 molecular layers thick. The volume of bound moisture in a soil layer is determined by the type of soil and varies widely from 2–3% for sand to 30–40% for clay and loess (Shutko, 1987). The bound water is inaccessible for plants and does not affect the salt regime of soils. Therefore, the monitoring system should identify the types of moisture in soils for these types to be considered in the models.

The soil moisture is expressed as a percentage of the weight of dry soils. The radiation models of moistening different types of soils taking into account soil density, its temperature, and salinity, have been studied in detail. In solving this problem, the wavelengths 2.25 cm, 18 cm, and 30 cm turned out to be most informative. The model of radio-thermal emission of the Earth's cover in conditions of non-uniformly moistened surface layer is based on the vertical gradient of the dielectric permeability coefficient $\varepsilon \cong (1 + 0.5\rho_s)^2$. The ε parameter is an informational indicator of the changing soil moisture. With growing soil moisture the ε value increases first slowly, if the soil was dry, then more rapidly. A weak dependence of ε at the stage of initial moistening is explained by the fact that the bound moisture is characterized by a low dielectric permeability. Diverse experimental dependences of ε on the types of soils and volumes of moistening has been discussed in Shutko (1987). The base of knowledge and an application of GIMS-technology permit solution of the problem of diagnostics of the water content W_s in the soil layer. Essentially, the problem is reduced to the retrieval of the function $W_s(z)$ for which the values $W_{si} = W_s(z_i)$ are known, being measured either remotely or *in situ*, meeting the conditions:

$$\partial W_s/\partial z\Big|_{z=0} \leq C \qquad 0 \leq W_s(z) \leq K_s(z) \tag{9.25}$$

where $K_s(z)$ is the prescribed function. The measurement errors follow the known law of distribution.

Practice has shown that regardless of the climatic zone, for an adequately accurate retrieval of the vertical profile of humidity in a 1-m layer, it is necessary to use microwave radiometers (10 cm and 30 cm) to measure the surface layer humidity (W_{s2}) and (100 cm) to measure (W_{s3}). Then an approximation $W_s^*(z) = az^3 + bz^2 + cz + d$ is assumed, whose unknown coefficients are found traditionally from the condition of minimum functional mean square deviation:

$$\sum_{i=1}^{3} [W_s(z_i) - W_s^*(z_i)]^2 = \min$$

Here it is necessary to observe the limitations of Equation (9.25).

This approach makes it possible from the data of microwave radiometric measurements, for instance, from the flying laboratory, to measure the water supply in a 1-m soil layer over large territories with an error not exceeding 0.03 g/cm^3 for the vegetation biomass <2 kg/m^2 and with an error of 0.07 g/cm^3 for biomass >2 kg/m^2.

Knowledge of the function $W_s(z)$ enables one to use the model of the water supply over the territory to retrieve the dynamic characteristics of soils and other elements of water balance as functions of geophysical and ecological parameters. An important specification of such a model is an experimental estimate of the upper soil layer thickness, at the lower boundary of which, from the estimates in Shutko (1987), the capillary bonds get broken after 6 days of watering. The characteristic thickness of this over-dried layer varies between 3–5 cm.

The choice of an optimal range of electromagnetic emission and synthesis, on the basis of an efficient method to measure the moisture content in soils require an additional study of the conditions of the formation of emission fields in the "atmosphere–soil–plant" system. Direct measurements should be combined with the use of data processing algorithms that include the models. An example of successful application of two algorithms to retrieve the soil moisture parameters differing in the sets of input data is given in Frate *et al.* (2003). The algorithms are based on the neural network and meant for processing remote radar measurements in *L*-range at two polarizations without preliminary information about the parameters of vegetation cover and surface roughness.

The monitoring of soil moisture is closely connected with the problem of soil salinity (Metternicht and Zinck, 2003). On a global scale, this process is recorded over 995 Mha of soils, with 77 Mha being salted – 58% referring to irrigated territories. Therefore, knowledge of only moisture content in a soil without data on its state is insufficient for reliable model estimates of the vegetation cover productivity and, as a result, for estimation of the capacity of the surface as a sink of CO_2. Finally, there is one more circumstance connected with the monitoring of soil moisture. The concept of GIMS-technology suggests that to estimate the soil moisture, one can use a respectively adjusted model of the territorial hydrological regime (Chapter 6), if the estimate of water content in snow cover serves as its input information (Magagi and Bernier, 2003).

References

Aagard, K. (ed.) (1998). *Proceedings of the ACSYS Conference on Polar Processes and Global Climate, Rosario, Argentina, 3–6 November 1997* (WCRP-106, WMO/TD No. 908). World Meteorological Organization, Geneva.

Abakumov, A.I. (1993). *Control and Optimization in the Models of Exploited Populations* (129 pp.). Dal'nauka, Vladivostok [in Russian].

Abstracts of the European Conference on Atmospheric UV Radiation (1998). Finnish Meteorological Institute, Helsinki (29 June–2 July, 1998, 127 pp.).

ACECHEM: Atmospheric Composition Explorer for CHEMistry and Climate Interaction (2001). European Space Agency, Nordwijk, The Netherlands (ESA SP-1257 (4), 103 pp.).

ACIA: Arctic Climate Impact Assessment (2000). Report on the 3rd Meeting of the Assessment Steering Commitee and a Scoping Workshop, 28 February–1 March 2000, Washington, DC (ACIA Rep. No. 1, 35 pp.). The Assessment Steering Commitee, New York.

ACIA Implementation Plan (2000). An assessment of consequences of climate variability and change and the effects of increased UV in the arctic region. The Assessment Steering Commitee, New York (35 pp.).

Adamenko, V.N. and Kondratyev, K.Ya. (1999). Global climate change and its empirical diagnostics. In: Yu.A. Izrael, Yu.P. Borisenkov, and K.Ya. Kondratyev (eds), *Anthropogenic Impact on Northern Nature and Its Consequences* (pp. 17–35). Kola Centre of the Russian Academy of Science, Apatity [in Russian].

Adamenko, V.N., Kondratyev, K.Ya., and Sinyakov, S.A. (1991a). Deposition of metals from the atmosphere at the North Pole compared to background results of the northwestern USSR. *Proceedings of the International Conference on the Role of the Polar Region in Global Change, Fairbanks, AL, 11–15 June 1990* (pp. 716–719). University of Alaska, Fairbanks.

Adamenko, V.N., Kondratyev, K.Ya., and Sinyakov, S.A. (1991b). Deposition from the atmosphere of metals at the North Pole as compared with background regions of north-western Europe. *Proceedings of the All-Union Geographical Society*, **123**(4), 316–323 [in Russian].

A Guide to World Resources 2000–2001. People and Ecosystems: The Frying Web of Life (2000). World Resources Institute, Washington, DC (389 pp.).

Aibulatov, N.A. (2000). *Cold War Echo in the Russian Arctic Seas* (307 pp.). GEOS, Moscow [in Russian].

Aitkenhead, M.J., McDonald, A.J.S., Dawson, J.J., Couper, G., Smart, R.P., Billett, M., Hope, D., and Palmer, S. (2003). A novel method for training neural networks for time-series prediction in environmental systems. *Ecological Modelling*, **162**, 87–95.

Akovetskiy, V.I., Zuev, V.E., Kondratyev, K.Ya., and Savinikh, V.P. (1991). What has happened to the planet? *International Life*, **10**, 72–77 [in Russian].

Albrecht, J. (2002). *Instruments for Climate Policy: Limited versus Unlimited Flexibility* (288 pp.). Edward Elgar, Cheltenham, UK.

Alcano, J., Leemans, R., and Kreileman, E. (eds) (2001). *Global Change Scenarios of the 21st Century* (232 pp.). Elsevier, Amsterdam.

Alexandrov, G. and Oikawa, T. (2002). TsuBiMo: A biosphere model of the CO_2 fertilization effect. *Climate Research*, **19**, 265–270.

Alexeev, G.V. (1998). *Arctic Climate Dynamics in the Global Environment* (World Climate Research Program 908, pp. 11–14). World Meteorological Organization, Geneva.

Alexeev, V.V., Zaitsev, S.I., and Kokorin, A.O. (1990). The growth of the rate of the atmosphere–ocean gas exchange in the case of rough seas. *Bulletin of the USSR Academy of Science*, **26**(3), 322–325 [in Russian].

Alexeev, V.V., Kryshev, I.I., and Sazykina, T.G. (1992). *Physical and Numerical Modeling of Ecosystems* (367 pp.). Gidrometeoizdat, St Petersburg [in Russian].

Allen, D.R., Bevilacqua, R.M., Nedoluha, G.E., Randall, C.E., and Manney, G.L. (2003) Unusual stratospheric transport and mixing during the 2002 Antarctic winter. *Geophys Res Lett*, **30**(12): article no. 1599 JUN 17.

Allison, I., Barry, R.G., and Goodison, B.E. (eds) (2001). *Climate and Cryosphere (CliC) Project Science and Co-Ordination Plan* (Version 1, WCRP-114, WMO/TD No. 1053), 73 pp.). World Meteorological Organization, Geneva.

Alverson, K. (2000). Long-term biogeophysical controls on the carbon cycle and their relevance to human concerns. *IHDP Update*, **3**, 1–4.

Alverson, K., Bradley, R., and Pedersen, T. (eds) (2002). *Paleoclimate, Global Change, and the Future* (400 pp.). Springer, Heidelberg.

Anderson, L.G., Olsson, K., and Chierici, M. (1998). A carbon budget for the Arctic Ocean. *Global Biogeochemical Cycles*, **12**(3), 455–466.

Anderson, R.P., Lew, D., and Peterson, A.T. (2003). Evaluating predictive models of species distributions: Criteria for selecting optimal models. *Ecological Modelling*, **162**(3), 211–232.

Andronova, N.G. and Schlesinger, M.E. (2001). Objective estimation of the probability density function for climate sensitivity. *Journal of Geophysics Research*, **106**(D19), 22605–22611.

Angell, J.K. (2000). Difference in radiosonde temperature trends for the period 1979–1998 of MSU data and the period 1959–1998 twice as long. *Geophysics Research Letters*, **27**(15), 2177–2180.

Angell, J.K. (2003). Effect of exclusion of anomalous tropical stations on temperature trends from a 63-station radiosonde network and comparison with other analyses. *Journal of Climate*, **16**(13), 2288–2295.

Aota, M., Shirasawa, K., Krapivin, V.F., and Mkrtchyan, F.A. (1993). A project of the Okhotsk Sea GIMS. *Proceedings of the 8th Internatiomal Symposium on Okhotsk Sea and Sea Ice and ISY/Polar Ice Extent Workshop, 1–5 February 1993* (pp. 498–500). Okhotsk Sea & Cold Ocean Research Association, Mombetsu, Japan.

Arsky, Yu.M., Zakharov, Yu.F., and Kalutskov, V.A. (1992). *Ecoinformatics* (520 pp.). Gidrometeoizdat, St Petersburg [in Russian].

Arsky, Yu.M., Krapivin, V.F., and Potapov, I.I. (1998). Forward to global ecoinformatics. *Problems of Environmental and Natural Resources*, **9**, 2–18 [in Russian].

Ashby W.R. (1956). *An Introduction to Cybernetics* (325 pp.). Chapman & Hall, London.

Aurela, M., Laurila, T., and Tuovinen, J.-P. (2001). Seasonal CO_2 balances of a subarctic mire. *Journal of Geophysics Research*, **106**(D2), 1623–1638.

Austin, J., Brimblecombe, P., and Sturges, W. (2002). *Air Pollution Science for the 21st Century* (350 pp.). Elsevier, Amsterdam.

Bacastow, R. (1981). Numerical evaluation of the evasion factor. In: *Carbon Cycle Modelling* (SCOPE-16, pp. 95–101). John Wiley & Sons, New York.

Bach, V., Krein, A., Berger, A., and Longhetto, A. (eds) (1987). *Carbon Dioxide in the Atmosphere* (530 pp.). Mir, Moscow [in Russian].

Bacmeister, J.T., Kuell, V., Offenmann, D., Riese, M., and Elkins, J.W. (1999). Intercomparison of satellite and aircraft observations of ozone, CFC-11, and NO_y using trajectory mapping. *Journal of Geophysics Research*, **104**(D12), 16375–16390.

Baldocchi, D., Falge, E., Gu, L., Olson, R., Hollinger, D., Running, S., Anthoni, P., Bernhofer, Ch., Davis, K., Evans, R. *et al.* (2001). FLUXNET: A new tool to study the temporal and special variability of ecosystem-scale carbon dioxide, water vapor, and energy flux densities. *Bulletin of the American Meteorological Society*, **82**(11), 2415–2434.

Baldwin, M., Hirooka, T., O'Neill, A., and Yoden, S. (2003) Major stratospheric warming in the southern hemisphere in 2002: Dynamical aspects of the ozone hole split. SPARC Newsletter, **20**, 24–26.

Baliunas, S. (2002). The Kyoto Protocol and global warming. *Imprimis*, **31**(.3), 1–7.

Ball, J.T. (1988). An analysis of stomatal conductance. Ph.D. Thesis, Stanford University, Stanford (89 pp.).

Bard, S.M. (1999). Global transport of anthropogenic contaminants and the consequences for the arctic marine ecosystems. *Marine Pollution Bulletin*, **38**(5), 356–379.

Barford, C.L., Wofsy, S.C., Goulden, M.L., Munger, J.W., Pyle, E.H., Urbanski, S.P., Hutyra, L., Saleska, J.R., Fitzjarrald, D., and Moore K. (2001) Factors controlling long- and short-term sequestration of atmospheric CO_2 in a mid-latitude forest. *Science*, **294**, 1688–1691.

Barker, H.W. and Stephens, G.L. (2001). An intercomparison of 1-D solar radiative transfer codes: From simple to complex cloudy atmospheres. *GEWEX News*, **11**(3), 7–10.

Barker, P.A., Street-Perrott, F.A., Leng, M.J., Greenwood, P.B., Swain, D.L., Perrott, R.A., Telford, R.J., and Ficken, K.J.A (2001). 14,000-year oxygen isotope record from diatom silica in two alpine lakes on Mt Kenya. *Science*, **292**, 2307–2310.

Barrie, L.A. (1986). Arctic air pollution: An overview of current knowledge. *Atmospheric Environment*, **20**(4), 643–663.

Barrie, L.A. (1996). Occurrence and trends of pollution in the Arctic troposphere. In: *Chemical Exchange between the Atmosphere and Polar Snow: Proceedings of NATO Advanced Research Workshop: Process of Chemical Exchange between the Atmosphere and Polar Snow, Ciocco, Italy, 19–23 March 1995* (pp. 95–129). Springer-Verlag, Berlin.

Bates, N.R. and Merlivat, L. (2001). The influence of short-term wind variability on air–sea CO_2 exchange. *Geophysics Research Letters*, **28**(17), 3281–3284.

Bates, N.R., Knap, A.H., and Michaels, A.F. (1998). Contribution of hurricanes to local and global estimates of air–sea exchange of CO_2. *Nature*, **395**(6697), 58–61.

Batijes, N.H. (1999). *Management Options for Reducing CO_2 Concentrations in the Atmosphere by Increasing Carbon Sequestration in Soil* (Report No. 410 200 031, 114 pp.). Dutch National Research Programme on Global Air Pollution and Climate Change, The Netherlands.

Battle, M., Bender, M.L., Tans, P.P., White, J.W.C., Ellis, J.T., Conway, T., and Francey, R.J. (2000). Global carbon sinks and their variability inferred from atmospheric O_2 and $\delta^{13}C$. *Science*, **287**(5462), 2467–2470.

Bazhin, N.M. (2000). Methane emission from a residual layer. *Second International Methane Mitigation Conference, 18–23 June 2000, Novosibirsk* (pp. 231–236). Novosibirsk University Press, Novosibirsk.

Bazilevich, N.I. and Rodin, L.E. (1967). The map-schemes of productivity and of the biological cycle of the most significant types of land vegetation. *Bulletin of the All-Union Geographical Society*, **99**(3), 190–194 [in Russian].

Bazzaz, F.A. (1986). Global CO_2 levels and the response of plants at the population and community levels. In: C. Rozenzweig and R. Dickinson (eds), *Climate–Vegetation Interactions* (pp. 31–35). UCAR, Boulder, MD.

Bazzaz, F.A. (1998). Tropical forests in a future climate: Changes in biological diversity and impact on the global carbon cycle. *Climate Change*, **39**(2–3), 317–336.

Beck, J.W., Richards, D.A., Edwards, R.L., Silverman, B.W., Smart, P.L., Donahue, D.J., Herrera-Osterheld, S., Burr, G.S., Calsoyas, L., Jull, A.J.T. *et al.* (2001). Extremely large variations of atmospheric ^{14}C concentration during the Last Glacial Period. *Science*, **292**(5526), 2453–2458.

Becker, E., Notholt, J., and Herber, A. (1999). Tropospheric aerosol measurements in the Arctic by FTIR-emission and star photometer extinction spectroscopy. *Geophysics Research Letters*, **26**(12), 1711–1714.

Beese, F.O., Hempel, G., Klemmer, P., Kruse-Graumann, L., Labitzke, K., Muhle, H., Schellnhuber, H.J., Simonis, U.E., Thoenes, H.-W., Velsinger, P., and Zimmermann, H. (1995). Scenario for the derivation of global CO_2 reduction targets and implementation strategies (39 pp.). German Advisory Council on Global Change, Dortmund.

Bekoriukov, V.I. and Fedorov, V.V. (1987). Empirical model of total ozone content above the southern hemisphere. *Meteorology and Hydrology*, **3**, 47–53 [in Russian].

Bender, M.L. (2000). Tracer from the sky. *Science*, **288**(5473), 1977–1978.

Bergamaschi, P., Hein, R., Brenninkmeijer, C.A.M., and Crutzen, P.J. (2000a). Inverse modeling of the global CO cycle. 2: Inversion of $^{13}C/^{12}C$ and $^{18}O/^{16}O$ isotope ratios. *Journal of Geophysics Research*, **105**(D2), 1929–1946.

Bergamaschi, P., Hein, R., Heimann, M., and Crutzen, P.J. (2000b). Inverse modeling of the global CO cycle. 1: Inversion of CO mixing ratios. *Journal of Geophysics Research*, **105**(D2), 1909–1928.

Berger, A., Loutre, M.F., and Crucifix, M. (2003). The Earth's climate in the next hundred thousand years (100 KYR). *Surviving Geophysics*, **24**(2), 117–138.

Berkowicz, R., Olesen, H.R., and Torp, U. (1986). The Danish Gaussian air pollution model: Description, test and sensitivity analysis in view of regulatory applications. In: C. De Wispelaere, F.A. Schiermeier, and N.V. Gillani (eds), *Air Pollution Modeling and Its Application* (Vol. 5, pp. 453–481). Plenum Press, New York.

Berresheim, H., Huey, J.W., Thorn, R.P., Eisele, F.L., Tanner, D.J., and Jefferson, A. (1998). Measurements of dimethyl sulfide, dimethyl sulfoxide, dimethyl sulfone, and aerosol ions at Palmer Station, Antarctica. *Journal of Geophysics Research*, **103**(D1), 1629–1637.

Bettinger, P., Johnson, D.L., and Johnson, K.N. (2003). Spatial forest plan development with ecological and economic goals. *Ecological Modelling*, **169**(2–3), 215–236.

Beven, K. (2002). Towards a coherent philosophy for modeling the environment. *Proceedings of the Royal Society London A*, **458**(2026), 2465–2484.

Bhatt, P.P., Remsberg, E.E., Gordley, L.L., McInerney, J.M., Brackett, V.G., and Russell, J.M. (1999). An evaluation of the quality of Halogen Occultation Experiment ozone profiles in the lower stratosphere. *Journal of Geophysics Research*, **104**(D8), 9261–9276.

Bichele, I., Moldau, X., and Ross, Yu. (1980). *A Sub-model of the Distribution of Assimilates and of the Growth of Plants in Conditions of Water Deficit* (Preprint No. A-5, 22 pp.). The Tartu Astrophysical Observatory, Tartu, Russia [in Russian].

Bilek, R.S., Tyler, S.C., Kurshara, M., and Yagi, K. (2001). Investigation of cattle methane production and emission over a 24-hour period using measurements of $\delta^{13}C$ and δD of emitted CH_4 and rumen water. *Journal of Geophysics Research*, **106**(D14), 15405–15413.

Bjorkstrom, A.A. (1979). A model of CO_2 interaction between atmosphere, ocean and land biota. In: *Global Carbon Cycle* (SCOPE-13, pp. 403–458). John Wiley & Sons, New York.

Blanchet, J.-P. (1989). Toward estimation of climatic effects due to Arctic aerosols. *Atmospheric Environment*, **23**(11), 2609–2626.

Blanchet, J.-P. and List, R. (1983). Estimation of optical properties of arctic haze using a numerical model. *Atmosphere and Ocean*, **21**(4), 444–465.

Bobylev, L.P., Kondratyev, K.Ya., and Johannessen, O.M. (eds) (2003). Arctic Environment Variability in the Context of Global Change. Springer-Praxis, Chichester, UK, 471 pp.

Boden, T.A., Kaiser, D.P., Speranski, R.J., and Stoss, F.W. (1994). *TRENDs '93: A Compendium of Data on Global Change* (176 pp.). Carbon Dioxide Informational Analysis Center, Oak Ridge National Laboratory, Oak Ridge, TN, 176 pp.

Bodenbender, J., Wassmann, R., Papen, H., and Rennenberg, H. (1999). Temporal and spatial variation of sulfur-gas-transfer between coastal marine sediments and the atmosphere. *Atmospheric Environment*, **33**(21), 3487–3502.

Boehmer-Christiansen, S.A. (2000). Who determines the policy concerning climate change and how is it done? *Bulletin of the Russian Geographical Society*, **132**(3), 6–22 [in Russian].

Boehmer-Christiansen, S.A. (2002). Keywords investing against climate change: Why failure remains possible. *Environmental Politics*, **11**(3), 1–30.

Boehmer-Christiansen, S.A., and Kellow, A. (2002). *International Environmental Policy: Interests and the Failure of the Kyoto Protocol* (214 pp.). Edward Elgar, Cheltenham, UK.

Bogatyrev, B.G. (1988). Modelling the transition processes on the border of vegetation zones in conditions of anthropogenic climate changes. Ph.D. thesis, Moscow State University, Moscow (114 pp.) [in Russian].

Bojkov, R.D. (1999). International assessment of ozone depletion: 1998. *WMO Bulletin*, **48**(1), 35–44.

Bojkov, R.D. and Visconti, G. (1998). *Atmospheric Ozone: Proceedings of the XVIII Quadrennial Ozone Symposium, 12–21 September 1996, L'Aquila, Italy* (Vol. 1, pp. 450–451). Parco Scientifico e Technologico, Abruzzo, Italy.

Bojkov, R.D., Balis, D., and Zerefos, C.S. (1998). Characteristics of the ozone decline in the Northern polar and middle latitudes during the winter–spring. *Meteorological and Atmospheric Physics*, **2**, 458–462.

Bolin, R. and Sukumar, R. (2000). Global perspective. In: R.T. Watson , I.R. Noble, B. Bolin, N.H. Ravindranath, D.J. Verardo, and D.J. Dokken (eds), *Land Use, Land-Use Change, and Forestry* (pp. 23–51). Cambridge University Press, Cambridge, UK.

Borell, P., Varotsos, C.A., and Kondratyev, K.Ya. (1998). Key problems of chemistry and photochemistry of the troposphere: The completion of the first phase of the EUROTRAC Project and future perspectives. *Ecological Chemistry*, **7**(1), 1–12 [in Russian].

Borisenkov, E.P. and Pichugin, Yu.A. (2001). Possible negative scenarios of biospheric dynamics as a result of anthropogenic activity. *Proceedings of the Russian Academy of Science*, **378**(6), 812–814 [in Russian].

Bornstein, R., Klotz, S., Pechinger, U., Salvador, R., Street, R., Shieh, L.J., Ludwig, F., and Miller, R. (1986). Application of linked three-dimensional PBL and dispersion models to New York City. In: C. De Wispelaere, F.A. Schiermeier, and N.V. Gillani (eds), *Air Pollution Modeling and Its Application* (Vol. 5, pp. 543–564). Plenum Press, New York.

Borodin, L.F., Krapivin, V.F., Golfeld, G.B., and Nazaryan, N.A. (1997). Search and identification of exceptional ecological and technological situations. *Problems of the Environmental and Natural Resources*, **10**, 2–20 [in Russian].

Borodin, L.F., Krapivin, V.F., Berezin, Yu.V., Levshin, I.P., and Tshernikov, A.A. (1998). Ideal and sub-ideal receptors of compound coded phase-manipulated signals. *Achievements of Present Radioelectronics*, **8**, 15–22 [in Russian].

Bortnik, V.N., Lopatina, S.A., and Krapivin, V.V. (1994). Simulation system for the study of hydrophysical fields of the Aral Sea. *Meteorology and Hydrology*, **9**, 102–106 [in Russian].

Boshetti, L., Brivio, P.A., and Gregoire, J.M. (2003). The use of Meteosat and GMS imagery to detect burned areas in tropical environments. *Remote Sensing of the Environment*, **85**(1), 78–91.

Bottenheim, J.W. and Barrie, L.A. (1996). Chemical reactions in the polar troposphere relevant to C, S, and N compounds. *Proceedings of the NATO Advanced Research Workshop: Process of Chemical Exchange between the Atmosphere and Polar Snow, Il Ciocco, Italy, 19–23 March 1995* (pp. 201–224). Springer-Verlag, Berlin.

Boucher, O. (2002). Aerosol radiative forcing and related feedbacks: How do we reduce uncertainties? *IGACtivities Newsletter*, **26**, 8–12.

Bounoua, L., Defries, R., Collatz, G.J., Sellers, P., and Khan, H. (2002). Effects of land cover conversion on surface climate. *Climate Change*, **52**(1–2), 29–64.

Bousquet, P., Peylin, P., Ciais, P., Le Quéré, C., Friedlingstein, P., and Tans, P.P. (2000). Regional changes in carbon dioxide fluxes of land and oceans since 1980. *Science*, **290**(5495), 1342–1346.

Boysen, M. (ed.) (2000). *Biennial Report 1998 & 1999* (130 pp.). Potsdam Institute for Climate Impact Research, Potsdam, Germany.

Bras, R.L. (1990). *Hydrology* (643 pp.). Addison-Wesley, New York.

Brasseur, G.P., Hauglustaine, D.A., Walters, S., Rasch, P.J., Muller, J.-F., Granier, C., and Tie, X.X. (1998). MOZART: A global chemical transport model for ozone and related chemical tracers, Part 1: Model description. *Journal of Geophysics Research*, **103**(D21), 28265–28290.

Brasseur, G.P., Prinn, R.G., and Pszenny, A.A.P. (eds) (2002). *Atmospheric Chemistry in a Changing World* (330 pp.). Springer-Verlag, Berlin.

Braswell, B.H., Schimel, D.S., Privette, J.L., Moore, B., Emery, W.J., Sulzman, E.W., and Hudak, A.T. (1996). Extracting ecological and biophysical information from AVHRR optical data: An integrated algorithm based on inverse modeling. *Journal of Geophysics Research*, **101**(D18), 23335–23348.

Braswell, B.H., Schimel, D.S., Linder, E., and Moore, B. (1997). The response of global terrestrial ecosystems to interannual temperature variability. *Science*, **278**, 870–872.

Bratus, A., Mescherin, A., and Novozhilov, A. (2001) Mathematical models of interaction between pollutant and environment. *Herald of the Moscow State University: Computational Mathematics and Cybernetics*, **15**(1), 23–28 [in Russian].

Braunlich, M., Aballain, O., Marik, T., Jockel, P., Brenninkmeijer, C.A.M., Chappellaz, J., Barnola, J.-M., Mulvaney, R., and Sturges, W.T. (2001). Changes in the global atmospheric methane budget over the last decades inferred from ^{13}C and D isotopic analysis of Antarctic firn air. *Journal of Geophysics Research*, **106**(D17), 20465–20482.

Brenninkmeijer, C.A.M. and Röckmann, T. (1998). Using isotope analysis to improve atmospheric CO budget calculations. *Proceedings of the International Symposium "Isotopes Technical Study into Past and Current Environmental Changes in the Hydrosphere and the Atmosphere, Vienna, 14–18 April 1997* (pp. 69–77). IIASA Press, Vienna.

Brenninkmeijer, C.A.M. and Röckmann, T. (1999). Anomalous (or not strictly mass dependent) isotope variations observed in important atmospheric trace gases. *IGACtivities Newsletter*, **16**, 13–18.

Breon, F.-M., Tanre, D., and Generoso, S. (2002). Aerosol effect on cloud droplet size monitored from satellite. *Science*, **295**(5556), 834–838.

Brombach, T., Hunziker, J.C., Chiodini, G., Cardellini, C., and Marini, L. (2001). Soil diffuse degassing and thermal energy fluxes from the southern Lakke Plain, Nisyros (Greece). *Geophysics Research Letters*, **28**(1), 69–72.

Bryden, H.L., McDonagh, E.L., and King, B.A. (2003). Changes in ocean water mass properties: Oscillations or trends? *Science*, **300**(5628), 2086–2088.

Bui Ta Long (1998). Data processing in the geoinformation monitoring systems in Vietnam. PhD thesis, Institute of Radioengineering and Electronics, Moscow (354 pp.) [in Russian].

Bukatova, I.L. and Makrusev, V.V. (2003). *Present Informational Technologies of Management* (252 pp.). Russian Custom Academy, Moscow.

Bukatova, I.L., Michasev, Yu.I., and Sharov, A.M. (1991). *Ecoinformatics: Theory and Experience of Evolutional Modelling* (205 pp.). Science Press, Moscow [in Russian].

Burroughs, W.J. (2001). *Climate Change: A Multidisciplinary Approach* (298 pp.). Cambridge University Press, Cambridge, UK.

Byakola, T. (2000). Technological options and policy measures for methane mitigation in Uganda – Possibilities and limitations. *2nd International Methane Mitigation Conference, 18–23 June 2000* (pp. 95–100). Novosibirsk University Press, Novosibirsk.

Byutner, E.K. (1986). *Planetary Gas Exchange* (240 pp.). Hydrometeoizdat, Leningrad [in Russian].

Cai, W. and Whellton, P.H. (2002). Evidence for a time-varying pattern of greenhouse warming in the Pacific Ocean. *Geophysics Research Letters*, **27**(16), 2577–2580.

Caldeira, K. and Duffy, P.B. (2000). The role of the Southern Ocean in uptake and storage of anthropogenic carbon dioxide. *Science*, **287**, 620–622.

Cannel, M.G.R., Milne, R., Hargreaves, K.J., Brown, T.A.W., Cruickshank, M.M., Bradley, R.I., Hope, D., Billett, M.F., Adger, W.N., and Subak, S. (1999.) National inventories of terrestrial carbon sources and sinks: The UK experience. *Climatic Change*, **42**(2), 505–530.

Carvalho, J.A., Jr, Costa, F.S., Gurgel Veras, C.A., Sandberg, D.V., Alvarado, E.C., Gielow, R., Serra, A.M., Jr, and Santos, J.C. (2001). Biomass fire consumption and carbon release rates of rainforest-clearing experiments conducted in northern Mato Grosso, Brazil. *Journal of Geophysics Research*, **106**(16), 17877–17887.

Caspersen, J.P., Pacala, S.W., Jenkins, J.C., Hurtt, G.C., Moorcroft, P.R., and Birdsey, R.A. (2000). Contributions of land-use history to carbon accumulation in US forests. *Science*, **290**(5494), 1148–1151.

Chahine, M.T. (1992). The hydrological cycle and its influence on climate. *Nature (UK)*, **359**(6394), 373–380.

Chandra, S., Varotsos, C., and Flynn, L.E. (1996). The mid-latitude ozone trends in the Northern Hemisphere. *Geophysics Research Letters*, **23**(5), 555–558.

Chanin, M.-L. (2001). Reports of the eighth session of the SPARC Scientific Steering Group and the SPARC 2000 General Assembly. *SPARC Newsletter*, **16**, 1–10.

Charette, M.A., Moran, S.B., Pike, S.M., and Smith, J.N. (2001). Investigating the carbon cycle in the Gulf of Main using the natural tracer thorium 234. *Journal of Geophysics Research*, **106**(C6), 11553–11580.

Charlton, A.J., O'Neill, A., Lahoz, W.A., Berrisford, P., van Velthoven, P., Eskes, H., and Kelder, H. (2003) The splitting of the stratospheric polar vortex in the southern hemisphere in September 2002: Dynamical evolution and impacts. *J. Atmos. Sci.* (submitted).

Chase, T.N., Pielke, R.A., Sr, Kittel, T.G.F., Nemani, R.R., and Running, S.W. (2000). Simulated impacts of historical land cover changes on global climate in northern winter. *Climate Dynamics*, **16**, 93–105.

Chase, T.N., Herman, B., Pielke, R.A., Sr, Zeng, X., and Leuthold, M. (2002). A proposed mechanism for the regulation of minimum midtropospheric temperatures in the Arctic. *Journal of Geophysics Research*, **107**(D14), ACU2/1–ACU2/7.

Chate, D.M. and Kamra, A.K. (1997). Collection efficiencies of large water drops collecting aerosol particles of various densities. *Atmospheric Environment*, **31**(11), 1631–1635.

Chen, J.M., Leblanc, S.G., Cihlar, J., Desjardins, R.L., and MacPherson, J.I. (1999). Extending aircraft- and tower-based CO_2 flux measurements to a boreal region using a Landsat thematic mapper land cover map. *Journal of Geophysics Research*, **104**(D14), 16859–16878.

Chen, J.M., Liu, J., Leblanc, S.G., Lacaze, R., and Roujean, J.-L. (2003). Multi-angular optical remote sensing for assessing vegetation structure and carbon absorption. *Remote Sensing of the Environment*, **84**(4), 516–525.

Chen, W., Chen, J., and Cihlar, J. (2000). An integrated terrestrial ecosystem carbon-budget model based on changes in disturbance, climate, and atmospheric chemistry. *Ecological Modelling*, **135**(1), 55–79.

Chisholm, S.W., Falkowski, P.G., and Cullen, J.J. (2001). Dis-crediting ocean fertilization. *Science*, **294**(5541), 309–311.

Chobadian, A., Goddard, A.J.H., and Gosman, A.D. (1985). Numerical simulation of coastal internal boundary layer developments and a comparison with simole models. In: C. De Wispelaere (ed.), *Air Pollution Modeling and Its Application* (Vol. 4, pp. 343–358). Plenum Press, New York.

Chock, D.P. and Winkler, S.L. (2000). A trajectory-grid approach for solving the condensation and evaporation equations of aerosols. *Atmospheric Environment*, **34**(18), 2957–2973.

Chou, M.-D., Chan, P.-K., and Wang, M. (2002). Aerosol radiative forcing derived from SeaWiFS-retrieved aerosol optical properties. *Journal of Atmospheric Science*, **59**(3), 748–757.

Choudhury, B. and Kukla, G. (1979). Impact of CO_2 on cooling of snow and water surfaces. *Nature*, **280**(5724), 668–671.

Christian, H.J., Blakeslee, R.J., Boccippio, D.J., Boeck, W.L., Buechler, D.E., Driscoll, K.T., Goodman S.J., Hall, J.M., Koshak, W.J., Mach, D.M. *et al.* (2003). Global frequency and distribution of lightning as observed from space by the Optical Transient Detector. *Journal of Geophysics Research*, **108**(D1), 4/1–4/15.

Christopher, S.A. and Zhang, J. (2002). Daytime variation of shortwave direct radiative forcing of biomass burning aerosols from GOES imager. *Journal of Atmospheric Science*, **59**(3), Part 2, 681–691.

Christy, J.R. and Spencer, R.W. (2003). Reliability of satellite data sets. *Science*, **301**(5636), 1046–1047.

Christy, J.R., Spencer, R.W., and Lobl, E.S. (1998). Analysis of the merging procedure for the MSU daily temperature time series. *Journal of Climate*, **11**, 2016–2041.

Christy, J.R., Spencer, R.W., Norris, W.B., Braswell, W.D., and Parker, D.E. (2003). Error estimates of Version 5.0 of MSU-AMSU bulk atmospheric temperatures. *Journal of Atmospheric and Oceanic Technology*, **20**, 613–629.

Ciais, P. (1999a). Isotopomers of CO_2 and their use in understanding the global carbon cycle. *IGACtivities Newsletter*, **16**, 18–19.

Ciais, P. (1999b). Restless carbon pools. *Nature*, **398**(6723), 111–112.

Ciais, P. (2002). Taking the pulse of the carbon cycle. *UHDP Update*, **2**, 11–12.

Cihlar, J., Denning, S., Ahern, F., Arino, O., Belward, A., Bretherton, F., Cramer, W., Dedieu, G., Field, C., Francey, R. *et al.* (2002). Initiative to quantify terrestrial carbon sources and sinks. *Earth Observation from Space Transactions*, **83**(1), 6–7.

Clark, M.P., Serreze, M.C., and Barry, R.G. (1996). Characteristics of Arctic Ocean climate based on COADS data, 1980–1993. *Geophysics Research Letters*, **23**(15), 1953–1956.

Clerbaux, C., Hadji-Lazaro, J., Hauglustaine, D., Megie, G., Khattatov, B., and Lamarque, J.-F. (2001). Assimilation of carbon monoxide measured from satellite in a three-dimensional chemistry-transport model. *Journal of Geophysics Research*, **106**(14), 15385–15394.

Coakley, J.A., Jr and Walsh, C.D. (2002). Limits to the aerosol indirect radiative effect derived from observations of ship tracks. *Journal of Atmospheric Science*, **59**(3), Part 2, 668–680.

Coconea, G. (2000). Methane gas emissions from the Romanian natural gas transport system. *2nd International Methane Mitigation Conference, 18–23 June 2000, Novosibirsk* (pp. 297–302). Novosibirsk University Press, Novosibirsk.

Cohard, J.-M., Pinty, J.-P., and Suhre, K. (2001). On the parametrization of activation spectra from cloud condensation nuclei microphysical properties. *Journal of Geophysics Research*, **105**(D9), 11753–11766.

Collatz, G.J., Berry, J.A., Farquhar, J.A., and Pierce, J. (1990). The relationship between the rubisco reaction mechanism and models of leaf photosynthesis. *Plant Cell Environment*, **13**, 219–225.

Collatz, G.J., Ball, G.J., Grivet, J.T., and Berry, J.A. (1991). Physiological and environmental regulation of stomatal conductance, photosynthesis and transpiration: A model that includes a laminar boundary layer. *Agriculture Forestry Meteorology*, **54**, 107–136.

Collatz, G.J., Ribas-Carbo, M., and Berry, J.A. (1992). Coupled photosynthesis stomatal conductance model for leaves of C_4 plants. *Australian Journal of Plant Physiology*, **19**, 519–538.

Collatz, G.J., Bounoua, L., and Los, S.O. (2000). A mechanism for the influence of vegetation on the response of the diurnal temperature range to changing climate. *Geophysics Research Letters*, **27**(20), 3381–3384.

Connors, V.S., Gormsen, B.B., Nolf, S., and Reichle, H.G., Jr (1999). Spaceborne observations of the global distribution of carbon monoxide in the middle troposphere during April and October 1994. *Journal of Geophysics Research*, **104**(D17), 21455–21470.

Cox, P., Friedlingstein, P., and Rayner, P. (2002). Modelling climate – carbon cycle feedbacks: A cross disciplinary collaboration priority. *Global Change Newsletter*, **49**, 12–14.

Cox, P.M., Betts, R.A., Jones, C.D., Spall, S.A., and Totterdell, I.J. (2000). Acceleration of global warming due to carbon-cycle feedbacks in a coupled climate model. *Nature*, **408**, 184–187.

Cox, R.A. (1999). Ozone and peroxy radical budgets in the marine boundary layer: Modeling the effect of NO_X. *Journal of Geophysics Research*, **104**(D7), 8047–8056.

Coyle, M., Fowler, D., and Ashmore, M. (2003). New directions: Implications of increasing tropospheric background ozone concentrations for vegetation. *Atmospheric Environment*, **37**(1), 153–154.

Craig, S.G., Holmén, K.J., Bonan, G.B., and Rasch, P.J. (1998). Atmospheric CO_2 simulated by the National Center for Atmospheric Research Community Climate Model, 1: Mean fields and seasonal cycles. *Journal of Geophysics Research*, **103**(D11), 13213–13235.

Cramer, W., Bondeau, A., Woodward, F.I., Prentice, I.C., Betts, R.A., Brovkin, V., Cox, P.M., Fisher, V., Foley, J.A., Friend, A.D. *et al.* (2001). Global response of terrestrial ecosystem structure and function to CO_2 and climate change: Results from six dynamic global vegetation models. *Global Change Biology*, **7**, 357–373.

Crane, K., Galasso, J., Brown, C., Cherkashov, G., Ivanov, G., Petrova, V., and Vanstayan, B. (2001). Northern ocean inventories of organochlorine and heavy metal contamination. *Marine Pollution Bulletin*, **43**(1–6), 28–60.

Crutzen, P.S. (1970). The influence of nitrogen oxides on the atmospheric ozone content. *Quarterly Journal of the Royal Meteorological Society*, **96**, 320–325.

Crutzen, P.S. (1998). The Bulletin interview. *WMO Bulletin*, **47**(2), 111–123.

Crutzen, P.S. and Ramanathan, V. (2000). The ascent of atmospheric sciences. *Science*, **290**(5490), 299–304.

Curry, J.A., Schramm, J.L., and Ebert, E.E. (1993). Impact of clouds on the surface radiation budget of the Arctic Ocean. *Meteorogical and Atmospheric Physics*, **57**, 197–217.

Curry, J.A., Rossow, W.B., Randall, D., and Schramm, J.L. (1996). Overview of Arctic cloud and radiation properties. *Journal of Climate*, **9**, 1731–1764.

Curry, J.A., Hobbs, P.V., King, M.D., Randall, D.A., Minnis, P., Isaac, G.A., Pinto, J.O., Uttal, T., Bucholtz, A., Cripe, D.G. *et al.* (2000). FIRE Arctic Clouds Experiment. *Bulletin of the American Meteorological Society*, **81**(1), 5–29.

Danilov-Danilyan, V.I. and Losev, K.S. (2000). *Ecological Challenge and Sustainable Development* (416 pp.). Progress, Moscow [in Russian].

De Laat, A.T.J., Zachariasse, M., Roelofs, G.J., van Velthoven, P., Dickerson, R.R., Rhoads, K.P., Oltmans, S.J., and Lelieveld, J. (1999). Tropospheric O_3 distribution over the Indian Ocean during spring 1995 evaluated with a chemistry-climate model. *Journal of Geophysics Research*, **104**(11), 13881–13895.

Del Frate, F., Ferrazzoli, P., and Schiavon, G. (2003.) Retrieving soil moisture and agricultural variables by microwave radiometry using neural networks. *Remote Sensing of the Environment*, **84**(2), 174–183, 214–233.

De Lucia, E.H., Hamilton, J.G., Naidu, S.L., Thomas, R.B., Andrews, J.A., Finzi, A., Lavine, M., Matamala, R., Mohan, J.E., Hendrey, G.R., and Schlesinger W.H. (1999). Net primary production of a forest ecosystem with experimental CO_2 enrichment. *Science*, **284**(5417), 1177–1179.

Delmas, R.J. and Legrand, M. (1998). Trends recorded in Greenland in relation with Northern Hemisphere anthropogenic pollution. *IGACtivities Newsletter*, **14**, 19–22.

Dementjeva, T.V. (2000). Emission of gases from peat-bog ecosystems. *Second International Methane Mitigation Conference, 18–23 June 2000, Novosibirsk* (pp. 223–226). Novosibirsk University Press, Novosibirsk.

Demirchian, K.S. and Kondratyev, K.Ya. (1998). Development of energetics and the environment. *Bulletin of the Russian Academy of Science: Energetics*, **6**, 3–27 [in Russian].

Demirchian, K.S. and Kondratyev, K.Ya. (1999). Scientific validity of predicted impacts of energetics on climate. *Bulletin of the Russian Academy of Science: Energetics*, **6**, 3–46 [in Russian].

Demirchian, K.S., Demirchian, K.K., Danilevich, Ya.B., and Kondratyev, K.Ya. (2002). Global climate warming, energetics and geopolicy. *Annals of Russian Academy of Science: Energetics*, **3**, 221–235 [in Russian].

De Rosnay, P., Bruen, M., and Polcher, J. (2000). Sensitivity of surface fluxes to the number of layers in the soil model used in GCMs. *Geophysics Research Letters*, **27**(20), 3329–3332.

Derwent, R.G., Collins, W.J., Johnson, C.E., and Stevenson, D.S. (2001). Transient behaviour of tropospheric ozone precursor in a global 3D CTM and their indirect greenhouse effects. *Climate Change*, **49**(4), 463–487.

Despres, A., Rancillac, F., and Bouville, A. (1986). First results of the data processing of the Vist European compaign on remote-sensing of air pollution. In: C. De Wispelaere, F.A. Schiermeier, and N.V. Gillani (eds) *Air Pollution Modeling and Its Application* (Vol. 5, pp. 371–382). Plenum Press, New York.

Dickson, R. (1999). All change in the Arctic. *Nature*, **397**, 389–391.

Dimitroulopoulou, C. and Marsh, A.R.W. (1997). Modelling studies of NO_3 night-time chemistry and its effects on subsequent ozone formation. *Atmospheric Environment*, **31**(18), 3041–3057.

Dolman, H., Nabuurs, G.J., Kuikman, P., Kruijt, B., and Brinkman, S. (2001). *Terrestrial Carbon Sinks and the Kyoto Protocol: The Scientific Issues* (Global Change Report No. 410 200 068, 31 pp.). Dutch National Research Programme on Global Air Pollution and Climate Change, Copenhagen.

Dong, J., Kaufmann, R.K., Myneni, R.B., Tucker, C.J., Kauppi, P.E., Liski, J., Buermann, W., Alexeyev, V., and Hughes, M.K. (2003). Remote sensing estimates of boreal and temperate forests woody biomass: Carbon pools, sources, and sinks. *Remote Sensing of the Environment*, **84**(3), 393–410.

Dong, X., Mace, G.G., Minnis, P., and Young, D.F. (2001). Arctic stratus cloud properties and their effect on the surface radiation budget: Selected cases from FIRE ACE. *Journal of Geophysics Research*, **106**(D14), 15297–15312.

Doskey, P.V. and Gaffney, J.S. (1992). Non-methane hydrocarbons in the arctic atmosphere at Barrow, Alaska. *Geophysics Research Letters*, **19**(4), 381–384.

Eastman, J.L., Coughenour, M.B., and Pielke, R.A., Sr (2001). The regional effects of COg and landscape change using a coupled plant and meteorological model. *Global Change Biology*, **7**, 797–815.

Egan, W.G., Hogan, A.W., and Zhu, H. (1991). Physical variation of water vapor, and the relation with carbon dioxide. *Geophysics Research Letters*, **18**(12), 2245–2248.

Elderfield, H. (2002). Carbonate mysteries. *Science*, **296**(5573), 1618–1621.

Eliasson, B., Riemer, P., and Wokaun, A. (1999). *Greenhouse Gas Control Technologies*. Elsevier, Amsterdam, 1095 pp.

Elliott, S., Lackner, K.S., Ziock, H.J., Dubley, M.K., Hanson, H.P., Barr, S., Ciczkowski, N.A., and Blake, D.R. (2001). Compensation of atmospheric COg buildup through engineered chemical sinkage. *Geophysics Research Letters*, **28**(7), 1235–1238.

Ellsaesser, H.W. (ed.) (1992). *Global 2000 Revisited: Mankind's Impact on Spaceship Earth* (436 pp.). Paragon House, New York.

Ellsaesser, H.W. (ed.) (1994). The unheard arguments: A rational view of stratospheric ozone. *21st Century Science and Technology*, **7**(23), 37–45.

Ellsaesser, H.W. (2002). The current status of global warming. *Energy and Environment*, **13**(1), 125–129.

Engelen, R.J., Denning, A.S., Gurney, K.R., and Stephens, G.L. (2001). Global observations of the carbon budget, 1: Expected satellite capabilities for emission spectroscopy in the EOS and NPOESS eras. *Journal of Geophysics Research*, **106**(D17), 20055–20068.

Ericksen, J.A., Schorran, D.E., Johnson, D.W., Lindberg, S.E., and Coleman, J.S. (2003). Accumulation of atmospheric mercury in forest foliage. *Atmospheric Environment*, **37**(12), 1613–1622.

Eriksson, E. (1963). The yearly circulation of sulfur in nature. *Journal of Geophysics Research*, **68**, 401–403.

Ernst W.G. (ed.) (2000). *Earth Systems: Processes and Issues* (566 pp.). Cambridge University Press, Cambridge, UK.

Ervin, D., Kahn, J.R., and Livingston, M.L. (eds) (2002). *Does Environmental Policy Work? The Theory and Practice of Outcomes Assessments* (256 pp.). Edward Elgar, Cheltenham, UK.

Eskes, H., Segers, A., and van Velthoven, P. (2003) Ozone forecasts of the stratospheric polar vortex splitting event in September 2002. *J. Atmos. Sci.* (submitted).

Essenhigh, R.H. (2001). Does CO_2 really drive global warming? *Chemistry and Innovation*, **31**(5), 44–46.

Esteve, P. (1998). Tropical dry ecosystems modelling and monitoring from space. *Ecological Modelling*, **108**(1–3), 175–188.

Everett, J.T. and Fitzharris, B.B. (eds) (2001). The Arctic and Antarctic. *The Regional Impacts of Climate Change* (Chapter 3, IPCC Report) (`http://www.grida.no/climate/ipcc/regional/042.htm`).

Falkowski, P., Scholes, R.J., Boyle, E., Canadell, D., Elser, J., Gruber, N., Hibbard, K., Högberg, P., Linder, S., Mackenzie, F.T. *et al.* (2000). The global carbon cycle: A test of our knowledge of Earth as a system. *Science*, **290**, 291–296.

Faluvegi, G.S., Alapaty, K., Reichle, H.G., Jr, Mathur, R., Raman, S., and Connors, V.S. (1999). Simulation of carbon monoxide transport during April 1994. *Journal of Geophysics Research*, **104**(D17), 21471–21485.

Fan, S., Gloor, M., Mahlman, J., Pacala, S., Sarmiento, J., Takahashi, T., and Trans, P. (1998). A large terrestrial carbon sink in North America implied by atmospheric and oceanic carbon dioxide data and models. *Science*, **282**, 442–446.

Fan, S.-M., Wofsy, S.C., Bakwin, P.S., Jacob, D.J., Anderson, S.M., Kebabian, P.L., McManus, J.B., Kolb, C.E., and Fitzjarrald, D.R. (1992). Micrometeorological measurements of CH_4 and CO_2 exchange between the atmosphere and subarctic tundra. *Journal of Geophysics Research*, **97**(D15), 16627–16644.

Fan, S.-M., Sarmiento, J.L., Gloor, M., and Pacala, S.W. (1999). On the use of regularization techniques in the inverse modeling of atmospheric carbon dioxide. *Journal of Geophysics Research*, **104**(D17), 21503–21512.

Fang, J., Chen, A., Peng, C., Zhao, S., and Ci, L. (2001). Changes in forest biomass carbon storage in China between 1949 and 1998. *Science*, **292**(5525), 2320–2322.

Farquhar, G.D., von Caemmerer, S., and Berry, J.A. (1980). A biochemical model of photosynthetic CO_2 fixation in leaves of C_3 species. *Planta*, **149**, 78–90.

Feingold, G., Remer, L., Ramaprasad, J., and Kaufman, Y.J. (2001). Analysis of smoke impact on clouds in Brazilian biomass burning region: An extension of Twomey's approach. *Journal of Geophysics Research*, **106**(D19), 22907–22922.

Feister, U. and Grewe, R. (1995). Higher UV radiation inferred from low ozone levels at northern mid-latitudes in 1992 and 1993. *Global and Planetary Change*, **11**(1), 25–34.

Feng, W., Chipperfield, M.P., Roscoe, H.K., Remedios, J.J., and Waterfall, A.M. (2003) Three-dimensional model study of the Antarctic ozone hole in 2002 and comparison with 2000. *J. Atmos. Sci.* (submitted).

Fenn, M.A., Browell, E.V., Butler, C.F., Grant, W.B., Kooi, S.A., Clayton, M.B., Gregory, G.L., Newell, R.E., Zhu, Y., Dibb, J.E. *et al.* (1999). Ozone and aerosol distributions and air mass characteristics over the South Pacific during the burning season. *Journal of Geophysics Research*, **104**(D13), 16197–16212.

Fergusson, A. and Wardle, D.I. (1998). *Arctic Ozone: The Sensitivity of the Ozone Layer to Chemical Depletion and Climate Change* (27 pp.). Minister of Public Works and Government Services, Toronto.

Ferm, M. and Hultberg, H. (1999). Dry deposition and internal circulation of nitrogen, sulfur and base cations to a coniferous forest. *Atmospheric Environment*, **33**(27), 4421–4430.

Fichmann, K.-U., Bramstedt, K., Weber, M., Rosanov, V.V., Hoogen, R., and Burrows, J.P. (1999). O_3 profiles from GOME satellite data-II: Observations in the Arctic spring 1997 and 1998. *Physics and Chemistry on Earth (C)*, **24**(5), 453–457.

Field, C.B. (2001). Sharing the garden. *Science*, **294**(5551), 2490–2491.

Field, C.B. and Fung, I.Y. (1999). Biogeochemical cycles: The not-so-big US carbon sink. *Science*, **285**(5427), 544–545.

Fioletov, V.E., Kerr, J.B., Hare, E.W., Labow, G.J., and McPeters, R.D. (1999). An assessment of the world ground-based total ozone network performance from the comparison with satellite data. *Journal of Geophysics Research*, **104**(D1), 1737–1748.

Flossman, A.I. (1998). Interaction of particles and clouds. *Journal of Atmospheric Science*, **55**(3), 879–887.

Folland, C.K., Karl, T.R., and Salinger, M.J. (2002). Observed climate variability and change. *Weather*, **57**(3), 269–278.

Fonteyn, D., Berrisford, P., Bonjean, S., Chabrillat, S., Charlton, A., Daerden, F., Errera, Q., and Lahoz, W. (2003) Chemical and dynamical data assimilation for the Antarctic winter 2002. *J. Atmos. Sci.* (submitted).

Forest, C.E., Stone, P.H., Sokolov, A.P., Allen, M.R., and Webster, M.D. (2002). Quantifying uncertainties in climate system properties with the use of recent climate observations. *Science*, **295**(5552), 113.

Forrester, J.W. (1971). *World Dynamics* (145 pp.). Wright-Allen Press, Cambridge, MA.

Foster, K.L., Plastridge, R.A., Bottenheim, J.W., Shepson, P.B., Finlayson-Pitts, B.J., and Spicer, P.W. (2001). The role of Br_2 and BrCl in surface ozone destruction at polar sunrise. *Science*, **291**, 472–474.

Frederick, K.D. (1994). Integrated assessment of the impacts of climate change on natural resources: An introductory essay. *Climate Change*, **28**(102), 1–14.

Fridborg, T., Hansen, B.U., Mordstroem, C., and Soegaard, H. (1999). Controls of the greenhouse gas exchange of a high-arctic ecosystem. *Geografiska Tidsskrifter*, **99**, 19–26.

Friedlingstein, P., Bopp, L., Ciais, P., Dufresne, J.L., Fairhead, L., LeTreut, H., Monfray, P., and Orr, J. (2001). Positive feedback between future climate change and the carbon cycle. *Geophysical Research Letters*, **28**, 1543–1546.

Friend, A.D. (1998). Parametrization of a global daily weather generator for terrestrial ecosystem modelling. *Ecological Modelling*, **109**(2), 121–140.

Fung, I. (2000). Variable carbon sinks. *Science*, **290**(5495), 1313–1314.

Gale, J. and Freund, P. (2000). Reducing methane emissions to combat global climate change: The role Russia can play. *2nd International Methane Mitigation Conference, 18–23 June 2000, Novosibirsk* (pp. 73–80). Novosibirsk University Press, Novosibirsk.

Galitskiy, V.V. (1985). Horizontal structure and dynamics of a single-age plant community: Numerical modelling. In: Yu. A. Izrael (ed.), *Mathematical Modelling of the Biogeocenotic Processes* (pp. 59–69). Science Press, Moscow [in Russian].

Gbondo-Tugbawa, S.S. and Driscoll, C.T. (2002). Evaluation of the effect of future controls on sulfur dioxide and nitrogen oxide emissions on the acid-base status of a northern forest ecosystem. *Atmospheric Environment*, **36**(10), 1631–1643.

Gedney, N. and Valdes, P.J. (2000). The effect of Amazonian deforestation on the northern hemisphere circulation and climate. *Geophysics Research Letters*, **27**(19), 3053–3056.

Germeyer, Yu.B. (1974). *Games with Non-antagonistic Interests* (289 pp.). Science Press, Moscow [in Russian].

Gerstengarbe, F.-W. (2002). *Angewandte Statistik* (PIK Report No. 75, 100 pp.). Potsdam Institute for Climate Impact Research, Potsdam, Germany [in German].

Gildor, H. and Tziperman, E. (2001). Physical mechanisms behind biogeochemical glacial–interglacial CO_2 variations. *Geophysics Research Letters*, **28**(12), 2421–2424.

Gillett, N.P., Allen, M.R., McDonald, R.E., Senior, G.A., Shindell, D.T., and Schmidt, G.A. (2002). How linear is the Arctic Oscillation response to greenhouse gases? *Journal of Geophysics Research*, **107**(D3), ACL 1/1–ACL 1/7.

Gillon, J. and Yakir, D. (2001). Influence of carbonic anhydrase activity in terrestrial vegetation on the ^{18}O content of atmospheric CO_2. *Science*, **291**(5513), 2584–2587.

Glantz, M.H. (2003). *Climate Affairs: A Primer* (184 pp.). Island Press, Boulder, CO.

Gleick, P.H. (1993). *A Guide to the World's Fresh Water Resources* (473 pp.). Oxford Scientific, Oxford, UK.

Gloor, M., Bakwin, P., Hurst, D., Lock, L., Draxler, R., and Tans, P. (2001). What is the concentration footprint of a tall tower? *Journal of Geophysics Research*, **106**(16), 17831–17840.

Goddéris, Y., François, L.M., and Veizer, J. (2001). The early Paleozoic carbon cycle. *Earth and Planetary Science Letters*, **190**(3–4), 181–196.

Goldstein, M.I., Corson, M.S., Lacher, T.E., and Grant, W.E. (2003). Managed forests and migratory bird populations: Evaluating spatial configurations through simulation. *Ecological Modelling*, **162**(1–2), 155–175.

Goody, R. (2002). Observing and thinking about the atmosphere. *Annual Review of Energy and the Environment*, **27**, 1–20.

Goody, R., Anderson, J., and North, G. (1998). Testing climate models: An approach. *Bulletin of the American Meteorological Society*, **79**, 2541–2549.

Gorshkov, V.G. (1990). *Biospheric Dynamics and Environmental Sustainability: Scientific and Technical Results: Theoretical and General Problems of Geography* (237 pp.). VINITI Press, Moscow [in Russian].

Gorshkov, V.G. (1995). *Physical and Biological Basis of Life Stability* (472 pp.). VINITI Press, Moscow [in Russian].

Gorshkov, V.G. and Makarieva, A.M. (1999). Impact of virgin and domesticated biota on the global environment. *Earth Research from Space*, **5**, 3–11 [in Russian].

Gorshkov, V.G., Gorshkov, V.V., and Makarieva, A.M. (2000). *Biotic Regulation of the Environment* (364 pp.). Springer-Praxis, Chichester, UK.

Gorshkov, V.G., Makarieva, A., Mackey, B., and Gorshkov, V.V. (2002). Biological theory and global change science. *Global Change Newsletter*, **48**, 11–14.

Gorstko, A.B. (1979). *Mathematical Model of the Azov Sea Ecosystem* (64 pp.). Znanie Press, Moscow [in Russian].

Gorstko, A.B. and Ougolnitskiy, G.A. (1990). *Introduction to the Modelling of Ecologo-economic Systems* (111 pp.). Rostov on Don State University Press, Rostov on Don, Russia [in Russian].

Goudrian, J., van Keulen, H., and van Laar, H.H. (eds) (1990). *The Greenhouse Effect and Primary Productivity in European Agro-ecosystems* (90 pp.). Pudoc, Wageningen, The Netherlands.

Grant, R.F., Gouldren, M.L., Wofsy, S.C., and Berry, J.A. (2001). Carbon and energy exchange by a black spruce–moss ecosystem under changing climate: Testing the mathematical model ECOSYS with data from the BOREAS experiment. *Journal of Geophysics Research*, **106**(D24), 33605–33621.

Grigoryev, Al.A. and Kondratyev, K.Ya. (2001a). *Ecological Catastrophes* (661 pp.). The St Petersburg Scientific Centre of the Russian Academy of Science, St Petersburg [in Russian].

Grigoryev, Al.A. and Kondratyev, K.Ya. (2001b). *Ecological Disasters* (206 pp.). The St Petersburg Scientific Centre of the Russian Academy of Science, St Petersburg [in Russian].

Grogan, P., Illeris, L., Michelsen, A., and Jonasson, S. (2001). Respiration of recently-fixed plant carbon dominates mid-winter ecosystem CO_2 production in subarctic heath tundra. *Climate Change*, **50**(1–2), 129–142.

Grooss, J.-U., Konopka, P., and Mueller, R. (2003) Ozone chemistry during the 2002 Antarctic vortex split. *J. Atmos. Sci.* (submitted).

Grossman, D. (2001). Dissent in the maelstrom. *Scientific American*, **285**(5), 38–39.

Guaduong, F.Q.C. and Masao, M. (2001). The carbon cycle of sandy lands in China and its global significance. *Climate Change*, **48**(4), 535–549.

Guirlet, M., Kibane-Dawe, I., and Harris, N. (1998). Third European Stratospheric Experiment on Ozone (THESEO). *SPARC Newsletter*, **10**, 24–25.

Guirlet, M., Chipperfield, M.P., Pyle, J.A., Goutail, F., Pommereau, J.P., and Kyrö, E. (2000). Modelled Arctic ozone depletion in winter 1997/1998 and comparison with previous winters. *Journal of Geophysics Research*, **105**(D17), 22185–22200.

Gyalistras, D. (2002). *How Uncertain Are Regional Climate Change Scenarios? Examples for Europe and the Alps* (Report No. 75, pp. 85–100). Potsdam Institute for Climate Research, Potsdam.

Haan, D., Zuo, Y., Gros, V., and Brenninkmeijer, C.A.M. (2001). Photochemical production of carbon monoxide in snow. *Journal of Atmospheric Science*, **40**(3), 217–230.

Habib, S. (2000). Technology thrusts for future Earth science applications. *Proceedings of the SPIE*, **2150**, 124–137.

Hackler, J.L., Houghton, R.A., and Lawrence, K.T. (1999). The US carbon budget: Contributions from land-use change. *Science*, **285**(5427), 574–577.

Hadley, A. and Toumi, R. (2002). A simple model which predicts some non-linear features between atmospheric sulphur and sulphur emissions. *Environmental Pollution*, **119**(3), 365–374.

Hadley, A. and Toumi, R. (2003). Assessing changes to the probability distribution of sulphur dioxide in the UK using a lognormal model. *Atmospheric Environment*, **37**(11), 1461–1474.

Hakola, H., Tarvainen, V., Laurila, T., Hiltunen, V., Hellen, H., and Keronen, P. (2003). Seasonal variation of VOC concentrations above a boreal coniferous forest. *Atmospheric Environment*, **37**(12), 1623–1634.

Hales, J. (1996). Scientific background for AMS Policy Statement on Atmospheric Ozone. *Bulletin of the American Meteorological Society*, **77**(6), 1245–1253.

Han, Q., Rossow, W.B., Zeng, J., and Welch, R. (2002). Three different behaviors of liquid water path of water clouds in aerosol–cloud interactions. *Journal of Atmospheric Science*, **59**(3), 726–735.

Hanna, E. and Braithwaite, R. (2003). The Greenland ice sheet: A global warming signal? *Weather*, **58**(9), 351–357.

Hansell, D.A. and Carlson, C.A. (eds) (2002). *Biogeochemistry of Marine Dissolved Organic Matter* (469 pp.). Elsevier, Amsterdam.

Hansen, G. and Chipperfield, M.P. (1999). Ozone depletion at the edge of the Arctic polar vortex 1996/1997. *Journal of Geophysics Research*, **104**(D1), 1846–1897.

Hansen, J.E. and Sato, M. (2001). Trends of measured climate forcing agents. *Proceedings of the National Academy of Science USA*, **98**(26), 14778–14783.

Hansen, J.E., Sato, M., and Ruedy, R. (1997). Radiative forcing and climate response. *Journal of Geophysics Research*, **102**, 6831–6864.

Hansen, J.E., Sato, M., Ruedy, R., Lacis, A., and Oinas, V. (2000). Global warming in the twenty-first century: An alternative scenario. *Proceedings of the National Academy of Science USA*, **97**(18), 9875–9880.

Harnik, N., Scott, R.K., and Perlwitz, J. (2003) Wave reflection and focusing prior to the major stratospheric warming of September 2002. *J. Atmos. Sci.* (submitted).

Harris, N.R.P., Kibane-Dawe, I., and Amanatidis, G.T. (eds) (1998). *Polar Stratospheric Ozone 1997: Proceedings of the 4th European Symposium, 22–26 September 1997, Schliersee, Bavaria, Germany* (Air Pollution Research Report No. 66, EUR 18032, 772 pp.). European Community, Brussels.

Harriss, R.C., Sachse, G.W., Hill, G.F., Wade, L., Bartlett, K.B., Collins, J.E., Steele, P., and Novelli, P. (1992). Carbon monoxide and methane in the North American Arctic and subarctic troposphere: July–August 1988. *Journal of Geophysics Research*, **97**(D15), 16589–16600.

Harshvardan, Schwartz, S.E., Benkovitz, C.V., and Guo, G. (2002). Aerosol influence on cloud microphysics examined by satellite measurements and chemical transport modeling. *Journal of Atmospheric Science*, **59**(2), Part 2, 714–725.

Hartmann, D.L. (2002). Tropical surprises. *Science*, **295**(5556), 811–812.

Harvey, L.D.D. (2001). A quasi-one-dimensional coupled climate-change cycle model, 2: The carbon cycle component. *Journal of Geophysics Research*, **106**(15), 22355–22372.

Harvey, L.D.D. and Huang, Z. (2001). A quasi-one-dimensional coupled climate-change cycle model, 1: Description and behavior of the climate component. *Journal of Geophysics Research*, **106**(C10), 22339–22353.

Haszpra, L., Bercza, P.S., Bakwin, P.S., Berger, B.W., Davis, K.J., and Weidinger, T. (2001). Measuring system for the long-term monitoring of biosphere/atmosphere of carbon dioxide. *Journal of Geophysics Research*, **106**(D3), 3057–3069.

Hauglustaine, D. (2002). Trace gas radiative forcing and related climate feedbacks: How do we reduce the uncertainties? *IGACtivities Newsletter*, **26**, 20–26.

Heans, K.A. (2001). Assessment of pre-industrial carbon dioxide content in the atmosphere using hydro-chemical data. *Proceedings of 1st International Conference on Global Warming and the Next Ice Age, 19–24 August 2001, Dalhousie University* (pp. 140–144). Dalhousie University Press, Nova Scotia.

Hellebrandt, H.J., Kern, J., and Scholz, V. (2003). Long-term studies on greenhouse gas fluxes during cultivation of energy crops on sandy soils. *Atmospheric Environment*, **37**(12), 1635–1644.

Hemming, D., Switsur, V.R., Waterhouse, J.S., and Heaton, T.H.E. (1998). Carbon stable isotope response of three tree species to recent climate and atmospheric changes. *Proceed-*

ings of the International Symposium: "Changes in the Hydrosphere and Atmosphere", 14–18 April 1997, Vienna (pp. 800–802).

Hensen, A., van den Bulk, W.C.M., Vermeulen, A. T., and Wyers, G.P. (1998). *CO_2 Exchange between Grassland and the Atmosphere: Results over a 4-year Period of CO_2 Measurements at Cabauw, The Netherlands.* Dutch National Research Programme on Global Air Pollution and Climate Change, Amsterdam, Report No. 410200020, 53 pp.

Herman, G. and Goody, R. (1976). Formation and persistence of summertime arctic status clouds. *Journal of Atmospheric Science*, **33**, 1537–1553.

Hernández, P.A., Notsu, K., Salazar, J.M., Mori, T., Natale, G., Okada, H., Virgili, G., Shimoke, Y., Sato, M., and Pérez, N.M. (2001). Carbon dioxide degassing by advective flow from Usu Volcano, Japan. *Science*, **292**(5514), 83–87.

Hidy, G.M. (2003). Snowpack and precipitation chemistry at high altitudes. *Atmospheric Environment*, **37**(9–10), 1231–1242.

Hitchman, M.H., Buker, M.L., and Tripoli, G.J. (1999). Influence of synoptic water on column ozone during Arctic summer 1997. *Journal of Geophysics Research*, **104**(D21), 26547–26563.

Hobbs, P.V., Rangno, A.L., Shupe, M.D., and Uttal, T. (2001). Airborne studies of cloud structures over the Arctic Ocean and comparisons with deductions from ship-based 35-GHz radar measurements. *Journal of Geophysics Research*, **106**, 15029–15044.

Holland, D.M. (2001). Explaining the Weddell Polynya – A large ocean eddy shed at Maud Rise. *Science*, **292**, 1697–1700.

Holloway, G. and Sou, T. (2001). Is Arctic sea ice rapidly thinning? *Ice and Climate News*, **1**, 2–5.

Hoppel, K., Bevilacqua, R., Allen, D., Nedoluha, G., and Randall, C. (2003) POAM III observations of the anomalous 2002 Antarctic ozone hole. *Geophys. Res. Lett.*, **30**, 1394, doi:10.1029/2003GL016899.

Hotuntsev, Yu.L. (2001). *Man, Technology, Environment* (224 pp.). Stable World, Moscow [in Russian].

Houghton, J.T., Calander, B.A., and Varney, S.K. (eds) (1992). *Climate Change 1992* (7 pp.). Cambridge University Press, Cambridge, UK.

Houghton, J.T., Filho, G.M., Calander, B.A., Harris, N., Kattenberg, A., and Maskell, K. (1996). *Climate Change 1995: The Science of Climate Change* (Intergovernmental Panel on Climate Change, 572 pp.). Cambridge University Press, Cambridge, UK.

Houghton, J.T., Ding, Y., Griggs, D.J., Noguer, M., Van der Linden, P.J., Dai, X., Masskell, K., and Johnson, C.A. (2001). *Climate Change 2001: The Scientific Basis* (Contribution of Working Group I to the Third Assessment Report of the Intergovernmental Panel Group on Climate Change, 881 pp.). Cambridge University Press, Cambridge, UK.

Houghton, R.A. (1999). The annual net flux of carbon to the atmosphere from changes in land use 1850–1990. *Tellus*, **518**, 298–313.

Houghton, R.A., Skole, D.L., Nobre, C.A., Hackler, J.L., Lawrence, K.T., and Chomentowski, W.H. (2000) Annual fluxes of carbon from deforestation and regrowth in the Brazilian Amazon. *Nature*, **403**(6767), 301–304.

Høv, Ø. (ed.) (1997). *Tropospheric Ozone Research. Tropospheric Ozone in the Regional and Sub-regional Context.* Springer for Science, Amsterdam, 455 pp.

Howarth, R.W., Stewart, J.W.B., and Ivanov, M.V. (eds). (1992). *Sulphur Cycling on the Continents: Wetlands, Terrestrial Ecosystems, and Associated Water Bodies* (372 pp.). John Wiley & Sons, Chichester, UK.

Huesemann, M.N., Skillman, A.D., and Crecelius, E.A. (2002). The inhibition of marine nitrification by ocean disposal of carbon dioxide. *Marine Pollution Bulletin*, **44**(2), 142–148.

Hughes, G.B., Giegengack, R., and Kritikos, H.N. (1999). Spectral indications of unexpected contributors to atmospheric CO_2 variability? *International Journal of Climatology*, **19**(8), 813–820.

Hughes, T.P., Baird, A.H., Bellwood, D.R., Card, M., Connoly, S.R., Folke, C., Grosberg, R., Hoegh-Gulberg, O., Jackson, J.B., Kleypas, J. *et al.* (2003). Climate change, human impacts, and the resilience of coral reefs. *Science*, **301**, 529–533.

Hulst, R. van (1979). The dynamics of vegetation: Succession in model communities. *Vegetatio*, **39**(2), 85–96.

IASC Project Catalogue (2001). The International Arctic Science Committee, Oslo (57 pp.).

Indermühle, A., Stocker, T.F., Joos, F., Fischer, H., Smith, H.J., Wahlen, M., Deck, B., Mastroianni, D., Tschumi, J., Blunier, T. *et al.* (1999). Holocene carbon-cycle dynamics based on CO_2 trapped in ice at Taylor Dome, Antarctica. *Nature*, **398**(6723), 121–126.

IPCC (1996). Climate change impacts on forests. In: R.T. Watson, M.C. Zinyowera, and R.H. Moss (eds), *Climate Change 1995: Impacts, Adaptations and Mitigation of Climate Change: Scientific–Technical Analyses* (Contribution of Working Group II to the Second Assessment Report of the Intergovernmental Panel on Climate Change, 879 pp.). Cambridge University Press, Cambridge, UK.

IPCC (2001a). *Climate Change 2001: Impacts, Adaptation, and Vulnerability* (Contribution of Working Group II to the Third Assessment Report of the Intergovernmental Panel on Climate Change, 1032 pp.). Cambridge University Press, Cambridge, UK.

IPCC (2001b). *Climate Change 2001: Mitigation* (Contribution of Working Group III to the Third Assessment Report of the Intergovernmental Panel on Climate Change, 752 pp.). Cambridge University Press, Cambridge, UK.

IPCC (2001c). *Third Assessment Report, Vol. 1: Climate Change 2001: The Scientific Basis* (881 pp.). Cambridge: Cambridge University Press, UK.

IPCC (2001d). The carbon cycle and atmospheric CO_2. *Third Assessment Report* (Working Group I, Chapter 3, pp. 183–237). Cambridge University Press, Cambridge, UK.

Irion, R. (2001). Fathoming the chemistry of the deep blue sea. *Science*, **293**(5531), 790–793.

Isaksen, I.S.A. (1998). *Tropospheric Ozone. Regional and Global Scale Interactions* (425 pp.). Reidel, Dordrecht, The Netherlands.

Isidorov, V.A. and Kondratyev, K.Ya. (2001). On global carbon cycle problems. *Problems of Environmental and Natural Resources*, **5**, 56–83.

Ito, A. and Oikawa, T. (2002). A simulation model of the carbon cycle in land ecosystems (Sim-CYCLE): A description based on dry-matter production theory and plot-scale validation. *Ecological Modelling*, **151**(2–3), 143–176.

Itoh, H. (2002). True versus apparent Arctic Oscillation. *Geophysics Research Letters*, **19**(8), 109/1–109/4.

Ivanov, A. (1978). *Introduction to Oceanography* (392 pp.). Mir, Moscow [in Russian].

Ivanov-Rostovtsev, A.G., Kolotilo, L.G., Tarasiuk, Yu.F., and Sherstiankin, P.P. (2001). *Self-organization and Self-regulation of Natural Systems* (216 pp.). Russian Geographical Society, St Petersburg [in Russian].

Iversen, T. (1989). Numerical modeling of the long-range atmospheric transport of sulphur dioxide and particulate sulphate to the Arctic. *Atmospheric Environment*, **23**(11), 2571–2596.

Iversen, T. (1996). Atmospheric transport pathways for the Arctic. In: E.W. Wolff and R.C. Bales (eds), *Chemical Exchange between the Atmosphere and Polar Snow* (NATO ASI Series Vol. 143, pp. 71–92). Springer-Verlag, Berlin.

Jacob, D.J., Leovy, C., Schoeberl, M., McNeal, R.J., and Gleason, J. (1995). Atmospheric trace gas measurements for the year 2000 and beyond. *The Earth Observer*, **7**, 31–38.

Jadin, E.A. and Vargin, P.N. (2003). Interannual variations of the Antarctic ozone hole, stratospheric circulation and sea surface temperature in the South Oceans. *J. Atmos. Sci.* (submitted).

Jaeglé, L., Jacob, D.J., Brune, W.H., Tan, D., Faloona, I.C., Weinheimer, A.S., Ridley, B.A., Campos, T.L., and Sachse, G.W. (1998a). Sources and chemistry of HO_x and production of ozone in the upper troposphere over the United States. *Geophysics Research Letters*, **25**(10), 1705–1708.

Jaeglé, L., Jacob, D.J., Brune, W.H., Tan, D., Faloona, I.C., Weinheimer, A.S., Ridley, B.A., Campos, T.L., and Sachse, G.W. (1998b). Sources of HO_x and production of ozone in the upper troposphere over the United States. *Geophysics Research Letters*, **25**(10), 1709–1712.

Jaeshke, W., Salkowski, T., Dierssen, J.P., Trümbach, J.V., Krischke, U., and Günther, A. (1999). Measurements of trace substances in the Arctic troposphere as potential precursors and constituents of Arctic haze. *Journal of Atmospheric Chemistry*, **34**, 291–319.

Jagovkina, S.V., Karol, I.L., Zubov, V.A., Lagun, V.E., Reshemikov, A.I., and Rosanov, E.V. (2000). The estimation of gas deposit leakage input into the total methane flux from the West Siberian region. *Proceedings of 2nd International Methane Mitigation Conference, 18–23 June 2000, Novosibirsk* (pp. 263–267). Novosibirsk University Press, Novosibirsk, Russia [in Russian].

Jaworowski, Z. (2003). Sun rules the climate. *21st Century Science and Technology*, Winter 2003/2004 (in print).

Jenkins, R., Barton, J., Bartzokas, A., Hesselberg, J., and Knutsen, H.M. (2002). *Environmental Regulation in the New Global Economy* (360 pp.). Edward Elgar, Cheltenham, UK.

Jepma, C., Meijer, H., Uiterkamp, T.S., and Weesie, P. (2001). Editorial. *Energy and Environment*, **12**(5–6), iii–ix.

Jeuken, A.B.M., Eskes, H.J., van Velthoven, P.F.J., Kelder, H.M., and Hølm, E.V. (1999). Assimilation of total ozone satellite measurements in a three-dimensional tracer transport model. *Journal of Geophysics Research*, **104**(D5), 5551–5563.

Johnson, D.E. and Ulyatt, M.I. (2000). Variations in the proportion of methane of total greenhouse gas emissions from US and NZ dairy production systems. *Proceedings of 2nd International Methane Mitigation Conference, 18–23 June 2000, Novosibirsk* (pp. 249–254). Novosibirsk University Press, Novosibirsk, Russia [in Russian].

Jones, A., Roberts, D.L., Woodage, M.J., and Johnson, C.E. (2001). Indirect sulphate aerosol forcing in a climate model with an interactive sulphur cycle. *Journal of Geophysics Research*, **106**(D17), 20293–20310.

Jones, C.D. and Cox, P.M. (2001). Modelling the volcanic signal in the atmospheric CO_2 record. *Global Biogeochemical Cycles*, **15**(2), 453–466.

Jones, G.S., Tett, S.F.B., and Stott, P.A. (2003). Causes of atmospheric temperature change 1960–2000: A combined attribution analysis. *Geophysics Research Letters*, **30**(5), 1228–1231.

Joos, F., Plattner, G.-K., Stocker, T.F., Marchal, O., and Schmittner, A. (1999). Global warming and marine carbon cycle feedbacks on future atmospheric CO_2. *Science*, **284**(5413), 464–468.

Kabat, P. and Nobre, C.A. (2000). Brazilia – European carbon research studies in LBA. *BAMS News*, **8**, 15–18.

Kaiser, J. (2000). Soaking up carbon in forests and fields. *Science*, **290**(5493), 922.

Kalabin, G.V. (2000). *Ecodynamics of the Anthropogenic Environment Province of the North* (294 pp.). Kola Scientific Centre, Russian Academy of Science, Apatity, Russia [in Russian].

Kamall, R. (2000). Capture and storage of CO_2. *Greenhouse Issues*, **51**, 2–4.

Kaminski, T., Rayner, P.J., Heimann, M., and Enting, I. (2001). On aggregation errors in atmospheric transport inversions. *Journal of Geophysics Research*, **106**(D5), 4703–4715.

Kane, D.L. and Reeburgh, W.S. (1998). Introduction to special section: Land–Air–Ice Interactions (LAII) Flux Study. *Journal of Geophysics Research*, **103**(28), 28911–28915.

Karl, T. and Gleckler, P.J. (2001). Tracking changes in AMIP model performance. *Proceedings of the 8th Scientific Assembly of IAMAS, 10–18 July 2001, Innsbruck* (p. 8). IAMAS Press, Vienna.

Karley, M.J., Beven, K.J., and Oliver, H.R. (1993). A method for predicting spatial distribution of evaporation using simple meteorological data. *Proceedings of International Symposium "Land Surface Range Space and Time Scales", 13–16 July 1993, Yokohama* (pp. 212, 619–626). IAHS, Yokohama, Japan.

Karol′, I.L. (2000). Impact of flights by world transport aircraft on the ozonosphere and climate. *Meteorology and Hydrology*, **7**, 17–32 [in Russian].

Kashapov, R.Sh. (2002). On the balance of organic carbon in the nature–economic system of Bashkortostan. *Bulletin of the Russian Geographical Society*, **134**(3b), 39–42 [in Russian].

Kashulina, G. and Reimann, C. (2002). Sulphur in the Arctic environment, (2): Results of multi-medium regional mapping. *Environmental Pollution*, **116**, 337–350.

Kashulina, G., Reimann, C., and Banks, D. (2003). Sulphur in the Arctic environment, (3): Environmental impact. *Environmental Pollution*, **124**(1), 151–171.

Katul, G.G. and Albertson, J.D. (1999). Modeling CO_2 sources, sinks, and fluxes within a forest canopy. *Journal of Geophysics Research*, **104**(D6), 6081–6092.

Katz, M.E., Pak, D.K., Dickens, G.R., and Miller, K.G. (1999). The source and fate of massive carbon input during the latest Paleocene thermal maximum. *Science*, **286**(5444), 1531–1533.

Katz, R.W. (2002). Techniques for estimating uncertainty in climate change scenarios and impact studies. *Climatic Research*, **20**, 167–185.

Katzenstein, A.S., Doezema, L.A., Simpson, I.J., Blake, D.R., and Rowland, F.S. (2003). Extensive regional atmospheric hydrocarbon pollution in the southwestern United States. *Proceedings of the National Academy of Sciences*, **100**(21), 11975–11979.

Kawamura, K., Legrand, M., and Cachier, H. (1998). Organic matter in polar aerosol, snow, and ice. *IGACtivities Newsletter*, **14**, 12–15.

Kawamura, K., Yokoyama, K., Fujii, Y., and Watanabe, O. (1999). Implication of azelaic acid in a Greenland ice core for oceanic and atmospheric changes in high latitudes. *Geophysics Research Letters*, **26**(7), 871–874.

Kaye, J.A. (1998). Summary of Total Ozone Mapping Spectrometer (TOMS) science team meeting. *The Earth Observer*, **10**(3), 15–22.

Kaye, J.A. and Rind, D. (1998). Summary of chemistry/climate modelling meeting. *The Earth Observer*, **10**(3), 27–36.

Keeling, C.D. and Bacastow, R.B. (1977). Impact of industrial gases on climate. In: R. Revelle, and W. Munk (eds) *Energy and Climate*. National Academy of Sciences, Geophysical Research Board, Geophysics Study Commitee, Washington, pp. 72–95.

Keeling, R.F. and Visbeck, M. (2001). Antarctic stratification and glacial CO_2. *Nature*, **412**(6847), 605–606.

Keeling, R.F., Manning, A.C., McElroy, E.M., and Shertz, S.R. (1998). Methods for measuring changes in atmospheric O_2 concentration and their application in southern hemisphere air. *Journal of Geophysics Research*, **103**(D3), 3381–3397.

Keller, M., Rocha, H., Trumbora, S., and Kruijt, B. (2001). Investigating the carbon cycle of the Amazon forests. *Global Change Newsletter*, **45**, 14–19.

Kelley, J.J. (1987). Carbon dioxide in the Arctic environment. *Journal of Earth Science*, **35**(2), 341–354.

Kelley, J.J., Krapivin, V.F., and Popovich, P.R. (1999a). Problems in Arctic environment monitoring. *Problems of Environmental and Natural Resources*, **6**, 32–40 [in Russian].

Kelley, J.J., Krapivin, V.F., and Popovich, P.R. (1999b). Problems in Arctic monitoring. *Problems of Environmental and Natural Resources*, **1**, 2–6 [in Russian].

Kennedy, D. (2001). An unfortunate U-turn on carbon. *Science*, **291**(5513), 2515.

Kennedy, M.J., Pevear, D.R., and Hill, R.J. (2002). Mineral surface control of organic carbon in black shale. *Science*, **295**(5555), 657–660.

Kerminen, V.-M., Hillamo, R. E., Mäkelä, T., Jaffrezo, J.-L., and Maenhut, W. (1998). The physicochemical structure of the Greenland summer aerosol and its relation to atmospheric processes. *Journal of Geophysics Research*, **103**(D5), 5661–5670.

Kerr, R.A. (2001). How grasses got the upper hand. *Science*, **293**, 1572–1573.

Kerr, R.A. (2002). Reducing uncertainties of global warming. *Science*, **295**(5552), 29–31.

Khalil, M.A.K., Rasmussen, R.A., Ren, L., Wang, M.X., Shearer, M.J., Dalluge, R.W., and Duan, C.-L. (2000). Methane emissions from rice fields. *Proceedings of 2nd International Methane Mitigation Conference, 18–23 June 2000, Novosibirsk* (pp. 13–30). Novosibirsk University Press, Novosibirsk, Russia [in Russian].

Kheshgi, H.S., Jain, A.K., Kotamarthi, V.R., and Wuebbles, D.J. (1999a). Future atmospheric methane concentrations in the context of the stabilization of greenhouse gas concentrations. *Journal of Geophysics Research*, **104**(D16), 19183–19190.

Kheshgi, H.S., Jain, A.K., and Wuebbles, D.S. (1999b). Model-based estimation of the global carbon budget and its uncertainty from carbon dioxide and carbon isotope records. *Journal of Geophysics Research*, **104**(D24), 31127–31143.

Khudoshina, M.Yu. (1996). Automatisation of investigations on the control of resources of sea coastal regions. PhD thesis, Institute of Radioengineering and Electronics, Moscow (268 pp.) [in Russian].

Khvorostyanov, L.I. (1995). Mesoscale processes of cloud formation, cloud-radiation interaction and their modelling with explicit cloud microphysics. *Atmospheric Research*, **39**, 1–67.

Khvorostyanov, L.I., Curry, J.A., Pinto, J.O., Shupe, M.D., Baker, B., and Sassen, K. (2001). Modeling with explicit spectral water and ice microphysics of a two-layer cloud system of altostratus and cirrus observed during the FIRE Arctic Clouds Experiment. *Journal of Geophysics Research*, **106**, 15099–15112.

Kim, W., Arai, T., Kanae, S., Oki, T., and Musiake, K. (2001) Application of the Simple Biosphere Model (SiB2) to a paddy field for a period of the growing season in GAME-Tropics. *Journal of the Meteorological Society of Japan*, **79**(1B), 387–400.

Kirilenko, A.P. (1990). Numerical modeling of the production process and the water cycle of forest ecosystems. PhD thesis, Computing Center of the Russian Academy of Science, Moscow (151 pp.) [in Russian].

Kiseleva, S.V. (1990). Study of the processes of CO_2 transport under different conditions for the water–air interface. PhD thesis, Moscow State University (171 pp.) [in Russian].

Klubov, S.V., Grigorian, S.V., and Davydova, S.L. (2000). Natural and technogenic distributions of the elements of the pollutants within the marine coastal zone. *Ecological Systems and Devices*, **5**, 42–45 [in Russian].

Koch, W. (1996). *Solarer Strahlungstransport in arktischen Cirrus* (Report No. 60, 99 pp.). German Institute for Coastal Research, Geesthacht.

Kohler, M.A. and Richards, M.A. (1962). Multicapacity basin accounting for predicting runoff from storm precipitation. *Journal of Geophysics Research*, **67**(13), 5187–5197.

Kojima, T. (1998). *The Carbon Dioxide Problem: Integrated Energy and Environmental Policies for the 21st Century* (240 pp.). Gordon & Breach, London.

Kok, M.T.J. and Verweij, W. (1999). *The Kyoto Protocol: Implications for Research: Report of the NRP Workshop: The Road after Kyoto, 20 January 1998* (Report No. 410 200 018, 22 pp.). Dutch National Research Programme on Global Air Pollution and Climate Change, Amsterdam University Press, The Netherlands.

Kolbe, A.E. (1999). Geographic extension of an uneven-aged, multi-species matrix growth model for northern hardwood forests. *Ecological Modelling*, **121**(2–3), 235–253.

Kolber, Z.S., Plumley, F.G., Lang, A.S., Beatty, J.T., Blankenship, R.E., van Dover, C.L., Vetriani, C., Koblizek, M., Rathgeber, C., and Falkowski, P.G. (2001). Contribution of aerobic photoheterotrophic bacteria to the carbon cycle in the ocean. *Science*, **292**(5526), 2492–2495.

Komhyr, W.D., Oltmans, S.J., Grass, R.D., and Leonard, R.K. (1991) Possible influence of long-term sea surface temperature anomalies in the tropical Pacific on global ozone. *Canadian J. Physics*, **69**, 1093–1102.

Kondragunta, S., Flynn, L.E., Neuendorffer, A., Miller, A.J., Long, C., Nagatani, R., Zhou, S., Beck, T., Beach, E., McPeters, R. *et al.* (2003) Vertical structure of the anomalous 2002 Antarctic ozone hole. *J. Atmos. Sci.* (submitted).

Kondratyev, K.Ya. (1989). *Ozone Dynamics* (212 pp.). VINITI, Moscow [in Russian].

Kondratyev, K.Ya. (1990). *Key Problems of Global Ecology* (454 pp.). VINITI, Moscow [in Russian].

Kondratyev K.Ya. (1991a). *Aerosol and Climate* (542 pp.). Hydrometeoizdat, Leningrad [in Russian].

Kondratyev, K.Ya. (1991b). Priorities in global ecology. *Bulletin of USSR Academy of Science, Geography*, **6**, 21–30 [in Russian].

Kondratyev, K.Ya. (1992a). Aerosol cloud climate interaction, Part 1: Aerosols; Part 2: Clouds. *Optics of the Atmosphere*, **5**(3), 317–325.

Kondratyev, K.Ya. (1992b). *Global Climate* (359 pp.). Science, St Petersburg [in Russian].

Kondratyev, K.Ya. (1993). Ecology and politics. *Bulletin of Russian Geographical Society*, **125**(2), 78–91 [in Russian].

Kondratyev, K.Ya. (1996). Global changes and demographic dynamics. *Bulletin of Russian Geographical Society*, **128**(3), 1–12 [in Russian].

Kondratyev, K.Ya. (1997). Key issues of global change at the end of the second millennium. *Newsletter of the European Geophysical Society*, **63**, 4–8.

Kondratyev, K.Ya. (1998a). Ecological risk: Real and hypothetical. *Bulletin of the Russian Geographical Society*, **130**(3), 13–24 [in Russian].

Kondratyev, K.Ya. (1998b). *Multidimensional Global Change* (771 pp.). Wiley-Praxis, Chichester, UK.

Kondratyev, K.Ya. (1999a). *Atmospheric Ozone Variability* (592 pp.). Springer-Praxis, Chichester, UK.

Kondratyev, K.Ya. (1999b). *Climatic Effects of Aerosols and Clouds* (264 pp.). Springer-Praxis, Chichester, UK.

Kondratyev, K.Ya. (1999c). *Ecodynamics and Geopolicy, Vol. 1: Global Problems* (1036 pp.). St Petersurg Scientific Center of the Russian Academy of Science, St Petersurg [in Russian].

Kondratyev, K.Ya. (2000a). Global changes on the border of two millennia. *Herald of the Russian Academy of Science*, **70**(9), 788–796 [in Russian].

Kondratyev, K.Ya. (2000b). Studying the Earth from space: The EOS scientific plan. *Earth Research from Space*, **3**, 82–91 [in Russian].

Kondratyev, K.Ya. (2001a). Impact of NATO's bombardment on the environment of the former Yugoslavia. *Bulletin of the Russian Academy of Science*, **1**, 63–67 [in Russian].

Kondratyev, K.Ya. (2001b). Key issues of global change at the end of the second millenium. *Our Fragile World: Challenges and Opportunities for Sustainable Development: EOLSS Vorrunner*, **1**, 147–165.

Kondratyev, K.Ya. (2001c). On the problem of global carbon cycle. *Problems of the Environment and Natural Resources*, **6**, 56–83 [in Russian].

Kondratyev, K.Ya. (2002a). Arctic environment, 1: Conceptual aspects. *Proceedings of the Russian Geographical Society*, **134**(5), 1–10 [in Russian].

Kondratyev, K.Ya. (2002b). Arctic environment, 2: Perspectives of future studies. *Proceedings of the Russian Geographical Society*, **134**(6), 1–9 [in Russian].

Kondratyev, K.Ya. (2002c). Global climate changes: Reality, hypotheses, and fantasies. *Earth Research from Space*, **1**, 3–23 [in Russian].

Kondratyev, K.Ya. (2003a). High-latitude environmental dynamics in the context of global change. *Időjàràs*, **107**(1), 1–29.

Kondratyev, K.Ya. (2003b). Radiative forcings due to aerosols. *Optics of the Atmosphere and Ocean*, **16**(1), 5–18.

Kondratyev, K.Ya. and Binenko, V.I. (eds) (1981). *Polar Aerosols, Extended Cloudiness, and Radiation* (150 pp.). Gidrometeoizdat, Leningrad.

Kondratyev, K.Ya. and Binenko, V.I. (2000). On the radiative forcing of clouds and aerosols. *Meteorology and Hydrology*, **1**, 33–41 [in Russian].

Kondratyev, K.Ya. and Demirchian, K.S. (2000). Global climate change and carbon cycle. *Bulletin of the Russian Geographical Society*, **132**(4), 1–20 [in Russian].

Kondratyev, K.Ya. and Demirchian, K.S. (2001a). Climate of the Earth and the Kyoto Protocol. *Herald of the Russian Academy of Science*, **71**(11), 100–107 [in Russian].

Kondratyev, K.Ya. and Demirchian, K.S. (2001b). Global climate and the Kyoto Protocol. *Problems of the Environment and Natural Resources*, **6**, 2–15 [in Russian].

Kondratyev, K.Ya. and Galindo, I. (1997). *Volcanic Activity and Climate* (382 pp.). A. Deepak, Hampton, VA.

Kondratyev, K.Ya. and Galindo, I. (2001). *Global Change Situations: Today and Tomorrow* (164 pp.). Universidad de Colima, Colima, Mexico,

Kondratyev, K.Ya. and Grigoryev, Al.A. (2000). Natural and anthropogenic ecological disasters. *Earth Research from Space*, **4**, 3–19 [in Russian].

Kondratyev, K.Ya. and Grigoryev, Al.A. (2002). *Environmental Disasters: Anthropogenic and Natural* (484 pp.). Springer-Praxis. Chichester, UK.

Kondratyev, K.Ya. and Isidorov, V.A. (2001). Global carbon cycle. *Optics of the Atmosphere and Ocean*, **14**(1), 1–10 [in Russian].

Kondratyev, K.Ya. and Johannessen, O.M. (1993). *The Arctic and Climate* (140 pp.). PROPO, St Petersburg [in Russian].

Kondratyev, K.Ya. and Krapivin, V.F. (2001a). An expert system to control the greenhouse effect. *Problems of Environmental and Natural Resources*, **6**, 23–39 [in Russian].

Kondratyev, K.Ya. and Krapivin, V.F. (2001b). Biocomplexity and the global ecological monitoring. *Earth Research from Space*, **1**, 1–8 [in Russian].

Kondratyev, K.Ya. and Krapivin, V.F. (2001c). Land ecosystems and their global dynamics. *Problems of Environmental and Natural Resources*, **5**, 24–38 [in Russian].

Kondratyev, K.Ya. and Krapivin, V.F. (2001d). Global dynamics of basic land ecosystems. *Earth Research from Space*, **4**, 3–12 [in Russian].

Kondratyev, K.Ya. and Krapivin, V.F. (2003a). Climate change and carbon cycle. *Problems of Environmental and Natural Resources*, **1**, 2–26 [in Russian].

Kondratyev, K.Ya. and Krapivin, V.F. (2003b). Global changes: Real and expected in the future. *Earth Research from Space*, **4**, 3–12 [in Russian].

Kondratyev, K.Ya. and Krapivin, V.F. (2003c). Global carbon cycle and climate. *Earth Research from Space*, **1**, 3–15 [in Russian].

Kondratyev, K.Ya. and Losev, K.S. (2002). Present problems of global civilization development and its possible perspectives. *Earth Research from Space*, **2**, 3–23 [in Russian].

Kondratyev, K.Ya. and Varotsos, C.A. (2000). *Atmospheric Ozone Variability: Implications for Climate Change, Human Health, and Ecosystems* (758 pp.). Springer-Praxis, Chichester, UK.

Kondratyev, K.Ya. and Varotsos, C.A. (2001). Global tropospheric ozone dynamics, Part 2: Numerical modelling of tropospheric ozone variability. *Environmental Scientific and Pollution Research*, **8**(2), 113–119.

Kondratyev, K.Ya. and Zhvalev, V.F. (eds) (1981). *First Global GARP Experiment, Vol. 2: Polar Aerosols, Extended Cloudiness, and Radiation* (150 pp.). Gidrometeoizdat, Leningrad [in Russian].

Kondratyev, K.Ya., Vasiliev, O.B., Ivlev, L.S., Nikolskiy, G.A., and Smoktiy, O.I. (1973). *Influence of Aerosol on the Radiation Transport: Possible Climatic Consequences* (266 pp.). Leningrad State University Press, Leningrad [in Russian].

Kondratyev, K.Ya., Grigoryev, Al.A., Pokrovskiy, O.M., and Shalina, E.V. (1983a). *Space Remote Sensing of Aerosol* (216 pp.). Hydrometeoizdat, Leningrad [in Russian].

Kondratyev, K.Ya., Moskalenko, N.I., and Pozdnyakov, D.V. (1983b). *Atmospheric Aerosol* (224 pp.). Hydrometeoizdat, Leningrad [in Russian].

Kondratyev, K.Ya., Loginov, V.F., and Kravchuk, E.G. (1990). Solar radiation trends as indicators of changes due to anthropogenic aerosols. *Proceedings USSR Academy of Science*, **315**(2), 341–344 [in Russian].

Kondratyev, K.Ya., Bondarenko, V.G., and Khvorostyanov, V.I. (1992a). A three-dimensional numerical model of cloud formation and aerosol transport in the orographically inhomogeneous atmospheric boundary layer. *Boundary-Layer Meteorology*, **61**(3), 265–285.

Kondratyev, K.Ya., Ortner, J., and Preining, O. (1992b). Priorities of global ecology now and in the next century. *Space Policy*, **8**(1), 39–48.

Kondratyev, K.Ya., Johannessen, O.M., and Melentyev, V.V. (1996). *High-latitude Climate and Remote Sensing* (200 pp.). Wiley-Praxis, Chichester, UK.

Kondratyev, K.Ya., Moreno Pena, F. and Galindo, I. (1997). *Sustainable Development and Population Dynamics* (128 pp.). Universidad de Colima, Colima, Mexico.

Kondratyev, K.Ya., Krapivin, V.F., and Pshenin, E.S. (2000). Conception of regional geoinformation monitoring. *Earth Research from Space*, **6**, 1–8 [in Russian].

Kondratyev, K.Ya., Demirchian, K.S., Baliunas, S., Adamenko, V.N., Bohmer-Christiansen, S., Idso, Sh.B., Postmentier, E.S., and Soon, W. (2001a). *Global Climate Changes: Conceptual Aspects* (125 pp.). St Petersburg Scientific Center of the Russian Academy of Science, St Petersburg [in Russian].

Kondratyev, K.Ya., Krapivin, V.F., and Borodin L.F. (2001b). Reconstruction of two-dimensional images in the problem of the Aral–Caspian aquageosystem monitoring. *Earth Research from Space*, **5**, 19–26 [in Russian].

Kondratyev, K.Ya., Losev, K.S., Ananicheva, M.D., and Chesnokova, I.V. (2001c). Some problems of the landscape history and ecology in the context of biotic regulation. *Bulletin of Russian Geographical Society*, **133**(5), 22–29 [in Russian].

Kondratyev, K.Ya., Krapivin, V.F., and Phillips, G.W. (2002a). *Global Environnmental Change: Modelling and Monitoring* (319 pp.). Springer-Verlag, Berlin.

Kondratyev, K.Ya., Krapivin, V.F., and Phillips, G.V. (2002b). *Problems of High-latitude Environmental Pollution* (280 pp.). St Petersburg State University, St Petersburg [in Russian].

Kondratyev, K.Ya., Krapivin, V.F., and Savinikh, V.P. (2003a). Global ecoinformatics and education. *Present Education*, **9**, 26–32 [in Russian].

Kondratyev, K.Ya., Krapivin, V.F., and Savinikh, V.P. (2003b). *Perspectives of Civilization Development: Multidimensional Analysis* (572 pp.). Logos, Moscow [in Russian].

Kondratyev, K.Ya., Krapivin, V.F., and Varotsos, C.A. (2003c). *Global Carbon Cycle and Climate Change* (368 pp.). Springer-Praxis, Chichester, UK.

Kondratyev, K.Ya., Losev, K.S., Ananicheva, M.D., and Chesnokova, I.V. (2003d). Price of Russian ecological service. *Herald of Russian Academy of Science*, **73**(1), 3–10 [in Russian].

Konopka, P., Grooß, J.-U., and Steinhorst, H.-M. (2003) Mixing and chemical ozone loss during and after the Antarctic polar vortex major warming in September 2002. *J. Atmos. Sci.* (submitted).

Kontorovich, A.E., Dobretsov, N.L., Lanerov, I.P., Korzhubayev, A.G., and Livshits, V.R. (1999). The energy strategy of Russia in the 21st century. *Herald of the Russian Academy of Science*, **69**(9), 771–789 [in Russian].

Korner, C. (1998). Tropical forests in a CO_2-rich world. *Climate Change*, **39**(2–3), 297–315.

Kozoderov, V.V. (1994). Problems of global changes in the context of space-borne observations and modeling the biogeophysical processes. *Earth Research from Space*, **6**, 51–67 [in Russian].

Kozoderov, V.V. (1998). Development of the theory of the biospheric global changes from data of the space-borne observations of the Earth. *Earth Research from Space*, **6**, 25–39 [in Russian].

Kraabol, A.G. and Stordal, F. (2000). Modelling chemistry in aircraft plumes, 2: The chemical conservation of NO_x to reservoir species under different conditions. *Atmospheric Environment*, **34**(23), 3951–3962.

Kraabol, A.G., Konopka, P., Stordal, F., and Schlager, H. (2000). Modelling chemistry in aircraft plumes, 1: Comparison with observations and evaluation of a layered approach. *Atmospheric Environment*, **34**(23), 3939–3950.

Kramer, H.J. (1995). *Observation of the Earth and Its Environment* (832 pp.). Springer-Verlag, Berlin.

Krapivin, V.F. (1969). About some Kotelnikov integral estimations. *Radioelectronics*, **XII**(7), 749–754 [in Russian].

Krapivin, V.F. (1978). *On the Theory of the Complex System Survivability* (248 pp.). Soviet Radio, Moscow [in Russian].

Krapivin, V.F. (1993). Mathematical model for global ecological investigations. *Ecological Modelling*, **67**(2–4), 103–127.

Krapivin, V.F. (1996). The estimation of the Peruvian current ecosystem by a mathematical model of biosphere. *Ecological Modelling*, **91**(1), 1–14.

Krapivin, V.F. (2000a). Radiowave ecological monitoring. In: V.V. Klyuev (ed.), *Ecological Diagnostics* (pp. 295–312). Znanie, Moscow.

Krapivin, V.F. (2000b). The greenhouse effect and the global biogeochemical cycle of carbon dioxide. *Problems of Environmental and Natural Resources*, **2**, 2–18 [in Russian].

Krapivin, V.F. and Kondratyev, K.Ya. (2002). *Global Changes of the Environment: Ecoinformatics* (724 pp.). St Petersburg University, St Petersburg [in Russian].

Krapivin, V.F. and Marenkin, D.S. (1999). Co-evolution of man and nature. *Problems of Environmental and Natural Resources*, **1**, 2–10 [in Russian].

Krapivin, V.F. and Nazarian, N.A. (1997). A numerical model to study the global sulfur cycle. *Numerical Modeling*, **9**(8), 36–50 [in Russian].

Krapivin, V.F. and Phillips, G.W. (2001). A remote sensing-based expert system to study the Aral–Caspian aquageosystem water regime. *Remote Sensing of the Environment*, **75**, 201–215.

Krapivin, V.F. and Potapov, I.I. (2002). *Methods of Ecoinformatics* (496 pp.). VINITI, Moscow [in Russian].

Krapivin, V.F. and Shutko, A.M. (2002). Investigations in the microwave monitoring of land covers. *Problems of Environmental and Natural Resources*, **4**, 44–53 [in Russian].

Krapivin, V.F. and Vilkova, L.P. (1990). Model estimation of excess CO_2 distribution in biosphere structure. *Ecological Modelling*, **50**, 57–78.

Krapivin, V.F., Svirezhev, Yu.M., and Tarko, A.M. (1982). *Mathematical Modeling of the Global Biospheric Processes* (272 pp.). Science, Moscow [in Russian].

Krapivin, V.F., Vilkova, L.P., Rochon, G.L. and Hicks, D.R. (1996). Model estimation of the role of urban areas in global CO_2 dynamics. *Proceedings of Eco-Informa '96, 4–7 November 1996, Florida* (pp. 17–22). MIT Press, Cambridge, Massachusetts.

Krapivin, V.F., Cherepenin, V.A., Nazaryan, N.A., Phillips, G.W., and Tsang, F.Y. (1997) Simulation model of radionuclides transport in the river system Angara/Yenisey. *Problems of Environmental and Natural Resources*, **2**: 41–48 [in Russian].

Krapivin, V.F., Cherepenin, V.A., Phillips, G.W., August, R.A., Pautkin, A.Yu., Harper, M.J. and Tsang, F.Y. (1998). An application of modelling technology to the study of radionuclear pollutants and heavy metals dynamics in the Angara–Yenisey river system. *Ecological Modelling*, **111**(1), 121–134.

Krapivin, V.F., Shutko, A.M., and Ivanov, D.S. (2002). A new technology for the synthesis of environmental monitoring systems. *Problems of Environmental and Natural Resources*, **4**, 2–7 [in Russian].

Krueger, K., Naujokat, B., and Labitzke, K. (2003) The unusual midwinter warming in the southern hemisphere stratosphere 2002: a comparison to northern hemisphere phenomena. *J. Atmos. Sci.* (submitted).

Krupa, S.V. (2003). Effects of atmospheric ammonia (NH_3) on terrestrial vegetation: A review. *Environmental Pollution*, **124**(2), 179–221.

Krupchatnikov, V.N. (1998). Simulation of CO_2 exchange processes in the atmosphere–surface biome system by the climate model ECSSib. *Russian Journal of Numerical, Analytical and Mathematical Modelling*, **13**(6), 479–492 [in Russian].

Kuck, L.R., Jr, Smith, T., Balsley, B.B., Helmig, D., Conway, T.J., Tans, P.P., Davis, K., Jensen, M.L., Bognar, J.A., Arrieta, R.V. *et al.* (2000). Measurements of landscape-scale fluxes of carbon dioxide in the Peruvian Amazon by vertical profiling through the atmospheric boundary layer. *Journal of Geophysics Research*, **105**(D17), 22137–22146.

Kukla, G.J. (2000). The last interglacial. *Science*, **287**, 987–988.

Kukla, G. and Went, E. (eds) (1992). *Start of Glacial* (NATO ASI Ser. I, No. 3, 353 pp.). NATO, New York.

Kump, L.R., Brantley, S.L., and Arthur, M.A. (2000). Chemical weathering, atmospheric CO_2, and climate. *Annual Reviews of Earth and Planetary Sciences*, **287**, 611–667.

Kushner, P.J. and Polvani, L.M. (2003) A very large, spontaneous stratospheric sudden warming in a simple AGCM: A prototype for the Southern-Hemisphere warming of 2002? *J. Atmos. Sci.* (submitted).

Kutsenogii, K.P. and Kutsenogii, P.K. (2000). Aerosols of the Siberia. Results of seven-year researches. *Siberian Ecological Journal*, **VII**(1), 11–20 [in Russian].

Kuzmin, P.O. (1957). Hydrological investigations of land waters. *International Association for Science and Hydrology*, **3**, 468–478.

La Deau, S. and Clark, J.S. (2001). Rising CO_2 levels and the fecundity of forest trees. *Science*, **292**(5514), 95–97.

Lafleur, P.M. (1999). Growing season energy and CO_2 exchange at a subarctic boreal woodland. *Journal of Geophysics Research*, **104**(D8), 9571–9580.

Lafleur, P.M., Roulet, N.T., and Admiral, S.W. (2001). Annual cycle of CO_2 exchange at a bog peat-land. *Journal of Geophysics Research*, **106**(D3), 3071–3082.

Lai, Ch.T., Katul, G., Oren, R., Ellsworth, D., and Schfer, K. (2000). Modeling CO_2 and water vapor turbulent flux distributions within a forest canopy. *Journal of Geophysics Research*, **105**(D21), 26333–26351.

Lal, M. and Yarasawa, H. (2001). Future climate change scenarios for Asia as inferred from selected coupled atmosphere–ocean global climate models. *Journal of the Meteorological Society of Japan*, **79**(1), 219–227.

Lal, R. (2001). Potential of desertification control to sequester carbon and mitigate the greenhouse effect. *Climate Change*, **51**, 35–72.

Lal, R., Follett, R.F., and Stewart, B.A. (eds) (1998). *Soil Processes and the Carbon Cycle* (609 pp.). Springer-Verlag/CRC Press, Berlin.

Landers, D.H., Bangay, G., Sisula, H., Golborn, T., and Liljelund, L.-E. (1995). Airborn contaminants in the Arctic: What we need to know. *The Science of the Total Environment*, **160/161**, 841–848.

Langmann, B. (2000). Numerical modelling of regional scale transport and photochemistry directly together with meteorological processes. *Atmospheric Environment*, **34**(21), 3585–3598.

Laurila, T., Plathan, P., Hakol, H., Koskinen, T., and Lättilä, H. (1995). Pallas, a site for studies of photochemically-active trace species in the European Arctic. *EUROTRAC Newsletter*, **15**, 2–4.

Lavorel, S., Lambin, E.F., Flannigan, M., and Scholes, M. (2002). Fires in the Earth System: The need for integrated research. *Global Change Newsletter*, **48**, 7–10.

Lazarev, A.I. and Savinykh, V.P. (1996). *Achievements of Domestic Astronautics in the Study of the Environment* (192 pp.). Hydrometeoizdat, St Petersburg [in Russian].

Leck, C., Bigg, E.K., Covert, D.S., Heitzenberg, J., Maenhant, W., Nilsson, E.D., and Wiedensohler, A. (1996). Overview of the atmospheric research program during the International Arctic Ocean Expedition of 1991 (IAOE-91) and its scientific results. *Tellus*, **488**(2), 136–155.

Leck, C. and Persson, C. (1996). The central Arctic Ocean as a source of dimethyl sulfide seasonal variability in relation to biological activity. *Tellus*, **488**(2), 156–177.

Ledley, T.S., Sundquist, E.T., Schwartz, S.E., Hall, D.K., Fellows, J.D., and Killeen, T.L. (1999). Climate change and greenhouse. *Earth Observations from Space*, **80**(39), 453–474.

Lee, K., Wannikhof, R., Takahashi, T., Doney, S.C., and Feely, R.A. (1998). Low interannual variability in recent oceanic uptake of atmospheric carbon dioxide. *Nature*, **396**(6707), 155–159.

Lee, X., Fuentes, J.D., Staebler, R.M., and Neumann, H.H. (1999). Long-term observation of the atmospheric exchange of CO_2 with a temperate deciduous forest in southern Ontario, Canada. *Journal of Geophysics Research*, **104**(D13), 15975–15984.

Legendre, L. and Krapivin, V.F. (1992). Model for vertical structure of phytoplankton community in Arctic regions. *Proceedings of the 7th International Symposium on the Okhotsk Sea and Sea Ice, 2–5 February 1992, Mombetsu, Japan* (pp. 314–316). Okhotsk Sea and Cold Ocean Research Association, Mombetsu.

Legendre, P. and Legendre, L. (1998). *Numerical Ecology* (853 pp.). Elsevier, Amsterdam.

Leimbach, M. and Toth, F.L. (2003). Economic development and emission control over the long term: The ICLIPS aggregated economic model. *Climate Change*, **56**(1–2), 139–165.

Lelieveld, J. and Dentener, F.J. (2000). What controls tropospheric ozone? *Journal of Geophysics Research*, **105**(D3), 3531–3552.

Lenz, R.J.M. and Beuttler, A. (2003). Experiences with GIS-based planning tool for spatial eco-balances. *Environmental Modelling and Software*, **18**(6), 581–585.

Leroux, M. (1998). *Dynamic Analysis of Weather and Climate* (356 pp.). Springer-Praxis, Chichester, UK.

Leslie, P.H. (1945). On the use of matrices in certain population mathematics. *Biometrika*, **33**(3), 183–212.

Levesque, J. and King, D.J. (2003). Spatial analysis of radiometric fractions from high-resolution multispectral imagery for modelling individual tree crown and forest canopy structure and health. *Remote Sensing of the Environment*, **84**(4), 589–602.

Levitus, S., Antonov, J.I., Wang, J., Delworth, T.L., Dixon, K.W., and Broccoli, A.J. (2001). Anthropogenic warming of Earth's climate system. *Science*, **292**(5515), 267–270.

Lieth, H. (1985). A dynamic model of the global carbon flux through the biosphere and its relations to climatic and soil parameters. *International Journal of Biometeorology*, **29**(2), 17–31.

Lin, D.-L., Sakoda, A., Shibasaki, R., Goto, N., and Suzuki, M. (2000). Modelling a global biogeochemical nitrogen cycle in terrestrial ecosystems. *Ecological Modelling*, **135**(1), 89–110.

Lin, X., Trainer, M., and Hsie, E.-Y. (1998). A modeling study of tropospheric species during the North Atlantic Regional Experiment (NARE). *Journal of Geophysics Research*, **103**(D11), 13593–13613.

Lindzen, R.S., Chou, M.-D., and Hou, A.Y. (2001). Does the Earth have an adaptive infrared iris? *Bulletin of the American Meteorological Society*, **82**, 417–432.

Lindzen, R.S. and Giannitsis, C. (2003). Reconciling observations of global temperature change. *Geophysics Research Letters*, **29**(10), X1–X3.

Liping, G., Erda, L., and Zhongpei, L. (2000). Methane emission flux and mitigation options and its relationship with N_2O emission from paddy soils. *Proceedings of 2nd International Methane Mitigation Conference, 18–23 June 2000, Novosibirsk* (pp. 217–222). Novosibirsk University Press, Novosibirsk.

Lischke, H., Lotter, A.F., and Fischlin, A. (2002). Untangling a Holocene pollen record with forest model simulations and independent climate data. *Ecological Modelling*, **150**(1–2), 1–21.

Lockwood, J.G. (2001). Abrupt and sudden climatic transitions and fluctuations: A review. *International Journal of Climatology*, **21**(9), 1153–1180.

Logan, J.A. (1999). An analysis of ozonesonde data for the troposphere: Recommendations for testing three-dimensional models and development of a gridded climatology for tropospheric ozone. *Journal of Geophysics Research*, **104**(D13), 16115–16149.

Logofet, D.O. (2002). Matrix population models: construction, analysis and interpretation. *Ecological Modelling*, **148**(3), 307–310.

Lohmann, K., Feichter, J., Penner, J., and Leatich, R. (2000). Indirect effect of sulfate and carbonaceous aerosols: A mechanistic treatment. *Journal of Geophysics Research*, **105**(D10), 12193–12206.

Losev, K.S. (2001). *Ecological Problems and Perspectives of Sustainable Development of Russia in the 21st Century* (400 pp.). Kosmosinform, Moscow [in Russian].

Lovett, R.A. (2002). Rain might be leading carbon sink factor. *Science*, **296**(5574), 1787.

Lu, F. and Li, Z. (2003). A model of ecosystem health and its application. *Ecological Modelling*, **170**(1), 55–59.

Lueker, T.J., Keeling, R.F., and Dubey, M.K. (2001). The oxygen to carbon dioxide ratios observed in emissions from a wildfire in Northern California. *Geophysics Research Letters*, **28**(12), 2413–2416.

Mabuchi, K., Sato, Y., and Kida, H. (2000). Numerical study of the relationships between climate and the carbon dioxide cycle on a regional scale. *Journal of the Meteorological Society of Japan*, **78**(1), 25–46.

Macdonald, A.M., O'Neil Baronger, M., Lee, K., Wallace, D., and Wanninkhof, R. (2002a). Subtropical Atlantic carbon transport. *International WOCE Newsletter*, **42**, 14–19.

Macdonald, A.M., Wanninkhof, R., O'Neil Baringer, M., Robbins, P.E., and Wallace, D.W.R. (2002b). Oceanic biogeochemical fluxes: A summary of the JGOFS portion of the WOCE/JGOFS Transport Workshop, 25–29 June 2001, Southampton. *International WOCE Newsletter*, **42**, 20–21.

MacKenzie, I.A., Harwood, R.S., Stott, P.A., and Watson, G.C. (1999). Radiative-dynamic effects of the Antarctic ozone hole and chemical feedback. *Quarterly Journal of the Royal Meteorological Society*, **125**, Part B(558), 2171–2203.

Maduro, R.A. and Schauerhammer, R. (1992). *The Holes in the Ozone Scare: The Scientific Evidence that the Sky Isn't Falling In* (356 pp.). 21st Century Science Associates/John Wiley & Sons, Washington, DC.

Magagi, R. and Bernier, M. (2003). Optimal conditions for wet snow detection using RADARSAT SAR data. *Remote Sensing of the Environment*, **84**(2), 221–233.

Majorivicz, J., Safanda, J., and Skinner, W. (2002). East to west retardation in the onset of the recent warming across Canada inferred from inversions of temperature logs. *Journal of Geophysics Research*, **107**(B10), ETG6/11–ETG6/12.

Maksudova, L.G., Savinykh, V.P., and Tshvetkov, V.Ya. (2000). Integration of sciences about the environment within the framework of geoinformatics. *Earth Research from Space*, **1**, 46–50 [in Russian].

Malinnikov, V.A., Savinykh, V.P., Sladkoptsev, S.A., and Tsypiba, E.M. (2000). *Geography from Space*. Moscow University of Geodesy and Cartography, Moscow [in Russian].

Mann, M.E. and Jones, P.D. (2003). Global surface temperatures over the past two millennia. *Geophysics Research Letters*, **30**(15), 1820–1821.

Manney, G.L., Sabutis, J.L., Allen, D.R., Lahoz, W.A., Scaife, A.A., Randall, C.E., Pawson, S., Swinbank, R., and Naujokat, B. (2003) Simulations of dynamics and transport during the September 2002 Antarctic Major Warming. *J. Atmos. Sci.* (submitted).

Mannion, A.M. (1997). *Global Environmental Change* (2nd edn, 387 pp.). Longman, London.

Marchuk, G.I. and Kondratyev, K.Ya. (1992). *Priorities in Global Ecology* (264 pp.). Science, Moscow [in Russian].

Marchuk, G.I., Kondratyev, K.Ya., Kozoderov, V.V., and Khvorostyanov, V.I. (1986). *Clouds and Climate* (512 pp.). Gidrometeoizdat, Leningrad [in Russian].

Marenkin, D.S. (2002). Modelling the energy fluxes in the atmosphere–plant–soil system. *Problems of Environmental and Natural Resources*, **4**, 77–87 [in Russian].

Marland, G., McCarl, B.A., and Schneider, U. (2001). Soil carbon: Policy and economics. *Climate Change*, **51**(1), 101–117.

Martens, P. (1998). *Health and Climate Change: Modelling the Impacts of Global Warming and Ozone Depletion* (224 pp.). Earthscan, London.

Matishov, D.G. and Matishov, G.G. (2001). *Radiation and Ecological Oceanology* (419 pp.). Kola Science Centre, Russian Academy of Science, Apatity [in Russian].

Matsumura, H. (2000). *Japan and the Kyoto Protocol Conditions for Ratification* (Energy and Environment Programme, 66 pp.). The Royal Institute of International Affairs, London.

McCarl, B.A. and Schneider U.A. (2001). Greenhouse gas mitigation in US agriculture and forestry. *Science*, **294**(5551), 2481–2482.

McCauley, L.L. and Meier, M.F. (eds) (1991). *Arctic System Science: Land/Atmosphere/Ice Interactions* (48 pp.). ARCUS, Fairbanks, AL.

McCormick, M.P. (1998). Summary of the SAGE III (Stratospheric Aerosol and Gas Experiment III) science team meeting. *The Earth Observer*, **10**(3), 7–14.

McGillis, W.R., Edson, J.B., Hare, J.E., and Fairrall, C.W. (2001). Direct covariance air–sea CO_2 fluxes. *Journal of Geophysics Research*, **106**(C8), 16729–16746.

McIntyre, A.D. (1999). The environment and the oil companies. *Marine Pollution Bulletin*, **38**(3), 155–156.

McKenzie, C., Schiff, S., Aravena, R., Kelly, C.J.T., and Lons, V. (1998). Effect of temperature on production of CH_4 and CO_2 from peat in a natural and flooded boreal forest wetlands. *Climate Change*, **40**(2), 247–266.

McNulty, S.G. (2002). Hurricane impacts on US forest carbon sequestration. *Environmental Pollution*, **116**, S17–S24.

Meadows, D.H., Meadows, D.L., Randers, J., and Behrens, W.W. (1972). *The Limits to Growth* (208 pp.). Potomak Associates, New York.

Meadows, M. (2000). United Kingdom methane emissions: Trends, projections and mitigation options. *Proceedings of 2nd International Methane Mitigation Conference, 18–23 June 2000, Novosibirsk* (pp. 37–44). Novosibirsk University Press, Novosibirsk.

Meijer, E.W. and Velthoven, P. (1997). The effect of the conversion of nitrogen oxides in aircraft exhaust plumes in global models. *Geophysics Research Letters*, **24**(23), 3013–3016.

Mejer, H.A.J. (2001). The science of greenhouse gases: Uncertainties in sources and sinks, and implications for verification. *Energy and Environment*, **12**(5–6), 425–446.

Mendelsohn, R. and Rosenberg, N.J. (1994). Framework for integrated assessment of global warming impacts. *Climate Change*, **28**(1–2), 15–44.

Menon, S., Del Genio, A.D., Koch, D., and Tselioudis, G. (2002). GCM simulations of the aerosol indirect effect. Sensitivity to cloud parametrization and aerosol burden. *Journal of Atmospheric Science*, **59**(3), Part 2, 692–743.

Metternicht, G.I. and Zinck, J.A. (2003). Remote sensing of soil salinity: potentials and constraints. *Remote Sensing of the Environment*, **85**(1), 1–20.

Metting, F.B., Smith, J.L., Amthor, J.S., and de Izaurral, R.C. (2001). Science needs and new technology for increasing soil carbon sequestration. *Climate Change*, **51**(1), 11–34.

Metzger, R.A. and Benford, G. (2001). Sequestering of atmospheric carbon through permanent disposal of crop residue. *Climatic Change*, **49**(1–2), 11–19.

Michaels, P.J., Singer, S.F., Knapperberger, O.P.E., Kerr, J.B., and McElroy, C.T. (1994). Analyzing ultraviolet-B radiation: Is there a trend? *Science*, **264**, 1341–1343.

Michalowski, B.A., Francisco, J.S., Li, S.-M., Barrie, L.A., Bottenheim, J.W., and Shepson, P.B. (2000). A computer model study of multiphase chemistry in the Arctic boundary layer during polar sunrise. *Journal of Geophysics Research*, **105**(D12), 15131–15146.

Milford, C., Hargreaves, K.J., Sutton, M.A., Loubet, B., Cellier, P. (2001). Fluxes of CO_2 over upland moorland in the vicinity of agricultural land. *Journal of Geophysics Research*, **106**(D20), 24169–24181.

Mintzer, I.M. (1987). *A Matter of Degrees: The Potential for Controlling the Greenhouse Effect* (Research Rep. No. 15, 70 pp.). World Resources Institute, New York.

Mirkin, B.M., Naumova, L.G., and Solomesch, A.I. (2002). *Present Science about Vegetation* (263 pp.). Logos Press, Moscow [in Russian].

Mischnenko, M., Penner, J., and Anderson, D. (2002). Global Aerosol Climatology Project. *Journal of Atmospheric Science*, **59**(1), 249–253.

Moe, A. and Tanger, K. (2000). *The Kyoto Mechanisms and Russian Climate Politics* (Energy and Environment Programme, 110 pp.). The Royal Institute of International Affairs, London.

Moiseev, N.N. (1990) *Man and Noosphere* (352 pp.). Young Guards, Moscow [in Russian].

Molemaker, M.J. and de Arellano, J.V.-G. (1998). Control of chemical reactions by convective turbulence in the boundary layer. *Journal of Atmospheric Science*, **55**(4), 568–579.

Molina, M.J. and Rawland, F.S. (1974). Predicted present stratospheric abundances of chlorine species from photodissociation of carbon tetrachlorine. *Geophysics Research Letters*, **1**, 309–315.

Monahan, E.C. and Dam, H.G. (2001). Bubbles: An estimate of their role in the global oceanic flux of carbon. *Journal of Geophysics Research*, **106**(C5), 9377–9384.

Monin, A.S. and Krasitsky, V.P. (1985). *Phenomena at the Ocean Surface* (375 pp.). Gidrometeoizdat, Leningrad [in Russian].

Monin, A.S. and Obukhov, A.M. (1954). Spatial characteristics of the turbulence in the upper atmospheric layer. *Annals of the Soviet Union Academy of Science*, **93**, 223–226 [in Russian].

Monnin, E., Indermühle, A., Dällenbach, A., Flückiger, J., Stauffer, B., Stocker, T.F., Raynaud, D., and Barnola, J.-M. (2001). Atmospheric CO_2 concentrations over the last glacial termination. *Science*, **291**(5501), 112–114.

Monod, A. and Carlier, P. (1999). Impact of clouds on the tropospheric ozone budget: Direct effect of multiphase photochemistry of soluble organic compounds. *Atmospheric Environment*, **33**, 4431–4446.

Morgan, M.G. (2000). Managing carbon from the bottom up. *Science*, **289**(5480), 2285.

Moritz, R.E., Bitz, C.M., and Steig, E.J. (2002). Dynamics of recent climate change in the Arctic. *Science*, **297**(5586), 1497–1501.

Morris, G.A., Kawa, S.R., Douglas, A.R., Schoeberg, M.R., Froidevaux, L., and Waters, J. (1998). Low-ozone pockets explained. *Journal of Geophysics Research*, **103**(D3), 3595–3610.

Muller, F. and Steinhardt, U. (2003). Landscape modelling and landscape analysis. *Ecological Modelling*, **168**(3), 215–216.

Müller, R., Schmidt U., Engel, A., McKenna, D.S., and Profitt, M.H. (2001). The O_3–N_2O relation from balloon-borne observations as a measure of Arctic ozone loss in 1991/92. *Quarterly Journal of the Royal Meteorological Society*, **127**(574), Part B, 1389–1412.

Munro, R., Siddans, R., Reburn, W.S., and Kerridge, B.S. (1998). Direct measurement of tropospheric ozone distributions from space. *Nature*, **392**, 168–171.

Nabuurs, G.J., Dolman, A.J., Verkaik, E., Whitmore, A.R., Daamen, W.P., Oenema, O., Kabat, P., and Mohren, G.M.J. (2000). *Resolving Issues on Terrestrial Biospheric Sinks in the Kyoto Protocol* (Report No. 410 200 030, 10 pp.). Dutch National Research Programme on Global Air Pollution and Climate Change, Amsterdam, The Netherlands,

Nabuurs, G.J. and Sikkema, R. (2001). International trade in wood products: Its role in the land use change and forestry carbon cycle. *Climate Change*, **49**(4), 377–395.

Nadelhoffer, K.L., Emmet, B.A., Gundersen, P., Kjønaas O.J., Koopmans, C.J., Schleppe, P., Tietem, A., and Wright, R.F. (1999). Nitrogen deposition makes a minor contribution to carbon sequestration in temperate forests. *Nature*, **398**(6723), 145–148.

Nagel, D., Herber, A., Thomason, L.W., and Leiterer, U. (1998). Vertical distribution of the spectral aerosol optical depth in the Arctic from 1993 to 1996. *Journal of Geophysics Research*, **103**(D2), 1857–1870.

Nagurny, A.P. and Maistrova, V.V. (2002). Long-term temperature trends for the free atmosphere in the Arctic. *Doklady RAS*, **170**(2), 345–349 [in Russian].

Nair, H., Allen, M., Froidevaux, L., and Zurek, R.W. (1998). Localized rapid ozone loss in the northern winter stratosphere: An analysis of UARS observations. *Journal of Geophysics Research*, **103**(D1), 1555–1571.

Nakicenovic, N., Grübler, A., and McDonald, A. (eds) (1998) *Global Energy Perspectives* (299 pp.). Cambridge University Press, Cambridge, UK.

Naoum, S. and Tsanis, I.K. (2003). Hydroinformatics in evapotranspiration estimation. *Environmental Modelling*, **18**(3), 261–271.

NASA/NOAA (2002a) Ozone Hole Press Release, 30 September, 2002 (`http://toms.gsfc.nasa.gov`)

NASA/NOAA (2002b) Ozone Hole Press Release, 6 December, 2002 (`http://www.gsfc.nasa.gov/topstory/20021206ozonehole.html`)

Nedoluha, G.E., Bevilacqua, R.M., Fromm, M.D., Hoppel, K.W., and Allen, D.R. (2003) POAM measurements of PSCs and water vapor in the 2002 Antarctic vortex. *Geophys. Res. Lett.*, **30**(15), 1796, 10.1029/2003GL017577.

Nefedova, E.I. (1994). Mathematical modeling of the global carbon cycle in the atmosphere–ocean system. PhD thesis, Computing Center of the Russian Academy of Science, Moscow (112 pp.) [in Russian].

Nefedova, E.I. and Tarko, A.M. (1993). Study of the global carbon cycle using the zonal model in the atmosphere–ocean system. *Annals of the Russian Academy of Science*, **333**(5), 645–647 [in Russian].

Nepstad, D., McGrath, D., Barros, A.C., Carvalho, G., Santilli, M., Vera Diaz, M., and Del, C. (2002). Frontier governance in Amazonia. *Science*, **295**(5555), 621–631.

Newell, R.E. and Pizer, W. (2002). Discounting the benefits of climate change policies using uncertain rates. *Resources*, **146**, 15–20.

Newell, R.E., Browell, E.V., Davis, D.D., and Liu, Sh.E. (1997). Western Pacific tropospheric ozone and potential vorticity: Implications for Asian pollution. *Geophysics Research Letters*, **24**(22), 2733–2736.

Newman, P.A. (1986) The final warming and polar vortex disappearance during the Southern Hemisphere spring. *Geophys. Res. Lett.*, **13**, 1228–1231.

Newman, P.A. and Nash, E.R. (2003) The unusual Southern Hemisphere stratosphere winter of 2002. *J. Atmos. Sci.* (submitted).

Nezlin, N.P. and McWilliams, J.C. (2003). Satellite data, empirical orthogonal functions, and the 1997–1998 El Niño off California. *Remote Sensing of the Environment*, **84**(2), 234–254.

Ni, J. (2001). Carbon storage in terrestrial eco-systems of China: Estimates at different spatial resolutions and their responses to climate change. *Climate Change*, **49**(3), 339–358.

Nicolis, C. and Nicolis, G. (1995). From short-scale atmospheric variability to global climate dynamics: Toward a systematic theory of averaging. *Journal of Atmospheric Science*, **52**(11), 1903–1913.

Nielsen, T.T. (1999). Characterization of fire regimes in the Experiment for Regional Sources and Sinks of Oxidants (EXPRESSO) study area. *Journal of Geophysics Research*, **104**(D23), 30713–30723.

Nikolo, N. and Zeller, K. (2003). Modeling coupled interactions of carbon, water, and ozone exchange between terrestrial ecosystems and the atmosphere, I: Model description. *Environmental Pollution*, **124**(2), 231–246.

Niles, J.O. and Schwarze, R. (2001). The value of careful carbon accounting in wood products: Editorial. *Climate Change*, **49**(4), 371–376.

Nilsson, S., Jonas, M., and Obersteiner, M. (2002). COP-6: A healing shock? An Editorial Essay. *Climate Change*, **52**(1–2), 25–28.

Nitu, C., Krapivin, V.F., and Bruno, A. (2000A). *Intelligent Techniques in Ecology* (150 pp.). Printech, Bucharest.

Nitu, C., Krapivin, V.F., and Bruno, A. (2000b). *System Modelling in Ecology* (260 pp.). Printech, Bucharest.

Nordhaus, W.D. (2001). Global warming economics. *Science*, **294**, 1283–1284.

Novozhilov, A.S. (2002). Identification of the parameters of a dynamic system that simulates the interaction between pollution and the environment. *Bulletin of the Russian Academy of Science, Series on the Theory of Systems and Management*, **3**, 123–127 [in Russian].

Oechel, W.C., Vourlitis, G.L., Hastings, S.J., Zulneta, R.C., Hinzman, L., and Kane, D. (2000). Acclimation of ecosystem CO_2 exchange in the Alaskan Arctic in response to climate warming. *Nature*, **406**(6799), 978–981.

O'Neill, B.C. and Oppenheimer, M. (2002). Dangerous climate impacts and the Kyoto Protocol. *Science*, **296**(5575), 1971–1972.

Orr, J.C., Maier-Reimer, E., Mikolajewicz, U., Monfray, P., Sarmiento, J.L., Toggweiler, J.R., Taylor, N.K., Palmer, J., Gruber, N., Sabine, C.L. *et al.* (2001). Estimates of anthropogenic carbon uptake from four three-dimensional global ocean models. *Global Biogeochemical Cycles*, **15**, 43–60.

Orsolini, Y.J., Randall, C.E., Manney, G.L., and Allen, D.R. (2003) An observational study of the final breakdown of the southern hemisphere stratospheric vortex in 2002. *J. Atmos. Sci.* (submitted).

Ougolnitsky, G.A. (1999). *Management of Ecologo–Economic Systems* (132 pp.). High School Books, Moscow [in Russian].

Ougolnitsky, G.A., Demianenko, Ya.M., and Dubrov, D.V. (2000). Methods of hierarchical optimization in forestry problems. In: G.A. Ougolnitsky (ed.), *Computer Modeling: Ecology* (pp. 79–89). High School Books, Moscow [in Russian].

Overland, J.E. and Adams, J.M. (2001). On the temporal character and regionality of the Arctic Oscillation. *Geophysics Research Letters*, **28**(14), 2811–2814.

Pacala, S.W., Hurtt, G.C., Baker, D., Peylin, P., Houghton, R.A., Birdsey, R.A., Heath, L., Sundquist, E.T., Stallard, R.F., Ciais, P., *et al.* (2001). Consistent land- and atmosphere-based US carbon sink estimates. *Science*, **292**, 2316–2320.

Painter, T.H., Dozier, J., Davis, R.E., and Green, R.O. (2003). Retrieval of subpixel snow-covered area and grain size from imaging spectrometer data. *Remote Sensing of the Environment*, **85**(1), 64–77.

Papakyriakou, T.N. and McCaughey, J.H. (1991). An evolution of evapotranspiration for a mixed forest. *Canadian Journal of Forestry Research*, **21**(11), 1622–1631.

Parkinson, C.L. (2003). Aqua: An Earth-observing satellite mission to examine water and other climate variables. *IEEE Transactions on Geoscience and Remote Sensing*, **41**(2), 173–183.

PARCS develops two updated research goals. (2000). *Witness the Arctic*, **9**(1), 8–9.

Patin, S. (1999). *Environmental Impact of the Offshore Oil and Gas Industry* (425 pp.). Eco-Monitor, New York.

Paulsen, C.A. (1970). The mathematical representation of wind and temperature profiles in the unstable atmospheric layer. *Journal of Applied Meteorology*, **9**, 857–861.

Peng, C. (1999). Modelling the response of net primary productivity (NPP) of boreal forest ecosystems to changes in climate and tire disturbance regime. *Ecological Modelling*, **122**(3), 175–193.

Peng, C., Jiang, H., Apps, M.J., and Zhang, Y. (2002a). Effects of harvesting regimes on carbon and nitrogen dynamics of boreal forests in central Canada: A process model simulation. *Ecological Modelling*, **155**(2–3), 177–189.

Peng, C., Liu, J., Dang, O., Apps, M.J., and Jiang H. (2002b). TRIPLEX: A generic hybrid model for predicting forest growth and carbon and nitrogen dynamics. *Ecological Modelling*, **153**(1–2), 109–130.

Penman, H.L. (1948). Natural evaporation from open water, bare soil and grass. *Proceedings of the Royal Society (London)*, **193**, 120–145.

Penner, J.E., Lister, D.H., Griggs, D.J., Dokken, D.J., and McFarland, M. (eds) (1999). *Aviation and the Global Atmosphere* (A special report of IPCC WG I and III, 373 pp.). Cambridge University Press, Cambridge, UK,

Pennise, D.M., Smith, H.R., Kithinji, J.P., Rezende, M.A., Raad, T.J., Zhang, J., and Fan, C. (2001). Emissions of greenhouse gases and other airborne pollutants from charcoal making in Kenya and Brazil. *Journal of Geophysics Research*, **106**(D20), 24143–24155.

Peritunen, J. (1998). LIGNUM: A model combining the structure and the functioning of trees. *Ecological Modelling*, **108**(1–3), 189–198.

Pervaniuk, V.S. (2001). A spatial numerical model of the global biogeochemical cycles of carbon and nitrogen in the atmosphere–ocean system. PhD thesis, Computing Center of the Russian Academy of Science (118 pp.) [in Russsian].

Philander, G. (2001). Why global warming is controversial. *Science*, **294**(5549), 2105–2106.

Phillips, G.W., August, R.A., Cherepenin, V.A., Harper, M.J., King, S.E., Krapivin, V.F., Pautkin, A.Yu., and Tsang, F.Y. (1997). Radionuclear pollutants in the Angara and Yenisey rivers of Siberia. *Radioprotection Colloques*, **32**, 299–304.

Pielke, R.A., Sr (2001a). Carbon sequestration: The need for an integrated climate system approach. *Bulletin of the American Meteorological Society*, **82**(11), 20–21.

Pielke, R.A., Sr (2001b). Earth system modeling An integrated assessment tool for environmental studies. In: T. Matsuno and H. Kida (eds), *Present and Future of Modeling Global Environmental Change: Toward Integrated Modeling* (pp. 311–337). Terrapub, Tokyo.

Pielke, R.A., Sr (2002). Overlooked issues in the US national climate and IPCC assessments. *Climate Change*, **52**(1–2), 1–11.

Pilegaard, K., Hummelshoj, P., and Jensen, N.O. (1998). Fluxes of ozone and nitrogen dioxide measured by eddy correlation over a harvested wheat field. *Atmospheric Environment*, **32**(5), 1167–1177.

Pirjola, L., Aaksonen, A., Aalto, P., and Külmälä, M. (1998). Sulphate aerosol formation in the Arctic boundary layer. *Journal of Geophysics Research*, **103**(D7), 8309–8322.

Pittock, A.B. (2002). What we know and don't know about climate change: Reflections on the IPCC TAR. *Climate Change*, **53**(4), 393–411.

Podgorny, I.A. and Ramanathan, V. (2001). A modeling study of the direct effect of aerosols over the tropical Indian Ocean. *Journal of Geophysics Research*, **106**(D20), 24097–24105.

Polishchuk, Yu.M. (1992). *Simulation Modelling of the Systems with Natural Components* (228 pp.). Science, Novosibirsk [in Russian].

Polissar, A.V., Hopke, P.K., Malm, W.C., and Sisler, J.F. (1998a). Atmospheric aerosol over Alaska, 1: Spatial and seasonal variability. *Journal of Geophysics Research*, **103**(15), 19035–19044.

Polissar, A.V., Hopke, P.K., Panteero, P., Malm, W.C., and Sisler, J.F. (1998b). Atmospheric aerosol over Alaska, 2: Elemental composition and sources. *Journal of Geophysics Research*, **103**(D15), 19045–19057.

Popovicheva, O.B., Starik, A.M., and Favorsky, O.N. (2000). Impacts of aviation on the gas and aerosol composition of the atmosphere. *Bulletin of the Russian Academy of Science: FAO*, **2**, 163–176 [in Russian].

Porte, A. and Bartelink, H.H. (2002). Modelling mixed forest growth: A review of models for forest management. *Ecological Modelling*, **150**(1–2), 141–188.

Post, W.M., Izaurralde, R.C., Mann, L.K., and Bliss, N. (2001). Monitoring and verifying changes of organic carbon in soil. *Climate Change*, **51**(1), 73–99.

Potosnak, M.J., Wofsy, S.C., Denning, A.S., Conway, T.J., Munger, J.W., and Barnes, D.H. (1999). Influence of biotic exchange and combustion sources on atmospheric CO_2 concentrations in New England from observations at a forest flux tower. *Journal of Geophysics Research*, **104**(D8), 9561–9569.

Potter, C.S. and Klooster, S.A. (1999). Detecting a terrestrial biosphere sink for carbon dioxide: Interannual ecosystem modeling for the mid-1980s. *Climate Change*, **42**(3), 489–503.

Potter, C.S., Klooster, S.A., de Carvalho, C.R., Genovese, V.B., Torregrosa, A., Dungan, J., Bobo, M., and Coughlan, J. (2001). Modeling seasonal and interannual variability in ecosystem carbon cycling for the Brazilian Amazon region. *Journal of Geophysics Research*, **106**(D10), 10423–10446.

Prather, M., Derwent, R., Ehhalt, D., Fraser, P., Sanhaeza, E., and Zhou, X. (1994). Other trace gases and atmospheric chemistry. In: J.-T. Houghton, L.G.M. Filho, J.P. Bruce, H. Lee, B.A. Callander, and E.F. Haites (eds), *Climate Change 1994* (pp. 73–126). Cambridge University Press, Cambridge, UK.

Prather, M., Midgley, P., Rowland, S.R., and Stolarski, R. (1996). The ozone layer: The road not taken. *Nature*, **381**(6583), 551–554.

Preller, R.H. and Cheng, A.B.E. (1999). Modeling the transport of radiactive contaminants in the Arctic. *Marine Pollution Bulletin*, **38**(2), 71–91.

Prentice, I.C., Farquhar, G.D., Fasham, M.J.R., Goulden, M.L., Heimann, M., Jaramillo, V.J., Kheshgi, H.S., Le Quéré, C., Scholes, R.J., and Wallace, D.W.R. (2001). The carbon cycle and atmospheric carbon dioxide. In: J.T. Houghton and D. Yihmi (eds), *Climate Change 2001: The Scientific Basis* (pp. 183–237). Cambridge University Press, Cambridge, UK.

Priestley, C.H.B. and Taylor, R.J. (1972). On the assessment of surface heat flux and evapotranspiration using large scale parameters. *Monthly Weather Review*, **100**, 81–92.

Prinn, R., Jacoby, H., Sokolov, A., Wang, C., Xiao, X., Yang, Z., Eckaus, R., Stone, P.H., Ellerman, A.D., Melillo, J. *et al.* (1999). Integrated Global System model for climate policy assessment: Feedbacks and sensitivity studies. *Climate Change*, **41**(304), 469–546.

Przybylak, R. (2002). *Variability of Air Temperature and Atmospheric Precipitation in the Arctic* (352 pp.). Kluwer, Hingham, MA.

Pszenny, A. and Brasseur, G. (1997). Tropospheric ozone: An emphasis on IGAC research. *Global Change Newsletter*, **30**, 2–10.

Qi, F., Guoduong, C., and Masao, M. (2001). The carbon cycle of sandy lands in China and its global significance. *Climate Change*, 48(4), 535–549.

Qi, Y., Xu, M., and Wu, J. (2002). Temperature sensitivity of soil respiration and its effects on ecosystem carbon budget: Nonlinearity begets surprises. *Ecological Modelling*, **153**(1–2), 131–142.

Rabotnov, T.A. (1987). *Experimental Phytocenology* (160 pp.). Moscow State University, Moscow [in Russian].

Ragas, A.M.J. and Leuven, R.S.E.W. (1999). Modelling of water quality-based emission limits for industrial discharges in rivers. *Water Science and Technology*, **39**(4), 185–192.

Rahn, K. (1981). Relative importances of North America and Eurasia as sources of arctic aerosol. *Atmospheric Environment*, **15**(8), 1447–1455.

Räisänen, J. (2001). CO_2-induced climate change in GMIP2 experiments: Quantification of agreement and role of internal variability. *Journal of Climate*, **14**(9), 2088–2104.

RAISE Plan focuses on ship-based research in Russia (2001). *Witness the Arctic*, **9**(1), 7–8.

Ramanathan, V. and Coakley, J.A. (1978). Climate modelling through radiative–convective models. *Reviews into Geophysical Space Physics*, **16**, 465–489.

Randall, C.E., Bevilacqua, R.M., Lumpe, J.D., Hoppel, K.W., Rusch, D.W., and Shettle, E.P. (2000). Comparison of Polar Ozone and Aerosol Measurement (POAM) II and Stratospheric Aerosol and Gas Experiment (SAGE) II aerosol measurements from 1994 to 1996. *Journal of Geophysics Research*, **105**(D3), 3929–3942.

Randall, C.E., Manney, G.L., Allen, D.R., Bevilacqua, R.M., Hornstein, J., Trepte, C., Lahoz, W.A., Ajtic, J., and Bodeker, G. (2003) Reconstruction and simulation of stratospheric ozone distributions during the 2002 Austral winter. *J. Atmos. Sci.* (submitted).

Rangno, A.L. and Hobbs, P.V. (2001). Ice particles in stratiform clouds in the Arctic and possible mechanisms for the production of high ice concentrations. *Journal of Geophysics Research*, **106**, 15065–15075.

Raputa, V.F., Smolyakov, B.S., Kutsenogii, K.P., Smirnova, A.I., and Yaroslavtseva, T.V. (2000). Analysis of temporal dynamics for the change in the composition of atmospheric aerosol in the north of the West Siberia. *Siberian Ecological Journal*, **VII**(1), 97–102 [in Russian].

Raupach, M.R. (2000). Equilibrium evaporation and the convective boundary layer. *Boundary-Layer Meteorology*, **96**, 107–141.

Raupach, M.R., Barrett, D., Briggs, P., and Kirby, M. (2001). Water, carbon and nutrient cycles on the Australian continent. *Global Change Newsletter*, **48**, 20–22.

Rau, G.H. and Caldeira, K. (2002). Minimizing effects of CO_2 storage in oceans. *Science*, **295**, 275–276.

Ravishankara, A.R., Hancock, G., Kawasaki, M., and Matsumi, Y. (1998). Photochemistry of ozone: Surprises and recent lessons. *Science*, **280**, 60–61.

Reichle, H.G., Jr, Anderson, B.E., Cannors, V.S., Denkins, T.C., Forbes, D.A., Gormsen, B.B., Langenfelds, R.L., Neil, D.O., Nolf, S.R., Novelli, P.C. *et al.* (1999). Space shuttle based global CO measurements during April and October 1994, MAPS instrument, data reduction, and data validation. *Journal of Geophysics Research*, **104**(D17), 21443–21454.

Reid, P.C. (2001). Climate change and the Continuous Plankton Recorder survey. *Marine Observation*, **71**(353), 118–123.

Reilly, J., Stone, P.H., Forest, C.E., Webster, M.D., Jacoby, H.D., and Prinn, R.G. (2001). Climate change: Uncertainty and climate change assessments. *Science*, **293**(5529), 430.

Remer, L.A., Kaufman, Y.J., Levin, Z., and Ghan, S. (2002). Model assessment of the ability of MODIS to measure top-of-atmosphere direct radiative forcing from smoke aerosols. *Journal of Atmospheric Science*, **59**(3), Part 2, 657–667.

Report from the Arctic Climate Impact Assessment Modeling and Scenarios Working Group (2001). *Workshop, 29–31 January 2001, Stockholm* (28 pp.). ACIA Secretariat, Fairbanks, AL.

Riedlinger, S.H. and Preller, R.H. (1991). The development of a coupled ice-ocean model for forecasting ice conditions in the Arctic. *Journal of Geophysics Research*, **96**, 16955–16977.

Riedo, M., Gyalistras, D., and Fuhrer, J. (2000). Net primary production and carbon stocks in differently managed grasslands: Simulation of site-specific sensitivity to climate change. *Ecological Modelling*, **134**(2–3), 207–227.

Rinsland, C.P., Salawitch, R.J., Gunson, M.R., Solomon, S., Zander, R., Mahieu, E., Goldman, A., Newchurch, M.J., Irion, F.W., and Chang, A.Y. (1999). Polar stratospheric descent of NO_y and CO and Arctic denitrification during winter 1992–1993. *Journal of Geophysics Research*, **104**(D1), 1847–1861.

Rochon, G.L., Krapivin, V.F., Watson, M., Fauria, S., Tsang, F.Y., and Fernandez, M. (1996). Remote characterization of the landsea interface: A case study of the Viet-Nam/South China Sea coastal zone. *Proceedings of the 2nd Ho Chi Minh City Conference on Mechanics, 24–25 September 1996, Ho Chi Minh City* (pp. 72–74). Institute of Applied Mechanics, Ho Chi Minh, Vietnam.

Röckmann, T., Brenninkmeijer, C.A.M., and Crutzen, P.J. (1999). Short-term variations in the $^{13}C/^{12}C$ ratio of CO as a measure of Cl activation during tropospheric ozone depletion events in the Arctic. *Journal of Geophysics Research*, **104**(D1), 1691–1697.

Rojstaczer, S., Sterling, S.M., and Moore, N.J. (2001). Human appropriation of photosynthesis products. *Science*, **294**, 2549–2552.

Romankevich, E.A. and Vetrov, A.A. (2001). *Carbon Cycle in the Russian Arctic Seas* (302 pp.). Science Press, Moscow [in Russian].

Romashkin, P.A., Hurst, D.F., Elkins, J.W., Dutton, G.S., and Wamsley, P.R. (1999). Effect of the tropospheric trend on the stratospheric tracer–tracer correlations: Methyl chloroform. *Journal of Geophysics Research*, **104**(D21), 26643–26652.

Roo-Zeilinska, E. and Solon, J. (1997). Effect of geographical location on species composition, vegetation diversity and phytoindicative characteristics in pine forests. *Environmental Pollution*, **98**(3), 347–360.

Rosa, L.P. and Ribeiro, S.K. (2001). The present, past, and future contributions to global warming of CO_2 emissions from fuels. *Climate Change*, **48**(2–4), 289–308.

Roscoe, H.K., Shanklin, J.D., and Colwell, S.R. (2003) Has the Antarctic vortex split before? *J. Atmos. Sci.* (submitted).

Rosenberg, N.J. and Izaurralde, R.C. (2001). Storing carbon agricultural soils to help head-off global warming. *Climate Change*, **51**(1), 1–10.

Rovinskiy, F.Ya., Chernogaeva, G.M., and Paramonov, S.G. (1995). The roles of river runoff and atmospheric transport in the sea pollution of the Russian North. *Meteorology and Hydrology*, **9**, 22–29 [in Russian].

Rowland, F.S. (1991). Stratospheric ozone depletion. *Annual Reviews of Physical Chemistry*, **42**, 731–740.

Royer, D.L., Wing, S.L., Beerling, D.J., Jolley, D.W., Koch, P.L., Hockey, L.J., and Berner, R.A. (2001). Paleobotanical evidence for near present-day levels of atmospheric CO_2 during part of the tertiary. *Science*, **292**(5525), 2310–2313.

Rudels, B., Larsson, A.-M., and Sehistendt, P.-I. (1991). Stratification and water mass formation in the Arctic Ocean. *Polar Research*, **10**(1), 19–31.

Russel, J.M. and Lastovicka, J. (1998). The IAGA/ICMA Symposium Session 2.18 on Middle Atmosphere Trends and Low Frequency Variability (14 August 1997, Uppsala, Sweden). *SPARC Newsletter*, **10**, 25–26.

Sabine, C.L. and Feely, R.A. (2001). Comparison of recent Indian Ocean anthropogenic CO_2 estimates with a historical approach. *Global Biogeochemical Cycles,* **15**, 31–42.

Safonov, V.S., Sobolevskaya, Yu.S., and Ivanova, N.A. (1999). Emission of CO_2 by conifers with increased concentrations of O_3 and CO_2. *Optics of the Atmosphere and Ocean,* **12**(12), 1157–1160 [in Russian].

Sands, R.D. and Leimbach, M. (2003). Modeling agriculture and land use in an integrated assessment framework. *Climate Change*, **56**(1–2), 185–210.

Santer, B.D., Sausen, R., Wigley, T.M.L., Boyle, J.S., Achuta Rao, K., Doutriauz, C., Hansen, J.E., Meehl, G.A., Roeckner, E., Ruedy, R. *et al.* (2002). Behavior of tropopause height and atmospheric temperature in models, re-analysis, and observations: Decadal changes. *Journal of Geophysics Research*, **108**(D1), 1/1–1/22.

Santer, B.D., Wehner, M.F., Wigley, T.M.L., Sausen, R., Meehl, G.A., Taylor, K.E., Ammann, C., Arblaster, I., Washington, W.M., Boyle, J.S. *et al.* (2003). Contributions of anthropogenic and natural forcing to recent tropopause height changes. *Science*, **301**(5632), 479–483.

Sarancha, D.A. (1995) *Biomodelling* (138 pp.). Computer Center of the Russian Academy of Science Press, Moscow [in Russian].

Sarma, K.M. (1998). Protection of the ozone layer: A success story of UNEP. *Linkages Journal*, **3**(3), 6–10.

Sarma, V.V.S.S., Kumar, M.D., and Manerikar, M. (2001). Emission of carbon dioxide from a tropical estuaria system, Goa, India. *Geophysics Research Letters*, **28**(7), 1239–1242.

Sarmiento, J.L. and Wofsy, S.C. (1999). *A US Carbon Cycle Science Plan* (Report of the Carbon and Climate Working Group, 69 pp.). US Global Change Research Program, Washington, DC.

Savinykh, V.P. (1990). Monitoring today: The Russian point of view. *Ecological Science*, **1**, 68–70 [in Russian].

Savinykh, V.P. (1991). *Visual–instrumental Investigations of the Earth with the Piloted Space Complex* (109 pp.). Bowels Press, Moscow [in Russian].

Savinykh, V.P. (1993). *Russian Cosmos: The Problems to the Whole World* (Vol. 5, pp. 3–8). Power Press, Moscow [in Russian].

Savinykh, V.P. and Tshvetkov, V.Ya. (2000). Integration of GIS technologies and systems of remote sensing data processing. *Earth Research from Space*, **2**, 83–86 [in Russian].

Savinykh, V.P. and Tshvetkov, V.Ya. (2001). *Geoinformation Analysis of Remote Sensing Data* (228 pp.). Map-Geocentre-Geodes, Moscow [in Russian].

Savinykh, V.P. and Tshvetkov V.Ya. (2002). Geoinformatics and synergetics. *Earth Research from Space*, **4**, 41–47 [in Russian].

Scaife, A.A., Jackson, D.R., Swinbank, R., Butchart, N., Thornton, H.E., Keil, M., and Henderson, L. (2003) Stratospheric vacillations and the major warming over Antarctica in 2002. *J. Atmos. Sci.* (submitted).

Scanlon, T.M. and Albertson, J.D. (2001). Turbulent transport of carbon dioxide and water vapor within a vegetation canopy during unstable conditions: Identification of episodes using wavelet analysis. *Journal of Geophysics Research*, **106**(D7), 7251–7262.

Schafer, A. and Victor, D.G. (1999). Global passenger travel: Implications for carbon dioxide emissions. *Energy*, **24**(8), 657–679.

Schimel, D.S. (1995). Terrestrial biogeochemical cycles: Global estimates with remote sensing. *Remote Sensing and the Environment*, **51**(1), 49–56.

Schimel, D.S. (1998). The carbon equation. *Nature*, **393**(6682), 208–209.

Schimel, D.S., Alves, D., Enting, I., Heimann, M., Joos, F., Raynand, D., and Wigley, T. (1996). CO_2 and carbon cycle. *Climate Change* (pp. 76–86). Cambridge University Press, Cambridge, UK.

Schimel, D.S., House, J.I., Hibbard, K.A., Bousquet, P., Ciais, P., Peylin, P., Braswell, B.H., Apps, M.J., Baker, D., Bondeau, A. *et al.* (2001). Recent patterns and mechanisms of carbon exchange by terrestrial ecosystems. *Nature*, **414**, 169–172.

Schindler, D.W. (1999). The mysterious missing sink. *Nature*, **398**(6723), 105–107.

Schlesinger, W.H. (1999). Carbon and agriculture: Carbon sequestration in soils. *Science*, **284**(5423), 2095–2096.

Schmidt, K.S. and Skidmore, A.K. (2003). Spectral discrimination of vegetation types in a coastal wetland. *Remote Sensing of the Environment*, **85**, 92–108.

Scholes, R.J. and Noble, I.R. (2001). Climate change: Storing carbon on land. *Science*, **294**(5544), 1012–1014.

Schönwiese C.-D. (2002). Klima in der Diskussion. *AFZ/Wald*, **57**(8), 386–389 [in German].

Schrope, M. (2001). Consensus science, or consensus politics? *Nature*, **412**(6843), 112–114.

Schubert, S.D. and Muntean, M.-J. (1998). An analysis of tropopause pressure and total ozone correlations. *Monthly Weather Review*, **116**, 569–582.

Schulte, R.P.O. (2003). Analysis of the production stability of mixed grasslands, II: A mathematical framework for the qualification of production stability in grassland ecosystems. *Ecological Modelling*, **159**(1), 71–99.

Schulte, R.P.O., Lantinga, E.A. and Struik, P.C. (2003). Analysis of the production stability of mixed grasslands, I: A conceptual framework for the qualification of production stability in grassland ecosystems. *Ecological Modelling*, **159**(1), 43–69.

Schultz, M.G., Jacob, D.J., Wang, Y., Logan, J.A., Atlas, E.L., Blake, D.R., Blake, N.J., Bradshaw, J.D., Browell, E.V., Fenn, M.A. *et al.* (1999). On the origin of tropospheric ozone and NO_x over the tropical South Pacific. *Journal of Geophysics Research*, **104**(D5), 5829–5843.

Schulze, E.-D., Wirth, C., and Heimann, M. (2000). Managing forests after Kyoto. *Science*, **289**, 2058–2059.

Scientific Assessment of Ozone Depletion: 1998 (1999). In: D.L. Albritton, R.T. Watson, P.J.Aucamp, and G. Megie (eds), *WMO Global Ozone Research and Monitoring* (Project Report No. 44, 625 pp.). World Meteorological Organization, Geneva.

Scorer, R.S. (1990). *Meteorology of Air Pollution* (160 pp.). Ellis Horwood, New York.

SEDAAR (1996). *The Strategic Environmental Distributed Active Archive Resources*. Macromedia Inc., Washington (http://www.ciesin.org).

Sedjo, R.A. (2001). Forest sinks as a tool for climate-change policy-making: A look at the advantages and challenges. *Resources*, **143**, 21–23.

Seibel, B.A. and Walsh, P.J. (2001). Potential impacts of CO_2 injections on deep-sea biota. *Science*, **294**(5541), 319–321.

Seidov, D.G. (1987). Numerical models for ocean circulation. *Earth and Universe*, **5**, 28–34 [in Russian].

Sellers, P.J. (1985). Canopy reflectance, photosynthesis and transpiration. *International Journal of Remote Sensing*, **6**, 1335–1372.

Sellers, P.J. (1987). Canopy reflectance, photosynthesis and transpiration, II: The role of biophysics in the linearity of their interdependence. *Remote Sensing of the Environment*, **21**, 143–183.

Sellers, P.J., Berry, J.A., Collatz, G.J., Field, C.B., and Hall, F.G. (1992) Canopy reflectance, photosynthesis, and transpiration, III: A reanalysis using improved leaf models and a new canopy integration scheme. *Remote Sensing of the Environment*, **42**, 187–216.

Sellers, P.J., Meeson, B.W., Hall, F.G., Asrar, G., Murphy, R.E., Schiffer, R.A., Bretherton, F.P., Dickinson, R.E. *et al.* (1995). Remote sensing of the land surface for studies of global change: Models–algorithms–experiments. *Remote Sensing of the Environment*, **51**(1), 3–26.

Sellers, P.J., Bounoua, L., Collatz, G.L., Randall, D.A., Dazlich, D.A., Los, S.O., Berry, J.A., Fung, I., Tucker, C.J., Field, C.B. *et al.* (1996a). Comparison of radiative and physiological effects of doubled atmospheric CO_2 on climate. *Science*, **271**, 1402–1406.

Sellers, P.J., Los, S.O., Tucker, C.J., Justice, C.O., Dazlich, D.A., Collatz, G.J., and Randall D.A. (1996b) A revised land surface parametrization (SiB2) for atmospheric GCMs, Part II: The generation of global fields of terrestrial biophysical parameters from satellite data. *Journal of Climate*, **9**(4), 708–737.

Sellers, P.J., Randall, D.A., Collotz, G.J., Berry, J.A., Field, C.B., Dazlich, D.A., Zhang, C., Collelo, G.D., and Bounoua, L. (1996c). A revised land surface parametrization (SiB2) for atmospheric GCMs, Part 1: Model formulation. *Journal of Climate*, **9**(4), 676–705.

Sellers, P.J., Dickinson, R.E., Randall, D.A., Betts, A.K., Hall, F.G., Berry, J.A., Collatz, G.J., Denning, A.S., Mooney, H.A., Nobre, C.A., Sato, N., Field, C.B., and Henderson-Sellers, A. (1997). Modeling the exchanges of energy, water, and carbon between continents and the atmosphere. *Science*, **275**, 502–509.

Seppelt, R. and Voinov, A. (2002). Optimization methodology for land use pattern using spatially explicit landscape models. *Ecological Modelling*, **151**(2–3), 125–142.

Sergin, S.Ya. (ed.) (1974). *Study of the Glaciers–Ocean–Atmosphere System* (138 pp.). USSR Far-East Scientific Centre, Vladivostok [in Russian].

Sharma, S., Barrie, L.A., Plummer, D., McConnell J.C., Brickell, P.C., Levasseur, M., Gosselin, M., and Bates, T.S. (1999). Flux estimation of oceanic dimethyl sulfide around North America. *Journal of Geophysics Research*, **104**(D17), 21327–21342.

Shaw, G.E. (1988). Antarctic aerosols: A review. *Review of Geophysics*, **26**, 89–112.

Shaw, G.E. (1993). The Arctic Haze phenomenon. *Bulletin of the American Meteorological Society*, **76**(112), 2403–2413.

Shimji Morimoto, Nakazawa, T., Higurachi, K., and Aoki, S. (2000). Latitudinal distribution of atmospheric CO_2 sources and sinks inferred by $\delta^{13}C$ measurements from 1985 to 1991. *Journal of Geophysics Research*, **105**(D19), 24315–24326.

Shindell, D.T., Rind, D., and Lonergan, P. (1998). Increased polar stratospheric ozone losses and delayed eventual recovery owing to increasing greenhouse-gas concentration. *Nature*, **392**, 589–592.

Shupe, M.D., Uttal, T., Matrosov, S.Y., and Frisch, A.S. (2001). Cloud water contents and hydrometeor sizes during the FIRE Arctic Clouds Experiment. *Journal of Geophysics Research*, **106**, 15015–15028.

Shutko, A.M. (1987). *Microwave Radiometry of the Water Surface and Soils* (190 pp.). Science, Moscow [in Russian].

Si, F.C. and Hai, L.D. (1997). Control of environmental pollution by the industrial and domestic waste in the key region of economical development. *Proceedings of the Workshop on Environmental Technology and Management in Ho Chi Minh City, 28–29 May 1997, Ho Chi Minh* (pp. 32–42). Institute of Applied Mechanics, Ho Chi Minh, Vietnam [in Vietnamese].

Siani, G., Paterne, M., Michel, E., Sulpizio, R., Sbrana, A., Arnold, M., and Haddad, G. (2001). Mediterranean Sea surface radiocarbon reservoir age changes since the last glacial maximum. *Science*, **294**(5548), 1917–1920.

Sid'ko, A.F. and Shevyrnogov, A.P. (1998) Study of seasonal dependence for spectral brightness of agricultural crops on the chlorophyll content and plant physiological parameters. *Earth Research from Space*, **3**, 96–105 [in Russian].

Simmons, A., Hortal, M., Kelly, G., McNally, A., Untch, A., and Uppala, S. (2003) ERA-40 Project Report Series – Analyses and forecasts of stratospheric winter polar vortex break-up: September 2002 in the Southern Hemisphere and related events from ECMWF operations and ERA-40. *J. Atmos. Sci.* (submitted).

Singer, S.F. (1997). *Hot Talk. Cold Science* (x + 110 pp.). Independent Institute, Oakland, CA.

Sinik, N., Loncar, E., and Vidic, S. (1985). The use of field data in average wet deposition modeling. In: C. De Wispelaere (ed.), *Air Pollution Modeling and Its Application* (Vol. IV, pp. 155–161). Plenum Press, New York.

Sinnhuber, B-M., Weber, M., Amankwah, A., Burrows, J.P. (2003) Total ozone during the unusual Antarctic winter of 2002. *Geophys. Res. Lett.* **30**(11), 1580, doi:10.1029/2002GL016798, 2003.

Skiles, J.W. (1998). A stochastic weather generation model for Alaska. *Ecological Modelling*, **110**(3), 211–232.

Sleep, N.H and Zahnle, K. (2001). Carbon dioxide cycling and implications for climate on ancient Earth. *Journal of Geophysics Research*, **106**(E1), 1373–1399.

Smirnov, A., Holben, B.N., Dubovik, O., Frouin, R., Eck, T.F., and Slutsker, I. (2003). Marine component in aerosol optical models derived from Aerosol Robotic Network data. *Journal of Geophysics Research*, **108**(D1), 4033–4044.

Smith, K.A., Thomson, P.E., Clayton, H., McTaggart, I.P., and Conen, F. (1998). Effects of temperature, water content and nitrogen fertilisation on emissions of nitrous oxide by soils. *Atmospherical Environment*, **32**(19), 3301–3309.

Smith, S.V. (2002). Carbon–nitrogen–phosphorus fluxes in the coastal zone: The global approach. *Global Change Newsletter*, **49**, 7–11.

Solomon, S. (2004) The hole truth. *Nature*, **427**, 289–291 (22 January, 2004); doi:10.1038/427289a.

Soon, W., Baliunas, S., Idso, C., Idso, S., and Legates, D.R. (2003). Reconstructing climatic and environmental changes of the past 1000 years: A re-appraisal. *Energy and Environment*, **14**(2–3), 233–296.

Sorokhtin, O.G. and Ushakov, S.A. (1999). *Global Evolution of the Earth* (446 pp.). Moscow State University, Moscow [in Russian].

Spero, H.J. and Lea, D.W. (2002). The cause of carbon isotope minimum events on glacial terminations. *Science*, **296**(5567), 522–525.

Spieles, D.J. and Mitsch, W.J. (2003). A model of macroinvertebrate trophic structure and oxygen demand in freshwater wetlands. *Ecological Modelling*, **161**(3), 181–192.

Staudt, A.C., Jacob, D.J., Logan, J.A., Bachiochi, D., and Krishnamurti, T.N. (2001). Continental sources, transoceanic transport, and interhemispheric exchange of carbon monoxide over the Pacific. *Journal of Geophysics Research*, **106**(D23), 32571–32589.

Steffen, W. and Tyson, P. (eds) (2001). *Global Change and the Earth System: A Planet under Pressure* (Vol. 4, 32 pp.). IGBP Science, Stockholm.

Steffen, W., Jager, J., Carson, D.J., and Bradshaw, C. (2002). *Challenges of a Changing Earth* (216 pp.). Springer-Verlag, Berlin,

Stein, A.F. and Lamb, D. (2000). The sensitivity of sulphur wet deposition to atmospheric oxidants. *Atmospherical Environment*, **34**(11), 1681–1690.

Stempell, D. (1985). *Weltbevölkerung* (205 pp.). Springer-Verlag, Berlin.

Stephen, S.M., Frankling, S.C., Stagg, R.M., and Brown, J.A. (2000). Sub-lethal effects of exposure of juvenile turbot to oil produced water. *Marine Pollution Bulletin*, **40**(11), 928–937.

Stephens, B.B. and Keeling, R.F. (2000). The influence of Antarctic sea ice on glacial-interglacial CO_2 variations. *Nature*, **404**(6774), 171–174.

Stern, L.A., Baisden, W.T., Amundson, R., and Wang, Y. (1998). The influence of soils on oxygen isotope ratio of atmospheric CO_2. *Proceedings of the International Symposium Isotopes Technnical Study of the Past and Current Environment: Changes in the Hydrosphere and the Atmosphere, 14–18 April 1997, Vienna* (pp. 27–36). IAEA, Vienna, Austria.

Sternberg, L. and De Angells, D. (2002). Carbon isotope composition of ambient CO_2 and recycling: A matrix model. *Ecological Modelling*, **154**(1–2), 179–192.

Stevenson, D.S., Johnson, C.E., Collins, W.J., Derwent, R.G., and Edwards, J.M. (2000). Future estimates of tropospheric ozone radiative forcing and methane turnover: The impact of climate change. *Geophysics Research Letters*, **27**(14), 2073–2076.

Stockwell, D.Z., Giannakopoulos, C., Plantevin, P.H., Carver, G.D., Chipperfield, M.P., Law, K.S., Pyle, J.A., Shallcross, D.E., and Wang, K.Y. (1999). Modelling NO_x from lightning and its impact on global chemical fields. *Atmospheric Environment*, **33**(27), 4477–4493.

Stolarski, R.S., McPeters, R.D., and Newman, P.A. (2003) The 2002 Antarctic Ozone. *J. Atmos. Sci.* (submitted).

Stone, D.A. and Weaver, A.J. (2002). Daily maximum and minimum temperature trends in a climate model. *Geophysics Research Letters*, **29**(9), 70:1–70:4, 10.1029/2001GL014556.

Stone, R.S., Dutton, E.G., Harris, S.M., and Longenecker, D. (2002). Earlier spring snowmelt in northern Alaska as an indicator of climate change. *Journal of Geophysics Research*, **107**(D10), ACL 10/1–ACL 10/15.

Straub, C.P. (1989). *Practical Handbook of Environmental Control* (537 pp.). CRC Press, Boca Raton, FL.

Streets, D.G., Jiang, K., Hu, X., Sinton, J.E., Zhang, Z.-Q., Xu, D., Jacobson, M.Z., and Hansen, H.E. (2001). Recent reductions in China's greenhouse gas emissions. *Science*, **294**(5548), 1835–1837.

Strich, S. (2002). Ergebnisse der Klimakonferenz von Marrakesch. *AFZ/Wald*, **57**(4), 278–279.

Sturm, M., Racine, C., and Tape, K. (2001). Increasing shrub abundance in the Arctic. *Nature*, **411**, 546.

Summer, M., Michael, K., Bradshaw, C.J.A., and Hindell, M.A. (2003). Remote sensing of Southern Ocean sea surface temperature: Implications for marine biophysical models. *Remote Sensing of the Environment*, **84**(2), 196–213.

Sun, V., Balyunas, S., Demirchyan, K.S., Kondratyev, K.Ya., Idso, S.B., and Postmentier, E.S. (2001). Influence of anthropogenic emissions of CO_2 on the climate: Unsolved problems. *Proceedings of the Russian Geographic Community*, **133**(2), 1–19 [in Russian].

Svirezhev, Yu.M. (1987). *Nonlinear Waves, Dissipative Structures and Disasters in Ecology* (366 pp.). Science Press, Moscow [in Russian].

Svirezhev, Yu.M. (2002). Simple spatially distributed model of the global carbon cycle and its dynamic properties. *Ecological Modelling*, **155**(1), 53–69.

Svirezhev, Yu. and Zavalishin, N. (2003). Forest–grass global vegetation model with forest age structure. *Ecological Modelling*, **160**(1–2), 1–12.

Svirezhev, Yu.M., Krapivin, V.F., and Logofet, D.O. (1985). Mathematical ecology: Forecast, assessment and optimization of impacts on natural ecosystems. In: A. Sydov, M. Thoma, and R. Vichnevetsky (eds), *Systems Analysis and Simulation* (Vol. 1, pp. 114–124). Akademie-Verlag, Berlin.

Svirezhev, Yu.M., Krapivin, V.F., Vilkova, L.P., and Tarko, A.M. (1995). Mathematical model of carbon cycle in the biosphere with consideration of spatial distribution of the land biomes. In: Yu.M. Svirezhev (ed.), *Mathematical Modelling the Biogeocenotical Processes* (pp. 6–14). Science, Moscow [in Russian].

Svirezhev, Yu.M., Steinborn, W.H., and Pomaz, U.L. (2003). Energy of solar radiation: Global scale. *Ecological Modelling*, **169**(2–3), 339–346.

Tabazadeh, A., Santee, M.L., Danilin, M.Y., Pumphrey, H.C., Newman, P.A., Hamill, P.J., and Mergenthaler, J.L. (2000). Quantifying denitrification and its effect on ozone recovery. *Science*, **288**, 1407–1411.

Takemura, T. (2002). *A study of Aerosol Distribution and Optical Properties with a Global Climate Model* (Report No. 16, 113 pp.). CCSR, University of Tokyo.

Takigawa, M., Takahashi, M., and Akigosh, H. (1999). Simulation of ozone and other chemical species using a Center for Climate System Research/National Institute for Environmental Studies atmospheric GCM with coupled stratospheric chemistry. Journal of Geophysics Research, **104**(D11), 14003–14018.

Tangley, L. (2001). High CO_2 levels may give fast-growing trees an edge. *Science*, **292**(5514), 36–37.

Taupin, F.G., Bessafi, M., Baldy, S., and Bremaud, P.J. (1999). Tropospheric ozone above the southwestern Indian Ocean is strongly linked to dynamical conditions prevailing in the tropics. *Journal of Geophysics Research*, **104**(D7), 8057–8066.

ten Brink, H.M., van Dorland, R., and Lelieveld, J. (1999). *AEROSOL: Cycle and Influence on the Radiation Balance: Measurements and Modeling of the Reduction of Radiation by Aerosol (MEMORA)* (Report No. 410200025, 48 pp.). Dutch National Research Programme on Global Air Pollution and Climate Change, Amsterdam, The Netherlands,

Terziev, F.S., Zatuchnoy, B.M. and Gershanovich, D.E. (eds) (1993). *The Okhotsk Sea* (168 pp.). Gidrometeoizdat, St Petersburg [in Russian].

Tett, S.F.B., Jones, G.S., Stott, P.A., Hill, D.S., Mitchell, J.F.B., Allen, M.R., Ingram, W.J., Johnes, T.C., Johnson, C.E., Jones, A. *et al.* (2002). Estimation of natural and anthropogenic contributions to twentieth century temperature change. *Journal of Geophysics Research*, **107**(D16), ACL 10/1–ACL 10/24.

Thebault, J.M. and Rabouille, S. (2003). Comparison between two mathematical formulations of the phytoplankton specific growth rate as a function of light and temperature, in two simulation models (ASTER and YOYO). *Ecological Modelling*, **163**, 145–151.

The US Climate Change Science Program: Vision for the Program and Highlights of the Science Strategic Plan (2003). A Report by the Climate Change Science Program and the Subcommittee on Global Change Research, Washington, DC (July 2003, 34 pp.).

Thompson, D.W.J., Baldwin, M.P., and Solomon, S. (2003): Stratospheric precursors of tropospheric climate anomalies in the Southern Hemisphere: 1979–2002. *J. Atmos. Sci.* (submitted).

Tianhong, L., Yanxin, S., and An, X. (2003). Integration of large scale fertilizing models with GIS using minimum unit. *Environmental Modelling*, **18**(3), 221–229.

Tie, X., Brasseur, G., Emmons, L., Horowitz, L., and Kinnison, D. (2001. Effects of aerosol on tropospheric oxidants: A global model study. *Journal of Geophysics Research*, **106**(D19), 22931–22964.

Tiezen, L.L. (1973). Photosynthesis and respiration in arctic tundra grasses: Field light intensity and temperature responses. *Arctic and Alpine Research*, **5**, 239–251.

Tikhonov, A.N. (1952). Systems of differential equations containing small parameters under the derivations. *Mathematical Collection*, **31**(3), 72–87 [in Russian].

Tikunov, V.S. (1997). *Classifications in Geography* (364 pp.). Institute of the Arts, Moscow [in Russian].

Tiller, M.N. (1986). *System Theory Application in Multimedia Risk Assessment* (pp. 413–420). CER Council ASCE, Taipei.

Timmreck, C., Graf, H.-F., and Kirchner, I. (1999). A one-and-half-year interactive MA/ECHAM4 simulation of Mount Pinatubo aerosol. *Journal of Geophysics Research*, **104**(D8), 9337–9359.

Timoshevskii, A., Yeremin, V., and Kalkuta, S. (2003). New method for ecological minitoring based on the method of self-organising mathematical models. *Ecological Modelling*, **162**(2–3), 1–13.

Tol, R.S.J. (2000). International climate policy: An assessment. *IHDP Newsletter Update*, **3**, 11–12.

Tsushima, Y. and Manabe, S. (2001). Influence of cloud feedback on annual variation global mean surface temperature. *Journal of Geophysics Research*, **106**(D19), 22646–22655.

Tuomi, J., Augner, M, and Leimar, O. (1999). Fitness interactions among plants: Optimal defence and evolutionary game theory. In: T.O. Vuorisalo and P.K. Mutikainen (eds), *Life History Evolution in Plants* (pp. 63–83). Kluwer Academic Press, Dordrecht, The Netherlands.

UNEP Scientific Assessment of Ozone Depletion (1994). WMO Global Ozone Research and Monitoring Project Report No. 37. World Meteorological Organization, Geneva.

Urquizo, N., Brook, J.R., Walmsley, J.L., and Burrows, W.R. (2003). Estimation of sulphate concentration in high elevation fog in Northeastern North America. *Atmospheric Environment*, **37**(8), 1087–1100.

US National Academy of Science (2001) *Climate Change Science: An Analysis of Some Key Issues* (A report, 28 pp.). US National Academy of Science, Washington, DC.

Valiela, I. and Bowen, J.L. (2002). Nitrogen sources to watersheds and estuaries: Role of land cover mosaics and losses within watersheds. *Environmental Pollution*, **118**(2), 239–248.

Varotsos, C.A. (ed.) (1997). *Atmospheric Ozone Dynamics Observations in the Mediterranean Region* (NATO ASI Series, 336 pp.). Springer-Verlag, Berlin,

Varotsos, C. (2002a) Atmospheric pollution and remote sensing: Implications for the Southern Hemisphere ozone hole split in 2002 and the Northern mid-lattitude ozone trend. COSPAR Symposium on Natural Hazards Using Space Technology, Houston, USA, October 9–19, 2002.

Varotsos, C. (2002b) The Southern Hemisphere ozone hole split in 2002. *ESPR Environ. Sci. and Pollut. Res.*, **9**, 375–376.

Varotsos, C. (2003a) Major sudden warming and strange twist of the ozone hole over Antarctica in 2002. *European Physics News*, **34**(2), 66–67.

Varotsos, C. (2003b): What is the lesson from the unprecedented event over Antarctica in 2002? *ESPR Environ. Sci. and Pollut. Res.*, **10**, 80–81.

Varotsos, C. (2003c) Why did a "no-ozone-hole" episode occur in Antarctica. *EOS*, **84**(19), 183.

Varotsos, C. (2004a) Atmospheric pollution and remote sensing: Implications for the southern hemisphere ozone hole split in 2002 and the Northern mid-latitude ozone trend. *Advances in Space Research*, **33**(3) 249–253.

Varotsos, C. (2004b) A review on the extraordinary events of the major, sudden stratospheric warming, the diminutive Antarctic Ozone Hole, and its split, in 2002. *ESPR Environ. Sci. and Pollut. Res.* (submitted).

Varotsos, C. and Kondratyev, K. Ya. (2003) The unusual split in the Antarctic ozone hole in September 2002. *Studying the Earth from Space*, **1**, 92–93.

Varotsos, C.A. and Kondratyev, K.Ya. (1998). Dynamics of total ozone in the mid-latitudes of the Northern Hemisphere. *Annals of the Russian Academy of Science*, **359**(6), 821–822 [in Russian].

Vasilevich, V.I. (1983). *Essays on Theoretical Phytocenology* (226 pp.). Science, Leningrad [in Russian].

Vasiliev, A.S. (1991). Hydrodynamic model for seasonal variability of physical fields in the sea of Okhotsk. *Proceedings of the 6th International Symposium on the Okhotsk Sea and Sea Ice, 3–6 February 1991, Mombetsu, Japan* (pp. 294–297). Okhotsk Sea and Sea Ice Association, Mombetsu, Japan.

Vega, M., Pardo, R., Barrado, E., and Deban, L. (1998). Assessment of seasonal and polluting effects of the quality of river water by exploratory data analysis. *Water Research*, **32**(12), 3581–3593.

Vernadsky, V.I. (1944). A few words about the noosphere. *Present Biology Successes*, **18**(2), 49 93 [in Russian].

Victor, B.G., Raustiala, K., and Skolnikoff, E.B. (eds) (1998). *The Implementation and Effectiveness of International Environmental Commitments: Theory and Practice* (737 pp.). MIT Press, Cambridge, MA.

Victor, D.G. (2001). *The Collapse of the Kyoto Protocol and the Struggle to Slow Global Warming* (178 pp.). Princeton University Press, Princeton, NJ.

Vilkova, L.P., Novichikhin, E.P., Santalov, N.P., and Yakovleva, G.D. (1998). Estimation of the parameters of the biogeochemical cycle of carbon on global and regional scales. *Problems of Environmental and Natural Resources*, **7**, 24–37 [in Russian].

Vinogradov, B.V. (1983). A quantitative estimation of the potential of soil moisture remote sounding. *Annals of USSR Academy of Science*, **272**(1), 247–250 [in Russian].

Vinogradov, M.E., Krapivin, V.F., Fleishman, B.S., and Shushkina, E.A. (1975). Mathematical model use for analysis of ocean ecosystem behavior in the pelagic region. *Oceanology (Moscow)*, **15**(2), 313–320 [in Russian].

Von Storch, H. and Flöser G. (eds) (1999). *Anthropogenic Climate Change* (351 pp.). Springer-Verlag, Berlin,

Vörösmarty, C.J., Hinzman, L.D., Peterson, B.J., Bromwich, D.H., Hamilton, L.C., Morison, J., Romanovsky, V.E., Sturm, M., and Webb, R.S. (2001). *The Hydrological Cycle and Its Role in Arctic and Global Environmental Change: A Rationale and Strategy for Synthesis Study* (84 pp.). Arctic Research Consortium of the U.S., Fairbanks, AL.

Wacker, U. (1995). Competition of precipitation particles in a model with parametrized cloud microphysics. *Journal of Atmospheric Science*, **52**(14), 2577–2589.

Wagner, F., Muller, D., and Ansmann, A.A. (2001). Comparison of the radiative impact of aerosols derived from vertically resolved (lidar) and vertically integrated (Sun photometer) measurements: Example of an Indian aerosol plume. *Journal of Geophysics Research*, **106**(D19), 22861–22870.

Wallace, J.M. and Thompson, D.W.J. (2002). Annual models and climate prediction. *Physics Today*, **55**(2), 28–33.

Walter, H. and Breckle, S.-W. (1985). *Ecological Systems of the Geosphere* (242 pp.). Springer-Verlag, Berlin.

Wang, D.Y., von Clarmann, T., Fisher, H., Funke, B., Garcia-Comas, M., Gil-Lopez, S., Glatthor, N., Grabowski, U., Hopfner, M., Kellmann, S., *et al.* (2003) Longitudinal variations of temperature and ozone profiles observed by MIPAS during the Antarctic stratosphere sudden warming of 2002. *J. Atmos. Sci.* (submitted).

Wang, F., Chen, J., and Forsling, W. (1997). Surface properties of natural aquatic sediments. *Water Research*, **31**(7), 1796–1800.

Wang, X.L. and Zwiers, F.W. (2001). Using redundancy analysis to improve dynamical seasonal mean 500-hPa geopotential forecast. *International Journal of Climatology*, **21**, 637–654.

Waple, A.M. and Lawrimore, J.H. (eds) (2003). State of the climate in 2002. *Bulletin of the American Meteorological Society*, **84**(6), S1–S68.

Watson, R.T., Noble, I.R., Bolin, B., Ravindranath, N.H., Verardo, D.J, and Dokken, D.J. (eds) (2000). *Land Use, Land-use Change, and Forestry* (377 pp.). Cambridge University Press, Cambridge, UK.

Wauben, W.M.F., Van Velthoven, P.F.J., and Kelder, H. (1997). 3D chemistry transport model study of changes in atmospheric ozone due to aircraft – NO emissions. *Atmospheric Environment*, **31**(12), 1819–1836.

Weaver, C., Ginoux, P., Chou, M.-D., and Joiner, J. (2002). Radiative forcing of Saharan dust: GOCART model simulations compared with ERBE data. *Journal of Atmospheric Science*, **59**(3), Part 2, 736–747.

Weber, M., Dhomse, S., Wittrock, F., Richter, A., Sinnhuber, B-M., and Burrows, J.P. (2003) Dynamical control of NH and SH winter/spring total ozone from GOME observations in 1995–2002. *Geophys. Res. Lett.*, **30**(11), 1583, doi:10.1029/2002GL016799.

Weisenstein, D.K., Ko, M.K.W., Dyominov, I.G., Pitarui, G., Riccardully, L., Visconti, G., and Bekki, S. (1998). The effect of sulfur emissions from HSCT aircraft: A 2D model intercomparison. *Journal of Geophysics Research*, **103**(ND1), 1527–1547.

Weller, G. and Lange, M. (eds) (1999). *Impacts of Global Climate Change in the Arctic Region* (59 pp.). International Arctic Science Committee, Fairbanks, AL.

White, A., Cannel, M.G.R., and Friend, A.D. (1999). Climate change impacts on ecosystems and the terrestrial carbon sink: A new assessment. *Global Environmental Change*, **9**(1001), S21–S30.

Wigley, T.M.L. (1998). The Kyoto Protocol: CO_2, CH_4 and climate implications. *Geophysics Research Letters*, **25**(13), 2285–2288.

Wigley, T.M.L. (1999). *The Science of Climate Change: Global and US Perspectives* (48 pp.). Pew Center on Global Climate Change, Arlington, VA.

Wigley, T.M.L. and Raper, S.C.B. (2001). Interpretation of high projections for global-mean warming. *Science*, **293**(5529), 451–455.

Winslow, J.C., Hunt, E.R., and Piper, S.C. (2003). The influence of seasonal water availability on global C_3 versus C_4 grassland biomass and its implications for climate change research. *Ecological Modelling*, **163**(1–2), 153–173.

Wirtz, K.W. (2000). Second order up-scaling: Theory and an exercise with a complex photosynthesis model. *Ecological Modelling*, **126**(1), 59–71.

Witness the Arctic (1994). ARCUS, Fairbanks, AL (Vol. 2, No. 1, 15 pp.).

Wofsy, S.C. (2001). Climate change: Where has all the carbon gone? *Science*, **292**(5525), 2261–2262.

Wofsy, S.C., Goulden, M.L., Munger, J.W., Pyle, E.H., Urbanski, S.P., Hutyra, L., Saleska, S.R., Fitzjarrald, D., and Moore, K. (2001). Factors controlling long- and short-term sequestration of atmospheric CO_2 in a mid-latitude forest. *Science*, **294**(5547), 1688–1691.

Wolff, E.W. and Cachier, H. (1998). Concentrations and seasonal cycle of black carbon in aerosol at a coastal Antarctic Station. *Journal of Geophysics Research*, **103**(D9), 11033–11041.

Wolff, E.W. and Martin, S. (1998). Sea ice as a source of sea salt aerosol in polar regions. *IGACtivities Newsletter*, **14**, 10–12.

World Meteorological Organization (WMO) (2002) Antarctic ozone hole splits in two. Press Release No. 681, October 1, 2002.

Wuebbles, D. (2002). Oversimplifying the greenhouse: An Editorial essay. *Climate Change*, **52**(4), 3240–3245.

Yacoubi, S.E., Jai, A.E., Jacewicz, P., and Pausas, J.G. (2003). LUCAS: An original tool for landscape modelling. *Environmental Modelling and Software*, **18**(5), 429–437.

Yakimov, S.P. (1996). Algorithms for the assessment of forest fire danger using remote sensing data. PhD thesis, State Technical University, Krasnoyarsk (155 pp.) [in Russian].

Yakovlev, O.I. (2001). *Space Radio Science* (320 pp.). Taylor & Francis, London.

Yang, X., Wang, M., Huang, Y., and Wang, Y. (2002). A one-compartment model to study soil carbon decomposition rate at equilibrium situation. *Ecological Modelling*, **151**(1), 63–73.

Yemshanov, D. and Perera, A.H. (2002). A spatially explicit stochastic model to simulate boreal forest cover transitions: General structure and properties. *Ecological Modelling*, **150**(1–2), 189–209.

Yokozawa, M. (1998). Effects of competition mode on spatial pattern in plant communities. *Ecological Modelling*, **106**(1), 1–16.

Young, O.R. and Wefering, D.M. (2002). The Global Carbon Project. *IHDP Newsletter Update*, **1**, 20–21.

Zaezdny, A., Tabak, D., and Wulich, D. (1989). *Engineering Applications of Stochastic Processes* (509 pp.). John Wiley & Sons, New York.

Zaitsev, S.I. (1988). Laboratory modeling of the processes of the ocean–atmosphere CO_2 exchange. PhD thesis, Moscow State University (170 pp.) [in Russian].

Zarco-Tejada, P.J., Rueda, C.A., and Ustin, S.L. (2003). Water content estimation in vegetation with MODIS reflectance data and model inversion methods. *Remote Sensing of the Environment*, **85**(1), 109–124.

Zavarzin, G.A. (ed.) (1999). *The Carbon Cycle in Russia* (329 pp.). Russian Academy of Science, Moscow [in Russian].

Zavarzin, G.A. and Kolotilova, N.N. (2001). *Introduction to Natural Microbiology* (256 pp.). Moscow State University, Moscow [in Russian].

Zerchaninova, I.L. and Potapov, I.I. (2001). Study of technologies by the Intergovernmental Group of Experts with respect to climate change in the context of progress achieved in Russia. *Problems of Environmental and Natural Resources*, **4**, 3–22 [in Russian].

Zerefos, C., Meleti, C., Balis, D., Tourpali, K., and Bais, A.F. (1998). Quasibiennial and longer-term changes in clear-sky UV-B solar irradiance. *Geophysics Research Letters*, **25**(23), 4345–4348.

Zhang, H., Henderson-Sellers, A., and McGuffie, K. (2001). The compounding effects of tropical deforestation and greenhouse warming on climate. *Climate Change*, **49**, 309–338.

Zhang, L., Moran, M.D., Makar, P.A., Brook, J.R., and Gong, S. (2002). Modelling gaseous dry deposition in AURAMS: A unified regional air-quality modelling system. *Atmospheric Environment*, **36**(2), 537–560.

Zlatev, Z., Christensen, J., and Høv, Ø. (1992). A Eulerian air pollution model for Europe with nonlinear chemistry. *Journal of Atmospheric Chemistry*, **15**(1), 1–37.

Zonneveld, C. (1998). A cell-based model for the chlorophyll a to carbon ratio in phytoplankton. *Ecological Modelling*, **113**(1–3), 55–70.

Zufall, M.J., Dai, W., Davidson, C., and Etyemezian, V. (1999). Dry deposition of particles to wave surface, 1: Mathematical modeling. *Atmospheric Environment*, **33**, 4273–4281.

Index

Printing: Mercedes-Druck, Berlin
Binding: Stein + Lehmann, Berlin

MIX
Papier aus verantwortungsvollen Quellen
Paper from responsible sources
FSC® C105338

If you have any concerns about our products,
you can contact us on
ProductSafety@springernature.com

In case Publisher is established outside the EU,
the EU authorized representative is:
Springer Nature Customer Service Center GmbH
Europaplatz 3, 69115 Heidelberg, Germany

Printed by Libri Plureos GmbH
in Hamburg, Germany